CLINICAL DISORDERS OF MEMBRANE TRANSPORT PROCESSES

CLINICAL DISORDERS
OF MEMBRANE
TRANSPORT PROCESSES

Edited by

Thomas E. Andreoli, M.D.
University of Texas Medical School
Houston, Texas

Joseph F. Hoffman, Ph.D.
Yale University School of Medicine
New Haven, Connecticut

Darrell D. Fanestil, M.D.
University of California, San Diego
La Jolla, California

and

Stanley G. Schultz, M.D.
University of Texas Medical School
Houston Texas

PLENUM MEDICAL BOOK COMPANY
New York and London

Library of Congress Cataloging in Publication Data

Physiology of membrane disorders. Selections.
    Clinical disorders of membrane transport processes.

    "This volume is a reprint with minor modifications of part VI of Physiology of
membrane disorders, second edition, published by Plenum Medical Book Company
in 1986" — T.p. verso.
    Includes bibliographies and index.
    1. Membrane disorders. I. Andreoli, Thomas E., 1935–   . II. Title. [DNLM: 1.
Biological Transport. 2. Membranes — physiopathology. 3. Metabolic Diseases. QS
532.5.M3 P5782c]
RB113.P492   1987   616   87-20247
ISBN-13: 978-0-306-42699-5   e-ISBN-13: 978-1-4684-1286-4
DOI: 10.1007/978-1-4684-1286-4

This volume is a reprint with minor modifications of Part VI of
Physiology of Membrane Disorders, Second Edition, published by
Plenum Medical Book Company in 1986.

© 1986, 1987 Plenum Publishing Corporation
233 Spring Street, New York, N.Y. 10013

Plenum Medical Book Company is an imprint of Plenum Publishing Corporation

Contributors

THOMAS E. ANDREOLI, M.D.
Edward Randall III Professor and Chairman
Department of Internal Medicine
Professor of Physiology and Cell Biology
University of Texas Medical School
Houston, Texas 77225

STANLEY H. APPEL, M.D.
Professor and Chairman
Program in Neuroscience
Department of Neurology
Baylor College of Medicine
Houston, Texas 77030

LOUIS V. AVIOLI, M.D.
Professor of Medicine
Department of Medicine
Washington University School of Medicine at The Jewish Hospital of
    St. Louis
St. Louis, Missouri 63110

LEE R. BERKOWITZ, M.D.
Department of Medicine
University of North Carolina
    School of Medicine
Chapel Hill, North Carolina 27514

STANLEY J. BIRGE, M.D.
Associate Professor of Medicine
Department of Medicine
Washington University School of Medicine at The Jewish Hospital of
    St. Louis
St. Louis, Missouri 63110

ROLAND C. BLANTZ, M.D.
Professor of Medicine
Department of Medicine
University of California, San Diego
School of Medicine
La Jolla, California 92093
Chief of Nephrology
Veterans Administration Medical Center
San Diego, California 92161

JOSEPH V. BONVENTRE, M.D., PH.D.
Assistant Professor of Medicine
Department of Preventive Medicine
    and Clinical Epidemiology
Massachusetts General Hospital
Department of Medicine

Harvard Medical School and
Massachusetts General Hospital
Boston, Massachusetts 02114

JOSEPH Y. CHEUNG, M.D.
Research Fellow in Medicine
Department of Preventive Medicine
    and Clinical Epidemiology
Massachusetts General Hospital
Department of Medicine
Harvard Medical School and
Massachusetts General Hospital
Boston, Massachusetts 02114

R. MICHAEL CULPEPPER, M.D.
Assistant Professor of Medicine
Division of Nephrology
University of Texas Medical School
Houston, Texas 77225

RALPH A. DeFRONZO, M.D.
Associate Professor of Medicine
Department of Medicine
Yale University School of Medicine
New Haven, Connecticut 06510

JOHN M. DIETSCHY, M.D.
Professor of Medicine
Department of Internal Medicine
Southwestern Medical School
University of Texas Health Science Center
Dallas, Texas 75235

ROBERT F. GILMOUR, JR., PH.D.
Assistant Professor of Pharmacology
    and Medicine
Krannert Institute of Cardiology
Departments of Medicine and Pharmacology
Indiana University School of Medicine
Indianapolis, Indiana 46223

ALBERT M. GORDON, PH.D.
Professor
Department of Physiology and Biophysics
University of Washington School of Medicine
Seattle, Washington 98195

STEVEN C. HEBERT, M.D.
Assistant Professor of Medicine
Division of Nephrology

University of Texas Medical School
Houston, Texas 77225
Present address
Department of Internal Medicine
Brigham and Women's Hospital
Boston, Massachusetts 02115

LEONARD R. JOHNSON, PH.D.
Professor of Physiology and Cell Biology
Department of Physiology and Cell Biology
University of Texas Medical School
Houston, Texas 77225

ALEXANDER LEAF, M.D.
Professor of Medicine
Ridley Watts Professor of Preventive Medicine
Department of Preventive Medicine
    and Clinical Epidemiology
Massachusetts General Hospital
Department of Medicine
Harvard Medical School and
Massachusetts General Hospital
Boston, Massachusetts 02114

ANTHONY D. C. MACKNIGHT, M.D., PH.D.
Department of Physiology
University of Otago Medical School
Dunedin, New Zealand

JOHN A. MANGOS, M.D.
Professor of Physiology and Pediatrics
Chairman, Department of Pediatrics
University of Texas Health Science Center
San Antonio, Texas 78284

JOSEPH PALMISANO, M.D.
Assistant Professor of Medicine
Division of Nephrology
Department of Medicine
University of Connecticut School of Medicine
Farmington, Connecticut 06032

JOHN C. PARKER, M.D.
Professor of Medicine
Department of Medicine
University of North Carolina
    School of Medicine
Chapel Hill, North Carolina 27514

JUAN C. PELAYO, M.D.
Assistant Professor of Medicine
Department of Medicine
University of California, San Diego
School of Medicine
La Jolla, California 92093
Veterans Administration Medical Center
San Diego, California 92161

GORDON A. PLISHKER, PH.D.
Assistant Professor of Neurology
Program in Neuroscience
Department of Neurology
Baylor College of Medicine
Houston, Texas 77030

BENGT RIPPE, M.D.
Department of Physiology
University of South Alabama
    College of Medicine
Mobile, Alabama 36688

ROBERT L. RUFF, M.D., PH.D.
Assistant Professor
Department of Neurology
Cleveland Veterans Administration Hospital and Case Western Reserve University
Cleveland, Ohio 44106

JERRY A. SCHNEIDER, M.D.
Professor of Pediatrics
Metabolic Diseases Division
Department of Pediatrics
University of California at San Diego
La Jolla, California 92093

JOSEPH D. SCHULMAN, M.D.
Director
Genetics and IVF Institute
Department of Obstetrics and Gynecology
Fairfax Hospital
Fairfax, Virginia 22031

PHILIP R. STEINMETZ, M.D.
Professor of Medicine
Chairman, Division of Nephrology
Department of Medicine
University of Connecticut School of Medicine
Farmington, Connecticut 06032

AUBREY E. TAYLOR, PH.D.
Professor and Chairman
Department of Physiology
University of South Alabama
    College of Medicine
Mobile, Alabama 36688

SAMUEL O. THIER, M.D.
Professor and Chairman
Department of Medicine
Yale University School of Medicine
New Haven, Connecticut 06510

HENRIK WESTERGAARD, M.D.
Assistant Professor of Medicine
Department of Internal Medicine
Southwestern Medical School
University of Texas Health Science Center
Dallas, Texas 75235

DOUGLAS P. ZIPES, M.D.
Professor of Medicine
Director of Cardiovascular Research
Krannert Institute of Cardiology
Department of Medicine
Indiana University School of Medicine
Roudebush Veterans Administration
    Medical Center
Indianapolis, Indiana 46223

Preface

Clinical Disorders of Membrane Transport Processes is a softcover book containing a portion of *Physiology of Membrane Disorders (Second Edition)*. The parent volume contains six major sections that deal with general aspects of the physiology of transport processes and specific aspects of transport processes in cells and in organized cellular systems, namely epithelia. This text contains the last section, which deals with the application of the physiology of transport processes to the understanding of clinical disorders.

We hope that this smaller volume will be helpful to individuals particularly interested in clinical derangements of membrane transport processes.

THOMAS E. ANDREOLI
JOSEPH F. HOFFMAN
DARRELL D. FANESTIL
STANLEY G. SCHULTZ

Preface to the Second Edition

The second edition of *Physiology of Membrane Disorders* represents an extensive revision and a considerable expansion of the first edition. Yet the purpose of the second edition is identical to that of its predecessor, namely, to provide a rational analysis of membrane transport processes in individual membranes, cells, tissues, and organs, which in turn serves as a frame of reference for rationalizing disorders in which derangements of membrane transport processes play a cardinal role in the clinical expression of disease.

As in the first edition, this book is divided into a number of individual, but closely related, sections. Part V represents a new section where the problem of transport across epithelia is treated in some detail. Finally, Part VI, which analyzes clinical derangements, has been enlarged appreciably.

THE EDITORS

Contents

CHAPTER 14: **Renal Tubular Defects in Phosphate and Amino Acid Transport**

RALPH A. DeFRONZO and SAMUEL O. THIER

CHAPTER 15: **Pulmonary Edema**

AUBREY E. TAYLOR and BENGT RIPPE

CLINICAL DISORDERS OF MEMBRANE TRANSPORT PROCESSES

CHAPTER 1

The Cellular Basis of Ischemic Acute Renal Failure

Alexander Leaf, Anthony D. C. Macknight, Joseph Y. Cheung, and Joseph V. Bonventre

1. Introduction

It has long been recognized that obstruction of the blood supply to the normothermic kidney for periods of greater than 1 hr will almost invariably result in tubular necrosis and the clinical picture of acute renal failure. As the duration of arterial obstruction increases from 30 to 60 min, the proportion of kidneys that suffer damage increases, as does the extensiveness of damage. In the past few years there has been heightened interest in the nature of the lesion(s) caused by ischemia which results in irreversible cell injury even if blood flow to the kidney is restored. Identifying the critical change or changes that doom the cell to certain death will not only add to our knowledge of cellular physiology but possibly provide the basis for therapeutic interventions aimed at preventing or delaying the onset of the irreversible change(s). It is this latter hope that has done much to stimulate research.

As so often happens in medicine, empirical manipulations have already resulted in claims of protection from ischemic injury before the nature of the injury itself is understood. Table I is a list of pharmacologic agents for which claims of protection from ischemic injury have been made in either the clinical or the experimental setting of renal ischemia. The differences in the known pharmacologic properties of this diverse group of compounds are so great that the existence of any common denominator to their mode of protection seems highly improbable.

The consequences of ischemia to the many essential activities of cells are also multiple. Among the potentially damaging effects imposed by ischemia on the kidneys are: depletion of ATP, lactic acidosis, peroxidation injury, calciphylaxis, membrane changes, activation of autolytic systems (e.g., phospholipases, lysosomal enzymes), and mitochondrial injury. These effects of ischemia are, of course, not mutually exclusive and they probably act in ensemble to cause irreversible cell injury. Whether effects of the numerous agents listed in Table I, for which therapeutic claims have been made, bear any relation to the possible multiple deleterious effects of ischemia on cells is not evident. The temptation is to assume that there must be one critical change in an important structure or function of the cell which renders ischemic cell injury irreversible. At present, however, support for such a logical assumption is lacking.

A considerable literature has accumulated recently about the cellular effects of noxious agents or conditions that compromise organ functions. The effects of various tissue poisons have been examined, but ischemic injury, which remains clinically the most important cause of acute kidney and heart failure, has been studied most extensively. It is the purpose of this chapter to review the state of our understanding of the nature of the cellular insult in acute ischemic renal failure. We will borrow heavily from studies on other tissues, especially the heart and liver. This is justifiable, since a common cellular response to the same injurious factor seems likely. Although many substances are toxic to the kidneys—including many medications in common use to heavy metals, mushroom toxins, and other known tissue poisons—mechanisms of their toxic action have recently been reviewed[25] and will not be considered here.

2. Ischemic Injury

All life depends ultimately upon a continuous supply of energy and, in higher forms, on availability of ATP, the cellular energy currency. As Schrodinger stated,[26] life depends upon a continuous flow of negative entropy. Ischemia disrupts the flow of oxygen and metabolizable substrate to cells and thus stops the essential provision of energy upon which the life of the cell

Alexander Leaf, Joseph Y. Cheung, and Joseph V. Bonventre • Department of Preventive Medicine and Clinical Epidemiology, Massachusetts General Hospital, and Department of Medicine, Harvard Medical School and Massachusetts General Hospital, Boston, Massachusetts 02114. Anthony D. C. Macknight • Department of Physiology, University of Otago Medical School, Dunedin, New Zealand.

Table I. Agents Reported as Protective in Ischemic Acute Renal Failure[1]

ATP-MgCl$_2$[2-4]
Bis-tris propane[5]
Bradykinin[6]
Chlorpromazine[7]
Clonidine[8]
Demeclocycline[8]
Furosemide[6,9,10]
Glucose, amino acids[11]
Inosine[12,13]
Lithium[8]
Mannitol[6,10,14-17]
Methylprednisolone[18]
Polyethylene glycol[19]
Propranolol[20]
Prostaglandin E[21]
Saralasin[22]
Verapamil[23,24]

depends. It also prevents the removal of the waste products of metabolism which accumulate in the tissue and which probably contribute to the cellular injury.

Although acute renal failure, that is, acute suppression of kidney function, can result from primary obstruction of blood flow to the kidneys or from tubular obstruction, it is the necrosis of renal tubule cells which results from ischemic injury that is the focus of this review. Isolated or cultured cells are particularly useful in determining whether agents which protect cells from ischemic injury *in vivo* actually are effective in sustaining the survival of cells directly during the ischemic period or are acting via improvement in blood flow and thus lessening the actual ischemic insult. With the kidney there is the additional possibility that the protective agent prevents tubular obstruction and thus protects against acute renal failure. To date it has not been easy to obtain reliable preparations of renal tubule cells for study and thus most investigators have utilized isolated hepatocytes or cardiac myocytes.

Ischemia is only incompletely simulated in experimental studies on the survival of isolated or cultured cells which are exposed to hypoxia and a bathing medium devoid of metabolic substrates. One realizes that such experimental conditions lack the stasis and small volume of extracellular fluid associated with ischemia *in vivo* which may contribute importantly to cellular injury. Accumulation of metabolic waste products in an ischemic tissue may contribute to cellular dysfunction. For example, the rise in extracellular K$^+$ due to its leakage from some cells will serve to accelerate depolarization of neighboring cells which may contribute to leakiness of cell membranes. Other cellular constituents leaking from injured cells may also adversely affect neighboring cells. The large volume of bathing medium relative to the volume of isolated or cultured cells eliminates the element of stasis. Few investigators have attempted to include the feature of stasis in their design of studies of hypoxia in isolated or cultured cells. Where such conditions have been imposed on rabbit retina *in vitro*, effects of neighboring cells on tissue injury have been observed.[27] Nevertheless, a considerable amount can be learned about the nature of ischemic cell injury even if all the consequences of ischemia are not imposed on the isolated cells.

2.1. Criteria of Cell Viability

In order to study the effects of ischemia or noxious agents on cell viability, it is necessary to adopt clear criteria that distinguish between a cell which is still viable and one which is irreversibly damaged. However simple the problem may seem, this issue has bedeviled investigators and at present no single, simple, but rigorous criterion of survivability of cells or tissues is available. The uncertainty increases as the "point of no return" is approached. It is a sobering fact that in this day of sophisticated biology, we cannot measure some feature of an injured cell and from the measurement know whether the cell can recover from the injury or not. Instead, a number of criteria, both morphologic and functional, are used to determine cell viability.

2.1.1. Morphologic Criteria of Viability

There have been numerous attempts to correlate morphologic alterations in renal tubular cells with the sequence of pathophysiologic events that occur in acute ischemic renal injury.[28-36] Trump and associates[31-34] have performed the most detailed studies and have described a sequence of stages that occur during the ischemic and postischemic periods. They are part of a continuum, with different time courses in different animal models, but are quite similar no matter how the "ischemia" is produced. There are many similarities to changes seen in other organs, such as heart[37] and liver,[38] and to changes in isolated cells[39] or cultured cells.[40]

The changes that occur in individual cells can be divided into an initial reversible phase, in which removal of the ischemic insult will result in a return of normal morphology and function, followed by an irreversible phase in which the cell proceeds to the necrotic stage whether or not the ischemic insult is removed. During the reversible phase the ultrastructural changes are, in sequence: (1) nuclear chromatin is clumped and mitochondrial matrix granules are lost; (2) the endoplasmic reticulum (both rough and smooth) becomes dilated, "blebs" form on the cell surface, and the cell appears swollen (interorganelle distance increases and electron density is reduced); (3) mitochondria pass into the condensed configuration, and the mitochondria outer compartment increases; (4) there is then a mixed population of mitochondria with some retaining the condensed configuration while others appear swollen; and (5) finally, all mitochondria are swollen (high-amplitude swelling).[41] Up to this point, recovery of both function and structure is possible if the ischemic insult is removed. If the ischemic insult is continued beyond the point when mitochondria show high-amplitude swelling, then cellular function does not recover and necrosis ensues. The first ultrastructural changes associated with this transition are the appearance of flocculent densities in the inner compartment of mitochondria and the disruption of cellular membrane systems.

The first event observed in cells exposed to ischemia, the clumping of heterochromatin, appears to be associated with changes in cell pH[42] and decreased RNA synthesis. The early loss of the mitochondrial matrix granules with ischemia may be associated with the loss of bound calcium from mitochondria.[43] Thus, it appears that an early event in the ischemic insult is the mobilization of sequestered calcium. There is, in fact, a net loss of calcium from the mitochondria, since inner mitochondrial membrane potential is dissipated in anaerobiosis and calcium uptake by this organelle is inhibited. Since the changes are reversible, the actual mobilization of the calcium is not critical in itself. However, continued high cellular concentration of ionized calcium can have serious consequences, as will be discussed later.

The second stage is characterized by cell swelling, the enlargement of the endoplasmic reticulum, and the appearance of plasma membrane blebs or "colliculi."[44] The cell swelling is the expected response to depletion of ATP, with inhibition of the Na+,K+-dependent ATPase, leading to the dissipation of ionic gradients between cells and extracellular fluid and consequent osmotic swelling due to unbalanced Donnan effects.[45,46] Whether blebbing is a part of the process of cell swelling representing herniations of plasma membranes at weak points at the cell surface or whether specific effects loosen the attachment of the plasma membrane to the underlying cytoskeleton is not clear. Cytochalasins B and D and phalloidin, which interact directly with cytoskeletal structures, cause severe blebbing[47] but may also have other effects on cell viability. Recently, evidence has been presented that calcium plays a role in this blebbing[47]; the concentration of calcium is a known moderator of cytoskeletal structures.[48]

Whatever the molecular mechanisms, blebbing is a distinctive feature of the ischemic kidney[19,30,31,44,49] and of other ischemic organs. It is also readily seen in isolated cell preparations and cultured cells exposed to anoxia.[50,51] We have been studying this phenomenon morphologically by direct microscopy of isolated cells subjected to oxygen and substrate depletion. Two cell systems have been examined in our laboratory: cells derived from a mixed population of kidney tubule explants maintained in culture for 5 to 7 days[51] and primary cultures of cells derived from proximal convoluted tubules.[52] Both of these cell populations, when exposed to anoxic conditions in a continuous-flow preparation, go through a distinct blebbing stage within 2 to 3 hr. The blebs appear at different times on different cells, expand rapidly, and can become quite large. As reported by Trump et al.,[42] the blebs contain cytoplasm with an apparently lower viscosity than the rest of the cytoplasm, since, when living cells are viewed with phase contrast or Nomarski optics, Brownian motion is much more evident in the bleb. These blebs show reduced electron density in electron micrographs, and there appears to be a sharp separation between the cytoplasm of the bleb and of the cell proper as the organelles and filaments, so readily seen in the cell cytoplasm, are not found in the blebs.[19,30,44,53] The blebs thus appear in regions where there is an alteration in the state of attachment between cytoskeletal elements and the plasma membrane[42] possibly reflecting an elevation in cytosolic calcium concentrations.

When cells are subject to sublethal injury, they recover normal volume, resume active transport, and reassemble microtubules and microfilaments.[29,32,37] The reversibility of bleb formation, however, is not clear. In intact kidneys the cells actually may shed the blebs. With reflow the tubule lumina are often filled with bleblike debris even within the loop of Henle or within Bowman's space which are not the usual places for bleb formation. After temporary ischemia the proximal tubule cells appear denuded of microvilli.[32] This would be a natural consequence of the shedding of blebs but Glaumann et al.[54] and Venkatachalam et al.[53] have demonstrated loss of microvilli, at least in part, due to internalization of microvilli. It may be that once bleb formation occurs, the surviving cell cannot reverse this deformation. Glaumann et al.[32] reported that blebs persisted on proximal tubule cells up to 3 hr after reinitiation of

blood flow. The fact that cells can have blebs for many hours but do not all take up trypan blue whether during reflow in the kidney or in culture, suggests that these membrane perturbations may not be directly linked to the cell death that occurs rapidly after 60 to 120 min of ischemia.[32,55,56] Blebs in isolated rat hepatocytes caused by incubation with cytochalasin B for 2 hr likewise are not associated with trypan blue staining.[57]

The third cellular event associated with ischemic exposure is the change in appearance of mitochondria. These changes have been studied intensively. The ability of the mitochondrial ATP-generating system to recover rapidly after a period of ischemia is obviously critical for cell survival. The first ultrastructural change is a transition from the orthodox to the condensed state. Normal isolated mitochondria are in the orthodox configuration which is associated with State 4 respiration.[58,59] but in intact cells it might be more accurate to say the mitochondria are in a state midway between States 3 and 4.[60,61] The biochemical correlate of the condensed configuration is State 3 respiration associated with an increased ADP/ATP ratio and enhanced ATP production. The fact that the mitochondria go to the condensed state provides evidence that the ion pumps and respiratory chain proteins of the mitochondrial inner membrane are intact and functioning at this point. If the ischemic insult is reversed, the mitochondria return to the orthodox state. If, however, the insult is continued, the mitochondria become swollen and reach the stage described as high-amplitude swelling.[41,54] In some cases the cells show a mixture of condensed and swollen mitochondria,[41] clearly indicating the progressive nature of these alterations.

The swelling of the inner compartment appears to continue until the mitochondria are disrupted.[41] If the ischemic insult is reversed prior to the disruption of the membrane, then even this much distortion of cellular ultrastructure is reversible and compatible with cell survival. If ischemia is continued beyond this point, the cell enters the irreversible stage. What one sees at the ultrastructural level is essentially everything described up to this point, with the exception that the mitochondria have all passed to the stage of high-amplitude swelling.

Trump et al.[41] have reported that a key additional change seen at this time, at least in kidney tubule cells, is the appearance of flocculent or "fluffy" densities within the mitochondrial inner compartment. These densities are quite different in appearance from the very-electron-dense intramitochondrial granules associated with the presence of divalent cations.[43,62,63] The "fluffy" densities have a protein component and may be the morphologic manifestation of denatured proteins, possibly those of the mitochondrial inner membrane. If indeed this is the case, it still may not be the critical change within the cell, since it is clear that even denaturation of proteins is reversible (renaturation), at least in some cases.[61] If these densities do indeed come from the mitochondrial inner membrane, then the key event may be a disruption in the critical spatial relationships of this membrane. Boine et al.[64] have reported that ischemia results in a decrease in the amount of cardiolipin, an important phospholipid component of the mitochondrial membrane; mitochondrial membrane proteins are also affected. If the alteration in mitochondrial membrane cannot be rapidly corrected, damage becomes irreversible. The damaged mitochondria cannot produce adequate amounts of ATP to repair the already serious but not yet "lethal" damage done to other cellular components. The cell then essentially continues in an energy-deprived state and proceeds to the necrotic stage.

2.1.2. Functional Criteria of Viability

Many different techniques assessing cellular function have been employed to determine cell viability. The very numbers of such techniques underline the fact that none is completely acceptable. The gold standard of cell viability could be reproductive capability. If a cell reproduces, it is alive. There are, however, many difficulties with this criterion. Conditions for cell growth in culture media may not be optimal for cell reproduction. There are also certain types of cells that so far have not been found to proliferate *in vitro,* and there are tissues that, when mature, exhibit no further cell duplication even *in vivo.*

The single most frequently used test of viability is the dye exclusion test. Generally, dye molecules do not permeate cells that are alive and have intact membranes, whereas injured cells take up the dye and can easily be distinguished from normal cells in this manner. A number of dyes have been used for such studies, including methylene blue,[65] eosin,[66] erythrocin B,[67] nigrosin,[68] and lissamine.[69] Of such dyes, however, trypan blue has been most commonly used to determine cell viability.

It is assumed in the use of trypan blue that uptake of the dye and staining of the nucleus indicate death of the cell. It has been shown, however, by Castellot *et al.*[70] that kidney cells from baby hamsters can be made reversibly permeable to trypan blue when they are incubated in a hypertonic medium. Nevertheless, such hypertonically treated cells are capable of synthesizing protein, RNA, and DNA when supplied with the appropriate substrates and cofactors. Furthermore, Baur *et al.*[71] have found that isolated liver cells can be stained reversibly with trypan blue by decreasing the pH of the medium bathing the cells.

A variant of the dye staining technique to monitor cell viability is the use of fluorescein diacetate. Live cells with intact plasma membranes will take up fluorescein diacetate which is hydrolyzed within the cell to yield free fluorescein to which the plasma membrane is normally impermeable. Free fluorescein, therefore, is retained within the living cell.[72,73] Dead cells are still able to hydrolyze fluorescein diacetate but are not able to retain the free fluorescein and thus will not fluoresce. Staining with fluorescein has been closely correlated with exclusion of trypan blue when both techniques have been compared.[51,74]

Other tests of viability that depend upon the permeability properties of plasma membranes have been used. The release of the cellular enzymes creatine phosphokinase and lactic acid dehydrogenase into the bathing medium has been used in studies of heart cells,[75] and lactate dehydrogenase release has been used in following the viability of liver cells. Although the release of enzymes from cells must indicate severe membrane damage, it is not certain that it necessarily indicates the death of the cell from which the enzyme is released. Since the blebs formed on the surface of the cell exposed to anoxic or ischemic injury can be shed by the cell, it is possible that enzyme release from cells may occur via the subsequent rupture of such blebs.

It has been suggested that the magnitudes of the plasma membrane potential and the intracellular Na^+ and K^+ concentrations are the most sensitive criteria of viability for isolated liver cells.[71] Although these may be the best criteria for cell quality or condition, they cannot be taken as criteria for cell death. It is well known that changes in membrane potential and intracellular elemental composition can result from incubation of cells and tissues in the cold[76–78] or with ouabain.[79] These changes are reversed rapidly, however, when the cells are re-

turned to 37°C or the ouabain is removed. More valid as a measure of viability than reductions in intracellular K^+ and increases in intracellular Na^+, therefore, is the ability of the cell to recover normal membrane potential or normal intracellular elemental composition.

Monitoring of cellular surface charge density with 8-anilinonaphthyl-1-sulfonic acid has been used to measure cell viability.[80] Other investigators[81,82] have used the cellular uptake of the nonmetabolized amino acid γ-aminoisobutyric acid, an active process, as a measure of membrane function.

Many cellular metabolic properties have also been used as criteria of viability. Though stimulation of respiration with exogenous substrates might be expected to indicate intactness of cell preparations, Mapes and Harris[83] showed that the oxidation of succinate by isolated liver cells was carried out primarily by the damaged cells in the preparation. Plasma membranes of many cells are impermeable to the di- and tricarboxylic acids of the citric acid cycle and only with damage to the cell membrane do these substrates bearing multiple electrical charges gain access to the mitochondria. Utilization of oxygen and generation of ATP in the presence of exogenous substrate have also been used as criteria for mitochondrial function. A "dead cell," as judged by increased membrane permeability, however, may show a respiratory response to added substrate if its mitochondria are functional. Vogt and Farber[84] have shown that mitochondria can recover a normal oxygen uptake and coupling of oxidative phosphorylation even when obtained from ischemic kidney tissue the cells of which were already doomed to necrosis.

Others have utilized a variety of complicated chemical sequences, such as the synthesis of nucleic acids or proteins as criteria for cell viability. Baur *et al.*[71] found that incorporation of uridine into trichloroacetic acid-precipitable material is a very sensitive indicator of cell damage. This technique, however, is difficult to interpret since it is not clear that a decreased capability to incorporate uridine is synonymous with cell death. Castellot *et al.*[70] used DNA synthesis and protein synthesis as measures of viability. It is well known,[85,86] however, that transient blockage of protein synthesis, of itself, does not necessarily result in irreversible injury to cells.

Low cellular levels of ATP have been correlated with cell death.[87,88] In brain,[89,90] kidney,[84] and liver,[91] however, it has been demonstrated that marked depletion of ATP can occur with minimal irreversible injury. Low levels of ATP may persist for as long as 48 hr in surviving liver cells.[92] Reperfusion of tissues whose ATP levels have fallen after a short period of ischemia results in reestablishment of ATP levels. The absolute level of ATP, therefore, is not an indicator of irreversible damage; rather, it is the ability to synthesize ATP after ischemic insult that determines the final outcome. Recovery of ATP levels within viable cells may require prolonged periods if the total nucleotide pool within the cell has fallen as a result of the ischemic exposure, or if the plasma membranes are permeable to the synthesized ATP.

There is, therefore, no simple and unequivocal means of distinguishing early the reversible from the irreversible stages of ischemic injury to cells. Were conditions for incubating cells today sufficient to assure prolonged survival of cells *in vitro,* then the needed distinction between viability and death could be made with certainty by prolonged incubation. Since this is not yet the case, most workers use several criteria of reversibility. The likely error from these methods is that cells that still have the potential for survival will be interpreted as dead. Despite all the limitations, many workers use trypan blue exclusion as a simple, routine test of cell viability. But conclusions based on such data must be regarded as somewhat uncertain.

2.2. Cellular Metabolism in Ischemia and Anoxia

2.2.1. Carbohydrate Metabolism

Both ischemia and anoxia cause profound disturbances in cellular metabolism and function, but with distinctly different patterns of metabolism. Both severely reduce or abolish oxidative phosphorylation, normally the major source of energy in mammalian tissues. With anoxia, glycolysis is stimulated severalfold in order to compensate for the loss of oxidative phosphorylation, but with ischemia, glycolysis is markedly inhibited,[93] owing to impaired washout of metabolites. Substrate availability is not a limiting factor to ATP formation in ischemia, therefore, as it is in anoxia. Rather, in ischemia it is the interruption of the delivery of oxygen and the stagnation of blood flow which are the primary factors preventing the utilization of carbohydrates and the synthesis of ATP.

In contrast to the sustained acceleration of glycolysis in anaerobic hearts, ischemia causes a transient increase in glycolytic flux lasting only a few minutes. This is followed by marked inhibition of glycolysis.[93] The initial increase in glycolysis is due to accelerated glycogenolysis in the heart.[94,95] Subsequent inhibition occurs despite increased levels of ADP, AMP, and inorganic phosphate. Ischemia causes both a rapid production and a slow washout of lactate from affected tissues. This accumulation of lactate and H^+ [94] inhibits lactate dehydrogenase so that NADH can no longer be oxidized to NAD^+ in the cytoplasm, and pyruvate is not converted to lactate. The resultant high levels of cytosolic NADH inhibit glyceraldehyde-3-phosphate dehydrogenase[96] and the decrease in intracellular pH affects phosphofructokinase.[97] These changes effectively shut off glycolysis and ATP production. In fact, the rate of glycolysis in the ischemic heart is about 1/10th the rate measured in the anaerobic heart.[94] Under ischemic conditions, provision of exogenous glucose without concomitant improvement in the washout of retained metabolites may even be "toxic" to cells. This is because ATP depletion would be accelerated by its consumption in the production of glucose-6-phosphate and fructose-1,6-diphosphate without net ATP synthesis since glycolysis is blocked at the glyceraldehyde-3-phosphate dehydrogenase step.

2.2.2. Lipid Metabolism

Although fatty acid metabolism is normally the major source of energy for the kidney,[98] it ceases in both anoxia and ischemia. Fatty acid metabolism is entirely dependent upon oxidative processes within the mitochondria which stop with absence of oxygen. With decrease in the supply of oxygen, uptake as well as oxidation of fatty acids declines sharply.[99,100] Oxidation is inhibited at the level of β-oxidation due to increased NADH and $FADH_2$ in mitochondria. As a result, long-chain acyl-CoA increases threefold in amount and long-chain acylcarnitine increases sixfold in the ischemic heart.[99] Most of the increase of acyl-CoA occurs in the mitochondria whereas most of the long-chain acyl-carnitine is confined in the cytosol.[101] The rise in cytosolic acyl-carnitine leads to an increase in cytosolic acyl-CoA. Accumulation of cytosolic acyl-CoA inhibits acyl-CoA synthetases,[102,103] resulting in higher tissue levels of free fatty acids. In addition, free fatty acids will increase as a result of the increased phospholipase activity in ischemia.[64,104,105] The increased intracellular fatty acid levels seem to account for the observed decrease in fatty acid uptake in the absence of oxygen.

Accumulation of free fatty acids and of intermediates of fatty acid metabolism interferes with a number of cell functions. High levels of fatty acyl-CoA inhibit adenine nucleotide translocase[106,107] and acyl-CoA synthetase activity.[102,103] Acylcarnitine inhibits mitochondrial respiration.[108] Free fatty acids also inhibit Na^+,K^+-ATPase[109] and calcium efflux from subcellular stores.[110] In addition to their adverse effect on enzymatic functions, fatty acids and their intermediate metabolites are active detergents which bind extensively to membranes.[111] Insertion of these detergent molecules into membranes contributes to the altered permeability and disruption of membrane structures seen in hypoxic tissues[112] and may also contribute to the uncoupling of mitochondrial respiration.[113–115]

2.2.3. Protein Metabolism

Little is known about the effects of ischemia on protein metabolism in kidneys, but this has been extensively studied in heart muscle. With the depression of ATP levels during ischemia, both protein synthesis and degradation are inhibited. Both chain initiation and elongation are inhibited by anoxia in heart muscle,[116] whereas amino acid uptake by cells, amino acid levels within cells, and the formation of aminoacyl-tRNA are all similar to those found in aerobic hearts.[116,117]

Protein degradation is inhibited by both anoxia and ischemia[118] and this may result from the accumulation of lactate, H^+, and fatty acid metabolites in the tissue.[108]

2.2.4. Adenine Nucleotide Metabolism

Early in ischemia, there is a fall in ATP and ADP levels that is associated with a commensurate increase in AMP and inorganic phosphate.[119,121] After 60 min of ischemia, cellular AMP levels are also decreased. The net loss from the cellular adenine nucleotide pool is secondary to the efflux from the cells of nucleoside breakdown products of AMP.[122,123] In addition to resulting in loss of substrate for purine salvage, nucleosides released from cells may have other effects on the ischemic organ. Elevated extracellular levels of adenosine have been implicated in the postocclusive reduction in renal blood flow seen in the rat kidney.[124,125] Adenosine and inosine are further catabolized locally to hypoxanthine and xanthine.[125,126] The oxidation of hypoxanthine to xanthine and then to urate is catalyzed by xanthine oxidase with formation of superoxides. Superoxides, if not promptly degraded enzymatically, are highly reactive and will damage vital cellular constituents, as discussed later.

The return of ATP levels to normal is delayed when ischemia is reversed because of the total reduction of the cellular pool of adenine nucleotides and nucleosides. In heart it is estimated to take 4 days before de novo synthesis of adenine nucleotides reestablishes normal cellular levels.[127] On the other hand, provision of exogenous adenosine during reperfusion restores preischemic ATP levels in 5 hr.[128] The return of ATP levels to normal may also be delayed or prevented after reversal of ischemia by high cytosolic concentrations of free fatty acids and of Ca^{2+} which will uncouple oxidative phosphorylation.

2.3. Acidosis

Tissue made ischemic rapidly accumulates H^+ and, in spite of intracellular buffers, an acidosis develops. In the globally ischemic rat heart, intracellular pH dropped from 7.05 to 6.2 within 13 min.[120] The intracellular pH after 60 min of anoxia in rat kidneys perfused with saline was 6.56 as measured by nuclear magnetic resonance.[5] In many experiments involving ischemic injury, it is unclear to what extent the acidosis contributes to the morphologic and functional abnormalities observed. As discussed previously in this chapter, low cell pH would be expected to result in inhibition of enzymes of the glycolytic pathway, and in particular phosphofructokinase.[97] In addition, acidosis may result in inhibition of the malate–aspartate cycle which is involved in the transport of NADH from cytoplasm to mitochondria for reoxidation.[129]

Williamson[129] compared the response of the heart to ischemia with the response to perfusion of the coronary arteries with acidotic, fully oxygenated fluid. He found that acidosis resulted in decreased left ventricular function, decreased high-energy phosphate stores, increased lactate levels and lactate/pyruvate ratios. As these changes were similar to those seen in ischemia, Williamson therefore attributed many of the adverse effects of ischemia to acidosis. In the kidney, Bore et al.[5] demonstrated a significant improvement in survival and renal function if renal ischemia were preceded by intrarenal infusion of bis-tris propane buffer. Buffer-pretreated kidneys had a fall in intracellular pH to 6.9 rather than 6.5 as observed in saline-pretreated ischemic kidneys. The authors conclude that the maintenance of intracellular pH was the protective factor.

There is evidence, however, that acidosis may actually protect cells against anoxic injury. Bing et al.[130] found that, while hypoxic heart muscle was further depressed by acid pH, upon reoxygenation there was enhanced recovery of the tissue that had been incubated at low pH. Nayler et al.,[131] employing respiratory rather than metabolic acidosis, also found improved recovery when the tissue was exposed to pH 6.9 rather than 7.5 during hypoxia. In their experiments, this protective effect of acidosis did not appear to be related to decreased contractility, and therefore energy consumption, during the hypoxic period. Vogel and Sperelakis[132] found that metabolic acidosis blocked the slow Ca^{2+} inward current in isolated perfused embryonic chick ventricles. Altschuld et al.[133] found similar inhibition of Ca^{2+} entry across the sarcolemma of isolated myocardial cells when these were incubated in hypoxic media at low pH. Sperelakis and Schneider[134] proposed that the decreased Ca^{2+} influx might be protective during periods of ischemia.

Calcium accumulation in mitochondria extracted from heart perfused at pH 6.9 was less than in mitochondria isolated from hearts perfused at pH 7.4 or 6.6.[131] Thus, since increased mitochondrial calcium may be detrimental to ATP generation, a mildly acidotic pH (6.9) may result in better preservation of mitochondrial function in vivo by preventing mitochondrial calcium intoxication.[131]

It is not only in excitable tissue that acidosis has proved to be protective of hypoxic cell function. Penttila and Trump[135] demonstrated that acidosis protected Ehrlich ascites tumor cells and kidney cortical cells in vitro against anoxic injury. We have recently found that isolated hepatocytes exposed to 50 min of anoxia have enhanced regeneration of ATP levels after recovery if the anoxic incubation is carried out at pH 6.9 rather than pH 7.5. This protection is also reflected in decreased release of lactate dehydrogenase into the media from the mildly acidotic,

protected cells. Measurements of intracellular pH have confirmed that the cells exposed to anoxia in an extracellular milieu of pH 6.9 have a lower intracellular pH than cells exposed to an extracellular pH of 7.5. Incubation under well-oxygenated conditions even at pH as low as 6.6 had no adverse effect on cellular ATP levels or enzyme release into the media. Hinnen et al.[136] demonstrated that decreased extracellular pH resulted in decreased Ca^{2+} uptake into Ehrlich ascites tumor cells. Struder and Borle[137] found that both respiratory and metabolic acidosis decreased total cell calcium and each of three exchangeable cellular pools of calcium. Furthermore, acidosis decreases flux between the exchangeable pools. Van der Saag et al.[138] have demonstrated that lowering the extracellular pH from 7.5 to 6.5 markedly decreases plasma membrane microviscosity in neuroblastoma cells.

2.4. Lysosomal Changes

In heart[139] and liver,[6,135,140] ischemia results in a marked increase in the percentage of lysosomal hydrolases that appear in the soluble phase of homogenized tissue. Whether these enzymes are released in situ preceding and contributing to loss of cell viability, or whether the release and activation of lysosomal enzymes is a postmortem effect is not known. Van Lancker[141] observed no change in the ratios of bound to free acid phosphatase and β-glucuronidase in rat kidney made ischemic for 1 hr. There was no release of hydrolytic enzymes during this time. Therefore, in the kidney, it appears that such acid hydrolase release is not an early event in cell death. In any case, the late irreversible changes in ischemic tissues in which varying degrees of autolysis are detected indicate that extensive proteolysis occurs and that proteolytic activity has taken precedence over the reduced rate of protein degradation observed as an earlier response to ischemia.

2.5. Mitochondrial Damage

Reduced oxygen delivery to tissues associated with ischemia inhibits oxidative phosphorylation in mitochondria. After 60 min of ischemia, mitochondria of the posterior papillary muscle of the canine heart show 98% reduction in substrate-stimulated respiration and complete loss of respiratory control, indicating a severe depression of electron transport and uncoupling of oxidative phosphorylation.[142] After 3 hr of global ischemia, mitochondria isolated from rat liver show a complete loss of respiratory control, a loss of adenine nucleotide translocase activity, decreases in the heme portions of cytochromes aa_3 and of $C + C_1$, and a decrease in dinitrophenol-activated ATPase.[143] In addition, such mitochondria are incapable of generating a membrane potential, lose 80% of their potassium and magnesium, and their sodium content increases 10-fold.[144] In the porcine heart, 2 hr of ischemia results in a 33% decrease in NADH-coenzyme Q reductase activity, the initial enzyme complex involved in electron transport.[145] Also, mitochondrial ATPase (F_1) activity decreases by one-half.[145] These results suggest damage from ischemia to enzymes involved in electron transport and phosphorylation. Damage to mitochondrial enzymes in ischemia may be mediated by increased cellular calcium,[146,147] swelling of mitochondria which may disturb critical spatial arrangements of enzymes essential for sequential activity,[148] or accumulation of free fatty acids and their metabolites which inhibit adenine nucleotide translocase and act as detergents.[106,107,112]

Increase in intracellular phosphate may also play a role in the mitochondrial changes associated with ischemia. Schwartz et al.[149] demonstrated that mitochondria isolated from rabbit heart showed a marked drop in State 3 respiration when medium phosphate was increased from 2 to 5 mM. Furthermore, the higher phosphate concentration resulted in swelling of mitochdondria[149] as well as accelerated loss of adenine nucleotides from the matrix space.[150] Depletion of matrix adenine nucleotides decreases ATP transport out of the mitochondria[151] and hence further reduces ATP production.

2.6. Superoxide Damage

It is well known that ionizing radiations and certain chemicals can result in the formation of highly reactive peroxides which are destructive to cells. Unsaturated lipids, which abound in cell membranes and which play an important role in determining their permeability and fluidity, are highly susceptible to peroxidation. Oxidation at the unsaturated carbon–carbon double bond changes the physical properties of these important lipids. Cholesterol, another important constituent of cell membranes, is also adversely affected by free radical reactions.[152] Free radical formation results in increased permeability of plasma membranes and gross cellular swelling followed by destruction of the cell. Examples of cell damage by this mechanism abound. High partial pressures of oxygen cause hemolysis of red blood cells.[153] The damage to lungs from inspiring very high concentrations of oxygen and some of the adverse effects of hyperbaric oxygen are other examples of lipid peroxidation resulting in cellular and tissue injury.[154] The ''aging pigments'' lipofucsin and ceroid are other examples of lipid peroxidation,[155] but these are formed slowly over the years, and their accumulation in certain tissues, namely brain and liver, provides histologic hallmarks of aging. They are presumed not to adversely affect the function of the cells in which they accumulate, but when they are present intracellularly in large amounts it is hard to accept that they do not interfere with cell function.

Since lipid peroxidation is often associated with high oxygen tensions, how can it occur in association with hypoxia or ischemia? As toxic lipid peroxides occur spontaneously even with normal concentrations of oxygen in tissues, there are a variety of cellular mechanisms to prevent lipid peroxidation. Enzymes, superoxide dismutase and catalases, have evolved to eliminate the superoxides and other free radicals in tissues. With ischemia there may be impairment or loss of function of these scavenger enzyme systems[156] so that, with reflow of oxygen-bearing blood to the ischemic tissue, superoxides and hydrogen peroxide are formed, causing unopposed lipid peroxidation and damage to cell membranes. Progressive destruction of cells can occur in experimental ischemia if blood flow is restored to the kidney,[44] heart,[157] or intestine.[206] During the anoxia and ischemia there is degradation of the adenine nucleotide pool with increase in hypoxanthine. On reoxygenation the hypoxanthine will react with xanthine oxidase to produce toxic superoxide (and hydrogen peroxide) with resultant tissue damage.[207] Although other reasons for such progressive cellular damage with reflow may be postulated, the sequence is consistent with damage due to possible lipid peroxide formation. Demopoules et al.[158] point out that if free radicals are generated in tissues by radiation or other means, sufficient oxygen tensions persist even in ischemic tissues to allow superoxide formation. Recknagel and Glende[154] have provided a detailed review of this topic.

2.7. Calciphylaxis

Eukaryotic cells maintain the concentration of free Ca^{2+} within the cytosolic compartment in the range of 10^{-7} to 10^{-8} M despite the facts that in their external milieu the Ca^{2+} activity is in the millimolar range (10^{-3} M) and the membrane potential is negative inside the cell. This large electrochemical gradient is maintained across the plasma membrane by energy-dependent Ca^{2+} extrusion mechanisms[159,160] assisted by the low permeability of plasma membranes to Ca^{2+}.

Anoxia and ischemia will cause increases in intracellular Ca^{2+} content and in cytosolic free Ca^{2+} by a series of interrelated effects. With oxygen deprivation, cellular levels of ATP drop rapidly, as discussed above. Early in the anoxic insult, cytosolic Ca^{2+} may be elevated at a time when total cell Ca^{2+} is unchanged.[166] This reflects an inhibition of both accumulation and retention of Ca^{2+} by cellular organelles, such as mitochondria and endoplasmic reticulum, related, at least in part, to the depletion of ATP. The fall in ATP also inhibits plasma membrane Na^+,K^+-ATPase, allowing an increase of intracellular Na^+. But elevation in extramitochondrial Na^+ results in release of Ca^{2+} from mitochondria from renal medulla and, to a lesser extent, from renal cortex.[167] The elevation of cytosolic Na^+ results in a decreased Na^+–Ca^{2+} antiport efflux of Ca^{2+} across plasma as well as mitochondrial membranes[159] and this will contribute to the increase in cytosolic Ca^{2+}.

This initial phase will be followed by a gain in total cell Ca^{2+} as plasma membrane permeability increases. The initial increase of cytosolic Ca^{2+} will activate membrane-bound phospholipases.[169] This activation of phospholipases will be associated with alterations in membrane enzyme activity[170] and membrane permeability. This in combination with possible effects of cell swelling and of the detergent effects of cytosolic free fatty acids[111] will further contribute to membrane damage. This damage will in turn increase membrane permeability, allowing the high extracellular concentrations of Ca^{2+} to flood the cytosol. Superoxide or free radical formation may contribute further to the leakiness of the plasma membranes. All these effects summate to raise cytosolic free Ca^{2+} to potentially toxic levels.

Ca^{2+} is a key regulator of multiple cell functions.[161] High cytosolic Ca^{2+} levels have four deleterious effects upon the cell: they increase ATP consumption, decrease ATP production, damage cell membranes, and disaggregate cytoskeletal structures.

The elevated cytosolic Ca^{2+} will stimulate plasma and endoplasmic Ca^{2+}-ATPases and cause activation of contractile proteins in cells. The leakiness of cell membranes, described above, raises intracellular Na^+ concentration which stimulates Na^+,K^+-ATPase. These actions add to the consumption of ATP. These effects occur when anoxia and ischemia have already reduced or suppressed ATP production. But the elevated cytosolic Ca^{2+} will increase mitochondrial Ca^{2+} uptake at the expense of what mitochondrial ATP synthesis remains or resumes with relief from anoxia or ischemia. Oxidative phosphorylation is uncoupled from ATP synthesis by the elevated cytosolic Ca^{2+}.[203]

Membrane-bound phospholipases will be activated by the increased intracellular free Ca^{2+}.[168] The activation of phospholipases will be associated with changes in membrane phospholipids resulting in alterations in membrane enzyme activity[170] and membrane permeability. Furthermore, the activation of phospholipases results in increased cytosolic free fatty acids which act as detergents,[111] further contributing to membrane damage. The free fatty acids, furthermore, can produce a rapid and extensive release of Ca^{2+} from kidney mitochondria.[168]

Elevated cytosolic Ca^{2+} would be expected to activate gelsolin, resulting in dissolution of the cellular actin networks.[171] Elevated cell Ca^{2+} would also be expected to disaggregate microtubules acting via the Ca^{2+}-binding protein, calmodulin.

With these known deleterious effects of elevated cytosolic Ca^{2+} on the cell, it is not surprising that many investigators, since McLean et al.,[162] have considered that an increase in cytosolic free Ca^{2+} may be a primary cause of irreversible cell injury. In a series of papers, Farber and associates[143,144,169,194–196] have reported the structural and biochemical changes which follow ischemia in rat liver which they attribute to elevated cytosolic free Ca^{2+}. Given that an increase in intracellular free Ca^{2+} has great potential for producing chaos in a cell struggling to survive an ischemic insult, much attention has been directed to trying to prevent that increase. Two general approaches have been taken. In one, extracellular Ca^{2+} is removed and the cell viability is compared to that in the presence of extracellular Ca^{2+}. In the second, drugs which block specific plasma membrane channels are employed.

Several examples of the first approach can be cited. Schanne et al.[163] exposed hepatocyte cultures to membrane-active toxins in the presence and absence of extracellular Ca^{2+}. They reported that the cells bathed in Ca^{2+}-free solution were markedly protected from anoxic injury. In contrast, Smith et al.[173] found that freshly isolated hepatocytes were more susceptible to toxic injury in the absence of Ca^{2+} than in its presence. In addition, we have found that freshly isolated hepatocytes are also more susceptible to anoxic injury in the absence of extracellular Ca^{2+} (J. V. Bonventre, unpublished observations). We have also examined the effect of extracellular Ca^{2+} on anoxic cell injury using freshly isolated myocytes from rat ventricles.[174] The myocytes were isolated from adult rat ventricles and remained rod-shaped in appearance and normal by electron microscopy when bathed with a Krebs–Hensleit bicarbonate buffer containing 1.25 mM Ca^{2+}. After a 45-min exposure to buffer containing less than 10 μM Ca^{2+}, intracellular K^+ had decreased and Na^+ had increased markedly (Fig. 1). If the cells were made anoxic, the absence of Ca^{2+} resulted in greater changes in electrolyte composition than in the population of anoxic cells incubated with 1.25 mM Ca^{2+}. Likewise, the incorporation of [14C]phenylalanine into cell protein was markedly impaired by the decreased extracellular Ca^{2+} (Fig. 2). The changes in cellular electrolytes are likely related to increases in membrane permeability to monovalent cations. Similar changes in membrane permeability associated with reduction in extracellular Ca^{2+} have been found in human lymphocytes,[174] squid giant axon,[175] frog lens,[1 cat pancreas,[177] and rat liver.[178,179] In addition to changes in membrane permeability, active transport of γ-aminoisobutyric acid is inhibited in hepatocytes incubated in Ca^{2+}-free media.[178] Therefore, since incubation of cells in Ca^{2+}-free solutions in itself can be deleterious, experiments utilizing such conditions cannot be used to establish the role of extracellular Ca^{2+} in cell death.

The second experimental approach to the study of the role of Ca^{2+} entry from the extracellular milieu, as a primary etiologic factor in cell injury, involves the use of the slow-Ca^{2+}-channel blocking agents, such as verapamil and nifedipine.

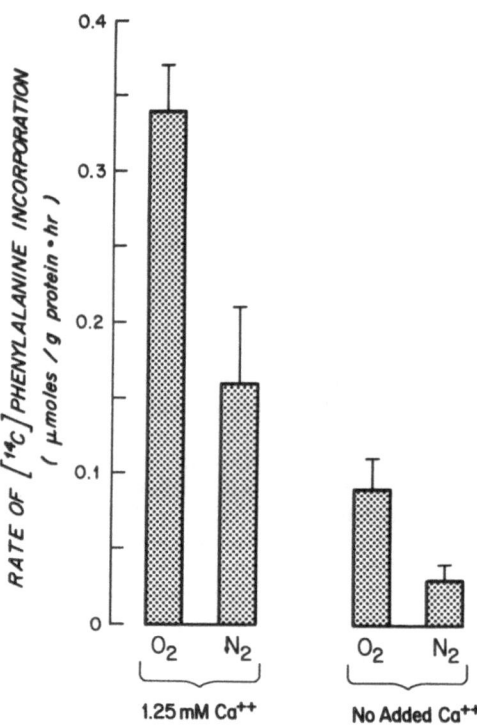

Fig. 1. Effect of Ca^{2+} removal from the extracellular milieu and anoxia on intracellular Na^+ and K^+ in isolated myocytes from adult rat ventricles. Intracellular Na^+ and K^+ were measured after 45 min of incubation in media containing either 1.25 mM Ca^{2+} or no added Ca^{2+} under aerobic or anaerobic conditions. The "no added Ca^{2+}" bathing solution had a Ca^{2+} concentration of 10 μM. From Ref. 174 with permission.

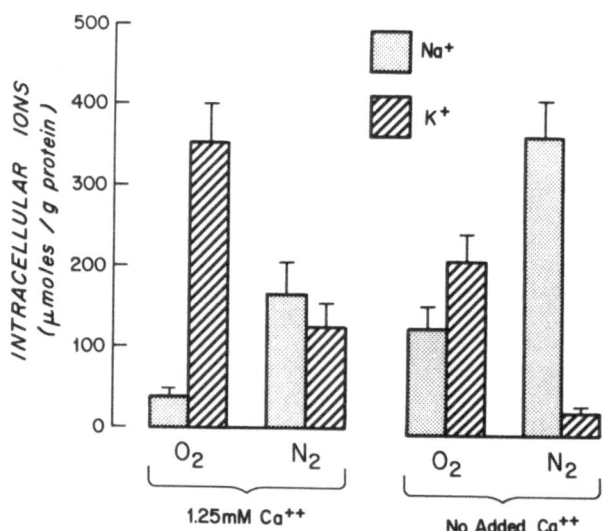

Fig. 2. Effect of Ca^{2+} removal from the extracellular milieu and anoxia on rate of [14C]phenylalanine incorporation into protein in myocytes isolated from adult rat ventricles. Cells were incubated with [14C]phenylalanine after 45 min of aerobic or anaerobic incubations. From Ref. 174 with permission.

Cells or tissues are bathed or perfused with media containing these agents in the presence of extracellular Ca^{2+}. It must be appreciated, however, that these agents act on the cell membrane in a very specific way. There is, a priori, therefore, no reason to assume that every pathway for Ca^{2+} entry into the cell will be blocked by these agents. Furthermore, in response to anoxia, the organization of the cell membrane breaks down with new pathways for Ca^{2+} permeability becoming established and these would not necessarily be blocked by the slow-Ca^{2+}-channel blocking agents. Nayler et al.[166] found that although verapamil delays the onset of the increase in resting tension induced by hypoxia in myocardial tissue, it does not alter the increase in total tissue Ca^{2+} seen with hypoxia. Bourdillon and Poole-Wilson[180] found that verapamil administration during reperfusion of rabbit myocardial tissue postischemia did not reduce $^{47}Ca^{2+}$ influx. In our laboratory,[181] we have found that verapamil and nifedipine prevent anoxic injury in isolated adult rat myocardial cells by decreasing the cellular contractile activity. When the cells were paced at a rate of 300 per min, to mimic the normal rat heart rate, the verapamil- and nifedipine-treated cells contracted at a slower rate than nontreated cells. This was associated with a protection of cellular high-energy phosphate levels and retention of the rod-shaped appearance. At lower stimulation frequencies, the control and the verapamil- and nifedipine-treated cells contracted at the same pace and no change in rate of decline of high-energy phosphate levels or in rate of decrease in number of rod-shaped cells was observed between the groups.

In nonexcitable cells, the effectiveness of the slow-Ca^{2+}-channel blocking agents in altering Ca^{2+} fluxes is not as well established as it is in excitable tissue.[182] In kidney, while preliminary reports suggested that verapamil blocked Ca^{2+} entry,[183,184] Gordon and Ferris[185] found that neither verapamil nor methoxyverapamil influenced the net influx or efflux of Ca^{2+} from urine to body fluids across renal tubules in vitro. Borle[186] also found no effect on Ca^{2+} influx in kidney cells and argued that kidney cells may have no voltage-dependent Ca^{2+} channels. We have studied the effects of verapamil in experimental models of acute renal failure in the rat.[23] We found no protection of renal function in a 40-min clamp model of acute renal failure as measured by changes in glomerular filtration rate 3 or 48 hr post-renal artery occlusion. Verapamil pretreatment in a norepinephrine model of acute renal failure protected renal function secondary to preservation of blood flow. Verapamil treatment postnorepinephrine had no beneficial effect on renal function.

We conclude, therefore, that although Ca^{2+} entry into the cell during anoxic insult may be an important contributor to ultimate cell death, it is not the sole mediator of such death, nor is it an essential prerequisite for cell death while tissue remains ischemic or anoxic. Such entry, however, may contribute importantly to the failure of cells to recover their normal function when oxygenated medium is provided following a period of ischemia or anoxia.

2.8. Cell Volume Regulation

It is well known that cell volume regulation is dependent upon metabolic supplies of energy. In the absence of ATP, the Na^+,K^+-ATPase of cell membranes can no longer pump Na^+ out of or K^+ into cells. Na^+ that enters cells down its electrochemical gradient accumulates and K^+ is lost. The resulting depolarization of cell membranes allows Cl^- to enter the cell.

The net gain of Na^+ and Cl^- chloride must exceed the loss of K^+. The increased solute content of the cell draws in water, and the cell swells.[45,46] Since this is one of the early and inevitable consequences of inhibition of metabolism, we examined the effect of preventing cell swelling during or shortly after an ischemic insult upon cell function and viability. In both dog heart[187,188] and rat kidney,[17,19,189] prevention of cell swelling was associated with protection of cells, at least temporarily, from ischemic injury.

In these studies, cell swelling was prevented by adding to the extracellular fluid a solute that penetrates cell membranes poorly. A number of substances have been found to be effective: mannitol, glucose, polyethylene glycol (PEG), and dextrans all serve this purpose. Since viability was protected to some degree by each, it seems that the effect is nonspecific and most likely due to the osmotic properties of the compounds, although they are capable of other effects such as "structuring water,"[190] and perhaps most importantly they may affect membrane permeability and stability.

Could cell swelling per se be a factor in the irreversible cell damage produced by prolonged ischemia? There is no direct evidence to support such a possibility. Even short periods of anoxia or ischemia are associated with cell swelling, but if blood flow and oxygen return before too long a delay, the swollen cells will promptly extrude the accumulated Na^+, Cl^-, and water, returning their volume to normal.[45] At the end of a 30- to 60-min period of clamping of the renal pedicle, all cells of the ischemic rat kidney appear swollen.[44] With reflow, many cells recover their normal appearance, but spotty necrosis of cells becomes apparent after 2 to 3 hr. The same pattern of increased cell death occurs after primary renal tubule cell cultures are returned to optimal culture conditions following 2 to 3 hr of anoxia.[51] Though it is not possible in the in vivo situation to follow the fate of a single swollen cell after correcting the ischemic insult in order to ascertain whether that particular cell survives or dies, DiBona and Powell,[37] by careful, quantitative morphometry, showed in the canine heart that the proportion of cells that became swollen during an ischemic period was closely related to the proportion of cells that later became necrotic. But cell swelling and cell death may simply be temporally associated phenomena consequent to some other primary effect, rather than the latter being caused by the former.

In studies utilizing isolated cardiac myocytes, we have found (unpublished observations) that, when suspended in an anoxic medium containing 6% PEG, the cells failed to stain with trypan blue, whereas similar cells incubated in a medium devoid of poorly penetrating solute, nearly all stained with trypan blue. This observation suggested that the poorly penetrating solute was protecting the viability of the isolated cardiac myocytes. However, if at the end of the period of anoxia the myocytes were resuspended in an oxygenated medium containing no PEG or other poorly penetrating solute, they promptly stained with trypan blue. Observations such as this suggest that the action of the poorly penetrating solute may be to preserve membrane stability and permeability rather than simply to prevent cells from swelling. It may be this effect on membrane stability and permeability that is common to those agents which have been found to prevent cell swelling and at the same time apparently to protect the cells from ischemic injury. This lack of importance of cell swelling on viability would be consistent with the often observed fact that cells can be made to swell by being placed in the cold but recover promptly when incubated at 37°C, or when placed in moderately hypotonic medium and then allowed to resume normal volume. Under both conditions the cells may undergo considerable swelling but appear to suffer no untoward consequences. Thus, the swelling that is so commonly an accompaniment of anoxia or ischemia, may in itself not be deleterious to the viablity of the cell. Other consequences of the anoxic or ischemic exposure must be primary.

This interpretation seems most likely to us at the present time, although it is possible that cell swelling could affect cell viability in several ways, and the cell swelling associated with exposure to anoxic or ischemic conditions may differ from the nonlethal cell swelling associated with exposure of cells to cold or to hypoosmotic media. Thus, it is possible that the distended plasma membrane around a swollen ischemic cell is more permeable than when in its normal configuration and that this increased permeability is not associated with exposure to cold or to hypotonic solutions. The increased membrane permeability would allow more rapid ingress of Na^+ and loss of K^+ from the cell. In addition to altering the intracellular environment of critical enzymes, this should increase the activity of the plasma membrane-bound Na^+,K^+-ATPase, which in turn would hasten the exhaustion of ATP in the cells. Ca^{2+}, or other potentially noxious substances in the extracellular fluids, would gain access to the cell, while important intracellular solutes might be lost through leaky plasma membranes. Anoxic or ischemic cell swelling may be associated with disruption of the cytoskeletal structures. If the cell swelling disrupts spatial relationships between membrane-bound enzymes that participate in sequential synthetic, degradative, or energy-producing chemical reactions required for the life of the cell, the possibility of recovering the essential normal flow of reactants from one fixed site to the next may be irreversibly lost and the fate of the cell irrevocably sealed.

At the present time, however, the observed protection of cell viability by nonpermeable solutes which prevent cell swelling seems more likely to be due to their effect in preserving normal cell membrane structure and permeability rather than through their effect in preventing cell swelling. In this context it should be mentioned that PEGs are used at high concentrations, usually greater than 30%, to fuse isolated cells together. This ability has been attributed by some investigators to a capability of PEG to order the structure of water.[191] If, at the concentrations of PEG used in our studies (4–8%), which are much below the concentration generally employed to fuse cells, the solute were creating a layer of structured water encasing cells, the low permeability of normal cell membranes might be preserved despite the ischemia. Since mannitol, glucose, and dextran all share with PEG the structure of ''poly-ols,'' this mode of protection may be common to all such substances. The permeability of ice to water molecules is some six orders of magnitude less than the self-diffusion rate of water in water.[192] The effect of such a layer of structured water about the ischemic cell could be considerable. Since mannitol is a known scavenger of OH^- radicals,[205] the protective action of this group of ''poly-ols'' may have another basis.

2.9. Recovery from Ischemia or Anoxia

As we have discussed, the cellular changes associated with ischemia or anoxia eventually become irreversible, but the precise nature of the changes which prevent recovery of reoxygenated and reperfused tissue has yet to be defined. *In vivo* the matter is further complicated in that organ function may remain severely damaged after a period of metabolic inhibition despite

restoration of cellular metabolism following reoxygenation. This situation may occur in acute ischemic renal failure, in which blockage of tubules by cellular debris may impair renal function despite restoration of circulation even though the epithelial cells have apparently recovered when the ischemic episode was of brief duration.

The extent of recovery upon reoxygenation is influenced by the duration of the metabolic duration, as well as by the nature of the insult; as discussed, the metabolic derangements and changes in cellular composition differ between ischemic and anoxic tissue. Different tissues, too, may respond differently, some being more sensitive to metabolic inhibition than others. Even within the same tissue, different groups of cells may respond differently. For example, following renal ischemia the ionic composition of proximal tubular epithelial cells deteriorates more rapidly than that of the distal tubular cells.[193] Evidence of necrosis also occurs most prominently in the straight portion of the proximal tubules in the subcortical zone of the kidney.[17,44,53] In addition, though cells may initially recover their composition when metabolism is restored, this recovery may not be sustained and evidence of impaired function may appear hours later.[193]

All these factors complicate the analysis of available data and caution is required in attempting to develop unifying hypotheses. Attention has been directed in this chapter largely to the known damaging effects to cells that would occur during the period of anoxia or ischemia. However, it is common experience to observe further tissue damage during the relief of anoxia or the return of circulation to cells or organs. In the kidney, at the end of 40 to 60 min of ischemia, virtually all cells are observed to be swollen but only after a few hours of reflow do necrotic cells appear.[44] The same has been observed in myocardium[37] and even with anoxic cultured kidney cells the proportion that are seen to take up trypan blue is considerably greater after 2 hr of reincubation in oxygenated medium than is the case at the end of the anoxic exposure.[51] What we do not know is whether these further changes upon oxygenation or renewed circulation simply represent late manifestations of injuries acquired by the cells during the anoxic or ischemic period or whether they represent continued further damage that may even depend upon the return of oxygenation or of blood flow. Both factors most likely occur as there are several noxious effects on cells which may be expected to result from or be accentuated by the return of oxygen or of blood flow.

The damaging effects of the uncoupling of oxidative phosphorylation by elevated cytosolic free fatty acid and free Ca^{2+} levels may not become manifest until oxygen and substrates become available. Cytosolic free Ca^{2+} will become high during anoxia or ischemia due to leakage through plasma membranes of the high extracellular content of Ca^{2+}. Only when oxidative metabolism resumes, however, will the high cytosolic Ca^{2+} result in accumulation of Ca^{2+} within mitochondria in preference to generation of ATP. The high cytosolic phosphate levels which also develop during anoxia or ischemia will contribute to the intramitochondrial accumulation of Ca^{2+} and the uncoupling of oxidative phosphorylation from ATP synthesis. This interference with the ability to generate high-energy phosphate bonds when oxygen has returned to the tissue may be even more detrimental to survival of cells than the period of cellular ''metabolic inactivity'' during the duration of anoxia or ischemia.

It seems quite unlikely that it is the return of more Ca^2 with reflow to cells partially damaged by ischemia that causes the progression of tissue damage following reflow to ischemic tissues. A simple calculation indicates that with extracellular fluids comprising 20% of tissue, sufficient Ca^{2+} is available to raise intracellular Ca^{2+} concentrations by 10- to 1000-fold over normal cytosolic concentrations. Were oxidative metabolism and ATP available to the cells, their mitochondria and endoplasmic reticulum could conceivably sequester such large amounts of intracellular Ca^{2+} and still preserve the essential low cytosolic concentrations of Ca^{2+}. But during anoxia or ischemia the needed ATP and mitochondrial H^+ gradients would not be available to accumulate the excessive intracellular Ca^{2+} in subcellular organelles.

Damage to cells from free radical formation that may increase with return of oxygen could also account for continued lethal injury during reoxygenation. Superoxides will cause lipid peroxidation affecting unsaturated lipids, both fatty acids and cholesterol, altering permeability properties of plasma and other cell membranes, as discussed. Conversion by proteases of xanthine dehydrogenase to xanthine oxidase in ischemic tissues[210] will increase the rate of formation of free radicals upon return of oxygen to the tissues with resulting increased tissue damage.[154,156,201] Reports of protection of bowel[207] and of myocardium[208,209] from ischemic injury by administration of superoxide dismutase following an ischemic insult have appeared.

Whereas the quantity of Ca^{2+} in the extracellular fluids contained in ischemic tissues will not be a limiting factor in the rise of cytosolic Ca^{2+} to toxic levels, this may not be true for other noxious agents. The Na^+ and water available in the extracellular fluids could limit the degree of cell swelling which might increase with reflow. Inorganic phosphate might increase in cells when more was delivered to them during reflow. The increased oxygen tensions with reflow would increase toxic superoxides and peroxides, as discussed. The initial removal with reflow from the tissue of the nucleotide breakdown products that had leaked from the cells during ischemia might further retard the reestablishment of the essential intracellular purine and pyrimidine nucleotide pool. Undoubtedly other deleterious effects may be accentuated during early reperfusion of tissue, the cells of which have lost much of their normal impermeability to movements of solutes from extracellular to intracellular fluids and vice versa. The fact that isolated or cultured cells bathed in large volumes of media may show further deterioration following relief of anoxia[51] makes somewhat less likely that deficiencies or excesses of extracellular substances, except for oxygen, modify the rate of recovery during reflow. This is not to say that the small volume of extracellular fluid may not importantly affect the cellular damage occurring during ischemia as compared with changes in isolated cells or tissues suspended in large volumes of anoxic media.[27]

In summary, one can, at our present state of understanding, construct a hypothesis which gives to Ca^{2+} a central role in the irreversible ischemic injury to cells. Ischemia inhibits oxidative metabolism with a prompt fall in ATP levels. This lack of ATP results in an accumulation in the cytosol of stores of intracellular Ca^{2+} released from mitochondria and endoplasmic reticulum. The elevated cytosolic free Ca^{2+} activates phospholipases which increase the permeability of cell membranes. The increased permeability of the plasma membrane allows further inundation of the cytosol by Ca^{2+} from extracellular fluids. If the period of cellular anoxia is relatively short (about 25 to 30 min or less), the changes in plasma membrane permeability need not be too great, cytosolic Ca^{2+} activity will not have increased too much, and reoxygenation will allow restoration of the nor-

mally low membrane permeability to Ca^{2+} with net extrusion from the cell of the Ca^{2+} accumulated when metabolism was inhibited. Though the mitochondria will initially accumulate some Ca^{2+} when metabolism is restored, this accumulation will be insufficient to impair oxidative phosphorylation to a significant extent. Following more prolonged periods of ischemia, however, the increases in permeability of plasma membranes will admit more Ca^{2+} into the cell and cytosolic free Ca^{2+} will rise even further. With resumption of oxidative metabolism, oxidative phosphorylation will be uncoupled from ATP synthesis because of Ca^{2+} accumulation in mitochondria. But if the ischemic or anoxic episode is sufficiently prolonged, mitochondria will have deteriorated to such an extent that they are no longer capable of generating a proton gradient across their inner membranes upon reoxygenation. Under these conditions, Ca^{2+} will not be accumulated by mitochondria[202] and no ATP can be generated. This persistence of deficient ATP synthesis will prevent recovery of cells and death will follow. But these damaging effects of high cytosolic Ca^{2+} levels are aided and abetted by many other parallel, associated, regressive changes in cellular structure and metabolism, as discussed, that are the consequences of lack of available free energy (ATP). Thus, high cytosolic Ca^{2+} levels are only part of the saga of cellular injury which begins and ends with the lack of ATP caused by ischemia or anoxia.

3. Summary

We have reviewed several aspects of ischemic injury in hopes of providing a focus for further research on critical aspects of the problem. Although a considerable amount has been learned, there are many important leads that need to be followed. The major difficulty to date has been determining which changes in cells exposed to ischemia or anoxia are primary noxious antemortem effects from those changes which are postmortem. This difficulty has plagued the interpretation of many experimental studies. The ambiguities in defining cell viability contribute to the experimental problem. It is likely that the irreversible stage in ischemic cell injury results from a combination of noxious effects and these will most likely include some of those we have discussed as well as others of which we have not yet conceived.

ACKNOWLEDGMENTS. This work was supported in part by NIH Grants HL-06664 and AM-27064. J.Y.C. was supported by National Research Service Fellowship Award HL-06378. J.V.B. was supported by NIH New Investigator Research Award AM-27957. A.D.C.M. received support from the Medical Research Council of New Zealand.

References

1. Bonventre, J. V. 1984. Cell response to ischemia. In: *Acute Renal Failure: Correlations between Morphology and Function.* K. Solez and A. Whetton, eds. Dekker, New York. pp. 195–217.
2. Chaudry, I. H., M. G. Clemens, and A. E. Baue. 1981. Alterations in cell function with ischemia and shock and their correction. *Arch. Surg.* **116**:1309–1317.
3. Gaudio, K. M., M. R. Taylor, I. H. Chaudry, M. Kashgarian, and N. J. Siegel. 1982. Accelerated recovery of single nephron function by the post-ischemic infusion of ATP–MgCl$_2$. *Kidney Int.* **22**:13–20.
4. Siegel, N. J., W. B. Glazier, I. H. Chaudry, K. M. Gaudio, B. Lytton, A. E. Baue, and M. Kashgarian. 1980. Enhanced recov-

ery from acute renal failure by the postischemic infusion of adenine nucleotides and magnesium chloride in rats. *Kidney Int.*
17:338–349.

5. Bore, P. J., P. Sehr, L. Chan, K. Thulborn, B. D. Ross, and G. K.
Radda. 1981. The importance of pH in renal preservation. *Transplant. Proc.* **13**:707–708.

6. Patak, R. V., S. Z. Fadem, M. D. Lifschitz, and J. H. Stein. 1979.
Study of factors which modify the development of norepinephrine-
induced acute renal failure in the dog. *Kidney Int.* **15**:227–237.

7. Dahlager, J. I., and T. Bilde. 1979. Renographic evaluation of
kidney preservation with chlorpromazine. *J. Nucl. Med.* **20**:18–
25.

8. Solez, K., T. Ideura, C. B. Silvia, B. Hamilton, and H. Saito.
1980. Clonidine after renal ischemia to lessen acute renal failure
and microvascular damage. *Kidney Int.* **18**:309–322.

9. DeTorrente, A., P. D. Miller, R. E. Cronin, P. E. Paulsen, A. L.
Erickson, and R. W. Schrier. 1978. Effects of furosemide and
acetylcholine in norepinephrine-induced acute renal failure. *Am.
J. Physiol.* **235**:F131–F136.

10. Hanley, M. J., and K. Davidson. 1981. Prior mannitol and
furosemide infusion in a model of ischemic acute renal failure.
Am. J. Physiol. **241**:F556–F560.

11. Abel, R. M., C. H. Beck, W. M. Abbott, J. A. Ryan, Jr., G. O.
Barnett, and J. E. Fisher. 1973. Improved survival from acute
renal failure after treatment with intravenous essential L-amino
acids and glucose. *N. Engl. J. Med.* **288**:695–699.

12. Fernando, A. R., D. M. G. Armstrong, J. R. Briffiths, W. F.
Hendry, E. P. N. O'Donaghue, J. P. Ward, L. E. Watkinson, and
J. E. A. Wickham. 1976. Enhanced preservation of the ischemic
kidney with inosine. *Lancet* **1**:555–557.

13. Marberger, M., R. Gunther, P. Alken, W. Rumpf, and M. Ranc.
1980. Inosine: Alternative of adjunct to regional hypothermia in
the prevention of post-ischemic renal failure. *Eur. Urol.* **6**:95–
102.

14. Boyer, D., R. D. Green, G. M. Collins, and N. A. Halasz. 1979.
Pharmacologic protection of rabbit kidneys from normothermic
ischemia. *Curr. Surg.* **36**:365–367.

15. Burke, T. J., R. E. Cronin, K. L. Bunchin, L. Peterson, and R. W.
Schrier. 1980. Ischemia and tubule obstruction during acute renal
failure in dogs: Mannitol in protection. *Am. J. Physiol.* **238**:F305–
F314.

16. Cronin, R. E., A. DeTorrente, P. D. Miller, R. E. Bulger, T. J.
Burke, and R. W. Schrier. 1978. Pathogenic mechanisms in early
norepinephrine-induced acute renal failure: Functional and histological correlates of protection. *Kidney Int.* **14**:115–125.

17. Flores, J., D. R. DiBona, C. H. Beck, and A. Leaf. 1972. The role
of cell swelling in ischemic renal damage and the protective effect
of hypertonic solute. *J. Clin. Invest.* **51**:118–126.

18. Toledo-Pereyra, L. H., V. R. Ramakrishnan, and M. Zammit.
1979. Study of the protective effect of methylprednisolone,
furosemide, and mannitol on ischemically damaged kidneys. *Eur.
Surg. Res.* **11**:179–184.

19. Frega, N. S., D. R. DiBona, and A. Leaf. 1979. The protection of
renal function from ischemic injury in the rat. *Pfluegers Arch.*
381:159–164.

20. Solez, K., R. J. D'Agostini, L. Stawowy, M. T. Freedman, W.
W. Scott, Jr., S. S. Siegelman, and R. H. Heptinstall. 1977.
Beneficial effect of propranolol in a histologically appropriate
model of post-ischemic acute renal failure. *Am. J. Pathol.*
88:163–192.

21. Mauk, R. H., R. V. Patak, S. Z. Fadem, M. D. Lifschitz, and J.
H. Stein. 1977. Effect of prostaglandin E administration in a
nephrotoxic and a vasoconstrictor model of acute renal failure.
Kidney Int. **12**:122–130.

22. Huland, H., H. J. Augustin, and T. Engels. 1981. The influence of
angiotensin II antagonist postischemic acute renal failure. *Urol.
Int.* **36**:15–22.

23. Bonventre, J. V., C. D. Malis, J. Y. Cheung, and A. Leaf. 1982.
Mechanisms of protection of verapamil in norepinephrine-induced
acute renal failure. *Clin. Res.* **30**:538a.

24. Burke, T. J., P. E. Arnold, and R. W. Schrier. 1981. A role for
intracellular calcium in the pathogenesis of norepinephrine (NE)-
induced acute renal failure (ARF). *Clin. Res.* **29**:457a.

25. Mudge, G. H., and G. G. Duggin. 1980. The symposium on drug
effect on the kidney. *Kidney Int.* **18**:539–711.

26. Schrodinger, E. 1956. *What Is Life?* Cambridge University Press,
London, 1944; Doubleday, New York.

27. Ames, A., III, and F. B. Nesbett. 1983. Pathophysiology of ischemic cell death. I. Time of onset of irreversible damage; importance
of different components of the ischemic insult. *Stroke* **14**:219–
226.

28. Coleman, S. E., J. Duggan, and R. L. Hackett. 1974. Freeze-
fracture studies of changes in nuclei isolated from ischemic rat
kidney. *Tissue Cell* **6**:521–534.

29. Cuppage, F. E., D. R. Neagoy, and A. Tate. 1967. Repair of the
nephron following temporary occlusion of the renal pedicle. *Lab.
Invest.* **17**:660–674.

30. Donahoe, J. F., M. A. Venkatachalam, D. B. Bernard, and N. G.
Levinsky. 1978. Tubular leakage and obstruction after renal ischemia: Structure–function correlations. *Kidney Int.* **13**:208–222.

31. Ginn, F. L., J. D. Shelbourne, and B. F. Trump. 1968. Disorders
of cell volume regulations. I. Effects of inhibition of plasma membrane adenosine triphosphatase with ouabain. *Am. J. Pathol.*
53:1041–1071.

32. Glaumann, B., H. Glaumann, I. K. Berezesky, and B. F. Trump.
1977. Studies on cellular recovery from injury. II. Ultrastructural
studies on the recovery of the pars convoluta of the proximal tubule
of the rat kidney from temporary ischemia. *Virchows Arch. B* **24**
:1–18.

33. Kreisberg, J. I., R. E. Bulger, B. F. Trump, and R. B. Nagle.
1976. Effect of transient hypotension on the structure function of
rat kidney. *Virchows Arch. B* **22**:121–133.

34. Mergner, W. J., S. H. Chang, and B. F. Trump. 1976. Studies on
the pathogenesis of ischemic cell injury. V. Morphologic changes
of the pars convoluta (P_1 and P_2) of the proximal tubule of rat
kidney made ischemic *in vitro*. *Virchows Arch. B* **21**:211–228.

35. Reimer, K. A., C. E. Ganote, and R. B. Jennings. 1972. Alterations in renal cortex following ischemic injury. III. Ultrastructure
of proximal tubules after ischemia or autolysis. *Lab. Invest.*
26:347–363.

36. Trump, B. F., J. M. Strum, and R. E. Bulger. 1974. Studies on the
pathogenesis of ischemic cell injury. I. Relation between ion and
water shifts and cell ultrastructure in rat kidney slices during swelling at 0–4°C. *Virchows Arch. B* **16**:1–34.

37. DiBona, D. R., and W. J. Powell. 1980. Quantitative correlation
between cell swelling and necrosis in myocardial ischemia in
dogs. *Circ. Res.* **47**:653–665.

38. Lemasters, J. J., S. Ji, and R. G. Thurman. 1981. Centrilobular
injury following low-flow hypoxia in isolated, perfused rat liver.
Science **213**:661.

39. Laiho, K. A., J. D. Shelbourne, and B. F. Trump. 1971. Observations on cell volume, ultrastructure, mitochondrial conformation
and vital dye uptake in Ehrlich ascites tumor cells. *Am. J. Pathol.*
65:203–229.

40. Kim, S. V. 1975. Brain hypoxia studied in mouse central nervous
system cultures. I. Sequential cellular changes. *Lab. Invest.*
33:658–669.

41. Trump, B. F., K. A. Laiho, W. J. Mergner, and A. V. Arstila.
1974. Studies on the subcellular pathophysiology of acute lethal
cell injury. *Beitr. Pathol.* **152**:243–271.

42. Trump, B. F., I. K. Berezesky, Y. Collan, M. W. Kahng, and W.
J. Mergner. 1976. Recent studies on the pathophysiology of ischemic cell injury. *Beitr. Pathol. Bd.* **158**:363–388.

43. Trump, B. F., W. J. Mergner, M. W. Kahng, and A. J. Saladino.
1976. Studies on the subcellular pathophysiology of ischemia.
Circulation **53**(Suppl. 1):I17–I26.

44. Frega, N. S., D. R. DiBona, B. Guertler, and A. Leaf. 1976.
Ischemic renal injury. *Kidney Int.* **10**:S17–S25.

45. Leaf, A. 1956. On the mechanism of fluid exchange of tissues *in
vitro*. *Biochem. J.* **62**:241–248.

46. Leaf, A. 1959. Maintenance of concentration gradients and regulation of cell volume. *Ann. N.Y. Acad. Sci.* **72**:396–404.

47. Jewell, S. A., G. Bellomo, H. Thor, S. Orrenius, and M. T. Smith. 1982. Bleb formation in hepatocytes during drug metabolism is caused by disturbances in thiol and calcium ion homeostasis. *Science* **217**:1257–1258.

48. Dedman, J. R., B. R. Brinkley, and A. R. Means. 1979. Regulation of microfilaments and microtubules by calcium and cyclic AMP. *Adv. Cyclic Nucleotide Res.* **11**:131–174.

49. Frega, N. S., D. R. DiBona, and A. Leaf. 1980. Enhancement of recovery from experimental ischemic acute renal failure. In: *Renal Pathophysiology—Recent Advances.* A. Leaf, G. Giebisch, L. Bolis, and S. Gorini, eds. Raven Press, New York. pp. 203–212.

50. Zollinger, H. E. 1948. Cytologic studies with the phase microscope. I. The formation of 'blisters' on cell in suspension (potocytosis) with observations on the nature of the cellular membrane. *Am. J. Pathol.* **24**:545–567.

51. Kreisberg, J. I., J. W. Mills, J. A. Jarrell, C. A. Rabito, and A. Leaf. 1980. Protection of cultured renal tubular epithelial cells from anoxic cell swelling and cell death. *Proc. Natl. Acad. Sci. USA* **77**:5445–5447.

52. Horster, M. 1979. Primary culture of mammalian nephron epithelia: Requirements for cell outgrowth and proliferation from defined explanted nephron segments. *Pfluegers Arch.* **383**:209–215.

53. Venkatachalam, M. A., D. B. Bernard, J. F. Donahoe, and N. G. Levinsky. 1978. Ischemic damage and repair in the rat proximal tubules: Differences among the S_1, S_2 and S_3 segments. *Kidney Int.* **14**:31–49.

54. Glaumann, B., H. Glaumann, I. K. Berezesky, and B. F. Trump. 1975. Studies on the pathogenesis of ischemic cell injury. II. Morphological changes of the pars convoluta (P_1 and P_2) of the proximal tubule of the rat kidney made ischemic in vivo. *Virchows Arch. B* **19**:281–302.

55. Glaumann, B., and B. F. Trump. 1975. Studies on the pathogenesis of ischemic cell injury. III. Morphological changes of the proximal pars recta tubules of the rat kidney made ischemic *in vivo. Virchows Arch. B* **19**:303–323.

56. Trump, B. F., and K. A. Laiho. 1975. Studies of cellular recovery from injury. I. Recovery from anoxia in Ehrlich ascites tumor cells. *Lab. Invest.* **33**:706–711.

57. Gravela, E., G. Poli, E. Albano, and M. V. Dianzanni. 1977. Studies on fatty liver with isolated hepatocytes. *Exp. Mol. Pathol.* **27**:339–352.

58. Hackenbrock, C. R. 1966. II. Electron transport-linked ultrastructural mechanical activity in mitochondria. I. Reversible ultrastructural changes with change in metabolic steady state in isolated liver mitochondria. *J. Cell Biol.* **30**:269–297.

59. Hackenbrock, C. R. 1968. II. Electron transport-linked ultrastructural transformations in mitochondria. *J. Cell Biol.* **37**:345–369.

60. Hackenbrock, C. R., T. G. Rehn, E. C. Weinbach, and J. J. Lemasters. 1971. Oxidative phosphorylation and ultrastructural transformation in mitochondria in the intact ascites tumor cell. *J. Cell Biol.* **51**:123–137.

61. Lehninger, A. L. 1975. *Biochemistry,* 2nd ed. Worth Publishing, New York.

62. Somlyo, A. P., A. V. Somlyo, and H. Shuman. 1981. Electron probe analysis of vascular smooth muscle: Composition of mitochondria, nuclei and cytoplasm. *J. Cell Biol.* **81**:316–335.

63. Sutfin, L. V., M. E. Holtrop, and R. E. Ogilvie. 1971. Microanalysis of individual mitochondrial granules with diameters less than 1000 angstroms. *Science* **174**:947–949.

64. Boine, I., E. E. Smith, and F. E. Hunter. 1970. The role of fatty acids in mitochondrial changes during liver ischemia. *Arch. Biochem. Biophys.* **139**:425–443.

65. Schrek, R. 1936. A method for counting the viable cells in normal and in malignant cell suspension. *Am. J. Cancer* **28**:389–392.

66. Hanks, J. H., and J. H. Wallace. 1958. Determination of cell viability. *Proc. Soc. Exp. Biol. Med.* **98**:188–192.

67. Philips, H. J., and J. E. Terryberry. 1957. Counting actively metabolizing tissue cultured cells. *Exp. Cell Res.* **13**:341–347.

68. Kaltenbach, J. P., M. H. Kaltenbach, and W. B. Lyons. 1958. Nigrosin as a dye for differentiating live and dead ascites cells. *Exp. Cell Res.* **15**:112–117.

69. Holmberg, B. 1961. On the permeability of lissamine green and other dyes in the course of cell injury and cell death. *Exp. Cell Res.* **22**:406–414.

70. Castellot, J. J., Jr., M. R. Miller, and A. B. Pardee. 1978. Animal cells reversibly permeable to small molecules. *Proc. Natl. Acad. Sci. USA* **75**:351–355.

71. Baur, H., S. Kasperek, and E. Pfaff. 1975. Criteria of viability of isolated liver cells. *Hoppe-Seylers Z. Physiol. Chem.* **356**:827–838.

72. Rotman, B., and B. W. Papermaster. 1966. Membrane properties of living mammalian cells as studied by enzymatic hydrolysis of fluorogenic ester. *Proc. Natl. Acad. Sci. USA* **55**:134–141.

73. Jarnagin, J. L., and D. W. Luchsinger. 1980. The use of fluorescein diacetate and ethidium bromide as a stain for evaluating viability of mycobacteria. *Stain Technol.* **55**:253–258.

74. Schanne, F. A. X., A. B. Kane, E. E. Young, and J. C. Farber. 1977. Calcium dependence of toxic cell death: A final common pathway. *Science* **206**:700–702.

75. Acosta, D., M. Pucketts, and R. McMillin. 1978. Ischemic myocardial injury in cultured heart cells: Leakage of cytoplasmic enzymes from injured cell. *In Vitro* **14**:728–732.

76. Conway, E. J., and H. Geoghegan. 1955. Molecular concentration of kidney cortex slices. *J. Physiol. (London)* **130**:438–445.

77. Conway, E. J. 1957. Nature and significance of concentration relation of potassium and sodium ions in skeletal muscle. *Physiol. Rev.* **37**:84–132.

78. Steinbach, H. B. 1952. Sodium and potassium balance of muscle and nerve. In: *Modern Trends in Physiology and Biochemistry: Woods Hole Lectures Dedicated to Memory of Leonor Michaelis.* E. S. G. Barron, ed. Academic Press, New York. pp. 173–192.

79. Schatzmann, H. J. 1953. Herzglykoside als Hemmstoffe für den aktiven kaliumund natrium transport durch die erythrocytenmembran. *Helv. Physiol. Pharmacol. Acta* **11**:346–354.

80. Pfaff, E., B. Schuler, H. Krell, and H. Hoke. 1980. Viability control and special properties of isolated rat hepatocytes. *Arch. Toxicol.* **44**:3–21.

81. Dickson, J. A. 1970. The uptake of non-metabolizable amino acids as an index of cell viability in vitro. *Exp. Cell Res.* **61**:235–245.

82. Edmondson, J. W., and N. U. Bang. 1981. Deleterious effects of calcium deprivation on freshly isolated hepatocytes. *Am. J. Physiol.* **241**:C3–C8.

83. Mapes, J. P., and R. A. Harris. 1975. On the oxidation of succinate by parenchymal cells isolated from rat liver. *FEBS Lett.* **51**:80–83.

84. Vogt, M. T., and E. Farber. 1968. On the molecular pathology of ischemic cell death: Reversible and irreversible cellular and mitochondrial metabolic alterations. *Am. J. Pathol.* **53**:1–26.

85. Shelburne, J. D., A. V. Arstila, and B. F. Trump. 1973. Studies on cellular autophagocytosis: The relationship of autophagocytosis to protein synthesis and to energy metabolism in rat liver and flounder kidney tubules *in vitro. Am. J. Pathol.* **73**:641–670.

86. Verbin, R. S., P. J. Goldblatt, and E. Farber. 1969. The biochemical pathology of inhibition of protein synthesis *in vivo. Lab. Invest.* **20**:529–536.

87. Laiho, K. V., and B. F. Trump. 1974. The relationship between cell viability and changes in mitochondrial ultrastructure, cellular ATP, ion and water content following injury of Ehrlich ascites tumor cells. *Virchows Arch. B* **15**:267–277.

88. Jennings, R. B., H. K. Hawkins, J. E. Lowe, M. L. Hill, S. Klotman, and K. A. Reimer. 1978. Relation between high energy phosphate and lethal injury in myocardial ischemia in the dog. *Am. J. Pathol.* **92**:187–207.

89. Kleihues, P., K. A. Hossmann, A. E. Pegg, K. Kobayashi, and V. Zimmermann. 1975. Resuscitation of the monkey brain after one

hour complete ischemia. III. Indication of metabolic recovery. *Brain Res.* **95**:61–73.

90. Kleihues, P., K. Kobayashi, and K. A. Hossmann. 1974. Purine nucleotide metabolism in the cat brain after one hour of complete ischemia. *J. Neurochem.* **23**:417–425.

91. Farber, E. 1973. ATP and cell integrity. *Fed. Proc.* **32**:1534–1539.

92. Van Lancker. 1976. *Molecular and Cellular Mechanisms in Disease.* Springer-Verlag, Berlin.

93. Kubler, W., and P. G. Spieckerman. 1970. Regulation of glycolysis in the ischemic and the anoxic myocardium. *J. Mol. Cell. Cardiol.* **1**:351–377.

94. Rovetto, M. J., J. T. Whitmer, and J. R. Neely. 1973. Comparison of the effects of anoxia and whole heart ischemia on carbohydrate utilization in isolated working rat heart. *Circ. Res.* **32**:699–711.

95. Wollenberger, A., and E. G. Krause. 1968. Metabolic control characteristics of the acutely ischemic myocardium. *Am. J. Cardiol.* **22**:349–359.

96. Furfine, C. S., and S. F. Velick, 1965. The acylenzyme intermediate and the kinetic mechanism of the glyceraldehyde-3-phosphate dehydrogenase reaction. *J. Biol. Chem.* **240**:844–855.

97. Trivedi, B., and W. H. Danforth. 1966. Effects of pH on the kinetics of frog muscle phosphofructokinase. *J. Biol. Chem.* **241**:4110–4114.

98. Wiedemann, M. J., and H. A. Krebs. 1969. The fuel of respiration of rat kidney cortex. *Biochem. J.* **112**:149–166.

99. Whitmer, J. T., J. A. Idell-Wenger, M. J. Rovetto, and J. R. Neely. 1978. Control of fatty acid metabolism in ischemic and hypoxic hearts. *J. Biol. Chem.* **253**:4305–4309.

100. Scheuer, J., and N. Brachfeld. 1966. Myocardial uptake and fractional distribution of palmitate-1-C^{14} by the ischemic dog heart. *Metabolism* **15**:945–954.

101. Idell-Wenger, J. A., L. W. Grotyohann, and J. R. Neely. 1978. Coenzyme A and carnitine distribution in normal and ischemic hearts. *J. Biol. Chem.* **253**:4310–4318.

102. Pande, S. V. 1973. Reversal by CoA of palmityl-CoA inhibition of long chain acyl-CoA synthetase activity. *Biochim. Biophys. Acta* **306**:15–20.

103. DeJong, W. C., and W. C. Hulsmann. 1970. A comparative study of palmityl-CoA synthetase activity in rat liver, heart, and gut mitochondrial and microsomal preparations. *Biochim. Biophys. Acta* **197**:127–135.

104. Smith, M. W., Y. Collan, M. W. Kahng, and B. F. Trump. 1980. Changes in mitochondrial lipids of rat kidney during ischemia. *Biochim. Biophys. Acta* **618**:192–201.

105. Farber, J. L., and E. E. Young. 1981. Accelerated phospholipid degradation in anoxic rat hepatocytes. *Arch. Biochem. Biophys.* **211**:312–320.

106. Bricknell, O. L., and L. H. Opie. 1978. Effects of substrate on tissue metabolic changes in the isolated rat heart during underperfusion and on release of lactate dehydrogenase and arrhythmias during reperfusion. *Circ. Res.* **43**:102–114.

107. Chua, B. H., and E. Shrago. 1977. Reversible inhibition of adenine nucleotide translocation by long chain acyl-CoA esters in bovine heart mitochondria and inverted submitochondrial particles. *J. Biol. Chem.* **252**:6711–6714.

108. Neely, J. R., and D. Feuvray. 1981. Metabolic products and myocardial ischemia. *Am. J. Pathol.* **102**:282–291.

109. Lamens, J. M. J., and W. C. Hulsmann. 1977. Inhibition of (Na$^+$-K$^+$)-stimulated ATPase of heart by fatty acids. *J. Mol. Cell. Cardiol.* **9**:343–346.

110. Katz, A. M., P. Nash-Adler, J. Micele, F. Messineo, and C. F. Louis. 1979. Inhibition of Ca influx from the sarcoplasmic reticulum by free fatty acids. *Circulation* **60**(Suppl. II):12.

111. Sheetz, M. P., R. G. Paitner, and S. J. Singer. 1976. Biological membranes as bilayer couples. III. Compensatory shape changes induced in membranes. *J. Cell Biol.* **70**:193–203.

112. Jennings, R. B. 1969. Early phase of myocardial ischemic injury and infarction. *Am. J. Cardiol.* **24**:753–765.

113. Pressman, B. C., and H. A. Lardy. 1956. Effect of surface active agents on the latent ATPase of mitochondria. *Biochim. Biophys. Acta* **21**:458–466.

114. Borst, D., J. A. Loos, E. J. Christ, and E. C. Slater. 1962. Uncoupling activity of long-chain fatty acids. *Biochim. Biophys. Acta* **62**:509–518.

115. Chefurka, W. 1966. Oxidative phosphorylation in *in vitro* aged mitochondria. I. Factors controlling the loss of the dinitrophenol-stimulated adenosine triphosphatase activity and respiratory control in mouse liver mitochondria. *Biochemistry* **5**:3887–3903.

116. Kao, R., D. E. Rannels, and H. E. Morgan. 1976. Effects of anoxia and ischemia on protein synthesis in perfused rat hearts. *Circ. Res.* **38**(Suppl. I):I124–I130.

117. Jefferson, L. S., E. B. Wolpert, K. E. Giger, and H. E. Morgan. 1971. Regulation of protein synthesis in heart muscle. III. Effect of anoxia on protein synthesis. *J. Biol. Chem.* **246**:2171–2178.

118. Chua, B., R. L. Kao, D. E. Rannels, and H. E. Morgan. 1979. Inhibition of protein degradation by anoxia and ischemia in perfused rat hearts. *J. Biol. Chem.* **254**:6617–6623.

119. Gerlach, E., B. Deuticke, and R. H. Driesbach. 1963. Zum verhalten von nucleotiden und ihren dephosphorylierten abbauprodukten in der niere bei ischamie und Kurzzeitiger postischamischer wiederdurchblutung. *Pfluegers Arch.* **278**:296–315.

120. Garlick, P. B., G. K. Radda, and P. J. Seeley. 1979. Studies of acidosis in the ischemic heart by phosphorus nuclear magnetic resonance. *Biochem. J.* **184**:547–554.

121. Kahng, M. W., I. K. Berezesky, and B. F. Trump. 1978. Metabolic and ultrastructural response of rat kidney cortex to *in vitro* ischemia. *Exp. Mol. Pathol.* **29**:183–198.

122. Katori, M., and R. M. Berne. 1966. Release of adenosine from anoxic hearts: Relationship to coronary flow. *Circ. Res.* **19**:420–425.

123. Schutz, W., J. Schrader, and E. Gerlach. 1981. Different sites of adenosine formation in the heart. *Am. J. Physiol.* **240**:4963–4970.

124. Osswald, H., H. J. Schmitz, and R. Kemper. 1977. Tissue content of adenosine, inosine and hypoxanthine in the rat kidney after ischemia and postischemic recirculation. *Pfluegers Arch.* **371**:45–49.

125. Sakai, K., M. Akima, and H. Nabata. 1979. A possible purinergic mechanism for reactive ischemia in isolated, cross-circulated rat kidney. *Jpn. J. Pharmacol.* **29**:235–242.

126. Jennings, R. B., K. A. Reimer, M. L. Hill, and S. E. Mayer. 1981. Total ischemia in dog hearts, *in vitro*. I. Comparison of high energy phosphate production, utilization, and depletion, and of adenine nucleotide catabolism in total ischemia *in vitro* vs. severe ischemia *in vivo*. *Circ. Res.* **49**:892–900.

127. Reimer, K. A., M. L. Hill, and R. B. Jennings. 1981. Prolonged depletion of ATP and of the adenosine nucleotide pool due to delayed resynthesis of adenine nucleotides following reversible myocardial ischemic injury in dogs. *J. Mol. Cell. Cardiol.* **13**:229–239.

128. Reibel, D. K., and M. J. Rovetto. 1979. Myocardial adenosine salvage rates and restoration of ATP content following ischemia. *Am. J. Physiol.* **237**:H247–H252.

129. Williamson, J. R., S. W. Schaffer, C. Ford, and B. Safer. 1976. Contribution of tissue acidosis to ischemic injury in the perfused rat heart. *Circ. Res.* **53**(Suppl. 1):i3–i14.

130. Bing, O. H. L., W. W. Brooks, and J. W. Messer. 1973. Heart muscle viability following hypoxia: Protective effect of acidosis. *Science* **180**:1297–1298.

131. Nayler, W. B., R. Ferrari, P. A. Poole-Wilson, and C. E. Yepez. 1979. A protective effect of a mild acidosis on hypoxic heart muscle. *J. Mol. Cell. Cardiol.* **11**:1053–1071.

132. Vogel, S., and N. Sperelakis. 1977. Blockade of myocardial slow inward current at low pH. *Am. J. Physiol.* **233**:C99–C103.

133. Altschuld, R. A., J. R. Hostetler, and G. P. Brierley. 1981. Response of isolated rat heart cells to hypoxia re-oxygenation, and acidosis. *Circ. Res.* **49**:307–316.

134. Sperelakis, N., and J. A. Schneider. 1976. A metabolic control mechanism for calcium ion influx that may protect the ventricular myocardial cell. *Am. J. Cardiol.* **37**:1079–1085.

135. Penttila, A., and B. F. Trump. 1974. Extracellular acidosis protects Ehrlich tumor cells and rat renal cortex against anoxic injury. *Science* **185**:272–278.

136. Hinnen, R., H. Miyamoto, and E. Racker. 1979. Ca^{2+} translocation in Ehrlich ascites tumor cells. *J. Membr. Biol.* **49**:309–324.

137. Struder, R. K., and A. B. Borle. 1979. Effect of pH on the calcium metabolism of isolated rat kidney cells. *J. Membr. Biol.* **48**:325–341.

138. Van der Saag, P. T., A. Feyen, W. Miltenburg-Vonk, and S. W. DeLatt. 1981. Plasma membrane-mediated effects of extracellular pH on the growth of neuroblastoma cells. *Exp. Cell Res.* **136**:351–358.

139. Osswald, H., H.-J. Schmitz, and R. Kemper. 1977. Tissue content of adenosine, inosine and hypoxanthine in the rat kidney after ischemia and postischemic recirculation. *Pfluegers Arch.* **371**:45–49.

140. Parce, J. W., C. C. Cunningham, and M. Waite. 1978. Mitochondrial phospholipase A_2 activity and mitochondrial aging. *Biochemistry* **17**:1634–1639.

141. Van Lancker. 1976. *Molecular and Cellular Mechanisms in Disease*. Springer-Verlag, Berlin.

142. Jennings, R. B., and C. E. Ganote. 1976. Mitochondrial structure and function in acute myocardial ischemic injury. *Circ. Res.* **38** (Suppl. 1):180–191.

143. Mittnacht, S., Jr., S. C. Sherman, and J. L. Farber. 1979. Reversal of ischemic mitochondrial dysfunction. *J. Biol. Chem.* **254**:9871–9878.

144. Mittnacht, S., Jr., and J. L. Farber. 1981. Reversal of ischemic mitochondrial dysfunction. *J. Biol. Chem.* **256**:3199–3206.

145. Rouslin, W., and R. W. Millard. 1981. Mitochondrial inner membrane enzyme defects in porcine myocardial ischemia. *Am. J. Physiol.* **240**:H308–H313.

146. Dhalla, N. S., P. K. Das, and G. P. Sharma. 1978. Subcellular basis of cardiac contractile failure. *J. Mol. Cell. Cardiol.* **10**:363–385.

147. Shen, A. C., and R. B. Jennings. 1972. Myocardial calcium and magnesium in acute ischemic injury. *Am. J. Pathol.* **67**:417–440.

148. Leaf, A. 1970. Regulation of intracellular fluid volume and disease. *Am. J. Med.* **49**:291–295.

149. Schwartz, A., G. Grupp, R. W. Millard, I. L. Grupp, D. A. Lathrop, M. A. Matlib, P. L. Vaghy, and J. R. Valle. 1981. Calcium-channel blockers: Possible mechanisms of protective effects on the ischemic myocardium. In: *New Perspectives on Calcium Antagonists*. G. B. Weiss, ed. American Physiological Society, Bethesda. pp. 191–210.

150. Meisner, H., and M. Klingenberg. 1968. Efflux of adenine nucleotides from rat liver mitochondria. *J. Biol. Chem.* **243**:3631–3639.

151. LaNoue, K. F., J. A. Watts, and C. D. Koch. 1981. Adenine nucleotide transport during cardiac ischemia. *Am. J. Physiol.* **241**:H663–H671.

152. Doleidin, F. H., S. R. Fahrenholtz, A. A. Lamola, and A. M. Trozzolo. 1974. Reactivity of cholesterol and some fatty acids toward singlet oxygen. *Photochem. Photobiol.* **20**:519–521.

153. Mengel, C. E., and H. E. Kann, Jr. 1966. Effect of *in vivo* hyperoxia of erythrocytes. III. *In vivo* peroxidation of erythrocytes lipid. *J. Clin. Invest.* **45**:1150–1157.

154. Recknagel, R. O., and E. A. Glende, Jr. 1977. Lipid peroxidation: A specific form of cellular injury. *Handbook of Physiology*, Section 9. Waverly Press/American Physiological Society, Bethesda. pp. 591–601.

155. Tappel, A. L. 1972. Vitamin E and free radical peroxidation of lipids. In: *Vitamin E and Its Role in Cellular Metabolism*. P. P. Nair and H. J. Kayden, eds. New York Academy of Sciences, New York. pp. 12–28.

156. Deneke, S. M., and B. L. Fanburg. 1980. Normobaric oxygen toxicity of the lung. *N. Engl. J. Med.* **303**:76–86.

157. Kloner, R. A., C. E. Ganote, D. Whalen, and R. B. Jennings. 1974. Effect of a transient period of ischemia on myocardial cells. II. Fine structure during the first few minutes of reflow. *Am. J. Pathol.* **74**:399–422.

158. Demopoules, H. B., E. S. Flamm, D. D. Pietronigro, and M. I. Sligman. 1980. The free radical pathology and microcirculation in the major central nervous system disorders. *Acta Physiol. Scand. Suppl.* **492**:91–119.

159. Baker, P. F., M. P. Blaustein, A. L. Hodgkin, and R. A. Steinhardt. 1969. The influence of calcium on sodium efflux in squid axons. *J. Physiol. (London)* **200**:431–458.

160. Schatzman, H. J. 1966. ATP-dependent Ca^{++}-extrusion from human red cells. *Experientia* **22**:364–365.

161. Rasmussen, H. 1981. *Calcium and cAMP as Synarchic Messengers*. Wiley–Interscience, New York.

162. McLean, A. E. M., E. McLean, and J. D. Judah. 1965. Cellular necrosis in the liver induced and modified by drugs. *Int. Rev. Exp. Pathol.* **4**:127–157.

163. Schanne, F. A. X., A. B. Kane, E. E. Young, and J. C. Farber. 1979. Calcium dependence of toxic cell death: A final common pathway. *Science* **206**:700–702.

164. Campbell, A. K., R. A. Daw, and J. P. Luzio. 1979. Rapid increase in intracellular free Ca^{2+} induced by antibody plus complement. *FEBS Lett.* **107**:55–60.

165. Campbell, A. K., and J. P. Luzio. 1981. Intracellular free calcium as a pathogen in cell damage initiated by the immune system. *Experientia* **37**:1110–1112.

166. Nayler, W. G., P. A. Poole-Wilson, and A. Williams. 1979. Hypoxia and calcium. *J. Mol. Cell. Cardiol.* **11**:683–706.

167. Haworth, R. A., D. R. Hunter, and H. A. Berkoff. 1980. Na^+ releases Ca^{2+} from liver, kidney and lung mitochondria. *FEBS Lett.* **110**:216–218.

168. Roman, I., P. Gmaj, C. Nowicka, and S. Angielski. 1979. Regulation of Ca^{2+} efflux from kidney and liver mitochondria by unsaturated fatty acids and Na^+ ions. *Eur. J. Biochem.* **102**:615–623.

169. Chien, K. R., J. Abrams, A. Serroni, J. T. Martin, and J. L. Farber. 1978. Accelerated phospholipid degradation and associated membrane dysfunction in irreversible, ischemic liver cell injury. *J. Biol. Chem.* **253**:4809–4817.

170. Coleman, R. 1973. Membrane-bound enzymes and membrane ultrastructure. *Biochim. Biophys. Acta* **300**:1–30.

171. Yin, H. L., and T. P. Stossel. 1979. Control of cytoplasmic actin gel-sol transformation by gelsolin, a calcium-dependent regulatory protein. *Nature (London)* **281**:583–586.

172. Dedman, J. R., B. R. Brinkley, and A. R. Means. 1979. Regulation of microfilaments and microtubules by calcium and cyclic AMP. *Adv. Cyclic Nucleotide Res.* **11**:131–174.

173. Smith, M. W., Y. Collan, M. W. Kahng, and B. F. Trump. 1980. Changes in mitochondrial lipids of rat kidney during ischemia. *Biochim. Biophys. Acta* **618**:192–201.

174. Quastel, M. R., G. B. Segel, and M. A. Lichtman. 1981. The effect of calcium chelation on lymphocyte monovalent cation permeability, transport, and concentration. *J. Cell. Physiol.* **107**:165–170.

175. Adelman, W. J., and J. W. Moore. 1961. Action of external divalent ion reduction on sodium movement in the squid giant axon. *J. Gen. Physiol.* **45**:93–103.

176. Delamere, N. A., and Paterson, C. A. 1978. The influence of calcium-free EGTA solution upon membrane permeability in the crystalline lens of the frog. *J. Gen. Physiol.* **71**:581–593.

177. Schulz, I., and K. Heil. 1979. Ca^{2+}-control of electrolyte permeability in plasma membrane vesicles from cat pancreas. *J. Membr. Biol.* **46**:41–70.

178. Edmondson, J. W., and N. U. Bang. 1981. Deleterious effects of calcium deprivation on freshly isolated hepatocytes. *Am. J. Physiol.* **241**:C3–C8.

179. Gordon, L. M., R. D. Sauerheber, and J. A. Esgate. 1978. Spin

label studies on rat liver and heart plasma membranes: Effects on temperature, calcium, and lanthanum on membrane fluidity. *J. Supramol. Struct.* **9**:299–326.

180. Bourdillon, P. D., and P. A. Poole-Wilson 1982. The effects of verapamil, quiescence and cardioplegia on calcium exchange and mechanical function in ischemia rabbit myocardium. *Circ. Res.* **50**:360–368.

181. Cheung, J. Y., A. Leaf, and J. V. Bonventre. 1982. Mechanism of protection by verapamil and nifedipine from anoxic injury in working isolated rat cardiac myocytes. *Clin. Res.* **30**:480a.

182. Williamson, J. R., R. H. Cooper, and J. B. Hoek. 1981. Role of calcium in the hormonal regulation of liver metabolism. *Biochim. Biophys. Acta* **639**:243–295.

183. Gemba, M., H. Obana, and Y. Matsushima. 1980. Divalent cation transport in kidney slices. III. Inhibitory action of verapamil on magnesium gain. *Jpn. J. Pharmacol.* **30**:389–391.

184. Matsushima, Y., M. Gemba, and H. Obana. 1979. Calcium transport in kidney slices: Effects of verapamil, diltiazem, and trimetazidine. *Jpn. J. Pharmacol.* **29**:A145.

185. Gordon, E. E., and R. K. Ferris. 1977. Stimulation of renal gluconeogenesis of verapamil and D-600. *Biochem. Pharmacol.* **26**:1089–1091.

186. Borle, A. B. 1981. Calcium transport by kidney cells. In: *Calcium and Phosphate Transport across Biomembranes.* F. Bronner and M. Peterlik, eds. Academic Press, New York. pp. 193–198.

187. Willerson, J. T., W. J. Powell, Jr., T. E. Guiney, J. J. Stark, C. A. Sanders, and A. Leaf. 1972. Improvement in myocardial function and coronary blood flow in ischemic myocardium after mannitol. *J. Clin. Invest.* **51**:2989–2998.

188. Powell, W. J., Jr., D. R. DiBona, J. Flores, N. Frega, and A. Leaf. 1976. Effects of hyperosmotic mannitol in reducing ischemic cell swelling minimizing myocardial necrosis. *Circulation* **53**(Suppl. 1):145–149.

189. Flores, J., D. R. DiBona, N. Frega, and A. Leaf. 1972. Cell volume regulation and ischemic tissue damage. *J. Membr. Biol.* **10**:331–343.

190. Blow, A. M. J., G. M. Botham, D. Fisher, A. H. Goodall, C. P. S. Tolcock, and J. A. Lucy. 1978. Water and calcium ions in cell fusion induced by polyethylene glycol. *FEBS Lett.* **94**:305–310.

191. Tilcock, C. P. S., and D. Fisher. 1979. Interaction of phospholipid membranes with poly(ethylene)glycols. *Biochim. Biophys. Acta* **577**:53–61.

192. Kuhn, W., and M. Thurkauf. 1958. Isotopentrennung beim Gefrieren von Wasser und Diffusionskonstantrem von D und ^{18}O im Eis. *Helv. Chim. Acta* **41**:938–971.

193. Mason, J., F. Beck, A. Dorge, R. Rick, and K. Thurau. 1981. Intracellular electrolyte composition following renal ischemia. *Kidney Int.* **20**:61–70.

194. Chien, K. R., J. Abrams, R. G. Pfau, and J. L. Farber. 1977. Prevention by chlorpromazine of ischemic liver cell death. *Am. J. Pathol.* **88**:539–558.

195. Chien, K. R., and J. L. Farber. 1977. Microsomal membrane dysfunction in ischemic rat liver cells. *Arch. Biochem. Biophys.* **180**:191–198.

196. Farber, J. L., J. T. Martin, and K. R. Chien. 1978. Irreversible ischemic cell injury: Prevention by chloropromazine of the aggregation of the intramembranous particles of rat liver plasma membranes. *Am. J. Pathol.* **92**:713–732.

197. Hearse, D. J. 1977. Reperfusion of the ischemic myocardium. *J. Mol. Cell. Cardiol.* **9**:605–616.

198. Ganote, C. E., R. Seabra-Gomes, W. G. Nayler, and R. B. Jennings. 1975. Irreversible myocardial injury in anoxic perfused rat hearts. *Am. J. Pathol.* **80**:419–450.

199. Grinwald, P. M., and W. G. Nayler. 1981. Calcium entry in the calcium paradox. *J. Mol. Cell. Cardiol.* **13**:867–880.

200. Chien, K. R., R. G. Pfau, and J. L. Farber. 1979. Ischemic myocardial cell injury: Prevention by chlorpromazine of an accelerated phospholipid degradation and associated membrane dysfunction. *Am. J. Pathol.* **97**:505–530.

201. Fridovich, I. 1978. Superoxide radicals, superoxide dismutases, and the aerobic lifestyle. *Photochem. Photobiol.* **28**:787–801.

202. Jennings, R. B., and H. K. Hawkins. 1980. Ultrastructural changes of acute myocardial ischemia. In: *Degradative Processes in Heart and Skeletal Muscle.* K. Wildenthal, ed. Elsevier/North-Holland, Amsterdam. pp. 295–346.

203. Rossi, C. S., and Lehninger, A. L. 1964. Stoichiometry of respiratory stimulation, accumulation of Ca^{++} and phosphate, and oxidative phosphorylation in rat liver mitochondria. *J. Biol. Chem.* **239**:3971–3980.

204. Vasdex, S. C., G. P. Biro, N. Narbatiz, and K. J. Kako. 1980. Membrane changes induced by early myocardial ischemia in the dog. *Can. J. Biochem.* **58**:1112–1119.

205. Grankvist, K., S. Marklund, J. Sehlin, and I. B. Taldejal. 1979. Superoxide dismutase catalase and scavengers of hydroxyl radical protect against the toxic action of alloxan on pancreatic islet cells *in vitro. Biochem. J.* **182**:17–25.

206. Haglund, U., and O. Lundgren. 1978. Intestinal ischemia and shock factors. *Fed. Proc.* **37**:2729–2733.

207. Granger, D. N., G. Ritili, and J. M. McCord. 1981. Superoxide radicals in feline intestinal ischemia. *Gastroenterology* **81**:22–29.

208. Bailie, M. B., S. R. Jolly, and B. R. Luckhesi. 1982. Reduction of myocardial ischemic injury by superoxide dismutase plus catalase. *Fed. Proc.* **41**:1736.

209. Shlafer, M., P. Kane, U. Wiggins, and M. Kirsh. 1982. Cytotoxic oxygen metabolites in pathogenesis of ischemic cardiac damage: Oxygen radical-metabolizing enzymes improve hypothermic cardioplegia. *Fed. Proc.* **41**:1765.

210. Roy, R. S., and J. M. McCord. 1982. Ischemia-induced conversion of xanthine dehydrogenase to xanthine oxidase. *Fed. Proc.* **41**:767.

CHAPTER 2

Genetic Variants Affecting the Structure and Function of the Human Red Cell Membrane

John C. Parker and Lee R. Berkowitz

1. Introduction

Investigators interested in the plasma membrane have long appreciated the red cell as an object of study. Red cells can be obtained in abundance and easily freed of contamination by other cell types. Mammalian red cells have no membranes other than the plasmalemma, which is easily isolated and purified. The "extracellular space" problem that plagues studies of cell solute and water content is virtually nonexistent. The techniques of making "resealed ghosts" and "inside-out vesicles" have provided ingenious approaches to problems that involve asymmetrical membrane properties. Our understanding of the ultrastructure and organization of lipids and proteins in biological membranes derives largely from the study of human erythrocytes. Much of what we know about the state of water and solutes in cytoplasm, the Na^+,K^+ pump, the Ca^{2+} pump, and some coupled, passive transport systems is based on work in red cell preparations.

The focus of this review is on inherited conditions that in one way or another affect the red cell membrane. Some of the topics to be discussed deal with conditions that cause shortened red cell survival; other sections discuss systemic illnesses in which abnormalities in the red cell membrane have been claimed to represent altered function in membranes of other cell types, e.g., nerve, skeletal muscle, or the media of arterioles. We have tried to focus on conditions that reveal something of the structure and function of the normal red cell membrane. Thus, we have omitted discussing glycolytic enzyme defects[452] and other inherited red cell disorders, such as the Heinz body anemias, in which the pathogenesis of membrane involvement is either not

understood or is not clearly related to the properties of normal cells.

Large portions of this paper appeared in *Physiological Reviews*.[R29] We are indebted to the editorial staff of that journal and to the Publications Committee of the American Physiological Society for permission to use the ideas, the figure, and many of the tables from that article in the present context.

2. Intrinsic Membrane Abnormalities

The past decade was important in advancing our understanding of the molecular architecture of the red cell membrane. Investigators from many disciplines have been stimulated to correlate the new information about membrane structure with ideas regarding solute and water transport, surface antigens, cell shape, cell mechanical properties, and the determinants of red cell survival and aging. Some current views of these structure–function relationships have derived from work on genetic variants. The burgeoning literature on red cell membrane composition and organization has been the subject of several excellent reviews.[58,187,280,281,283,289] We briefly summarize this field, concentrating on features related to inherited traits.

Current models of membrane structure begin with the concept that the major permeability barrier between the cytoplasm and the exterior is a lipid sheet two molecules thick.[107,186] Each layer of the sheet consists of phospholipid molecules packed tightly together with their polar or charged ends facing the aqueous phases on either side of the membrane, while the hydrocarbon fatty acid moieties point toward the midplane at the junction of the two laminae. Several classes of phospholipids are represented in the membrane and are distributed asymmetrically.[290,386,454] The external leaf of the bilayer is relatively rich in phosphatidylcholine, sphingomyelins, and glycolipids, whereas the inner half contains most of the phos-

John C. Parker and Lee R. Berkowitz • Department of Medicine, University of North Carolina, School of Medicine, Chapel Hill, North Carolina 27514.

19

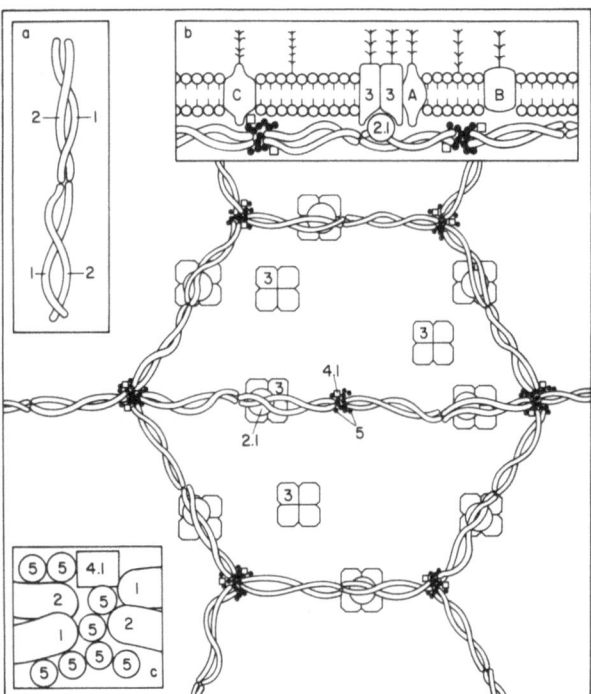

Fig. 1. Relationships among lipid bilayer, integral membrane proteins, and membrane skeleton (not to scale). Lattice on inner membrane surface is arbitrarily drawn in hexagonal array, with cytoplasmic portions of some band 3 tetramers linked via ankyrin (band 2.1) to spectrin tetramers (bands 1 and 2). Other band 3 tetramers are not connected to membrane skeleton. Inset a: spectrin bands 1 and 2 bound side to side to form dimers that in turn join end to end to form a tetramer. Band 2 is phosphorylated at the COOH-terminal end[201] on head of molecule at dimer–dimer binding site.[282,320,R34] Inset b: edge-on view of membrane featuring linkages among spectrin, ankyrin (band 2.1), and band 3. Glycophorin C ("glycoconnectin" or SGβ) shown interacting with band 4.1.[324] Glycophorin A (SGα) is shown closely related to band 3.[335] Branched structures on certain phospholipid molecules, band 3, and glycophorins represent carbohydrate chains. Inset c: spectrin tetramers associated at tail ends with actin (band 5) and band 4.1.

phatidylserine, phosphatidylethanolamine, and phosphatidylinositol. Cholesterol is present in the bilayer in a 1 : 1 molar ratio with phospholipids. Proteins comprise about half the membrane mass and are divided into two classes, integral and peripheral.[417] Integral membrane proteins are inserted in the bilayer and retain their position there by virtue of strongly hydrophobic amino acid sequences. Some integral proteins pass through the bimolecular lipid sheet and are in contact with the aqueous phases on either side. Peripheral membrane proteins comprise a reticular network on the inner or cytoplasmic face of the membrane; although they may be associated with integral membrane proteins or lipids, they are not thought to be hydrophobically bound within the bilayer itself.

Some features of membrane organization are shown in Fig. 1. The proteins of the membrane are designated by their conventional numbers, reflecting their relative mobilities on SDS–polyacrylamide gel electrophoresis.[423] The most external structures of the membrane are branched polysaccharide chains that are covalently linked either to phospholipids or to integral membrane proteins.[22,198] Sialic acid residues in these

chains confer a negative surface charge on the cell that is referred to as the zeta potential.[139] Most of the antigenic diversity of red cells is due to genetic variations in the carbohydrate and amino acid sequences of the glycolipids and proteins on the outer membrane surface. The ABH and Ii antigens, for example, are identified with carbohydrate chains located on band 3 and glycolipids.[149,198] The MN and Ss systems reflect differences in amino acid sequences on the NH_2-terminal, externally located ends of integral membrane sialoglycoproteins known as glycophorins[22] (see Table III). The Rh system is partly associated with the band 3 protein.[457]

Band 3 (M_r 93,000), the most abundant integral membrane protein, spans the bilayer and performs the important function of anion exchange. Current views of band 3 include a number of concepts. (1) Unlike the glycophorins, the NH_2-terminal end of band 3 is internal. (2) It exists in the membrane as a tetramer. (3) Of the 1 million copies of this molecule in each cell, 20–40% are tethered to the network of integral membrane proteins at the inner membrane surface.[90,200,244,282,468] Among other integral proteins in the red cell membrane are the Na^+,K^+ pump,[154,245] the Ca^{2+} pump,[246] and the glucose transporter.[230,236] These comprise such a small proportion of the membrane mass that their ultrastructural relationships have not yet been characterized as well as those of band 3.

The major mass of the membrane skeleton is composed of spectrin, or bands 1 and 2 (M_r 240,000 and 220,000, respectively) (Fig. 1).[289] Bands 1 and 2 are both elongated, worm-shaped structures about 1000 Å long and 50 Å wide.[416] In the membrane, one band 1 molecule binds side to side with one band 2 molecule to form a linear heterodimer.[322] Two such heterodimers join head to head to form a tetramer roughly 2000 Å long. There are 100,000 spectrin tetramers per cell. The tetramers can become associated with other tetramers (tail to tail) in the presence of band 4.1 (M_r 78,000) and oligomers of band 5 (M_r 60,000), a protein similar to muscle actin.[60,449] At a point 200–250 Å from the midpoint of the tetramer there is a binding site for band 2.1 (M_r 180,000–200,000) or ankyrin, which in turn binds to the cytoplasmic domain of band 3.[35,232,449,R4] In this way, 20–40% of band 3 tetramers are thought to be linked to the membrane skeleton, since there are only about 100,000 ankyrin molecules per cell. Band 3 has associations with other proteins, such as glyceraldehyde-3-phosphate dehydrogenase, band 4.2, aldolase,[175,243,446,494] and possibly hemoglobin[135,395] which are not depicted in Fig. 1. A second point of articulation of the membrane skeleton with integral proteins of the bilayer is the suggested association between glycophorin C ["glycoconnectin," periodic acid–Schiff (PAS), 2, sialoglycoprotein β (SGβ)] and band 4.1 at the junction of the spectrin tetramers; the evidence for this is preliminary.[324] Not represented in Fig. 1 are two recent suggestions about the configuration of membrane skeletal proteins: Spectrin can form oligomers containing more than four molecules[R24]; and actin polymers may arrange themselves in long, linear arrays parallel to the membrane surface.[R2]

The membrane skeleton thus interacts with the bilayer and its contents; the lateral mobility in the plane of the membrane of integral proteins, such as band 3, is restrained by the connections with the spectrin lattice on the inner surface.[336] Moreover, the asymmetry of the phospholipids in the bilayer may partly depend on the integrity of spectrin.[196,197]

The shape and mechanical properties of the cell are thought to be determined by the microarchitecture of the membrane and its skeleton. It is possible to prepare hemoglobin-free membranes of red cells that retain the normal biconcave disk appearance. But when the skeletal proteins are dissected away from such membranes by treatment in solutions of low ionic strength, the bilayer lipids and integral proteins fragment into small, spherical vesicles.[423] Conversely, when cells or ghosts are extracted with a nonionic detergent such as Triton X-100, the bilayer lipids and integral proteins become solubilized and the membrane skeletal assembly is left intact, retaining the configuration of the original cells.[493] Studies of stress–strain relationships suggest that the viscoelastic and viscoplastic properties of the cell are influenced by the membrane skeleton. Relatively small deformations of the cell over short time periods are completely reversible, whereas prolonged or extreme distortions result in permanent morphological changes.[R19] Such plastic behavior would not be expected of a simple bilayer and is thought to reflect the properties of the underlying skeletal assembly that can be molded or rearranged by the application of stress over time.[138,254,321] What determines normal cell shape is disputable. The asymmetry of lipids in the bilayer has been mentioned in both as a cause and an effect of the membrane curvature.[196,338,413] There is evidence, for example, that cholesterol may be heterogeneously distributed in the dimple and edge regions.[328] The same debate about cause and effect may be raised in regard to the membrane skeletal proteins that retain the shape of the cells from which they were prepared. The concept that metabolism or energy is somehow involved in maintenance of cell shape was proposed when partial ATP depletion was found to cause a reversible shape change.[329] The discovery that some membrane proteins, especially band 2, are phosphorylated by kinases in the cell suggested a role for ATP in determining membrane shape. This concept, however, has recently been questioned, and thus the role of ester-bonded phosphates on membrane proteins remains unclear.[143,356,R1]

Although many genetic variations in the red cell membrane have been described, very few are understood in molecular detail. In some instances it is only known that the shape or the permeability of the membrane is abnormal, but the specific structural component of the membrane responsible for this deviation is often not identified. Thus, it is difficult to construct a meaningful classification of these phenotypic entities. The discussion is organized according to the salient phenomenological characteristics that have focused attention on atypical cells.

2.1. Disorders of Red Cell Shape

Among the most instructive membrane defects is a heterogeneous group of conditions characterized by elongated and sometimes fragmented red cells in the peripheral blood. These conditions include elliptocytosis, ovalocytosis, poikilocytosis, and pyropoikilocytosis.[283] All are congenital, but the modes of inheritance are not equally clear; some may be fresh mutations. One form of hereditary elliptocytosis is transmitted as an autosomal dominant with at least one genetic determinant on chromosome 1 near the Rh locus.[93] There is a wide range in the clinical severity of these disorders: some propositi have a normal red cell life span, whereas others may have a lethal hemolytic anemia. Abnormal cell shape ties this group together; results from several laboratories indicate that the shape abnormality is sometimes traceable to specific alterations in the membrane skeleton.[90,282] The persistence of shape abnormalities, not only in hemoglobin-free ghosts but also in Triton-extracted cells or membranes in which all but the skeletal network proteins have

been dissolved, provides circumstantial evidence for this idea.[281,283] Normal red cells fragment when heated to 49–50°C, but this tendency is exaggerated in some of these variants. In pyropoikilocytosis the cells are particularly vulnerable to heat, and fragmentation occurs at 45–46°C or sometimes with prolonged incubation at body temperature in vitro.[350,499,501] The observation that purified spectrin from heat-sensitive cells also has an abnormal tendency to heat denaturation indicates that heat instability is caused by an abnormality of membrane proteins.[281,283] In some variants of this syndrome, whether heat sensitive or not, there appears to be a defect in the spectrin structure itself that impedes the normal association of heterodimers into tetramers.[265–267,R22] Abnormal tryptic fragments found in the spectrin of certain patients with this syndrome seem to support this possibility.[89,257] This abnormality would interfere with the ability of the membrane skeleton to form a cohesively linked lattice at the inner surface (see Fig. 1). In another variant of the elliptocytosis–poikilocytosis syndrome, Tchernia et al.[148,434] suggest a clear-cut absence of band 4.1 in three sisters, whereas the parents and another sister have about half-normal amounts of this protein. The red cell structure of the homozygotes was grossly abnormal and spherocytic, whereas in the heterozygotes only mild elliptocytosis was seen. The observations on this family firmly support other evidence that band 4.1 is a key part of the membrane skeletal assembly. The red cells of an unrelated patient with elliptocytosis and band 4.1 deficiency had increased extractability of glycophorin C by Triton X-100. This observation supports the notion that band 4.1 and glycophorin C interact to form part of the linkage between the bilayer and skeletal proteins.[324]

Quite a different story has evolved from the investigations of Agre et al.[7] of two unrelated families with a variant of the elliptocytosis syndrome. In the affected individuals the lesion has been identified as a partial deficiency of high-affinity ankyrin-binding sites on the inner membrane surface. It is not certain whether the underlying abnormality is in the cytoplasmic portion of band 3 to which ankyrin binds, but the studies of these patients reinforce the concept that ankyrin binding plays a stabilizing role in the maintenance of normal red cell shape. The finding of normal anion fluxes in these cells is evidence of the integrity of the transport function of band 3, despite the possibility of an abnormal structural relationship in its cytoplasmic or ankyrin-binding segment.

Phenomenological observations on these cells (i.e., mechanical instability, rigidity, leakiness to Ca^{2+} and other cations, heat sensitivity, and reduced survival in the circulation) probably reflect membrane skeletal abnormalities. Inasmuch as the red cells in these disorders appear to acquire their abnormal shape after they leave the marrow,[401] one can speculate that the deformation they encounter in the microvasculature is less reversible than in normal cells and that they become molded into the configurations observed in peripheral blood because of altered viscoelastic properties of the membrane skeletal structure. Many hereditary elliptocytes are normal in terms of the findings described, and further studies will undoubtedly disclose other specific defects in the array of proteins that function together to stabilize the shape of the red cell.

Hereditary spherocytosis (HS) is a relatively common heterogeneous condition identifiable by dense, round cells in the peripheral blood and by varying degrees of hemolysis and anemia.[283] The illness is ameliorated but not cured by splenectomy.[80,475] Inheritance is autosomal dominant in most cases, and genetic determinants may be on autosome 8 or 12.[240] There is no agreement about the molecular defect in this disease, although most investigators believe HS represents an alteration in the structure of the red cell membrane.[451,495] The circulating red cells, and possibly the reticulocytes, of HS patients have normal volume but a reduced surface area and low lipid content per cell.[97,373] The round shape and the abnormally high intracellular hemoglobin concentration render the cells poorly deformable.[330] *In vitro* the cells show an increased Na^+ permeability[40,476] and an increased tendency to lose surface lipids on metabolic depletion.[94,373]

Beyond these observations there is much controversy.[451,495] The following ideas are some that have been asserted and refuted: (1) HS cells show reduced phosphorylation of membrane proteins[11,42,49,50,189,298,437,438,488]; (2) fatty acids in the HS membrane lipids are abnormal[228,250,268,496]; (3) microviscosity of the membrane is abnormal[18,95,229]; and (4) active Ca^{2+} transport is diminished in HS cells.[497] Most patients with HS show a normal pattern of membrane proteins on SDS–polyacrylamide gel electrophoresis, but two families with reduced band 4.2 have been found.[202,339] Qualitative membrane protein abnormalities relating to solubility,[137,222] extractability,[205,411] and ATPase activity[241] have been reported, but the relationship of these observations to current models of membrane ultrastructure is not clear. Recent reports from two independent laboratories show that the red cells of some people with HS have a lesion on one of the spectrin chains that results in a decreased binding affinity between spectrin and band 4.1.[R15,R39]

A well-characterized autosomal-recessive spherocytic hemolytic anemia has been described in mice with homozygotes showing absence of both bands of spectrin in SDS–polyacrylamide gel separations of red cell membrane proteins. When normal mouse spectrin is incorporated into the spherocytes by hypotonic hemolysis and resealing, the deficient ghosts acquire some of the properties of normal mouse red cells. Although the spherocytic shape persists, the tendency of the membrane to fragment and fuse in lysolecithin-containing hypertonic media is reversed, and the time course of cell swelling in response to hypotonic stress is normalized.[414] These observations support the idea that spectrin may play a role in stabilizing the membrane. It would be interesting to see whether the increased lateral mobility of integral membrane proteins in mouse spherocytes[412] is normalized by spectrin incorporation. A human family with a recessively transmitted variant of HS has recently been described in which the homozygote has a deficiency of both bands of spectrin. Physiological studies on these cells should give further insight into the consequences of spectrin deficiency.[6]

The underlying defect in the common variety of HS is still under investigation, but studies of cation transport in these patients' cells suggest that the determinants for active and passive Na^+ permeability in the red cell may be genetically linked. This interesting observation is discussed in connection with the red cell permeability variants. Comprehensive and up-to-date reviews of red cell membrane skeletal disorders authored by workers in the field were collected and edited by Palek.[R28]

2.2. Cation Transport Defects

An apparent primary derangement of red cell Na^+–K^+ transport resulting in varying degrees of hemolytic anemia has been reported in about 25 patients.[9,44,45,78,86,128,177,213,224, 253,270,308,312,344,365,406,415,481,482,502] Freshly drawn red cells

from these individuals are high in Na^+ and low in K^+. In some variants the increment in Na^+ exceeds the deficit in K^+, and the cells are consequently overhydrated (hydrocytosis). In other instances the sum of $Na^+ + K^+$ content is normal[213] or low,[9,86,177,365,502] and cell water is variably decreased (xerocytosis). These syndromes are often called hereditary stomatocytosis because of the mouth-shaped central pallor of the red cells on dried blood smears, although not all the variants have this morphological feature. This subject has been reviewed in detail.[180,283,331,355,477]

The functional defect in all these cases appears to be an increase in the passive permeability of the cells to Na^+ and K^+. The Na^+,K^+ pump rate is increased in every case, partly because of a rise in cell Na^+ concentration[344] and partly because of an increased number of pumps per cell.[482] The fate of the red cells *in vivo* depends somewhat on the relative magnitude of the Na^+ and K^+ permeabilities plus the pump's ability to compensate for the increased passive ion movements. Hydrocytes gain cations and water as they circulate in the bloodstream,[344] whereas in xerocytes the normal tendency of red cells to lose K^+ and H_2O with age is accelerated.[206] Other factors may determine red cell survival in these patients, since a close correlation between cation permeability and severity of hemolysis is not always observed.

Passive net movements of monovalent cations in red cells do not occur exclusively by simple diffusion.[R12] There are various downhill transport mechanisms for Na^+ and K^+ that, if increased, might account for the abnormal steady-state ion and H_2O contents in these red cell variants (Tables I and VII). In most cases no attempt has been made to analyze the various permeability routes that might be altered. In contrast, Wiley[477] found in a patient with xerocytosis that much of the abnormal passive Na^+ and K^+ flux was both saturable and inhibited by furosemide; the residual or diffusion permeability was also increased. Chailley *et al.*[78] studied a family of stomatocytosis patients whose cells showed a disproportionate increase of furosemide-sensitive Na^+ flux that was interpreted to represent a departure from the 1 Na^+ : 1 K^+ stoichiometry characteristic of the Na^+–K^+ cotransport pathway in normal red cells. Wiley[477] studied a patient with hydrocytosis in whom furosemide-sensitive K^+ flux was completely absent from red cells, thus suggesting that the massive increase in K^+ permeability was via an electrodiffusion pathway.

It has long been known that increased concentration of Ca^{2+} in the red cell may cause a large and selective increase in K^+ permeability.[263] Wiley[477] and Mentzer *et al.*[306] have considered whether the relatively high K^+ leak in xerocytes might be due to an increase in cell Ca^{2+} and have concluded that this is unlikely. Ca^{2+}-mediated K^+ leak does appear to play a role in the K^+ loss observed in red cells from individuals with pyruvate kinase deficiency.[176,249,305] On the other hand, in a patient with hemolytic anemia with massively increased Ca^{2+} permeability, cell Na^+ and K^+ contents were normal[477,483]

A variety of 1 : 1 exchange pathways can be demonstrated by measuring the fluxes of radioactive cations across the red cell membrane. Some K^+/K^+ exchange occurs under normal circumstances,[181] and Na^+/Na^+ exchange can also be measured (see Table VII).[181] These fluxes do not cause a change in the Na^+ or K^+ content of the cells, since neither pathway conducts net movements across the membrane. Nevertheless, in some of the discussed syndromes, an increase in Na^+/Na^+ exchange has been reported.[86,128,312]

The increase in pump flux in some of these disorders is

Table I. Pathways Capable of Net Passive Na$^+$ and/or K$^+$ Transport across the Human Red Cell Membrane

Pathway	Pharmacological inhibitors	Alterations in disease
KCl cotransport[127]a	Furosemide	
Na$^+$-K$^+$ cotransport[480]a	Furosemide	Hereditary stomatocytosis[477]
Ca^{2+}-dependent K$^+$ leak[263]	Quinidine, cetiedil[37]	Pyruvate kinase deficiency[305]
Heinz body-induced K$^+$ leak[343]a	Furosemide[341]	? Thalassemia[192]
"Ground" Na$^+$ and K$^+$ leaks, electrodiffusion		Hereditary stomatocytosis[477]

aCl$^-$-dependent pathways; whether these pathways involve Cl$^-$ cotransport is not clear.[83,127,341]

greater than can be accounted for by either the increased cell Na$^+$ concentration or the relative youth of the cell population; the number of pumps per cell is apparently increased.[344,482] These observations led Wiley[477] to hypothesize that the genes for pump and leak may be linked, so that a mutation leading to a greater passive Na$^+$–K$^+$ permeability may, *pari passu*, cause a higher density of active-transport sites. Wiley[476] concluded from extensive observations on patients with HS that the patients who had the greatest values for Na$^+$ influx also had the highest membrane Na$^+$,K$^+$-ATPase activities. Wiley also showed that this had nothing to do with the rate of red cell destruction *in vivo*. Cumberbatch and Morgan[99] recently observed the same relationship between active and passive Na$^+$ movements in normal individuals and speculated that the red cell contains "pump-leak units" that may be coded for by closely associated genes.

The structural basis for these permeability aberrations remains obscure. Many stomatocytosis patients have an increase in total lipid per cell, and in some the proportion of phosphatidylcholine is elevated.[224,344,415,477,502] Red cells of most patients have a normal protein pattern on SDS–polyacrylamide gels, although a deficiency in band 7 was observed in one patient.[306,R21] Bienzle et al.[44,45] discovered an abnormal protein (M_r 25,000) by two-dimensional electrophoresis. Banga et al.[31] reported a missing protein (M_r 29,000) in the red cell membranes of a patient with exercise-induced hemolysis; this patient may have a variant of the xerocytosis syndrome.[306,R21] A nonspecific but large decrease in membrane protein phosphorylation was observed in a patient with hydrocytosis.[308] Exhaustive and careful work in a group of patients with xerocytosis showed a number of minor lipid and protein alterations that reflected cell immaturity, since the same changes were observed in the red cells of other patients with high reticulocyte counts.[142,399,400,420] Grossly high Na$^+$ and low K$^+$ contents in the red cells of spectrin-deficient mice are evidence that ultrastructural lesions can influence Na$^+$ and K$^+$ movements.[467] Preliminary studies indicate that these spherocytes are leaky to cations (C. H. Joiner, personal communication). Further evidence that a structural basis exists for the permeability alterations in human hereditary stomatocytosis is provided by the remarkable observation of Mentzer et al.[306,307] that virtually all the transport abnormalities in a patient with hydrocytosis were normalized by treating the cells with a variety of cross-linking agents, the most effective of which was dimethyl adipimidate. Unfortunately, no characteristic association of membrane constituents was observed in the patient's cells that would give a clue to the site of the basic molecular defect.[306]

An apparently inherited variation in the structure of band 3, the principal anion transport protein, has been reported.[323]

About 6–7% of normal humans have a slightly longer intracytoplasmic NH$_2$-terminal segment of this membrane constituent, but their red cell anion transport is not affected. If this protein were functionally deficient, the patient might have a chronic respiratory alkalosis, because the removal of CO$_2$ from the respiratory control center would be impaired.[474]

The influence of cell age on various properties of the cell pervades this discussion. Young red cells may have mitochondria, they have an increased surface area, and both passive and active transport pathways may be increased. Most patients with hemolytic disease have a high proportion of young red cells in the circulation. Investigators have usually been careful to control for this consideration. In a recent essay, Wiley[479] points out that the normal shift of surface antigens of the Ii system may provide a useful adjunctive index of cell age in connection with studies on patients with accelerated or ineffective erythropoiesis.

2.3. Surface-Antigen Variants

A number of interesting red cell variants have been identified by their unusual antigenic expression. Although none of these provide detailed information regarding structure-function relationships, they will probably prove instructive when they are better understood. Three variants are particularly interesting.

2.3.1. McLeod Erythrocyte

The Kell system of antigens on the surface of red cells, neutrophils, and monocytes appears to be coded for by two genes, one on the X chromosome and the other on an autosome. Full expression of Kell antigens requires the products of both genes to be on the cell surface. The antigen coded for on the X chromosome is called Kx, and the gene that determines Kx is called *Xk*. There are four known alleles of *Xk*; two cause an absence of Kx on the red cell. Red cells lacking the Kx antigen are called McLeod cells[17] and have a number of abnormal features, including reduced survival *in vitro*.[293,427,473] McLeod cells tend to assume an acanthocytic shape with irregular excrescences or protuberances that are easily recognized on blood smears.[163,427,473] Two other abnormalities appear to be linked to the *Xk* gene: one is an inherited defect of neutrophilic leukocytes called chronic granulomatous disease (CGD) that renders the cells incapable of killing certain types of ingested bacteria,[294] and the other is a mild form of muscular dystrophy characterized by a high level of creatine kinase (CK) in plasma and abnormal skeletal muscle histology.[292] The relationships among these *Xk*-associated conditions are shown in Table II. Not

Table II. Variants of the *Xk* Gene[a]

| *Xk* allele | Kx antigen | | Phenotype |
	Red cells	Neutrophils	
X1k	+	+	Normal
X2k	−	−	McLeod RBC, CGD, normal CK
X3k	+	−	Normal RBC, CGD, normal CK
X4k	−	+	McLeod RBC, no CGD, high CK

[a]CGD, chronic granulomatous disease; CK, creatine kinase; +, presence of Kx antigen; −, absence of Kx antigen.

Table III. Nomenclature of Sialoglycoproteins or Glycophorins and Their Presence or Absence[a]

Nomenclature				
Sialoglycoprotein (SG)	Sgα	SGδ	SGβ	SGγ
Glycophorin (Gp)	GpA	GpB	GpC	
Genetic variants				
Normal	+	+	+	+
En(a⁻)	−	+	+	+
MkMk	−	−	+	+

[a]+, presence of protein in En(a⁻) and MkMk variants; −, absence of protein in En(a⁻) and MkMk variants.

all patients with CGD have an absence of Kx on their neutrophils; indeed, not all forms of CGD are X-linked. Although Duchenne's dystrophy is X-linked and associated with a high CK and variably abnormal red cells, not one person with this serious illness has been found to have McLeod red cells. Males with McLeod red cells have 30–60% acanthocytes in their blood and a chronic hemolytic anemia. Their mothers have fewer acanthocytes and normal red cell survival, suggesting inactivation of a portion of their abnormal *Xk*-bearing chromosomes by the Lyon effect, resulting in two populations of cells, some normal and some with the McLeod defect.[473]

In addition to their abnormal shapes, McLeod cells have reduced osmotic water permeability, normal Na^+–K^+ transport, and normal glycolytic enzymes.[163] Their lipid content is normal in some experiments and reduced in others but with normal proportions of various phospholipid classes.[163,427] Because acanthocytes are also found in inherited disorders of plasma lipoproteins (abetalipoproteinemia), the possibility that McLeod cells result from an abnormality in the plasma was tested and found unlikely.[163] The membrane proteins of McLeod cells have a normal pattern on SDS–polyacrylamide gel electrophoresis, but studies with ghosts and intact cells show a striking increase in labeling with radioactive phosphate in both spectrin and band 3,[431] although the controls for these studies did not have a comparably young red cell population. The precise glycoprotein or glycolipid that bears the Kx antigen has not yet been determined, but evidence suggests that this genetic variant involves a structural element that determines red cell shape, water permeability, and perhaps metabolism of membrane phosphorus without affecting the Na^+,K^+ pump. A striking suggestion about McLeod cells is that their band 3 molecules may exist in the membrane as dimers rather than tetramers.[R14]

2.3.2. En(a⁻) and MkMk Cells

The sialoglycoproteins comprise about 2% of the red cell membrane proteins and are identified on SDS–polyacrylamide gels by PAS staining. The four species of sialoglycoproteins (α, β, γ, δ) are rich in carbohydrates and contain 70–80% of the sialic acid groups on the red cell surface. Several antigen systems are carried by the sialoglycoproteins, among which the most prominent are MN and Ss. The nomenclature of these membrane constituents varies among investigators and is complicated because one protein may be present in two or more electrophoretic bands due to the formation of homo- and heterodimers. Some features of the four sialoglycoproteins are given in Table III.[22,158]

Sialoglycoprotein α (SGα) is the most abundant (0.5–2 million copies per cell) and most studied of the group; its prima-

ry structure is known.[440] The NH_2-terminal segment is rich in carbohydrate and located on the outer cell surface, where it is recognized as the M or N antigen depending on the amino acids at positions 1 and 5. A hydrophobic portion of the molecule passes through the membrane, and the COOH-terminal end constitutes a cytoplasmic domain. Sialoglycoprotein is tentatively thought to be associated in the membrane with band 3, the anion transport protein.[335] SGδ contains the same NH_2-terminal 26-amino-acid sequence as the N-antigenic form of SGα but thereafter diverges to express the Ss and U antigens.[158] The genes for SGα and SGδ are apparently located close together on autosome 4.[53] The minor sialoglycoproteins, SGβ and SGγ, are not as well characterized. Mueller and Morrison[324] recently suggested that SGβ (glycophorin C or "glycoconnectin") participates in the linkage of integral and peripheral membrane proteins.

Individuals with deficiencies of one or more red cell sialoglycoproteins have been found. The En(a⁻) cells, first recognized because of their lack of MN antigens, have no SGα.[159,161,432] The MkMk cells with no MN or Ss reactivity lack both SGα and SGδ.[23,217,439] The existence of these recessively transmitted variants offers an opportunity to investigate the structural and physiological role of the sialoglycoproteins. Both En(a⁻) and MkMk red cells have 30–45% of the normal sialic acid content, with a consequent reduction in surface charge.[139,161,439] Surprisingly, no hematological abnormality has been discovered in these individuals. The red cells look normal and appear to have a normal membrane structure in freeze–fracture electron micrographs.[29] Their osmotic fragility is normal.[439] Band 3 is more heavily glycosylated than normal in both variants,[433,439] but there is no clinical evidence of altered Cl^-/HCO_3^- exchange. The distribution of phospholipids in the membrane of En(a⁻) cells is normal.[455] These observations led to reconsideration of the role of sialic acid loss in the process of red cell aging. There is evidence that as red cells come to the end of their life span they lose 10–15% of their sialic acid.[91,279] Enzymatic cleavage of sialic acid from red cells causes them to be removed rapidly from the circulation.[225] A "physiological autoantibody" believed to be involved in the removal of senescent red cells binds to both old and enzymatically desiala red cells.[237] The normal function and survival of En(a⁻) and MkMk cells, however, suggest that simple quantitative sialic acid deficiency must not be relevant to the process of red cell aging and removal from the circulation. Because the sialoglycoproteins are integral membrane proteins and because at least one of them, SGα, spans the bilayer, they might affect transport. We have found no published studies of solute and water movements in these cells. The lack of impairment in survival *in vivo* does not necessarily mean that some functions of the cell might not be instructively altered in En(a⁻) or MkMk erythrocytes. A possible example is the recent observation that cells lacking glycophorin A are resistant to infection by *Plasmodium falciparum*.[R30,R36]

2.3.3. Rh Null Cells

Individuals who lack all Rh-antigenic activity on their red cells (Rh null) have chronic hemolytic anemia.[425] The Rh complex comigrates with band 3[457] and other, smaller proteins.[318] It can be modified by SH-reactive compounds and requires phospholipid for full antigenic expression.[188,R13] The Rh sites (10,000–30,000 per cell[256,260]) are distributed over the surface of the red cell membrane in a totally random pattern[333] or in couplets.[248] Cross-linking studies suggest that the sites normally have considerable mobility in the plane of the membrane.[458] Some features of Rh null cells include stomato- and spherocytic morphology[425]; altered expression of Ss, U, and Fy antigens[260,425]; increased osmotic fragility and autohemolysis[425]; normal glycolytic enzymes[425]; normal membrane proteins on SDS–polyacrylamide gel[418]; normal proportions of phospholipids with minor alterations in fatty acid composition[419,425]; and an altered response to treatment with phospholipases.[418] An interesting finding is the increased active and passive Na^+ and K^+ fluxes, with increased Na^+,K^+-ATPase activity and ouabain binding,[256,259,R3] although these observations have been criticized on the grounds that they may reflect nothing more than an immature red cell population as revealed by the increased proportion of i to I antigens.[478] The Rh complex is coded for by a gene on autosome 1 near the locus for a variant of hereditary elliptocytosis.[93] Individuals with Rh null cells inherit a null gene from each of their parents, but whether Rh null cells lack the gene product or whether they possess it in altered or disguised form[364] is not known. Clearly, more must be known about the variant before convincing conclusions can be drawn about structure–function relationships of these abnormal cells.

3. Hemoglobinopathies

3.1. Sickle-Cell Disease

The formation of hemoglobin S is caused by an abnormal nucleotide sequence on autosome 11, which results in the substitution of valine for glutamate at position 6 from the NH_2-terminus of the β-globin chain.[235,258] The red cells of homozygotes contain only S β-chains, whereas the cells of heterozygotes contain about 35% of this abnormal gene product. With deoxygenation, hemoglobin S polymerizes to form a gel, resulting in gross deformation of the red cell, hemolysis, and occlusion of small blood vessels.[108,337] Alterations in the cell membrane occur with sickling, which may be important in the pathogenesis of the disease and may also provide information about normal membrane characteristics. For this discussion we consider these phenomena under three headings: (1) cell volume homeostasis; (2) cell Ca homeostasis; and (3) membrane organization. This separation is obviously artificial, inasmuch as the three areas are so closely related.

Throughout this section an entity known as the irreversibly sickled cell (ISC) is referred to frequently. ISCs, once deformed by deoxygenation, fail to return to a normal biconcave shape on reexposure to oxygen.[108] Some ISCs are found in

the blood of most patients with sickle-cell disease and can be isolated and quantified by virtue of their increased density.[R10] Cells similar to ISCs can be produced *in vitro* by prolonged incubation under various conditions.[41,85] The role of the ISC in the pathogenesis of the symptoms and signs of sickle-cell disease is disputed.[81,203,407,408] Note that ISCs retain their distorted shape even though the hemoglobin within them is depolymerized. The evidence suggests that irreversible shape change is due to alterations in structure of the cell membrane, specifically in the arrangement of its skeletal or peripheral proteins.

3.1.1. Cell Volume Homeostasis

Perutz and Mitchison[359] were the first to describe a potential abnormality in the water content of the sickle erythrocyte. From calculations of hemoglobin S crystallization they predicted that sickling would be associated with a 20% loss of cell water. Since their prediction, numerous attempts at measuring the water content of the sickle erythrocyte have yielded conflicting results. Tosteson *et al.*[444] showed that, during 30 min of deoxygenation, sickle erythrocytes lose K^+ and gain Na^+ without a change in water content. These findings have been confirmed.[144,179,252] Masys *et al.*[295] however, using radiolabeled albumin as a plasma marker, showed dilution of the albumin when sickle erythrocytes were deoxygenated, thus suggesting water loss from cells. Hargens *et al.*[199] showed that on deoxygenation, concentrated solutions of hemoglobin S lose osmotic pressure. This would probably cause water loss in the intact sickle erythrocyte. Kaul *et al.*,[140] employing a continuous density gradient system, showed that with deoxygenation, normal erythrocytes became less dense and had a decrease in mean corpuscular hemoglobin concentration. This can be explained by an influx of Cl^- and osmotically obliged water as hemoglobin becomes less negatively charged with deoxygenation.[223] When sickle erythrocytes were deoxygenated in this system, however, their density increased, indicating that water was lost.

Gulley *et al.*[342] reported that even well-oxygenated sickle cells have a water content that, although normal in absolute terms, may be suboptimal for deformability. With the ektacytometer, deformability was measured as a function of cell water content. Normal cells show maximum deformability at normal water content; when they are experimentally dehydrated, the deformability diminishes abruptly, presumably due to increased cytoplasmic viscosity. When normal red cells are experimentally overhydrated, their deformability falls again, due to the assumption of an increasingly spheroid shape. Well-oxygenated homozygous sickle cells usually have a normal water content. In contrast to normal cells, however, the deformability of sickle cells shows an initial increase as they are caused to gain water. Only after substantial swelling of the cell does the deformability of sickle cells begin to fall. These results suggest that even well-oxygenated sickle cells may be suboptimally hydrated.

Although controversy persists regarding the water shifts that occur acutely with deoxygenation of whole populations of sickle cells, there is general agreement that ISCs are dehydrated[178,179]; indeed, because of this feature they are separable as the densest cells in a sickle-cell patient's blood.[85] Production of ISCs *in vitro* was found to be inhibited by manipulations that prevent loss of water from the cells.[85,179] This observation supports the notion that loss of cell water associated with the gelation of hemoglobin S *in vivo* plays a key role in the irreversible deformation of the cell membrane characteristic of ISCs.

The controversy concerning the water content of sickle erythrocytes is more than theoretically important. Investigation of the kinetics of hemoglobin S gelation has suggested a role for cell water that may have implications for therapy.[337] Hofrichter *et al.*[209] deoxygenated hemoglobin S under low-temperature conditions that inhibited gelation. On rewarming the deoxyhemoglobin, gelation occurred with a delay time inversely proportional to the 30th power of the hemoglobin S concentration. Similar gelation kinetics have been found in the intact sickle erythrocyte,[92] thus suggesting that a small dilution of the cell contents might result in considerable retardation of sickling. A study by Zarkowsky and Hochmuth[500] supports this idea. An application of this concept is the report of Rosa *et al.*[376] that when sickle-cell patients are made chronically hyponatremic, their red cells swell, the cellular hemoglobin concentration is reduced, and the patients have fewer episodes of painful vascular occlusion, although there is no improvement in their anemia.

In summary, observations regarding the acute effects of sickling on cell water content are controversial. ISCs are depleted of water, and dehydration may be a prerequisite for their formation. Cell water content has a marked effect on the kinetics of gelation of hemoglobin S.

3.1.2. Ca^{2+} Homeostasis

Eaton *et al.*[130] found that sickle cells have a Ca^{2+} content eight times the value for normal erythrocytes. The highest levels of Ca were seen in the dense fraction of ISC. Radioactive Ca uptake was abnormally high in oxygenated sickle cells and rose to even higher levels with deoxygenation, a condition that did not influence Ca influx in normal cells. Other reports have confirmed these findings.[346,347] Accumulation of Ca in sickle cells probably occurs partly because of increased passive Ca entry.[277] Larsen *et al.* have advanced the concept that cell shape change (induced in their experiments by shearing force) may increase passive Ca leak.[255] Others have suggested that sickle cells have impaired active, outward Ca^{2+} transport.[51,R26] Membrane-bound Ca^{2+},Mg^{2+}-ATPase in sickle cells may have an abnormally low response to the cytoplasmic activator protein calmodulin.[115,185,277]

There are two potential consequences of increased intracellular Ca, and both may contribute to the pathogenesis of sickling. (1) Increased Ca may lead to decreased water content of the sickle erythrocyte, a change that markedly affects the kinetics of gelation of hemoglobin S and is probably a prerequisite for ISC formation. (2) Increased Ca may alter the membrane constituents of the sickle erythrocyte and thus lead to irreversible shape changes that are independent of cell volume.

Gardos[173] first observed the ability of Ca to alter the water content of the erythrocyte. He noted that ATP-depleted erythrocytes became selectively permeable to K^+. Because K^+ is the major intracellular cation, the erythrocytes lost K^+ and water. This process was inhibited if a Ca chelator was present in the medium. The accepted explanation for the "Gardos effect" is that ATP depletion leads to a failure of the outwardly directed, energy-dependent Ca^{2+} pump. Ca^{2+} levels then rise in the erythrocyte, making the membrane selectively permeable to K^+.[263,396]

There is considerable controversy regarding the role of the Gardos reaction in the pathogenesis of sickle-cell anemia. Glader and Nathan[179] found that deoxygenation and ATP depletion of sickle erythrocytes *in vitro* led to K^+ loss, dehydration, and the appearance of ISC-like cells provided Ca was present in the incubation medium. No ISC-like cells were seen when a Ca chelator was present. Likewise, if sickle erythrocytes were placed in a medium with sufficient K^+ to eliminate the gradient for K^+ loss, the formation of ISC-like cells did not occur.[179] These observations were confirmed by Clark *et al.*[85] Steinberg *et al.*[424] provided further evidence that the Gardos effect may mediate K^+ loss in sickle cells. They measured K^+ loss from deoxygenated red cells of a group of patients with the homozygous condition and found that cells containing the least K^+ after incubation in nitrogen had the highest rates of radioactive Ca uptake.

Berkowitz and Orringer[37,38] reported that the mechanism of action of cetiedil, a potent *in vitro* antisickling agent, may be explained by the drug's ability to inhibit the Gardos effect. Previous investigators showed that cetiedil has no effect on the solubility or oxygen affinity of hemoglobin S.[27,34] At an *in vitro* concentration of 0.1 mM (point of maximum antisickling action), selective K^+ loss secondary to Ca accumulation is inhibited by cetiedil.

Bookchin and Lew[52,262] argued against a role for Ca^{2+}-mediated K^+ loss in the pathogenesis of sickling with reports that Ca-loaded sickle cells do not show selective increase in K^+ permeability unless depleted of ATP or exposed to the Ca ionophore A23187. Furthermore, ATP levels are not decreased in either the reversibly sickled erythrocyte or the ISC.[88,178] The idea that Ca may be compartmentalized within intracellular inside-out vesicles has been offered as an explanation for a refractory state of the Ca^{2+}-dependent K^+ channel in sickle cells.[R6]

Other effects of Ca on the erythrocyte membrane, independent of volume changes, have been described. In normal erythrocytes, Dreher *et al.*[121] evaluated the separate effects of ATP depletion, dehydration, K^+ loss, and Ca accumulation on rigidity as determined by the deformation pressure in a micropipette system. Increased rigidity was not seen in ATP-depleted dehydrated erythrocytes that had lost K^+. Increased rigidity was observed only when Ca accumulation was added to these variables. In another study, ATP-depleted cells had low deformability, which returned to normal if agents (e.g., chlorpromazine, papaverine) were used that displace membrane-bound Ca^{2+}.[375] Dzandu and Johnson[129] reported abnormal phosphorylation of membrane polypeptides in sickle erythrocytes. The same abnormal pattern of phosphorylation can be created in the normal erythrocyte with Ca loading. Palek *et al.*[348,349] have demonstrated that introduction of Ca (> 0.5 mM) into either sickle or normal erythroyctes leads to transamidative cross-linking of membrane proteins. However, *in vivo* ISCs did not show such cross-linking unless subjected to ATP depletion and Ca accumulation *in vitro*. Experimental manipulations leading to elevation of intracellular Ca were also reported to cause membrane vesicle loss[12,15] and to promote adsorption of hemoglobin to the membrane.[16] The significance of these studies for *in vivo* sickle erythrocytes is not clear. The ionophore A23187 was most often used to increase intracellular Ca, and many of these effects were seen at Ca levels far above the range reported for sickle cells or ISCs in fresh blood samples.

A summary of these data indicates that the sickle erythrocyte clearly has an elevated Ca content that is related to deoxygenation. The role of intracellular Ca^{2+} in causing an increased K^+ permeability via the Gardos mechanism is controversial. The magnitude of the Ca elevation in the freshly drawn sickle cell is below the level at which alterations in membrane structure have been produced under experimental circumstances *in vitro*.

3.1.3. Membrane Composition

There is no evidence to suggest a coexistent genetic abnormality in the membrane of the sickle erythrocyte responsible for the characteristic shape changes that occur with gelation of hemoglobin S.[183] However, repeated episodes of sickling lead to irreversible membrane damage.[1,13,14] Clark and Shohet[87] demonstrated this effect by enclosing hemoglobin S within normal red cell membranes, using a resealed-ghost technique. These "hybrid" cells sickled on deoxygenation and underwent irreversible shape changes with prolonged hypoxia.

Sustained association of hemoglobin with the membrane may explain the influence of hemoglobin S on red cell shape. Denatured hemoglobin S has been demonstrated on the membranes of freshly drawn sickle cells.[26] *In vitro* experiments at pH < 7.3 and very low ionic strength show that hemoglobin S binds to normal red cell membranes more avidly than does hemoglobin A and that the binding sites involve band 3 and possibly phospholipids.[410] The relevance of these observations to *in vivo* conditions is not clear, however.

Binding of hemoglobin to the membrane, particularly when associated with oxidative denaturation, may explain why K^+ and water loss occur in sickle cells. Normal red cells briefly exposed to acetylphenylhydrazine acquire Heinz bodies and show a K^+ leak that is independent of the presence of Ca^{2+}, inhibited by furosemide, and Cl^- dependent (Table I).[341,343] Sickle cells are low in vitamin E, spontaneously generate oxygen radicals, and are subject to lipid peroxidation.[105,R18,R20] These observations suggest that some of the membrane alterations seen in sickle cells are caused by oxidative damage.

An alternative explanation for the influence of hemoglobin S on red cell shape does not involve permanent hemoglobin–membrane binding. Lux *et al.*[285] exposed ISCs to Triton X-100, forming spectrin–actin networks free of hemoglobin, lipids, and integral membrane proteins. The "Triton shells" retained the ISC shape. Further work with these irreversibly sickled membrane skeletons has revealed their normal protein composition. Conditions that inhibit the formation of spectrin–actin complexes, such as exposure to hyper- or hypotonic media or zinc salts, cause a reversal of the ISC shape to a spherical configuration. These changes were independent of the ATP or Ca^{2+} content of the cells.[284] The absence of hemoglobin in these irreversibly sickled membrane skeletons militates against a role for hemoglobin binding in the maintenance of the distorted cell shape. The irreversible change in membrane configuration seen in ISCs may be secondary to rearrangements in the spectrin–actin network, perhaps induced by prolonged deformation of the cell.

Lubin *et al.*[275] showed a striking reversible change in the orientation of membrane phospholipids with sickling. Well-oxygenated sickle cells have the same phospholipid asymmetry as normal cells. On sickling, however, the external leaflet of the bilayer becomes enriched in phosphatidylethanolamine and phosphatidylserine, and the amount of phosphatidylcholine decreases. On reoxygenation these changes are almost completely normalized, except for ISCs. There is circumstantial evidence that these alterations are determined by rearrangements of spectrin, since oxidation of spectrin in normal cells is associated with similar changes in lipid orientation.[197] The possibility remains, however, that the oxidative effects are due to changes in membrane constituents other than spectrin.

Abnormal phospholipid distribution with sickling may also explain the work of Hebbel *et al.*,[203] which demonstrates that sickled red cells are more adherent to endothelial cells than normal erythrocytes. Explanations for increased adherence include a clustering of negative charges at the cell surface and the influence of unidentified factors in plasma from patients with sickle-cell disease.[203,R23] A redistribution of phospholipids may also contribute to the adherence phenomenon, inasmuch as a shift in the orientation of phosphatidylserine to the outer membrane leaflet leads, in the presence of Ca, to increased cell–cell contact, cell fusion, and acceleration of blood clotting.[84,487,R32]

In summary, the changes in monovalent cation permeability, Ca flux, and membrane skeletal configuration that accompany the deformation of sickled cells have furnished some insights into the properties of the red cell membrane. (1) Cell shape may influence passive cation permeability. (2) Ca entry into the cell may not lead to abrupt increase in K^+ permeability in ATP-replete cells. (3) Important structural and functional relationships probably exist between the membrane skeleton and the lipid bilayer with its integral proteins. Further study of this common genetic variant will undoubtedly add to our knowledge about the organization of the red cell membrane.

3.2. Hemoglobin CC Disease

The formation of hemoglobin C is due to an alteration in the nucleotide sequence of autosome 11 which leads to the substitution of lysine for glutamate at position 6 from the NH_2-terminus of the β-globin chain.[R9] The fact that hemoglobin C is more positively charged than hemoglobin A, plus the observation that red cells from CC patients have reduced contents of Na^+, K^+, and water, have generated interest in the transport properties of these cells.[R25]

Brugnara *et al.* have shown that the Cl^- ratio is normal in CC cells. In addition, they have reported that the maximum rate of the Na^+,K^+ pump and the ouabain/furosemide-insensitive fluxes of Na^+ and K^+ are increased compared to normal red cells. K^+ efflux was stimulated by cell swelling and by a reduction of cytoplasmic pH, effects not seen in control red cells. Quinine, Mg^{2+}, trifluoperazine, and Ca^{2+} chelation had no effect on K^+ efflux.[R7,R8]

Berkowitz and Orringer investigated the effects of deoxygenation and Ca^{2+} loading on CC red cells.[R5,R27] Unlike SS cells, CC cells showed no increase in passive Na^+ or K^+ movements upon deoxygenation. However, when CC cells were exposed to a series of concentrations of Ca^{2+} in media containing the ionophore A23187, they showed a twofold increase in the sensitivity of the Ca^{2+}-activated K^+ channel. It took half as much Ca^{2+} to trigger K^+ loss in CC cells as it did in AA cells. Furthermore, measurement of Ca concentration in freshly drawn CC cells showed values (50 μmoles/liter RBC) that were 5 times those of AA controls.

The mechanism by which hemoglobin C alters the function

of the cell membrane may be related to recent findings by Reiss *et al.* that under physiological conditions *in vitro*, hemoglobin C binds to the membrane. Since the binding is competitively inhibited by the enzyme glyceraldehyde-3-phosphate dehydrogenase,[243] it was postulated that hemoglobin C binds to the cytoplasmic portion of band 3.[R31]

4. Endocrine Disorders

In this and the following sections we discuss several genetic disorders with clinical manifestations in tissues other than the red cell. In each condition, alterations were reported in red cell structure or function, although in no case are the aberrant properties prejudicial to red cell survival or oxygen transport. Biopsy of red cells is easily performed, and information gained from studying their properties may be useful in diagnosis and genetic counseling. Furthermore, to the extent that the red cell membrane has structural components or transport properties in common with other tissues, clinically important features of disease may be disclosed. Because we are emphasizing the membrane, we ignore conditions such as galactosemia, hypoxanthine-guanine ribosyltransferase deficiency, and acute intermittent porphyria in which systemic metabolic lesions may be diagnosed and understood by performing assays in erythrocytes.

The action of parathormone on target tissues includes binding of the hormone to receptors on the cell surface and subsequent generation in the cell of cAMP by adenylate cyclase. The interaction between the receptor and adenylate cyclase is mediated by a "coupling protein" in the plasma membrane called N protein, which binds guanine nucleotides and can be ADP-ribosylated in the presence of cholera toxin. Although parathormone receptors have not been reported in the red cell, and although the presence of adenylate cyclase in mature human erythrocytes is controversial, N protein appears to be an integral red cell membrane protein with a cytoplasmic domain. Its monomer has a molecular weight of 42,000. Pseudohypoparathyroidism, a trait arguably assigned to the X chromosome, is characterized by an unresponsiveness of target tissues in bone and kidney to parathormone. Some patients with this disease fail to generate cAMP in the presence of adequate hormone levels, and the red cell membranes of these individuals have about half the normal content of N protein.[146,R35] This finding supports the clinical observation that many patients with pseudohypoparathyroidism have impaired function of other organs, such as the thyroid and the renal collecting duct; responses to hormones in both depend on the generation of cAMP. Presumably, the deficiency of N protein is generalized to include the plasma membrane of many cell types, and thus the action of hormones such as thyrotropin and ADH may be impaired. The relationship of these findings to a putative effect of parathormone on red cell Ca^{2+} transport is unclear.[48]

Another genetically determined endocrine disorder, diabetes mellitus, is often characterized by a decrease in insulin binding to physiologically important cells in the liver and adipose tissue. Reduced insulin binding is also found in blood mononuclear leukocytes of diabetics with chronically increased insulin levels.[340] Human erythrocytes also bind insulin.[164] Red cells from adult-onset diabetics show a 10–15% reduction in insulin binding, although the binding affinity is normal.[111,374] The functional importance of insulin to the red cell is unclear, but the binding of hormone to this cell apparently reflects meaningful changes in other tissues. Increased insulin binding was found in red cells of anorexia nervosa patients[332,459] and of individuals treated with the biguanide class of antidiabetic drugs.[212] Patients with renal failure, on the other hand, showed diminished insulin binding correctable by hemodialysis.[165]

5. Manic-Depressive Disease

Cade's demonstration[69] of the effectiveness of orally administered lithium salts in the management of patients with manic psychosis led to trials of this drug in a variety of psychiatric illnesses. Lithium is currently the treatment of choice in the management of certain patients with bipolar manic-depressive disease. Measurements of Li^+ concentration in red cells and plasma during chronic therapy with lithium salts have shown that each person achieves a cell-to-plasma ratio that is stable over time; the value of that ratio, however, varies widely among different individuals. Subsequent studies have indicated several membrane transport routes for Li^+ in red cells. One of these pathways, the Li^+/Na^+ exchanger, shows great quantitative variation among red cells from various donors, and this heterogeneity appears to be partly determined by genetic factors. Several hypotheses have been advanced about the relationship between red cell Li^+ transport and psychiatric illness, although the information on this subject remains controversial.[133,442]

5.1. Modes of Li^+ Transport across the Red Cell Membrane

In the steady state (i.e., in blood of a patient on chronic, constant-dose lithium therapy), Li^+ is not equilibrated across the red cell membrane. The cell-to-plasma concentration ratio varies from 0.2 to 0.9, far from the 1.3–1.4 value predicted from the 10-mV potential across the red cell membrane. The disequilibrium of the Li^+ is explicable in terms of recent data on Li^+ transport.[123,133,193,351,353,397,442]

5.1.1. Na^+,K^+ Pump

Although Li^+ can be carried in both directions across the membrane by the ouabain-sensitive Na^+,K^+ pump, the affinities of this system are such that at physiological concentrations of cell Na^+ and plasma K^+, no Li^+ is transported by that mechanism.

5.1.2. Li^+/Na^+ Exchange

A 1 Li^+ : 1 Na^+ exchange operates in either direction across the red cell membrane. The mechanism can bind either Li^+ or Na^+ on one side of the membrane and exchange the transported species for either Li^+ or Na^+ on the opposite side. Net movement of Li^+ across the membrane can be accomplished by having a low Na^+ concentration on the side from which Li^+ is transported and a high Na^+ concentration on the opposite side. Both sides have an affinity 15- to 18-fold higher for Li^+ than for Na^+. Under physiological conditions the plasma Na^+ is 14–15 times higher than the cell Na^+. Consequently, Li^+ tends to be transported out of the cell against its electrochemical gradient in exchange for Na^+, which moves into the cell down its gradient. The energy that fuels the active extrusion of Li^+ derives from the disequilibrium of Na^+ across the membrane, which in turn results from the activity of the ATP-dependent Na^+,K^+ pump. The Li^+/Na^+ exchanger is inhibited by phloretin, is resistant to ouabain, and is independent of

ATP.[353,397] This system accounts for the disequilibrium of Li$^+$ across the red cell membrane in the blood of patients taking the drug. There are genetically determined differences in the V_{max} of the Li$^+$/Na$^+$ exchanger that account for the variations in the cell-to-plasma Li$^+$ ratio of different individuals on lithium therapy. Under normal conditions in people not receiving lithium, the transport pathway operates as a Na$^+$/Na$^+$ exchanger[394] of unknown physiological importance.

5.1.3. LiCO$_3^-$ Pathway

In the presence of HCO$_3^-$ solutions, Li$^+$ and, to a lesser extent, Na$^+$ form small amounts of LiCO$_3^-$ and NaCO$_3^-$. These are rapidly transported across the membrane, in exchange for Cl$^-$ or HCO$_3^-$, by the anion exchanger identified with band 3. Movements of Li$^+$ by this route are passive, downhill, and under physiological circumstances in a patient on lithium therapy would account for about half of the influx of Li$^+$ into the cell.[157]

5.1.4. Electrodiffusion

Li$^+$ can move across the membrane by simple passive diffusion of the ion down its electrochemical potential energy gradient.

5.1.5. Cotransport

Recent evidence suggests that Li$^+$ can cross the membrane by a passive system that normally carries Na$^+$ and K$^+$ together in the same direction. The magnitude of Li$^+$ flow by this route under physiological conditions is not clear.[442]

5.2. Genetic Heterogeneity in Li$^+$ Transport

The differences in cell-to-plasma concentration ratios seen among patients on lithium therapy can be explained by variations in the V_{max} of the pathway of Li$^+$/Na$^+$ exchange as studied in vitro.[122,345,351] Investigations of twins and large families, by both in vivo and in vitro methods, show that the rate of Li$^+$/Na$^+$ exchange in human red cells is influenced by genetic factors.[116,118,119,219,345,351,392] The values show a higher concordance for monozygotic than for dizygotic twins. When large families are compared, the variance between families is significantly greater than the variance within families. Individuals with unusually low values for Li$^+$/Na$^+$ exchange were found to have first-degree relatives of both sexes with similar values, suggesting in one pedigree an autosomal dominant mode of inheritance.[345] However, an extensive study of parent–child correlations versus sib–sib correlations did not distinguish whether this quantitative trait is inherited via single or multiple genes.[118] Although most authors agree that age, ethnic origin, and state of mania or depression do not correlate with the rate of Li$^+$-Na$^+$ exchange, there is some difference of opinion regarding the influence of sex; some observers find higher ratios in males,[219] others in females.[392]

5.3. Correlations with Manic-Depressive Disease

Mendels and Frazer[302] originally proposed that the steady-state ratio of cell-to-plasma Li$^+$ concentration correlated with the likelihood that patients with affective disorders would benefit from lithium therapy. Those individuals with high ratios

(and therefore a low activity of the Li$^+$/Na$^+$ exchanger) were most responsive, a suggestion both confirmed[76] and disputed.[77,133,392] Subsequently, it was suggested that the activity of the Li$^+$/Na$^+$ exchanger correlates with the presence of bipolar as opposed to unipolar affective illness,[155,247] although this has also been controversial.[10,77,110,122,304,391]

The combined evidence suggests that bipolar manic-depressive disease is a heterogeneous syndrome. Some patients with this diagnosis have a striking history of the disorder in first-degree relatives.[39,303,304] About 25% of manic-depressive patients have a low value for Li$^+$/Na$^+$ exchange in the red cell membrane that results in a high ratio of cell-to-plasma Li$^+$ concentration during chronic therapy with the drug. First-degree relatives of these patients may also have low values for Li$^+$/Na$^+$ exchange, but not all of these relatives have psychiatric illness.[119,122,286,345,352,390] The recent hypothesis that red cell Li$^+$/Na$^+$ exchange is somehow related to essential hypertension indicates that further investigation of individuals who have had red cell studies[eg., 118] with regard to their blood pressure history would be useful.

5.4. Miscellaneous Findings

Several observations on manic-depressive disease bear mention. Therapy with lithium salts results in a reversible fall in the affinity of the red cell Li$^+$/Na$^+$ exchanger for Li$^+$ [133,134] An irreversible consequence of lithium therapy is that red cell choline levels increase by as much as 10-fold owing to an inhibition of choline efflux. This change lasts for the life of the cell in the circulation.[231,264] An unconfirmed observation on some manic-depressive patients is that they have a Na$^+$,K$^+$ pump deficiency that is corrected by lithium therapy.[210,211]

6. Essential Hypertension

Essential hypertension is prevalent in populations with high salt intake. Studies of humans and experimental animals show that, in addition to diet, genetic factors predispose to development of high blood pressure. A possibility of recent interest is that studies of cation transport in blood cells might disclose alterations that would serve as genetic markers for hypertensive diathesis and perhaps also provide insight into the pathogenesis of the disease.

Many observations are focused on the mechanism by which Na$^+$ and K$^+$ metabolism affects blood pressure in humans and animals. In salt-dependent hypertension there is increased peripheral vascular resistance apparently mediated by a humoral factor with pressor and natriuretic properties. The walls of blood vessels show an increased Na$^+$ content in some models of the disease. Some studies suggest that the Na$^+$,K$^+$ pump or Na$^+$, K$^+$-ATPase is suppressed in the blood vessels of hypertensive individuals, whereas in other models the Na$^+$,K$^+$ pump rate is thought to be increased secondary to an increased passive Na$^+$ permeability of the plasma membrane of vascular smooth muscle. The humoral pressor-natriuretic factor in salt-sensitive hypertensive animals may itself be an inhibitor of the Na$^+$,K$^+$ pump.[184,185,195] The increased Na$^+$ content of the arterioles may contribute to increased peripheral resistance by causing edema of the vascular wall, by sensitizing the tissues to endogenous pressor hormones, or by leading to an elevated level of Ca^{2+} in the sarcoplasm with a resulting rise in arteriolar tone.[112,194]

Table IV. Studies of Na$^+$ Content in Blood Cells of Humans and Rats

	Na$^+$ content, concentration, or activity	
	Normal	Increased
Red cells		
Humans with essential hypertension and/or their offspring	56, 63, 73, 327, 405, 445, 461, 462, 464, 465, 469	102, 141, 174, 271–273, 450, 470–472
Rats with hypertension	33	36, 368
Leukocytes, humans with hypertension		19, 20, 25, 96, 132, 153, 371, 436

Table VI. Studies of Passive Unspecified Na$^+$ Movements in Blood Cells of Humans and Rats

	Passive Na$^+$ flux	
	Normal	Increased
Red cells		
Humans with essential hypertension	288,[a] 371	152, 204, 287, 288,[b] 370, 470–472
Rats with hypertension		33, 156, 368, 484
Leukocytes, humans with essential hypertension	371	

[a]Black patients.
[b]White patients.

The data on red cells are abundant and controversial. Because it is difficult to fit the observations into a coherent picture, the information is presented in a tabular, descriptive format.

6.1. Abnormalities in Cell Ion Content

A substantial alteration in either passive or active ion movements might lead to a measurable change in steady-state ion content, provided the abnormal transport pathway is capable of net ion transfer. Some investigators claim that freshly drawn blood cells from hypertensive humans and animals indeed contain more Na$^+$ than cells from normal controls (Table IV). Not all studies agree, however, insofar as red cells are concerned. It is remarkable that some groups of investigators draw blood from hypertensive patients for elaborate transport studies and yet do not report the ion contents of fresh cells.[101,167–172]

6.2. Abnormalities in Na$^+$,K$^+$ Pump

Numerous measurements reflecting the activity of the Na$^+$,K$^+$ pump in blood cells from hypertensive individuals have been reported. These include studies of the effect of ouabain or other cardiac glycosides on tracer Na$^+$ efflux and K$^+$ influx in cells with unaltered ion contents, measurements of net uphill ion movements in cells treated to raise Na$^+$ and lower K$^+$ content, assays for Na$^+$,K$^+$-ATPase, and determinations of ouabain binding (Table V). The results are disparate and conflicting. Most studies have been done on isolated, washed cells in synthetic media. If there is a circulating inhibitor of the Na$^+$,K$^+$ pump in essential hypertension, it would seem appropriate to conduct assays of that mechanism with plasma from

hypertensives. Poston et al.[371] recently demonstrated a highly significant inhibition of the ouabain-sensitive Na$^+$ efflux rate constant in leukocytes of normal individuals after preincubation in plasma from individuals with essential hypertension.

6.3. Unspecified Increases in Passive Na$^+$ Movements

Elevated rates of isotopic Na$^+$ flux through pathways other than the Na$^+$,K$^+$ pump were originally reported by Wessels et al.[470–472] and recently by Mahoney et al.[287,288] (Table VI). No attempt was made in these investigations to ascertain whether the increased Na$^+$ movements were due to Na$^+$/Na$^+$ exchange, Na$^+$–K$^+$ cotransport, or simple Na$^+$ leak. The following sections review efforts to determine the nature of the Na$^+$ pathways in red cells of hypertensives.

6.4. Abnormalities in Na$^+$/Na$^+$ and Na$^+$/Li$^+$ Exchange

Section 5 reviews evidence that human red cells have a system that can conduct Na$^+$/Na$^+$ or Na$^+$/Li$^+$ exchange or countertransport. The report of Canessa et al.[73] that red cells from hypertensives have a highly significant twofold elevation in this pathway stimulated a number of studies yielding conflicting results (see Table VII). Among the findings are that Li$^+$/Na$^+$ countertransport is elevated in hypertensive males but not females,[219] that it is high in hypertensive whites but not blacks,[72,74] and that it is simply not a good marker for hypertension.[124–126]

Table V. Studies of Na$^+$,K$^+$ Pump in Blood Cells of Humans and Rats

	Na$^+$,K$^+$ pump, Na$^+$,K$^+$-ATPase activity, or ouabain binding		
	Decreased	Normal	Increased
Red cells			
Humans with essential hypertension and/or their offspring	141, 461, 462	426, 463	167–169, 464–466, 491
Rats with hypertension	393	36	
Leukocytes, humans with hypertension	132, 153, 371, 436		

Table VII. Features of Ouabain-Insensitive, Monovalent Cation Countertransport and Cotransport Systems of Human Red Cells

	Countertransport Na$^+$(Li$^+$)/Na$^+$(Li$^+$)	Cotransport Na$^+$(Li$^+$)–K$^+$
Modes of transport	*trans* stimulation of Na$^+$ or Li$^+$ transport by Na$^+$ or Li$^+$	*cis* stimulation of Na$^+$ or Li$^+$ transport by K$^+$
Affinity for internal sites	Li$^+$ > Na$^+$	Na$^+$ > Li$^+$
Phloretin	inhibits	"uncouples"[166]
Furosemide	no effect	inhibits
Cl$^-$ removal	no effect	inhibits[83,127]
Activity in hypertensive patients and their offspring		
Increased	4, 70, 71, 73, 75, 100, 219,^{c} 261, 445,^{a} 490^c	4, 74, 75, 147
Normal	72,^{a} 74,^{a} 124–126, 167, 168, 219^b	106, 124–126, 426, 461, 462
Decreased	73,^{a} 74,^{a} 100, 101, 109, 166–172, 309	

^{a}Black patients.
^{b}Female patients.
^{c}Male patients.

6.5. Abnormal Na$^+$–K$^+$ Cotransport

Tosteson[441] recently reviewed the origins of the hypothesis that the human red cell membrane contains a mechanism for the coupled cotransport of Na$^+$ and K$^+$. The affinities of this system show that no net cation movements through the cotransporter occur at physiological cell and plasma cation concentrations. At abnormally high cell Na$^+$ concentrations, however, the cotransporter couples the outward movement of K$^+$ down its chemical gradient with the outward, uphill movement of Na$^+$. The cotransporter can thus work parallel to the Na$^+$,K$^+$ pump to extrude Na$^+$ from the cell, but the energy for outward Na$^+$ cotransport derives ultimately from the activity of the Na$^+$,K$^+$ pump in maintaining a steep chemical gradient for K$^+$ across the membrane. The stoichiometry of the cotransport system has been depicted as 1 Na$^+$: 1 K$^+$.[443] In osmotically swollen cells, however, the ratio may be changed so K$^+$ flux exceeds Na$^+$ flux.[3] In a variant of hereditary stomatocytosis, on the other hand, the ratio of Na$^+$: K$^+$ flux through the cotransporter was very high.[78] An absolute requirement for Cl$^-$ (or Br$^-$) constitutes an additional feature of the cotransport system.[83,127]

Movements of ions through the Na$^+$–K$^+$ cotransporter are inhibited by furosemide and some other diuretics. Just as ouabain sensitivity is often used as the criterion for assigning a given flux to the Na$^+$,K$^+$ pump, furosemide inhibition is used to identify fluxes through the cotransport pathway. But because furosemide also affects the Na$^+$,K$^+$ pump to some degree, the operational definition used by some workers for cotransport fluxes is "ouabain-insensitive, furosemide-sensitive." A conceptual problem with this definition is the presumption that Na$^+$–K$^+$ cotransport is uninfluenced by ouabain under all circumstances and that drug effects sufficiently define the system.

Furosemide is also an inhibitor of anion exchange and might therefore alter Na$^+$ movements (as the ion pair NaCO$_3^-$) under physiological conditions.[59,157] The most convincing demonstrations of Na$^+$–K$^+$ cotransport do not rely on the presence of inhibitors but rather on the manipulation of ion concentrations.

Garay and colleagues[101,109,166–172] reported that people with essential hypertension, normotensive individuals with two hypertensive parents, and hypertensive rats have a deficiency of ouabain-insensitive, furosemide-sensitive Na$^+$ and K$^+$ movements in their red cells. A study of 14 families suggests that the defect in cotransport and essential hypertension are inherited together as autosomal dominant traits.[309] The defect is demonstrated by preincubating cells in solutions with *p*-chloromercuribenzene sulfonate (PCMBS) to raise their Na$^+$ concentration. After reversing the PCMBS effect with cysteine, the cells are placed in a physiological salt solution and the rates of Na$^+$ and K$^+$ movements are measured in the presence and absence of ouabain and furosemide. Under these conditions the maximum rate of ouabain-sensitive fluxes is increased in the red cells of hypertensives. The authors believe this observation reflects a compensatory hypertrophy of the Na$^+$,K$^+$ pump due to the deficient Na$^+$–K$^+$ cotransporter. The studies of Wambach et al.,[465,466] which showed elevated Na$^+$,K$^+$-ATPase in red cell ghosts from hypertensives, are cited by the Garay group in support of this notion.

Walter and Distler[461,462] did not confirm these findings. Using different methods that avoid PCMBS preincubation, they observed a normal ouabain-insensitive, furosemide-sensitive Na$^+$ efflux and a somewhat diminished ouabain-sensitive Na$^+$ efflux rate constant in the red cells of their hypertensive patients. Duhm et al.,[124–126] Swarts et al.,[426] and Davidson et al.[106] were unable to confirm that the rate of red cell Na$^+$–K$^+$ cotransport correlates with hypertension.

A criticism of the cotransport studies is that some may not have been conducted at saturating concentrations of internal Na$^+$ ($K_{1/2}$ 13 mM). In these instances, variations may reflect different degrees of activation of the mechanism rather than its maximum velocity. Alternatively, there may be heterogeneity among individuals in the affinity of cotransporter for internal Na$^+$.[R19–R21]

6.6. Relationships between Countertransport and Cotransport Studies

Recently, two groups compared observations that hypertensive patients and their relatives have increased Li$^+$/Na$^+$ countertransport (Boston) and reduced Na$^+$–K$^+$ cotransport (Paris). These investigators have also begun to study similarities and differences in the two modes of transport (Table VII). The two transport systems appear to be discrete entities. Each can run maximally when the other is inhibited, and they have different affinities for Na$^+$ and Li$^+$. In individuals from different geographic areas, the level of each type of transport in red cells varies. For example, normal values for countertransport vary by a factor of 3 between the Boston and Paris studies. In both studies the hypertensive patients and some of their relatives have elevated countertransport levels; however, the cotransport system is reduced in Paris hypertensives and increased in Boston counterparts.[71,75] Canessa et al.[74] studied countertransport and cotransport in the red cells of Boston whites and Philadelphia blacks. Hypertension, or a strong family history of hypertension, was associated with increased countertransport and cotransport in Boston patients. The Philadelphia patients, however, showed reduced cotransport and normal countertransport.

6.7. Miscellaneous Observations on Red Cells of Hypertensives

Based on viscometric studies, Dintenfass and Girolani[113] thought that red cell rigidity contributed to the increased peripheral resistance of hypertensives. Abnormal binding of Ca^{2+} [369] and Na^+ [450] has been reported in membranes from red cells of hypertensive patients.

6.8. Summary

The conflicting data regarding Na^+ concentration, Na^+, K^+ pump activity, Li^+/Na^+ countertransport, and Na^+-K^+ cotransport in red cells of hypertensive patients seem inexplicable except on the basis of heterogeneous patient populations and differences in experimental protocol. Future work in this area should probably be conducted by standardized methods. Both countertransport and cotransport pathways show great interindividual variation, and the values for any individual's red cells are reproducible over time.[122,124,126] There is evidence from family and twin studies that the activities of both systems are inherited traits (see Section 5).[106,309] But new data suggest that these permeabilities may vary with diet,[319,445] body weight,[5] and pregnancy.[R41] A relatively unexplored phenomenon is the heterogeneity of red cell Na^+ concentrations in the black population.[30,136,274,327] This appears to be genetically determined and may be related to the possibly distinctive nature of hypertensive disease among blacks.[98,131,491,492] Recent large series[R33,R37,R38] suggest that in some regions of the world there may be, within the hypertensive population, subgroups of individuals, who have abnormal red cell cation transport.[R40] The usefulness of these findings for diagnosis, genetic counseling, therapy, or an understanding of the origin of high blood pressure remains to be determined.

Apparently, there is a defect in the Na^+,K^+ pump in the leukocytes of individuals with essential hypertension; this leads to a high cell Na^+ concentration. A circulating inhibitor of the Na^+,K^+ pump may be involved,[371] although lack of a plasma factor necessary for optimum pump activity seems possible.[82]

How do the findings in blood cells relate to the pathogenesis of hypertension? Blaustein[46] suggests that if arteriolar smooth muscle had a transport defect leading to high sarcoplasmic Na^+,[448] then the steady-state intracellular level of Ca^{2+} might be increased through the operation of a membrane Ca^{2+}/Na^+ exchanger. The high cell Ca^{2+} would in turn lead to an increase in myofibrillar tone, thus narrowing the lumina of arterioles and causing high peripheral vascular resistance. In support of a role for Ca^{2+}/Na^+ exchange in maintaining arteriolar tone, quinidine, which inhibits this transport pathway,[354] causes vasodilation and hypotension when acutely infused.[403] Recently, Blaustein and colleagues have provided confirmation of the hypothesis that a circulating Na^+,K^+ pump inhibitor can be demonstrated in the plasma of hypertensive patients.[R17] The origin of such an endogenous substance may be the central nervous system.[R16]

7. Inherited Neuromuscular Disorders

Red cell abnormalities have been reported in patients with Duchenne's dystrophy (DD), myotonic dystrophy (MyD), Hun-

Table VIII. General Features of Duchenne's Dystrophy, Myotonic Dystrophy, and Huntington's Chorea[a]

	Duchenne's dystrophy	Myotonic dystrophy	Huntington's chorea
Inheritance	XLR	AD	AD
Weakness	+	+	−
Chorea	−	−	+
Dementia	−	−	+

[a] XLR, X-linked recessive trait; AD, autosomal dominant trait; +, presence of feature; −, absence of feature.

tington's chorea (HC), congenital myotonia, and Friedreich's ataxia. Although none of these disorders is understood in molecular terms, observations on the involved neuromuscular tissues, plus the red cell findings, suggest that the underlying genetic abnormalities reside in the plasma membrane. Portions of the extensive and controversial literature in this area have been reviewed.[32,66,366,381,383,387,388,447] The following account emphasizes the most studied conditions (DD, MyD, and HC). Table VIII shows some general characteristics of these diseases.

Evidence implicating the sarcolemma in the pathogenesis of DD includes the finding in patients' plasma of enzymes from skeletal muscle cytoplasm, such as aldolase and creatine kinase; presumably, a membrane defect allows the enzymes to leak out. A decrease in sarcolemmal adenylate cyclase is also reported in this condition.[388] In MyD, certain electrophysiological abnormalities of muscle suggest a lesion in the sarcolemma. A myotonia-like illness can be induced in humans and experimental animals by treatment with an inhibitor of cholesterol synthesis that causes the accumulation of desmosterol, a compound not normally present, in all body membranes.[387] MyD is not only a disease of muscle: along with putative red cell abnormalities, many patients have cataracts and testicular atrophy. HC has not been linked to abnormality of neuronal plasma membranes; biochemically, however, the central nervous system becomes depleted of γ-aminobutyric acid and choline acetyltransferase.[66]

Almost everything that can be measured in red cells has been claimed abnormal in these diseases. The observations virtually all fall into two categories: either they are unconfirmed, or they are confirmed by some and not by others (Tables IX–XII). Some quantitative features are increased in one laboratory and decreased in others. All observers agree, however, that none of these neuromuscular syndromes is associated with reduced red cell survival.

7.1. Na^+-K^+ Transport

Brown et al.[61] were among the first to study red cells in patients with muscular dystrophy. Their claim that ouabain exerted a paradoxical stimulatory action on Na^+,K^+-ATPase in red cell membranes from patients with DD was promptly assailed because the Na^+ and K^+ concentrations used in the assay were unphysiologically low, and when appropriate Na^+ and K^+ concentrations were used, ouabain had its expected inhibitory effect.[242] Reports of a slight increase in red cell Na^+,K^+-ATPase in HC have come from two independent laboratories (Table IX), but no corresponding physiological studies of Na^+-K^+ transport in this condition are available. Hobbs et al.[207] disputed the claim that the coupling ratio of the Na^+,K^+ pump

Table IX. Studies of Na$^+$–K$^+$ Transport in Red Cells from Patients with Neuromuscular Disease

	Duchenne's dystrophy	Myotonic dystrophy	Huntington's chorea
Cell K$^+$			
Normal	114		
Cell Na$^+$			
Normal			68
Increased	120		
Na$^+$K$^+$ pump			
Normal	372, 389	207	
Increased	409		
Decreased	422		
Qualitatively abnormal		218	
Na$^+$K$^+$-ATPase			
Normal	208, 242, 387	378, 387, 389	
Increased			66, 67, 429
Decreased	314, 334, 357, 422	334	
Qualitatively abnormal	8, 61, 62, 299, 334, 357, 360	334	
Ouabain binding			
Qualitatively abnormal	314		

is altered (1 : 1 instead of 3 Na$^+$: 2 K$^+$) in MyD.[218] If a real abnormality of Na$^+$–K$^+$ transport existed in any of these conditions, one might expect it to be reflected in the Na$^+$ or K$^+$ content of the cells. Remarkably, this simple measurement has rarely been made. A confirmation of Dowben and Holley's early observation[120] of an increased red cell Na$^+$ content in DD seems necessary.

7.2. Ca^{2+} Transport and Ca^{2+}-Related Phenomena

Several reports of increased Ca^{2+},Mg^{2+}-ATPase and increased ATP-dependent Ca^{2+} transport in DD have appeared from diverse laboratories. In MyD red cells, Ca^{2+},Mg^{2+}-ATPase was reported both increased and decreased. The specific increase in K$^+$ permeability observed in normal red cells when Ca^{2+} is introduced into the cytosol[263] was reported normal, increased, and decreased in DD and MyD (Table X). Passive, inward Ca^{2+} movements in hypertonically shrunken cells incubated in a Na$^+$-free medium were elevated in MyD red cells.[367]

7.3. Cl$^-$ and Water Movements

Cl$^-$ exchange was reported 13% increased in red cells from patients with HC, a finding that the authors related to abnormalities in protein mobility as reported by ESR measurements.[43,64] Some but not all patients with DD had reduced

Table X. Studies of Ca^{2+} Movements, Ca^{2+}-Mediated K$^+$ Flux, and Other Permeabilities of Red Cells from Patients with Neuromuscular Disease

	Duchenne's dystrophy	Myotonic dystrophy	Huntington's chorea
Ca^{2+} pump			
Increased	316		
Qualitatively abnormal	316		
Ca^{2+},Mg^{2+}-ATPase			
Increased	208, 278, 389	389	
Decreased		278	
Ca^{2+}-mediated K$^+$ flux			
Increased	24, 114, 380, 381		
Decreased	428	24, 380	
Normal		191	
K$^+$ flux			
Increased	215		
Ca^{2+} flux			
Increased		367	
Cl$^-$ flux			
Increased			43
H$_2$O flux			
Decreased	28		
Cell Mg^{2+}			
Increased	47		

Table XI. Studies of Physical Properties of Red Cells in Neuromuscular Diseases

	Duchenne's dystrophy	Myotonic dystrophy	Huntington's chorea
Survival *in vivo*			
Normal	2		
Morphology			
Normal	311	460	
Qualitatively abnormal	214, 276, 296, 297, 313, 460	313	291
Membrane fluidity[a]			
Normal	65	32, 79, 160	66, 498
Qualitatively abnormal	66, 382, 398, 485	66	64, 66, 362
Surface charge			
Increased	55	55	
Qualitatively abnormal	429, 430		
Osmotic fragility			
Normal			68
Increased	150, 182, 238, 269		
Deformability	57, 358		
Decreased			

[a]Includes ESR and fluorescence probes of lipids and proteins.

water permeability in their red cells as determined by NMR spectroscopy (Table X).[28]

7.4. Morphology and Physical Properties

Patients with DD were reported to have abnormally high percentages of echinocytes[214,276,296,297] or stomatocytes[313] in their blood, but these findings have been disputed.[311] Stomatocytes were reportedly abundant in the blood of MyD and HC patients. The number of intramembranous particles per unit area of freeze-fractured red cell membranes was reported low in DD.

We have chosen not to review the large number of publications on the use of spin-label and fluorescent probes to study the fluidity or microviscosity of red cell membranes in neuromuscular disorders. Barchi[32] reviewed the controversial results critically and informatively and considers the information inconclusive but worthy of further study (Table XI). Table XI includes references to other claimed abnormalities. One confirmed observation is a slightly increased osmotic fragility of red cells in DD, perhaps signifying a low surface-to-volume ratio in these cells.

A finding worthy of confirmation is the report by Pickard *et al.*[363] that lymphocyte capping is diminished in blood samples from DD patients and their mothers. The authors interpret this finding as an indication of altered membrane fluidity in the cells.

7.5. Composition and Metabolism

Although the electrophoretic pattern of SDS-extracted membrane proteins is reportedly normal in DD, MyD, and HC, much controversial literature exists regarding the ability of membrane preparations to catalyze the phosphorylation of endogenous and exogenous proteins by ATP (protein kinase). Tsung and Palek[447] have perceptively reviewed the inconclusiveness and difficulty of these studies. Lipid composition, another feature that is fraught with experimental pitfalls, has been claimed to be normal and abnormal as reviewed by Plishker and Appel.[366] Other reports of metabolic and en-

zymatic properties of red cells in neuromuscular disease are summarized in Table XII.

7.6. Conclusions

Many of the abnormal features of red cell membranes in neuromuscular disease are difficult to measure. When ghosts are prepared from intact cells, some normal membrane constituents may be lost or degraded. Some cytoplasmic components may become associated with the membrane; various changes in stored preparations may occur over time. The membrane itself is thus not a pure preparation, and the measurement of enzymatic activities may be confounded by variable accessibility of substrates to reactive sites. Until more precise understanding of the structure–function relationships of the red cell membrane is achieved, judgment should be reserved concerning the existence of abnormalities on the red cells of patients with neuromuscular disease. The aforementioned provocative observations recently reported in patients with McLeod cells (i.e., X linkage, mild muscular disease, and clear-cut alterations in red cell morphology) suggest that this model may be of great interest to investigators seeking a correlation between red cell and sarcolemmal membranes.

8. Adenosine Deaminase Variants

It has long been suspected that red cells have access to metabolic substrates *in vivo* that are not abundant in peripheral venous plasma. The evidence for this has been circumstantial. Pig red cells, which cannot metabolize glucose, become rapidly depleted of ATP when incubated in their own plasma *in vitro*. Other substrates such as purine nucleosides, riboses, or trioses can be used by these cells and may be available in the hepatic circulation.[226,239,300] Human red cells can utilize glucose, but they cannot synthesize the adenine ring *de novo*. Their adenine nucleotide content falls slowly on incubation in plasma *in vitro*. Thus, McManus and Lambe[301] postulated that some influx of adenine or adenosine must occur as the cells circulate in the

Table XII. Biochemical Features of Red Cells in Neuromuscular Diseases

	Duchenne's dystrophy	Myotonic dystrophy	Huntington's chorea
Membrane proteins			
Normal	21, 182		498
Lipids			
Normal	182, 366	191, 435	
Qualitatively abnormal	216, 234, 251	190	
Glucose, purine, nucleoside metabolism			
Normal	220		
Increased	103	103, 325, 326	
Decreased	227		
Qualitatively abnormal	54, 421	421	
Cell ATP			
Normal	220		361
Increased	103, 104	103, 104	
Spermine			
Increased	315		
Protein kinase			
Normal	145, 151, 221		
Increased	383–385, 456	221	
Decreased	377–380, 489		
Other enzymes[a]			
Normal	162	378	498
Abnormal	299, 317, 388, 404		

[a]Adenylate cyclase, acetylcholinesterse, phospholipase, acetyl coenzyme A-lysolecithin transferase, methyltransferase, catalase.

bloodstream. Two genetic disorders, adenosine deaminase deficiency and adenosine deaminase excess, alter red cell ATP content and provide strong evidence that adenosine is a key metabolite for the human erythrocyte in the circulation.

Adenosine is metabolized by two enzymes in the red cell. Adenosine kinase (K_m 1.9 μM) phosphorylates adenosine to AMP, which in turn is converted to ADP and ATP. Adenosine deaminase (K_m 40 μM) deaminates adenosine to inosine, which is transported out of the cell.[301,310] Adenosine deaminase deficiency is a recessive trait determined by a gene on autosome 20; homozygotes have red cell adenine nucleotide (deoxyribo + ribonucleotide) levels that are about twice normal.[402,R11] Adenosine deaminase excess is an autosomal dominant trait in which the red cell enzyme is 45- to 70-fold increased; heterozygotes have half-normal red cell ATP levels and a chronic hemolytic anemia.[453] The consequences for red cell ATP of these two inborn errors of metabolism underscore the importance of adenosine as a physiological substrate *in vitro*.

References

1. Acquaye, C. T. A., M. Wilchek, and M. Gorecki. 1982. The effect of antisickling agents on the fluidity of erythrocyte membranes. *Biochem. Int.* 4:221–227.

2. Adornato, B. T., L. Corash, and W. K. Engel. 1977. Erythrocyte survival in Duchenne muscular dystrophy. *Neurology* 27:1093–1094.

3. Adragna, N., and D. C. Tosteson. 1984. Effect of volume change on ouabain-insensitive net onward cation movements in human red cells. *J. Membr. Biol.* 78:43–52.

4. Adragna, N., M. L. Canessa, H. Solomon, E. Slater, and D. C. Tosteson. 1982. Red cell lithium–sodium countertransport and sodium–potassium cotransport in patients with essential hypertension. *Hypertension* 4:795–804.

5. Adragna, N., C. Ellison, H. Solomon, S. Mathysse, D. C. Tosteson, and M. Canessa. 1982. Intrafamilial correlation coefficients for Na countertransport and cotransport in essential hypertension. *Clin. Res.* 30:333A.

6. Agre, P., E. P. Orringer, and V. Bennett. 1982. Deficient red cell spectrin in severe, recessively inherited spherocytosis. *N. Engl. J. Med.* 306:1155–1161.

7. Agre, P., E. P. Orringer, D. H. K. Chui, and V. Bennett. 1981. A molecular defect in two families with hemolytic poikilocytic anemia: Reduction of high affinity membrane binding sites for ankyrin. *J. Clin. Invest.* 68:1566–1576.

8. Akari, S., and S. Matwari. 1971. Ouabain and erythrocyte ghost adenosine triphosphatase: Effects in human muscular dystrophies. *Arch. Neurol.* 24:187–190.

9. Albala, M. M., N. L. Fortier, and B. E. Glader. 1978. Physiologic features of hemolysis associated with altered cation transport and 2,3-diphosphoglycerate content. *Blood* 52:135–141.

10. Albrecht, V. J., and B. Müller-Oerlinghausen. 1976. Zur klinische Bedeutung der intraerythrozyteeren lithium Konzentration: ergebnisse einer katamnestiche Studie. *Arzneim. Forsch.* 26:1145–1147.

11. Ali, S. A., E. C. Gordon-Smith, and H. S. Selhi. 1979. Kinetics of red cell membrane phosphorylation: Altered affinity of HS membrane protein acceptors. *Br. J. Haematol.* 42:225–230.

12. Allan, D., M. M. Billah, J. B. Finean, and R. H. Mitchell. 1976. Release of diacylglycerol-enriched vesicles from erythrocytes with increased intracellular [Ca^{2+}] *Nature (London)* 261:58–60.

13. Allan, D., A. R. Limbrick, P. Thomas, and M. P. Westerman. 1981. Microvesicles from sickle erythrocytes and their relation to irreversible sickling. *Br. J. Haematol.* 47:383–390.

14. Allan, D., A. R. Limbrick, P. Thomas, and M. P. Westerman. 1982. Release of spectrin-free vesicles on reoxygenation of sickled erythrocytes. *Nature (London)* 295:612–613.

15. Allan, D., and R. H. Mitchell. 1976. Production of 1,2-diacylglycerol in human erythrocyte membranes exposed to low concentrations of calcium ions. *Biochim. Biophys. Acta* 455:824–830.

16. Allen, D. W., and S. Cadman. 1979. Calcium-induced erythrocyte membrane changes: The role of adsorption of cytosol proteins and proteases. *Biochim. Biophys. Acta* **551**:1–9.

17. Allen, F. H., S. M. R. Krabbe, and P. A. Corcoran. 1961. A new phenotype (McLeod) in the Kell blood group system. *Vox Sang.* **6**:555–560.

18. Aloni, B., M. Sninitzky, S. Moses, and A. Livne. 1975. Elevated microviscosity in membrane of erythrocytes affected by hereditary spherocytosis. *Br. J. Haematol.* **31**:117–123.

19. Ambrosini, E., F. V. Costa, L. Montebugnoli, C. Borghi, and B. Magnani. 1981. Intralymphocytic sodium concentration as an index of response to stress and exercise in young subjects with borderline hypertension. *Clin. Sci.* **61**:25s–27s.

20. Ambrosini, E., F. Tartagni, L. Montebugnoli, and B. Magnani. 1979. Intralymphocytic sodium in hypertensive patients: A significant correlation. *Clin. Sci.* **57**:325s–327s.

21. Anand, R., and A. E. H. Emery. 1981. Erythrocyte spectrin in Duchenne muscular dystrophy. *Clin. Chim. Acta* **117**:345–354.

22. Anstee, D. J. 1981. The blood group MNSs-active sialoglycoproteins. *Semin. Hematol.* **28**:13–21.

23. Anstee, D. J., W. J. Mawby, and M. J. A. Tanner. 1979. Abnormal blood group Ss active sialoglycoprotein in the membranes of Miltenberger class III, IV, and V human erythrocytes. *Biochem. J.* **183**:193–203.

24. Appel, S., and A. D. Roses. 1976. Membrane biochemical studies in myotonic muscular dystrophy. In: *Membranes and Disease*. L. Bolis, J. F. Hoffman, and A. Leaf, eds. Raven Press, New York. pp. 189–195.

25. Araoye, M. A., I. M. Khatri, L. L. Y. Yao, and E. D. Fries. 1978. Intracellular sodium in hypertensive patients. *Clin. Res.* **26**:53A.

26. Asakura, T., K. Minakata, K. Adachi, M. O. Russell, and E. Schwartz. 1977. Denatured hemoglobin in sickle erythrocytes. *J. Clin. Invest.* **59**:633–640.

27. Asakura, T., S. T. Ohnishi, K. Adachi, M. Ozguc, K. Hashimoto, M. Singer, M. O. Russell, and E. Schwartz. 1980. Effect of cetiedil on erythrocyte sickling: New type of antisickling agent that may affect erythrocyte membranes. *Proc. Natl. Acad Sci. USA* **77**:2955–2959.

28. Ashley, D. L., and J. H. Goldstein. 1981. Nuclear magnetic resonance evidence for abnormal water transport in Duchenne muscular dystrophy erythrocytes. *Biochem. Biophys. Res. Commun.* **100**:364–369.

29. Bachi, T., K. Whiting, M. J. A. Tanner, M. N. Metaxas, and D. J. Anstee. 1977. Freeze-fracture electron microscopy of human erythrocytes lacking the major membrane sialoglycoprotein. *Biochim. Biophys. Acta* **464**:635–639.

30. Balfe, J. W., C. Cole, E. K. M. Smith, J. B. Graham, and L. G. Welt. 1968. A hereditary transport defect in the human red blood cell. *J. Clin. Invest.* **47**:4a.

31. Banga, J. P., J. C. Pinder, W. B. Gratzer, D. C. Linch, and E. R. Huens. 1979. An erythrocyte membrane protein abnormality in march hemoglobinuria. *Lancet* **2**:1048–1049.

32. Barchi, R. L. 1980. Physical probes of biological membranes in studies of the muscular dystrophies. *Muscle Nerve* **3**:82–97.

33. Ben-Ishay, D., A. Avriham, and R. Viskoper. 1975. Increased erythrocytes sodium efflux in genetic hypertensive rats of the Hebrew University strain. *Experientia* **31**:660–662.

34. Benjamin, L. J., G. Kokkini, and C. M. Peterson. 1980. Cetiedil: Its potential usefulness in sickle cell disease. *Blood* **55**:265–270.

35. Bennett, V., and P. J. Stenbock. 1980. Association between ankyrin and the cytoplasmic domain of band 3 isolated from human erythrocyte membranes. *J. Biol. Chem.* **255**:6424–6432.

36. Berglund, G., L. Sigstrom, S. Landin, B. E. Karlberg, and H. Herlitz. 1981. Intra-erythrocytic sodium and Na,K-ATPase concentration and urinary aldosterone excretion in spontaneously hypertensive rats. *Clin. Sci.* **60**:229–232.

37. Berkowitz, L. R., and E. P. Orringer. 1981. The effect of cetiedil, an *in vitro* antisickling agent, on erythrocyte membrane cation permeability. *J. Clin. Invest.* **68**:1215–1220.

38. Berkowitz, L. R., and E. P. Orringer. 1982. Effects of cetiedil on

39. monovalent cation permeability in the erythrocyte: An explanation for the efficacy of cetiedil in the treatment of sickle cell anemia. *Blood Cells* **8**:283–288.

39. Bertelsen, A., B. Havald, and M. Hauge. 1977. A Danish twin study of manic-depressive disorders. *Br. J. Psychol.* **130**:330–351.

40. Bertles, J. F. 1957. Sodium transport across the surface membrane of red blood cells in hereditary spherocytosis. *J. Clin. Invest.* **36**:816–824.

41. Bertles, J. F., and P. F. A. Milner. 1968. Irreversibly sickled erythrocytes: A consequence of the heterogeneous distribution of hemoglobin types in sickle cell anemia. *J. Clin. Invest.* **47**:1731–1741.

42. Beutler, E., E. Ginto, and C. Johnson. 1976. Human red cell protein kinase in normal subjects and patients with hereditary spherocytosis, sickle cell disease, and autoimmune hemolytic anemia. *Blood* **48**:887–898.

43. Bialas, W. A., W. R. Markesberry, and D. A. Butterfield. 1980. Increased chloride transport in erythrocytes in Huntington's disease. *Biochem. Biophys. Res. Commun.* **95**:1895–1900.

44. Bienzle, U., S. Bhadki, H. Knufferman, D. Niethammer, and E. Kleihauer. 1977. Abnormality of erythrocyte membrane protein in a case of congenital stomatocytosis. *Klin. Wochenschr.* **55**:569–572.

45. Bienzle, U., D. Niethammer, U. Keeberg, K. Ungefahr, K. Kohne, and E. Kleihauer. 1975. Congenital stomatocytosis and chronic hemolytic anemia. *Scand. J. Haematol.* **15**:339–346.

46. Blaustein, M. P. 1977. Sodium ions, calcium ions, blood pressure regulation, and hypertension: A reassessment and a hypothesis. *Am. J. Physiol.* **232**:C165–C173.

47. Boellinger, J. W., E. J. Olson, D. Frederickson, and E. R. Hughes. 1965. Plasma and erythrocyte magnesium in muscular dystrophy. *Am. J. Dis. Child.* **110**:172–175.

48. Bogin, E., S. G. Massry, J. Levi, M. Djaldeti, G. Bristol, and J. Smith. 1982. Effect of parathyroid hormone on osmotic fragility of human erythrocytes. *J. Clin. Invest.* **69**:1017–1025.

49. Boivin, P., J. Delaunay, and C. Galand. 1979. Altered erythrocyte membrane protein phosphorylation in an unusual case of hereditary spherocytosis. *Scand. J. Haematol.* **23**:251–255.

50. Boivin, P., and C. Galand. 1977. Erythrocyte membrane phosphorylation in hereditary spherocytosis. *Biomedicine* **27**:34–36.

51. Bookchin, R. M., and V. L. Lew. 1980. Progressive inhibition of the Ca pump and Ca:Ca exchange in sickle red cells. *Nature (London)* **284**:561–563.

52. Bookchin, R. M., and V. L. Lew. 1981. Effect of a "sickling pulse" on calcium and potassium transport in sickle cell trait red cells. *J. Physiol. (London)* **312**:265–280.

53. Bootsma, D., and P. J. MacAlpine. 1979. Report of the committee on the genetic constitution of chromosomes 2, 3, 4, and 5. *Cytogenet. Cell Genet.* **25**:21–31.

54. Bosia, A., G. P. Pescormina, and P. Arose. 1971. Red cell glycolysis in the myodystrophic child. *Eur. J. Clin. Invest.* **1**:413–420.

55. Bosmann, H. B., D. M. Gersten, R. C. Griggs, J. L. Howland, M. S. Hudecki, S. Katzare, and J. McLaughlin. 1976. Erythrocyte surface membrane alterations: Findings in human and animal muscular dystrophies. *Arch. Neurol.* **33**:135–138.

56. Bracharz, H., H. Laas, and G. Betzein. 1962. Uber die Wirkung von aldosteronantagonisten auf der erhohten Blutdruck. *Med. Klin.* **57**:233–238.

57. Brain, M. C., I. Kohn, A. J. McKomas, Y. F. Missirlis, M. P. Rathbone, and J. Vickers. 1978. Red cell instability in Duchenne's syndrome. *N. Engl. J. Med.* **298**:403.

58. Branton, D., C. M. Cohen, and J. Tyler. 1981. Interaction of cytoskeletal proteins on the human erythrocyte membrane. *Cell* **24**:24–32.

59. Brazy, P. C., and R. B. Gunn. 1976. Furosemide inhibition of chloride transport in human red blood cells. *J. Gen. Physiol.* **68**:583–599.

60. Brenner, S. L., and E. D. Korn. 1980. Spectrin/actin complex

isolated from sheep erythrocytes accelerates actin polymerization by simple nucleation. *J. Biol. Chem.* **255**:1670–1676.

61. Brown, H. D., S. K. Chattopadhyay, and A. B. Patel. 1967. Erythrocyte abnormality in human myopathy. *Science* **157**:1577–1578.

62. Brown, H. D., S. K. Chattopadhyay, and A. B. Patel. 1968. Diazacholesterol effect on membrane ATPase. *Metabolism* **17**: 555–559.

63. Burck, H. C. 1971. Der Elektrolytgehalt der Erythrozyten in rahmen der diagnostik der Herzinsuffizienz. *Verh. Dtsch. Ges. Inn. Med.* **77**:140–142.

64. Butterfield, D. A., M. L. Braden, and W. R. Markesbery. 1978. Erythrocyte membrane alterations in Huntington disease: Effects of gamma-amino butyric acid. *J. Supramol. Struct.* **9**:125–130.

65. Butterfield, D. A., D. B. Chestnut, S. H. Appel, and A. D. Roses. 1976. Spin label study of erythrocyte membrane fluidity in myotonic and Duchenne muscular dystrophy and congenital myotonia. *Nature (London)* **263**:159–161.

66. Butterfield, D. A., and W. R. Markesbery. 1980. Specificity of biophysical and biochemical alterations in erythrocyte membranes in neurological disorders. *J. Neurol. Sci.* **47**:261–271.

67. Butterfield, D. A., J. Q. Oeswein, M. E. Prenz, K. C. Hisle, and W. R. Markesbery. 1978. Increased Na,K-ATPase activity in erythrocyte membranes in Huntington's disease. *Ann. Neurol.* **4**:60–62.

68. Butterfield, D. A., M. J. Purdy, and W. R. Markesbery. 1979. Electron spin resonance, hematological, and deformability studies of erythrocytes from patients with Huntington's disease. *Biochim. Biophys. Acta* **551**:452–458.

69. Cade, J. F. 1949. Lithium salts in the treatment of psychotic excitement. *Med. J. Aust.* **2**:349–352.

70. Canali, M., E. Borghi, E. Sani, A. Curti, A. Montranari, A. Novarini, and A. Borghetti. 1981. Increased lithium–sodium countertransport in essential hypertension: Its relationship to family history of hypertension. *Clin. Sci.* **61**:13s–15s.

71. Canessa, M., N. Adragna, I. Bize, T. Connolly, H. Solomon, G. Williams, E. Slater, and D. C. Tosteson. 1980. Ouabain-insensitive cation transport in red cells of normotensive and hypertensive patients. In: *Intracellular Electrolytes and Arterial Hypertension.* H. Zumkley and H. Losse, eds. Thieme, Stuttgart. pp. 239–250.

72. Canessa, M., A. Spalvins, N. Adragna, and B. Falkner. 1984. Red cell sodium countertransport and cotransport in normotensive and hypertensive blacks. *Hypertension* **6**:344–351.

73. Canessa, M., N. Adragna, H. S. Solomon, T. M. Connolly, and D. C. Tosteson. 1980. Increased sodium–lithium countertransport in red cells of patients with essential hypertension. *N. Engl. J. Med.* **302**:772–776.

74. Canessa, M., N. Adragna, H. Solomon, D. C. Tosteson, B. Falkner, and C. Ellison. 1982. Red cell Na transport polymorphism and essential hypertension. *Clin. Res.* **30**:334A.

75. Canessa, M., I. Bize, H. Solomon, N. Adragna, D. C. Tosteson, G. Dagher, R. Garay, and P. Meyer. 1981. Na countertransport and cotransport in human red cells: Function, dysfunction, and genes in essential hypertension. *Clin. Exp. Hypertens.* **3**:783–796.

76. Casper, R. C., G. Pandey, L. Gosenfeld, and J. M. Davis. 1976. Intracellular lithium and chemical response. *Lancet* **2**:418–419.

77. Cazzulo, C. L., E. Smeraldi, E. Sacchetti, and S. Bottinelli. 1975. Intracellular lithium concentration and clinical response. *Br. J. Psychiatry* **126**:298–300.

78. Chailley, B., C. Feo, R. Garay, G. Dagher, R. Bruckdorfer, S. Fischer, J. P. Piau, and J. Delaunay. 1981. Evidence for imbalanced furosemide-sensitive Na + K cotransport in hereditary stomatocytosis. *Scand. J. Haematol.* **27**:365–373.

79. Chalikian, D. M., and R. L. Barchi. 1980. Fluorescent probe analysis of erythrocyte membranes in myotonic dystrophy. *Neurology* **30**:277–285.

80. Chapnam, R. G. 1968. Red cell life span after splenectomy in hereditary spherocytosis. *J. Clin. Invest.* **47**:2263–2267.

81. Chien, S., S. Usami, and J. F. Bertles. 1970. Abnormal rheology of oxygenated blood in sickle cell anemia. *J. Clin. Invest.* **49**:623–634.

82. Chipperfield, A. R. 1978. Stimulation of active transport in human erythrocytes by human plasma. *J. Physiol. (London)* **276**:29P–30P.

83. Chipperfield, A. R. 1981. Chloride dependence of furosemide- and phloretin-sensitive passive sodium and potassium fluxes in human red cells. *J. Physiol. (London)* **312**:435–444.

84. Chiu, D., B. Lubin, B. Roelofsen, and L. L. M. van Deenen. 1981. Sickled erythrocytes accelerate clotting in vitro: An effect of abnormal membrane lipid asymmetry. *Blood* **58**:398–401.

85. Clark, M. R., J. C. Guattelli, N. Mohandas, and S. B. Shohet. 1980. Influence of red cell water content on the morphology of sickling. *Blood* **55**:823–830.

86. Clark, M. R., N. Mohandas, V. Caggiano, and S. B. Shohet. 1978. Effect of abnormal cation transport on deformability of dessicocytes. *J. Supramol. Struct.* **8**:521–532.

87. Clark, M. R., and S. B. Shohet. 1976. Hybrid erythrocytes for membrane studies in sickle cell disease. *Blood* **47**:121–131.

88. Clark, M. R., R. C. Unger, and S. B. Shohet. 1978. Monovalent cation composition and ATP and lipid content of irreversibly sickled cells. *Blood* **51**:1169–1178.

89. Coetzer, T., and S. S. Zail. 1981. Tryptic digestion of spectrin in variants of hereditary elliptocytosis. *J. Clin. Invest.* **67**:1241–1248.

90. Cohen, C. M., and D. Branton. 1981. The normal and abnormal red cell cytoskeleton: A renewal search for molecular defects. *Trends Biochem. Sci.* **6**:266–268.

91. Cohen, N. S., J. E. Eckholm, M. G. Luthra, and D. J. Hanahan. 1976. Biochemical characterization of density-separated human erythrocytes. *Biochim. Biophys. Acta* **419**:229–242.

92. Coletta, M., J. Hofrichter, F. A. Ferrone, and W. A. Eaton. 1982. Kinetics of sickle cell hemoglobin polymerization in single red cells. *Nature (London)* **300**:194–197.

93. Cook, P. J. L., J. E. Noades, M. S. Newton, and R. De Mey. 1977. On the orientation of the Rh: Ell linkage group. *Ann. Hum. Genet.* **41**:157–162.

94. Cooper, R. A., and J. H. Jandl. 1969. The selective and conjoint loss of red cell lipids. *J. Clin. Invest.* **48**:906–914.

95. Cooper, R. A., W. H. Sawyer, M. H. Leslie, J. S. Hill, F. M. Gill, and J. S. Wiley. 1980. Normal fluidity of red cell membranes in hereditary spherocytosis. *Br. J. Haematol.* **46**:299–301.

96. Costa, F. V., E. Ambrosini, L. Montebugnoli, L. Paccaloni, L. Vasconi, and B. Magnani. 1981. Effects of a low-salt diet and of acute salt loading on blood pressure and intralymphocytic sodium concentration in young subjects with borderline hypertension. *Clin. Sci.* **61**:21s–23s.

97. Crosby, W. H. 1952. The pathogenesis of spherocytes and leptocytes (target cells). *Blood* **7**:261–274.

98. Cruikshank, J. K., and D. G. Beevers. 1982. Epidemiology of hypertension in blacks and whites. *Clin. Sci.* **62**:1–6.

99. Cumberbatch, M., and D. B. Morgan. 1981. Relations between sodium transport and sodium concentration in human erythrocytes in health and disease. *Clin. Sci.* **60**:555–564.

100. Cusi, D., C. Barlassina, M. Ferrandi, P. Palazzi, E. Celega, and G. Bianchi. 1981. Relationship between altered Na–K cotransport and Na–Li countertransport in the erythrocytes of "essential" hypertensive patients. *Clin. Sci.* **61**:33s–36s.

101. Dagher, G., and R. P. Garay. 1980. A Na–K cotransport assay for essential hypertension. *Can. J. Biochem.* **58**:1069–1074.

102. D'Amico, G. 1958. Red cell Na and K in congestive heart failure, essential hypertension and myocardial infarction. *Am. J. Med. Sci.* **236**:156–161.

103. Danon, M. J., W. E. Marshall, and A. Omachi. 1977. Erythrocyte metabolism in muscular dystrophy. *Neurology* **27**:398.

104. Danon, M. J., W. E. Marshall, and A. Omachi. 1978. Erythrocyte metabolism in muscular dystrophy. *Arch. Neurol.* **35**:592–595.

105. Das, S. K., and C. N. Rajagopalan. 1980. Superoxide dismutase,

glutathione peroxidase, catalase, and lipid peroxidation of normal and sickled erythrocytes. *Br. J. Haematol.* **44**:87–92.

106. Davidson, J. S., L. H. Opie, and B. Keding. 1982. Sodium–potassium cotransport activity as genetic marker in essential hypertension. *Br. Med. J.* **284**:539–541.

107. Davson, H., and J. F. Danielli. 1952. *The Permeability of Natural Membranes.* Cambridge University Press, London.

108. Dean, J., and A. N. Schechter. 1978. Sickle cell anemia: Molecular and cellular bases of therapeutic approaches. *N. Engl. J. Med.* **299**:752–763.

109. Demendonca, M., M. Grichois, R. P. Garay, J. Sassard, D. Ben-Ishay, and P. Meyer. 1980. Abnormal net Na and K fluxes in erythrocytes of three varieties of hypertensive rats. *Proc. Natl. Acad. Sci. USA* **77**:4283–4286.

110. Demisch, V. L., and H. J. Bochnik. 1976. Zur Verbesserung der Lithiumprophylaxe endogen phasischer Psychosen: Aspekte der parallelen Lithiumbestimmung in Serum und in Erythrozyten. *Arzneim. Forsch.* **26**:1149–1151a.

111. DePirro, R., A. Fusco, R. Lauro, I. Testa, F. Ferreti, and C. DeMartinis. 1980. Erythrocyte insulin receptors in non-insulin-dependent diabetes mellitus. *Diabetes* **29**:96–99.

112. De Wardener, H., and G. A. MacGregor. 1980. Further observations on Dahl's hypothesis that a saluretic substance may be responsible for a sustained rise in arterial pressure: Its possible role in essential hypertension. *Kidney Int.* **18**:1–9.

113. Dintenfass, L., and A. Girolani. 1978. Rigidity of red cells in essential hypertension. *Haemostasis* **7**:298–302.

114. Dise, C. A., D. B. P. Goodman, W. C. Lake, A. Hodson, and H. Rasmussen. 1977. Enhanced sensitivity to calcium in Duchenne muscular dystrophy. *Biochem. Biophys. Res. Commun.* **79**:1286–1290.

115. Dixon, E., and R. M. Winslow. 1981. The interaction between Ca,Mg-ATPase and the soluble activator (calmodulin) in erythrocytes containing haemoglobin S. *Br. J. Haematol.* **47**:391–397.

116. Dorus, E., G. N. Pandey, and J. M. Davis. 1975. Genetic determinants of lithium ion distribution: An *in vitro* and *in vivo* monozygotic–dizygotic twin study. *Arch. Gen. Psychiatry* **32**:1097–1102.

117. Dorus, E., G. N. Pandey, and A. Frazer. 1974. Genetic determinant of lithium ion distribution: An *in vitro* monozygotic–dizygotic twin study. *Arch. Gen. Psychiatry* **31**:463–465.

118. Dorus, E., G. N. Pandey, R. Shaughnessy, and J. M. Davis. 1980. Lithium transport across the RBC membrane: A study of genetic factors. *Arch. Gen. Psychiatry* **37**:80–81.

119. Dorus, E., G. N. Pandey, R. Shaughnessy, M. Gaviria, E. Val, S. Eriksen, and J. M. Davis. 1979. Lithium transport across red cell membrane: A cell membrane abnormality in manic-depressive patients. *Science* **205**:932–934.

120. Dowben, R. M., and K. R. Holley. 1959. Erythrocyte alterations in muscle disease. *J. Lab. Clin. Med.* **54**:867–870.

121. Dreher, K. L., J. W. Eaton, E. Berger, K. P. Breslawee, P. L. Blackshear, and J. G. White. 1980. Calcium-induced erythrocyte rigidity. *Am. J. Pathol.* **101**:543–556.

122. Duhm, J., and B. F. Becker. 1978. Studies on the lithium transport across the red cell membrane. IV. Interindividual variations in the Na-dependent Li countertransport system of human erythrocytes. *Pfluegers Arch.* **370**:211–219.

123. Duhm, J., F. Eisenried, B. F. Becker, and W. Greil. 1976. Studies on the lithium transport across the red cell membrane. I. Li uphill transport by the Na-dependent Li countertransport system of human erythrocytes. *Pfluegers Arch.* **364**:147–155.

124. Duhm, J., and B. O. Gobel. 1982. Sodium–lithium exchange and sodium–potassium cotransport in human erythrocytes. I. Evaluation of a simple uptake test to assess the activity of the two transport systems. *Hypertension* **4**:468–476.

125. Duhm, J., B. Gobel, R. Lorenz, and P. L. Weber. 1981. Na–K cotransport and Na–Li countertransport in erythrocytes from essential hypertensive patients. *Pfluegers Arch. Suppl.* **391**:R21.

126. Duhm, J., B. O. Gobel, R. Lorenz, and P. L. Weber. 1982. Sodium–lithium exchange and sodium–potassium cotransport in

human erythrocytes. II. A simple uptake test applied to normotensive and essential hypertensive individuals. *Hypertension* **4**:477–482.

127. Dunham, P. B., G. W. Steward, and J. C. Ellory. 1980. Chloride-activated passive potassium transport in human erythrocytes. *Proc. Natl. Acad. Sci. USA* **77**:1711–1715.

128. Dutcher, P. O., G. B. Segal, S. A. Feig, D. R. Miller, and M. R. Klemperer. 1975. Cation transport and its altered regulation in human stomatocytic erythrocytes. *Pediatr. Res.* **9**:924–927.

129. Dzandu, J. K., and R. M. Johnson. 1980. Membrane protein phosphorylation in intact normal and sickle cell erythrocytes. *J. Biol. Chem.* **255**:6382.

130. Eaton, J. W., T. D. Skelton, H. S. Swofford, C. E. Kolpin, and H. S. Jacob. 1973. Elevated erythrocyte calcium in sickle cell disease. *Nature (London)* **246**:105–106.

131. Editorial. 1980. Hypertension in blacks and whites. *Lancet* **2**:73–74.

132. Edmonson, R. P. S., R. D. Thomas, P. J. Hilton, J. Patrick, and N. F. Jones. 1975. Abnormal leukocyte composition and sodium transport in essential hypertension. *Lancet* **1**:1003–1005.

133. Ehrlich, B. E., and J. M. Diamond. 1980. Lithium, membranes, and manic-depressive illness. *J. Membr. Biol.* **52**:187–200.

134. Ehrlich, B. E., J. M. Diamond, W. Kaye, E. M. Ornita, and L. Gosenfeld. 1979. Lithium transport in erythrocytes from a pair of twins with manic disorder. *Am. J. Psychiatry* **136**:1477–1478.

135. Eisinger, J., J. Flores, and J. M. Salhany. 1982. Association of cytosol hemoglobin with the membrane in intact erythrocytes. *Proc. Natl. Acad. Sci. USA* **79**:408–412.

136. Elston, R. C. 1972. Pedigree analysis of quantitative traits. *Am. J. Hum. Genet.* **24**:37a.

137. Englehardt, R. 1976. Impaired reassembly of red blood cell membrane components in hereditary spherocytosis. In: *Membranes and Disease.* L. Bolis and J. F. Hoffman, eds. Raven Press, New York. pp. 75–80.

138. Evans, E. A., and R. M. Hochmuth. 1978. Mechanochemical properties of membranes. *Curr. Top. Membr. Transp.* **10**:1–65.

139. Eylar, E. H., M. A. Madoff, O. V. Brody, and J. L. Oncley. 1962. The contribution of sialic acid to the surface charge of the erythrocyte. *J. Biol. Chem.* **237**:1992–2000.

140. Kaul, D. K., M. E. Fabry, P. Windish, S. Baez, and R. L. Nagel. 1983. Erythrocytes in sickle cell anemia are heterogeneous in their rheological and hemodynamic characteristics. *J. Clin. Invest.* **72**:22–31.

141. Fadeke Aderounmu, A., and L. A. Salako. 1979. Abnormal cation composition and transport in erythrocytes from hypertensive patients. *Eur. J. Clin. Invest.* **9**:369–375.

142. Fairbanks, G. 1980. The red cell membrane in normal and abnormal states. In: *Red Blood Cell and Lens Metabolism.* S. K. Srivistava, ed. Elsevier, Amsterdam. pp. 191–212.

143. Fairbanks, G., V. Patel, and J. E. Dino. 1981. Biochemistry of ATP-dependent red cell membrane shape change. *Scand. J. Clin. Lab. Invest.* **41**(Suppl. 156):135–140.

144. Fales, F. W. 1978. Water distribution in blood during sickling of erythrocytes. *Blood* **51**:703–709.

145. Falk, R. S., D. Campion, D. Guthrie, R. S. Sparks, and C. F. Fox. 1979. Phosphorylation of red cell membrane proteins in Duchenne muscular dystrophy. *N. Engl. J. Med.* **300**:258–259.

146. Farvel, Z., A. A. Brickman, H. R. Kaslow, V. M. Brothers, and H. R. Bourne. 1980. Defect of receptor–cyclase coupling protein in pseudohypoparathyroidism. *N. Engl. J. Med.* **303**:237–242.

147. Feig, P., P. Mitchell, and R. I. Sha'afi. 1982. Further characterization of the sodium transport abnormality found in red blood cells of subjects with essential hypertension. *Kidney Int.* **21**:187.

148. Feo, C. J., S. Fischer, J. P. Piau, M. J. Grange, and G. Tchernia. 1980. Premier observation de l'absence d'une proteine de la membrane erythrocytaire (Bande 4.1) dans un cas d'anemie elliptocytaire familiale. *Nouv. Rev. Fr. Hematol.* **22**:315–325.

149. Finne, J. 1980. Identification of the blood group ABH-active glycoprotein components of human erythrocyte membranes. *Eur. J. Biochem.* **104**:181–189.

150. Fischer, E. R., E. Silvestri, J. W. Vester, S. Nolan, U. Ahmad, and T. S. Danowski. 1976. Increased erythrocyte osmotic fragility in pseudohypertrophic muscular dystrophy. *J. Am. Med. Assoc.* **236**:955.

151. Fischer, S. M. Tortolero, J. P. Piau, J. Delaunay, and G. Schapira. 1978. Protein kinase and adenylate cyclase of erythrocyte membrane from patients with Duchenne muscular dystrophy. *Clin. Chim. Acts* **88**:437–440.

152. Fitzgibbon, W. R., T. O. Morgan, and J. B. Meyers. 1980. Erythrocyte ^{22}Na efflux and urinary sodium excretion in essential hypertension. *Clin. Sci.* **59**:195s–197s.

153. Forrester, T. E., and G. A. O. Alleyne. 1980. Leukocyte electrolytes and sodium efflux rate constants in the hypertension of pre-eclampsia. *Clin. Sci.* **59**:199s–201s.

154. Fossel, E. T., and A. K. Solomon. 1981. Relation between red cell membrane Na,K-ATPase and band 3 protein. *Biochim. Biophys. Acta* **649**:557–571.

155. Frazer, A., J. Mendels, D. Brunswick, J. London, M. Pring, T. A. Ramsey, and J. Rybakowski. 1978. Erythrocyte concentrations of the lithium ion: Clinical correlates and mechanisms of action. *Am. J. Psychiatry* **135**:1065–1069.

156. Friedman, S. M., M. Nakashima, R. A. McIndoe, and C. L. Friedman. 1976. Increased erythrocyte permeability to Li and Na in the spontaneously hypertensive rat. *Experientia* **32**:476–478.

157. Funder, J., D. C. Tosteson, and J. O. Wieth. 1978. Effects of bicarbonate on lithium transport in human red cells. *J. Gen. Physiol.* **71**:721–746.

158. Furthmayer, H. 1978. Structural comparison of glycophorins and immunochemical analysis of genetic variants. *Nature (London)* **271**:519–524.

159. Furuhjelm, U., G. Myllyla, H. R. Nevanlinna, S. Nordling, A. Pirkola, A. Gooch, R. Sanger, and P. Tippett. 1969. The red cell phenotype En(a−) and the anti-Ena serological and physiochemical aspects. *Vox Sang.* **17**:256–278.

160. Gaffney, B. J., D. B. Drachman, D. C. Lin, and G. Tennekoon. 1980. Spin-label studies of erythrocytes in myotonic dystrophy: No increase in membrane fluidity. *Neurology* **30**:272–276.

161. Gahmberg, C. G., G. Myllyla, J. Leikola, A. Pirkola, and S. Nordling. 1976. Absence of the major sialoglycoprotein in the membrane of human En(a−) erythrocytes and increased glycosylation of band 3. *J. Biol. Chem.* **251**:6108–6116.

162. Galbraith, D. A., and D. C. Watts. 1980. Changes in some cytoplasmic enzymes from red cells fractionated into age groups by centrifugation in Ficoll/Triosil gradients: Comparison of normal humans and patients with Duchenne muscular dystrophy. *Biochem. J.* **191**:63–70.

163. Galey, W. R., A. P. Evan, P. S. van Nice, W. G. Dail, B. M. Wimer, and R. A. Cooper. 1978. Morphology and physiology of the McLeod erythrocyte. *Vox Sang.* **34**:152–161.

164. Gambhir, K. K., J. A. Archer, and C. J. Bradley. 1978. Characteristics of human erythrocyte insulin receptors. *Diabetes* **27**:701–708.

165. Gambhir, K. K., J. A. Archer, S. G. Nerurkar, I. A. Cruz, and M. Sanders. 1981. Erythrocyte insulin receptors in chronic renal failure. *Nephron* **28**:4–10.

166. Garay, R. P., N. Adragna, M. Canessa, and D. C. Tosteson. 1981. Outward Na and K cotransport in human red cells. *J. Membr. Biol.* **62**:169–174.

167. Garay, R. P., and G. Dagher. 1980. Erythrocyte Na and K transport systems in essential hypertension. In: *Intracellular Electrolytes and Arterial Hypertension.* H. Zumkley and H. Losse, eds. Thieme, Stuttgart. pp. 69–76.

168. Garay, R. P., G. Dagher, and P. Meyer. 1980. An inherited sodium ion–potassium ion cotransport defect in essential hypertension. *Clin. Sci.* **59**:191s–193s.

169. Garay, R. P., G. Dagher, M. G. Pernollet, M. A. Devynk, and P. Meyer. 1980. Inherited defect in a Na,K-cotransport system in erythrocytes from essential hypertensive patients. *Nature (London)* **284**:281–283.

170. Garay, R. P., M. Demendonca, J. L. Elghozi, M. A. Devynk, G.

Dagher, M. G. Pernollet, M. L. Grichois, D. Ben-Ishay, and P. Meyer. 1979. Clinical and pathological relevance of erythrocyte cation fluxes measurement in hypertension. *Clin. Sci.* **57**: 329s–332s.

171. Garay, R. P., J. L. Elghozi, G. Dagher, and P. Meyer. 1980. Laboratory distinction between essential and secondary hypertension by measurement of erythrocyte cation fluxes. *N. Engl. J. Med.* **302**:769–771.

172. Garay, R. P., and P. Meyer. 1979. A new test showing abnormal net Na and K fluxes in erythrocytes of essential hypertension patients. *Lancet* **1**:349–353.

173. Gardos, G. 1959. The role of calcium in the potassium permeability of human erythrocytes. *Acta Physiol. Acad. Sci. Hung.* **15**:121–125.

174. Gessler, U. 1962. Intra- und extrazellulare elektrolytveranderungen bei essentieller Hypertonie vor und nach Behandlung. *Z. Kreislaufforsch.* **51**:177–183.

175. Gillies, R. J. 1982. The binding site for aldolase and G3PDH in erythrocyte membranes. *Trends Biochem. Sci.* **7**:41–42.

176. Glader, B. E. 1976. Salicylate-induced injury of pyruvate kinase-deficient erythrocytes. *N. Engl. J. Med.* **294**:916–918.

177. Glader, B. E., N. Fortier, M. M. Albala, and D. G. Nathan. 1974. Congenital hemolytic anemia associated with dehydrated erythrocytes and increased potassium loss. *N. Engl. J. Med.* **291**:491–496.

178. Glader, B. E., S. E. Lux, A. Muller-Soyano, O. S. Platt, R. D. Propper, and D. G. Nathan. 1978. Energy reserve and cation composition of irreversibly sickled cells *in vivo. Br. J. Haematol.* **40**:527–532.

179. Glader, B. E., and D. G. Nathan. 1978. Cation permeability alterations during sickling: Relationship to cation composition and cellular hydration of irreversibly sickled cells. *Blood* **51**:983–989.

180. Glader, B. E., and D. W. Sullivan. 1979. Erythrocyte disorders leading to potassium loss and cellular dehydration. In: *Normal and Abnormal Red Cell Membranes.* S. E. Lux, V. T. Marchesi, and C. F. Fox, eds. Liss, New York. pp. 503–513.

181. Glynn, I. M., and S. J. D. Karlish. 1975. The sodium pump. *Annu. Rev. Physiol.* **37**:13–55.

182. Godin, D. V., M. A. Bridges, and P. J. M. MacLeod. 1978. Chemical compositional studies of erythrocyte membranes in Duchenne muscular dystrophy. *Res. Commun. Chem. Pathol. Pharmacol.* **20**:331–349.

183. Goldberg, M. A., A. T. Lalos, and H. F. Bunn. 1981. The effect of erythrocyte membrane preparations on the polymerization of sickle hemoglobin. *J. Biol. Chem.* **256**:193–197.

184. Gonick, H. C., H. J. Kramer, W. Paul, and E. Lu. Circulating inhibitor of sodium–potassium-activated adenosine triphosphatase after expansion of extracellular fluid volume in rats. *Clin. Sci. Mol. Med.* **53**:329–334.

185. Gopinath, R. M., and F. F. Vincenzi. 1979. (Ca + Mg)-ATPase activity of sickle cell membranes: Decreased activation by red blood cell cytoplasmic activator. *Am. J. Hematol.* **7**:303–312.

186. Gorter, E., and F. Grendel. 1925. On the bimolecular layers of lipids on the chromocytes of the blood. *J. Exp. Med.* **41**:439–443.

187. Gratzer, W. B. 1981. The red cell membrane and its cytoskeleton. *Biochem. J.* **198**:1–8.

188. Green, F. A. 1972. Erythrocyte membrane lipids and Rh antigen activity. *J. Biol. Chem.* **247**:881–887.

189. Greenquist, A. C., and S. B. Shohet. 1976. Phosphorylation in erythrocyte membranes from abnormally shaped cells. *Blood* **438**:877–886.

190. Grey, J. E., H. J. Gitelman, and A. D. Roses. 1980. Myotonic muscular dystrophy: Defective phospholipid metabolism in the erythrocyte plasma membrane. *J. Clin. Invest.* **65**:1478–1482.

191. Grey, J. E., H. J. Gitelman, and A. D. Roses. 1981. Comment on erythrocyte phospholipid metabolism in myotonic dystrophy. *Ann. Neurol.* **10**:494.

192. Gunn, R. B., D. N. Silvers, and W. F. Rosse. 1972. Potassium permeability in beta thalassemia minor red cells. *J. Clin. Invest.* **51**:1043–1050.

193. Haas, M., J. Schooler, and D. C. Tosteson. 1975. Coupling of lithium to sodium transport in human red cells. *Nature (London)* **258**:425–427.

194. Haddy, F. J. 1980. Mechanism, prevention, and therapy of sodium-dependent hypertension. *Am. J. Med.* **69**:746–758.

195. Haddy, F. J., M. B. Pamani, and D. L. Clough. 1979. Humoral factors and the sodium–potassium pump in volume expanded hypertension. *Life Sci* **24**:2105–2118.

196. Haest, C. W. M., G. Plasa, and B. Deuticke. 1981. Selective removal of lipids from the outer membrane layer of human erythrocytes without hemolysis. *Biochim. Biophys. Acta* **609**:701–708.

197. Haest, C. W. M., G. Plasa, D. Kamp, and B. Deuticke. 1978. Spectrin as a stabilizer of the phospholipid asymmetry in the human erythrocyte membrane. *Biochim. Biophys. Acta* **509**:21–32.

198. Hakamori, S. 1981. Blood group ABH and Ii antigens of human erythrocytes: Chemistry, polymorphism, and their developmental change. *Semin. Hematol.* **28**:39–62.

199. Hargens, A. R., L. J. Bowie, D. Lent, S. Caruthers, R. M. Peters, H. T. Hammel, and P. F. Scholander. 1980. Sickle-cell hemoglobin: Fall in osmotic pressure upon deoxygenation. *Proc. Natl. Acad. Sci. USA* **77**:4310–4312.

200. Hargreaves, W. R., K. N. Giedd, A. Verkleij, and D. Branton. 1980. Reassociation of ankyrin with band 3 in erythrocyte membranes and in lipid vesicles. *J. Biol. Chem.* **255**:11965–11972.

201. Harris, H. W., and S. E. Lux. 1980. Structural characterization of the phosphorylation sites of human erythrocyte spectrin. *J. Biol. Chem.* **255**:11512–11520.

202. Hayashi, S., R. Kootumo, S. Ishigami, G. Tsujino, S. Saeki, and T. Tanaka. 1974. Abnormality in a specific protein of the erythrocyte membrane in hereditary spherocytosis. *Biochem. Biophys. Res. Commun.* **57**:1038–1044.

203. Hebbel, R. P., O. Yamada, C. F. Moldow, H. S. Jacob, J. C. White, and J. W. Eaton. 1980. Abnormal adherence of sickle erythrocytes to cultured vascular endothelium: Possible mechanism for microvascular occlusion in sickle cell disease. *J. Clin. Invest.* **65**:154–160.

204. Henningsen, N. C., S. Mattsson, B. Nosslin, D. Nelson, and O. Ohlsson. 1979. Abnormal whole body and cellular (erythrocytes) turnover of ^{22}Na in normotensive relatives of probands with established essential hypertension. *Clin. Sci.* **57**:321s–324s.

205. Hill, J. S., W. H. Sawyer, G. J. Howlett, and J. S. Wiley. 1981. Hereditary spherocytosis of man: Altered binding of cytoskeletal components to the erythrocyte membrane. *Biochem. J.* **201**:259–266.

206. Hjelm, M. 1974. Methodological aspects of current procedures to separate erythrocytes into age groups. In: *Cellular and Molecular Biology of Erythrocytes*. H. Yoshikawa and S. Rapoport, eds. University Park Press, Baltimore. pp. 427–444.

207. Hobbs, A. S., R. A. Bromback, and B. W. Festoff. 1979. Monovalent cation transport in myotonic dystrophy: Na-K pump ratio in erythrocytes. *J. Neurol. Sci.* **41**:299–306.

208. Hodson, A., and D. Pleasure. 1977. Erythrocyte cation-activated adenosine triphosphatase in Duchenne muscular dystrophy. *J. Neurol. Sci.* **32**:361–369.

209. Hofrichter, J., P. D. Ross, and W. A. Eaton. 1974. Kinetics and mechanism of deoxyhemoglobin S gelation: A new approach to understanding sickle cell disease. *Proc. Natl. Acad. Sci. USA* **71**:4864–4868.

210. Hokin-Neaverson, M., W. A. Burchhardt, and J. W. Jefferson. 1976. Increased erythrocyte Na pump and Na,K-ATPase activity during lithium therapy. *Res. Commun. Chem. Pathol. Pharmacol.* **14**:117–126.

211. Hokin-Neaverson, M., D. A. Spiegel, and C. W. Lewis. 1975. Deficiency of erythrocyte sodium pump activity in bipolar manic-depressive psychosis. *Life Sci.* **15**:1739–1748.

212. Holle, A., W. Mangels, M. Dreyer, J. Kuhnau, and H. W. Rudiger. 1981. Biguanide treatment increases the number of insulin-receptor sites on human erythrocytes. *N. Engl. J. Med.* **305**:563–566.

213. Honig, G. R., P. S. Lacson, and H. S. Maurer. A new familial disorder with abnormal erythrocyte morphology and increased permeability of the erythrocytes to sodium and potassium. *Pediatr. Res.* **5**:159–162.

214. Howells, K. F. 1976. Structural changes of erythrocyte membranes in muscular dystrophy. *Res. Exp. Med.* **168**:213–217.

215. Howland, J. L. 1974. Abnormal potassium conductance associated with genetic muscular dystrophy. *Nature (London)* **251**:724–725.

216. Howland, J. L., and S. L. Iyer. 1977. Erythrocyte lipids in heterozygous carriers of Duchenne's muscular dystrophy. *Science* **198**:309–310.

217. Hudson, C., D. Lee, D. G. Cooper, C. M. Giles, E. W. Iken, J. Poole, D. Grimaldi, and D. J. Anstee. 1979. Mk in three generations of an English family. *J. Immunogenet.* **6**:391–401.

218. Hull, K., and A. D. Roses. 1976. Stoichiometry of sodium and potassium transport in erythrocytes of patients with myotonic muscular dystrophy. *J. Physiol. (London)* **254**:169–181.

219. Ibsen, K. K., H. A. Jensen, J. O. Wieth, and J. Funder. 1982. Essential hypertension: Sodium–lithium countertransport in erythrocytes from patients and from children having one hypertensive parent. *Hypertension* **4**:703–709.

220. Igisu, H., S. Mawatari, and Y. Kuroiwa. 1979. Erythrocyte ATP in Duchenne dystrophy: Effects of ouabain and propranolol. *Neurology* **29**:992–995.

221. Iyer, S. L., P. A. Hoenig, A. P. Sherblom, and J. L. Howland. 1977. Membrane function affected by genetic muscular dystrophy. 1. Erythrocyte ghost protein kinase. *Biochem. Med.* **18**:384–391.

222. Jacob, H. S. 1972. The abnormal red cell membrane in hereditary spherocytosis: Evidence for the crucial role of membrane microfilaments. *Br. J. Haematol.* **23**(Suppl.):35–44.

223. Jacobs, M. A., and D. R. Stewart. 1947. Osmotic properties of the erythrocyte. XII. Ionic and osmotic equilibrium with a complex external solution. *J. Cell. Comp. Physiol.* **30**:79–103.

224. Jaffe, E. R., and E. L. Gottfried. 1968. Hereditary nonspherocytic hemolytic disease associated with an altered phospholipid composition of the erythrocytes. *J. Clin. Invest.* **47**:1375–1381.

225. Jancik, J., R. Schauer, and H. Streicher. 1975. Influence of membrane-bound *N*-acetyl neuraminic acid on the survival of erythrocytes in man. *Hoppe-Seylers Z. Physiol. Chem.* **356**:1329–1331.

226. Jarvis, S. M., J. D. Young, M. Ansay, A. L. Archibald, R. A. Harkness, and R. J. Simmonds. 1980. Is inosine the physiological energy source of pig erythrocytes? *Biochim. Biophys. Acta* **597**:183–188.

227. Johnson, E. C., M. K. Young, P. A. Stacy, and C. H. Beatty. 1979. Erythrocyte glucose metabolism in Duchenne muscular dystrophy. *Clin. Chim. Acta* **98**:77–85.

228. Johnsson, R., and N. E. Saris. 1981. Plasma and erythrocyte lipids in hereditary spherocytosis. *Clin. Chim. Acta* **114**:263–268.

229. Johnsson, S., R. Johnsson, J. Gripenberg, and P. Vuopio. 1980. The fluidity gradient in erythrocyte membranes in hereditary spherocytosis: A spin label study. *Br. J. Haematol.* **46**:73–78.

230. Jones, M. N., and J. K. Nickson. 1981. Monosaccharide transport proteins of the human erythrocyte membrane. *Biochim. Biophys. Acta* **650**:1–20.

231. Jope, R. S., D. J. Jenden, B. E. Ehrlich, and J. M. Diamond. 1978. Choline accumulates in erythrocytes during lithium therapy. *N. Engl. J. Med.* **299**:833–834.

232. Junbu, Y., S. Sato, T. Nakao, and M. Nakao. 1982. Ankyrin is necessary for both drug-induced and ATP-induced shape change of human erythrocyte ghosts. *Biochem. Biophys. Res. Commun.* **104**:1087–1092.

234. Kalofoutis, A., G. Jullien, and V. Spanos. 1977. Erythrocyte phospholipids in Duchenne muscular dystrophy. *Clin. Chim. Acta* **74**:85–87.

235. Kan, Y. W., and A. M. Dozy. 1980. Evolution of the hemoglobin S and C genes in world populations. *Science* **209**:388–391.

236. Kasahara, M., and P. C. Hinkle. 1977. Reconstitution and pu-

rification of the D-glucose transporter from human erythrocytes. *J. Biol. Chem.* **252**:7384–7390.

237. Kay, M. M. B. 1978. Role of physiologic autoantibody in the removal of senescent red cells. *J. Supramol. Struct.* **9**:555–567.

238. Kim, H. D., M. G. Luthra, R. P. Watts, and L. Z. Stern. 1980. Factors influencing osmotic fragility of red blood cells in Duchenne's muscular dystrophy. *Neurology* **30**:726–731.

239. Kim, H. D., R. P. Watts, M. G. Luthra, C. R. Schwalbe, R. T. Connor, and K. Brendel. 1980. A symbiotic relationship of energy metabolism between a "nonglycolytic" mammalian red cell and the liver. *Biochim. Biophys. Acta* **589**:256–263.

240. Kimberling, W. J., R. A. Taylor, R. G. Chapman, and H. A. Lubs. 1978. Linkage and gene localization of hereditary spherocytosis (HS). *Blood* **52**:859–867.

241. Kirkpatrick, F. H., G. M. Woods, and P. L. Lacelle. 1975. Absence of one component of spectrin adenosine triphosphatase in hereditary spherocytosis. *Blood* **46**:945–954.

242. Klassen, G. A., and R. Blostein. 1969. Adenosine triphosphatase and myopathy. *Science* **163**:492–493.

243. Kliman, H. J., and T. L. Steck. 1980. Association of glyceraldehyde-3-phosphate dehydrogenase with the human red cell membrane. *J. Biol. Chem.* **255**:6314–6321.

244. Knauf, P. A. 1979. Erythrocyte anion exchange and the band 3 protein: Transport kinetics and molecular structure. *Curr. Top. Membr. Transp.* **12**:249–363.

245. Knauf, P. A., F. Proverbio, and J. F. Hoffman. 1974. Chemical characteristics and Pronase accessibility of the Na:K pump-associated phosphoprotein of human red blood cells. *J. Gen. Physiol.* **63**:305–323.

246. Knauf, P. A., F. Proverbio, and J. F. Hoffman. 1974. Electrophoretic separation of different phosphoproteins associated with Ca-ATPase and Na,K-ATPase in human red cell ghosts. *J. Gen. Physiol.* **63**:324–336.

247. Knorring, L., L. Oerland, and C. Ferris. 1976. Evaluation of the lithium RBC/plasma ratio as a predictor of the prophylactic effect of lithium in affective disorders. *Pharmakopsychiatr. Neuropsychopharmakol.* **9**:81–84.

248. Knox, E. G. 1977. Distribution of rhesus antigens on red cell surfaces. *Br. J. Haematol.* **37**:537–541.

249. Koller, C. A., E. P. Orringer, and J. C. Parker. 1979. Quinine protects pyruvate kinase deficient red cells from dehydration. *Am. J. Hematol.* **7**:193–199.

250. Kuiper, P. J. C., and A. Livne. 1972. Differences in fatty acid composition between normal erythrocytes and hereditary spherocytosis affected cells. *Biochim. Biophys. Acta* **260**:755–758.

251. Kunze, D., G. Reichmann, E. Egger, G. Leuschner, and E. Eckhardt. 1973. Erythrozyten lipide bei progressiver Muskeldystrophie. *Clin. Chim. Acta* **43**:333–341.

252. Kuratsin-Mills, J., M. Kudo, and S. K. Addae. 1974. Cation content and transport characteristics of the sickle cell erythrocyte and their relationship with the structural changes in the membrane. *Clin. Sci. Mol. Med.* **46**:679–692.

253. Lande, W., K. Cerrone, and W. Mentzer. 1979. Congenital anemia with abnormal cation permeability and cold hemolysis *in vitro*. *Blood* **54**:29a.

254. Lange, Y., R. A. Hardesman, and T. L. Steck. 1982. Role of the reticulum in the stability and shape of the isolated human erythrocyte membrane. *J. Cell Biol.* **92**:714–721.

255. Larsen, F. L., S. Katz, B. D. Roufogalis, and D. E. Brooks. 1981. Physiological shear stresses enhance the Ca permeability of human erythrocytes. *Nature (London)* **294**:667–668.

256. Lauf, P. K., and C. H. Joiner. 1976. Increased potassium transport and ouabain binding in human Rh null red blood cells. *Blood* **48**:457–468.

257. Lawler, J., S. C. Liu, J. Prchal, and J. Palek. 1982. The molecular defect of spectrin in hereditary pyropoikilocytosis. Alterations in the trypsin resistant domain involved in spectrin self-association. *Clin. Res.* **30**:322A.

258. Lebo, R. V., A. V. Carrano, K. Burkhart, A. M. Dozy, L.-C. Yu, and Y. W. Kan. 1979. Assignment of human beta, gamma, and delta globin genes to the short arm of chromosome 11 by chromosome sorting and DNA restriction enzyme analysis. *Proc. Natl. Acad. Sci. USA* **76**:5804–5808.

259. Lee, P., and M. M. Stevenson. 1974. Membrane permeability to sodium and potassium in Rh null red blood cells. *Proc. Int. Congr. Physiol. Sci., 11th, New Delhi.* **11**:16–18.

260. Levine, P., D. Tripodi, J. Struck, C. M. Zmijewski, and W. Pollack. 1973. Hemolytic anemia associated with Rh null but not with Bombay blood. *Vox Sang.* **24**:417–424.

261. Levy, R., R. Zimlichman, A. Keynan, and A. Livne. 1982. The erythrocyte membrane in essential hypertension: Modified temperature dependence of Li efflux. *Biochim. Biophys. Acta* **685**:214–218.

262. Lew, V. L., and R. M. Bookchin. 1980. A Ca-refractory state of the Ca-sensitive K permeability mechanism in sickle cell anemia red cells. *Biochim. Biophys. Acta* **602**:196–200.

263. Lew, V. L., and H. G. Ferreira. 1978. Calcium transport and the properties of a calcium-activated potassium channel in red cell membranes. *Curr. Top. Membr. Transp.* **10**:217–277.

264. Lingsch, C., and K. Martin. 1976. An irreversible effect of lithium administration to patients. *Br. J. Pharmacol.* **57**:323–327.

265. Liu, S., and J. Palek. 1980. Spectrin tetramer–dimer equilibrium and the stability of erythrocyte membrane skeletons. *Nature (London)* **285**:586–588.

266. Liu, S., J. Palek, and J. Prchal. 1981. Defective membrane skeletal assembly in hereditary elliptocytosis. In: *Erythrocyte Membranes: Recent Clinical and Experimental Advances*. W. C. Kruckeberg, J. W. Eaton, and G. J. Brewer, eds. Liss, New York. pp. 157–169.

267. Liu, S., J. Palek, J. Prchal, and R. P. Castleberry. 1981. Altered spectrin dimer–dimer association and red cell membrane instability in hereditary pyropoikilocytosis. *J. Clin. Invest.* **68**:597–605.

268. Livne, A., B. Aloni, S. Moses, and P. J. C. Kuiper. 1973. Linolenoyl sorbitol and the fragility of hereditary spherocytes. *Br. J. Haematol.* **25**:429–435.

269. Lloyd, S. J., and M. G. Nunn. 1978. Osmotic fragility of erythrocytes in Duchenne muscular dystrophy. *Br. Med. J.* **2**:252.

270. Lo, S. S., W. H. Hitzig, and H. R. Marti. 1970. Stomatozytose. *Schweiz. Med. Wochenschr.* **100**:1977–1979.

271. Losse, H., H. Wehmeyer, and F. Wessels. 1960. Der Wasser- und Elektrolytgehalt von Erythrozyten bei arterieller Hypertonie. *Klin. Wochenschr.* **38**:393–395.

272. Losse, H., W. Zidek, H. Zumkley, F. Wessels, and H. Vetter. 1981. Intracellular Na as a genetic marker of essential hypertension. *Clin. Exp. Hypertens.* **3**:627–640.

273. Losse, H., H. Zumkley, and H. Wehmeyer. 1962. Untersuchungen über den Elektrolyt- und Wassergehalt von Erythrozyten bei arterieller Hypertonie. *Z. Kreislaufforsch.* **51**:43–51.

274. Love, W. D., and G. E. Burch. 1953. Plasma and erythrocyte sodium and potassium concentrations in a group of southern white and Negro blood donors. *J. Lab. Clin. Med.* **41**:258–267.

275. Lubin, B., D. Chiu, J. Bastacky, B. Rodofson, and L. L. M. van Deenen. 1981. Abnormalities in membrane phospholipid organization in sickled erythrocytes. *J. Clin. Invest.* **67**:1643–1649.

276. Lumb, E., and A. E. H. Emery. 1975. Erythrocyte deformation in Duchenne muscular dystrophy. *Br. Med. J.* **3**:467–468.

277. Luthra, M. G., and D. A. Sears. 1982. Increased Ca, Mg, and Na + K ATPase activities in erythrocytes of sickle cell anemia. *Blood* **60**:1332–1336.

278. Luthra, M. G., L. Z. Stern, and H. D. Kim. 1979. Ca and Mg-ATPase of red cells in Duchenne and myotonic dystrophy: Effect of soluble cytoplasmic activator. *Neurology* **29**:835–841.

279. Lutz, H. U., and J. Fehr. 1979. Total sialic acid content of glycophorins during senescence of human red blood cells. *J. Biol. Chem.* **254**:11177–11180.

280. Lux, S. E. 1979. Dissecting the red cell membrane. *Nature (London)* **281**:426–428.

281. Lux, S. E. 1979. Spectrin–actin membrane skeleton of normal and abnormal red blood cells. *Semin. Hematol.* **16**:21–51.

282. Lux, S. E. 1982. The membrane skeleton of abnormal red blood cells: Workshop report. In: *Differentiation and Function of Hematopoietic Cell Surfaces.* V. T. Marchesi and R. Gallo, eds. Liss, New York. pp. 197–206.

283. Lux, S. E., and B. E. Glader. 1981. Disorders of the red cell membrane. In: *Hematology of Infancy and Childhood,* 2nd ed. D. G. Nathan and F. A. Oski, eds. Saunders, Philadelphia. pp. 456–565.

284. Lux, S. E., and K. John. 1978. The role of spectrin and actin in irreversibly sickled cells: Unsickling of "irreversibly" sickled ghosts by conditions which interfere with spectrin–actin polymerization. In: *Biochemical and Clinical Aspects of Hemoglobin Abnormalities.* W. S. Caughey, ed. Academic Press, New York. pp. 335–352.

285. Lux, S. E., K. John, and M. J. Karnovsky. 1976. Irreversible deformation of the spectrin–actin lattice in irreversibly sickled cells. *J. Clin. Invest.* **58**:955–963.

286. Lyttkins, L., V. Soderberg, and L. Wetterberg. 1973. Increased lithium erythrocyte–plasma ratio in manic-depressive psychosis. *Lancet* **1**:40.

287. Mahoney, J. R., N. L. Etkin, J. D. McSwigan, and J. W. Eaton. 1982. Assessment of red cell sodium transport in essential hypertension. *Blood* **59**:439–442.

288. Mahoney, J. R., N. L. Etkin, J. D. McSwigan, M. W. Forstoffel, J. R. Eckman, and J. W. Eaton. 1982. Racial differences in hypertension-associated RBC Na permeability. *Clin. Res.* **30**:338A.

289. Marchesi, V. T. 1979. Functional properties of the human red blood cell membrane. *Semin. Hematol.* **16**:3–20.

290. Marinetti, G. V., and K. Cattieu. 1982. Asymmetric metabolism of phosphatidylethanolamine in the human red cell membrane. *J. Biol. Chem.* **257**:245–248.

291. Markesbery, W. R., and D. A. Butterfield. 1977. Scanning electron microscopy studies of erythrocytes in Huntington's disease. *Biochem. Biophys. Res. Commun.* **78**:560–564.

292. Marsh, W. L., N. J. Marsh, A. Moore, W. A. Symmans, C. L. Johnson, and C. M. Redman. 1981. Elevated serum creatine phosphokinase in subjects with McLeod syndrome. *Vox Sang.* **40**:403–411.

293. Marsh, W. L., R. Oyen, and M. E. Nichols. 1976. Kx antigen, the McLeod phenotype, and chronic granulomatous disease. *Vox Sang.* **31**:356–362.

294. Marsh, W. L., R. Oyen, M. E. Nichols, and F. H. Allen. 1975. Chronic granulomatous disease and the Kell blood groups. *Br. J. Haematol.* **29**:247–262.

295. Masys, D. R., P. A. Bromberg, and S. P. Balcerzak. 1974. Red cells shrink during sickling. *Blood* **44**:885–889.

296. Matheson, D., and J. L. Howland. 1974. Erythrocyte deformation in human muscular dystrophy. *Science* **184**:165–166.

297. Matheson, D., and J. L. Howland. 1975. Erythrocytes in human muscular dystrophy [reply]. *Science* **187**:454.

298. Matsumoto, N., Y. Yawata, and H. S. Jacob. 1977. Association of decreased membrane phosphorylation with red blood cell spherocytosis. *Blood* **49**:233–239.

299. Mawatari, S., M. Schonberg, and M. Olarte. 1976. Biochemical abnormalities of erythrocyte membrane in Duchenne dystrophy. *Arch. Neurol.* **33**:489–493.

300. McManus, T. J. 1974. Alternate pathways for metabolism: A comparative view. In: *The Human Red Cell in Vitro.* T. J. Greenwalt and G. A. Jamieson, eds. Grune & Stratton, New York. pp. 49–61.

301. McManus, T. J., and C. Lambe. 1973. Species differences in nucleoside metabolism of red cells. In: *Erythrocytes, Thrombocytes, Leukocytes.* E. Gerlach, K. Moser, E. Deutsch, and W. Williams, eds. Thieme, Stuttgart. pp. 135–139.

302. Mendels, J., and A. Frazer. 1973. Intracellular lithium concentration and clinical response: Towards a membrane theory of depression. *J. Psychiatr. Res.* **10**:9–18.

303. Mendelwicz, J., and J. L. Fleiss. 1974. Linkage studies with X-chromosome markers in bipolar (manic-depressive) and unipolar (depressive) illness. *Biol. Psychiatry* **9**:261–294.

304. Mendlewicz, J., P. Verbanck, P. Linkowski, and J. Wilmotte. 1978. Lithium accumulation in erythrocytes of manic-depressive patients: An *in vitro* twin study. *Br. J. Psychiatry* **133**:436–444.

305. Mentzer, W. C., R. L. Bahner, H. Schmidt-Schöenbein, S. H. Robinson, and D. G. Nathan. 1971. Selective reticulocyte destruction in erythrocyte pyruvate kinase deficiency. *J. Clin. Invest.* **50**:688–699.

306. Mentzer, W. C., G. K. H. Lam, B. H. Lubin, A. Greenquist, S. L. Schrier, and W. Lande. 1978. Membrane effects of imidoesters in hereditary stomatocytosis. *J. Supramol. Struct.* **9**:275–288.

307. Mentzer, W. C., B. H. Lubin, and S. Emmons. 1976. Correction of the permeability defect in hereditary stomatocytosis by dimethyl adipimidate. *N. Engl. J. Med.* **294**:1200–1204.

308. Mentzer, W. C., W. B. Smith, J. Goldstone, and S. B. Shohet. 1975. Hereditary stomatocytosis: Membrane and metabolic studies. *Blood* **46**:659–669.

309. Meyer, P., R. P. Garay, C. Nazaret, G. Dagher, M. Bellet, M. Broyer, and J. Finegold. 1981. Inheritance of abnormal erythrocyte cation transport in essential hypertension. *Br. Med. J.* **282**:1114–1117.

310. Meyskins, F. L., and H. E. Williams. 1971. Adenosine metabolism in human erythrocytes. *Biochim. Biophys. Acta* **240**:170–179.

311. Miale, T. D., J. L. Frias, and D. L. Lawson. 1975. Erythrocytes in human muscular dystrophy. *Science* **187**:452.

312. Miller, D. R., F. R. Rickles, M. A. Lichtman, P. L. Lacelle, J. Bates, and R. I. Weed. 1971. A new variant of hereditary hemolytic anemia with stomatocytosis and erythrocyte cation abnormality. *Blood* **38**:184–204.

313. Miller, S. E., A. D. Roses, and S. H. Appel. 1976. Scanning electron microscopy studies in muscular dystrophy. *Arch. Neurol.* **33**:172–174.

314. Mishra, S. K., M. Hobson, and D. Desiah. 1980. Erythrocyte membrane abnormalities in human myotonic dystrophy. *J. Neurol. Sci.* **46**:333–340.

315. Mollica, F., S. L. Volti, A. Rapisarda, G. Loingo, L. Pavone, and A. Vanella. 1980. Increased erythrocyte spermine in Duchenne muscular dystrophy. *Pediatr. Res.* **14**:1196–1198.

316. Mollman, J. E., J. C. Cardenas, and D. E. Pleasure. 1980. Alteration of calcium transport in Duchenne erythrocytes. *Neurology* **30**:1236–1239.

317. Moore, R. B., and S. H. Appel. 1980. Methylation of membrane phospholipids in patients with myotonic and Duchenne muscular dystrophy. *Exp. Neurol.* **70**:380–391.

318. Moore, S., C. F. Woodrow, and D. B. L. McClelland. 1982. Isolation of membrane components associated with human red cell antigens Rh₀(D), (c), (E), and Fyᵃ. *Nature (London)* **295**:529–531.

319. Morgan, T., J. Meyers, and W. Fitzgibbon. 1981. Sodium intake, blood pressure, and red cell sodium efflux. *Clin. Exp. Hypertens.* **3**:641–654.

320. Morrow, J. S. 1982. Structure of the erythrocyte cytoskeleton: Workshop report. In: *Differentiation and Function of Hematopoietic Cell Surfaces.* V. T. Marchesi and R. C. Gallo, eds. Liss, New York. pp. 193–196.

321. Morrow, J. S., and V. T. Marchesi. 1981. Self-assembly of spectrin oligomers *in vitro*: A basis for a dynamic cytoskeleton. *J. Cell Biol.* **88**:463–468.

322. Morrow, J. S., D. W. Speicher, W. J. Knowles, C. J. Hsu, and V. T. Marchesi. 1980. Identification of functional domains of human erythrocyte spectrin. *Proc. Natl. Acad. Sci. USA* **77**:6592–6596.

323. Mueller, T. J., and M. Morrison. 1977. Detection of a variant of protein 3, the major transmembrane protein of the human erythrocyte. *J. Biol. Chem.* **252**:6573–6576.

324. Mueller, T. J., and M. Morrison. 1981. Glycoconnectin (PAS2), a membrane attachment site for the human erythrocyte cytoskeleton. In: *Erythrocyte Membranes: Recent Clinical and Ex-*

perimental Advances, Volume 2. W. C. Krukenberg, J. W. Eaton, and G. J. Brewer, eds. Liss, New York. pp. 95–112.

325. Müller, M. M., M. Frass, and B. Mamloi. 1979. Metabolism of adenine and adenosine in erythrocytes of patients with myotonic muscular dystrophy. *Adv. Exp. Med. Biol.* **122A**:183–188.

326. Müller, M. M., R. Kuzmits, M. Frass, and B. Mamoli. 1980. Purine metabolism of erythrocytes in myotonic dystrophy. *J. Neurol.* **223**:59–66.

327. Munro-Faure, A. D., D. M. Hill, and J. Anderson. 1971. Ethnic differences in human blood cell sodium concentration. *Nature (London)* **231**:457–458.

328. Murphy, J. R. 1965. Erythrocyte metabolism. VI. Cell shape and the localization of cholesterol in the erythrocyte membrane. *J. Lab. Clin. Med.* **65**:756–763.

329. Nakao, M., T. Nakao, and S. Yamazoe. 1960. Adenosine triphosphate and the maintenance of shape of the human red cell. *Nature (London)* **187**:945–946.

330. Nakashima, K., and E. Beutler. 1979. Erythrocyte cellular and membrane deformability in hereditary spherocytosis. *Blood* **53**:481–486.

331. Nathan, D. G., and S. B. Shohet. 1970. Erythrocyte ion transport defects and hemolytic anemia: Hydrocytosis and dessicytosis. *Semin. Hematol.* **7**:381–408.

332. Nerurkar, S. G., and K. K. Gambhir. 1979. Insulin receptor assay in human erythrocytes as an index to insulin sensitivity of body tissues. *Clin. Chim. Acta* **25**:1672–1673.

333. Nicholson, G. L., S. P. Masouredis, and S. J. Singer. 1971. Quantitative, 2-dimensional ultrastructural distribution of Rh(D) antigenic sites on human erythrocyte membranes. *Proc. Natl. Acad. Sci. USA* **68**:1416–1420.

334. Niebroj-Dobosz, I. 1976. Erythrocyte ghosts Na,K-ATPase activity in Duchenne's muscular dystrophy and myotonia. *J. Neurol.* **214**:61–69.

335. Nigg, E. A., C. Bron, M. Girardet, and R. J. Cherry. 1980. Band 3–glycophorin A association in erythrocyte membranes demonstrated by combining protein diffusion measurements with antibody-induced cross linking. *Biochemistry* **19**:1887–1893.

336. Nigg, E. A., and R. J. Cherry. 1980. Anchorage of a band 3 population at the erythrocyte cytoplasmic membrane surface: Protein rotational diffusion measurements. *Proc. Natl. Acad. Sci. USA* **77**:4702–4706.

337. Noguchi, C. T., and A. N. Schechter. 1981. Review: The intracellular polymerization of sickle hemoglobin and its relevance to sickle cell disease. *Blood* **58**:1057–1068.

338. Nordland, J. R., C. F. Schmidt, S. N. Dicken, and T. E. Thompson. 1981. Transbilayer distribution of phosphatidylethanolamine in large and small unilamellar vesicles. *Biochemistry* **20**:3237–3241.

339. Nozawa, Y., T. Noguchi, H. Iida, H. Fukushima, and Y. Ito. 1974. Erythrocyte membrane in hereditary spherocytosis: Alterations in surface ultrastructure and membrane proteins as inferred by scanning electron microscopy and SDS disc gel electrophoresis. *Clin. Chim. Acta* **55**:81–85.

340. Olefsky, J. M., and G. M. Reaven. 1974. Decreased insulin binding to lymphocytes from diabetic subjects. *J. Clin. Invest.* **54**:1323–1328.

341. Orringer, E. P. 1984. A further characterization of the selective K movements observed in human red blood cells following acetylphenylhydrazine exposure. *Am. J. Hematol.* **16**:355–366.

342. Gulley, M. L., D. W. Ross, C. Feo, and E. P. Orringer. 1982. The effect of cell hydration on the deformability of normal and sickle erythrocytes. *Am. J. Hematol.* **13**:283–291.

343. Orringer, E. P., and J. C. Parker. 1977. Selective increase of potassium permeability in red blood cells exposed to acetylphenylhydrazine. *Blood* **50**:1013–1021.

344. Oski, F. A., J. L. Naiman, S. F. Blum, H. S. Zarkowsky, J. Whaun, S. B. Shohet, A. Green, and D. G. Nathan. 1969. Congenital hemolytic anemia with high-sodium, low-potassium red cells: Studies of three generations of a family with a new variant. *N. Engl. J. Med.* **280**:909–916.

345. Ostrow, D. G., G. N. Pandey, J. M. Davis, S. W. Hurt, and D. C. Tosteson. 1978. A heritable disorder of lithium transport in erythrocytes of a subpopulation of manic-depressive patients. *Am. J. Psychiatry* **135**:1070–1078.

346. Palek, J. 1977. Red cell membrane injury in sickle cell anemia. *Br. J. Haematol.* **35**:1–9.

347. Palek, J. 1977. Red cell calcium content and transmembrane calcium movements in sickle cell anemia. *J. Lab. Clin. Med.* **89**:1365–1374.

348. Palek, J., S. C. Liu, and P. A. Liu. 1978. Spectrin assembly in irreversibly sickled cell membranes: Role of Ca and ATP. In: *Biochemical and Clinical Aspects of Hemoglobin Abnormalities*. W. S. Caughey, ed. Academic Press, New York. pp. 353–367.

349. Palek, J., S. C. Liu, and P. A. Liu. 1978. Crosslinking of the nearest membrane protein neighbors in ATP depleted, Ca enriched, and irreversibly sickled red cells. In: *Erythrocyte Membranes: Recent Clinical and Experimental Advances*. W. C. Krukeberg, J. W. Eaton, and G. J. Brewer, eds. Liss, New York. pp. 75–88.

350. Palek, J., S. Liu, P. Liu, J. Prchal, and R. P. Castleberry. 1981. Altered assembly of spectrin in red cell membranes in hereditary pyropoikilocytosis. *Blood* **57**:130–139.

351. Pandey, G. N., E. Dorus, J. M. Davis, and D. C. Tosteson. 1979. Lithium transport in human red blood cells: Genetic and clinical aspects. *Arch. Gen. Psychiatry* **36**:902–908.

352. Pandey, G. N., D. G. Ostrow, M. Haas, E. Dorus, R. C. Casper, J. M. Davis, and D. C. Tosteson. 1977. Abnormal lithium and sodium transport in erythrocytes of a manic-depressive patient and some members of his family. *Proc. Natl. Acad. Sci. USA* **74**:3607–3611.

353. Pandey, G. N., B. Sarkadi, M. Haas, R. B. Gunn, J. M. Davis, and D. C. Tosteson. 1978. Lithium transport pathways in human red blood cells. *J. Gen. Physiol.* **72**:233–247.

354. Parker, J. C. 1978. Sodium and calcium movements in dog red blood cells. *J. Gen. Physiol.* **71**:1–17.

355. Parker, J. C., E. P. Orringer, and T. J. McManus. 1978. Disorders of ion transport in red blood cells. In: *Physiology of Membrane Disorders*. T. E. Andreoli, J. F. Hoffman, and D. D. Fanestil, eds. Plenum Press, New York. pp. 773–800.

356. Patel, V. P., and G. Fairbanks. 1981. Spectrin phosphorylation and shape change of human erythrocyte ghosts. *J. Cell Biol.* **88**:430–440.

357. Pearson, T. W. 1978. Na,K-ATPase of Duchenne muscular dystrophy erythrocyte ghosts. *Life Sci.* **22**:127–132.

358. Pearcy, A. K., and M. E. Miller. 1975. Reduced deformability of erythrocyte membranes from patients with Duchenne muscular dystrophy. *Nature (London)* **258**:147–148.

359. Perutz, M., and J. M. Mitchison. 1950. State of haemoglobin in sickle cell anaemia. *Nature (London)* **166**:677–680.

360. Peter, J. B., M. Worsfold, and C. M. Pearson. 1969. Erythrocyte ghost ATPase in Duchenne dystrophy. *J. Lab. Clin. Med.* **74**:103–108.

361. Pettegrew, J. W., T. Glonek, and R. M. Stewart. 1979. Phosphorus-31 nuclear magnetic resonance studies in blood in Huntington's disease. *Trans. Am. Neurol. Assoc.* **104**:233–235.

362. Pettegrew, J. W., J. S. Nichols, and R. M. Stewart. 1980. Membrane studies in Huntington's disease: Steady state fluorescence studies of intact erythrocytes. *Ann. Neurol.* **8**:381–386.

363. Pickard, N. A., H. D. Gruemer, H. L. Verrill, E. R. Isaacs, M. Robinow, W. E. Nance, E. C. Myers, and B. Goldsmith. 1978. Systemic membrane defect in the proximal muscular dystrophies. *N. Engl. J. Med.* **299**:841–846.

364. Plapp, F. V., J. P. Evans, and L. L. Tilzer. 1981. Detection of $Rh_o(D)$ antigen on the inner surface of Rh negative erythrocyte membranes. *Fed. Proc.* **40**:208.

365. Platt, O. S., S. E. Lux, and D. G. Nathan. 1981. Exercise-induced hemolysis in xerocytosis: Erythrocyte dehydration and shear sensitivity. *J. Clin. Invest.* **68**:631–638.

366. Plishker, G. A. and S. H. Appel. 1980. Red blood cell alterations in muscular dystrophy: The role of lipids. *Muscle Nerve* **3**:70–81.

367. Plishker, G. A., H. J. Gitelman, and S. H. Appel. 1978. Myotonic muscular dystrophy: Altered calcium transport in erythrocytes. *Science* **200**:323–325.

368. Postnov, Y. U., S. Orlov, P. Gulak, and A. Shevchenko. 1976. Altered permeability of the erythrocyte membrane for sodium and potassium ions in spontaneously hypertensive rats *Pfluegers Arch.* **365**:257–264.

369. Postnov, Y. U., S. N. Orlov, and N. I. Pokudin. 1979. Decrease of calcium binding by the red blood cell membrane in spontaneously hypertensive rats and in essential hypertension. *Pfluegers Arch* **379**:191–195.

370. Postnov, Y. U., S. N. Orlov, A. Shevchenko, and A. M. Adler. 1977. Altered sodium permeability, calcium binding, and Na,K-ATPase activity in the red cell membrane in essential hypertension. *Pfluegers Arch.* **371**:263–269.

371. Poston, L., R. B. Sewell, S. P. Wilkinson, P. J. Richardson, R. Williams, E. M. Clarkson, G. A. MacGregor, and H. E. de Wardener. 1981. Evidence for a circulating sodium transport inhibitor in essential hypertension. *Br. Med. J.* **282**:847–849.

372. Probstfield, J. L., Y. Wang, and A. L. From. 1972. Cation transport in Duchenne muscular dystrophy erythrocytes. *Proc. Soc. Exp. Biol. Med.* **141**:479–481.

373. Reed, C. F., and S. N. Swisher. 1966. Erythrocyte lipid loss in hereditary spherocytosis. *J. Clin. Invest.* **45**:777–781.

374. Robinson, T. J., J. A. Archer, K. K. Gambhir, V. W. Hollis, L. Carter, and C. Bradley. 1979. Erythrocytes: A new cell type for the evaluation of insulin receptor defects in diabetic humans. *Science* **205**:200–202.

375. Rogausch, H. 1978. Influence of Ca on red cell deformability and adaptation to sphering agents. *Pfluegers Arch.* **373**:43–47.

376. Rosa, R. M., B. E. Bierer, R. Thomas, J. S. Stoff, M. Kruskall, S. Robinson, H. F. Bunn, and F. H. Epstein. 1980. A study of induced hyponatremia in the prevention and treatment of sickle cell crisis. *N. Engl. J. Med.* **303**:1138–1143.

377. Roses, A. D., and S. H. Appel. 1973. Protein kinase activity in erythrocyte ghosts of patients with myotonic muscular dystrophy. *Proc. Natl. Acad. Sci. USA* **70**:1855–1859.

378. Roses, A. D., and S. H. Appel. 1974. Muscle membrane protein kinase in myotonic muscular dystrophy. *Nature (London)* **250**:245–247.

379. Roses, A. D., and S. H. Appel. 1975. Phosphorylation of component a of the erythrocyte membrane in myotonic muscular dystrophy. *J. Membr. Biol.* **20**:51–58.

380. Roses, A. D., and S. H. Appel. 1978. Inherited membrane disorders of muscle: Duchenne muscular dystrophy and myotonic muscular dystrophy. In: *Physiology of Membrane Disorders*. T. E. Andreoli, J. F. Hoffman, and D. D. Fanestil, eds. Plenum Press, New York pp. 801–815.

381. Roses, A. D., S. H. Appel, D. A. Butterfield, S. E. Miller, and D. B. Chestnut. 1975. Specificity of biochemical and biophysical tests in Duchenne and myotonic muscular dystrophy, carrier states, and congenital myotonia. *Trans. Am. Neurol. Assoc.* **100**:131–134.

382. Roses, A. D., D. A. Butterfield, S. H. Appel, and D. B. Chestnut. 1975. Phenytoin and membrane fluidity in myotonic dystrophy. *Arch. Neurol.* **32**:535–538.

383. Roses, A. D., G. B. Hartwig, M. Mabry, Y. Nagano, and S. E. Miller. 1980. Red blood cell and fibroblast membranes in Duchenne and myotonic muscular dystrophy. *Muscle Nerve* **3**:36–54.

384. Roses, A. D., M. H. Herbstreith, and S. H. Appel. 1975. Membrane protein kinase alteration in Duchenne muscular dystrophy. *Nature (London)* **254**:350–351.

385. Roses, A. D., M. J. Roses, S. E. Miller, K. L. Hull, and S. H. Appel. 1976. Carrier detection in Duchenne muscular dystrophy. *N. Engl. J. Med.* **294**:193–198.

386. Rothman, J. E., and J. Lenard. 1977. Membrane asymmetry. *Science* **195**:743–753.

387. Rowland, L. P. 1976. Pathogenesis of muscular dystrophies. *Arch. Neurol.* **33**:315–321.

388. Rowland, L. P. 1980. Biochemistry of muscle membranes in Duchenne muscular dystrophy. *Muscle Nerve* **3**:3–20.

389. Ruitenbeek, W. 1979. Membrane-bound enzymes of erythrocytes in human muscular dystrophy. *J. Neurol. Sci.* **41**:71–80.

390. Rybakowski, J. 1977. Pharmacogenetic aspects of red blood cell lithium index in manic-depressive psychosis. *Biol. Psychiatry* **12**:425–429.

391. Rybakowski, J., M. Chlopocka, Z. Kapelski, B. Hernacka, Z. Szajnerman, and K. Kasprzak. 1974. Red blood cell lithium index in patients with affective disorders in the course of lithium prophylaxis. *Int. Pharmakopsychiatry* **9**:116–171.

392. Rybakowski, J., and W. Strzyżewski. 1976. Red blood cell lithium index and long term maintenance treatment. *Lancet* **1**:1408–1409.

393. Rygielski, D. B., D. L. Kropp, and W. N. Duran. 1981. Hypertension and the Na-K pump. *Fed. Proc.* **40**:611a.

394. Sachs, J. R. 1971. Ouabain-insensitive sodium movements in the human red blood cell. *J. Gen. Physiol.* **57**:259–282.

395. Salhany, J. M., and N. Shaklai. 1979. Functional properties of human hemoglobin bound to the erythrocyte membrane. *Biochemistry* **18**:893–899.

396. Sarkadi, B. 1980. Active calcium transport in human red cells. *Biochim. Biophys. Acta* **604**:159–190.

397. Sarkadi, B., J. K. Allimoff, R. B. Gunn, and D. C. Tosteson. 1978. Kinetics and stoichiometry of Na-dependent Li transport in human red blood cells. *J. Gen. Physiol.* **72**:249–265.

398. Sato, B., K. Nishikida, L. T. Samuels, and F. H. Tyler. 1978. Electron spin resonance studies of erythrocytes from patients with Duchenne muscular dystrophy. *J. Clin. Invest.* **61**:251–259.

399. Sauberman, N., G. Fairbanks, H. U. Lutz, N. L. Fortier, and L. M. Snyder. 1981. Altered red blood cell surface area in hereditary xerocytosis. *Clin. Chim Acta* **114**:149–161.

400. Sauberman, N., N. L. Fortier, G. Fairbanks, R. J. O'Connor, and L. M. Snyder. 1979. Red cell membrane in hemolytic disease: Studies of variables affecting electrophoretic analysis. *Biochim. Biophys. Acta* **556**:292–313.

401. Schartum-Hansen, H. 1935. Die Genese der Ovalocyten. *Acta Med. Scand.* **86**:348–360.

402. Schmalsteig, F. C., A. S. Goldman, G. C. Mills, T. M. Monahan, J. A. Nelson, and R. M. Goldblum. 1976. Nucleoside metabolism in adenosine deaminase deficiency. *Pediatr. Res.* **10**:393.

403. Schmidt, P. G., L. D. Nelson, A. L. Mark, D. D. Heistad, and F. M. Abboud. 1974. Inhibition of adrenergic vasoconstriction by quinidine. *J. Pharmacol. Exp. Ther.* **188**:124–143.

404. Schmitt, J., R. Royer, C. Schmidt, and R. Mucka. 1976. Approche, nosographie des myopathies hereditaires: Essai de classification et de diagnostic par les cholinesterases globulaires et les pseudocholinesterases seriques. *Rev. Neurol.* **132**:481–487.

405. Schroeder, E. 1968. Relationship between plasma renin, plasma sodium, and erythrocyte sodium in healthy persons and hypertensives. *Ger. Med. Mon.* **13**:384–389.

406. Schroter, W., and K. Ungefahr. 1976. Studies on the cation transport in high sodium and low potassium red cells in hereditary hemolytic anemia associated with stomatocytosis. In: *Membranes and Disease*. L. Bolis, J. F. Hoffman, and A. Leaf, eds. Raven Press, pp. 95–98.

407. Serjeant, G. R. 1975. Fetal haemoglobin in homozygous sickle cell disease. *Clin. Haematol.* **4**:109–122.

408. Serjeant, G. R., B. Serjeant, and P. F. A. Milner. 1969. The irreversibly sickled cell: A determinant of haemolysis in sickle cell anemia. *Br. J. Haematol.* **17**:527–533.

409. Sha'afi, R. I., S. B. Rodan, R. L. Hinz, S. M. Fernandez, and G. A. Rodan. 1975. Abnormalities in membrane microviscosity and ion transport in genetic muscular dystrophy. *Nature (London)* **254**:525–526.

410. Shaklai, N., V. S. Sharma, and H. M. Ranney. 1981. Interaction of sickle cell hemoglobin with erythrocyte membranes. *Proc. Natl. Acad. Sci. USA* **78**:65–68.

411. Sheehy, R., and G. B. Ralston. 1978. Abnormal binding of spec-

trin to the membrane of erythrocytes in some cases of hereditary spherocytosis. *Blut* **36**:145–148.

412. Sheetz, M. P., M. Schindler, and D. E. Koppel. 1980. Lateral mobility of integral membrane proteins is increased in spherocytic erythrocytes. *Nature (London)* **285**:510–512.

413. Sheetz, M. P., and S. J. Singer. 1974. Biological membranes as bilayer couples: A molecular mechanism of drug–erythrocyte interaction. *Proc. Natl. Acad. Sci. USA* **71**:4457–4461.

414. Shohet, S. B. 1979. Reconstitution of spectrin-deficient spherocytic mouse erythrocyte membranes. *J. Clin. Invest.* **64**:483–494.

415. Shohet, S. B., D. G. Nathan, B. M. Livermore, S. A. Feif, and E. R. Jaffe. 1973. Hereditary hemolytic anemia associated with abnormal membrane lipid. II. Ion permeability and transport abnormalities. *Blood* **42**:1–8.

416. Shotton, D. M., B. E. Burke, and D. Branton. 1979. The molecular structure of human erythrocyte spectrin: Biophysical and electron microscopic studies. *J. Mol. Biol.* **131**:303–329.

417. Singer, S. J., and G. L. Nicholson. 1972. The fluid mosaic model of the structure of cell membranes. *Science* **175**:720–731.

418. Smith, J. A., F. V. Lucas, A. P. Martin, D. A. Senhauser, and M. L. Vorbeck. 1973. Lipid–protein interactions of erythrocyte membranes: Comparison of normal O, Rh(D) positive with the rare O, Rh null. *Biochem. Biophys. Res. Commun.* **54**:1015–1023.

419. Smith, J. A., and A. J. Sinclair. 1977. Rh O and the erythrocyte membrane. *Blood* **49**:491–492.

420. Snyder, L. M., H. U. Lutz, N. Sauberman, J. Jacobs, and N. L. Fortier. 1978. Fragmentation and myelin formation in hereditary xerocytosis and other hemolytic anemias. *Blood* **52**:750–761.

421. Solomons, C. C., S. P. Ringel, E. I. Nwuke, and H. Suga. 1977. Abnormal adenine metabolism of erythrocytes in Duchenne and myotonic muscular dystrophy. *Nature (London)* **268**:55–56.

422. Souweine, G., J. C. Bernard, Y. Lasna, and J. Lachanat. 1978. The sodium pump of erythrocytes from patients with Duchenne muscular dystrophy: Effect of ouabain on the active sodium flux and on Na,K-ATPase. *J. Neurol.* **217**:287–294.

423. Steck, T. L. 1974. The organization of proteins in the human red cell membrane. *J. Cell Biol.* **62**:1–19.

424. Steinberg, M. H., J. W. Eaton, E. Berger, M. B. Coleman, and F. J. Oelshlegel. 1978. Erythrocyte calcium abnormalities and the clinical severity of sickling disorders. *Br. J. Haematol.* **40**:533–539.

425. Sturgeon, P. 1970. Hematological observations in the anemia associated with blood type Rh null. *Blood* **36**:310–320.

426. Swarts, H. G. P., S. L. Bonting, J. J. DePont, F. S. Steckhoven, T. A. Thien, and A. V. Laar. 1981. Cation fluxes and Na,K-activated ATPase activity in erythrocytes of patients with essential hypertension. *Hypertension* **3**:641–649.

427. Symmans, W. A., C. S. Shepard, W. L. Marsh, R. Oyen, S. B. Shohet, and B. J. Linehan. 1979. Hereditary acanthocytosis associated with the McLeod phenotype of the Kell blood group system. *Br. J. Haematol.* **42**:575–583.

428. Szentistvanyi, I., Z. Janka, and L. Heiner. 1980. Calcium-dependent potassium transport in progressive muscular dystrophy. *Eur. Neurol.* **19**:39–42.

429. Szibor, R., K. Redmann, H. Deike, and K. Muller. 1976. Der Einfluss von Ouabain auf die elektrophoretische Beweglichkeit der Erythrozyten bei 7 Patienten mit progressiver Muskeldystrophie typ Duchenne. *Helv. Paediatr. Acta* **31**:249–256.

430. Szibor, R., V. Steinbicker, K. Redmann, and E. Heuse. 1979. Lyon phenomenon in ouabain treated erythrocytes of Duchenne muscular dystrophy carriers as revealed by cell electrophoresis. *Clin. Genet.* **25**:475–479.

431. Tang, L. L., C. M. Redman, D. Williams, and W. L. Marsh. 1976. Biochemical studies on McLeod phenotype erythrocytes. *Biochem. J.* **153**:271–277.

432. Tanner, M. J. A., and D. J. Anstee. 1976. The membrane change in En(a⁻) human erythrocytes. *Biochem. J.* **153**:271–277.

433. Tanner, M. J. A., R. E. Jenkins, D. J. Anstee, and J. R. Clamp. 1976. Abnormal carbohydrate composition of the major penetrating membrane protein of En(a⁻) human erythrocytes. *Biochem. J.* **155**:701–703.

434. Tchernia, G., N. Mohandas, and S. B. Shohet. 1981. Deficiency of skeletal membrane protein band 4.1 in homozygous hereditary elliptocytosis: Implications for erythrocyte membrane stability. *J. Clin. Invest.* **68**:454–460.

435. Thomas, N. S. T., and P. S. Harper. 1978. Myotonic dystrophy: Studies of the lipid composition and metabolism of erythrocytes and skin fibroblasts. *Clin. Chim. Acta* **83**:13–23.

436. Thomas, R. D., R. P. S. Edmonson, P. J. Hilton, and N. F. Jones. 1975. Abnormal sodium transport in leukocytes from patients with essential hypertension and the effect of treatment. *Clin. Sci. Mol. Med.* **48**:169s–170s.

437. Thompson, S., and A. H. Maddy. 1981. The abnormal phosphorylation of spectrin in human hereditary spherocytosis. *Biochim. Biophys. Acta* **649**:31–37.

438. Thompson, S., and A. H. Maddy. 1981. The molecular basis of the defect in phosphorylation of spectrin in human hereditary spherocytosis. *Biochim. Biophys. Acta* **649**:38–44.

439. Tokunaga, E., S. Sasakawa, K. Tanaka, H. Kawamata, C. M. Giles, E. W. Iken, J. Poole, D. J. Anstee, W. Mawby, and M. J. A. Tanner. 1979. Two apparently healthy Japanese individuals of type MkMk have erythrocytes which lack both the blood group MN and Ss active sialoglycoproteins. *J. Immunogenet.* **6**:383–390.

440. Tomita, M., H. Furthmayer, and V. T. Marchesi. 1978. Primary structure of glycophorin A: Isolation and characterization of peptides and complete amino acid sequence. *Biochemistry* **17**:4756–4770.

441. Tosteson, D. C. 1981. Cation countertransport and cotransport in human red cells. *Fed. Proc.* **40**:1429–1433.

442. Tosteson, D. C. 1981. Lithium and mania. *Sci. Am.* **244**:164–174.

443. Tosteson, D. C., N. Adragna, I. Bize, H. Solomon, and M. Canessa. Membranes, ions, and hypertension. *Clin. Sci.* **61**:5s–10s.

444. Tosteson, D. C., E. Carlsen, and E. T. Dunham. 1952. The effects of sickling on ion transport. *J. Clin. Invest.* **31**:406–411.

445. Trevisan, M., R. Cooper, D. Ostrow, W. Miller, S. Sparks, Y. Leonas, A. Allen, M. Steinhauer, and J. Stamler. 1981. Dietary sodium, erythrocyte sodium concentration, sodium-stimulated lithium efflux, and blood pressure. *Clin. Sci.* **61**:29s–32s.

446. Tsai, I.-H., S. N. Prasanna-Murthy, and T. L. Steck. 1982. Effect of red cell membrane binding on the catalytic activity of glyceraldehyde-3-phosphate dehydrogenase. *J. Biol. Chem.* **257**:1438–1442.

447. Tsung, P., and J. Palek. 1980. Red cell membrane protein phosphorylation in hemolytic anemias and muscular dystrophies. *Muscle Nerve* **3**:55–69.

448. Tuck, M. L., R. P. Garay, and P. Meyer. 1982. Identification of the Na-K cotransport system in vascular smooth muscle cells: Effects of catecholamines on cation transport. *Clin. Res.* **30**:340A.

449. Tyler, J. M., B. N. Reinhardt, and D. Branton. 1980. Associations of erythrocyte membrane proteins: Binding of purified bands 2.1 and 4.1 to spectrin. *J. Biol. Chem.* **255**:7034–7039.

450. Urry, D. W., T. L. Trapane, K. S. Andrews, M. M. Long, H. W. Overbeck, and S. Oparil. 1980. NMR observation of altered sodium interaction with human erythrocyte membranes of essential hypertensives. *Biochem. Biophys. Res. Commun.* **96**:514–521.

451. Valentine, W. N. 1977. The molecular lesion of hereditary spherocytosis. *Blood* **49**:241–245.

452. Valentine, W. N., D. E. Paglia, and E. Beutler. 1980. The primary cause of hemolysis in enzymopathies of anaerobic glycolysis: A viewpoint and a commmentary. *Blood Cells* **6**:819–829.

453. Valentine, W. N., D. E. Paglia, and F. Gilsanz. 1977. Hereditary hemolytic anemia with increased red cell adenosine deaminase (45 to 70 fold) and decreased adenosine triphosphate. *Science* **195**:783–785.

454. Van Deenen, L. L. M. 1979. Structural organization and dynamics of phospholipids in red cell membranes. In: *Normal and Abnormal Red Cell Membranes*. S. E. Lux, V. T. Marcheso. amd C. F. Fox, eds. Liss, New York. 451–456.

455. Van Meer, G., C. G. Gahmberg, J. A. F. Op den Kamp, and L. L. M. van Deenen. 1981. Phospholipid distribution in human En(a⁻) red cell membranes which lack the major sialoglycoprotein, glycophorin A. *FEBS Lett.* **135**:53–55.

456. Vickers, J. D., A. J. McComas, and M. P. Rathbone. 1978. Alterations of membrane phosphorylation in erythrocyte membranes from patients with Duchenne muscular dystrophy. *Can. J. Neurol. Sci.* **5**:437–442.

457. Victoria, E. J., L. C. Mahan, and S. P. Masouredis. 1981. Anti-Rh₀(d) binds to band 3 glycoprotein of the human erythrocyte membrane. *Proc. Natl. Acad. Sci. USA* **78**:2898–2902.

458. Victoria, E. J., E. A. Muchmore, E. J. Sudora, and S. Masouredis. 1975. The role of antigen mobility in anti-Rh₀(D)-induced agglutination. *J. Clin. Invest.* **56**:292–301.

459. Wachslicht-Rodbard, H., H. A. Gross, D. Rodbard, M. H. Ebert, and J. Roth. 1979. Increased insulin binding to erythrocytes in anorexia nervosa: Restoration to normal with refeeding. *N. Engl. J. Med.* **300**:882–887.

460. Wakayama, Y., A. Hodson, E. Bunilla, D. Pleasure, and D. L. Schotland. 1979. Freeze–fracture studies of erythrocyte plasma membrane in human neuromuscular diseases. *Neurology* **29**:670–675.

461. Walter, U., and A. Distler. 1980. Effects of ouabain and furosemide on ATPase activity and sodium transport in erythrocytes of normotensives and of patients with essential hypertension. In: *Intracellular Electrolytes and Arterial Hypertension*. H. Zumkley and H. Losse, eds. Thieme, Stuttgart. pp. 170–181.

462. Walter, U., and A. Distler. 1982. Abnormal sodium efflux in erythrocytes of patients with essential hypertension. *Hypertension* **4**:205–210.

463. Wambach, G., and A. Helber. 1981. Na,K-ATPase in erythrocyte ghosts is not a marker for primary hypertension. *Clin. Exp. Hypertens.* **3**:663–674.

464. Wambach, G., A. Helber, G. Bonner, and W. Hummerich. 1978. Natrium-kalium-ATPase-aktivitat in Erythrozytenghost und Elektrolytkonzentration in erythrozyten von Patienten mit essentieller Hypertonie. *Verh. Dtsch. Ges. Inn. Med.* **84**:800–805.

465. Wambach, G., A. Helber, G. Bonner, and W. Hummereich. 1979. Natrium-kalium ATPase aktivitat in Erythrozytenghosts von Patienten mit essentieller Hypertonie. *Klin. Wochenschr.* **57**:169–172.

466. Wambach, G., A. Helber, G. Bonner, W. Hummereich, A. Konrads, and W. Kauffman. 1980. Sodium–potassium ATPase activity in erythrocyte ghosts of patients with primary and secondary hypertension. *Clin. Sci.* **59**:183s–185s.

467. Waymouth, C. 1973. Erythrocyte sodium and potassium levels in normal and anemic mice. *Comp. Biochem. Physiol. A* **44**:751–766.

468. Weinstein, R. S., J. K. Khodad, and T. L. Steck. 1980. The band 3 protein intramembrane particle of the human red blood cell. In: *Membrane Transport in Erythrocytes*. U. V. Lassen, H. H. Ussing, and J. O. Wieth, eds. Munksgaard, Copenhagen. pp. 35–50.

469. Weller, J. M. 1959. The acid–base balance and sodium distribution of the blood in essential hypertension. *J. Lab. Clin. Med.* **53**:553–556.

470. Wessels, F., G. Junge-Hülsing, and H. Losse. 1967. Untersuchungen zur Natriumpermeabilität der Erythrozyten bei hypertonikern und normotonikern mit familiarer Hochdruckbelastung. *Z. Kreislaufforsch.* **56**:374–389.

471. Wessels, F., and H. Losse. 1967. Beziehungen zwischen natriumstoffwechsel der Erythrozyten und Gefassreagibilitat bei normotonikern mit familiarer Hochdruckbelastung. *Klin. Wochenschr.* **45**:850–852.

472. Wessels, F., H. Zumkley, and H. Losse. 1970. Untersuchungen zur Frage des zusammenhangs zwischen kationenpermeabilitat der Erythrozyten und Hochdruckdisposition. *Z. Kreislaufforsch.* **59**:415–426.

473. White, W., E. D. Washington, B. H. Sabo, M. Stroup, J. McCreary, R. Oyen, and W. L. Marsh. 1980. Anti-Km in a transfused man with the McLeod syndrome. *Blood Transfusion Immunohematol.* **23**:305–317.

474. Wieth, J. O., and J. Brahm. 1980. Kinetics of bicarbonate exchange in human red cells: Physiological implications. In: *Membrane Transport in Erythrocytes*. U. V. Lassen, H. H. Ussing, and J. O. Wieth, eds. Munksgaard, Copenhagen. pp. 467–487.

475. Wiley, J. S. 1970. Red cell survival studies in hereditary spherocytosis. *J. Clin. Invest.* **49**:666–672.

476. Wiley, J. S. 1972. Coordinated increase of sodium leak and sodium pump in hereditary spherocytosis. *Br. J. Haematol.* **22**:529–542.

477. Wiley, J. S. 1977. Genetic abnormalities of cation transport in the human erythrocyte. In: *Membrane Transport in Red Cells*. J. C. Ellory and V. L. Lew, eds. Academic Press, New York. pp. 337–361.

478. Wiley, J. S. 1978. Cation fluxes in Rh null red cells. *Blood* **51**:555–556.

479. Wiley, J. S. 1981. Increased cation permeability in thalassemia and conditions of marrow stress. *J. Clin. Invest.* **67**:1094–1102.

480. Wiley, J. S., and R. A. Cooper. 1974. A furosemide-sensitive cotransport of sodium plus potassium in the human red blood cell. *J. Clin. Invest.* **53**:745–755.

481. Wiley, J. S., R. A. Cooper, K. Adachi, and T. Asakura. 1979. Hereditary stomatocytosis: Association of low, 2,3-diphosphoglycerate with increased cation pumping by the red cell. *Br. J. Haematol.* **41**:133–141.

482. Wiley, J. S., J. C. Ellory, M. A. Shumann, C. C. Shaller, and R. A. Cooper. 1975. Characteristics of the membrane defect in the hereditary stomatocytosis syndrome. *Blood* **46**:337–354.

483. Wiley, J. S., and F. M. Gill. 1976. Red cell calcium leak in congenital hemolytic anemia with extreme microcytosis. *Blood* **47**:197–210.

484. Wiley, J. S., J. S. Hutchinson, F. Mendelsohn, and A. E. Doyle. 1980. Increased sodium permeability of erythrocytes in spontaneously hypertensive rats. *Clin. Exp. Pharmacol. Physiol.* **7**:527–530.

485. Wilkerson, L. S., R. C. Perkins, R. Roelofs, L. Swift, L. R. Dalton, and J. H. Park. 1978. Erythrocyte membrane abnormalities in Duchenne muscular dystrophy monitored by saturation transfer electron paramagnetic resonance spectroscopy. *Proc. Natl. Acad. Sci. USA* **75**:838–841.

487. Wilschut, J., and D. Papahadapoulos. 1979. Ca-induced fusion of phospholipid vesicle monitored by mixing of aqueous content. *Nature (London)* **281**:690–692.

488. Wolfe, L. C., and S. E. Lux. 1978. Membrane protein phosphorylation of intact normal and hereditary spherocytic erythrocytes. *J. Biol. Chem.* **253**:3336–3342.

489. Wong, P., and A. D. Roses. 1979. Isolation of an abnormally phosphorylated erythrocyte membrane band 3 glycoprotein from patients with myotonic muscular dystrophy. *J. Membr. Biol.* **45**:147–166.

490. Woods, J. W., R. J. Falk, A. W. Pittman, P. J. Klemmer, B. S. Watson, and K. Namboodiri. 1982. Increased red cell sodium-lithium countertransport in normotensive sons of hypertensive parents. *N. Engl. J. Med.* **306**:593–595.

491. Woods, K. L., D. G. Beevers, and M. West. 1981. Familial abnormality of erythrocyte cation transport in essential hypertension. *Br. Med. J.* **282**:1186–1188.

492. Woods, K. L., D. G. Beevers, and M. J. West. 1981. Racial differences in red cell cation transport and their relationship to experimental hypertension. *Clin. Exp. Hypertens.* **3**:655–662.

493. Yu, J., D. A. Fischman, and T. L. Steck. 1973. Selective solubilization of proteins and phospholipids from red cell membrane by nonionic detergents. *J. Supramol. Struct.* **1**:233–248.

494. Yu, J., and T. L. Steck., 1975. Associations of band 3, the pre-

dominant polypeptide of the human erythrocyte membrane. *J. Biol. Chem.* **250**:9176–9184.

495. Zail, S. S. 1977. The erythrocyte membrane abnormality of hereditary spherocytosis. *Br. J. Haematol.* **37**:305–310.

496. Zail, S. S., and A. Pickering. 1979. Fatty acid composition in hereditary spherocytosis. *Br. J. Haematol.* **42**:399–402.

497. Zail, S. S., and K. Van den Hoek. 1976. Studies on calcium transport and calcium-dependent adenosine triphosphatase activity of erythrocyte membranes in hereditary spherocytosis. *Br. J. Haematol.* **34**:605–611.

498. Zanella, A., C. Izzo, G. Meola, M. Mariari, M. T. Colotti, V. Silani, G. Pellegara, and G. Scarlato. 1980. Metabolic impairment and membrane abnormality in red cells from Huntington's disease. *J. Neurol. Sci.* **47**:93–103.

499. Zarkowsky, H. S. 1979. Heat-induced erythrocyte fragmentation in neonatal elliptocytosis. *Br. J. Haematol.* **41**:515–518.

500. Zarkowsky, H. S., and R. M. Hochmuth. 1979. Sickling times of individual erythrocytes at zero PO$_2$. *J. Clin. Invest.* **56**:1023–1034.

501. Zarkowsky, H. S., N. Mohandas, C. B. Soeaker, and S. B. Shohet. 1975. A congenital hemolytic anemia with thermal sensitivity of the erythrocyte membrane. *Br. J. Haematol.* **29**:537–543.

502. Zarkowsky, H. S., F. A. Oski, R. Sha'afi, S. B. Shohet, and D. G. Nathan. 1968, Congenital hemolytic anemia with high sodium, low potassium red cells. *N. Engl. J. Med.* **278**:573–581.

R1. Anderson, J. M., and J. M. Tyler. 1980. State of spectrin phosphorylation does not affect erythrocyte shape or spectrin binding to erythrocyte membrane. *J. Biol. Chem.* **255**:1259–1265.

R2. Atkinson, A. L., J. S. Morrow, and V. T. Marchesi. 1982. The polymeric state of actin in the human erythrocyte cytoskeleton. *J. Cell. Biochem.* **18**:493–505.

R3. Ballas, S. K., M. R. Clark, N. Mohandas, H. F. Colfer, M. S. Caswell, M. O. Bergren, H. A. Perkins, and S. B. Shohet. 1984. Red cell membrane and cation deficiency in Rh null syndrome. *Blood* **63**:1046–1055.

R4. Bennett, V. 1982. The molecular basis for membrane–cytoskeleton association in human erythrocytes. *J. Cell. Biochem.* **18**:49–65.

R5. Berkowitz, L. R., and E. P. Orringer. 1984. Ion transport in hemoglobin CC red cells. *Blood* **64**:23a.

R6. Lew, V. L., A. Hockaday, M. Sepulveda, A. P. Somolyo, A. V. Somolyo, O. E. Ortiz, and R. M. Bookchin. 1985. Compartmentalization of sickle-cell calcium in endocytic vesicles. *Nature (London)* **315**:586–589.

R7. Brugnara, C., A. S. Kopin, H. F. Bunn, and D. C. Tosteson. 1985. Regulation of cation content and cell volume in erythrocytes from patients with homozygous Hemoglobin C. *J. Clinical Investigation,* **75**:1608–1617.

R8. Brugnara, C., A. Kopin, H. F. Bunn, and D. C. Tosteson. 1984. Cation transport in hemoglobin CC red cells. *J. Gen. Physiol.* **84**:33a.

R9. Bunn, H. F., B. G. Forget, and H. M. Ranney. 1977. *Human Hemoglobins.* Saunders, Philadelphia. pp. 222–225.

R10. Clark, M. R., N. Mohandas, S. H. Embury, and B. H. Lubin. 1982. A simple laboratory alternative to irreversibly sickled cell (ISC) counts. *Blood* **60**:659–662.

R11. Coleman, M. S., J. Donofrio, J. J. Hutton, L. Hahn, A. Daoud, B. Lampkin, and J. Dyminski. 1978. Identification and quantitation of adenine deoxynucleotides in erythrocytes of patients with adenosine deaminase deficiency and severe combined immunodeficiency. *J. Biol. Chem.* **253**:1619–1626.

R12. Ellory, J. C., P. B. Dunham, P. J. Logue, and G. W. Stewart. 1982. Anion-dependent cation transport in erythrocytes. *Philos. Trans. R. Soc. London Ser. B* **229**:483–495.

R13. Gahmberg, C. G. 1982. Molecular identification of the human Rh$_o$ (D) antigen. *FEBS Lett.* **140**:93–97.

R14. Glaubensklee, C. S., A. P. Evan, and W. R. Galey. 1982. Structural and biochemical analysis of the McLeod erythrocyte membrane. I. Freeze fracture and discontinuous polyacrylamide gel electrophoresis. *Vox Sang.* **42**:262–271.

R15. Goodman, S. R., K. A. Schiffer, L. A. Casoria, and E. Eyster. 1982. Identification of the molecular defect in the erythrocyte membrane skeleton of some kindreds with hereditary spherocytosis. *Blood* **60**:772–784.

R16. Halperin, J., R. Schaeffer, L. Galvez, and S. Malave. 1983. Ouabain-like activity in human cerebrospinal fluid. *Proc. Natl. Acad. Sci. USA* **80**:6101–6104.

R17. Hamlyn, J. M., R. Ringel, J. Schaeffer, P. D. Levinson, B. P. Hamilton, A. A. Kowarski, and M. P. Blaustein. 1982. A circulating inhibitor of Na + K ATPase associated with essential hypertension. *Nature (London)* **300**:650–652.

R18. Hebbel, R. P., J. W. Eaton, M. Balasingham, and M. H. Steinberg. 1982. Spontaneous oxygen radical generation by sickle erythrocytes. *J. Clin. Invest.* **70**:1253–1259.

R19. Hochmuth, R. M. 1982. Solid and liquid behavior or red cell membrane. *Annu. Rev. Biophys. Bioeng.* **11**:43–55.

R20. Jain, S. S., and S. B. Shohet. 1984. A novel phospholipid in irreversibly sickled cells: Evidence for *in vivo* peroxidative membrane damage in sickle cell disease. *Blood* **63**:362–367.

R21. Lande, W. M., P. V. W. Thiemann, and W. C. Mentzer. 1982. Missing band 7 membrane protein in two patients with high Na, low K erythrocytes. *J. Clin. Invest.* **70**:1273–1280.

R22. Liu, S., J. Palek, and J. T. Prchal. 1982. Defective spectrin dimer–dimer association in hereditary elliptocytosis. *Proc. Natl. Acad. Sci. USA* **79**:2072–2076.

R23. Mohandas, N., and E. Evans. 1984. Adherence of sickle erythrocytes to vascular endothelial cells: Requirement for both cell membrane changes and plasma factors. *Blood* **64**:282–287.

R24. Morrow, J. S., W. B. Haigh, and V. T. Marchesi. 1981. Spectrin oligomers: A structural feature of the erythrocyte cytoskeleton. *J. Supramol. Struct. Cell. Biochem.* **17**:275–287.

R25. Murphy, J. R. 1968. Hemoglobin CC disease: Rheological properties of erythrocytes and abnormalities in cell water. *J. Clin. Invest.* **47**:1483–1495.

R26. Niggli, V., E. S. Adunyah, B. F. Cameron, E. A. Bababunmi, and E. Carafoli. 1982. The Ca pump of sickle cell plasma membranes: Purification and reconstitution of the ATPase enzyme. *Cell Calcium* **3**:131–151.

R27. Orringer, E. P., C. Skrzynia, and L. R. Berkowitz. 1984. Quantification of the role of calcium in K loss in intact RBC. *Clin. Res.* **32**:318A.

R28. Palek, J. 1983. Blood cell cytoskeleton. I. Red cell membrane skeleton. *Semin. Hematol.* **20**:139–242.

R29. Parker, J. C., and L. R. Berkowitz. 1983. Physiologically instructive genetic variants involving the human red cell membrane. *Physiol. Rev.* **63**:261–313.

R30. Pasvol, G., J. S. Wainscoat, and D. J. Weatherall. 1982. Erythrocytes deficient in glycophorin resist invasion by the malarial parasite, *Plasmodium falciparum. Nature (London)* **297**:64–66.

R31. Reiss, G., H. M. Ranney, and N. Shaklai. 1982. Association of hemoglobin C with erythrocyte ghosts. *J. Clin. Invest.* **70**:946–952.

R32. Schwartz, R., N. Duzgunes, D. Chiu, and B. Lubin. 1983. Interaction of phosphatidylserine–phosphatidylcholine liposomes with sickle erythrocytes. *J. Clin. Invest.* **71**:1570–1580.

R33. Smith, J. B., K. O. Ash, S. C. Hunt, W. M. Hentschel, W. Sprowell, M. M. Dadone, and R. R. Williams. 1984. Three red cell sodium transport systems in hypertensive and normotensive Utah adults. *Hypertension* **6**:159–166.

R34. Speicher, D. W., J. S. Morrow, W. J. Knowles, and V. T. Marchesi. 1982. A structural model of human erythrocyte spectrin: Alignment of chemical and functional domains. *J. Biol. Chem.* **257**:9093–9101.

R35. Spiegel, A. M., M. A. Levine, S. J. Marx, and G. D. Auerbach. 1982. Pseudohypoparathyroidism: The molecular basis for hormone resistance—a retrospective. *N. Engl. J. Med.* **307**:679–680.

R36. Tanner, M. J. A. 1982. Red cell invasion by the malarial parasite. *Trends Biochem. Sci.* **7**:231.

R37. Trevisan, M., D. Ostrow, R. Cooper, K. Liu, S. Sparks, A. Okonek, E. Stevens, J. Marquardt, and J. Stamler. 1983. Abnormal red cell ion transport and hypertension: The Peoples Gas Company study. *Hypertension* **5**:363–367.

R38. Wiley, J. S., D. A. Clarke, L. A. Bonaquisto, J. D. Scarlett, S. B. Harrap, and A. E. Doyle. 1984. Erythrocyte cation cotransport and countertransport in essential hypertension. *Hypertension* **6**:360–368.

R39. Wolfe, L. C., K. M. John, J. C. Falcone, B. S. Byrne, and S. E. Lux. 1982. A genetic defect in the binding of protein 4.1 to spectrin in a kindred with hereditary spherocytosis. *N. Engl. J. Med.* **307**:1367–1374.

R40. Canessa, M. 1984. The polymorphism of red cell Na and K transport in essential hypertension: Findings, controversies, and perspectives. In: *Erythrocyte Membrane 3: Recent Clinical and Experimental Advances.* Liss, New York. pp. 293–315.

R41. Worley, R. J., W. M. Hentschel, C. Cormier, S. Nutting, G. Pead, K. Zelenkos, J. B. Smith, K. D. Ash, and R. R. Williams. 1982. Increased sodium-lithium countertransport in erythrocytes of pregnant women. *New Eng. J. Med.* **307**:412–416.

Inherited Membrane Disorders of Muscle

Duchenne Muscular Dystrophy and Myotonic Muscular Dystrophy

Gordon A. Plishker and Stanley H. Appel

1. Introduction

Myotonic (MyD) and Duchenne (DMD) muscular dystrophy are two of the more common and devastating inherited muscular dystrophies in man. In this chapter we will review MyD and DMD in terms of the pertinent clinical data and present a review of the recent electrophysiological, morphological, and biochemical findings in these two disorders.

The muscular dystrophies present as many different clinical syndromes. Each is a separate entity with a distinctive inborn error of metabolism and with clinical expression in several organ systems in addition to muscle. The specific metabolic defect has not been defined in any of the major dystrophies and, at present, we have no simple way to classify their clinical and probably biochemical heterogeneity.

Progress in understanding the pathogenesis of these human muscular dystrophies has been complicated by several factors. Animal models of dystrophy or myotonia do not present the clinical, physiological features of the human disorders. Neither myotonia or dystrophy alone can be defined as a disease process and there is no evidence that these conditions operate through similar mechanisms. Each of the muscular dystrophies may represent a class of genetic defects having similar phenotypic clinical expression. The genetic defects in these disorders have pleiotropic expressions in many organ systems.

The strategy in our own investigations has been to assume that the defect is in many different organ systems and that the

elucidation of the biochemical mechanisms responsible for abnormalities in red blood cells and cells grown in tissue culture will lead to the identification of the primary genetic defect which in muscle tissue causes the dystrophy and myotonia. It is clear that current recombinant DNA technology may resolve the genetic locus of the specific inborn errors in these diseases. Nevertheless, an understanding of the phenotypic expression of these genetic defects must include an explanation for the alterations in membrane structure and function in these disorders.

2. Myotonic Muscular Dystrophy

Myotonia is clinically manifested as muscle stiffness. It is most marked after periods of rest or exposure to cold, and lessens with exercise. In electrophysiological terms, it can be defined as an abnormal tendency of the muscle membrane to discharge trains of repetitive action potentials in response to either a willed contraction or to direct electrical or mechanical stimulation. This altered excitability persists following *d*-tubocurarine or nerve block, which helps differentiate this condition from neuromyotonia or Isaac's syndrome, in which a primary neural defect results in the increased release of acetylcholine from nerve terminals.

Several different diseases have been described with muscle myotonia as a prominent clinical finding. The most common of these disorders is MyD. This condition is inherited as an autosomal dominant trait, and although penetrance is high, the clinical presentation is extremely variable. Some individuals manifest a full constellation of skeletal, cardiac, and smooth muscle abnormalities as well as bony changes, cataracts, endocrine dysfunction, and personality defects. Other patients have

Gordon A. Plishker and Stanley H. Appel • Program in Neuroscience, Department of Neurology, Baylor College of Medicine, Houston, Texas 77030.

only cataracts or an individual symptom complex such as dysphagia.[1-15] Other myotonic disorders described in man include myotonia congenita, paramyotonia congenita and hyperkalemic periodic paralysis, myotonia levior, and myotonia combined with dwarfism, diffuse bone disease, and unusual ocular and facial abnormalities. The clinical syndromes of these inherited myotonic disorders in man are quite different from each other and suggest that the underlying primary biochemical defects may be unique for each disorder. The distinctive clinical features further suggest that physiological alterations giving rise to myotonia in MyD may be different from those giving rise to myotonia either in myotonia congenita or paramyotonia congenita. Most experimental and clinical data suggest that the myotonic disorders of man represent genetically induced primary alterations in the structure and function of cellular membranes. They are likely due to different inborn errors of metabolism and more specifically different defects of critical membrane ionic events. The specific metabolic defect has not been defined in any of these disorders.

2.1. Physiological Investigations of Myotonia

2.1.1. Goat Myotonia

The most thorough physiological investigations have been carried out in the inherited myotonia of goat. This disorder resembles human myotonia congenita and has been used as a model of the latter. Both may be inherited as an autosomal dominant trait and in both, clinical expression is almost entirely confined to skeletal muscles. In the muscles of the myotonic goat, the membrane resistance is increased, Cl^- conductance is decreased, K^+ conductance is increased, and Na^+ inactivation is slower.[16-19] Most of the remaining electrical parameters, including resting membrane potential, rate of the rise of the action potential, the overshoot of the action potential, the resting Na^+ permeability (P_{Na}), and the total fiber capacitance, are normal.[16,19,20] The low resting membrane Cl^- conductance is considered to be the major factor accounting for the myotonic excitability of these cells.

Adrian and Bryant[21] have used this basic abnormality to propose an interesting explanation of myotonia. It has been demonstrated that under normal circumstances, an action potential spreading into the transverse tubule system raises the K^+ concentration in the lumen of the tubules. The passage of several action potentials would increase the tubular K^+ and result in accumulative afterdepolarization which would be offset by the entry of Cl^- into the muscle fiber. Adrian and Bryant postulate that in goat myotonia the decreased Cl^- conduction cannot counterbalance the depolarizing effect of increased tubular K^+ and repetitive active potentials result.

A similar explanation could be invoked with the experimental myotonias produced in animals by the administration of monocarboxylic acids such as 2,4-dichlorophenoxyacetate. These agents appear to block Cl^- conductance by altering the natural ionic preference of the channels for Cl^-.[22,23] This has led to the proposal that in the animal myotonias a similar compound acts either to block the Cl^- channel or to change its selective permeability.[19,22-24]

The cation conductance abnormalities have been observed by voltage clamp in myotonic goat fibers. These abnormalities were in the direction of promoting myotonia but were insufficient to induce myotonia in the presence of a normal Cl^- conductance.[19,25-27] The slow rate of Na^+ inactivation observed could result in the myotonic fiber having a larger steady-state depolarizing Na^+ current in the threshold region. This would be expected to increase the chance of spontaneous firing. The increases detected in outward K^+ conductances would be expected to increase tubular K^+ accumulations. This would result in further membrane depolarizations. The magnitude of these changes was small compared to the decreases in Cl^- conductance. The Cl^- effect remains the principal underlying factor for the myotonia. The observation of abnormalities in additional ionic conductances has led to the speculation that there may be a membrane abnormality that is basic to all channel alterations.[19] Decreases in Cl^- conductances, increases in the electrogenic Na^+,K^+-ATPase, and possible alterations in Na^+ channel inactivation have been reported in animals treated with the cholesterol-lowering agent, diazacholesterol.[28-32] Although membrane lipid abnormalities may result from diazacholesterol administration to animals and in turn may be associated with the alterations in muscle membrane conductance, no similar biochemical abnormalities have been substantiated in either goat or human myotonic disorders (see Section 2.2.4).

2.1.2. Myotonia Congenita

In many respects, goat myotonia resembles human myotonia congenita. Both may be inherited as an autosomal dominant trait, and both have a clinical expression almost entirely confined to the skeletal muscle with no systemic involvement. The electrical parameters of intercostal muscle from patients with myotonia congenita exhibit similar alterations to those described for goat myotonia. Membrane resistance is increased and Cl^- conductance is dramatically decreased whereas fiber capacitance is normal.[33] However, the K^+ conductance of human myotonia congenita muscles is distinctively different from goat myotonic muscles. Lipicky and Bryant[34] noted that the K^+ conductance was decreased in some patients with myotonia congenita whereas it was twofold higher in goat myotonia. Muscles from two patients with myotonia congenita had normal Cl^- conductances and normal K^+ conductance. Clearly, further studies are necessary and it is possible that the human myotonia congenita may not be a homogeneous entity with disturbance in the Cl^- conductance as the sole abnormality. Nonetheless, the ability of impaired Cl^- conductances to lead to repetitive action potentials is characteristic of this type of human myotonia. Under normal circumstances, Cl^- conductance represents an important mechanism for preventing the repetitive action potentials in these muscles and when such a mechanism is impaired myotonia may ensue.

2.1.3. Paramyotonia Congenita

The repetitive discharges of paramyotonia congenita do not appear to be due to impaired Cl^- conductance. The myotonic activity of these muscle cells is worsened by cooling. Investigation of intercostal muscle fibers from patients with paramyotonia congenita has led to the suggestion that in these cells there is a temperature-dependent abnormality in the Na^+ conductance pathway which leads to the repetitive discharges.[35] Decreasing the temperature results in an abnormal increase in Na^+ permeability by decreasing the Na^+ channel inactivation, leading to membrane depolarization. Cl^- conductance does not appear to be abnormal. Thus, paramyotonia congenita may represent an example of an impaired Na^+ channel inactivation resulting in the myotonic activity. The impairment would produce abnormal electrical activity analogous to that produced by scorpion venom. Thus, as in paramyotonia congenita, application of scorpion toxin in the early stages results in repetitive spiking, whereas with prolonged application it leads to depolarization, steady-state inactivation of Na^+ channels, and failure of all electrical activity.[32]

2.1.4. MyD

Measurements of the passive membrane properties of intercostal biopsies from humans with MyD indicate that Cl^- abnormalities cannot explain the abnormal electrical and contractile behavior of this disease. Studies from several different laboratories have documented decreased resting membrane potential in MyD.[23,34,36] Lipicky and Bryant[34] noted an increased membrane capacitance with normal membrane resistance and conductance. Hofmann et al.[37] earlier reported no change in K^+ concentration gradients, which could explain the partial depolarization and spontaneous membrane potential oscillations. Thus, it is clear that defects in Cl^- conductance cannot explain the physiological alterations present in MyD. Preliminary voltage clamp studies suggest the Na^+ currents are normal in MyD.[38]

In our laboratory we have used tissue culture of human myotonic dystrophy biopsies to study the basic membrane electrical properties and possible abnormalities in MyD.[39] Our initial studies demonstrated four specific differences between MyD and control human myotubes: (1) decreased resting potential of approximately 10 mV; (2) decreased action potential afterhyperpolarization amplitude; (3) decreased outward-going (delayed) rectification in steady-state current–voltage plots; and (4) common appearance of depolarizing afterpotentials associated with repetitive firing of action potentials in response to a single step stimulus or slow depolarizing ramp.

The decreased resting potential did not appear to explain the repetitive firing. Measurements of the Na^+/K^+ permeability ratio in MyD myotubes were variable with no significant differences from control. Furthermore, in either cultured normal or MyD myotubes, there was no evidence of a Cl^- conductance. This may explain the markedly increased input resistance we observed in our studies of cultured human myotubes compared to rat myotubes. Depolarization of the resting potential of the control myotubes to the value of the MyD myotubes did not produce increased repetitive firing or depolarizing afterpotentials.

Our most recent experiments suggest that alterations in the action potential afterpotential appear to be the most significant abnormality in MyD myotubes.[40] However, the ionic parameters responsible for this afterpotential have not yet been defined in human myotubes. The hyperpolarizing afterpotential in cultured rat muscle is due to a Ca^{2+}-dependent K^+ conductance.[41,42] However, in cultured human muscle this relationship has not been established. Neither the blockage of Ca^{2+} entry by the application of nickel nor the chelation of intracellular Ca^{2+} by EGTA injection had any effect on the hyperpolarizing afterpotential of control myotubes. Extracellular application or intracellular injection of tetraethylammonium, a known blocker of delayed voltage-dependent K^+ conductance, significantly reduced the hyperpolarizing afterpotential. This suggests that the typical hyperpolarizing afterpotential observed in the control myotubes is dominated by the delayed voltage-dependent K^+ conductance.

Unexpectedly we found that the hyperpolarizing afterpotential of control myotubes could be converted to a depolarizing afterpotential by the injection of calcium.[40] This effect was observed immediately after the insertion of an electrode containing 0.5 M $CaCl_2$ and 0.5 M KCl. This depolarizing afterpotential persisted after the removal of the calcium-containing electrode and the impalement of the fiber with a normal KCl electrode. The return of the afterpotential to pre-calcium injection levels was observed an hour after the calcium-containing electrode had been removed. Similar conversions of the afterpotential control myotube occurred when extracellular Ca^{2+} concentration was increased from 2 to 10 mM. The depolarizing phase of the afterpotential was greatest immediately after the application of the high extracellular Ca^{2+} and became smaller after a few minutes. No significant change was seen in the more rapid phase of the action potential.

In contrast to the findings with control myotubes, MyD myotubes typically exhibited a depolarizing afterpotential, which in the presence of high extracellular Ca^{2+} converted to a hyperpolarizing afterpotential.[40] Voltage clamp measurements showed that a 2-sec outward current of the control myotubes was depressed after extracellular Ca^{2+} was raised from 2 to 10 mM. The depression of the outward current was most evident at depolarized potentials and shortly after solution changes. In contrast, in myotonic myotubes 10 mM extracellular Ca^{2+} enhanced this outward current. These data provide support for our working hypothesis that the defect in MyD is related to an abnormality in outward K^+ conductance. The hyperpolarizing afterpotential induced in myotonic muscle by high Ca^{2+} may be related to an activation of the Ca^{2+}-activated K^+ conductance either by an enhanced Ca^{2+} entry or by a shift in the Ca^{2+} dependence of the K^+ conductance. A number of K^+ conductances having discrete characteristics have been reported.[43–48] It is not clear which of these, if any, are responsible for the conductance observed in our studies. We have not ruled out the possibility that the enhancement of inward currents is responsible for the depolarizing afterpotential noted in normal Ca^{2+} values. In red blood cells from patients with MyD, we have found an increased passive influx of Ca^{2+}.[49] The myotonic red blood cells also have K^+ effluxes of reduced sensitivity to Ca^{2+}.[50]

Myotonia may thus occur in a number of disorders. The available experimental evidence suggests that myotonia congenita, paramyotonia congenita, and MyD appear to differ with respect to clinical expression, the organ systems involved, and the physiological mechanism producing myotonia. At least three potential candidates presently exist for producing repetitive action potentials in muscle: (1) impairment of the Cl^- conductance; (2) a delay in the inactivation of the Na^+ channel; and (3) an impairment in a K^+ conductance. Human myotonia congenita may be an example of impairment in Cl^- conductance, paramyotonia congenita an impairment in the Na^+ inactivation process, and MyD an impairment in a K^+ conductance mechanism. Alterations in a K^+ conductance mechanism may not only explain the muscle myotonia, but also the impairment of the cardiac conductile system, the presence of cataracts, the bone changes, the enhancement of insulin release in the presence of a glucose load, and the smooth muscle dysfunction.

2.2. Biochemical Investigations

2.2.1. Na$^+$,K$^+$-ATPase

Investigations of the enzymatic and transport activities of the Na^+,K^+-ATPase have provided conflicting results. Brown et al.[51] initiated the interest in the ouabain-sensitive Na^+,K^+-ATPase by reporting that ouabain stimulated, rather than inhibited, the hydrolysis of ATP in erythrocyte ghosts prepared from patients with DMD, MyD, and myotonia congenita. The ouabain stimulation was only detected in the presence of low concentrations of Na^+ and K^+. At more physiological Na^+ and K^+ concentrations, no alterations in the Na^+,K^+-ATPase could be demonstrated in either MyD or DMD cells.[52–55] In MyD, Hull and Roses[56] reported an alteration in the stoichiometry of the ouabain-sensitive Na^+/K^+ exchange in erythrocytes. This finding has not been confirmed by other investigators.[57] Studies of Desnuelle et al.[58] suggested a decreased number of ouabain-binding sites in MyD muscle.

2.2.2. Protein Kinase

Altered protein phosphorylation was observed both in erythrocytes and in muscle biopsies obtained from patients with MyD.[59–61] The predominant effect was a decrease in the protein phosphorylation in the membranes as compared to controls. In fresh erythrocyte ghosts, the predominant decrease in phosphorylation on SDS–polyacrylamide gels was in the 90,000–100,000 molecular weight region. Several integral membrane proteins, including the Na^+,K^+-ATPase, fall in this molecular weight range. The variable activities with frozen red cell preparations, and the normal phosphorylation of spectrin in fresh red cell membranes make it unlikely that an alteration in the enzyme is the primary genetic defect in MyD. Although the polypeptide substrate could be abnormal, it is equally plausible that the lipid microenvironments surrounding these proteins or the mobility of the proteins in the membrane could give rise to the altered phosphorylation patterns. The report of a discernible difference in the temperature response of membrane phosphorylation in MyD and the biophysical evidence of alterations in membrane viscosity are consistent with this possibility.[62–68]

2.2.3. Insulin Receptors

The nature of the insulin resistance that occurs in patients with myotonic dystrophy appears to be unique among the forms of insulin resistance that have been studied. Insulin resistance states such as obesity,[69] insulin-dependent diabetes mellitus,[70] ataxia telangiectasia,[71] congenital lipodystrophy,[72] and acanthosis nigricans with severe insulin resistance[73] are all characterized by endogenous hyperinsulinism and a reciprocal decrease in insulin receptor. All of these conditions are associated with an increase in the fasting insulin concentrations and an excessive insulin response to a glucose challenge. In MyD the majority of patients show a resistance to endogenous insulin but this is not always associated with an increase in the fasting insulin concentrations.[74–82]

The reports on monocyte insulin binding in patients with MyD show little agreement. Kobayashi et al.[76] found normal insulin binding to monocytes and no correlation between monocyte receptors and the fasting insulin concentrations in their seven patients. Festoff and Moore[74] reported a dramatic decrease in insulin binding to monocytes in six patients even though five had normal endogenous insulin levels. The contradictions between the findings of Festoff and Moore[74] and those of Kobayashi et al.[76] are especially striking in view of the fact that three of Festoff and Moore's subjects had been included in Kobayashi and colleagues' earlier study. Tevaarwerk and Hudson[75] studied monocyte insulin binding in 12 patients and found a normal number of receptors, but with decreased affinities. They found that insulin binding was proportional to the responsiveness to exogenous insulin. Our own

work[82] concurs with that of Tevaarwerk and Hudson.[75] We found a modest decrease in overall binding accompanied by substantial decreases in receptor affinity. In our study of 12 patients with MyD, we found resistance to exogenous insulin in nine. Only three of our patients had increased fasting insulin concentrations, but eight had an excessive insulin response to a glucose challenge. Insulin resistance was correlated with a decreased monocyte insulin receptor affinity. The mechanism for this increase in receptor affinity in insulin resistance remains obscure. No insulin antibodies, insulin receptor antibodies, or other circulating insulin-binding inhibitors were found in the serum. This affinity decrease may somehow be related to the compensatory nature of hyperinsulinism seen in this condition. It could also represent an alteration that is intrinsic to the receptor protein or is secondary to some other general membrane defect.

2.2.4. Lipid Composition

Theoretically, the changes in enzyme activities, membrane biophysics, and ion transport processes in MyD could be explained by a primary defect in lipid metabolism. Many enzymes are either dependent on or influenced by the membrane lipids. Increases in the membrane fluidity and myotonia can be induced in animals with drugs that interfere with lipid metabolism.[83,84] Although some studies have suggested specific lipid abnormalities in MyD, alterations in the major lipid constituents of the membrane have not been established. The studies of the phospholipid composition of erythrocyte membranes in MyD have not revealed any alteration in this autosomal disease. Ruitenbeek's[85] report of a reduction in palmitoleic acid (16:1) has not been substantiated either by McLaughlin and Engel[86] or by us.[55] The report by Grey et al.[87] that Ca^{2+}-dependent phosphatidic acid synthesis in erythrocytes was markedly impaired in MyD was not substantiated by us[88] or by the original observers.[89] We have found an alteration in the methylation of phosphatidylethanol.[90] Since this is only observed in female MyD patients, it is unlikely that this is a primary metabolic defect.

The major neutral lipid in the cell is cholesterol. In the erythrocyte it represents over 80% of this class of lipids that are weakly absorbed to silica gel and that can be eluted with pure chloroform. It has been suggested that in MyD there is a primary abnormality involving the enzyme Δ24-reductase which converts desmosterol to cholesterol. This hypothesis was based on the observation that myotonia may develop in animals or man following the administration of drugs such as diazacholesterol and triparanol which block cholesterol biosynthesis (particularly the Δ24-reductase reaction).[91–95] It was also based on the report of an atypical sterol accumulation in the blood of patients with MyD.[94]

The administration of diazacholesterol induced myotonia in man,[93] goat,[89,93] and rat.[92,93] In man, alterations in the sterol composition of plasma membranes from erythrocytes and from one muscle sample were associated with the clinical and electromyographic myotonia.[95] Total sterols were reduced in the plasma and the plasma membrane of erythrocytes. Both fractions showed increases in desmosterol, the immediate precursor of cholesterol. Discontinuation of the diazacholesterol therapy was accompanied by the disappearance of both clinical and EMG manifestations of myotonia and the abnormalities in sterol composition. In the goat,[91,95] similar observations were made following therapy with diazacholesterol. In the rat,[92,93] diazacholesterol and triparanol caused an accumulation of desmosterol in plasma and muscle and resulted in myotonia, cataracts, an enhancement of red blood cell Na^+,K^+-ATPase activity, and an increase in muscle membrane resistance. Recent studies suggest that an accumulation of desmosterol in the sarcolemma causes alterations in several ionic conductances.[29,31,32] A decreased Cl^- conductance which reduces the resting membrane potential causes repetitive firing in fibers having large inward Na^+ currents and short-duration outward K^+ currents.

The observation of elevated serum desmosterols in MyD by Wakamatsu et al.[96] was the first and only evidence to link the drug-induced myotonia to the hereditary disease. Nine patients with MyD were shown to have 13.4 ± 4.6 mg/100 ml desmosterol while 11 control subjects showed a level of 5.8 ± 2.1 mg/100 ml, as determined by the colorimetric method of Frantz et al.[97] Several investigators have been unable to substantiate this observation of raised desmosterol levels in patients with MyD. Andiman et al.[98] examined the red blood cell membranes and plasma from 10 patients with MyD and two patients with myotonia congenita for alterations in sterol composition. Desmosterol was not detected in either the sera or the red cell ghosts of the myotonic patients using gas–liquid chromatography. Peter et al.[93] also found no accumulations of desmosterol in either plasma of red cells of patients with MyD or myotonia congenita. Thomas and Harper[99] found no elevation in the desmosterol level in plasma or in red cell ghosts of five patients with MyD. They also studied cultured fibroblasts from myotonic dystrophy patients and from normal individuals. Both normal and MyD cells had similar growth patterns in delipidated sera media and similar patterns of [^{14}C]acetate incorporation into the various lipid fractions. They found no evidence of desmosterol accumulations in MyD fibroblasts. These negative results raise a question about the accuracy of the colorimetric identification of desmosterol by Wakamatsu et al.[96] It is important to note that the Frantz et al.[97] method used by Wakamatsu et al. reacts with substances other than desmosterol. Thus, it would appear that there is no established correlation between MyD and abnormal accumulations of desmosterol.

3. Duchenne Muscular Dystrophy

DMD is a progressive, lethal disease that is inherited as an X-linked trait. Boys become symptomatic early in motor development and diagnosis is usually made between 3 and 6 years of age. Symptoms of proximal muscle weakness predominate; a waddling, "ducklike" gait and a lordic posture are common and elevated serum enzymes are the rule.[100] Patients are usually unable to walk by age 10 and frequently die before age 20.[101] Approximately 10% of the mothers of affected sons are symptomatic primarily with complaints of proximal muscle weakness. Many more are asymptomatic but have slight elevations in serum enzymes, especially creatinine kinase and lactate dehydrogenase isoenzyme-5.[100]

3.1. Morphological Investigations

The leakage of enzymes as well as creatinine from DMD muscle fibers prompted morphological study of the muscle's plasma membrane or sarcolemma. Focal defects in the plasma membrane of nonnecrotic muscle fibers were observed by electron microscopy.[102–104] These lesions readily permitted the entry of peroxidase- and procaine yellow-containing extracellular fluid. The myofibrils were highly constricted in the area near the membrane lesion. These findings appeared to be specific since they were only rarely observed in control specimens or in biopsies of patients with other muscular dystrophies.

Freeze–fracture studies demonstrated changes in both the distribution and the number of intramembranous particles in plasma membranes of muscle cells from patients with DMD, MyD, and facioscapulohumeral muscular dystrophy.[105–107] In DMD there was a significant diminution in the number of particles as well as the number of aggregated particles or orthogonal arrays. These orthogonal arrays refer to aggregates of four or more 60- to 70-Å particles. Both freeze–fracture faces of muscle plasma membrane contained openings or caveolae, which were significantly increased in DMD. The loss of orthogonal arrays was not specific for DMD. Significant decreases were also observed in facioscapulohumeral muscular dystrophy and in myositis. In MyD there was one report of an increase in the density of intramembranous particles in muscle sarcolemma. Freeze–fracture analysis of cultured muscle cells from DMD patients have revealed no significant differences from control.[108,109]

The physiological function associated with these different morphological structures has not been established. Studies in red blood cells suggest that one of the constituents of these intramembranous particles is a 95,000-dalton protein band which is involved with anion movements across the plasma membrane.[110] It has been shown that the caveolae in vertebrate muscle correspond to the T-tubule system openings.[111,112] In dystrophic chickens, increases in caveolar density were associated with an extensive proliferation of the T-tubule system.[113–115] It has not yet been established whether the increase in caveolar density noted in DMD is also associated with a proliferation of the T-tubule system.

Thompson et al.[116] studied seven patients with DMD and demonstrated that cultured DMD cells showed impaired contact inhibition compared to control cells. This was observed during the proliferation and migration of mononuclear cell suspensions prepared from muscle biopsies. The DMD cultures showed less contact inhibition and the presence of many small clumps of cells. As myoblast fusion proceeded, the developing myotubes proliferated between the clustered cells. At the ultrastructural level the DMD mononuclear cells contained abnormal vacuoles. Unpublished studies from our laboratory have not substantiated this alteration in contact inhibition.

The application of scanning electron microscopy to erythrocytes has demonstrated an abnormality in the shape of dystrophic erythrocytes following fixation.[117,118] In the fresh, unfixed blood from patients with DMD and MyD, there were no differences. Following fixation there were quantitative differences. Studies of blood from patients with MyD and patients and carriers of DMD all showed an abnormal number of cup-shaped erythrocytes (stomatocytes).[117] The occurrence of stomatocytes was influenced by many variables such as the medium, the number of washes, and the pH at the time of fixation. These factors increased the percentage of stomatocytes in samples from patients with muscular dystrophy as compared to normal controls.

Erythrocytes that are spherical and whose surface bears uniformly spaced projections or spicules are called echinocytes. These cells are found in vitro in blood from both normal subjects and patients and carriers of DMD. The number of echinocytes found is a function of the time interval between blood drawing and cell fixation. The longer the interval, the greater is the number of echinocytes. As compared to normals, a higher percentage of cells from patients and carriers of DMD are transformed.[118] The variability and the nonspecific character of the stomatogenesis and echinogenesis make these morphological tests totally impractical as a diagnostic tool, but these results do support the idea of a general membrane defect in these disorders.

3.2. Biochemical Investigations

3.2.1. Ca^{2+}-ATPase

Hodson and Pleasure[54] found that the apparent affinity of the Ca^{2+}-stimulated Mg^{2+}-dependent ATPase (Ca^{2+}-ATPase) for its Mg^{2+}-ATP substrate was higher in DMD red blood cells. The Ca^{2+}-ATPase activity of DMD erythrocyte ghosts was normal at the physiological substrate concentration. In detergent-treated cells, Luthra et al.[119] found slightly higher Ca^{2+}-ATPase activity in DMD and lower activity in MyD. However, using hypertonically prepared ghosts, they did not find any differences in activity. The addition of crude hemolysates stimulated Ca^{2+}-ATPase activities to a greater extent in DMD and to a lesser degree in MyD. Since the effect they observed was the same with hemolysates from either patients or control subjects, they concluded that the difference in activation involved inherent properties of the membrane. Mollman et al.[120] reported that Ca^{2+} transport into resealed inside-out erythrocyte vesicles was increased in DMD.

Calmodulin is an important activator of this enzyme. It increases both the overall activity and the Ca^{2+} affinity of this ATPase. Unfortunately, endogenous measurements of calmodulin in the membrane were not performed in either study. It is not known if differences in the calmodulin level contributed to the observations in DMD and MyD.

This Ca^{2+}-ATPase can also be activated by membrane phospholipids and by free fatty acids.[121] Phosphatidylserine produces a half-maximal stimulation at about 7 nmoles phospholipid/μg protein. Cardiolipin caused a half-maximal stimulation at 3 nmoles/μg protein. We found that phosphatidylinositol 4,5-bisphosphate was about 10 times as effective as phosphatidylserine in stimulating the erythrocyte Ca^{2+}-ATPase.[122] Phosphatidylinositol 4-phosphate was slightly less effective than phosphatidylinositol 4,5-bisphosphate. The latter caused stimulation of the ATPase activity which was greater than that produced by saturating levels of calmodulin. The concentration of this lipid required for maximal stimulation was about 2 μM. The low amount of phosphatidylinositol 4,5-bisphosphate required for this response argues for potential physiological regulation of the Ca^{2+}-ATPase by this lipid. Alterations in the metabolism of this lipid could contribute to the observations seen in DMD.

3.2.2. Protein Kinase

Studies by Roses et al.[123–125] demonstrated an increased protein phosphorylation in DMD. The most significant increases in the rate and degree of phosphorylation (approximately 20%) were in the 220,000-dalton polypeptide known as spectrin. Subsequently, conflicting reports on spectrin phosphorylation have appeared in the literature. Iyer et al.[126] and Fisher et al.[127] did not confirm the original observation. Falk et al.[128] showed an increase in the phosphorylation of the spectrin band of approximately 18%. Recently, Roses et al.[129,130] reported that the increased phosphorylation is present in peptide fractions pre-

pared by either tryptic or cyanogen bromide cleavage of purified spectrin. No unique phosphorylated peptides are present in DMD material. The results to date require confirmation. Furthermore, it is not clear whether the enzyme, the substrate, or the other membrane constituents were responsible for the altered phosphorylation. Since the altered activities did not affect all endogenous membrane substrates uniformly, it is unlikely that disturbances in the protein kinase were primarily responsible. Although the spectrin or the protein kinase could be abnormal, it is equally plausible that the lipid microenvironment surrounding these proteins or the mobility of the proteins in the membrane could give rise to the altered phosphorylation patterns.

3.2.3. Adenylate Cyclase

The synthesis of cAMP from ATP is controlled by adenylate cyclase. This provides the mechanism for extracellular control of cellular protein phosphorylation. Specific plasma membrane receptors permit the regulation of adenylate cyclase by monoamines and peptides. In skeletal muscle, only a single receptor regulates the adenylate cyclase. This receptor is a β_2-adrenergic receptor that is activated by epinephrine. In skeletal muscle, cAMP phosphorylation affects the activity of phosphorylase kinase, glycogen synthetase, and protein phosphatases. Alterations in adenylate cyclase have been observed in homogenates of DMD muscle,[131–136] cultured DMD muscle cells,[137] and in erythrocyte membranes.[138–140] The major changes were an increased basal activity and a reduction in epinephrine sensitivity.

3.2.4. Phospholipid and Fatty Acid Composition

In theory, changes in morphology and enzyme activities in DMD could be explained by primary defects in lipid metabolism. Quantitatively, phospholipids represent the most important fraction of lipids in the plasma membrane. In 1973, Kunze et al.[141] reported an abnormal phospholipid composition of red blood cell membranes in patients they classified as having either an X chromosomal or autosomal malignant muscular dystrophy. This observation initiated other investigations of phospholipid composition in MyD and DMD. The initial observation of Kunze et al. was a significant increase in sphingomyelin and a decrease in phospholipid fraction that was thought to be composed of phosphatidylethanolamine and phosphatidylserine. Since that initial observation there have been some conflicting results from several laboratories. Kalofoutis et al.[142] reported a moderately significant increase in sphingomyelin. Three other studies did not find any significant differences in sphingomyelin.[99,143,144] Since sphingomyelin is the only phospholipid that does not appear to be affected by autoxidation and since precautions were not taken in the Kunze et al. study, it appears likely that there is no abnormality in the phospholipid composition in DMD. The fatty acid composition of the principal phospholipids in erythrocytes has been examined in DMD by several investigators, and no significant alterations have been found.[85,141,143,145]

Alterations in the fatty acid composition of neutral lipid fractions have been reported in DMD by two laboratories. Howland and Iyer[146] and Ruitenbeek[85] found that palmitoleic acid (16:1) was lower in DMD. The Howland and Iyer study examined total lipid extracts from patients and carriers. The major alteration was in carriers. The Ruitenbeek study involved a more extensive evaluation of lipid fractions, and he found palmitoleic to be lower in the di- and triglyceride fractions. These findings have not been confirmed by McLaughlin and Engel[86] or by us.[55] There are several reasons to view the palmitoleic acid abnormality with caution. In general, palmitoleic acid represents 1% or less of the total fatty acid composition of normal erythrocytes. Howland and Iyer's control value was significantly higher at 4.9%. The Howland and Iyer and the Ruitenbeek studies did not mention an attempt to sex and age match patients and controls, and we have found differences between adults and children. Children have higher percentages of oleic acid and lower percentages of arachidonic, palmitic, and palmitoleic acid. Since Ruitenbeek did not detect phosphatidylinositols, this suggests that his lipid extractions may have been incomplete. Finally, precautions to prevent autoxidation were not mentioned in either study.

4. Summary

Electrophysiological, morphological, and biochemical studies of MyD and DMD have revealed a large number of abnormalities. Unfortunately, the critical questions remain to be answered. The molecular defects responsible for the dystrophic processes and the myotonia have not been elucidated. Many membrane alterations have been delineated in both DMD and MyD, but it is not clear whether any of these are the primary expression of the inborn error of metabolism, or a secondary effect. However, regardless of whether membrane perturbations are a primary or secondary expression of the genetic defect, they contribute significantly to the alteration of muscle structure and function and merit intensive investigation. Further advances in our understanding of normal membranes may ultimately provide important clues to both of these debilitating muscular dystrophies.

References

1. Thomasen, E. 1948. *Myotonia in Thomsen's Disease (Myotonia Congenita), Paramyotonia and Dystrophia Myotonia.* Universitetsforlaget, Aarhus.
2. Caughey, J. E., and N. C. Myriathopoulos. 1963. *Dystrophia Myotonia and Related Disorders.* Thomas, Springfield, Ill.
3. Decourt, J. 1966. Les troubles endocrineins de la dystrophie myetonique (Maladie de Steinert). In: *Progressive Muskeldystrophie, Myotonie, Myasthenie.* E. Kuhn, ed. Springer-Verlag, Berlin. pp. 265–280.
4. Drucker, W. D., L. P. Rowland, K. Sterling, and N. P. Cristy. 1961. On the function of the endocrine glands in myotonic muscular dystrophy. *Am. J. Med.* **31**:941–950.
5. Kuhl, W. J., I. S. Harper, and R. M. Dowben. 1961. Thyroxine and triiodothyronine turnover studies in dystrophia myotonica. *J. Clin. Endocrinol.* **31**:1592–1595.
6. Gorden, P., R. C. Griggs, S. R. Nissley, J. Roth, and W. K. Engel. 1969. Studies of plasma insulin in myotonic dystrophy. *J. Clin. Endocrinol.* **29**:684–690.
7. Church, S. C. 1967. The heart in myotonia atrophica. *Arch. Intern. Med.* **119**:176–181.
8. Bache, R. J., and G. A. Sarosi. 1968. Myotonia atrophica: Diagnosis in patient with complete heart block and Stokes–Adams syncope. *Arch. Intern. Med.* **121**:369–372.
9. Kohn, N. H., J. S. Faires, and T. Rodman. 1964. Unusual manifestations due to involvement of involuntary muscle in dystrophia myotonica. *N. Engl. J. Med.* **271**:1179–1183.
10. Hughes, D. T. D., J. C. Swann, J. A. Gleeson, and F. I. Lee.

11. 1965. Abnormalities in swallowing associated with dystrophia myotonica. *Brain* **88**:1037–1042.
11. Kuhn, V. E., J. Schaaf, W. Wenz, and W. Stein. 1965. Untersuchungen am Verdauungstrokt bei myotonischer dystrophie. *Schweiz. Med. Wochenschr.* **95**:1263–1266.
12. Harvey, J. C., D. H. Sherbourne, and C. E. Siegel. 1965. Smooth muscle involvement in myotonic dystrophy. *Am. J. Med.* **39**:81–90.
13. Wochner, D. R., G. Drews, W. Strober, and T. A. Waldmann. 1966. Accelerated breakdown of immunoglobulin G (IgG) in myotonic dystrophy: A hereditary error of immunoglobulin catabolism. *J. Clin. Invest.* **45**:321–329.
14. Lindsley, D. B., and E. C. Curnen. 1936. An electromyographic study of myotonia. *Arch. Neurol. Psychiatry* **35**:253–269.
15. Denny-Brown, D., and S. Nevin. 1941. The phenomenon of myotonia. *Brain* **64**:1–18.
16. Bryant, S. H. 1973. The electrophysiology of myotonia with a review of congenital myotonia of goats. In: *New Developments in Electromyography and Clinical Neurophysiology,* Volume 1. J. E. Desmedt, ed. Karger, Basel, pp. 420–450.
17. Bryant, S. H. 1979. Myotonia in the goat. *Ann. N.Y. Acad. Sci.* **317**:314–325.
18. Bryant, S. H. 1977. The physiological basis of myotonia. In: *Pathogenesis of Human Muscular Dystrophies.* L. P. Rowland, ed. Excerpta Medica, Amsterdam, pp. 715–728.
19. Bryant, S. H. 1982. Physical basis of myotonia. In: *Disorders of the Motor Unit.* D. L. Shotland, ed. Wiley, New York. pp. 381–389.
20. DeCoursey, T. E., S. H. Bryant, and K. M. Owenburg. 1981. Dependence of membrane potential on extracellular ionic concentrations in myotonic goats and rats. *Am. J. Physiol.* **240**:C56–C63.
21. Adrian, R. H., and S. H. Bryant. 1974. On the repetitive discharge in myotonic muscle fibers. *J. Physiol. (London)* **240**:505–515.
22. Furman, R. E., and R. L. Barchi. 1977. The pathophysiology of myotonia produced by aromatic carboxylic acids. *Ann. Neurol.* **4**:357–365.
23. Palade, P. T., and R. L. Barchi. 1978. On the inhibition of muscle membrane chloride conductance by aromatic carboxylic acids. *J. Gen. Physiol.* **69**:879–896.
24. Bryant, S. H., and A. Morales-Aguilera. 1971. Chloride conductance in normal and myotonic muscle fibers and the action of monocarboxylic aromatic acids. *J. Physiol. (London)* **219**:367–383.
25. DeCoursey, T. E., and S. H. Bryant. 1979. Sodium current in normal and myotonic mammalian skeletal muscle fibers. *Biophys. J.* **25**:69a.
26. Bryant, S. H., and T. E. DeCoursey. 1980. Sodium currents in cut skeletal muscle fibers from normal and myotonic goats. *J. Physiol. (London)* **307**:31P–32P.
27. Valle, R., and S. H. Bryant. 1980. Potassium conductance in myotonic and normal mammalian skeletal muscle fibers. *Fed. Proc.* **39**:2073.
28. Furman, R. E., and R. Barchi. 1981. 20,25-Diazacholesterol myotonia: An electrophysiological study. *Ann. Neurol.* **10**:251–260.
29. D'Alonzo, A. J., and J. J. McArdle. 1982. An evaluation of fast and slow-twitch muscle from rats treated with 20,25-diazacholesterol. *Exp. Neurol.* **78**:46–66.
30. Peter, J. B., and W. Feihn. 1975. Diazacholesterol myotonia: Accumulation of desmosterol and increased adenosine triphosphatase activity of sarcolemma. *Science* **179**:910–912.
31. D'Alonzo, A. J., and J. J. McArdle. 1982. Effects of 20,25-diazacholesterol treatment on the decay of end-plate currents. *Exp. Neurol.* **76**:681–683.
32. Barchi, R. L. 1982. A mechanistic approach to the myotonic syndromes. *Muscle Nerve* **5**:560–563.
33. Lipicky, R. J. 1977. Studies in human myotonic dystrophy. In: *Pathogenesis of Human Muscular Dystrophies.* L. P. Rowland, ed. Elsevier/North-Holland, Amsterdam. pp. 729–738.

34. Lipicky, R. J., and S. H. Bryant. 1973. A biophysical study of the human myotonias. In: *New Developments in Electromyography and Clinical Neurophysiology,* Volume 1. J. E. Desmedt, ed. Karger, Basel. pp. 451–463.

35. Lehmann-Horn, F., R. Rudel, R. Denglen, H. Lorkovic, A. Haass, and K. Ricker. 1981. Membrane defects in paramyotonia congenita with and without myotonia in a warm environment. *Muscle Nerve* 4:396–406.

36. McComas, A. J., and K. Mrozek. 1968. The electrical properties of muscle fiber membranes in dystrophic myotonica and myotonia congenita. *J. Neurol. Neurosurg. Psychiatry* 31:441–447.

37. Hofmann, W. W., W. Alston, and G. Rowe. 1966. A study of individual neuromuscular junctions and myotonia. *Electroencephalogr. Clin. Neurophysiol.* 21:521–537.

38. DeCoursey, T. E., S. H. Bryant, and R. J. Lipicky. 1982. Sodium currents in human skeletal muscle fibers. *Muscle Nerve* 5:614–618.

39. Merickel, M., R. Gray, P. Chauvin, and S. H. Appel. 1981. Cultured muscle from myotonic muscular dystrophy patients: Altered membrane electrical properties. *Proc. Natl. Acad. Sci. USA* 78:648–652.

40. Appel, S. H., M. Merickel, R. Gray, and R. B. Moore. 1984. Membrane abnormalities in the myotonic disorders. In: *Neuromusc. Dis.* G. Serratrice, ed. Raven Press, New York. pp. 167, 172.

41. Barrett, J. N., E. F. Barrett, and L. B. Dribin. 1981. Calcium-dependent slow potassium conductance in rat myotubes. *Dev. Biol.* 82:258–266.

42. Merickel, M. B., and R. Gray. 1979. Membrane properties of rat skeletal muscle in primary tissue culture. *Biophys. J.* 25:200a.

43. Thompson, S. H. 1977. Three pharmacologically distinct potassium channels in molluscan neurones. *J. Physiol. (London)* 265:465–488.

44. Duval, A., and C. Leoty. 1978. Ionic currents in mammalian fast skeletal muscle. *J. Physiol. (London)* 278:403–423.

45. Connor, J. A. 1980. The fast K^+ channel and repetitive firing. In: *Molluscan Nerve Cells: From Biophysics to Behavior.* J. Koester and J. H. Byrne, ed. Cold Spring Harbor Laboratory, Cold Spring Harbor, N.Y.

46. Brown, D. A., and P. R. Adams, 1980. Muscarinic suppression of a novel voltage-sensitive K^+ current in a vertebrate neurone. *Nature (London)* 283:673–676.

47. Almers, W., and P. T. Palade. 1981. Slow calcium and potassium currents across frog muscle membrane: Measurements with a Vaseline-gap technique. *J. Physiol. (London)* 312:159–176.

48. Barrett, J. N., K. L. Magleby, and B. S. Pallotta. 1982. Properties of single calcium-activated potassium channels in cultured rat muscle. *J. Physiol. (London)* 331:211–230.

49. Plishker, G. A., H. J. Gitelman, and S. H. Appel. 1978. Myotonic muscular dystrophy: Altered calcium transport in erythrocytes. *Science* 200:323–325.

50. Roses, A. D., and S. H. Appel. 1978. Inherited membrane disorders of muscle: Duchenne muscular dystrophy and myotonic muscular dystrophy. In: *Physiology of Membrane Disorders.* T. Andreoli, J. Hoffman, and D. D. Fanestil, eds. Plenum Press, New York. pp. 801–815.

51. Brown, H. D., S. K. Chattopadhyay, and A. B. Patel. 1967. Erythroycte abnormality in human myopathy. *Science* 157:1577–1578.

52. Klassen, G. A., and R. Blostein. 1969. Adenosine triphosphatase and myopathy. *Science* 163:492–493.

53. Souweine, G., J. C. Bernard, Y. Lasne, and J. Lachanat. 1978. The sodium pump of erythrocytes from patients with Duchenne muscular dystrophy: Effect of ouabain on the active sodium flux and on (Na^+, K^+) ATPase. *J. Neurol.* 217:287–294.

54. Hodson, A., and D. Pleasure. 1977. Erythrocyte cation-activated adenosine triphosphatases in Duchenne muscular dystrophy. *J. Neurol. Sci.* 32:361–369.

55. Plishker, G. A., and S. H. Appel. 1980. Red blood cell alterations in muscular dystrophy: The role of lipids. *Muscle Nerve* 3:70–81.

56. Hull, K. I., and A. D. Roses. 1966. Stoichiometry of sodium and potassium transport in erythrocytes from patients with myotonic muscular dystrophy. *J. Physiol. (London)* 254:169–181.

57. Hobbs, A. S., R. A. Brumback, and B. W. Festoff. 1979. Monovalent cation transport in myotonic dystrophy: Na-K pump ratio in erythrocytes. *J. Neurol. Sci.* 41:299–306.

58. Desnuelle, C., A. Lombet, G. Serratrice, and M. Lazdunski. 1982. Sodium channel and sodium pump in normal and pathological muscles from patients with myotonic muscular dystrophy and lower motor neuron impairment. *J. Clin. Invest.* 69:358–367.

59. Roses, A. D., and S. H. Appel. 1973. Protein kinase activity in erythrocyte ghosts of patients with myotonic muscular dystrophy. *Proc. Natl. Acad. Sci. USA* 70:1855–1859.

60. Roses, A. D., and S. H. Appel. 1974. Muscle membrane protein kinase in myotonic muscular dystrophy. *Nature (London)* 250:245–247.

61. Roses, A. D., and S. H. Appel. 1975. Phosphorylation of a component of the human erythrocyte membrane in myotonic muscular dystrophy. *J. Membr. Biol.* 20:51–58.

62. Vickers, J. D., A. J. McComas, and M. P. Rathbone. 1979. Myotonic muscular dystrophy: Abnormal temperature response of membrane phosphorylation in erythrocyte membranes. *Neurology (Minneapolis)* 29:791–796.

63. Butterfield, D. A., A. D. Roses, M. L. Cooper, S. H. Appel, and D. B. Chesnut. 1974. A comparative electron spin resonance study of the erythrocyte membrane in myotonic muscular dystrophy. *Biochemistry* 13:5078–5082.

64. Butterfield, D. A., D. B. Chesnut, A. D. Roses, and S. H. Appel. 1974. Electron spin resonance studies of erythrocytes from patients with myotonic muscular dystrophy. *Proc. Natl. Acad. Sci. USA* 71:909–913.

65. Butterfield, D. A., D. B. Chesnut, S. H. Appel, and A. D. Roses. 1976. Spin label study of erythrocyte membrane fluidity in myotonic and Duchenne muscular dystrophy and congenital myotonia. *Nature (London)* 263:159–161.

66. Butterfield, D. A., A. D. Roses, S. H. Appel, and D. B. Chesnut. 1976. Electron spin resonance studies of membrane proteins in erythrocytes in myotonic muscular dystrophy. *Arch. Biochem. Biophys.* 177:226–234.

67. Butterfield, D. A. 1977. Electron spin resonance investigations of membrane proteins in erythrocytes in muscle diseases. *Biochim. Biophys. Acta* 470:1–7.

68. Butterfield, D. A. 1981. Myotonic muscular dystrophy: Time-dependent alterations in erythrocytes membrane fluidity. *J. Neurol. Sci.* 52:61–67.

69. Wigand, J. P., and W. B. Blackard. 1979. Downregulation of insulin-receptors in obese man. *Diabetes* 28:281–287.

70. Olefsky, J. M., and G. M. Reaven. 1977. Insulin binding in diabetes: Relationships with plasma insulin levels and insulin sensitivity. *Diabetes* 26:680–688.

71. Bar, R. S., W. R. Levis, M. M. Rechler, L. C. Harrison, C. Siebert, J. Podskalny, J. Roth, and M. Muggeo. 1978. Extreme insulin resistance in ataxia telangiectasia: Defect in affinity of insulin receptors. *N. Engl. J. Med.* 298:1164–1171.

72. Oseid, S., H. Beck-Nielsen, O. Pedersen, and O. Svik. 1977. Decreased binding of insulin to its receptor in patients with congenital generalized lipodystrophy. *N. Engl. J. Med.* 296:245–248.

73. Kahn, C. R., J. S. Flier, R. S. Bar, J. A. Archer, P. Gorden, M. M. Martin, and J. Roth. 1976. The syndromes of insulin resistance and acanthosis nigricans: Insulin-receptor disorders in man. *N. Engl. J. Med.* 294:739–745.

74. Festoff, B. W., and W. V. Moore. 1979. Evaluation of the insulin-receptor in myotonic dystrophy. *Am. Neurol.* 6:60–65.

75. Tevaarwerk, G. J. M., and A. J. Hudson. 1977. Carbohydrate metabolism and insulin resistance in myotonia dystrophica. *J. Clin. Endocrinol. Metab.* 44:491–498.

76. Kobayashi, M., J. C. Meek, and E. Streib. 1977. The insulin-receptor in myotonic dystrophy. *J. Clin. Endocrinol. Metab.* 45:821–824.

77. Poffenbarger, P. L., T. Pozefsky, and J. S. Soeldnerl. 1976. The direct relationship of proinsulin–insulin hypersecretion to basal serum levels of cholesterol and triglyceride in myotonic dystrophy. *J. Lab. Clin. Med.* **87**:384–396.

78. Cudworth, A. G., and B. A. Walker. 1975. Carbohydrate metabolism in dystrophia myotonica. *J. Med. Genet.* **12**:157–161.

79. Barbosa, J., F. Q. Nuttall, W. Kennedy, and F. Goetz. 1974. Plasma insulin in patients with myotonic dystrophy and their relatives. *Medicine (Baltimore)* **53**:307–323.

80. Gorden, P., R. C. Griggs, S. P. Nissley, J. Roth, and W. K. Engel. 1969. Studies of plasma insulin in myotonic dystrophy. *J. Clin. Endocrinol. Metab.* **29**:684–690.

81. Huff, T. A., E. S. Horton, and H. E. Lebovitz. 1967. Abnormal insulin secretion in myotonic dystrophy. *N. Engl. J. Med.* **277**:837–841.

82. Stuart, C. A., R. M. Armstrong, S. A. Provow, and G. A. Plishker. 1983. Insulin resistance in patients with myotonic dystrophy. *Neurology* **33**:679–685.

83. Butterfield, D. A., and W. E. Watson. 1977. Electron spin resonance studies of an animal model of human congenital myotonia: Increased erythrocyte membrane fluidity in rats with 20,25-diazacholesterol induced myotonia. *J. Membr. Biol.* **32**:165–176.

84. Chalikian, D. M., and R. L. Barchi. 1982. Sarcolemmal desmosterol accumulation and membrane physical properties in 20,25-diazacholesterol myotonia. *Muscle Nerve* **5**:118–124.

85. Ruitenbeek, W. 1978. The fatty acid composition of various lipid fractions isolated from erythrocytes and blood plasma of patients with Duchenne and congenital myotonic muscular dystrophy. *Clin. Chim. Acta* **89**:99–110.

86. McLaughlin, J., and W. K. Engel. 1979. Lipid composition of erythrocytes: Findings in Duchenne's muscular dystrophy and myotonic atrophy. *Arch. Neurol.* **36**:351–354.

87. Grey, J. E., H. J. Gitelman, and A. D. Roses. 1980. Myotonic muscular dystrophy: Defective phospholipid metabolism in the erythrocyte plasma membrane. *J. Clin. Invest.* **65**:1478–1482.

88. Moore, R. B., S. H. Appel, and G. A. Plishker. 1981. Myotonic dystrophy: Calcium-dependent phosphatidic acid synthesis in erythrocytes. *Ann. Neurol.* **10**:491–493.

89. Grey, J., H. Gitelman, and A. D. Roses. 1981. Comment on erythrocyte phospholipid metabolism in myotonic muscular dystrophy. *Ann. Neurol.* **10**:494.

90. Moore, R. B., and S. H. Appel. 1980. Methylation of erythrocyte membrane phospholipids in patients with myotonic and Duchenne muscular dystrophy. *Exp. Neurol.* **70**:380–391.

91. Burns, T. W., H. E. Dale, and P. L. Langley. 1965. Normal and myotonic goats receiving diazacholesterol. *Am. J. Physiol.* **209**:1227–1232.

92. Peter, J. B., R. M. Andiman, R. L. Bowman, and T. Nagatomo. 1973. Myotonia induced by diazacholesterol: Increased (Na⁺ K⁺)-ATPase activity of erythrocyte ghosts and development of cataracts. *Exp. Neurol.* **41**:738–744.

93. Peter, J. B., S. H. Dromgoole, D. S. Champion, K. E. Stempel, and R. L. Bowman. 1975. Experimental myotonia and hypocholesterolemic agents. *Exp. Neurol.* **49**:115–122.

94. Winer, N., D. M. Klachko, R. D. Baer, P. I. Langley, and T. W. Burns. 1966. Myotonic response induced by inhibitors of cholesterol biosynthesis. *Science* **153**:312–313.

95. Winer, N., J. M. Martt, J. E. Somers, L. Wolcott, H. E. Dale, and T. W. Burns. 1965. Induced myotonia in man and goat. *J. Lab. Clin. Med.* **66**:758–769.

96. Wakamatsu, H., H. Nakamura, K. Ito, W. Anazawa, S. Okajima, S. Okamoto, K. Shigeno, and Y. Goto. 1970. Serum desmosterol and other lipids in myotonic dystrophy. *Keio J. Med.* **19**:145–149.

97. Frantz, I. D., Jr., M. L. Mobberley, and G. J. Schroepeer, Jr. 1959. Effects of MER-29 on the intermediary metabolism of cholesterol. *Prog. Cardiovasc. Dis.* **2**:511–518.

98. Andiman, R. M., J. B. Peter, and G. Dhopeshwarkar. 1974. Myotonic dystrophy and myotonia congenita: ATPase and lipid composition of erythrocyte membranes and serum lipids with special reference to desmosterol. *Neurology (Minneapolis)* **24**:352.

99. Thomas, N. S. T., and P. S. Harper. 1978. Myotonic dystrophy: Studies on the lipid composition and metabolism of erythrocytes and skin fibroblasts. *Clin. Chim. Acta* **83**:13–23.

100. Roses, A. D., M. J. Roses, G. A. Nicholson, and C. R. Roe. 1977. Lactate dehydrogenase isoenzyme 5 (LDH-5) in Duchenne muscular dystrophy. *Neurology (Minneapolis)* **27**:414–421.

101. Appel, S. H., and A. D. Roses. 1977. The muscular dystrophies. In: *The Metabolic Basis of Inherited Disease*, 4th ed. J. B. Stanbury, J. B. Wyngaarden, and D. S. Frederickson, eds. McGraw–Hill, New York.

102. Mokri, B., and A. G. Engel. 1975. Duchenne dystrophy: Electron microscopic findings pointing to a basic or early abnormality in the plasma membrane of the muscle fiber. *Neurology (Minneapolis)* **25**:111–120.

103. Schmalbruch, H. 1975. Segmental fiber breakdown and defects of the plasmalemma in diseased human muscle. *Acta. Neuropathol.* **33**:129–141.

104. Bradley, W. B., and J. J. Fulthorpe. 1978. Studies of sarcolemmal integrity in myopathic muscle. *Neurology (Minneapolis)* **28**:670–677.

105. Schotland, D. L., E. Bonilla, and M. VanMeter. 1977. Duchenne dystrophy: Alteration in plasma membrane structure. *Science* **196**:1005–1007.

106. Schotland, D. L., E. Bonilla, and Y. Wakayama. 1979. Pathogenesis of muscle cell damage in the dystrophies: Morphologic aspects including freeze fracture studies. In: *Current Topics in Nerve and Muscle Research.* A. J. Aguayo and G. Karpati, eds. Excerpta Medica, Amsterdam. pp. 29–38.

107. Bonilla, E., D. L. Schotland, and Y. Wakayama. 1982. Freeze–fracture studies in human muscular dystrophy. In: *Disorders of the Motor Unit.* D. L. Schotland, ed. Wiley, New York. pp. 475–487.

108. Osame, M. 1981. Morphological abnormalities of muscle plasma membrane in Duchenne muscular dystrophy: A review for freeze fracture microscopic studies. *Clin. Neurol.* **21**:1074–1076.

109. Osame, M., A. G. Engel, C. J. Rebouche, and R. E. Scott. 1982. Freeze–fracture electron microscopic study of cultured muscle cells in Duchenne dystrophy. In: *Disorders of The Motor Unit.* D. L. Schotland, ed. Wiley, New York. pp. 489–501.

110. Gunn, R. B., and R. B. Kirk. 1976. Anion transport and membrane morphology. *J. Membr. Biol.* **27**:265–282.

111. Franzini-Armstrong, C., L. Landmesser, and G. Pilar. 1975. Size and shape of transverse tubule openings in frog twitch muscle fibers. *J. Cell Biol.* **64**:493–497.

112. Rayns, D. G., F. O. Simpson, and W. S. Bertaud. 1968. Surface features of striated muscle. II. Guinea pig skeletal muscle. *J. Cell Sci.* **3**:475–482.

113. Costello, B. R., and S. A. Shafiq. 1979. Freeze fracture study of muscle plasmalemma in normal and dystrophic chickens. *Muscle Nerve* **2**:191–201.

114. Malouf, N. N., and J. R. Sommer. 1976. Chicken dystrophy: The geometry of the transverse tubules. *Am. J. Pathol.* **84**:299–316.

115. Beringer, T. 1978. Stereologic analysis of normal and dystrophic avian myofibers. *Exp. Neurol.* **61**:380–394.

116. Thompson, E. J., R. Yasin, G. VanBeet, K. Nurse, and S. Ali-Ani. 1977. Myogenic defect in human muscular dystrophy. *Nature (London)* **268**:241–243.

117. Miller, S. E., A. D. Roses, and S. H. Appel. 1976. Scanning electron microscopy studies in muscular dystrophy. *Arch. Neurol.* **33**:172–174.

118. Grassi, E., B. Lucci, C. Marchini, S. Ottonello, M. Parma, R. Reggiani, G. L. Rossi, and J. Tagliavini. 1978. Deformed erthrocytes in muscular dystrophies. *Neurology (Minneapolis)* **28**:842–844.

119. Luthra, M. G., I. Z. Stern, and H. D. Kim. 1979. (Ca⁺⁺ + Mg⁺⁺)-ATPase of red cells in Duchenne and myotonic dystrophy: Effect of soluble cytoplasmic activator. *Neurology (Minneapolis)* **29**:835–841.

120. Mollman, J. E., J. C. Cardenas, and D. E. Pleasure. 1980. Altera-

tions of calcium transport in Duchenne erythrocytes. *Neurology* **30**:1236–1239.

121. Niggli, V., E. S. Adunyah, and E. Carafoli. 1981. Acidic phospholipids, unsaturated fatty acids, and limited proteolysis mimic the effect of calmodulin on the purified Ca^{2+}-ATPase. *J. Biol. Chem.* **256**:8588–8592.

122. Choquette, D., G. Hakim, A. G. Filoteo, G. A. Plishker, J. R. Bostwick, and J. T. Penniston. 1985. Regulation of plasma membrane Ca^{2+} ATPase by lipids of the phosphatidylinositol cycle. *Biochem. Biophys. Res. Comm.* **125**:908–915.

123. Roses, A. D., M. H. Herbstreith, and S. H. Appel. 1975. Membrane protein kinase alterations in Duchenne muscular dystrophy. *Nature (London)* **254**:350–351.

124. Roses, A. D., and S. H. Appel. 1976. Erythrocyte spectrin peak II phosphorylation in Duchenne muscular dystrophy. *J. Neurol. Sci.* **29**:185–193.

125. Roses, A. D., M. H. Herbstreith, B. Metcalf, and S. H. Appel. 1976. Increased phosphorylated components of erythrocyte membrane spectrin band II with reference to Duchenne muscular dystrophy. *J. Neurol. Sci.* **30**:167–178.

126. Iyer, S. L., P. A. Hoenig, A. P. Sherblom, and J. L. Howland. 1977. Membrane function affected by genetic muscular dystrophy. *Biochem. Med.* **18**:384–391.

127. Fisher, S., M. Tortolero, J. P. Piau, J. Delaunay, and G. Shapira. 1978. Protein kinase and adenylate cyclase of erythrocyte membrane from patients with Duchenne muscular dystrophy. *Clin. Chim. Acta* **88**:437–440.

128. Falk, R. S., D. Champion, and D. Futhrie. 1979. Phosphorylation of red cell membrane proteins in Duchenne muscular dystrophy. *N. Engl. J. Med.* **300**:258.

129. Roses, A. D., M. Herbstreith, and P. Shile. 1980. The isolation of abnormally (^{32}P)-phosphorylated cyanogen bromide cleavage products of erythrocyte spectrin in Duchenne muscular dystrophy. *Neurology (Minneapolis)* **30**:423.

130. Roses, A. D., M. E. Mabry, M. H. Herbstreith, P. V. Shile, and C. V. Balakrishnan. 1982. Increased ^{32}P-phosphorylation of spectrin peptides in Duchenne muscular dystrophy. In: *Disorders of the Motor Unit.* D. L. Schotland, ed. Wiley, New York. pp. 413–422.

131. Mawatari, S., A. Takagi, and L. P. Rowland. 1974. Adenyl cyclase in normal and pathologic human muscle. *Arch. Neurol.* **30**:96–102.

132. Canal, N., L. Frattola, and S. Smirne. 1975. The metabolism of cyclic 3'-5' adenosine monophosphate (cAMP) in diseased muscle. *J. Neurol.* **208**:259–265.

133. Susheela, A. K., R. D. Kaul, and K. Sachdeva. 1975. Adenyl cyclase activity in Duchenne dystrophic muscle. *J. Neurol.* **24**:361–363.

134. Khokhlov, A. P., and V. K. Malakhovsky. 1978. Characteristics of cyclic 3'-5' AMP turnover in patients with progressive muscular atrophy. *Vopr. Med. Khim.* **24**:754–758.

135. Willner, J. H., C. G. Cerri, and H. Somer. 1978. Adenyl cyclase: Abnormal in Duchenne carrier muscle. *IVth Int. Congr. Neuromusc. Dis. Abstr.* p. 478.

136. Takahaski, K., H. Takao, and T. Takae. 1978. Adenylate cyclase in Duchenne and Fukuyama type of dystrophy. *Kobe J. Med. Sci.* **24**:193–198.

137. Mawatari, S., A. Miranda, and L. P. Rowland. 1976. Adenyl cyclase abnormality in Duchenne muscular dystrophy: Muscle cells in culture. *Neurology* **26**:1021–1026.

138. Mawatari, S., M. Schonberg, and M. Olarte. 1976. Biochemical abnormalities of erythrocyte membrane in Duchenne dystrophy adenosine triphosphatase and adenyl cyclase. *Arch. Neurol.* **33**:489–493.

139. Wacholtz, M. D., S. G. Doible, and S. Jackowsky. 1979. Adenylate cyclase and ATPase activities in red cell membranes of patients and genetic carriers of Duchenne muscular dystrophy. *Clin. Chim. Acta* **96**:255–259.

140. Lane, R. J. M., P. Maskrey, and G. A. Nicholson. 1978. An evaluation of some carrier detection techniques in Duchenne muscular dystrophy. *IVth Int. Congr. Neuromusc. Dis.* p. 488.

141. Kunze, D., G. Reichmann, E. Egger, G. Leuschner, and H. Eckhardt. 1973. Erythrozytenlipide bei progressiver muskeldystrophia. *Clin. Chim. Acta* **43**:333–341.

142. Kalofoutis, A., G. Jullien, and V. Spanos. 1977. Erythrocyte phospholipids in Duchenne muscular dystrophy. *Clin. Chim. Acta* **74**:85–87.

143. Kobayashi, T., S. Mawatari, and Y. Kuroiwa. 1978. Lipids and proteins of erythrocyte membrane in Duchenne muscular dystrophy. *Clin. Chim. Acta* **85**:259–266.

144. Koski, C. L., F. B. Jungalwala, and E. H. Kolodny. 1978. Normality of erythrycyte phospholipids in Duchenne muscular dystrophy. *Clin. Chim. Acta* **85**:295–298.

145. Kohlschutter, A., U. N. Wiesmann, N. N. Herschkowitz, and E. Ferber. 1976. Phospholipid composition of cultivated skin fibroblasts in Duchenne's muscular dystrophy. *Clin. Chim. Acta* **70**:463–465.

146. Howland, J. I., and S. L. Iyer. 1977. Erythrocyte lipids in heterozygous carriers of Duchenne muscular dystrophy. *Science* **198**:309–310.

Disorders of Muscle

The Periodic Paralyses

Robert L. Ruff and Albert M. Gordon

1. Introduction

The periodic paralyses are characterized by episodic attacks of paralysis. They are usually classified as primary (familial) or secondary and, according to the associated change in serum potassium, as hypokalemic, hyperkalemic, or normokalemic. The pattern of electrolyte changes is usually similar from attack to attack in a given patient; however, rarely, a single patient will have paralysis associated with an elevated, depressed, or unchanged serum K^+ concentration.[1] The different forms of periodic paralysis share several features: (1) the attacks may last from minutes to days and occur sporadically, (2) the paralysis can be focal or generalized, (3) the tendon reflexes are depressed, (4) respiratory and cranial nerve-innervated muscles are relatively spared, (5) rest after exercise can provoke an attack whereas continuous activity can allay paresis, (6) cooling of muscles may produce weakness, (7) whereas early in the course of the disease the patient may have normal strength between attacks, progressive proximal weakness can develop after repeated attacks in the primary forms of periodic paralysis.[2]

In hypokalemic periodic paralysis the attacks are triggered by excessive influx of K^+ into muscle whereas hyperkalemic paralysis is associated with exaggerated efflux of K^+ from muscle. The relationships of the paralysis to the changes in muscle and serum K^+ have not been resolved completely. However, it is clear that the paralysis results from a defect in the muscle fibers as they do not contract with either nerve or direct stimulation during an attack.[3] Therefore, the paralysis must result from a defect in one or more of the steps of excitation–contraction coupling.[4] In the following sections we discuss the clinical

features of the different types of periodic paralysis, the pathophysiology of the serum and muscle K^+ changes, and the rationale for various forms of treatment.

2. Clinical Features

2.1. Hypokalemic Periodic Paralysis

2.1.1. Familial Hypokalemic Periodic Paralysis

This disease is transmitted as an autosomal dominant disorder with reduced penetrance in women resulting in a male/female ratio of 4 : 1.[5] Sporadic cases have also been noted.[6,7] The attacks usually begin in the first or second decade of life with the majority of patients having a paralytic attack before 16 years of age.[8] The attacks are usually infrequent during adolescence, become more frequent (occasionally occurring daily) during early adulthood, and may cease when the patient reaches middle age. During an attack, the serum K^+ falls, but not always below the normal range. The patients become oliguric and there is renal retention of K^+ and Na^+.[9–11] The serum K^+ returns to normal with recovery of muscle strength. Sinus bradycardia and other electrocardiographic signs of hypokalemia develop when the serum K^+ drops below the normal range.[12,13] The serum concentration of creatine kinase and its MB isoenzyme may rise during an attack.[14] The paralysis is usually flaccid but myotonia may be present as the weakness develops.[15] Common precipitating factors are: a high-carbohydrate meal, rest after exercise, cold, and emotional excitement.[6,8,16–18] Although patients usually have normal strength between attacks, repeated attacks may result in permanent proximal myopathy.[2,18] Rarely, mild weakness has been noted in adolescent patients who have had few attacks.[19] Occasionally, patients may develop cardiomyopathy in association with permanent extremity weakness.[20]

The diagnosis is suggested by a positive family history, hypokalemia during a paralytic attack, and a normal serum K^+

Robert L. Ruff • Department of Neurology, Cleveland Veterans Administration Hospital and Case Western Reserve University, Cleveland, Ohio 44106. Albert M. Gordon • Department of Physiology and Biophysics, University of Washington School of Medicine, Seattle, Washington 98195.

between attacks. The diagnosis can be confirmed by inducing hypokalemic paralysis in the hospital and restoring strength with K^+ replacement. The usual provocative test is to administer glucose, as an intravenous infusion, at 50 g/hr or orally 2 g/kg, combined with 10–20 U of regular insulin given subcutaneously.[5,8] This will usually induce hypokalemia and paralysis within 3 hr. If not, the test may be repeated after the patient has been subjected to heavy exercise and given 4–8 g of NaCl solution orally.[8] Occasionally, patients will be refractory to this test.[21] An alternative test was developed by Engel *et al.*[18] in which a very low dose of epinephrine (2 μg/min) is infused into the brachial artery for 5 min and the amplitude of the evoked compound action potential in hand muscles is recorded. A positive result is a greater than 30% reduction in the muscle action potential. Both of these tests must be carried out with extreme caution by experienced individuals. The glucose–insulin challenge can provoke hypoglycemia or extreme hypokalemia and excess intraarterial epinephrine can produce severe vasospasm.

2.1.2.   Thyrotoxic Periodic Paralysis

The presentation of thyrotoxic periodic paralysis closely resembles that of familial hypokalemic periodic paralysis. Patients have recurrent attacks of weakness that can be precipitated by a carbohydrate challenge with or without insulin,[22,23] muscle cooling,[22] or rest after exercise.[23] Serum K^+ is usually decreased during the attacks.[22,23] However, in some cases, the patients are normokalemic during attacks.[22] There are a number of differences between patients that develop thyrotoxic periodic paralysis and those that have familial hypokalemic periodic paralysis. Satoyoshi *et al.* reported that 8.9% of 492 patients with thyrotoxicosis had periodic paralysis,[22] and 82% of the patients became thyrotoxic before or coincident with the onset of paralytic attacks. Most cases of thyrotoxic periodic paralysis are sporadic. The age of onset is greater than 20 years of age in over 90% of cases of thyrotoxic periodic paralysis compared with 60% of patients with familial hypokalemic periodic paralysis having their first attack before 16 years of age. The male/female ratio is approximately 6 : 1 in thyrotoxic periodic paralysis compared with 4 : 1 in familial hypokalemic periodic paralysis.[8,22] More than 80% of the reported cases of thyrotoxic periodic paralysis have occurred in Orientals, whereas familial periodic paralysis is uncommon in Orientals.[22,24]

2.1.2a.   Relationship   to   Thyrotoxicosis.
Thyrotoxic periodic paralysis resolves in over 90% of cases with normalization of thyroid function.[22,24] Okihiro and Nordyke reported a Japanese man with thyrotoxic periodic paralysis that resolved with thyroidectomy, who had recurrence of attacks when treated with triiodothyronine.[25] Episodic attacks of weakness have resulted from administration of thyroid hormone in one Japanese and three Caucasian euthyroid patients.[26,27] The Japanese patient developed paralysis with doses of thyroid hormone which did not produce signs of thyrotoxicosis.[26] Conversely, iatrogenic thyrotoxicosis did not worsen the course of a patient with familial hypokalemic periodic paralysis. Interestingly, this patient had attacks of paralysis when thyroid treatment was withdrawn.[24] Hence, the relationship between thyroid status and weakness is markedly different in thryotoxic periodic paralysis and familial hypokalemic periodic paralysis.

Table I. Causes of Secondary Hypokalemic Periodic Paralysis

Renal K^+ wastage
    Primary hyperaldosteronism[29]
    Renal tubular acidosis[30]
    Fanconi's syndrome[31]
    Acute tubular necrosis[32]
    Bilateral ureterocolostomy[33]
    Secondary hyperaldosteronism[34]
    Mineralocorticoid treatment[28]
    Natural licorice[35]
    Ammonium chloride treatment[36]
Gastrointestinal K^+ loss
    Chronic diarrhea[37]
    Villous adenoma[38]
    Gastrin-secreting adenoma[39]
    Laxative abuse[40]
    Barium poisoning[41,164]

2.1.3.   Hypokalemic Periodic Paralysis Due to Excess Renal or Gastrointestinal K^+ Loss

Severe K^+ depletion usually produces generalized weakness, but occasionally patients will develop episodic paralysis in association with serum $K^+ < 3$ meq/liter. One patient had paralysis induced by insulin.[28] Hypokalemia between attacks and absence of a family history of episodic paralysis separate these patients from those with familial hypokalemic periodic paralysis. Some of the causes of secondary hypokalemic periodic paralysis are shown in Table I.

2.2.   Hyperkalemic Periodic Paralysis

2.2.1.   Primary Hyperkalemic Periodic Paralysis

This disorder, also referred to as adynamia episodica hereditaria, usually has an autosomal dominant inheritance[2,42–47] with sporadic cases occasionally reported.[48,49] In the familial disorder, attacks usually commence in childhood or adolescence.[2,42–45,47] At the onset of an attack, the patients may develop myalgia,[44] myotonia,[44,47] elevated serum concentration of creatine kinase and creatinuria,[50] and characteristically the serum K^+ rises though not always beyond the normal range.[2,42–45] The electrocardiogram shows T-wave elevation in association with the hyperkalemia.[47] Hypocalcemia during paralysis was found in three patients.[45,48] A diuresis with elevated K^+ excretion frequently accompanies the paralysis.[42] Paralysis may be precipitated by rest after exercise, oral ingestion of 1–2.5 g potassium, cold exposure, pregnancy, or administration of glucocorticoids[42–45,51] Initially, strength is usually normal between attacks, but a permanent proximal myopathy may develop after repeated attacks.[52] Patients frequently manifest myotonia[44,45,53] or abnormal electrical excitability of muscle[54] and peripheral nerves.[42,55]

Myotonic hyperkalemic periodic paralysis and paramyotonia congenita have many features in common: autosomal dominant inheritance, myotonia, myotonia and weakness provoked by exposure to cold, and weakness produced by rest after exercise. The similarities between these disorders suggested that they might be a single disease.[45,46] However, there are slight differences which distinguish paramyotonia and hyperkalemic periodic paralysis: (1) some patients with hyperkalemic periodic paralysis have no clinical or electromyographic evidence of myotonia[42,56] (2) K+ loading may not provoke weakness in patients with paramyotonia,[42,57] (3) repeated muscle contraction aggravates myotonia in paramyotonia and lessens myotonia in hyperkalemic periodic paralysis,[57] and (4) the attacks of weakness in paramyotonia may be associated with reduced serum K+.

2.2.2. Secondary Hyperkalemic Periodic Paralysis

Renal K+ retention due to adrenal insufficiency,[58–67] or inhibition of mineralocorticoid action[68] can produce persistent hyperkalemia and paralytic attacks. The attacks are induced by rest after exercise with a further elevation of the serum K+ above 7 meg/liter.

2.3. Normokalemic Periodic Paralysis

This autosomal dominant disorder is characterized by paralytic attacks starting in childhood and provoked by rest after exercise or cold exposure.[1,69,70] K+ loading induced paralysis in one family,[69] but not in another family.[70] Large doses of Na+ improve the weakness.[69,70] Usually there is no consistent pattern of electrolyte changes associated with paralysis. However, Chesson et al.[1] reported a patient who initially had episodes characteristic of familial normokalemic periodic paralysis and then developed spontaneously occurring and provokable episodes of both hypo- and hyperkalemic periodic paralysis superimposed on a persistent proximal myopathy.

There are reports of four other forms of periodic paralysis which are not associated with fluctuations in serum K+ levels.[71–74] These disorders differ from the previously discussed forms of periodic paralysis in that they are associated with other skeletal, cardiac, or developmental abnormalities. In one family, the paralysis appeared to result from a structural defect in the sarcoplasmic reticulum which may inhibit Ca^{2+} release or storage.[74]

2.4. Morphological Changes

The characteristic histopathological finding in periodic paralysis is a vacuolar myopathy[75] which has been found in the primary hypokalemic,[15,17,23,75–79] secondary hypokalemic,[28,30] thyrotoxic,[22,23,79–81] primary hyperkalemic,[8,47,79,82,83] and normokalemic[1,2,82] disorders. The vacuolation appears to be more prominent after repeated paralytic attacks and consequently has been associated with the permanent proximal myopathy,[8] but myopathy can occur without vacuolar changes.[84] The vacuoles are located in the center of the fiber, are usually lined with membrane, may contain PAS-reactive material (probably glycogen), and are frequently associated with swelling of the T-tubules and mitochondria, and dilation of the terminal cisternae of the sarcoplasmic reticulum (Fig. 1). Engel[8,78] studied the development of vacuoles in primary hypokalemic periodic paralysis and suggested the following sequence: (1) the vacuoles are formed by the coalescence of proliferating T-tubular and sarcoplasmic reticulum membrane; and (2) small vacuoles join to form larger vacuoles which may communicate with the extracellular space via the T-tubules. Why the vacuoles develop is still unclear. Engel[8] and Kao and Gordon[86] suggested that the vacuoles were a consequence of

Fig. 1. (Top) Longitudinal section of muscle from a patient with hypokalemic periodic paralysis.[77] The moderately stretched muscle has a portion of a myotonic vacuole in the middle right side of the micrograph. Fixed in osmium tetroxide with Price clamp. Stained with uranyl acetate and lead acetate. Original magnification ×17,500. (Bottom) Longitudinal section of muscle from a rat kept on a K+-deficient diet for 4 weeks.[86] The sarcoplasmic reticulum is dilated with prominent triads. Fixed with formaldehyde–glutaraldehyde. Stained with uranyl acetate and lead citrate. Original magnification ×12,000.

the sarcoplasmic and sarcolemmal abnormalities found in the periodic paralyses. The morphological changes may reflect intracellular K+ depletion as clinical[28,30,85] and experimental[86–88] K+ depletion produces vacuolar changes which closely resemble those seen in the periodic paralyses (Fig. 1).

2.5. Intracellular Electrolyte Changes

A common feature of muscle from patients with periodic paralysis is that the intracellular concentration of K+ is lower, and Na+ and Cl− are higher than normal values (Table II). The intracellular electrolyte changes are similar to those found with K+ depletion, supporting the proposal that the vacuolar changes in periodic paralysis result from an alteration in the intracellular electrolyte composition.[8,86]

Table II. Muscle Electrolytes in Patients with Primary Periodic Paralysis[a]

Disorder	State of patient	K^{+b}	Na^{+b}	Cl^{-b}	Water[c]	N^d
Hypokalemic periodic paralysis[17,18,77,89–95]	Paralyzed	78.1	42.9	29.3	777.3	18
	Not paralyzed	78.8	43.8	26.3	766.7	20
Thyrotoxic periodic paralysis[22,94]	Paralyzed	84.0	40.3	—	786	4
	Not paralyzed	86.1	43.3	25.2	787	15
Hyperkalemic periodic paralysis[46,53,96]	Not paralyzed	83.8	42.1	21.6	774.7	8
Controls[46]		95	35	21	782	12

[a]Average of literature values.
[b]Expressed as meq/kg tissue wet wt.
[c]Expressed as ml/kg tissue wet wt.
[d]Number of specimens included in average.

3. Pathophysiology of the Periodic Paralysis

3.1. Changes in the Distribution of Electrolytes

3.1.1. Hypokalemic Periodic Paralysis

3.1.1a. Familial Hypokalemic Periodic Paralysis. The etiology of the hypokalemia during paralytic attacks has been an intriguing question. Conn et al.[97] suggested that intermittent hyperaldosteronism produced the hypokalemia due to renal K^+ loss. This proposal was supported by the observation that hyperaldosteronism,[29] mineralocorticoid treatment,[28] and licorice ingestion[35] could produce hypokalemic weakness. Hyperaldosteronism has been observed during paralytic attacks.[97,98] However, aldosterone levels are usually not elevated during paralytic attacks[18,99,100] and there is frequently urinary retention of Na^+, K^+, and water.[8,18] The occasional finding of elevated aldosterone production may be due to a decrease in extracellular volume which sometimes accompanies paralytic attacks.[8]

Based on arterial and venous plasma measurements, Zierler and Andres[101] found that influx of potassium into muscle produced the hypokalemia in spontaneous or induced paralytic attacks. This finding was confirmed by Grob et al.[99] and Shizume et al.[102] The intracellular K^+ store is much larger than the quantity of extracellular K^+, and thus a shift of sufficient K^+ into muscle to lower the serum K^+ content to less than 1.5 meq/liter may produce very little change in the K^+ content of skeletal muscle.[8] Interestingly, red blood cell K^+ transport is normal[103] and red blood cells do not accumulate K^+ during paralytic attacks.[18] With recovery from the paralysis, K^+ moves from muscle into the extracellular space until the serum K^+ is nearly normal. K^+ is released from muscle to venous blood even when the recovery is produced by K^+ administration. The recovery is followed by a K^+ diuresis.[99,101,102]

The etiology of the episodic excessive influx of K^+ into muscle is not known. McArdle[16] suggested that a block in muscle hexose-phosphate metabolism could result in an excess of nondiffusible phosphorylated anions during glycogen breakdown or synthesis, which would result in an influx of cations and water into muscle cells. Shy et al.[17] modified this proposal by suggesting that muscle glycogen synthesis was impaired in hypokalemic periodic paralysis. These proposals could account for the induction of paralysis by insulin administration or carbohydrate meals. However, Engel et al.[104] studied carbohydrate metabolism and mitochondrial respiration in muscle from patients with hypokalemic periodic paralysis and found no abnor-

malities. In addition, Gordon et al.[77] demonstrated that paralytic attacks were not always associated with increased muscle intracellular water content.

Streeten and Speller[105] observed that administration of a variety of exogenous mineralocorticoids or ACTH, but not aldosterone or glucocorticoids, could precipitate paralytic attacks. The ACTH-induced paralysis could be prevented by metyrapone or SU-9055 (agents which inhibit adrenal steroidogenesis). Based on these findings, they suggested that elevated levels of 18-hydroxy-11-deoxycorticosterone might be responsible for the hypokalemia. The difficulties with this proposal are: (1) paralytic attacks are usually associated with urinary K^+ retention[8,18] and (2) mineralocorticoids do not appear to promote cellular uptake of K^+.[106] Thus, while mineralocorticoids can precipitate paralytic attacks due to their kaliopenic action, they do not mimic the electrolyte changes that occur during spontaneous attacks and are probably not the natural trigger.

Intracellular K^+ accumulation by muscle could result from excessive pumping of K^+ into muscle cells. Samaha[107] reported that basal Na^+,K^+-ATPase activity in muscle from patients with familial hypokalemic periodic paralysis was normal. Insulin and β-adrenergic hormones appear to stimulate the Na^+,K^+ pump in muscle[108–113]; but paralytic attacks are usually not associated with hypersecretion of insulin or epinephrine.[5,114,115] However, episodic acceleration of muscle Na^+–K^+ transport could result from increased response of the Na^+,K^+ pump to insulin or epinephrine stimulation. Hofmann et al.[116] reported that skeletal muscle from a single patient with hypokalemic periodic paralysis bound more insulin than normal. It was not clear if the increased insulin binding represented an increase in number or binding affinity of insulin receptors. However, the increased insulin binding could result in an exaggerated response of the Na^+,K^+ pump to insulin which could account for the induction of paralysis by insulin administration or high-carbohydrate meals. Unfortunately, there are no reported studies of hormone stimulation of Na^+–K^+ transport in muscle from patients with any form of periodic paralysis. Zierler and Andres have found that in patients with hypokalemic periodic paralysis, the nocturnal shift of K^+ into muscle cells[101] was exaggerated over that seen in normal subjects. This finding may explain why paralytic attacks often take place during sleep and in early morning.

3.1.1b. Thyrotoxic Periodic Paralysis. The fluid and electrolyte changes in this disorder are similar to those found in primary hypokalemic periodic paralysis. Muscle uptake of

K+ produces hypokalemia[102] and there is usually diminished renal K+ excretion during an attack.[117] Thyroid hormone directly stimulates Na+–K+ transport in mouse skeletal muscle[118] and also enhances muscle sensitivity to β-adrenergic stimulation.[119] Despite the increased Na+,K+ pump activity, thyroid treatment did not alter skeletal muscle Na+ or K+ content,[118] suggesting that passive K+ efflux and Na+ influx were also increased by thyroid hormone. In human red blood cells from thyrotoxic patients, both passive Na+ and K+ fluxes and active Na+–K+ transport were increased; however, passive fluxes exceeded the Na+–K+ pump rate, resulting in decreased intracellular K+ and elevated intracellular Na+ content.[120] This might explain why muscle from patients with thyrotoxic periodic paralysis has excess intracellular Na+ and is K+ depleted (Table II).[22,94] Enhanced β-agonist sensitivity of the Na+,K+ pump could explain why epinephrine induces hypokalemia. It is not known if there is also increased sensitivity to insulin-stimulated K+ uptake by muscle. However, insulin concentration is elevated during spontaneous paralytic attacks in thyrotoxic periodic paralysis.[121]

3.1.1c. Secondary Hypokalemic Periodic Paralysis.

Patients with this disorder have total body K+ depletion, and usually have a persistent myopathy. They can become paralyzed when the serum K+ falls below 2–3 meg/liter. In most of these patients, the further fall in K+ is triggered by aggravation of their underlying K+ wasting disorder. In the case of a man with mineralocorticoid-induced hypokalemia, insulin administration repeatedly produced hypokalemic paralysis.[28] Offerijns et al.[122] noted that in K+-depleted rats, insulin produced hypokalemia and diaphragmatic paralysis. Thus, in both clinical and experimental K+-depleted states, insulin can produce hypokalemic paralysis.

In K+-depleted mammals, the intracellular K+ content of skeletal muscle is decreased while the K+ content of heart, brain, and red blood cells is maintained at nearly normal values.[123] The density of ouabain-binding sites and the Na+–K+ transport capacity are diminished in muscle[123] and increased in red blood cells.[124] The selective suppression of muscle Na+–K+ transport may be mediated by α-adrenergic stimulation,[125] and muscle activity.[126] Despite the decreased density of Na+,K+ pumps, K+-depleted muscle is able to markedly increase Na+–K+ transport activity when appropriately stimulated.[127,128] Thus, it is possible that insulin stimulation of Na+–K+ transport could further lower serum K+ in K+-depleted patients and experimental animals.

3.1.2. Hyperkalemic Periodic Paralysis

3.1.2a. Primary Hyperkalemic Periodic Paralysis.

Measurement of arterial and venous plasma K+ concentrations showed that the hyperkalemia during paralytic attacks results from a net efflux of K+ from muscle.[42,44,46,51] Consequently, muscle K+ content decreases during attacks.[46,53,96] In spite of normal renal K+ clearance,[44,51] regulation of plasma K+ is markedly disturbed. Patients may have several episodes of hyperkalemia daily.[129] There may be a generalized impairment in K+ transport as red blood cells also have diminished K+ influx and net K+ loss during paralytic attacks.[130] The efflux of K+ from muscle is not caused by diminished secretion of insulin, catecholamines, or glucagon[131]) or elevated serum mineralocorticoid or glucocorticoid activity.[132] Basal Na+,K+-ATPase activity in muscle is

normal.[107] It is possible that Na+,K+ transport may be aberrantly triggered by hormonal stimuli.

3.1.2b. Secondary Hyperkalemic Periodic Paralysis.

Patients with adrenal insufficiency (Addison's disease) are prone to develop hyperkalemia for two reasons. First, the Addisonian patient cannot increase mineralocorticoid output to compensate for a K+ load. Second, diminished glucocorticoid activity reduces the sensitivity to adrenergic stimulation.[133] Consequently, the ability of muscle to take up K+ is impaired due to reduced β-adrenergic stimulation of the Na+,K+ pump.[109,110,113] A consequence of the decreased K+ uptake is that skeletal muscle becomes K+ depleted.[106]

In summary, the electrolyte shifts associated with weakness in the various forms of periodic paralysis are due to excessive K+ influx or efflux from skeletal muscle. The mechanisms for these abnormal K+ fluxes are not known.

3.2. Paralysis Results from Loss of Membrane Excitability

The flaccid weakness in the various forms of periodic paralysis is due to loss of muscle contractility. Nerve conduction and neuromuscular transmission are normal during hypokalemic paralysis attacks.[8,99] Consequently, the loss of contractility could result from (1) loss of surface membrane excitability, (2) impaired T-tubule excitation, (3) decreased release of Ca2+ from the sarcoplasmic reticulum, (4) defective myofibrillar Ca2+-stimulated ATPase activity, or (5) depletion of ATP and other high-energy phosphates. The currently available data suggest that mechanisms 3–5 are not the cause of paralysis. In patients with primary hypokalemic periodic paralysis, Engel and Lambert[134] studied the contraction of skinned fibers (sarcolemma mechanically removed) from paralyzed muscle in response to application of microdrops of Ca2+-containing solutions. They found that the Ca2+ sensitivity of the paralyzed muscle was normal. Engel et al.[135] also found ATP content in paralyzed muscle to be normal. Ca2+ uptake by the sarcoplasmic reticulum was impaired during paralytic attacks and normal between attacks in five patients with thyrotoxic periodic paralysis.[81,136] Diminished Ca2+ uptake by the sarcoplasmic reticulum would impair muscle relaxation but not necessarily affect the initiation of contraction.[4] Unfortunately, Ca2+ release from the sarcoplasmic reticulum has not been studied. T-tubule excitability and coupling to Ca2+ release also remain to be studied. The T-tubules are dilated in the various forms of periodic paralysis but the triadic connections to the terminal cisternae appear to be preserved.[76–79,82,83]

The available data suggest that sarcolemmal inexcitability causes the paralysis. During a paralytic attack in primary hypokalemic[77,137] or hyperkalemic[54] periodic paralysis, EMG recording shows a progressive decline in the number of fibers that can be electrically activated. Extracellular recording from single muscle fibers demonstrates a failure of the surface membrane to propagate an action potential.[137] The amount of tension that a muscle can generate is directly proportional to the number of muscle fibers that remain electrically excitable.[77] A difference between these two forms of periodic paralysis is that in hyperkalemic paralysis there may be a brief period of increased muscle fiber excitability with spontaneous action potentials at the onset of a paralytic attack.[54] Intracellular membrane potential recordings during paralysis usually demonstrated that the mus-

Table III. Muscle Membrane Potential Measurements in Periodic Paralysis[a]

	Not paralyzed (mV)[b]	Paralyzed (mV)[b]
Primary hypokalemia		
Creutzfeldt et al.[140]	−85.6 (−87.4)	−77.1
Hofmann and Smith[95]c	−60.2 (−75.8)	−50.5 (−79.5)
Riecker and Bolte[141]	−78 (−87.2)	−49.1
Rüdel et al.[144]c	−75 (−85.5)	−54 (−99.4)
Shy et al.[17]	−75.4	−71.2
Thyrotoxic		
Hofmann and Smith[95]c	−44.8 (−75.8)	−25.0 (−79.5)
Primary hyperkalemic		
Brooks[138]	−72 (−65)	−45
Creutzfeldt et al.[140]	−68.5 (−87.4)	−43.6, −51.5
Lehmann-Horn et al.[143]c	−83.0, −80.8 (−80.2)	−50, −64.3 (−72.5)
McComas et al.[142]	−66.3 (−83.6)	
Norris[139]	−63 (−70)	
Normokalemic		
McComas et al.[142]	−61.3 (−83.6)	

[a]The values shown are mean values of the membrane potentials of extremity muscle recorded in vivo (except as noted).
[b]Values in parentheses are from control subjects.
[c]From intercostal muscles studied in vitro.

cle fibers were depolarized (Table III) and electrically inexcitable.[95,138,143,144] Therefore, the muscle paralysis can largely be attributed to loss of membrane excitability due to depolarization-induced inactivation of the voltage-sensitive Na^+ channels which normally generate the action potential.[145] However, in some cases the depolarization may not have been of sufficient magnitude to completely explain the membrane inexcitability.[17,143]

It is also possible that voltage-dependent Na^+ channel gating may be altered in periodic paralysis. Rüdel et al.[144] noted that muscle fibers from patients with hypokalemic periodic paralysis studied in bathing medium of normal K^+ concentration had decreased excitability and markedly reduced action potential amplitude, despite having resting potentials between −70 and −80 mV. Excitability and action potential amplitude could be increased by hyperpolarizing the fibers. The muscle fibers from a patient with hyperkalemic periodic paralysis studied in high-K^+ medium were inexcitable at membrane potentials which would not normally be sufficiently positive to inactivate Na^+ channels.[143] These findings suggest that there was excessive Na^+ channel inactivation at or near the resting potential. In mammalian muscle fibers, there are at least three inactivation processes that have different kinetics and voltage dependence. The fastest inactivation process is able to close Na^+ channels on a millisecond timescale.[145−147] The slower inactivation processes act on a timescale of hundreds of milliseconds (slow inactivation) or minutes (ultraslow inactivation).[146] The slower inactivation processes are active at more negative membrane potentials than is true for fast inactivation. Preliminary studies in vitro in rat and human muscle suggested that more than half of the available Na^+ channels are inactivated at the resting potential due to the slower forms of inactivation.[146−149] Consequently, the reduced excitability of fibers from patients with hypokalemic or hyperkalemic periodic paralysis may result from slight depolarization of the membrane potential increasing the amount of inactivation due to the slow inactivation processes. In addition, there may also be a shift in the voltage dependence of the inactivation processes.[144]

While paralysis appears to be associated with significant membrane depolarization, it is less clear whether the fibers from patients with hypokalemic periodic paralysis may be depolarized to a lesser extent between attacks (Table III). Hofmann and Smith[95] and Rüdel et al.[144] noted depolarization of fibers in solutions with normal K^+ concentration; however, their measurements were made in vitro and may not accurately reflect the in vivo membrane potential.[149,150]

3.3. Origin of the Muscle Membrane Depolarization

The membrane potential, E_m, is determined by the intra- and extracellular concentrations of Na^+, K^+, and Cl^-; the permeability P_{Na}, P_K, and P_{Cl} of these ions; and an additional contribution, E_p, from electrogenic pumping the most significant of which is due to the Na^+,K^+ pump.[127,128] The modified Goldman equation provides the relationship between these factors:

$$E_m = \frac{RT}{F} \ln \frac{([K^+]_o + \alpha[Na^+]_o + \beta[Cl^-]_i)}{([K^+]_i + \alpha[Na^+]_i + \beta[Cl^-]_o)} + E_p$$

where R is Boltzmann's constant, T is the absolute temperature, F is Faraday's constant, α is P_{Na}/P_K, β is P_{Cl}/P_K, and the subscript i refers to intracellular and o refers to extracellular. The membrane depolarization in both hypo- and hyperkalemic paralysis has been attributed to increased P_{Na}, [8,95,139,140,142−144] decreased P_K,[77,151,152] or reduction in E_p.[138,153]

Layzer[153] proposed that in hypokalemic periodic paralysis, P_K was reduced and the membrane potential was maintained at a near-normal value by an increased E_p. During an attack, the $[K^+]_o$ could fall to a sufficiently low level to turn off the Na^+,K^+ pump, and the membrane would depolarize due to loss of E_p. Layzer[153] acknowledged that during an attack of hypokalemic paralysis, the serum K^+ content does not decrease sufficiently to inhibit the Na^+,K^+ pump; however, he suggested that the large flux of K^+ into muscle might deplete the T-tubules of K^+ so that $Na^+−K^+$ transport in the T-tubules would stop. This is an interesting proposal in light of the ability of the large inward K^+ current carried by the inward rectifier K^+ channel to deplete the T-tubules of K^+.[154] However, there appears to be few, if any, Na^+,K^+ pumps in the T-tubule membrane[155] so that accelerated Na^+-K^+ transport should not be influenced by the T-tubular K^+ concentration. In addition, Akaike[156] demonstrated that in the absence of external K^+, there is sufficient leakage of K^+ out of the muscle to maintain active Na^+-K^+ transport. Consequently, it appears unlikely that the Na^+,K^+ pump is blocked during attacks of hypokalemic paralysis.

The ability of either elevated or reduced extracellular K^+ to depolarize the sarcolemma can be partially explained by the passive membrane properties of skeletal muscle. The resting K^+ conductance is mediated by "inward rectifier" channels which allow K^+ to flow into the cell more easily than it passes out. Stanfield et al.[157] demonstrated that the K^+ conductance depends on the membrane potential and $[K^+]_o$. As the $[K^+]_o$ is lowered, the K^+ conductance at a given membrane potential decreases and the curve of conductance vs. membrane potential shifts toward more hyperpolarized potentials (Fig. 2). Given this behavior of the K^+ channels, why do normal muscle fibers hyperpolarize in solutions with low K^+ concentrations? The answer is probably that in normal muscle fibers the K^+ conductance is so much larger than the Na^+ conductance that even when the $[K^+]_o$ is lowered and consequently the K^+ conductance is lowered, the K^+ conductance still exceeds the Na^+ conductance and the membrane potential approaches the K^+ equilibrium potential and hyperpolarizes. Figure 3 shows that as the ratio P_{Na}/P_K is increased (proceeding from curve A to curve C), the inward rectifier characteristics become more prominent and the membrane will depolarize as the $[K^+]_o$ is reduced. Thus, the inward rectifying characteristic of the K^+ channel can

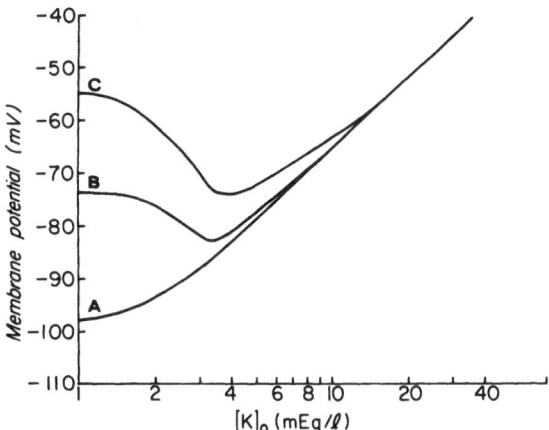

Fig. 3. The computed membrane potential is plotted as a function of the serum K^+ concentration assuming that $P_{Na} = 2.5 \times 10^{-8}$ cm/sec, $P_{Cl} = 10^{-6}$ cm/sec; the intracellular ion concentrations (meq/kg wet wt) were $[Na^+]_i = 50.6$, $[K^+]_i = 78.6$, $[Cl^-]_i = 5.1$ (77); Donnan values of 0.96 for cations and 1.04 for anions. For K^+ current, the values measured on frog muscle by Adrian and Freygang[183] were used with a K^+ conductance in the linear portion of the P_K vs. membrane potential relationship of (A) 2×10^{-5}/ohm-cm^2, (B) 10^{-5}/ohm-cm^2, (C) 2×10^{-6}/ohm-cm^2. Similar curves can be generated by increasing P_{Na} rather than decreasing P_K.

explain the depolarization in low-K^+ solutions if one also assumes that P_{Na}/P_K at the resting potential is greater than normal. Other consequences of increased P_{Na}/P_K are that the resting membrane potential would be less negative and that either high or low serum K^+ concentrations could produce depolarization. The resting membrane potentials of patients with hypokalemic periodic paralysis may be depolarized compared to normal values (Table III). The depolarizing capacity of elevated or reduced serum K^+ was seen in a patient suffering hypokalemic periodic paralysis who was given intravenous injections of K^+, without careful monitoring of the serum $[K^+]$. After an initial improvement, his condition became much worse than at the onset; he had developed hyperkalemia and paralysis.[158] Thus, K^+ treatment turned a hypokalemic paralysis into a hyperkalemic paralysis, both presumably due to sarcolemmal depolarization and inexcitability.

The possibility that increased P_{Na} caused the membrane depolarization in both hypo- and hyperkalemic periodic paralysis was suggested by measurement of a lower than normal input resistance in depolarized fibers[8,142] and that procaine partially repolarized the membrane potential of paralyzed muscle fibers.[95,139]

Rüdel et al.[144] presented some evidence that the paradoxical membrane depolarization seen in muscle fibers from patients with hypokalemic periodic paralysis is associated with an increased steady-state Na^+ conductance. The muscle fibers from three patients depolarized when bathed in low-K^+ solutions. The depolarization was produced by a persistent inward current which was probably due to an influx of Na^+ or Ca^{2+}. They did not attempt to distinguish between these possibilities. The slope of the membrane current–voltage relationship in paralyzed fibers was the same as in control fibers which implied that the Cl^- and K^+ conductance were not significantly altered but the inward rectifying property of the K^+ channels allowed this persistent inward current to be more effective in depolarizing the

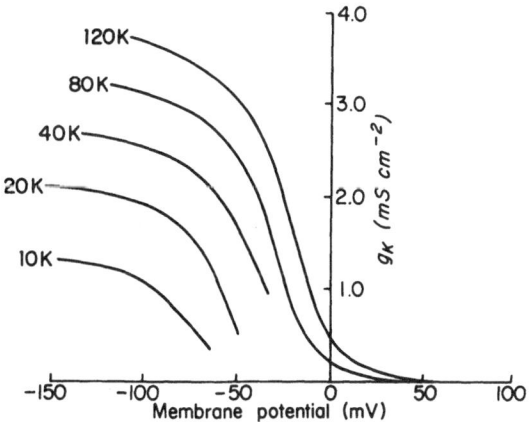

Fig. 2. Relation between K^+ channel conductance [g_K, mS/cm^2 = m/ohm per cm$^2 = I_m/(V - V_{equil})$] and membrane potential for frog skeletal muscle measured using the three-electrode voltage clamp technique by Stanfield et al.[157] for five different extracellular K^+ concentrations indicated by the curves (10, 20, 40, 80, and 120 mM [$K^+]_o$). To obtain these data they have subtracted off a constant conductance (a so-called "leak" conductance) seen at depolarized potentials so that the conductance decreases to zero (in contrast, see Fig. 4 where this has not been subtracted for the hypokalemic rat data). Note how the maximum $[K^+]_o$ decreases so that at a particular membrane potential, g_K falls dramatically as $[K^+]_o$ is decreased. Figure redrawn after Stanfield et al.[157] with permission.

patients' muscles. This inward current was not blocked by tetrodotoxin which suggests that the depolarization was not produced by the opening of the normal voltage-dependent Na^+ channels. In addition, whatever channel is responsible for this persistent inward current at depolarized membrane potentials, does not inactivate in the same voltage range as normal voltage-dependent Na^+ channels. Membrane depolarization and paralysis could also be induced by cooling the muscle fibers to 27°C which suggests that P_K/P_{Na} is temperature sensitive and may account for the clinically observed cold-induced weakness.

Hyperkalemic periodic paralysis also appears to be associated with increased Na^+ conductance. Lehmann-Horn et al. [143] studied muscle fibers from two patients with hyperkalemic periodic paralysis. The mechanism of paralysis appeared to be slightly different for the fibers of each patient. Muscle fibers from both patients developed an excessive depolarization compared to control fibers in response to elevating the $[K^+]_o$, but the first patient was more sensitive with the depolarization occurring from 3.5 mM to 7 mM (Table III). For the first patient, as the $[K^+]_o$ was increased, the muscle fibers were spontaneously active with membrane potentials between −60 and −70 mV. When the $[K^+]_o$ was increased to 7 mM, the fibers depolarized to −50 mV and were electrically inexcitable. The depolarization was associated with an increased tetrodotoxin-blockable Na^+ conductance. The fibers from the second patient depolarized without activity to a mean value of −64.3 mV in 7 mM K^+ solution and in 10 mM K^+ solution the mean potential was −55 mV which is 15 mV less negative than the predicted K^+ equilibrium potential. The fibers from the second patient became electrically inexcitable and paralyzed at resting potentials which were not sufficiently positive to inactivate normal muscle fibers. When the fibers from either patient were bathed in a 3.5 mM K^+ solution, lowering the temperature from 37°C to 27°C resulted in depolarization and paralysis, whereas fibers from normal subjects showed only slight changes in membrane potential or force-generating capacity. Fibers from the first patient fibrillated in response to cooling. The spontaneous activity was attributed to an increased Na^+ conductance. The membrane potentials of fibers from the second patient were studied at 27°C and interestingly the muscle was paralyzed when 60% of the fibers had resting potentials between −70 and −85 mV. They attributed the excessive depolarization in the first patient and the excitability in the second patient to defects in the Na^+ channels although other causes cannot be ruled out.

The weakness and myotonia in paramyotonia appear to result from increase in the muscle Na^+ and Cl^- conductance which are triggered by cooling the muscle. [159] At 37°C, paramyotonic muscle fibers had normal resting potentials and Na^+, K^+, and Cl^- conductances. At 27°C, the fibers were depolarized with increased Na^+ and Cl^- conductance. Depolarization could be blocked by tetrodotoxin. Hence, as the paramyotonic fibers were cooled, Na^+ conductance increased, the fibers became hyperexcitable at first and then inexcitable, presumably due to Na^+ channel inactivation. The temperature-dependent behavior of the paramyotonic muscle fibers was similar to that observed in the fibers from one of the patients with hyperkalemic periodic paralysis (patient 1) discussed above. It is not clear if the cold-induced myotonia seen in paramyotonia and some cases of hyperkalemic periodic paralysis results from a common membrane defect. However, a local anesthetic derivative, tocainide, is able to prevent cold-induced muscle stiffness and weakness in paramyotonia congenita, [160] and paramyotonia with hyperkalemic episodic paralysis. [161]

In summary, the work of Rüdel et al. provides evidence that the paralysis and myotonia in hypo- and hyperkalemic periodic paralysis and paramyotonia may be associated with abnormalities in Na^+ conductance. It is not clear whether the possible abnormalities in Na^+ conductance represent modifications in the normal Na^+ channel molecules or the incorporation of novel species of Na^+ channels with unique sensitivity to extracellular K^+ and temperature into the sarcolemma.

3.4. Mechanisms of Insulin-Induced Depolarization in Hypokalemic Periodic Paralysis

The administration of insulin to patients with hypokalemic periodic paralysis usually produces profound weakness. [5,8] Hofmann and Smith [95] demonstrated that insulin applied in vitro to muscle from patients with hypokalemic periodic paralysis could produce sarcolemmal inexcitability and paralysis. Hofmann et al. [116] subsequently demonstrated that muscle fibers from patients with hypokalemic periodic paralysis have an abnormally high sensitivity to insulin, and they suggested that insulin could depolarize muscle fibers by decreasing the P_K/P_{Na} ratio.

3.4.1. Studies on an Animal Model

The limited data available from human studies do not resolve the etiology of the membrane depolarization in response to insulin in hypokalemic periodic paralysis. In order to explore this problem more thoroughly, it has been useful to study an animal model of this disease. The K^+-depleted rat provides a reasonable model of hypokalemic periodic paralysis because its muscles share many of the important characteristics found in diseased human muscle:

1. Hypokalemia and insulin administration depolarize and paralyze muscle fibers [122,151,162,163] and the membrane repolarizes by increasing $[K^+]_o$. [151,162]
2. $[K^+]_i$ is lower and $[Na^+]_i$ is higher than normal fibers. [123,127,128]
3. Structural changes occur such as vacuolization, distension of the sarcoplasmic reticulum, and swelling of the T-tubules and mitochondria. [86,87]

The K^+-depleted rat differs from humans with primary hypokalemic paralysis by having basal hypokalemia, elevated muscle oxygen consumption, [163] and normal insulin binding. [116] However, the close parallel allows study of the mechanism of hypokalemic paralysis and the role of insulin in producing membrane depolarization. In vitro, the addition of insulin to solutions bathing muscle from K^+-deficient rats caused a reduction in the amplitude of the twitch if the bathing solutions had lower than normal K^+ concentrations. The size of the twitch returned to normal when the concentration of K^+ in the bath was elevated. [122,151,162] Insulin also produced weakness when muscles were studied in vivo. [163] The paralysis can be induced by zinc-free insulin, indicating that the weakness was a direct effect of insulin and was not due to zinc (normally present in commercially available insulin) reducing the K^+ conductance. [151] When the twitch response to direct electrical stimulation was nearly abolished in low-K^+ and insulin-containing solutions, the muscle's contractile response to caffeine or elevated $[K^+]_o$ was nearly normal. [151] In addition, using skinned fibers, the myofibrillar sensitivity to Ca^{2+} was normal (Ruff and Martyn, unpublished data). Caffeine acts by releasing Ca^{2+} from the sarcoplasmic reticulum.[4] A normal caffeine contracture and normal myofibrillar sensitivity to Ca^{2+} suggest that the Ca^{2+} content of the sarcoplasmic reticulum is probably normal. High $[K^+]_o$ presumably trigger Ca^{2+} release by depolarizing the T-tubules.[4] Therefore, a normal K^+ contracture suggests that inactivation of the coupling mechanism between membrane depolarization and sarcoplasmic reticulum release of Ca^{2+} is not the cause of weakness.

The paralysis results from loss of surface membrane excitability due to depolarization-induced Na^+ channel inactivation. The paralyzed fibers are depolarized[151,162,163] and the sarcolemma is inexcitable to direct electrical stimulation.[151,162] Membrane excitability can be restored by passing current to artificially hyperpolarize the membrane.[151] The membrane depolarization appears to result from a combination of increased P_{Na} and decreased P_K. Insulin is not exerting its action on the Na^+, K^+ pump in the animal model, since ouabain does not alter the depolarization of the muscle caused by insulin or alter the time course of hypokalemic and insulin-induced paralysis.[151] In low-K^+ solution, muscle fibers paralyzed by insulin had reduced input resistance (R_i), suggesting that the depolarization was associated with increased P_{Na}.[151] However, Kao and Gordon[151,152] argued that the increased P_{Na} resulted from the membrane depolarization, and that the depolarization was actually caused by reduced P_K. In Na^+- and Cl^--free solutions, depolarization of muscle fibers was prevented and R_i increased, indicating reduced P_K.

Chronic K^+ depletion in humans also alters the response of human muscle to insulin. Insulin produces a slight hyperpolarization of normal mammalian muscle.[110] However, in a K^+-depleted patient, insulin administration produced muscle paralysis and electrical inexcitability.[28]

Another clinical syndrome which suggests that reduced K^+ conductance can evoke hypokalemic paralysis is Ba^{2+} intoxication.[41,164] Patients developed extremity paralysis and hypokalemia. Experimentally, Ba^{2+} blocked K^+ channels and depolarized fibers bathed in low-K^+ solutions.[165]

3.4.2. The Action of Insulin

In our laboratory, Eleanor Bond has studied the effect of insulin on the inward rectifying K^+ channel in normal and K^+-depleted rats. These studies were performed with fibers bathed in high-$[K^+]_o$ solutions to enhance the current through the inward rectifier channel. Elevated $[K^+]_o$ has the additional effect of making it possible for the fibers from K^+-depleted rats to regain intracellular K^+ and lose excess Na^+. Consequently, muscle from control and K^+-depleted rats had similar intracellular ionic compositions. Under these circumstances, the voltage dependence of the inward rectifier was similar in control and K^+-depleted fibers. Insulin added to the bathing solution reduced the conductance in the low-conductance portion of the curve (Bond, unpublished data). Thus, insulin exaggerated the inward rectifying behavior of the K^+ channel (Fig. 4). Bond (unpublished data) also studied the threshold for Na^+ channel activation and found that the threshold was similar in K^+-depleted and control muscle fibers. Thus, insulin did not appear to alter the threshold for Na^+ channel activation. Kao and Gordon[151] demonstrated that the insulin effect could be mimicked by Con A, a lectin which can bind to and cross-link insulin receptors. Hence, agonist binding to the insulin receptor or cross-linking of insulin receptors was sufficient to alter K^+ conductance. Glucose uptake was not necessary, since insulin-induced paralysis of K^+-depleted muscles occurred without glucose in the bathing solution.[151]

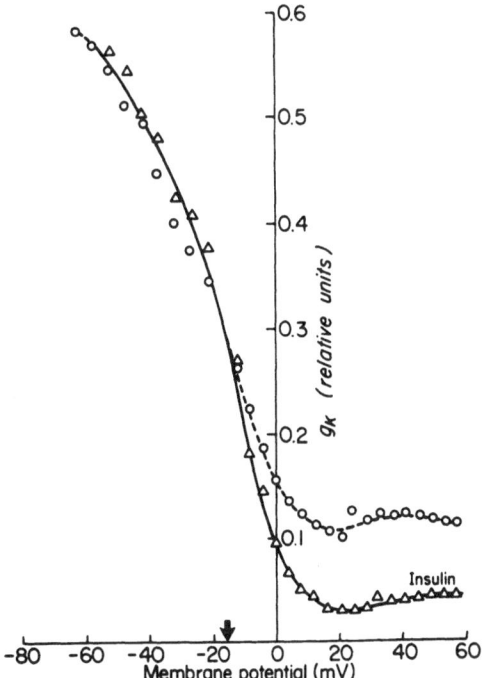

Fig. 4. K^+ conductance measured on muscle fiber from the medial omohyoid muscle from K^+-depleted rats in the presence and absence of zinc-free insulin (12 mU/ml). The membrane current–voltage relationship was measured using the three-electrode voltage clamp technique[157] at the end of muscle fibers. The external solution contained 80 mMK^+, 44 mM SO_4^{2-} (to decrease Cl^- currents), 6 mM added $CaSO_4$ (to compensate for bound $CaSO_4$ to keep free Ca^{2+} constant), and 3 × 10^{-8} M tetrodotoxin (to block voltage-sensitive Na^+ channels). The major remaining current is presumably due to K^+. The holding potential in these solutions was -18 mV, indicated by the arrow. The relative K^+ channel chord conductance is calculated from $g_K = I_m/(V - V_{equil})$ with $V_{equil} = 18.3$ mV. The same cell was recorded before ($\bigcirc$) and 120 min after ($\triangle$) the addition of insulin. Note that the curves with and without insulin overlap for membrane potentials in the hyperpolarizing direction but that insulin decreases the limiting conductance for polarized potentials. Data kindly provided by Eleanor Bond.

In conclusion, the muscle membrane inexcitability in hyper- and hypokalemic periodic paralysis appears to result from sarcolemmal depolarization producing Na^+ channel inactivation. The voltage-dependent gating of Na^+ channels may also be altered. Insulin may depolarize muscle fibers by reducing K^+ conductance. Figure 5 summarizes the possible mechanisms of acute paralysis and chronic weakness in hypokalemic periodic paralysis. Insulin and other factors, such as β-adrenergic stimulation and nocturnal rest, induce a shift of K^+ from extracellular fluid into cells, hypokalemia, and an increase in intracellular K^+. The hypokalemia increases Na^+ conductance, and insulin reduces K^+ conductance. The muscle membrane then depolarizes, the Na^+ channels inactivate, the sarcolemma becomes inexcitable, and the muscle paralyzed. Chronic changes in intracellular Na^+ and K^+ concentrations may cause a vacuolar change in the sarcotubular system and chronic weakness and alter the response of skeletal muscle to insulin. With this scheme, one can explain most clinical features and laboratory findings and rationalize therapeutic approaches.

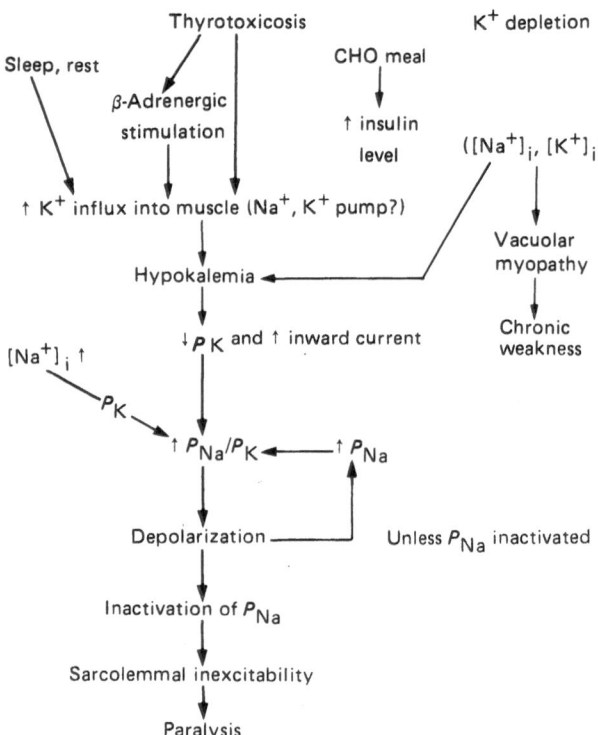

Fig. 5. Hypothesized scheme for mechanism of paralysis in hypokalemic periodic paralysis.

4. Therapy and Its Rationale

4.1. Hypokalemic Periodic Paralysis

4.1.1. Primary Hypokalemic Periodic Paralysis

Acute attacks are treated with oral K^+ (2–10 g KCl).[6] Intravenous K^+ is usually not required and if given care must be exercised to avoid hyperkalemia.[158] Relatively little K^+ is required to replace the extracellular K^+. Exercise or electrical stimulation of motor nerves will improve the strength of contracting muscles.[166] The beneficial effect of exercise is probably due to the accumulation of K^+ in the T-tubules[154] which can increase the conductance of inward rectifying K^+ channels located in the tubules.

Prophylactic treatment includes maintenance of a negative Na^+ balance with a low-Na^+ diet, diuretics, or spironolactone. Na^+ depletion may reduce intracellular Na^+ to more normal values, which may diminish the possible block of K^+ channels by intracellular Na^+.[167] A low-carbohydrate diet is sometimes beneficial and may act by minimizing postprandial insulin release. In a similar fashion, diazoxide may prove useful by reducing insulin secretion.[168] Acetazolamide, a carbonic anhydrase inhibitor, is successful in the majority of patients in preventing spontaneous and insulin-evoked attacks by eliminating insulin-induced hypokalemia.[169-173] The metabolic acidosis produced by acetazolamide is important for its action. Administration of $NaHCO_3$ eliminates the ability of acetazolamide to prevent insulin-induced hypokalemia,[172] and NH_4Cl potentiates the action of acetazolamide.[170] In normal individuals, metabolic acidosis produces a shift of K^+ out of muscle.[106] The

metabolic acidosis may also inhibit Na^+ channel inactivation.[174] Acetazolamide elicited attacks of paralysis in one patient who developed hypokalemia in response to the drug.[11]

4.1.2. Thyrotoxic Periodic Paralysis

Treatment is primarily directed at normalizing thyroid function. Until the patient becomes euthyroid, treatment of acute attacks and prophylactic measures are the same as for primary hypokalemic periodic paralysis. Acetazolamide is usually not effective in this disorder as it fails to prevent hypokalemia. Propranolol, a β-adrenergic blocker, can prevent attacks,[175,176] but, interestingly, in one case reserpine, which depletes catecholamines from nerve terminals, was not effective.[177]

4.1.3. Secondary Hypokalemic Periodic Paralysis

Therapy is K^+ replacement and correction of the underlying K^+ wasting disorder.

4.2. Hyperkalemic Periodic Paralysis

4.2.1. Primary Hyperkalemic Periodic Paralysis

Acute treatment is directed at lowering serum K^+ by facilitating cellular uptake of K^+ using glucose and insulin.[42,44,96] Epinephrine can quickly stop an attack.[44,53,96] Intravenous Ca^{2+} can abort attacks[2,42] presumably due to a shift in the voltage sensitivity of the Na^+ channel.[174] Acetazolamide can prevent attacks by stimulating muscle and red blood cell K^+ uptake[51,130,178] and insulin release.[179] As noted before, the metabolic acidosis may also shift the voltage dependence of Na^+ channel inactivation. The $β_2$ agonist salbutamol, which stimulates Na^+–K^+ active transport in muscle, is a very effective prophylactic agent.[180-182] Other preventative measures are frequent meals of high carbohydrate content, and low-dose diuretic therapy. The local anesthetic derivative tocainide may be useful in preventing myotonia and weakness induced by cold.[161]

4.2.2. Secondary Hyperkalemic Periodic Paralysis

Again, treatment is aimed at the primary disorder. Intravenous glucose and insulin may be used to acutely lower serum K^+.

4.3. Normokalemic Periodic Paralysis

In one family, large doses of NaCl or acetazolamide were effective in preventing attacks.[69] The mechanism of action of these agents was not clear.

5. Summary

The periodic paralyses are a group of syndromes characterized by episodic attacks of muscular paralysis, usually accompanied by changes in the K^+ content of serum in skeletal muscles. They are classified as primary or secondary and according to the associated changes in serum K^+ levels and clinical manifestations as (1) hypokalemic periodic paralysis, (2) hyperkalemic periodic paralysis, and (3) normokalemic periodic paralysis.

In hypokalemic periodic paralysis, the paralysis is associated with a fall in serum K^+ levels. Factors such as insulin, β-adrenergic stimulation, and nocturnal rest, which induce a shift in K^+ from extracellular fluid into muscles causing hypokalemia, can bring on the paralysis. Studies on patients and an animal model (the K^+-deficient rat) show that the paralysis is due to an inexcitability of the muscle cell membrane caused by depolarization. An increase in Na^+ conductance causes the membrane potential to depolarize when the $[K^+]_o$ falls. Studies on the animal model indicate that insulin also has a direct effect to reduce K^+ conductance. The result of these two factors is a depolarization and presumed inactivation of Na^+ channels, producing inexcitability and paralysis. It is shown that with this hypothesis, one can explain most of the clinical features and laboratory findings and rationalize therapeutic approaches. Morphological changes including vacuolization of muscle membranes can be associated with chronic weakness, but are not the causative feature of the electrolyte imbalance as they are observed in the animal model as well.

Thyrotoxic periodic paralysis is similar to hypokalemic periodic paralysis in that attacks are associated with decreased serum K^+ resulting from muscle uptake of K^+. The paralysis appears to result from depolarization-induced Na^+ channel inactivation.

In hyperkalemic periodic paralysis, the paralysis is associated with an elevated serum K^+ level due to a K^+ efflux from muscle cells. Muscle membranes are depolarized and inexcitable during the paralysis. An increase in Na^+ conductance due to an abnormally functioning Na^+ channel in these patients may contribute to the membrane depolarization. As with the hypokalemic periodic paralysis, there are morphological changes in muscle involving vacuolization of muscle membranes. It is not clear what role the structural changes play in the paralysis. They may be secondary to changes in muscle electrolytes.

Few cases of normokalemic periodic paralysis have been studied so that little can be said about the direct cause of the paralysis. There are similar morphological changes to those seen in the other two forms of periodic paralysis, but again there is no suggestion as to how these changes contribute to the paralysis.

ACKNOWLEDGMENT. Partially supported by USPHS Grants NS-00498 and NS-16696.

References

1. Chesson, A. L., S. S. Schochet, and B. H. Peter. 1979. Biphasic periodic paralysis. *Arch. Neurol.* **36**:700–704.
2. Pearson, C. M. 1964. The periodic paralysis: Differential features and pathological observations in permanent myopathic weakness. *Brain* **87**:341–358.
3. Hartwig, H. 1875. Über emien fall von intermittiereuder paralysis spinalis. *Zentralbl. Med. Wiss.* **13**:428.
4. Endo, M. 1977. Calcium release from the sarcoplasmic reticulum. *Physiol. Rev.* **57**:71–102.
5. Johnsen, T. 1981. Familial periodic paralysis with hypokalemia: Experimental and clinical investigations. *Dan. Med. Bull.* **28**:1–27.
6. Talbot, J. H. 1941. Periodic paralysis. *Medicine (Baltimore)* **20**:85–99.
7. Sagild, U. 1963. Hypoglycemia induced by potassium administration during attacks of periodic paralysis. *Acta Med. Scand.* **173**:329–332.
8. Engel, A. G. 1977. Hypokalemic and hyperkalemic periodic paralysis. In: *Scientific Approach to Clinical Neurology.* E. S. Gold-ensohn and S. H. Appel, eds. Lea & Febiger, Philadelphia. pp. 1742–1765.
9. Biermond, A., and A. P. Daniels. 1934. Familial periodic paralysis and its transition into spinal muscular atrophy. *Brain* **57**:91–109.
10. Allott, E. N., and B. McArdle. 1938. Further observations on familial periodic paralysis. *Clin. Sci. Mol. Med.* **3**:229–245.
11. Torres, C. F., R. C. Griggs, R. T. Moxley, and A. N. Bender. 1981. Hypokalemic periodic paralysis exacerbated by acetazolamide. *Neurology (New York)* **31**:1423–1428.
12. Van Buchem, F. S. P. 1957. The electrocardiogram and potassium metabolism. *Am. J. Med.* **23**:376–384.
13. Mathew, R., and P. S. Gupta. 1979. Periodic paralysis (report of six cases). *J. Assoc. Physicians India* **27**:663–667.
14. Wolf, P., J. Griffiths, J. Koett, and J. Howell. 1979. The presence of serum creatine kinase 2 (MB) in hypokalemic familial periodic paralysis. *Enzyme* **24**:197–199.
15. Resnick, J. S., and W. K. Engel. 1967. Myotonic lid lag in hypokalemic periodic paralysis. *J. Neurol. Neurosurg. Psychiatry* **30**:47–51.
16. McArdle, B. 1956. Familial periodic paralysis. *Br. Med. Bull.* **12**:226–229.
17. Shy, M., T. Wanko, P. T. Rowley, and A. G. Engel. 1961. Studies in familial periodic paralysis. *Exp. Neurol.* **3**:53–121.
18. Engel, A. G., E. H. Lambert, J. W. Rosevear, and W. N. Tauxe. 1965. Clinical and EMG studies in a patient with primary hypokalemic periodic paralysis. *Am. J. Med.* **38**:626–640.
19. Buruma, O. J. S., and G. T. A. M. Bots. 1978. Myopathy in familial hypokalemic periodic paralysis independent of paralytic attacks. *Acta Neurol. Scand.* **57**:171–179.
20. Buruma, O. J. S., J. J. Schipperheyn, and G. T. A. M. Bots. 1981. Heart muscle disease in familial hypokalemic periodic paralysis. *Acta Neurol. Scand.* **64**:12–21.
21. Chen, R. F. 1959. Familial periodic paralysis. *Arch. Neurol.* **1**:475–480.
22. Satoyoshi, E., K. Murakami, H. Kowa, M. Kinoshita, and Y. Mishiyama. 1963. Periodic paralysis in hyperthyroidism. *Neurology (Minneapolis)* **13**:746–752.
23. Engel, A. G. 1966. Electron microscopic observations in primary hypokalemic and thyrotoxic periodic paralysis. *Mayo Clin. Proc.* **41**:797–808.
24. Engel, A. G. 1961. Thyroid function and periodic paralysis. *Am. J. Med.* **30**:327–333.
25. Okihiro, M. M., and R. A. Nordyke. 1966. Hypokalemic periodic paralysis experimental precipitation with sodium lithyronine. *J. Am. Med. Assoc.* **198**:277–279.
26. Layzer, R. B., and E. Goldfield. 1974. Periodic paralysis caused by abuse of thyroid hormone. *Neurology (Minneapolis)* **24**:949–955.
27. Ober, K. P., and J. F. Hennessy. 1981. Jodbasedow and thyrotoxic periodic paralysis. *Arch. Intern. Med.* **141**:1225–1227.
28. Ruff, R. L. 1979. Insulin-induced weakness in hypokalemic myopathy. *Ann. Neurol.* **6**:139–140.
29. Conn, J. W. 1955. Presidential address. I. Painting background. II. Primary aldosteronism. *J. Lab. Clin. Med.* **45**:3–28.
30. Raskin, R. J., J. T. Tesan, and O. J. Lawless. 1981. Hypokalemic periodic paralysis in Sjogren's syndrome. *Arch. Intern. Med.* **141**:1671–1673.
31. Milne, M. D., S. W. Stanbury, and A. E. Thompson. 1952. Observations on Fanconi syndrome and renal hyperchloremic acidosis in adults. *Q. J. Med.* **21**:61–74.
32. Bull, G. M., A. M. Joekes, and K. G. Lowe. 1950. Renal function studies in acute tubular necrosis. *Clin. Sci.* **9**:379–385.
33. Sataline, L. R., and J. M. Simonelli. 1961. Potassium paresis following ureterosignoidostomy. *J. Urol.* **85**:559–564.
34. Cohen, T. 1959. Hypokalemic muscle paralysis associated with administration of chlorothiazide. *J. Am. Med. Assoc.* **170**:2083–2086.
35. Salassa, R. M., V. R. Mattox, and J. W. Rosevear. 1962. Inhibi-

tion of the 'mineralo-corticoid' activity by licorice by spironolactone. *J. Clin. Endocrinol. Metab.* **22**:1156–1159.

36. Goulon, M., M. Rapin, J. Lissac, J. J. Pocidolo, and O. Mantel. 1962. Quadriplegia with hypokalemia and hypokalemic acidosis secondary to the absorption of ammonium chloride over a three-year period. *Bull. Soc. Hop. Med. Paris* **113**:113–115.

37. Keye, J. D. 1952. Death in potassium deficiency. *Circulation* **5**:766–769.

38. Keyloun, V. E., and W. J. Grace. 1967. Villous adenoma of the rectum associated with severe electrolyte imbalance. *Am. J. Dig. Dis.* **12**:104–108.

39. Verner, V. J., and A. B. Morrison. 1958. Islet cell tumor and a syndrome of refractory watery diarrhea and hypokalemia. *Am. J. Med.* **25**:374–382.

40. Schwartz, W. B., and A. S. Relman. 1953. Metabolic and renal studies in chronic potassium depletion resulting from overuse of laxatives. *J. Clin. Invest.* **32**:258–265.

41. Lewi, Z., and Y. Bar-Khayim. 1964. Food poisoning from barium carbonate. *Lancet* **2**:342–344.

42. Gamstorp, I., M. Hauge, H. F. Helweg-Larsen, H. Mjönes, and U. Sagild. 1957. Adynamia episodica hereditaria. *Am. J. Med.* **23**:385–390.

43. Armstrong, F. B. 1962. Hyperkalemic familial periodic paralysis. *Ann. Intern. Med.* **57**:455–461.

44. McArdle, B. 1962. Adynamia episodica hereditaria and its treatment. *Brain* **85**:121–148.

45. Layzer, R. B., R. E. Lovelace, and L. P. Rowland. 1967. Hyperkalemic periodic paralysis. *Arch. Neurol.* **16**:455.

46. Klein, R., and J. E. Haddow. 1970. Arterial–venous differences in serum potassium concentrations in paramyotonia. *Pediatr. Res.* **4**:286–294.

47. Thomas, C. 1978. Recent findings on the geneology, clinical aspects, and histology of a family with periodic hyperkalemic paralysis. *Rev. Neurol.* **134**:45–58.

48. Dyken, M. L., and G. D. Timmons. 1963. Hyperkalemic periodic paralysis with hypocalcemic episode. *Arch. Neurol.* **9**:508–517.

49. Riggs, J. E., R. T. Moxley, R. C. Griggs, and F. A. Horner. 1981. Hyperkalemic periodic paralysis: An apparent sporadic case. *Neurology (New York)* **31**:1157–1159.

50. Hudson, A. J., K. P. Strickland, and A. J. Wilensky. 1967. Serum enzyme studies in familial hyperkalemic periodic paralysis. *Clin. Chim. Acta* **17**:331–336.

51. Streeten, D. H. P., T. G. Dalakos, and H. Fellerman. 1971. Studies on hyperkalemic periodic paralysis: Evidence of changes in plasma Na and Cl and induction of paralysis by adrenal glucocorticoids. *J. Clin. Invest.* **50**:142–155.

52. McDonald, R. D., N. B. Rewcastle, and J. G. Humphrey. 1968. The myopathy of hyperkalemic periodic paralysis. *Arch. Neuol.* **19**:274–279.

53. Carson, M. J., and C. M. Pearson. 1964. Familial hyperkalemic periodic paralysis with myotonic features. *J. Pediatr.* **64**:853–865.

54. Buchthal, F., L. Engback, and I. Gamstorp. 1958. Paresis and hyperexcitability in adynamia episodica hereditaria. *Neurology (Minneapolis)* **8**:347–351.

55. Segura, R. P., and J. H. Petajan. 1979. Neural hyperexcitability in hyperkalemic periodic paralysis. *Muscle Nerve* **2**:245–249.

56. Bradley, W. G. 1969. Adynamia episodica hereditaria. *Brain* **92**:345–378.

57. Garcin, R., P. Rondot, and M. Fardeau. 1964. Sur les accidents neuromusculaires et en particulier sur une 'myopathie vacuolaire' observés au cours d'un traitement prolongé par la chloroquine: Amélioration rapid apres arrêt due médicament. *Rev. Neurol.* **111**:117–126.

58. Bull, G. M., A. B. Carter, and K. G. Lowe. 1953. Hyperpotassaemic paralysis. *Lancet* **2**:60–63.

59. Marks, L. J., and E. Feit. 1953. Flaccid quadriplegia, hyperkalemia, and Addison's disease. *Arch. Intern. Med.* **91**:56–57.

60. Richardson, G. O., and J. C. Sibley. 1953. Flaccid quadriplegia

associated with hyperpotasseamia. *Can. Med. Assoc. J.* **69**:504–506.

61. Mollaret, P., M. Goulon, and M. Tornilhac. 1958. Contribution a l'étude des paralyses avel hyperkaliémie. I. Róle de l'insuffisance cortico-surrénalienne. *Rev. Neurol.* **98**:341–357.

62. Pollen, R. H., and R. H. Williams. 1960. Hyperkalemic neuromyopathy in Addison's disease. *N. Engl. J. Med.* **263**:273–278.

63. Faw, M. L., and R. V. Ewer. 1962. Intermittent paralysis and chronic adrenal insufficiency. *Ann. Intern. Med.* **57**:461–463.

64. Bell, H., W. L. Hayes, and J. Vosburgh. 1965. Hyperkalemic paralysis due to adrenal insufficiency. *Arch. Intern. Med.* **115**:418–420.

65. Daughaday, W. H., and D. Rendleman. 1967. Severe symptomatic hyperkalemia in an adrenalectomized woman due to enhanced mineralocorticoid requirement. *Ann. Intern. Med.* **66**:1197–1202.

66. Van Dellen, R. C., and D. C. Purnell. 1969. Hyperkalemic paralysis due to adrenal insufficiency. *Mayo Clin. Proc.* **44**:904–914.

67. Vilchez, J. J., A. Cabello, J. Bendito, and T. Villarroya. 1980. Hyperkalemic paralysis neuropathy and persistent motor neuron discharges at rest in Addison's disease. *J. Neurol. Neurosurg. Psychiatry* **43**:818–822.

68. Ucezne, E. O., and B. P. Harrold. 1980. Hyperkalemic paralysis due to spironolactone. *Postgrad. Med. J.* **56**:256.

69. Poskanzer, D. C., and D. N. S. Kerr. 1961. A third type of periodic paralysis with normokalemia and favourable response to sodium chloride. *Am. J. Med.* **31**:328–336.

70. Meyers, K. R., D. H. Gilden, C. F. Rinaldi, and J. L. Hansen. 1972. Periodic muscle weakness, normokalemia, and tubular aggregates. *Neurology (Minneapolis)* **22**:269–274.

71. Tyler, F. H., F. E. Stephens, F. D. Gunn, and G. T. Perkoff. 1951. Studies in disorders of muscle. VII. Clinical manifestations and inheritance of a type of periodic paralysis without hypopotassemia. *J. Clin. Invest.* **30**:492–502.

72. Stevens, J. R. 1954. Familial periodic paralysis, myotonia, progressive amyotrophy, and pes cavus in members of a single family. *A.M.A. Arch. Neurol. Psychiat.* **72**:726–734.

73. Klein, R. 1963. Periodic paralysis with cardiac arrhythmia. *J. Pediatr.* **62**:371–376.

74. Iannoccone, S. T., K. Bove, B. Nagy, and F. J. Samaha. 1982. Familial dystrophy associated with periodic paralysis and a unique deficit of the major protein of sarcoplasmic reticulum. *Ann. Neurol.* **12**:108–109.

75. Goldflam, S. 1895. Weitere Mittheilung uber die paroxysmale, familiare Lahmung. *Dtsch. Z. Nervenheilkd.* **7**:1–12.

76. Biczyskowa, W., A. Fidzianska, and H. Jedrzejowska. 1969. Light and electron microscopic study of the muscles in hypokalemic periodic paralysis. *Acta Neuropathol.* **12**:329–346.

77. Gordon, A. M., J. R. Green, and D. Lagonoff. 1970. Studies on a patient with hypokalemic familial periodic paralysis. *Am. J. Med.* **48**:185.

78. Engel, A. G. 1970. Evolution and contents of vacuoles in primary hypokalemic periodic paralysis. *Mayo Clin. Proc.* **45**:774–814.

79. Pellissier, J. F., M. C. Faugere, D. Gambarelli, and M. Toba. 1980. Histological, histochemical and ultrastructural studies in five cases. *Semin. Hop. Paris* **56**:1393–1404.

80. Norris, F. H., B. J. Panner, and J. M. Stormont. 1968. Thyrotoxic periodic paralysis: Metabolic and ultrastructural studies. *Arch. Neurol.* **19**:88–94.

81. Takaci, A., D. L. Schotland, S. DiMauro, and L. P. Rowland. 1973. Thyrotoxic periodic paralysis: Function of SR and muscle glycogen. *Neurology (Minneapolis)* **23**:100.

82. Bradley, W. G. 1969. Ultrastructural changes in adynamia episodica hereditaria and normokalemic familial periodic paralysis. *Brain* **42**:379–390.

83. Walter, G. F., and L. Auböck. 1980. Case report: Ultrastructural changes of muscles during attacks of adynamia episodic hereditaria. *Neuropathol. Appl. Neurobiol.* **6**:313–317.

84. Dyken, M., W. Zeman, and T. Rusche. 1969. Hypokalemic peri-

odic paralysis: Children with permanent myopathic weakness. *Neurology (Minneapolis)* **19**:691–699.

85. Wainwright, J., and D. Pirie. 1980. Hypokalemic myopathy. *S. Afr. Med. J.* **58**:81–84.

86. Kao, L. I., and A. M. Gordon. 1977. Alteration of skeletal muscle cellular structures by potassium depletion. *Neurology (Minneapolis)* **27**:855–860.

87. DeCosta, W. J. P. 1979. Experimental hypokalaemic ultrastructural changes in rat gastrocnemius muscle. *Acta Neurol.* **45**:79–82.

88. Corbett, A. J., and M. Pollock. 1981. Experimental potassium depletion myopathy. *J. Neurol. Sci.* **49**:193–206.

89. Vastola, E. F., and C. A. Bertrand. 1956. Intracellular water and potassium in periodic paralysis. *Neurology (Minneapolis)* **6**:523–530.

90. Conn, J. W., and D. H. Streeten. 1960. Intermittent aldosteronism in periodic paralysis. In: *Metabolic Basis of Inherited Disease.* J. B. Stanbury, J. B. Wyngaarden, and P. S. Fredrickson, eds. McGraw-Hill, New York. pp. 883–898.

91. Talso, P. J., M. F. Glenn, Y. T. Oester, and J. Fudema. 1963. Body composition in hypokalemic familial periodic paralysis. *Ann. N.Y. Acad. Sci.* **110**:994–1001.

92. Kreindler, A., V. Ionasesco, I. Petrovici, A. Petresco, and G. Calcaianu. 1964. Etudes cliniques et de laboratoire dans la paralysie periodique familiale avec hypokalémie. *Rev. Neurol.* **110**:377–385.

93. Niall, J. F., and R. K. Poy. 1966. Studies in familial hypokalemic periodic paralysis. *Aust. Ann. Med.* **15**:352–356.

94. Shishiba, Y., K. Shizume, M. Sakuma, H. Yamauchi, K. Nakao, and S. Okinaka. 1966. Studies on electrolyte metabolism in idiopathic and thyrotoxic periodic paralysis. III. Intra- and extracellular concentrations of potassium and sodium in muscle and their changes during induced attacks of paralysis. *Metabolism* **15**:153–162.

95. Hofmann, W. W., and R. A. Smith. 1970. Hypokalaemic periodic paralysis studied *in vitro. Brain* **93**:445–474.

96. Klein, R., T. Egan, and P. Usher. 1960. Changes in sodium, potassium and water in hyperkalemic familial periodic paralysis. *Metabolism* **9**:1005–1024.

97. Conn, J. W., S. S. Fajans, L. H. Louis, D. H. P. Streeten, and R. D. Johnson. 1957. Intermittent aldosteronism in periodic paralysis: Dependence of attacks on retention of sodium, and failure to induce attacks by restriction of dietary sodium. *Lancet* **1**:802–805.

98. Gordon, M. 1978. Periodic familial paralysis with hypokalemia. Hemodynamic and metabolic studies: favorable effect of acetazolamide. *Rev. Neurol.* **134**:655–672.

99. Grob, D., R. J. Johns, and Ä. Liljestrand. 1957. Potassium movement in patients with familial periodic paralysis: Relationship to defect in muscle function. *Am. J. Med.* **23**:356–375.

100. Sonkodi, S. 1980. Familial periodic paralysis (hypopotassaemic form). *Wien. Klin. Wochenschr.* **92**:610–613.

101. Zierler, K. L., and R. Andres. 1957. Movement of potassium into skeletal muscle during spontaneous attack in family periodic paralysis. *J. Clin. Invest.* **36**:730–737.

102. Shizume, K., Y. Shishiba, M. Sakuma, H. Yamauchi, K. Nakao, and S. Okinaka. 1966. Studies on electrolyte metabolism in idiopathic and thyrotoxic periodic paralysis. I. Arteriovenous differences of electrolytes during induced paralysis. *Metabolism* **15**:138–144.

103. Buruma, O. J., T. M. A. R. Dubbelman, A. W. de Bruyne, and J. van Steveninck. 1978. Erythrocyte membrane studies in familial hypokalemic periodic paralysis. *Arch. Neurol.* **35**:615–616.

104. Engel, A. G., C. S. Potter, and J. W. Rosevear. 1967. Studies on carbohydrate metabolism and mitochondrial respiratory activities in primary hypokalemic periodic paralysis. *Neurology (Minneapolis)* **17**:329–338.

105. Streeten, D. H. P., and P. J. Speller. 1974. The role of mineralocorticoids in the pathogenesis of hypokalemic periodic paralysis. *J. Clin. Endocrinol. Metab.* **39**:326–333.

106. Stearns, R. H., M. Cox, P. U. Feig, and I. Singer. 1984. Internal potassium balance and the control of the plasma potassium concentration. *Medicine (Baltimore)* **60**:339–354.

107. Samaha, F. J. 1969. Sodium–potassium adenosine triphosphate in diseased muscle: Studies on periodic paralysis, myasthenia gravis, and Eaton–Lambert syndrome. *Neurology (Minneapolis)* **19**:551–552.

108. Clausen, T., and J. A. Flatman. 1977. The effect of catecholamines on Na–K transport and membrane potential in rat soleus muscle. *J. Physiol. (London)* **270**:383–414.

109. Clausen, T., and P. G. Kohn. 1977. The effect of insulin on the transport of sodium and potassium in rat soleus muscle. *J. Physiol. (London)* **265**:19–42.

110. Flatman, J. A., and T. Clausen. 1979. Combined effects of adrenaline and insulin on active electrogenic Na–K transport in rat soleus muscle. *Nature (London)* **281**:580–581.

111. Moore, R. D., and J. L. Rabovsky. 1979. Mechanism of insulin action on resting membrane potential of frog skeletal muscle. *Am. J. Physiol.* **236**:C249–C254.

112. Erlij, D., and H. F. Schoen. 1981. Effects of insulin on alkalication movements across muscle and epithelial cell membranes. *Ann. N.Y. Acad. Sci.* **372**:272–290.

113. Clausen, T. 1982. Adrenergic control of $Na^+–K^+$ homeostasis. *Acta Med. Scand. (Suppl.)* **672**:111–115.

114. Viskoper, R. J., J. Fidel, T. Horn, D. Trivoni, and J. Chaco. 1973. On the beneficial action of acetazolamide in hypokalemic periodic paralysis: Study of carbohydrate metabolism. *Am. J. Med. Sci.* **226**:125–129.

115. Vallat, G. 1977. Hypokalemic familial periodic paralysis with basal electrocardiographic changes. *Semin. Hop. Paris* **53**:1163–1165.

116. Hofmann, W. W., B. T. Adornato, and H. Reich. 1983. The relationship of insulin receptors to hypokalemic periodic paralysis. *Muscle Nerve* **6**:48–51.

117. Shizume, K., Y. Shishiba, M. Sakuma, H. Yamauchi, K. Nakao, and S. Okinaka. 1966. Studies on electrolyte metabolism in idiopathic and thyrotoxic periodic paralysis. II. Total exchangeable sodium and potassium. *Metabolism* **15**:145–152.

118. Biron, R., R. Burger, A. Chinet, T. Clausen, and R. DuBois-Ferrierer. 1979. Thyroid hormones and the energetics of active sodium–potassium transport in mammalian skeletal muscles. *J. Physiol. (London)* **297**:47–60.

119. Sharma, V. K., and S. P. Banerjee. 1978. β-Adrenergic receptors in rat skeletal muscle: Effects of thyroidectomy. *Biochim. Biophys. Acta* **539**:538–542.

120. Smith, E. K. M., and P. D. Samuel. 1970. Abnormalities in the sodium pump of erythryoctes from patients with hypothyroidism. *Clin. Sci.* **38**:49–61.

121. Shishiba, Y., T. Shimizu, and T. Saito. 1972. Elevated immunoreactive insulin concentration during spontaneous attacks in thyrotoxic periodic paralysis. *Metabolism* **21**:285–290.

122. Offerijns, F. G. J., D. Westerink, and A. F. Willebrands. 1958. The relation of potassium deficiency to muscular paralysis by insulin. *J. Physiol. (London)* **141**:377–384.

123. Norgaard, A., K. Kjeldsen, and T. Clausen. 1981. Potassium depletion decreases the number of ^3H-ouabain binding sites and the active Na–K transport in skeletal muscle. *Nature (London)* **293**:739–741.

124. Erdmann, E., and W. Krawietz. 1977. Increased number of ouabain binding sites in human erythrocyte membranes in chronic hypokalemia. *Acta Biol. Med. Ger.* **36**:879–883.

125. Akaike, N. 1981. Sodium pump in skeletal muscle: Central nervous system-induced suppression by α-adrenoreceptors. *Science* **213**:1252–1254.

126. Clausen, T., K. Kjeldsen, and A. Norgaard. 1983. Effects of denervation on sodium, potassium and [^3H] ouabain binding in muscles of normal and potassium-depleted rats. *J. Physiol. (London)* **345**:123–134.

127. Akaike, N. 1975. Contribution of an electrogenic sodium pump to

membrane potential in mammalian skeletal muscle fibers. *J. Physiol. (London)* **245**:499–520.

128. Akaike, N. 1978. Resting and action potentials in white muscle of potassium deficient rats. *Comp. Biochem. Physiol.* **61A**:629–633.

129. Lewis, E. D., R. C. Griggs, and R. T. Moxley. 1979. Regulation of plasma potassium in hyperkalemic periodic paralysis. *Neurology (New York)* **29**:1131–1137.

130. Hoskins, B., T. H. Maren, and F. Q. Vroom. 1974. Acetazolamide and potassium flux in RBC's—Studies in patients with hyperkalemic periodic paralysis. *Arch. Neurol.* **31**:187–189.

131. Clausen, T., P. Wang, H. Orskov, and O. Kristensen. 1980. Hyperkalemic periodic paralysis: Relationships between changes in plasma water, electrolytes, insulin and catecholamine during attacks. *Scand. J. Clin. Lab. Invest.* **40**:211–220.

132. Streeten, D. H. P., T. G. Dalakos, and H. Fellerman. 1971. Studies on hyperkalemic periodic paralysis: Evidence of changes in plasma number, CL, and induction of paralysis by adrenal glucocorticoids. *J. Clin. Invest.* **50**:142–155.

133. Ruff, R. L. 1985. Endocrine myopathies (hyper- and hypofunction of adrenal, thyroid, pituitary, parathyroid glands, and iatrogenic steroid myopathy). In: *Myology*. B. Q. Banker and A. G. Engel, eds. McGraw-Hill, New York.

134. Engel, A. G., and E. H. Lambert. 1969. Calcium activation of electrically inexcitable muscle fibers in primary hypokalemic periodic paralysis. *Neurology (Minneapolis)* **19**:851–858.

135. Engel, A. G., C. S. Potter, and J. W. Rosevear. 1964. Nucleotides and adenosine monophosphate deaminase activity in muscle in primary hypokalemic periodic paralysis. *Nature (London)* **292**:670–672.

136. Au, K. S., and R. T. T. Yeung. 1972. Thyrotoxic periodic paralysis: Periodic variation in the muscle calcium pump activity. *Arch. Neurol.* **26**:543–548.

137. DeGrandis, D., A. Fiaschi, E. Tomelleri, and D. Orrico. 1978. Hypokalemic periodic paralysis: A single fiber electromyographic study. *J. Neurol. Sci.* **37**:107–112.

138. Brooks, J. E. 1969. Hyperkalemic periodic paralysis. *Arch. Neurol.* **20**:13–18.

139. Norris, F. B., Jr. 1962. Unstable membrane potential in human myotonic muscle. *Electroencephalogr. Clin. Neurophysiol.* **14**:197–201.

140. Creutzfeldt, O. D., B. C. Abbott, W. M. Fowler, and C. M. Pearson. 1963. Muscle membrane potentials in episodic adynamia. *Electroencephalogr. Clin. Neurophysiol.* **15**:508–519.

141. Riecker, G., and H. D. Bolte. 1966. Membranpotentiale eizelner skeletmuskelzellen bein kypkaliamischer periodischer muskelparalyse. *Klin. Wochenschr.* **44**:804–807.

142. McComas, A. J., K. Mrozek, and W. G. Bradley. 1968. The nature of the electrophysiological disorder in adynamia episodica. *J. Neurol. Neurosurg. Psychiatry* **31**:448–452.

143. Lehmann-Horn, F., R. Rüdel, K. Ricker, H. Lorkovic, R. Dengler, and H. C. Hopf. 1983. Two cases of adynamia episodica hereditaria in vitro investigation of muscle cell membrane and contraction parameters. *Muscle Nerve* **6**:113–121.

144. Rüdel, R., F. Lehmann-Horn, K. Ricker, and G. Küther. 1984. Hypokalemic periodic paralysis in vitro investigation of muscle fiber membrane parameters. *Muscle Nerve* **7**:110–120.

145. Hodgkin, A. L., and A. F. Huxley. 1952. A quantitative description of membrane current and its application to conduction and excitation in nerve. *J. Physiol. (London)* **117**:500–544.

146. Almers, W., P. R. Stanfield, and W. Stühmer. 1983. Slow changes in currents through sodium channels in frog muscle membrane. *J. Physiol. (London)* **339**:253–271.

147. Almers, W., W. M. Roberts, and R. L. Ruff. 1984. Voltage clamp of rat and human skeletal muscle: Measurements with an improved loose-patch technique. *J. Physiol. (London)* **347**:751–768.

148. Roberts, W. M., R. L. Ruff, and W. Almers. 1983. Sodium currents in human muscle under voltage clamp. *Soc. Neurosci. Abstr.* **9**:1278.

149. Ruff, R. L., W. Stühmer, and W. Almers. 1982. The effect of glucocorticoid treatment on mammalian muscle excitability. *Pfluegers Arch.* **395**:132–137.

150. Ruff, R. L., D. Martyn, and A. M. Gordon. 1982. Glucocorticoid treatment does not alter muscle membrane excitability. *Am. J. Physiol.* **243**:E512–521.

151. Kao, L. I., and A. M. Gordon. 1965. Mechanism of insulin-induced paralysis of muscles from potassium-depleted rats. *Science* **188**:740.

152. Gordon, A. M., and L. I. Kao. 1978. Disorders of muscle membranes: The periodic paralyses. In: *Physiology of Membrane Disorders*. T. E. Andreoli, J. F. Hoffman, and D. D. Fanestil, eds. Plenum Press, New York. pp. 817–829.

153. Layzer, R. B. 1982. Periodic paralysis and the sodium–potassium pump. *Ann. Neurol.* **11**:547–552.

154. Almers, W. 1979. Potassium concentration changes in the transverse tubules of vertebrate skeletal muscle. *Fed. Proc.* **39**:1527–1532.

155. Venosa, R. A., and P. Horwicz. 1981. Density and apparent location of the sodium pump in frog sartorius muscle. *J. Membr. Biol.* **59**:225–232.

156. Akaike, N. 1982. Hyperpolarization of mammalian skeletal muscle fibers in K-free media. *Am. J. Physiol.* **242**:C12–C18.

157. Stanfield, P. R., N. B. Standen, C. A. Leech, and F. M. Ashcroft. 1980. Inward rectification in skeletal muscle fibers. In: *Advances in Physiological Science*, Volume V. E. Varga, A. Kover, T. Kovacs, and L. Kovacs, eds. Pergamon Press, Elmsford, N.Y. pp. 247–262.

158. Sestoft, L. 1967. Direct transition from hypo- to hyperkalemic paralysis during potassium treatment of familial periodic paralysis. *Dan. Med. Bull.* **14**:157–160.

159. Lehmann-Horn, F., R. Rüdel, R. Dengler, H. Lorkovic, A. Haass, and K. Ricker. 1981. Membrane defects in paramyotonia congenita with and without myotonia in a warm environment. *Muscle Nerve* **4**:396–406.

160. Ricker, K., A. Haass, R. Rüdel, R. Böhlen, and H. G. Mertens. 1980. Successful treatment of paramyotonia congenita (Eulenberg): Muscle stiffness and weakness prevented by tocainide. *J. Neurol. Neurosurg. Psychiatry* **43**:812–826.

161. Ricker, K., R. Böhlen, and R. Rohkamm. 1983. Different effectiveness of tocainide and hydrochlorothiazide in paramyotonia congenita with hyperkalemic episodic paralysis. *Neurology (Cleveland)* **33**:1615–1618.

162. Osaku, M., and I. Ohtsuki. 1970. Mechanism of muscular paralysis by insulin with special reference to periodic paralysis. *Am. J. Physiol.* **219**:1173–1182.

163. Dengler, R., W. W. Hofmann, and R. Rüdel. 1979. Effect of potassium depletion and insulin on resting and stimulated skeletal rat muscle. *J. Neurol. Neurosurg. Psychiatry* **42**:818–826.

164. Huang, K. W. 1943. Pa-Ping: Transient stimulating family periodic paralysis. *Chin. Med. J.* **61**:305–312.

165. Henderson, E. G. 1974. Strophanthidin sensitive electrogenic mechanisms in frog sartorius muscles exposed to barium. *Pfluegers Arch.* **350**:81–95.

166. Campa, J. F., and D. B. Sanders. 1974. Familial hypokalemic periodic paralysis. Local recovery after nerve stimulation. *Arch. Neurol.* **31**:110–115.

167. Bezanilla, F., and C. M. Armstrong. 1972. Negative conductance caused by entry of sodium and cesium ions into the potassium channels of squid axon. *J. Gen. Physiol.* **60**:588–608.

168. Johnsen, T. 1977. Trial of the prophylactic effect of diazoxide in the treatment of familial periodic hypokalemia. *Acta Neurol. Scand.* **56**:525–532.

169. Griggs, R. C., W. K. Engel, and J. S. Resnick. 1970. Acetazolamide treatment of hypokalemic periodic paralysis. *Ann. Intern. Med.* **73**:39–48.

170. Viskoper, R. J., A. Licht, J. Fidel, and J. Chaco. 1973. On the beneficial action of acetazolamide treatment in hypokalemic periodic paralysis: A metabolic and electromyographic study. *Am. J. Med. Sci.* **266**:119–123.

171. Johnsen, T. 1977. Effect upon serum insulin, glucose and potassium concentrations of acetazolamide during attacks of familial periodic paralysis. *Acta Neurol. Scand.* **56**:533–541.
172. Gerard, J. M. 1979. Familial periodic paralysis with hypokalemia, hyperaldosteronism, and extracellular vacuolization. *Rev. Neurol.* **134**:761–772.
173. Riggs, J. E., R. C. Griggs, and R. T. Moxley. 1984. Dissociation of glucose and potassium arterial–venous differences across the forearm by acetazolamide. *Arch. Neurol.* **41**:35–38.
174. Campbell, D. T., and B. Hille. 1976. Kinetic and pharmacological properties of the sodium channel of frog skeletal muscle. *J. Gen. Physiol.* **67**:309–323.
175. Conway, M. J., J. A. Seibel, and R. P. Eaton. 1974. Thyrotoxicosis and periodic paralysis improvement with beta blockade. *Ann. Intern. Med.* **81**:332–336.
176. Yeung, R. T. F., and T. F. Tse. 1974. Thyrotoxic periodic paralysis: Effect of propranolol. *Am. J. Med.* **57**:584–590.
177. Resnick, J. 1969. Thyrotoxic periodic paralysis. *Am. J. Med.* **47**:831–838.
178. Riggs, J. E., R. C. Griggs, R. T. Moxley, and E. D. Lewis. 1981. Acute effects of acetazolamide in hyperkalemic periodic paralysis. *Neurology (New York)* **31**:725–729.
179. Hoskins, B. 1977. Studies on the mechanism of action of acetazolamide in the prophylaxis of hyperkalemic periodic paralysis. *Life Sci.* **20**:343–350.
180. Wang, P., and T. Clausen. 1975. Treatment of attacks in hyperkalaemic familial periodic paralysis by inhalation of salbutamol. *Lancet* **1**:221–223.
181. Wang, P., T. Clausen, and H. Oerskov. 1978. Salbutamol inhalations suppress attacks of hyperkalemia in familial periodic paralysis. *Monogr. Hum. Genet.* **10**:62–65.
182. Dahl-Jørgensen, K., and H. Michalsen. 1979. Adynamia episodica herediteria: Treatment with salbutamol. *Acta Paediatr. Scand.* **68**:583–585.
183. Adrian, R. H., and W. H. Freygang. 1962. Potassium conductance of frog muscle membrane under controlled voltage. *J. Physiol. (London)* **163**:104–138.

Pathophysiology of Cardiac Arrhythmias

Robert F. Gilmour, Jr. and Douglas P. Zipes

1. Introduction

Normal cardiac rhythm occurs when spontaneous electrical impulses generated in the sinoatrial (SA) node are transmitted via the specialized conducting pathways to working myocardium. This orderly progression of impulse formation and propagation ensures a synchronous sequence of contraction in atrial and ventricular myocardium. Myocardial disease, cardioactive drugs, neurotransmitters, and other interventions can disrupt cardiac rhythm by altering spontaneous activity in the sinus node and other potential pacemaker regions in the heart and by impairing impulse propagation. Significant abnormalities of automaticity or conduction precipitate rhythm disturbances that compromise cardiac function, often to the point of producing clinical symptoms or death.

The purpose of this chapter is to review present knowledge concerning cardiac electrophysiological alterations that provoke cardiac arrhythmias. We will consider separately abnormalities of impulse formation, abnormalities of impulse propagation, and selected instances where abnormal impulse formation and propagation interact. Recent experimental observations and the proposed mechanisms for these observations will be emphasized and relevant clinical examples given.

1.1. Abnormalities of Impulse Formation

Under appropriate circumstances, all types of cardiac cells can generate spontaneous activity. Normally SA and certain atrioventricular (AV) nodal cells and cells in the His–Purkinje system demonstrate diastolic or phase 4 depolarization that, if uninterrupted, brings the cell to threshold and initiates an action potential. Some experimental and pathological conditions may induce phase 4 depolarization and automaticity in tissues that are not normally automatic, such as atrial and ventricular muscle fibers. In addition, atrial and ventricular muscle, Purkinje fibers, and several other cardiac tissues can develop sustained rhythmic activity, distinct from automaticity, that is triggered by early or delayed afterdepolarizations. In the following section we will discuss abnormal impulse formation due to altered normal diastolic depolarization, abnormal automatic mechanisms, and triggered activity.

1.2. Altered Normal Automaticity

The spontaneous discharge rate of the SA node normally determines the cardiac rhythm, since SA nodal cells exhibit the most rapid rate of diastolic depolarization. Impulses generated in the SA node control the cardiac rhythm by activating more slowly depolarizing pacemakers in the heart before they reach threshold. [1,2] If the spontaneous discharge rate of the SA node slows or if impulses generated by the SA node are blocked before they exit the SA nodal area, or before they enter the AV node, His–Purkinje system, or ventricle, latent pacemakers can reach threshold (escape) and assume control of the heart rate.

Disease states, autonomic nervous system influences, and cardiac drugs such as digitalis can increase the spontaneous discharge rates of subsidiary pacemakers to rates exceeding that of the SA node, so that these subsidiary pacemakers usurp control of the cardiac rhythm. In this regard, data from several studies suggest that automaticity in subendocardial Purkinje fibers surviving acute myocardial infarction in the dog are responsible for the spontaneous arrhythmias that occur in these animals 24–48 hr after infarction. [3–5] Purkinje fibers excised from infarcted zones and studied *in vitro* exhibit enhanced automaticity that appears to be due to accelerated normal diastolic depolarization. The combination of enhanced Purkinje fiber au-

Robert F. Gilmour, Jr. • Krannert Institute of Cardiology, Departments of Medicine and Pharmacology, Indiana University School of Medicine, Indianapolis, Indiana 46223.   **Douglas P. Zipes** • Krannert Institute of Cardiology, Department of Medicine, Indiana University School of Medicine, and Roudebush Veterans Administration Medical Center, Indianapolis, Indiana.

tomaticity and entrance block in adjacent infarcted myocardium may allow spontaneous Purkinje fiber discharges to alter the ventricular rhythm *in vivo*.

The effects of cardiac disease, drugs, and neurotransmitters on automaticity of the SA node and subsidiary pacemaking tissues may vary, depending on the ionic currents responsible for diastolic depolarization and action potentials in these fibers. Normal automaticity in the SA and AV nodal areas arises from cells with maximum diastolic membrane potentials of -60 to -70 mV, whereas normal Purkinje fiber automaticity is generated by cells having maximum diastolic potentials of -80 to -95 mV. The ionic basis for diastolic depolarization in both types of tissue is still being investigated[2,6,7] (see also Chapter 29). However, it is clear that upon reaching threshold for an active response, the upstroke of the SA and AV nodal cell action potentials are mediated primarily by the slow inward Ca^{2+} current and can be blocked by calcium channel blocking agents such as verapamil.[8,9] In contrast, the action potential upstroke of the Purkinje cell is mediated by the fast inward Na^+ current and is suppressed by sodium-channel blocking agents such as quinidine[10] and tetrodotoxin,[11] but not by verapamil.

Acetylcholine suppresses and β-adrenergic agonists enhance diastolic depolarization in the SA and AV nodes and in Purkinje cells[2,12] (Chapter 29). Acetylcholine appears to act primarily by increasing potassium conductance, thereby increasing potassium efflux and counteracting the depolarizing effects of sodium or calcium influx during diastole. β-Adrenergic agonists indirectly affect the pacemaker current as well as the slow inward calcium current and accelerate attainment of threshold. Accordingly, sympathetic or parasympathetic stimulation can induce rhythm changes in the heart by altering normal automaticity in both calcium channel-dependent and sodium channel-dependent tissues.

1.3. Abnormal Automaticity

Although normally polarized atrial and ventricular muscle do not exhibit diastolic depolarization or automaticity, they may generate spontaneous activity if they are depolarized to membrane potentials in the range of -50 to -60 mV.[13, 14] Purkinje fibers that have maximum diastolic potentials in this range also discharge spontaneously, generally at faster rates than those produced at normal diastolic potentials[15] (Fig. 1). Spontaneous Purkinje fiber and myocardial action potentials generated at reduced membrane potentials are similar in some respects to action potentials normally present in the SA and AV nodes. For example, the spontaneous discharge rate and amplitude of depolarization-induced action potentials are increased by manipulations that increase the slow inward calcium current, such as catecholamines and elevated extracellular calcium concentration,[13–16] or decrease outward potassium current, such as Ba^{2+}.[17] Verapamil and low extracellular Ca^{2+} concentration can completely suppress depolarization-induced spontaneous activity, whereas fast sodium channel blockers and reduced extracellular Na^+ concentration have only a small inhibitory effect (Fig. 1).

Depolarization-induced automaticity can occur as a result of myocardial disease[18–21] or experimental manipulations such as exposure to Ba^{2+} or current injection.[13–17] Observations in diseased myocardium studied *in vitro* suggest that events such as myocardial ischemia may induce abnormal automaticity in previously quiescent tissue or increase the spontaneous discharge rate of latent pacemakers in the His–Purkinje system, thereby facilitating pacemaker escape and the induction of premature ventricular complexes or tachycardia. The same agents that suppress normal automaticity in Purkinje fibers may not suppress enhanced automaticity if the latter is mediated by calcium channel-dependent action potentials.

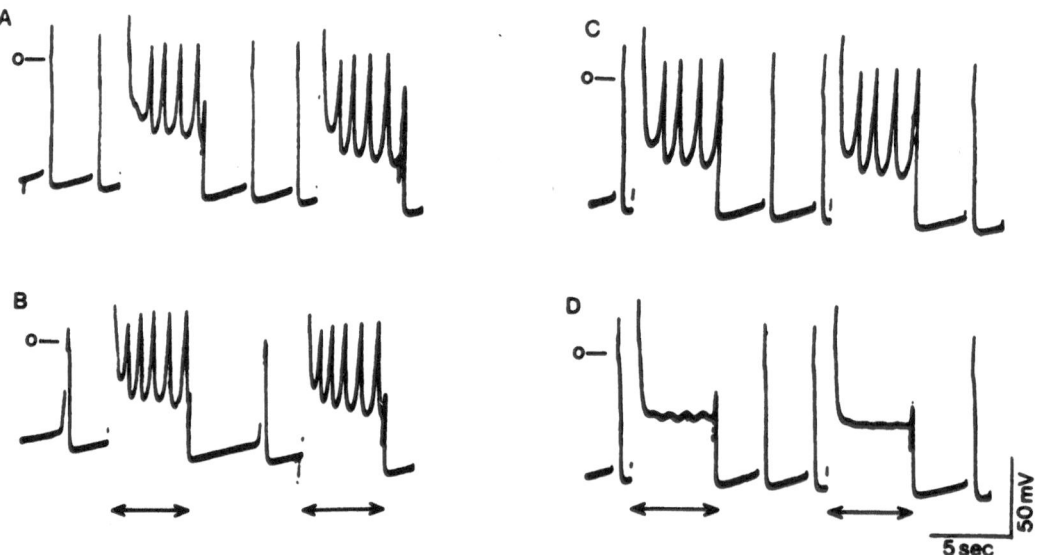

Fig. 1. Differential effects of lidocaine (3 mg/liter) and verapamil (3×10^{-6} M) on automaticity initiated at high and low levels of transmembrane potential. Depolarizing current pulses were delivered across a single sucrose gap at various times (indicated by the interval between the arrowheads) to depolarize the fiber to -50 to -55 mV. (A) Control; (B) during superfusion with lidocaine; (C) return to control after 30 min of washout; (D) during superfusion with verapamil. Lidocaine slowed the spontaneous discharge rate and reduced the amplitude of action potentials arising from the high level of membrane potential but did not alter action potentials or spontaneous rate at the low level of membrane potential. Verapamil suppressed automaticity at the low level of membrane potential without altering the spontanous discharge rate at the high level of membrane potential. Reproduced with permission from Elharrar and Zipes.[15]

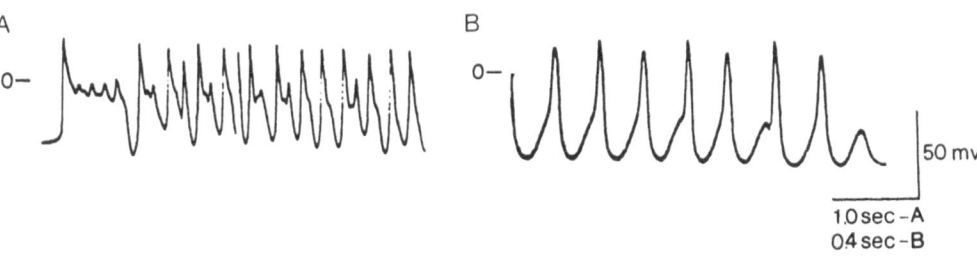

Fig. 2. Triggered sustained rhythmic activity recorded in rat myocardium damaged by epinephrine. (A) Triggered activity arising from early afterdepolarizations; (B) spontaneous activity in the terminal phase of a triggered burst which appears to be generated by delayed afterdepolarizations. The membrane gradually hyperpolarizes, a delayed afterdepolarization fails to reach threshold, and the spontaneous activity ceases. Reproduced with permission from Gilmour and Zipes.[21]

1.4. Triggered Activity

Another form of impulse formation found in both normal and diseased cardiac tissue is triggered sustained rhythmic activity.[22] Triggered activity occurs when a driven action potential is accompanied by a secondary voltage oscillation (afterdepolarization) that reaches the threshold for a regenerative response. The required antecedent action potential distinguishes triggered activity from automaticity, where threshold is attained by spontaneous diastolic depolarization. Afterdepolarizations may occur prior to complete repolarization of the triggering action potential (early afterdepolarizations) (Fig. 2A) or may follow complete repolarization (delayed afterdepolarizations) (Fig. 2B). Once triggered, the initial action potential may induce a burst or train of action potentials that continue in the absence of further driven impulses. Triggered activity can terminate spontaneously with a subthreshold afterdepolarization (Fig. 2B) or be terminated by a properly timed driven impulse or train of impulses.[22] In contrast to other forms of impulse formation, triggered activity often is accelerated, rather than suppressed, following a period of rapid overdrive pacing.

Triggered activity due to delayed afterdepolarizations occurs in apparently normal atrial myocardium,[18,23] upper pectinate muscles bordering the crista terminalis,[24] and the mitral valve.[25] Various experimental manipulations, such as exposure to digitalis,[26,27] induce triggered activity, and triggered action potentials have been recorded in diseased human atrium and ventricle studied in vitro[18,20] (Fig. 3). Digitalis-induced afterdepolarizations may be due to an inward current, possibly carried by sodium, that is augmented by elevated intracellular calcium ion concentration (the transient inward current).[28,29] The magnitude of this current usually is too small to contribute appreciably to the normal action potential contour. However, exposure to digitalis produces inhibition of the sodium-potassium pump and an increase in intracellular sodium ion concentration. Exchange of sodium for calcium is accelerated, the intracellular calcium concentration rises, and the transient inward current is increased. The resultant membrane depolarization may bring the cell to threshold and trigger an action potential. An increase in intracellular calcium or sodium concentration during periods of rapid pacing also may induce the transient inward current and explain, in part, overdrive acceleration of triggered activity.

As predicted from the above explanation, both sodium and calcium channel blocking agents suppress digitalis-induced de-

layed afterdepolarizations in vitro.[26,27] Presumably, the sodium channel blockers reduce the influx and subsequent accumulation of sodium, whereas the calcium channel blockers prevent an increase in intracellular calcium levels by limiting calcium entry. However, afterdepolarizations recorded in chronically diseased human atrium are suppressed by verapamil, but not by tetrodotoxin or lidocaine,[30] suggesting that the ionic events responsible for delayed afterdepolarizations and triggered activity may differ in some preparations, depending on the type of tissue and the experimental conditions.

The presence of triggered activity in diseased human atrium and ventricle studied in vitro suggests that triggered foci may precipitate atrial and ventricular arrhythmias in vivo. It has been proposed that arrhythmias due to triggered activity might be accelerated following a period of rapid pacing, whereas other forms of spontaneous activity should be suppressed. Recent clinical and experimental studies have used this approach with some success, but more data are needed and differentiation of this mechanism from others that cause arrhythmias still cannot be made with certainty.[31,32] The experimental data also indi-

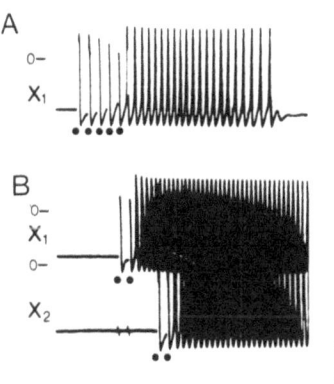

Fig. 3. Triggered sustained rhythmic activity in diseased human ventricle. (A) Spontaneous activity triggered by a series of driven action potentials (indicated by the dots) at a recording site X_1. Note the gradual increase in the size of the afterdepolarizations until the afterdepolarization reaches threshold and maintains sustained rhythmic activity following cessation of pacing. The sustained rhythmic activity terminates when the last afterdepolarization fails to reach threshold. (B) Initiation of triggered activity by intracellular current injection (indicated by dots beneath the respective action potential recordings) at sites X_1 and X_2 which lie along the same trabeculum. Although sites X_1 and X_2 were approximately 4 mm apart, triggered sustained rhythmic activity from one site did not propagate to the other site. For current pulses, cycle length = 2000 msec; pulse duration = 10 msec; pulse intensity = 200 nA. Vertical calibration, 50 mV. Horizontal calibration, 10 sec. Reproduced with permission from Gilmour et al.[20]

cate that the precipitation or termination of a tachycardia by a single premature complex cannot be used as the sine qua non for the presence of reentry, since a single stimulus can induce or terminate triggered activity.

2. Abnormalities of Impulse Propagation

Propagation of the cardiac impulse involves a complex interaction between active and passive membrane properties of individual cardiac cells. An action potential elicited in one cell induces the excitation of a neighboring cell through the flow of local circuit current. The ability of this current to bring the next cell to threshold depends on the intensity of the input, i.e., the amplitude and upstroke velocity of the initial action potential, the excitability of the neighboring cell, and the magnitude of intercellular current flow. The latter is determined mainly by the amount of current leak across the sarcolemmal membrane and the resistance to intercellular current flow produced by cell junctions.

Disruption of the normal activation sequence may lead to slow conduction and block. This in turn may be responsible for the generation of bradyarrhythmias, if the SA node impulse blocks in the SA or AV junction and a slow junctional or ventricular escape rhythm ensues, or tachyarrhythmias if reentry is precipitated. In the following section, we will discuss abnormalities of active membrane properties, excitability, and cell coupling, and their roles in the production of slow conduction, block, and reentrant excitation.

2.1. Active Membrane Properties and Excitability

Activation and reactivation of the fast inward sodium current occurs most rapidly at normal diastolic potentials (-80 to -95 mV) and less rapidly as the membrane becomes depolarized.[33–35] Therefore, loss of resting potential in myocardial and Purkinje cells causes partial inactivation of the fast inward Na$^+$ current and subsequent reductions of action potential amplitude and upstroke velocity, excitability, and conduction velocity. Recovery of excitability is retarded at low resting potentials, which prolongs refractoriness and contributes to rate-dependent conduction disturbances such as Wenckebach periodicity.[36]

Myocardial ischemia or infarction produce several environmental alterations that reduce transmembrane potential. These include elevated extracellular potassium concentration,[37,38] intracellular and extracellular acidosis,[37,39–41] hypoxia,[37,40] and release of sarcolemmal membrane components such as lysophosphatides.[42] As expected, the action potential amplitude and upstroke velocity of ischemic cells are depressed.[23,43] Excitability may be increased during the early phase of acute myocardial ischemia, as the membrane depolarizes to potentials near the threshold potential.[23,44] Prolonged ischemia decreases excitability, however, as sodium channels become progressively inactivated by sustained depolarization and the threshold potential shifts to less negative values.[23,44]

Cells along the border of an infarcted region of myocardium also may have reduced action potential amplitudes and upstroke velocities, even if their resting membrane potentials are within the normal range.[19,20] This may be due to electrotonically mediated attenuation of border-zone cell action potential upstrokes by the slowly rising action potentials in adjacent ischemic cells. In regions of mottled infarction, therefore, a

normal resting potential may not necessarily correlate with normal active membrane properties.

Under circumstances where membrane depolarization completely inactivates the fast inward sodium current, impulse propagation still may occur via calcium channel-dependent action potentials, called slow responses.[45] Experimentally, depolarization of myocardial and Purkinje cells to potentials in the range of -60 to -50 mV by elevated extracellular potassium largely inactivates the fast inward Na$^+$ current and induces inexcitability. Excitability can be restored by exposing the cells to β-adrenergic agonists such as isoproterenol. The resultant slow responses conduct slowly and have refractory periods that outlast terminal action potential repolarization. They are unaffected by sodium channel blockers, but are suppressed by verapamil and nifedipine.[45]

Since potassium[37,38] and catecholamines[46] are released into the extracellular space during ischemia, it has been proposed that slow responses occur in ischemic myocardium and mediate slow conduction.[45] Several lines of evidence fail to support this hypothesis. First, other metabolic alterations that occur in ischemic myocardium, such as hypoxia and acidosis,[47] suppress slow responses, making their appearance unlikely during acute ischemia. Second, if ischemia-induced slow conduction were due to the slow response, sodium-channel blockers would exert little, if any, effect on the slow conduction, whereas calcium-channel blockers would be expected to worsen the slow conduction or suppress it entirely. However, tetrodotoxin and lidocaine, which do not inhibit the slow response, suppress poorly conducting action potentials recorded in ischemic myocardium in vitro[48] or in situ,[23] but verapamil fails to suppress these action potentials[23] (Fig. 4). Fast-Na$^+$-channel blockers also exacerbate conduction delay and increase the incidence of ventricular arrhythmias during acute myocardial ischemia in vivo.[49,50] In contrast, calcium-channel blockers usually reduce conduction delay and ventricular arrhythmias during acute ischemia,[49,51] possibly by reducing the size of the infarct zone. These studies suggest that residual fast inward Na$^+$

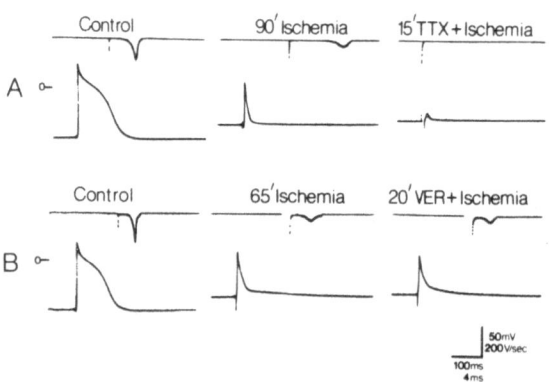

Fig. 4. Effects of ischemia and ischemia + tetrodotoxin (TTX) (A) or + verapamil (B) on V_{max} (upper recordings) and action potential configuration (lower recordings) of hamster atrial transplant cells. Following 90 min of ischemia, the transplant cell shown in (A) was exposed to TTX (10^{-5} M) for 15 min. TTX progressively decreased action potential amplitude and V_{max} and the cell became inexcitable. After 65 min of ischemia in another atrial transplant (B), action potential configuration was similar to that seen at 90 min of ischemia in (A). Exposure of this cell to verapamil (2×10^{-6} M) for 20 min did not produce inexcitability. Reproduced with permission from Gilmour and Zipes.[23]

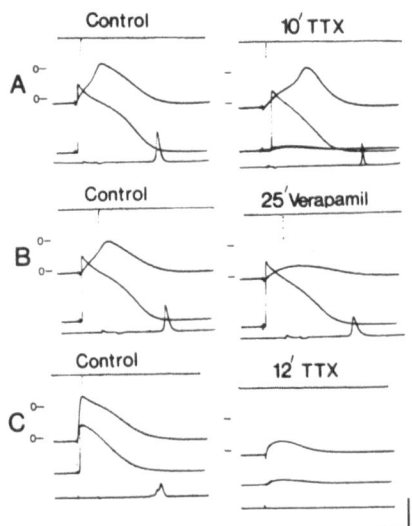

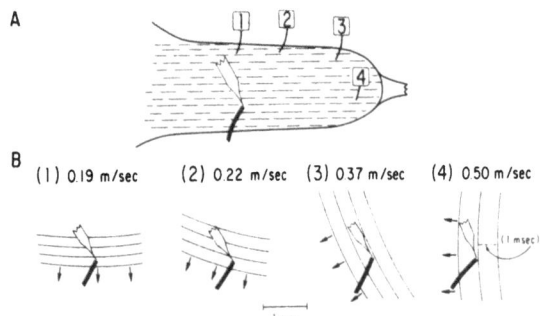

Fig. 6. Changes in propagation velocity with direction in the papillary muscle. (A), a schematic drawing of the preparation shows the locations of the stimulus electrodes (numbered) with respect to the recording site. (B) shows the direction of propagation (illustrated by isochrones) and the associated velocities from excitation initiated at each of the stimulus sites. The distance scale applies to (B) only, where the magnification is greater than in (A). Reproduced with permission from Spach *et al.*[58]

Fig. 5. Effects of tetrodotoxin (TTX) and verapamil on action potentials in diseased human ventricle, removed from a patient at the time of endocardial resection for recurrent ventricular tachycardia. (A) Action potentials and upstroke velocity recordings from an abnormal cell (upper action potential recording) and a relatively normal cell (lower action potential recording) before (left) and after (right) exposure to TTX for 10 min. $\dot{V}_{max}$ for the lower cell is shown in the bottom tracing. TTX produced activation delay and intermittent conduction block near the normal cell but had little effect on the action potential of the abnormal cell. Two consecutive cycles are superimposed in the right panel. In (B), after washout of TTX, the same two cells were exposed to verapamil for 25 min. Verapamil reduced the action potential amplitude of the abnormal cell without affecting its resting membrane potential and slightly reduced action potential amplitude and $\dot{V}_{max}$ in the normal cell (right panel). (C) The effects of TTX on a different piece of myocardium taken from the same patient. Control recordings are shown on the left. In these cells, TTX markedly reduced action potential amplitude and $\dot{V}_{max}$ while verapamil only slightly reduced action potential amplitude (not shown). Reproduced with permission from Gilmour *et al.*[20]

current mediates the action potential upstroke in ischemic cells, producing a so-called depressed fast response.

Although slow responses do not appear to mediate slow conduction during acute myocardial ischemia, they may contribute to abnormal impulse formation and propagation in chronically diseased myocardium. Several recent *in vitro* studies of chronically diseased human myocardium have found depressed spontaneous and driven action potentials that are suppressed by verapamil and low extracellular Ca^{2+} concentrations, but not by tetrodotoxin[19,20,52] (Figs. 5A, B). Other depressed action potentials are found in these preparations, however, that are suppressed by tetrodotoxin, but not by verapamil[20] (Fig. 5C).

At the present time, the relative roles of depressed fast responses and slow responses in the initiation and perpetuation of ventricular tachyarrhythmias in humans are unknown. However, preliminary data indicate that the present generation of calcium channel blocking agents are far less effective than sodium channel blocking agents in suppressing recurrent ventricular tachycardia.[53,54]

2.2. Cell Coupling

Cardiac cells are electrically coupled to one another by means of low-resistance intercellular junctions that permit the passive spread of current.[33–35] Decreased intercellular coupling, such as occurs during hypoxia and other interventions that increase intracellular calcium and hydrogen ion concentration,[55–57] also may contribute to slow conduction by retarding the passive spread of current between cells and the attainment of threshold. Cells that are completely uncoupled from one another may act as inexcitable islands that force the excitation wavefront to assume a more circuitous route, thus delaying propagation and increasing the likelihood of reentrant excitation. In contrast, uncoupling electrically abnormal areas of myocardium from normal surrounding cells may isolate aberrant spontaneous or driven action potentials within small regions of the heart and prevent the induction of arrhythmias. For example, Fig. 3 illustrates that triggered activity could be provoked in several human ventricular cells, yet this activity did not propagate more than 3–4 mm from its site of origin.

Recent studies by Spach *et al.*[58,59] suggest that the anatomical arrangement of cardiac cells may determine their junctional resistance, so that impulse propagation proceeds more rapidly in a direction that is parallel to muscle fiber orientation than in the transverse direction (Fig. 6). Therefore, impulses normally may conduct preferentially in one direction due to advantageous fiber orientation. If the impulse is diverted from the preferential path due to a zone of block or change in activation sequence, then the alternate route it travels may have higher coupling resistance. Hence, conduction will be slowed, despite the fact that the active electrical properties of cells in the alternate path are normal. In addition, disparities in conduction velocities dependent on the path traveled are amplified by interventions that alter upstroke velocity and action potential amplitude, such as elevated potassium concentration. Since myocardial ischemia may produce zones of block or variations in activation sequences due to latent pacemaker escape, these studies provide an additional mechanism whereby slow conduction, block, and reentrant excitation may occur.

2.3. Reentry

Reentrant excitation occurs when an impulse excites a region of myocardium, conducts slowly around an area of inexcitable tissue, and returns to reexcite the original region. Unidirectional block in one limb of the circular pathway, prolonged conduction time in the other limb, and a relatively brief refractory period in the previously excited region are required to prevent the circulating wavefront from impinging on refractory tissue and extinguishing itself.

The classic description of reentrant excitation by Schmitt and Erlanger[60] requires an anatomical or electrophysiological obstacle around which a slowly conducting impulse circles. However, experiments in atrial myocardium have demonstrated that the obstacle need not be fixed.[61] A premature stimulus can generate a propagating wavefront that establishes a vortex of conduction around a central area of electrotonically depolarized cells. These cells function as an inexcitable island, despite the fact that their electrophysiological properties are normal. The cycle length of this type of reentry is determined in large part by the refractory period of cells in the propagating wavefront. Shortening their refractory period shortens the tachycardia cycle length, until the leading edge of the propagating wavefront encounters the refractory tail of the previous wavefront and further propagation fails.

More recently, it has been demonstrated that a certain type of reentry, known as reflection, results from transmission of impulses back and forth along the same myocardial or Purkinje fiber, rather than around an obstacle.[62] Reflection occurs when two excitable segments of myocardium are separated by an inexcitable segment having an appropriate impedance (Fig. 7). The inexcitable segment acts as a conduit for passive current flow between the two excitable segments. Action potentials elicited in one of the excitable segments propagate to the junction with the inexcitable segment, at which point active transmission ceases and current flows passively into and across the inexcitable cells. Since no active responses occur in the inexcitable segment, the magnitude of the electrotonic potential tends to diminish with the distance traveled, as current dissipates among the cells and intercellular spaces. If sufficient current passes through the inexcitable segment, the distal segment is brought to threshold and active propagation resumes. Action potentials in the distal segment then generate an electrotonic potential that is directed back across the same inexcitable segment. If the proximal segment has had sufficient time to recover excitability, it can be reexcited, completing one cycle of a reentrant circuit that may become self-perpetuating.

3. Interactions between Abnormal Impulse Formation and Propagation

Since excitability, action potential amplitude, and maximum action potential upstroke velocity depend on the voltage at which activation occurs, impulse propagation into and out of spontaneously depolarizing regions may change significantly as diastolic depolarization proceeds. Conversely, changes in conduction may affect the degree of entrance and exit block surrounding an automatic focus and alter the ability of regenerative and subthreshold potentials to modulate pacemaker activity. Impulse conduction and automaticity also may interact due to the flow of injury current between regions of disparate electrophysiological properties. These and other possible interactions between abnormal impulse formation and propagation will be discussed in the following section.

3.1. Modulation of Impulse Propagation

In spontaneously discharging Purkinje fibers having maximum diastolic potentials more negative than −70 mV, action potentials elicited by premature stimuli late in diastole have greater conduction velocities than those elicited earlier in di-

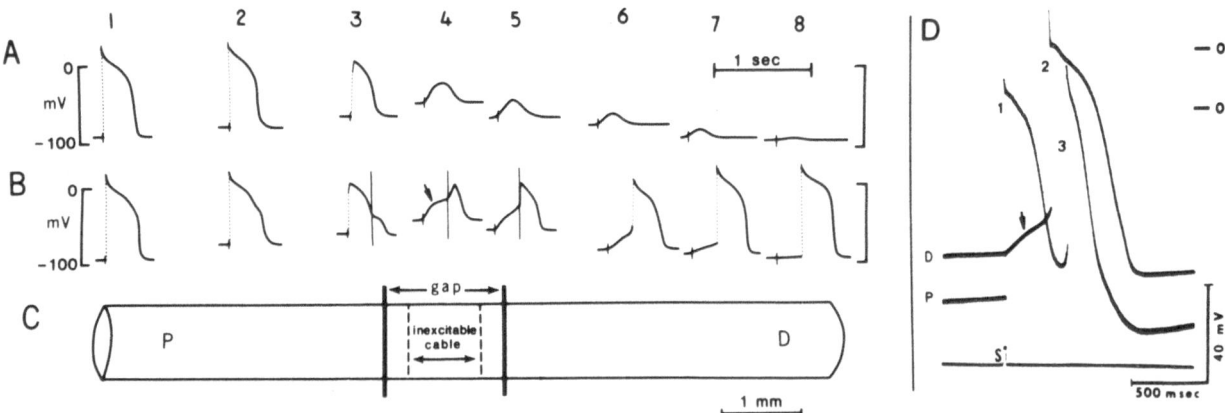

Fig. 7. Reflection. (C) Schematic representation of electrotonic transmission across an area of inexcitability in a Purkinje fiber. The central compartment (gap) of the tissue bath contains a solution which prevents active conduction in the central segment (inexcitable cable) of the Purkinje fiber. Action potentials from various locations along the Purkinje fiber are presented in (A) when transmission across the inexcitable area is unsuccessful, and in (B) when transmission is successful. In (A), the regenerative impulse encountering the region of block decays along the length of the blocked segment. In (B), the slowly rising electrotonic potential (arrow) resulting from the activity in the proximal tissue brings the distal segment to threshold and thus restores after a delay active propagation via electrotonic transmission. When the delay in transmission across the gap is sufficiently long, electrotonic transmission in the reverse direction over the same blocked segment can reexcite the proximal segment and generate a repetitive response. This is illustrated in (D). The two traces were recorded from the two active segments on the proximal (P) and distal (D) sides of the inexcitable gap. The bottom trace is the stimulus marker (S). Stimulation of the proximal segment (1) was transmitted to the distal segment (2) following a slowly developing electrotonic depolarization (arrow) that reaches the threshold of the tissue beyond the inexcitable gap after a delay of 300 msec. Electrotonic transmission in the reverse direction induced a closely coupled (404 msec) reflected response (3) in the proximal segment following expiration of its refractory period. Reproduced with permission from Antzelevitch et al.[62]

astole[63] since the cell is closer to threshold voltage and more easily excited. In this instance, the increased excitability overrides the effects of partial inactivation of the fast inward sodium current induced by slow diastolic depolarization. Thus, normal or "super-normal" conduction may occur, despite a 30% decrease in maximum upstroke velocity and reduced action potential amplitude.[63] These observations may explain the apparent improvement in conduction produced *in vivo* by late premature stimuli.[64]

In contrast to the situation cited above, premature stimuli delivered to spontaneously depolarizing cells with maximum diastolic potentials less than −70 mV may produce more slowly conducting action potentials as diastolic depolarization approaches threshold.[65] At maximum diastolic potentials in this range, significant inactivation of the fast inward sodium current and shift of the threshold potential to less negative values can occur, making the cells more difficult to excite. Spontaneous membrane depolarization induces further sodium or calcium channel inactivation and progressive reductions in action potential amplitude and maximum upstroke velocity that may not be counteracted by an appropriate increase in excitability. As a result, conduction velocity decreases. Under certain circumstances, however, stimuli delivered to a depolarized region of His-Purkinje tissue in late diastole produce slowly propagating action potentials in the absence of phase four depolarization.[66,67] This time-dependent decrease in excitability may be due to more complete reactivation of the transient outward current, rather than to progressive inactivation of sodium or calcium inward currents.[67]

The complete spectrum of excitability in depolarized Purkinje cells that do not demonstrate diastolic depolarization consists of low excitability and action potential amplitude during early diastole, maximal excitability in mid-diastole and lower excitability in late diastole.[66,67] If such a region of depolarized cells is coupled to more normally polarized Purkinje tissue, the biphasic changes of action potential amplitude and excitability in the depolarized region may interact with changes in excitability secondary to diastolic depolarization in surrounding tissue. As a result impulses generated in the depolarized area will block during early diastole, since action potential amplitude in the depolarized region is low and membrane potential in surrounding cells is far from threshold. In mid-diastole, action potential amplitude in the depolarized region is maximal, membrane potential in surrounding cells is closer to threshold and conduction will be more likely to occur. In late diastole conduction block may result from a decrease in action potential amplitude in the depolarized region, despite continued diastolic depolarization in surrounding cells. Finally, if the spontaneously depolarizing region is very close to threshold, even a low amplitude action potential generated by the depolarized region may propagate.

Biphasic changes in excitability in depolarized cells demonstrating no significant phase four depolarization may interact, therefore, with spontaneously depolarizing surrounding myocardium to produce several types of conduction disturbances. These include acceleration- and deceleration-dependent bundle branch block, supernormality and overdrive suppression and facilitation of conduction.[64,66,67]

3.2. Modulation of Impulse Formation

As discussed previously, automatic cells with the fastest rate of diastolic depolarization usually suppress more slowly

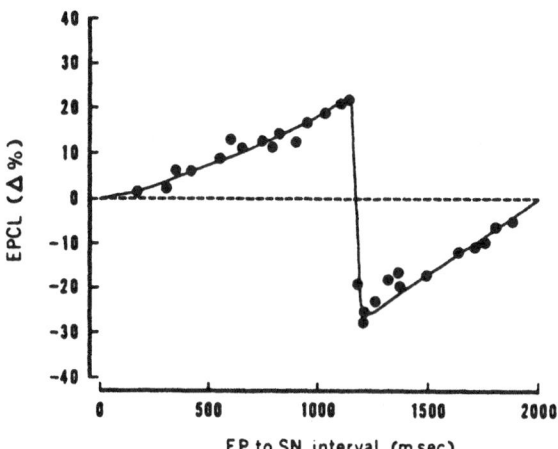

Fig. 8. Phase–response curve of a Purkinje fiber pacemaker mounted in a sucrose gap preparation. The *x* axis (EP to SN interval) represents the time after a spontaneous discharge of the pacemaker (EP) at which stimulated responses (SN) were evoked in the segment of fiber beyond the gap. The *y* axis (EPCL) represents the prolongation (first half) or abbreviation (second half) of the pacemaker cycle as a function of that time. EPCL = ectopic pacemaker cycle length; EP to SN interval = time after ectopic pacemaker discharge, EP, at which an evoked response, SN, was generated to stimulate a ventricular response of sinus nodal (SN) origin. Intrinsic cycle length of EP, in the absence of electrotonic influence, was about 2000 msec. Abrupt phase reversal occurred at about 58% of the intrinsic cycle length. Reproduced with permission from Jalife and Moe.[67]

depolarizing foci, unless block at some site between the dominant and latent pacemakers prevents resetting of the latent pacemaker. However, if entrance or exit block into or out of the automatic focus is incomplete, propagated impulses may induce subthreshold potentials distal to the site of block that predictably alter the discharge rate of a focus.[66,67] Subthreshold depolarizing impulses that arrive in the focal area early in diastole delay the subsequent spontaneous discharge, whereas impulses that arrive later in diastole accelerate the subsequent discharge. The complete phase-response curve for this type of interaction is shown in Fig. 8 and an example of pacemaker advance and delay in diseased human ventricular myocardium is given in Fig. 9.

Phase resetting by subthreshold stimuli probably involves several ionic phenomena. Subthreshold depolarizations that occur early in diastole may partially inactivate the pacemaker current. Later in diastole, as the cell spontaneously depolarizes to potentials closer to threshold, the subthreshold depolarization accelerates attainment of threshold and the initiation of an early spontaneous discharge. If subthreshold impulses are delivered to the focus at the same point in each spontaneous cycle, the pacemaker mechanism will be repeatedly reset or accelerated, depending on the timing of the impulses. As a result, the focus may become entrained at a slower or faster discharge rate than normal. By this mechanism parasystolic foci may interact with impulses (e.g., the normally propagating sinus impulses) that are transmitted electronically across an inexcitable region of myocardium and become "modulated," that is sped up, slowed down or entrained by the dominant rhythm. The theoretical and experimental implications of modulated parasystole have been considered in detail.[67–69]

In addition to advancing or delaying the next expected

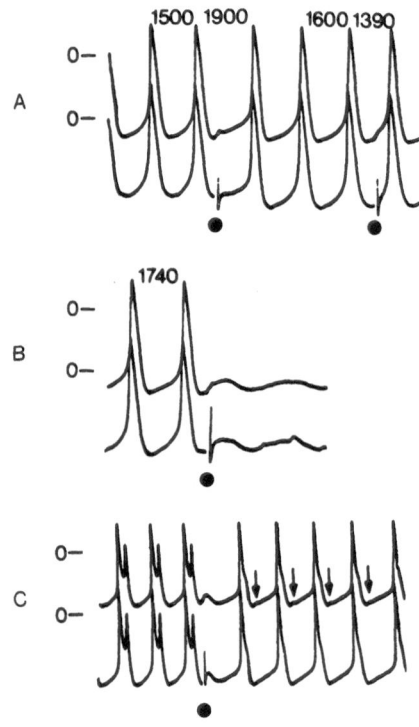

Fig. 9. Modulation of pacemaker activity by subthreshold current pulses in diseased human ventricle. (A) Two recording sites along the same trabeculum in a spontaneously active preparation. Current pulses (indicated by the dots) of 30-msec duration were injected through the lower microelectrode at various times. The interval between the spontaneous action potentials is given in milliseconds above each cycle. Injection of a subthreshold current pulse through the lower microelectrode relatively early in the spontaneous cycle (approximately 680 msec after initiation of the rapid portion of the preceding action potential upstroke) produced a subthreshold depolarization in the upper record and delayed the next spontaneous discharge by 400 msec to 1900 msec. This response would fall in the first half of the curve indicated in Fig. 8. A current pulse of the same intensity and duration delivered later in the spontaneous cycle (950 msec after the preceding upstroke) accelerated the next discharge by 210 msec to 1390 msec, relative to the previous two action potentials. The response to this current injection falls in the second half of the graph depicted in Fig. 8. In (B), a stimulus at a precise interval in the cardiac cycle (called the singular point; in this example 930 msec after the preceding action potential upstroke) abolished pacemaker activity. (C) illustrates that a single subthreshold pulse (dot) converted biphasic action potentials to action potentials with a single active component followed by delayed afterdepolarizations (arrows). Vertical calibration, 50 mV; horizontal calibration, 2 sec in (A) and (B), 4 sec in (C). Reproduced with permission from Gilmour *et al.*[20]

spontaneous discharge, subthreshold impulses that invade a focal area at a precise time, known as the singular point,[70] produce complete suppression, or annihilation of spontaneous activity[20,71] (Fig. 9B). Spontaneous activity may not resume until an external stimulus is applied. The ionic mechanism of annihilation in cardiac tissue has not been determined. Perhaps subthreshold stimuli delivered at the singular point interrupt the normal sequence of pacemaker current activation at a critical stage and arrest the membrane at an equilibrium potential, where the net current flow across the membrane is zero. There is theoretical and experimental evidence to support the pres-

ence of stable equilibrium potentials[72,73] in addition to the normal or usual resting membrane potential.

The complete suppression of spontaneous activity by a single subthreshold response and the reinitiation of such annihilated automaticity by a second stimulus further undermines the diagnostic specificity of tachycardia initiation and termination by single premature stimuli *in vivo*, generally considered diagnostic of reentry. Rather, the experimental demonstration of annihilation may be related to some clinical situations where ventricular tachycardia can be terminated by a single critically-timed premature stimulus that need not propagate throughout the ventricle.[74,75]

Single subthreshold responses also may alter the configuration of spontaneous action potentials, as illustrated in Fig. 9C. In this example, the introduction of a subthreshold depolarization converted action potentials accompanied by a suprathreshold early afterdepolarization to action potentials followed by a subthreshold delayed afterdepolarization (indicated by the arrows). Conceivably, such subthreshold stimuli may change the electrocardiographic manifestation of polymorphic ventricular tachycardia[76] from one configuration to another.

3.3. Injury Currents

Abnormal impulse formation and propagation also may interact via the production of injury currents.[77] Injury currents are induced when a region of myocardium is depolarized, relative to a neighboring region, establishing a voltage gradient between the two groups of cells. Several disparities in cellular electrical properties may produce such a gradient. These include: (1) reduced resting membrane potential in one group of cells, compared to that of neighboring cells; (2) reduced action potential amplitude, so that current flows from the cells with the higher amplitude (i.e., those that become more depolarized during the upstroke phase of the action potential) to cells with a lower amplitude; and (3) reduced action potential duration, which induces current flow from cells having the longer state of depolarization to those that return sooner to their resting potential. For a potential gradient to be maintained, however, the depolarized and nondepolarized regions must be separated by an inexcitable region to prevent equalization of membrane potential and action potential duration by electrotonic influences.[78]

Circular flow of injury current across the borders of ischemic or infarcted myocardium conceivably could initiate reentrant rhythms by reexciting normal myocardium.[77,79] Sustained injury current might initiate spontaneous depolarizations, perhaps by a mechanism similar to that illustrated in Fig. 1. Bursts of spontaneous activity could produce an automatic tachycardia, whereas a single spontaneous discharge might provide a premature impulse that establishes a reentrant circuit. However, the presence of an inexcitable zone and the production of spontaneous activity by injury currents *in vivo* remain to be demonstrated conclusively. Therefore, the initiation of arrhythmias by this mechanism presently is speculative.

4. Electrophysiological Mechanisms Responsible for Clinically Occurring Arrhythmias

As stated earlier, the genesis of cardiac arrhythmias generally is divided into categories of disorders of impulse formation, disorders of impulse conduction, or combinations of both.[12,80,81] It is important to realize, however, that our present diagnostic tools do not permit unequivocal determination of the electrophysiological mechanisms responsible for most clinically occurring arrhythmias and certainly do not allow convincing differentiation of the ionic mechanisms responsible for a particular arrhythmia. No electrocardiographic or electrophysiological criteria exist that unquestionably separate reentry from automaticity clinically nor are there tools, such as pacemakers, drugs, cardioverters, or surgery, that exert an effect specific enough to separate the two possibilities. At best, one can postulate only that a particular arrhythmia is "most consistent with" or "best explained by" one or the other electrophysiological mechanism. In the final analysis, *all* current clinical criteria established to support reentry can be mimicked by pacemaker activity and vice versa. Furthermore, some tachyarrhythmias may be started by one mechanism and perpetuated by another. For example, premature ventricular depolarization due to abnormal automaticity may precipitate a ventricular tachycardia sustained by reentry. Given these comments, mechanisms demonstrated to cause abnormal electrical activity in a variety of clinical situations will be considered.

4.1. Disorders of Impulse Formation

This category is defined as inappropriate discharge rate of the normal pacemaker, the sinus node (e.g., sinus rates too fast or too slow for the physiological needs of the patient), or discharge from an ectopic pacemaker that controls the atrial or ventricular rhythm for one complex or more. It is important to recall that such disorders of impulse formation can be due to a speeding or slowing of a *normal* pacemaker mechanism (e.g., phase 4 diastolic depolarization that is ionically normal for the sinus node or for an ectopic site such as a Purkinje fiber but occurs inappropriately fast or slow) or due to an ionically *abnormal* pacemaker mechanism. The patient with persistent sinus tachycardia at rest or sinus bradycardia during exertion exhibits inappropriate sinus nodal discharge rates, but the ionic mechanisms responsible for sinus nodal discharge still may be normal. Conversely, when a patient experiences ventricular tachycardia during an acute myocardial infarction, abnormal ionic mechanisms probably are operative to generate this tachycardia. Although pacemaker activity generally is not found in ordinary working myocardium, myocardial ischemia conceivably can bestow abnormal pacemaker properties on cells such as ventricular muscle fibers, permitting them to depolarize automatically. Based on the rate response to catecholamine administration of isolated fibers exhibiting normal phase 4 diastolic depolarization and on *in vivo* studies during stellate ganglion stimulation, it is likely that rates much in excess of 200 bpm are not due to enhanced normal automaticity.[45,81,82] Rhythms due to automaticity may be slow atrial, junctional, and ventricular escape rhythms, certain types of atrial tachycardias (such as those produced by digitalis), accelerated junctional (nonparoxysmal junc-

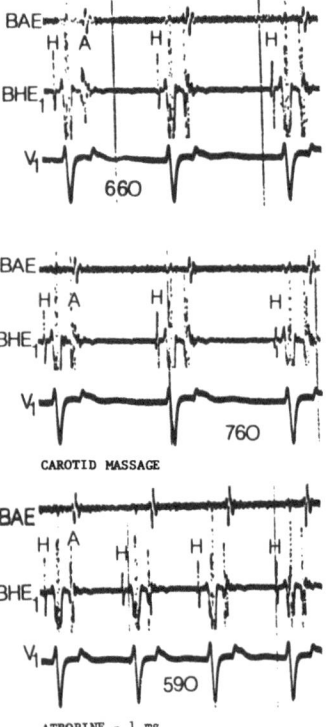

Fig. 10. Nonparoxysmal AV junctional tachycardia. (A)–(C) illustrate control, response to carotid sinus massage, and response to atropine (1 mg intravenously). Note that His bundle depolarization is the earliest recordable electrical activity in each cycle. The atria are depolarized retrogradely (low right atrial activity recorded in BHE precedes high right atrial activity recorded in BAE). Note also that carotid sinus massage slows the junctional discharge rate while atropine speeds it. From these tracings alone one could not distinguish the rhythm from other types of supraventricular tachycardias (for example, see Fig. 15). However, the onset and termination of this tachycardia were typical of nonparoxysmal AV junctional tachycardia. Reproduced with permission from Zipes.[80]

tional tachycardia) (Fig. 10), and idioventricular rhythms and parasystole.

Electrophysiological concepts about parasystole have been drastically revised in the last several years.[68,69,83] Classically, parasystole has been considered similar to a fixed-rate, asynchronously discharging pacemaker: its timing is not altered by the dominant rhythm, it produces depolarization when the myocardium is excitable, and the intervals between discharges are multiples of a basic interval (Fig. 11). Complete entrance block, constant or intermittent,[84] insulates and protects the parasystolic focus from surrounding electrical events and accounts

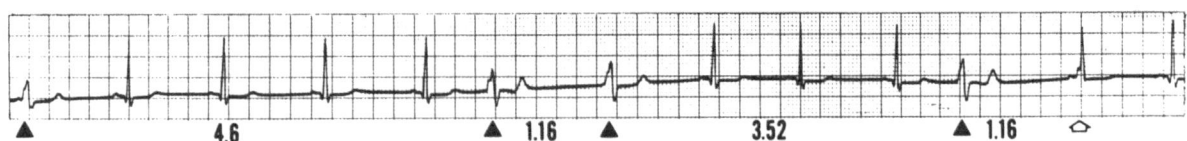

Fig. 11. Ventricular parasystole. Parasystolic discharge of ventricular origin (solid triangles) occurs at a constant interectopic interval of 1.16 sec or multiple thereof. Variable coupling interval between the normally conducted QRS complexes and parasystolic complexes along with fusion beats (open arrow) are also present. Reproduced with permission from Zepes.[158]

for such activity. Occasionally, the focus may exhibit exit block, during which it may fail to depolarize excitable myocardium.[85] As indicated earlier, data from recent experiments have shown that, in fact, the dominant cardiac rhythm may modulate parasystolic discharge to speed up or slow down its rate via electrotonic interactions with the dominant rhythm across an area of depressed excitability. Complex interactions of complete silence, concealed or manifest bigeminy, trigeminy, quadrigeminy, and periods of more complex group beating may occur owing to the entraining effects of the dominant rhythm on the ectopic focus. Exactly how these observations will alter the "rules" established to diagnose parasystole electrocardiographically is still being established. However, there appear to be clinical[86] examples to support these experimental observations.

4.2. Disorders of Impulse Conduction

Reentry probably is the cause of many tachyarrhythmias, including various kinds of supraventricular and ventricular tachycardias, flutter, and fibrillation. It is important to remember, however, that initiation or termination of tachycardia by pacing stimuli, the demonstration of electrical activity bridging diastole, fixed coupling, and a variety of other clinically used techniques, while consistent with reentry, do not constitute proof of its existence.

4.2.1. Atrial Flutter and Fibrillation

Mapping studies[87,88] in animals and mapping and stimulation studies in man[89,90] provide some evidence to suggest that many forms of atrial flutter are due to reentry.

Much indirect evidence supports reentry as a cause of fibrillation[91,92] but this is difficult to prove beyond question. Several studies have established that a critical mass of myocardium is required to maintain fibrillation,[93–96] lending support to Moe's hypothesis that multiple wavelets of reentry, influenced by the mass of the tissue, refractory periods, and conduction velocity, maintain fibrillation.[97]

4.2.2. Sinus and Atrial Reentry

Reentry in parts of the atrium has been reported to occur in several experimental models as well as in humans. The sinus node shares with the AV node electrophysiological features such as the potential for dissociation of conduction, i.e., an impulse can be made to conduct in some nodal fibers but not in others.[98] For example, premature stimulation can produce slow propagation in the sinus node, block in some areas, and the development of repetitive responses most likely due to reentry. Evidence from a number of studies in man[99,100] and animals[98,101] supports the concept that sustained reentry in the sinus node can occur and cause supraventricular tachycardia (Fig. 12). Allessie and Bonke[101] have shown that the reentrant circuit in the isolated rabbit preparation was located entirely within the sinus node and that the atrium was not an essential link. Reentry within the atrium, unrelated to the sinus node, has been shown experimentally to occur in the rabbit[61,102] and dog atrium,[103] with or without an anatomical obstacle, and may be a cause of supraventricular tachycardia in man. Examples of supraventricular tachycardia purported to be due to atrial reentry have been cited,[99,104,105] but their relative infrequency in the published literature suggests that this is not a commonly recog-

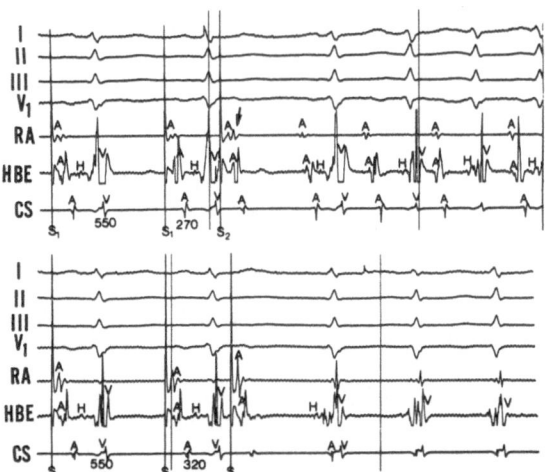

Fig. 12. Sinoatrial nodal reentry. (A) Premature stimulation of the high right atrium at an S_1–S_2 interval of 270 msec initiates an atrial tachycardia with an activation sequence similar to that occurring during high right atrial pacing. The premature P wave blocks proximal (arrow) to the His bundle but the tachycardia is still initiated. In (B), premature stimulation of the high right atrium at an S_1–S_2 interval of 320 msec results in a prolonged AH interval and initiation of AV nodal reentry. Retrograde low right atrial activation recorded in the HBE lead occurs before ventricular activation and is followed by left atrial (CS) and high right atrial activation (arrow). This is in sharp contrast to the top panel in which the atrial activation sequence begins in the high right atrium, then the low right atrium (HBE), and finally the left atrium (CS). Reproduced with permission from Zipes.[80]

nized cause of supraventricular tachycardia in man. Distinguishing atrial tachycardia due to automaticity from atrial tachycardia sustained by reentry over very small areas, i.e., microreentry, is very difficult, and therefore conclusions regarding clinical electrophysiological mechanisms of this supraventricular tachycardia based on currently available information must be accepted cautiously.

4.2.3. AV Nodal Reentry

It has been known for over 75 years that stimulation of the ventricle could produce responses that propagated to the atrium and then back to the ventricles,[106] sometimes being sustained to cause a reciprocating rhythm.[107,108] Scherf and Shookoff[109] ascribed this phenomenon to "longitudinal functional dissociation" in the AV junction, suggesting that some fibers of the AV conduction system had shorter refractory periods and were able to conduct in one direction whereas conduction in the other fibers with longer refractory periods was blocked. The impulse was then able to reexcite the previously refractory fibers. Mendez et al.[110] demonstrated that an impulse traveling from ventricle to atrium, if timed properly, did not prevent another impulse traveling simultaneously from atrium to ventricle from reaching the His bundle. For this to occur, the impulses traveling in opposite directions must have been conducting in different AV nodal pathways.

Microelectrode studies on isolated rabbit AV nodal preparations provide further evidence to support these concepts of longitudinal AV nodal dissociation and reentry.[111] Cells in the upper portion of the AV node can be dissociated during propaga-

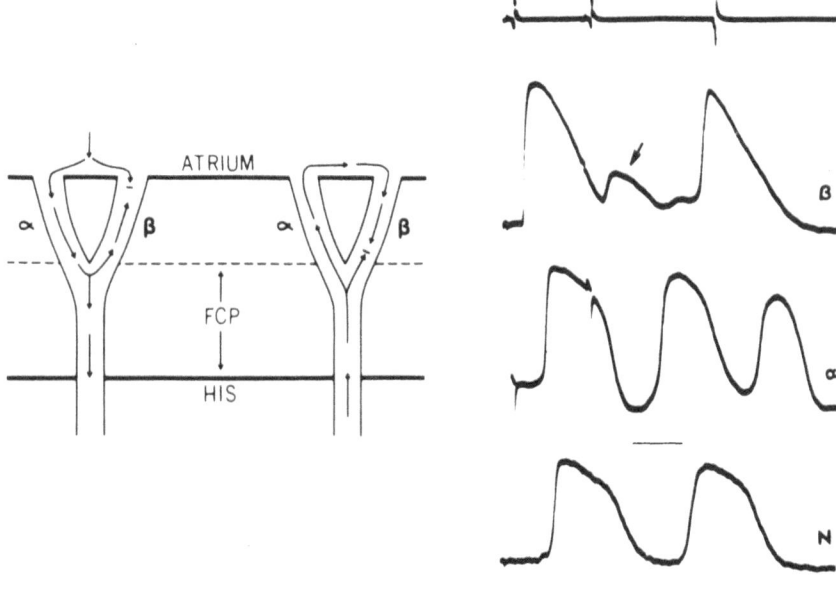

Fig. 13. Atrial echoes. (Left) Schematic representation of intranodal dissociation responsible for an atrial echo. A premature atrial response fails to penetrate the β pathway, which exhibits unidirectional block but propagates anterogradely through the α pathway. Once the final common pathway (FCP) is engaged, the impulse may return to the atrium via the now-recovered β pathway to produce an atrial echo. The neighboring diagram illustrates the pattern of propagation during generation of a ventricular echo. A premature response in the His bundle traverses the final common pathway, encounters a refractory β pathway, and returns through a now-recovered β pathway to produce a ventricular echo. (Right) Actual recordings from the atrium (top tracing), cells impaled in the β region (second tracing), α region (third tracing), and N portion of the AV node (bottom tracing) in an isolated rabbit preparation. The basic response to A_1 activated both α and β pathways and the N cell (first tier of action potentials). The premature atrial response, A_2, caused only a local response in the β cell (arrow), was delayed in transmission to the α cell, and was further delayed in propagation to the N cell. Following the α response, a retrograde spontaneous response occurred in the β cell and propagated to the atrium (E). This atrial response represents an atrial echo. The echo returned to stimulate the α cell but was not propagated to the N cell. Reproduced with permission from Mendez and Moe.[113]

tion of premature stimuli, so that one group of cells, called α, can discharge in response to a premature stimulus at a time when another group of cells, called β, fails to discharge.[112,113] The impulse then can propagate to the mid and lower portions of the AV node and turn around (without needing to activate the His bundle) to reexcite the β group of cells and produce an atrial echo (Fig. 13). Experiments[114] using multiple simultaneous microelectrode recordings and episodes of sustained AV nodal tachycardia in isolated rabbit atria[115] largely confirmed these observations.

One of the first studies employing premature electrical stimulation to initiate and terminate a paroxysmal supraventricular tachycardia *in situ* was performed in a dog.[116] Subsequent investigations in man support the concepts[117,118] that reentry, initiated by a critical degree of AV nodal conduction delay,[119,120] can be localized to the AV node, possibly occurring over functionally distinct dual AV nodal pathways[121] (Fig. 14). The premature atrial response can block anterogradely in one AV nodal pathway that conducts more rapidly (fast pathway, or β pathway) but has a longer refractory period than a second pathway (slow pathway, or α pathway). The premature atrial response travels to the ventricle over the slow (α) pathway and back to the atrium over the fast (β) pathway[122] (Fig. 15). Less commonly, the premature atrial response can block in the slow pathway and travel in the fast pathway anterogradely, using the slow pathway retrogradely.[123] A plot of the A_1–A_2 interval versus the A_2–H_2 or H_1–H_2 interval shows a "break" in the curve when the A_2–H_2

interval suddenly prolongs, presumably as conduction now travels over the slow pathway. Tachycardia usually begins at that point (Fig. 16).

In patients who have dual AV nodal physiology, both pathways exhibit electrophysiological responses in the anterograde direction characteristic of AV nodal fibers. However, in some patients, the electrophysiological features of the retrogradely conducting pathway differ greatly from those of the anterogradely conducting pathway, in response not only to atrial or ventricular pacing but also to various drugs.[105,122,124–126] Thus, it is not clear whether the retrogradely conducting pathway is extranodal,[127] is intranodal but insulated from the rest of the AV node, or includes only a small amount of AV nodal tissue.

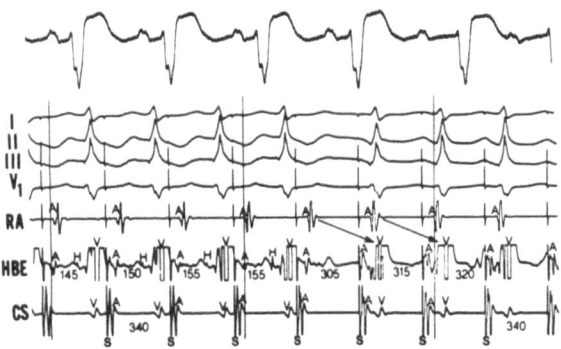

Fig. 14. Two sets of PR intervals. In the top panel, the PR interval during spontaneous sinus rhythm lengthens suddenly from 240 msec to 400 msec with minimal change in the PP interval. In the bottom panel, during coronary sinus pacing at a constant cycle length of 340 msec, the AH interval suddenly increases from 155 to 305 msec, consistent with anterograde block in the fast nodal pathway. Subsequently, each paced P wave precedes the next QRS complex slightly but conducts with a long AH interval to the following QRS complex (arrows). Reproduced with permission from Gilmour and Zipes.[159]

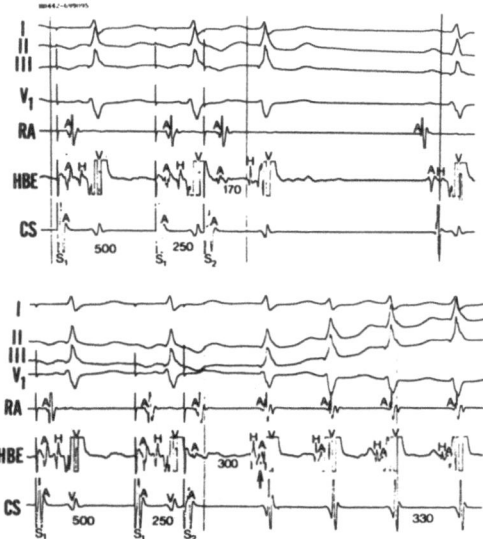

Fig. 15. Initiation of AV nodal reentrant tachycardia in a patient with dual AV nodal pathways. Upper and lower panels show the last two paced beats of a train of stimuli delivered to the coronary sinus at a pacing cycle length of 500 msec. The results of premature atrial stimulation at an S_1–S_2 interval of 250 msec on two occasions are shown. In the upper panel, S_2 was conducted to the ventricle with an AH interval of 170 msec and then was followed by a sinus beat. In the lower panel, S_2 was conducted with an AH interval of 300 msec and initiated AV nodal reentry. Note that the retrograde atrial activity occurs (arrow) prior to the onset of ventricular septal depolarization and is superimposed with the QRS complex. Retrograde atrial activity begins first in the low right atrium (HBE lead) and then progresses to the high right atrium (RA) and coronary sinus (CS) recordings. Reproduced with permission from Zipes.[80]

4.2.4. Preexcitation Syndrome

The accumulated anatomical and electrophysiological data support a reentrant mechanism to explain tachycardias related to an accessory pathway more than they support that mechanism for any other kind of tachycardia.[128–131] Electrophysiological studies have demonstrated that, in most patients who have reciprocating tachycardia associated with the Wolff–Parkinson–White syndrome, the accessory pathway conducts more rapidly than does the normal AV node but takes a longer time to recover excitability, i.e., the anterograde refractory period of the accessory pathway exceeds that of the AV node.[129,132] Consequently, a premature atrial complex that occurs sufficiently early blocks anterogradely in the accessory pathway and continues to the ventricle over the normal AV node and His bundle. After the ventricles have been excited, the impulse is able to enter the accessory pathway retrogradely and return to the atrium. A continuous conduction loop of this kind establishes the circuit for the tachycardia. Although exceptions occur, the usual activation wave during such a reciprocating tachycardia in a patient with an accessory pathway occurs in this fashion: anterogradely over the normal AV node–His–Purkinje system and retrogradely over the accessory pathway, resulting in a normal QRS complex (Fig. 17). In some patients, the accessory pathway may be capable only of retrograde conduction,[130,133,134] but the circuit and mechanism of tachycardia remain the same. Less commonly, the accessory pathway may conduct only anterogradely.[130,135]

In addition to the response of the reciprocating tachycardia to premature stimulation, one of the strongest lines of evidence supporting a reentrant mechanism is that interruption of the presumed reentrant loop at widely separated points (by surgically

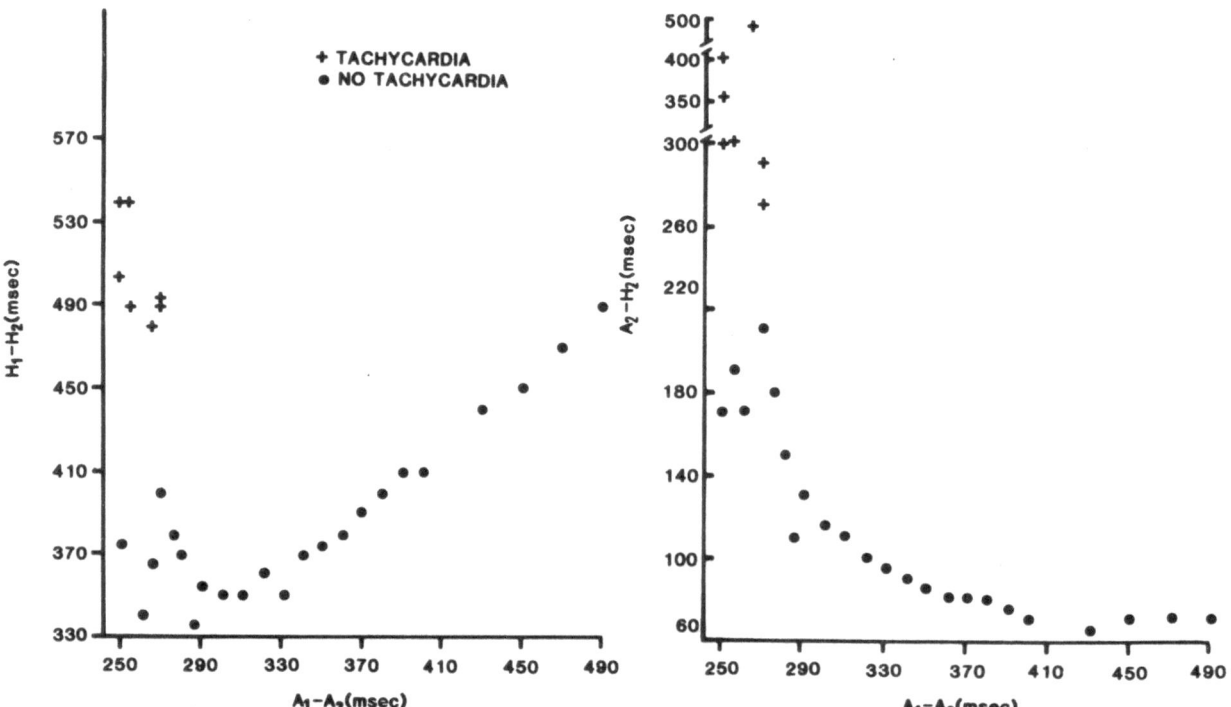

Fig. 16. Discontinuous AV nodal curves. At a critical A_1–A_2 interval the H_1–H_2 interval and the A_2–H_2 intervals increase markedly. At the break in the curves, AV nodal reentrant tachycardia is initiated. Reproduced with permission from Zipes.[80]

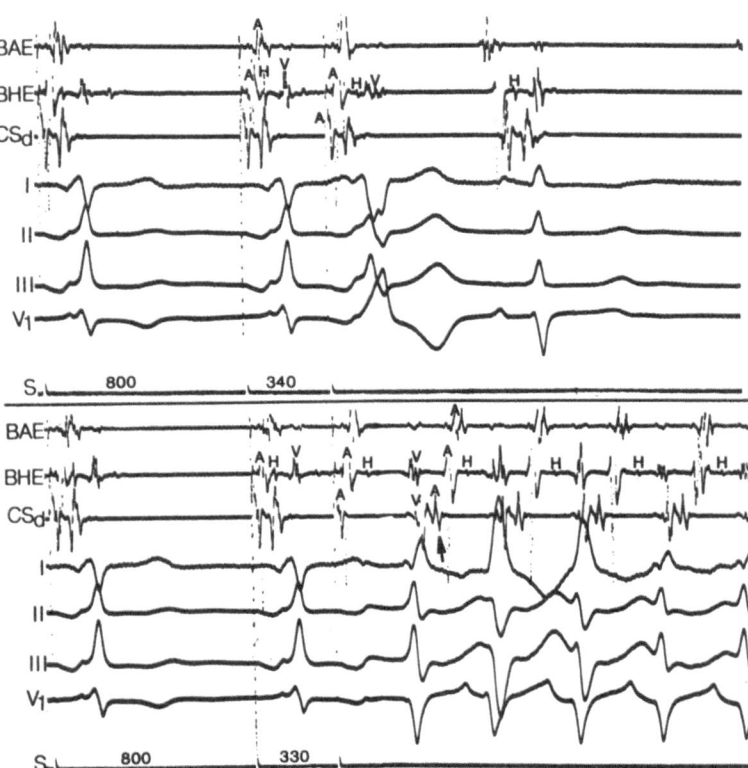

Fig. 17. Initiation of reciprocating tachycardia in Wolff–Parkinson–White syndrome. The last two beats of a regular train at a cycle length of 800 msec and the premature stimulus (stimuli) are shown. In the top recording, premature left atrial stimulation at a cycle length of 340 msec was followed by conduction over the accessory pathway to the ventricle and full preexcitation. In the bottom panel, shortening the premature interval by 10 msec resulted in anterograde block over the accessory pathway, conduction over the normal AV node–His bundle route, and loss of ventricular preexcitation (slight functional aberration and HV prolongation occur). Initiation of a reciprocating tachycardia follows. Note that during reciprocating tachycardia, atrial activation is recorded earliest in the coronary sinus lead, followed by low and high right atrial activation, and is consistent with a left-sided accessory pathway (arrow). Reproduced with permission from Zipes.[80]

cutting the normal AV node–His bundle pathway *or* the accessory pathway) eliminates the ability to develop supraventricular tachycardia.

Other pathways may constitute the circuit for reciprocating tachycardias in patients who have some form of the Wolff–Parkinson–White syndrome. Conduction may proceed anterogradely over the accessory pathway and retrogradely over the AV node–His bundle, so-called "antidromic tachycardia." Two accessory pathways may form the circuit. In the Lown–Ganong–Levine syndrome (short PR interval and normal QRS complex),[136] conduction over a James fiber[137] that connects the atrium to the distal portion of the AV node and His bundle, has been a postulated pathway although the presence of this entity as a distinct syndrome is unlikely.[138]

For some patients, electrophysiological findings may be consistent with communications between the AV node and the ventricle (nodoventricular) or the His bundle–bundle branches and the ventricle (fasciculoventricular).[139] Tachycardia in patients with nodoventricular fibers may be due to reentry using these fibers as the anterograde pathway and the His–Purkinje fibers and a portion of the AV node retrogradely. No direct relationship may exist between fasciculoventricular fibers and tachycardia.

4.2.5. Ventricular Reentry

Reentry within ventricular muscle with or without participation of Purkinje fibers probably is responsible for many

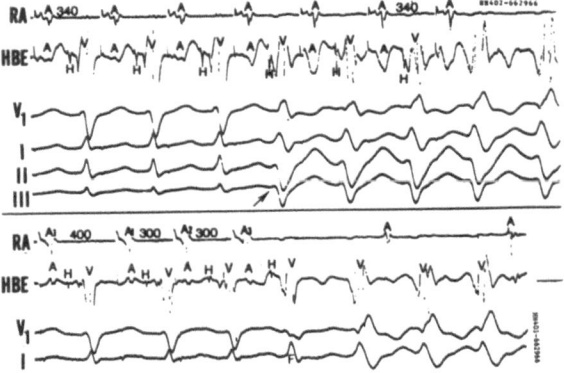

Fig. 18. During continuous right atrial pacing at a cycle length of 340 msec (top panel), a ventricular tachycardia with right bundle branch block and left axis deviation emerges (arrow) at a cycle length of 335 msec. Note the shortening of the HV interval during ventricular tachycardia. His bundle activation occurs after the onset of the QRS complex and finally can no longer be seen in the last two QRS complexes. Premature atrial stimulation also induces ventricular tachycardia in this patient (bottom panel). The last cycle of a basic train of eight stimuli (A_1–A_1 cycle length 400 msec) is displayed. Two premature atrial complexes (A_2, A_3) are introduced at cycle lengths of 300 msec. The first premature atrial complex (A_2) conducts normally. However, A_3 conducts with a fusion QRS complex (F), indicating that the ventricular tachycardia begins after A_2. Reproduced with permission from Zipes *et al.*[31]

BCL 900 ; S₁ S₂ 391

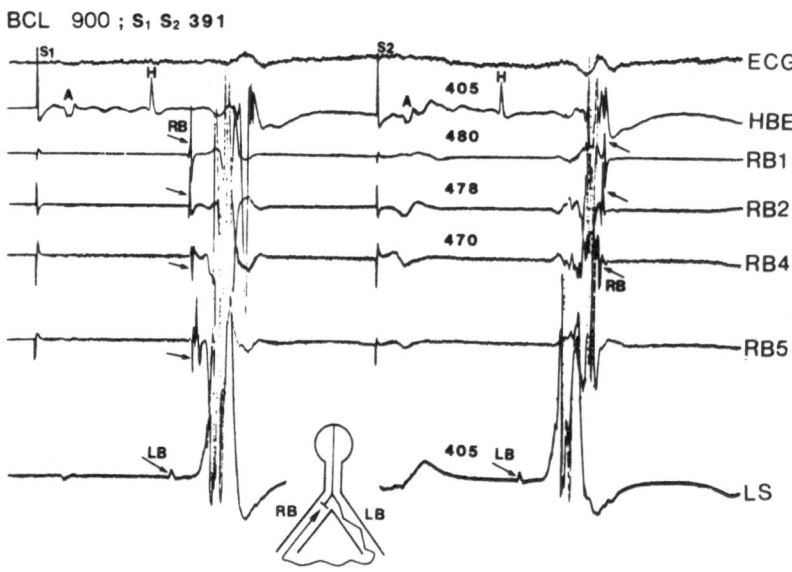

Fig. 19. Reentry in the bundle branches of the dog heart. Recordings were made with an electrode in the His bundle area (HBE), a multipolar plaque electrode sewn on the right bundle branch portraying activation from base to apex direction in electrograms RB_1 to RB_5 and from the left bundle branch (LB) in the left septal lead (LS). During the last basic cycle, S_1 applied to the atrium causes conduction to proceed from the His bundle through the left and right bundle branches to the ventricle. Note orderly progression of conduction from RB_1 to RB_5. At a premature right atrial interval of 391 msec, conduction is blocked distal to His, travels to the ventricle over the left bundle branch, and then activates the right bundle branch retrogradely from RB_4 to RB_2 to RB_1. Retrograde right bundle branch deflection in RB_5 cannot be seen. Inset shows activation sequence in bundle branches following S_2. Reproduced with permission from Glassman and Zipes.[142]

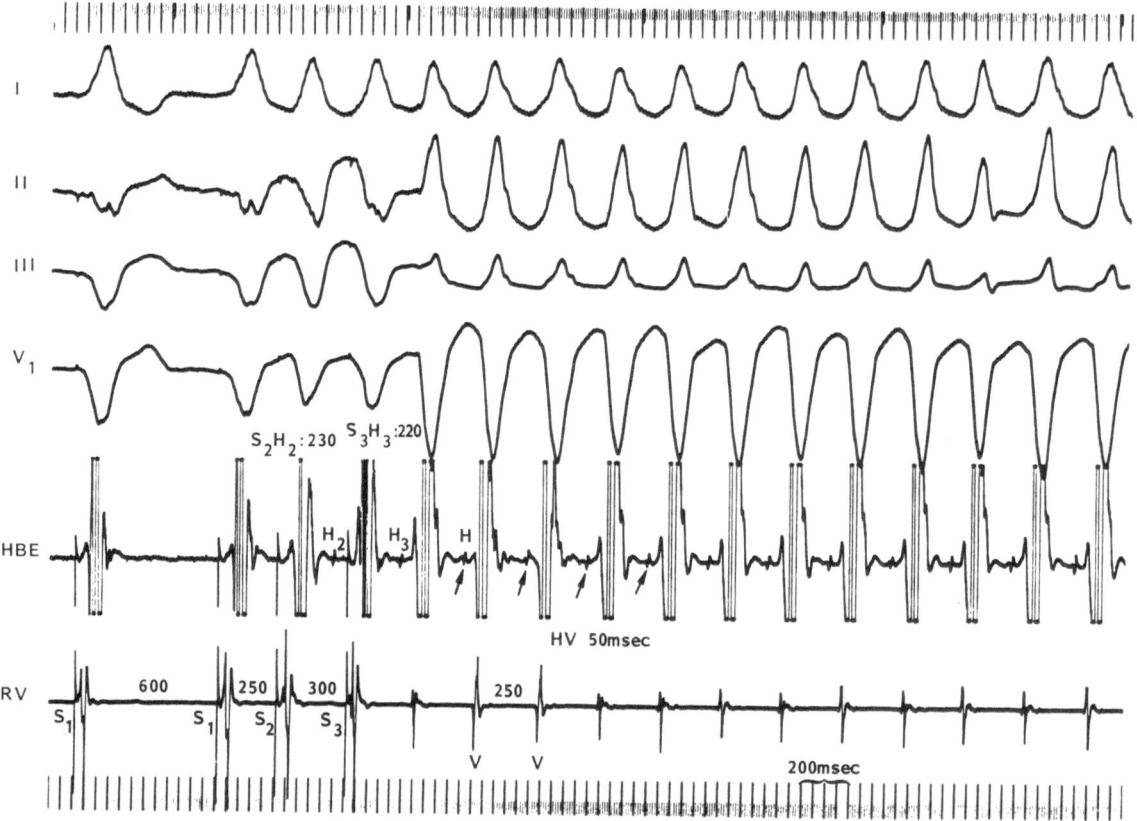

Fig. 20. Bundle branch reentrant ventricular tachycardia. The last two stimuli (S_1–S_1) of a train of eight are shown, along with a premature stimulus (S_2) delivered at 250 msec and another premature stimulus (S_3) delivered at an interval of 300 msec. After the S_3 stimulus, the retrograde His (H_3) delays to 220 msec and is followed by a ventricular tachycardia (cycle length 250 msec) in which each ventricular depolarization is preceded by His bundle deflection at an HV interval of 50 msec, comparable to that in sinus rhythm. The patient has atrial fibrillation (not well seen) effectively eliminating participation of the atrium in this tachycardia. Reproduced with permission from Lloyd *et al*.[146]

ventricular tachycardias (Fig. 18). In addition, reentry has been demonstrated to occur over the bundle branches in the *in situ* dog heart[140–142] (Fig. 19) and probably in man as well[130,143] (Fig. 20). However, bundle branch reentry does not appear to be a common cause of sustained ventricular tachycardia in the dog or in man.[31,144–146] Reentry within ventricular muscle is difficult to prove because the large muscle mass makes complete mapping of the reentrant loop and interruption of the pathway difficult. Fontaine *et al.*[147] have shown that an appropriately placed ventriculotomy incision interrupted sustained ventricular tachycardia in man. In general, experimental studies provide only supportive evidence of reentry.[148]

Ischemia, both acute and chronic, has been a favorite model to investigate because it generates fragmented, delayed activation that may be conducive to the generation of reentrant excitation.[149–151] However, even the demonstration of continuous electrical activity spanning diastole that is linked temporally and perhaps causally to the generation of sustained ventricular arrhythmia[145,152,153] constitutes only circumstantial proof. For example, such continuous electrical activity could be produced by oscillatory pacemaker activity conducting with delay to the surrounding myocardium and could be unrelated to reentry. Several recent animal studies using an extensive array of extracellular electrodes[154–157] have provided important information concerning the sequence of activation during ventricular tachyarrhythmias, strongly suggesting in some of the studies that reentry is the responsible mechanism.

5. Summary

Cardiac arrhythmias occur due to abnormalities of impulse formation, impulse propagation, or a combination of the two disorders. Abnormal impulse formation may involve accelerated or decelerated normal phase 4 depolarization in the SA and AV nodes and in the His–Purkinje system, abnormal automatic mechanisms, such as slow response automaticity in depolarized Purkinje fibers and atrial and ventricular myocardium, and triggered sustained rhythmic activity consequent to early or delayed afterdepolarizations.

Clinically, altered normal automaticity can be responsible for sinus bradycardia or tachycardia and slow junctional and ventricular escape rhythms. Altered normal automaticity in Purkinje fibers that survive myocardial infarction or abnormal automaticity in diseased Purkinje fibers and atrial or ventricular myocardium may generate idioventricular rhythms, parasystole, and atrial or ventricular tachyarrhythmias. Potentially, triggered activity could cause atrial and ventricular tachyarrhythmias, but the signficance and incidence of triggered arrhythmias remain to be determined.

Abnormal impulse propagation results from depressed active or passive membrane properties. Reduced action potential amplitude, upstroke velocity, excitability, and cell-to-cell coupling can produce slow conduction, block, and reentrant excitation. Reentry involves unidirectional block and slow conduction around a fixed anatomical or electrophysiological obstacle, a vortex of conduction around a central area of unexcited myocardium ("leading circle"), or electrotonic current flow to and fro across an inexcitable gap (reflection).

Slow conduction and block can permit latent pacemaker escape and precipitate various reentrant tachyarrhythmias. There is good evidence to suggest that dual AV nodal pathways and accessory pathways in patients with Wolff–Parkinson-White syndrome sustain reentrant excitation and that reentry can occur within the atrium and ventricle. Bundle branch and sinus node reentry appear to be less common clinically.

Abnormal impulse formation and propagation interact with one another to alter the conduction velocity of diastolic impulses initiated in a spontaneously depolarizing region of myocardium and to accelerate, decelerate, and entrain parasystolic foci and annihilate spontaneous activity. In addition, microreentry and spontaneous complexes that initiate an automatic or reentrant tachycardia may be induced by injury currents.

The clinical consequences of stimuli delivered to depolarized foci during early, mid- and late diastole may be acceleration and deceleration dependent bundle branch block, and supernormal conduction. Parasystolic rhythms may be modulated by changes in the sinus rate and terminated by a single subthreshold stimulus.

As new mechanisms for the development of cardiac arrhythmias are identified, the tests used to identify previously described mechanisms may need to be revised. For example, initiation and termination of a presumed reentrant tachycardia by a single premature stimulus does not exclude triggered activity or initiation and annihilation of pacemaker activity as possible mechanisms. Therefore it is still not possible, given present discriminating techniques, to be unequivocably certain about the mechanisms responsible for most clinically-occurring cardiac arrhythmias.

ACKNOWLEDGMENTS. Supported in part by the Herman C. Krannert Fund, Indianapolis, Indiana; by Grants HL-06308, HL-07182, and Young Investigator Award HL-24851, all from the National Heart, Lung and Blood Institute of the National Institutes of Health, Bethesda, Maryland; by the Attorney General of Indiana Public Health Trust and by the Roudebush Veterans Administration Medical Center, Indianapolis, Indiana; and by a Grant-In-Aid from the American Heart Association, Indiana Affiliate, Inc., Indianapolis, Indiana.

References

1. Vassalle, M. 1977. The relationship among cardiac pacemakers: Overdrive suppression. *Circ. Res.* **41**:269–277.
2. Vassalle, M. 1982. Cardiac automaticity and its control. In: *Excitation and Neural Control of the Heart*. M. L. Levy and M. Vassalle, eds. American Physiological Society, Washington, D.C. pp. 59–77.
3. Friedman, P. L., J. F. Stewart, J. J. Fenoglio, and A. L. Wit. 1973. Survival of subendocardial Purkinje fibers after extensive myocardial infarction in dogs: *In vitro* and *in vivo* correlations. *Circ. Res.* **33**:597–611.
4. Lazzara, R., N. El-Sherif, and B. J. Scherlag. 1973. Electrophysiological properties of Purkinje cells in one day old myocardial infarction. *Circ. Res.* **33**:722–734.
5. Horowitz, L. N., J. F. Spear, and E. N. Moore. 1975. Subendocardial origin of ventricular arrhythmias in 24-hour old experimental myocardial infarction. *Circulation* **53**:56–63.
6. Brown, H. F. 1982. Electrophysiology of the sinoatrial node. *Physiol. Rev.* **62**:505–530.
7. DiFrancesco, D. 1981. A new interpretation of the pacemaker current in calf Purkinje fibers. *J. Physiol. (London)* **314**:359–376.
8. Zipes, D. P., and C. Mendez. 1973. Action of manganese ions and tetrodotoxin on atrioventricular nodal transmembrane potentials in isolated rabbit hearts. *Circ. Res.* **32**:447–454.
9. Zipes, D. P., and J. C. Fischer. 1974. Effects of agents which inhibit the slow channel on sinus node automaticity and atrioventricular conduction in the dog. *Circ. Res.* **34**:184–192.

10. Weidmann, S. 1955. Effects of calcium ions and local anesthetics on electrical properties of Purkinje fibres. *J. Physiol. (London)* **129**:568–582.

11. Dudel, J., K. Peper, R. Rudel, and W. Trautwein. 1967. The effect of tetrodotoxin on the membrane current in cardiac muscle (Purkinje fibers). *Pfluegers Arch.* **295**:213–226.

12. Hoffman, B. F., and P. F. Cranefield. 1960. *Electrophysiology of the Heart*. McGraw–Hill, New York.

13. Katzung, B. G. 1975. Effects of extracellular calcium and sodium on depolarization-induced automaticity in guinea pig papillary muscle. *Circ. Res.* **37**:118–127.

14. Surawicz, B., and S. Imanishi. 1976. Automatic activity in depolarized guinea pig ventricular myocardium: Characteristics and mechanisms. *Circ. Res.* **39**:751–759.

15. Elharrar, V., and D. P. Zipes. 1980. Voltage modulation of automaticity in cardiac Purkinje fibers. In: *The Slow Inward Current and Cardiac Arrhythmias*. D. P. Zipes, J. C. Bailey, and V. Elharrar, eds. Martinus Nijhoff, The Hague. pp. 357–373.

16. Surawicz, B. 1980. Depolarization-induced automaticity in atrial and ventricular myocardial fibers. In: *The Slow Inward Current and Cardiac Arrhythmias*. D. P. Zipes, J. C. Bailey, and V. Elharrar, eds. Martinus Nijhoff, The Hague. pp. 375–396.

17. Antoni, H., and I. Oberdisse. 1965. Elektrophysiologische Untersuchungen uber die Barium-induzierte Schrittnache-Aktivitat um isolierten Saugetiermyocard. *Pfluegers Arch.* **284**:259–272.

18. Hordof, A. J., R. Edie, J. R. Malm, B. F. Hoffman, and M. R. Rosen. 1976. Electrophysiologic properties and response to pharmacologic agents of fibers from diseased human atria. *Circulation* **54**:774–779.

19. Singer, D. H., C. M. Baumgarten, and R. E. Jeneick. 1981. Cellular electrophysiology of cellular and other dysrhythmias: Studies on diseased and ischemic heart. *Prog. Cardiovasc. Dis.* **24**:97–156.

20. Gilmour, R. F., Jr., J. J. Heger, E. N. Prystowsky, and D. P. Zipes. 1983. Cellular electrophysiological abnormalities of diseased human ventricular myocardium. *Am. J. Cardiol.* **51**:137–144.

21. Gilmour, R. F., Jr., and D. P. Zipes. 1980. Electrophysiological characteristics of rodent myocardium damaged by adrenaline. *Cardiovasc. Res.* **14**:582–589.

22. Cranefield, P. F. 1977. Action potentials, afterpotentials and arrhythmias. *Circ. Res.* **41**:415–423.

23. Gilmour, R. F., Jr., and D. P. Zipes. 1982. Electrophysiological response of vascularized hamster cardiac transplants to ischemia. *Circ. Res.* **50**:599–614.

24. Saito, T., M. Otoguro, and T. Matsubara. 1978. Electrophysiological studies of the mechanism of electrically induced sustained rhythmic activity in the rabbit right atrium. *Circ. Res.* **42**:199–206.

25. Wit, A. L., and P. F. Cranefield. 1976. Triggered activity in cardiac muscle fibers of the simian mitral valve. *Circ. Res.* **38**:85–98.

26. Ferrier, G. R. 1977. Digitalis arrhythmias: Role of oscillatory afterpotentials. *Prog. Cardiovasc. Dis.* **19**:459–490.

27. Rosen, M. R., and P. Danilo, Jr. 1980. Effects of tetrodotoxin, lidocaine, verapamil and AHR-2666 on ouabain-induced delayed afterdepolarizations in canine Purkinje fibers. *Circ. Res.* **46**:117–124.

28. Lederer, W. J., and R. W. Tsien. 1976. Transient inward current underlying arrhythmogenic effects of cardiotonic steroids in Purkinje fibres. *J. Physiol. (London)* **263**:73–100.

29. Vassalle, M., and A. Mugelli. 1981. An oscillatory current in sheep cardiac Purkinje fibers. *Circ. Res.* **48**:618–631.

30. Mary-Rabine, L., A. J. Hordof, P. Danilo, Jr., J. R. Malm, and M. R. Rosen. 1980. Mechanisms for impulse initiation in isolated human atrial fibers. *Circ. Res.* **47**:267–277.

31. Zipes, D. P., P. R. Foster, P. J. Troup, and D. H. Pedersen. 1979. Atrial induction of ventricular tachycardia: Reentry versus triggered automaticity. *Am. J. Cardiol.* **44**:1–8.

32. Rosen, M. R., C. Fisch, B. F. Hoffman, P. Danilo, Jr., D. E. Lovelace, and S. B. Knoebel. 1980. Can accelerated atrioventricular junction escape rhythms be explained by delayed afterdepolarizations? *Am. J. Cardiol.* **45**:1272–1284.

33. Weidmann, S. 1955. The effect of the cardiac membrane potential on the rapid availability of the sodium carrying system. *J. Physiol. (London)* **127**:213–224.

34. Noble, D. 1979. *The Initiation of the Heartbeat*, 2nd ed. Oxford University Press, London.

35. Fozzard, H. A. 1977. Cardiac muscle: Excitability and passive electrical properties. *Prog. Cardiovasc. Dis.* **19**:343–359.

36. Zipes, D. P., C. Mendez, and G. K. Moe. 1983. Some examples of Wenckebach periodicity in cardiac tissues, with an appraisal of mechanisms. In: *Frontiers of Cardiac Electrophysiology*. M. Rosenbaum and M. V. Elizari, eds. Nijhoff, The Hague. pp. 357–375.

37. Downar, E., M. J. Janse, and D. Durrer. 1977. The effect of "ischemic" blood on transmembrane potentials of normal porcine ventricular myocardium. *Circulation* **55**:455–462.

38. Hill, J. L., and L. S. Gettes. 1980. Effect of acute coronary artery occlusion on extracellular K$^+$ activity in swine. *Circulation* **61**:678–778.

39. Neeley, J. R., J. T. Whitmer, and M. J. Rovetto. 1975. Effect of coronary blood flow on glycolytic flux and intracellular pH in isolated rat hearts. *Circ. Res.* **37**:733–741.

40. Steenbergen, C., G. Deleeuw, T. Rich, and J. R. Williamson. 1977. Effects of acidosis and ischemia on contractility and intracellular pH of rat heart. *Circ. Res.* **41**:849–858.

41. Poole-Wilson, P. A. 1978. Measurement of intracellular pH in pathological states. *J. Mol. Cell. Cardiol.* **10**:511–526.

42. Corr, P. B., M. E. Cain, F. X. Witkowski, D. A. Price, and B. E. Sobel. 1979. Potential arrhythmogenic electrophysiological derangements in canine Purkinje fibers induced by lysophosphoglycerides. *Circ. Res.* **44**:822–832.

43. Downar, E., M. J. Janse, and D. Durrer. 1977. The effect of acute coronary artery occlusion on subepicardial transmembrane potentials in the intact porcine heart. *Circulation* **56**:217–224.

44. Elharrar, V., P. R. Foster, T. L. Jirak, W. E. Gaum, and D. P. Zipes. 1977. Alterations in canine myocardial excitability during ischemia. *Circ. Res.* **40**:98–105.

45. Cranefield, P. F. 1975. *The Conduction of the Cardiac Impulse: The Slow Response and Cardiac Arrhythmias*. Futura Publishers, Mt. Kisco, N.Y.

46. Ceremuzynski, L., J. Staszewska-Barczak, and K. Herbaczynska-Cedro. 1969. Cardiac rhythm disturbances and the release of catecholamines after coronary occlusion in dogs. *Cardiovasc. Res.* **3**:190–197.

47. Sperelakis, N., and J. A. Schneider. 1976. A metabolic control mechanism for calcium ion influx that may protect the ventricular myocardial cell. *Am. J. Cardiol.* **37**:1079–1085.

48. Lazzara, R., R. R. Hope, N. El-Sherif, and B. J. Scherlag. 1979. Effects of lidocaine on hypoxic and ischemic cardiac cells. *Am. J. Cardiol.* **41**:872–879.

49. Elharrar, V., W. E. Gaum, and D. P. Zipes. 1977. Effect of drugs on conduction delay and incidence of ventricular arrhythmias induced by acute coronary occlusion in dogs. *Am. J. Cardiol.* **39**:544–549.

50. Ruffy, R., L. V. Rosenshtraukh, V. Elharrar, and D. P. Zipes. 1979. Electrophysiological effects of ethmozin on canine myocardium. *Cardiovasc. Res.* **13**:354–363.

51. Peter, T., T. Fujimoto, H. Hamamoto, A. McCullen, and W. J. Mandel. 1982. Electrophysiologic effects of diltiazem in canine myocardium, with special reference to conduction delay during ischemia. *Am. J. Cardiol.* **49**:602–605.

52. Spear, J. F., L. N. Horowitz, A. B. Hodess, H. MacVaugh, III, and E. N. Moore. 1979. Cellular electrophysiology of human myocardial infarction. I. Abnormalities of cellular activation. *Circulation* **59**:247–256.

53. Zipes, D. P., and R. F. Gilmour, Jr. 1982. Calcium antagonists and their potential role in the prevention of sudden coronary death. *Ann. N.Y. Acad. Sci.* **382**:258–288.

54. Wellens, H. J. J., J. Farré, and F. W. Bär. 1980. The role of the slow inward current in the genesis of ventricular tachyarrhythmias in man. In: *The Slow Inward Current and Cardiac Arrhythmias.* D. P. Zipes, J. C. Bailey, and V. Elharrar, eds. Martinus Nijhoff, The Hague. pp. 507–514.

55. DeMello, W. C. 1982. Intercellular communication in cardiac muscle. *Circ. Res.* **51**:1–9.

56. Wojtczak, J. 1979. Contracture and increase in internal longitudinal resistance of cow ventricular muscle induced by hypoxia. *Circ. Res.* **44**:88–95.

57. Weingart, R. 1977. The actions of ouabain on intercellular coupling and conduction velocity in mammalian ventricular muscle. *J. Physiol. (London)* **264**:341–365.

58. Spach, M. S., W. T. Miller, III, D. B. Geselowitz, R. C. Barr, J. M. Kootsey, and E. A. Johnson. 1981. The discontinuous nature of propagation in normal canine cardiac muscle: Evidence for recurrent discontinuities of intracellular resistance that affect the membrane currents. *Circ. Res.* **48**:39–54.

59. Spach, M. S., J. M. Kootsey, and J. D. Sloan. 1982. Active modulation of electrical coupling between cardiac cells of the dog: A mechanism for transient and steady state variations in conduction velocity. *Circ. Res.* **51**:347–362.

60. Schmitt, F. O., and J. Erlanger. 1928. Directional differences in the conduction of the impulse through heart muscle and their possible relation to extrasystolic and fibrillatory contractions. *Am. J. Physiol.* **87**:326–347.

61. Allessie, M. A., F. I. M. Bonke, and F. J. G. Schopman. 1977. Circus movement in rabbit atrial muscle as a mechanism of tachycardia. III. The "leading circle" concept: A new model of circus movement in cardiac tissue without the involvement of an anatomical obstacle. *Circ. Res.* **41**:9–18.

62. Antzelevitch, C., J. Jalife, and G. K. Moe. 1980. Characteristics of reflection as a mechanism of reentrant arrhythmias and its relationship to parasystole. *Circulation* **61**:182–191.

63. Peon, J., G. R. Ferrier, and G. K. Moe. 1978. The relationship of excitability of conduction velocity in canine Purkinje tissue. *Circ. Res.* **43**:125–135.

64. Rosenbaum, M. B., J. O. Lazzari, and M. V. Elizari. 1976. The role of phase 3 and phase 4 block in clinical electrocardiography. In: *The Conduction System of the Heart.* H. J. J. Wellens, K. I. Lie, and M. J. Janse, eds. Lea & Febiger, Philadelphia. pp. 126–142.

65. Singer, D. H., R. Lazzara, and B. F. Hoffman. 1967. Interrelationship betewen automaticity and conduction in Purkinje fibers. *Circ. Res.* **21**:537–558.

66. Jalife, J., C. Antzelevitch, V. Lamanna, and G. K. Moe. 1983. Rate-dependent changes in excitability of depressed cardiac Purkinje fibers as a mechanism of intermittant bundle branch block. *Circulation* **67**:912–922.

67. Gilmour, R. F., J. J. Salata, and D. P. Zipes. 1985. Rate-related suppression and facilitation of conduction in isolated canine cardiac Purkinje fibers. *Circ. Res.* **57**:35–45.

68. Jalife, J. and G. K. Moe. 1976. Effect of electrotonic potentials on pacemaker activity of canine Purkinje fibers in relation to parasystole. *Circ. Res.* **39**:801–809.

69. Moe, G. K., J. Jalife, W. J. Mueller, and B. Moe. 1977. A mathematical model of parasystole and its application to clinical arrhythmias. *Circulation* **56**:968–979.

70. Winfree, A. T. 1970. Integrated view of resetting a circadian clock. *J. Theor. Biol.* **28**:327–374.

71. Jalife, J., and C. Antzelevitch. 1979. Phase resetting and annihilation of pacemaker activity in cardiac tissue. *Science* **206**:695–697.

72. Best, E. N. 1979. Null space in the Hodgkin–Huxley equations. *Biophys. J.* **27**:87–104.

73. Gadsby, D. C., and P. F. Cranefield. 1977. Two levels of resting potential in cardiac Purkinje fibers. *J. Gen. Physiol.* **70**:725–746.

74. Wellens, H. J. J., K. I. Lie, and D. Durrer. 1974. Further observations on ventricular tachycardia as studied by electrical stimulation of the heart: Chronic recurrent ventricular tachycardia and ven-

tricular tachycardia during acute myocardial infarction. *Circulation* **49**:647–653.

75. Ruffy, R., K. J. Friday, and W. F. Southworth. 1983. Termination of ventricular tachycardia by single extrastimulation during ventricular effective refractory period. *Circulation* **67**:457–459.

76. Josephson, M. E., L. N. Horowitz, A. Farshidi, S. R. Spielman, E. L. Michelson, and A. M. Greenspan. 1979. Recurrent sustained ventricular tachycardia. IV. Pleomorphism. *Circulation* **59**:459–468.

77. Janse, M. J., F. J. L. van Capelle, H. Morsink, A. G. Kléber, F. Wilms-Schopman, R. Cardinal, C. N. d'Alnoncourt, and D. Durrer. 1980. Flow of "injury" current and patterns of excitation during early ventricular arrhythmias in acute regional myocardial ischemia in isolated porcine and canine hearts: Evidence for two different arrhythmogenic mechanisms. *Circ. res.* **47**:151–165.

78. Moe, G. K., and C. Mendez. 1973. Physiological basis of premature beats and sustained tachycardia. *N. Engl. J. Med.* **288**:250–254.

79. Wit, A. L., P. F. Cranefield, and B. F. Hoffman. 1972. Slow conduction and reentry in the ventricular conducting system. II. Single and sustained circus movements in networks of canine and bovine Purkinje fibers. *Circ. Res.* **30**:11–22.

80. Zipes, D. P. 1984. Genesis of cardiac arrhythmias: Electrophysiologic considerations. In: *Heart Diseases: A Textbook of Cardiovascular Medicine.* E. Braunwald, ed. Saunders, Philadelphia, pp. 605–647.

81. Hoffman, B. F., and M. R. Rosen. 1981. Cellular mechanisms for cardiac arrhythmias. *Circ. Res.* **49**:1–15.

82. Randall, W. C. 1977. Sympathetic control of the heart. In: *Neural Regulation of the Heart.* W. C. Randall, ed. Oxford University Press, London.

83. Jalife, J., C. Antzelevitch, and G. K. Moe. 1983. Models of parasystole and reflection. In: *Frontiers of Cardiac Electrophysiology.* M. B. Rosenbaum and M. V. Elizari, eds. Martinus Nijhoff, The Hague, pp. 217–238.

84. Cohen, H., R. Langendorf, and A. Pick. 1973. Intermittent parasystole: Mechanism of protection. *Circulation* **48**:761–774.

85. Nau, G. J., A. E. Aldariz, R. S. Acunzo, P. A. Chiale, M. V. Elizari, and M. B. Rosenbaum. 1981. Concealed ventricular parasystole uncovered in the form of ventricular escapes of variable coupling. *Circulation* **64**:199–207.

86. Nau, G. J., A. E. Aldariz, R. S. Acunzo, M. S. Halpern, J. M. Davidenko, M. V. Elizari, and M. B. Rosenbaum. 1982. Modulation of parasystolic activity by non-parasystolic beats. *Circulation* **66**:462–469.

87. Boineau, J. P., C. R. Mooney, R. D. Hudson, D. G. Hughes, R. A. Erdin, Jr., and A. C. Wilds. 1977. Observations on reentrant excitation pathways and the refractory period distributions in spontaneous and experimental atrial flutter in the dog. In: *Reentrant Arrhythmias.* H. E. Kulbertus, ed. University Park Press, Baltimore. pp. 72–98.

88. Pastelin, G., C. Mendez, and G. K. Moe. 1978. Participation of atrial specialized conduction pathways in atrial flutter. *Circ. Res.* **42**:386–393.

89. Josephson, M. E., and S. F. Seides. 1979. *Clinical Cardiac Electrophysiology.* Lea & Febiger, Philadelphia. p. 68.

90. Waldo, A. L., W. A. H. MacLean, R. B. Karp, N. T. Kouchoukos, and T. N. Janes. 1977. Entrainment and interruption of atrial flutter with atrial pacing: Studies in man following open heart surgery. *Circulation* **56**:737–744.

91. Zipes, D. P. 1975. Electrophysiological mechanisms involved in ventricular fibrillation. *Circulation* **51–52**(Suppl. III):120–130.

92. Schecter, D. C. 1979. Flashbacks: Ventricular fibrillation. Part I and II. *Pace* **2**:490 and 648.

93. McWilliam, J. A. 1887. Fibrillar contraction of the heart. *J. Physiol. (London)* **8**:296–310.

94. Garrey, W. E. 1914. The nature of fibrillary contraction of the heart: Its relation to tissue mass and form. *Am. J. Physiol.* **33**:397–414.

95. Porter, W. T. 1894. On the results of ligation of the coronary arteries. *J. Physiol. (London)* **15**:121–138.

96. Zipes, D. P., J. Fischer, R. M. King, A. D. Nicoll, and W. W. Jolly. 1975. Termination of ventricular fibrillation in dogs by depolarizing a critical amount of myocardium. *Am. J. Cardiol.* **36**:37–44.

97. Moe, G. K., and J. A. Abildskov. 1969. Atrial fibrillation as a self-sustaining arrhythmia independent of focal discharge. *Am. Heart J.* **58**:59–70.

98. Han, J., A. M. Malozzi, and G. K. Moe. 1968. Sinoatrial reciprocation in the isolated rabbit heart. *Circ. Res.* **22**:355–362.

99. Wu, D., F. Amat-y-Leon, P. Denes, R. C. Dhingra, R. J. Pietras, and K. M. Rosen. 1975. Demonstration of sustained sinus and atrial reentry as a mechanism of paroxysmal supraventricular tachycardia. *Circulation* **51**:234–243.

100. Weisfogel, G. M., W. P. Batsford, K. L. Paulay, M. E. Josephson, J. B. Ogunkelu, M. Akhtar, S. F. Seides, and A. N. Damato. 1975. Sinus node reentrant tachycardia in man. *Am. Heart J.* **90**:295–304.

101. Allessie, M. A., and F. I. M. Bonke. 1979. Direct demonstration of sinus nodal reentry in the rabbit heart. *Circ. Res.* **44**:557–568.

102. Allessie, M. A., F. I. M. Bonke, and F. J. G. Schopman. 1976. Circus movement in rabbit and atrial muscle as a mechanism of tachycardia. II. The role of nonuniform excitability in the occurrence of unidirectional block, as studied with multiple microelectrodes. *Circ. Res.* **39**:168–177.

103. Boineau, J. P., R. B. Schuessler, C. R. Mooney, C. B. Miller, A. C. Wylds, R. D. Hudson, J. M. Bouemans, and C. W. Brockus. 1980. Natural and evoked atrial flutter due to circus movement in dogs. *Am. J. Cardiol.* **45**:1167–1181.

104. Coumel, P., D. Flammang, P. Attuel, and J. F. Leckercq. 1979. Sustained intra-atrial reentrant tachycardia: Electrophysiologic study of 20 cases. *Clin. Cardiol.* **2**:176–178.

105. Wu, D., P. Denes, F. Amat-y-Leon, R. Dhingra, C. R. C. Wyndham, R. Bauernfeind, P. Latif, and K. M. Rosen. 1978. Clinical electrocardiographic and electrophysiologic observations in patients with paroxysmal supraventricular tachycardia. *Am. J. Cardiol.* **41**:1045–1051.

106. Hering, H. E. 1907. Ueber die Automatie des Saugethierherzens. *Arch. Ges. Physiol.* **116**:143–164.

107. Mines, G. R. 1913. On dynamic equilibrium in the heart. *J. Physiol. (London)* **46**:350–383.

108. Mines, G. R. 1914. On circulating excitations in heart muscles and their possible relation to tachycardia and fibrillation. *Trans. R. Soc. Can. Sect. 4* **8**:43–52.

109. Scherf, D., and C. Shookoff. 1926. Experimenteele Untersuchungen uber die "Umkehr-Extrasystole" (reciprocating beat). *Wien. Arch. Inn. Med.* **12**:501–509.

110. Mendez, C., J. Han, P. D. Garcia de Jalon, and G. K. Moe. 1965. Some characteristics of ventricular echoes. *Circ. Res.* **16**:562–581.

111. Watanabe, Y., and L. S. Dreifus. 1965. Inhomogeneous conduction in the AV node: A model for reentry. *Am. Heart J.* **70**:505–514.

112. Mendez, C., and G. K. Moe. 1966. Some characteristics of transmembrane potentials of AV nodal cells during propagation of premature beats. *Circ. Res.* **19**:993–1010.

113. Mendez, C., and G. K. Moe. 1966. Demonstration of dual AV conduction system in the isolated rabbit heart. *Circ. Res.* **19**:378–393.

114. Janse, M. J., F. J. L. VanCapelle, G. E. Freud, and D. Durrer. 1971. Circus movement within the AV node as a basis for supraventricular tachycardia as shown by multiple microelectrode recording in the isolated rabbit heart. *Circ. Res.* **28**:403–414.

115. Wit, A. L., B. N. Goldreyer, and A. N. Damato. 1971. An *in vitro* model of paroxysmal supraventricular tachycardia. *Circulation* **43**:862–875.

116. Moe, G. K., W. Cohen, and R. L. Vick. 1963. Experimentally induced paroxysmal AV nodal tachycardia in the dog. *Am. Heart J.* **65**:87–92.

117. Schuilenburg, R. M., and D. Durrer. 1968. Atrial echo beats in the human heart elicited by induced atrial premature beats. *Circulation* **37**:680–693.

118. Hunt, N. C., F. R. Cobb, M. B. Waxman, H. J. Zeft, R. H. Peter, and J. J. Morris, Jr. 1968. Conversion of supraventricular tachycardias with atrial stimulation: Evidence for reentry mechanism. *Circulation* **38**:1060–1065.

119. Bigger, J. T., and B. N. Goldreyer. 1970. The mechanism of supraventricular tachycardia. *Circulation* **42**:673–688.

120. Goldreyer, B. N., and A. N. Damato. 1971. Essential role of atrioventricular conduction delay in the initiation of paroxysmal supraventricular tachycardia. *Circulation* **43**:679–687.

121. Denes, P., D. Wu, R. C. Dhingra, R. Chuquimia, and K. M. Rosen. 1973. Demonstration of dual AV nodal pathways in patients with paroxysmal supraventricular tachycardia. *Circulation* **48**:549–555.

122. Wu, D. 1982. Dual atrioventricular nodal pathways: A reappraisal. *Pace* **5**:72–81.

123. Wu, D., P. Denes, F. Amat-y-Leon, C. R. C. Wyndham, R. Dhingra, and K. M. Rosen. 1977. An unusual variety of atrioventricular nodal reentry due to retrograde dual atrioventricular nodal pathways. *Circulation* **56**:50–59.

124. Spurrell, R. A. J., D. M. Krikler, and E. Sowton. 1974. Effects of verapamil on electrophysiological properties of anomalous atrioventricular connection in Wolff–Parkinson–White syndrome. *Br. Heart J.* **36**:256–264.

125. Wellens, H. J. J., S. L. Tan, F. W. H. Bär, D. R. Duren, K. I. Lie, and H. M. Dohmen. 1977. Effects of verapamil studied by programmed electrical stimulation of the heart in patients with paroxysmal reentrant supraventricular tachycardia. *Br. Heart J.* **39**:1058–1066.

126. Rinkenberger, R. L., E. N. Prystowsky, J. J. Heger, P. J. Troup, W. M. Jackman, and D. P. Zipes. 1980. Effects of intravenous and chronic oral verapamil administration in patients with supraventricular tachyarrhythmias. *Circulation* **62**:996–1010.

127. Denes, P., R. Kehoe, and K. M. Rosen. 1973. Multiple reentrant tachycardias due to retrograde conduction of dual atrioventricular bundles with atrioventricular nodal-like properties. *Am. J. Cardiol.* **44**:162–170.

128. Wolff, L., J. Parkinson, and P. D. White. 1950. Bundle branch block with short PR interval in healthy young people prone to paroxysmal tachycardia. *Am. Heart J.* **5**:685–704.

129. Durrer, D., L. Schoo, R. M. Schuilenburg, and H. J. J. Wellens. 1967. The role of premature beats in the initiation and the termination of supraventricular tachycardia in the Wolff–Parkinson–White syndrome. *Circulation* **36**:644–662.

130. Zipes, D. P., R. L. DeJoseph, and D. A. Rothbaum. 1974. Unusual properties of accessory pathways. *Circulation* **49**:1200–1211.

131. Gallagher, J. J. 1978. The pre-excitation syndromes. *Prog. Cardiovasc. Dis.* **20**:285–327.

132. Wellens, H. J. J. 1976. The electrophysiologic properties of the accessory pathway in the Wolff–Parkinson–White syndrome. In: *The Conduction System of the Heart.* H. J. J. Wellens, K. I. Lie, and M. H. Janse, eds. Lea & Febiger, Philadelphia. pp. 567–587.

133. Coumel, P., and P. Attuel. 1974. Reciprocating tachycardia in overt and latent preexcitation: Influence of functional bundle branch block on the rate of the tachycardia. *Eur. J. Cardiol.* **1**:423–436.

134. Barold, S. S., and P. Coumel. 1977. Mechanisms of atrioventricular junctional tachycardia: Role of renentry and concealed accessory bypass tracts. *Am. J. Cardiol.* **39**:97–106.

135. Hammill, S. C., E. L. C. Pritchett, G. J. Klein, W. M. Smith, and J. J. Gallagher. 1980. Accessory atrioventricular pathways that conduct only in the antegrade direction. *Circulation* **62**:1335–1348.

136. Lown, B., W. F. Ganong, and S. A. Levine. 1952. The syndrome of short PR interval, normal QRS complex and paroxysmal rapid heart action. *Circulation* **5**:693–706.

137. James, T. N. 1970. The Wolff–Parkinson–White syndrome:

Evolving concepts of its pathogenesis. *Prog. Cardiovasc. Dis.* **13**:159–189.

138. Jackman, W. M., E. N. Prystowsky, G. V. Naccarelli, & N. S. Finebers, G. T. Rahilly, J. J. Heger, and D. P. Zipes. 1983. Reevaluation of enhanced atrioventricular nodal conduction: Evidence to suggest a continuum of normal atrioventricular nodal physiology. *Circulation* **67**:441–448.

139. Gallagher, J. J., W. M. Smith, J. H. Kasell, D. W. Benson, Jr., R. Sterba, and A. O. Grant. 1981. Role of Mahaim fibers in cardiac arrhythmias in man. *Circulation* **64**:176–189.

140. Moe, G. K., C. Mendez, and J. Han. 1965. Aberrant AV impulse propagation in the dog heart: A study of functional bundle branch block. *Circ. Res.* **16**:261–286.

141. Zipes, D. P. 1975. Reentry in the ventricles: Recent advances in ventricular conduction. *Adv. Cardiol.* **14**:51.

142. Glassman, R. D., and D. P. Zipes. 1981. Site of antegrade and retrograde functional right bundle branch block in the intact canine heart. *Circulation* **64**:1277–1286.

143. Akhtar, M., C. Gilbert, F. G. Wolfe, and D. H. Schmidt. 1978. Reentry within the His–Purkinje system: Elucidation of reentrant circuit using right bundle branch and His bundle recordings. *Circulation* **58**:295–304.

144. Spurrell, R. A. J., E. Sowton, and D. C. Deuchar. 1973. Ventricular tachycardia in four patients evaluated by programmed electrical stimulation of the heart and treated in two patients by surgical division of anterior radiation of left bundle branch. *Br. Heart J.* **35**:1014–1025.

145. Josephson, M. E., L. N. Horowitz, A. Farshidi, and J. A. Kastor. 1978. Recurrent sustained ventricular tachycardia. I. Mechanisms. *Circulation* **57**:431–440.

146. Lloyd, E. A., D. P. Zipes, J. J. Heger, and E. N. Prystowsky. 1982. Sustained ventricular tachycardia due to bundle branch block reentry. *Am. Heart J.* **104**:1095–1097.

147. Fontaine, G., G. Guiraudon, R. Frank, F. Fillette, C. Cabrol, and Y. Grosgogeat. 1982. Surgical management of ventricular tachycardia unrelated to myocardial ischemia or infarction. *Am. J. Cardiol.* **49**:397–410.

148. Wallace, A. G., and R. J. Mignone. 1966. Physiologic evidence concerning the reentry hypothesis for ectopic beats. *Am. Heart J.* **72**:60–70.

149. Scherlag, B. J., R. H. Helfant, J. I. Haft, and A. N. Damato. 1970. Electrophysiology underlying ventricular arrhythmias due to coronary ligation. *Am. J. Physiol.* **219**:1665–1671.

150. Boineau, J. P., and J. L. Cox. 1973. Slow ventricular activation in acute myocardial infarction: A source of reentrant premature ventricular contractions. *Circulation* **48**:703–713.

151. Waldo, A. L., and G. A. Kaiser. 1973. Study of ventricular arrhythmias associated with acute myocardial infarction in the canine heart. *Circulation* **47**:1222–1228.

152. El-Sherif, N., B. J. Scherlag, R. Lazzara, and R. R. Hope. 1977. Reentrant ventricular arrhythmias in the late myocardial infarction period. I. Conduction characteristics in the infarction zone. *Circulation* **55**:686–701.

153. El-Sherif, N., B. J. Scherlag, R. Lazzara, and R. R. Hope. 1977. Reentrant ventricular arrhythmias in the late myocardial infarction period. II. Patterns of initiation and termination of reentry. *Circulation* **55**:702–719.

154. El-Sherif, N., R. A. Smith, and K. Evans. 1981. Canine ventricular arrhythmias in the late myocardial infarction period. 8. Epicardial mapping of reentrant circuits. *Circ. Res.* **49**:255–265.

155. Wit, A. L., M. A. Allessie, F. I. M. Bonke, W. Tammers, J. Smeets, and J. J. Fenoglio, Jr. 1982. Electrophysiologic mapping to determine the mechanism of experimental ventricular tachycardia initiated by premature impulses: Experimental approaches and initial results demonstrating reentrant excitation. *Am. J. Cardiol.* **49**:166–185.

156. Ideker, R. E., G. J. Klein, L. Harrison, W. Smith, J. Kasell, K. Reimer, A. G. Wallace, and J. J. Gallagher. 1981. The transition to ventricular fibrillation induced by reperfusion after acute ischemia in the dog: A period of organized epicardial activation. *Circulation* **63**:1371–1379.

157. Klein, G. J., R. E. Ideker, W. M. Smith, L. A. Harrison, J. Kasell, A. G. Wallace, and J. J. Gallagher. 1979. Epicardial mapping of the onset of ventricular tachycardia initiated by programmed stimulation in the canine heart with chronic infarction. *Circulation* **60**:1375–1384.

158. Zipes, D. P. 1983. Arrhythmias. In: *Comprehensive Cardiac Care.* K. G. Andreoli, V. K. Fowkes, D. P. Zipes, and A. G. Wallace, eds. Mosby, St. Louis, pp. 167–333.

159. Gilmour, R. F., Jr., and D. P. Zipes. 1984. Basic Electrophysiologic mechanisms for the development of arrhythmias. Clinical Application. *Med. Clin. N. Amer.* **68**:795–818.

Pathophysiology of Peptic Ulcer Disease

Leonard R. Johnson

1. Introduction

Gastric and duodenal ulcer are both peptic ulcer diseases. The term *peptic* refers to the fact that acid and pepsin are necessary to produce the break in the mucosa which is called an ulcer. In humans the best evidence for this statement occurs at the opposite ends of the secretory range. Individuals whose parietal cells are continuously stimulated to secrete acid by gastrin, being released from a tumor (gastrinoma or Zollinger–Ellison syndrome), develop ulcers in all but 7% of the cases.[1] Those having the achlorhydria of pernicious anemia rarely develop ulcers.

The relative contributions of acid and pepsin to the development of an ulcer are difficult to assess. Obviously, some acid must always be present, since it is needed to convert pepsinogen to the active enzyme. To further complicate matters, the presence of pure hydrochloric acid in the stomach stimulates pepsin secretion via a local cholinergic reflex.[2,3] Literally hundreds of studies have been published involving the production of experimental ulcers with acid and pepsin.[4] Some investigators have stressed the role of acid as the ulcerogenic agent,[5–7] while others have emphasized the importance of pepsin.[8–10]

Although Goldberg et al.[11] studied the production of esophagitis, they conducted one of the few quantitative evaluations of the roles of acid and pepsin in ulcer formation. A 1-hr infusion of HCl (pH 1.0) into the esophagus of an anesthetized cat produced severe esophagitis, which was not exacerbated by the addition of pepsin. Apparently, pepsin has little effect on the digestion of most protein at this pH,[12] and it actually may be denatured at pH 1.0.[12,13] At pHs 1.6 and 2.0, erosions were produced only when pepsin was added to the acid. At these pHs, pepsin probably causes erosions by protein digestion. The importance of pepsin was further demonstrated by the finding that

no erosions were produced after a pepsin inhibitor was added to an acid–pepsin solution known to produce esophagitis.[11] The authors concluded that although high concentrations of acid alone may cause erosions, in a more physiological range (pH 1.6–2.0) the major function of acid was to activate pepsin.

Pathophysiologically, an ulcer develops when the destructive factors acting on the mucosa overwhelm those processes serving to protect it. While there are other factors which may act to injure the mucosa, acid and pepsin are certainly the most important. Those processes which protect the mucosa from damage, however, are less easily defined. Each of the following has been considered by some investigators to be an important part of the resistance to mucosal injury: mucus, alkaline secretion, mucosal blood flow, cell membrane constituents, prostaglandin levels, mucosal cell turnover and replication. The balance between destructive factors and mucosal resistance is the important concept in ulcerogenesis. An ulcer may occur when mucosal resistance decreases below normal just as one may occur when acid and pepsin secretion increase above normal levels. This is best emphasized by the lack of correlation between peak acid output and severity of ulcer.[14] Furthermore, gastric ulcers commonly develop in individuals who secrete less than normal amounts of acid.[15] Although as a group duodenal ulcer patients secrete higher than average amounts of acid, a large number of patients with duodenal ulcer have normal or lower than normal secretory rates.[15]

The role of decreased mucosal resistance in the development of an ulcer becomes all the more apparent when one considers that an ulcer is a localized lesion. Although a large area of mucosa is exposed to acid, pepsin, and other damaging agents, an ulcer usually occurs in a well-defined area. This is true even in cases of experimental ulcer when large amounts of gastric juice or acid and pepsin are introduced into the gut and in cases of gastrinoma where acid secretory rates may be extremely high. Thus, a local defect in the resistance of the mucosa must be postulated in order to explain the focal development of the lesion. Theoretically, this defect could be due to impaired blood

Leonard R. Johnson • Department of Physiology and Cell Biology, University of Texas Medical School, Houston, Texas 77225.

flow, low rate of cell replication, or perhaps a decreased rate of alkaline secretion at this point.

There are many other factors in addition to acid–pepsin and mucosal resistance which play important roles in the pathogenesis of ulcer. These include heredity, duodenogastric reflux, pancreatic secretion, psychological factors, and geographic and socioeconomic factors. However, within the scope of this volume on membrane disorders, I will concentrate on the factors which lead to hypersecretion of H^+ across the parietal cell membrane and those which alter cell membranes or other components of mucosal resistance allowing acid and pepsin to damage the cells.

2. Gastric vs. Duodenal Ulcer

Although they are both called peptic ulcers, some investigators consider gastric and duodenal ulcers to have different pathogeneses. This concept is based on a number of findings. First, patients with gastric ulcer on the average secrete lower than normal amounts of acid, while those with duodenal ulcer secrete higher than normal amounts.[16] Increased acid secretion in duodenal ulcer patients is seen in the basal state, in response to food, sham-feeding, insulin, 2-deoxyglucose, gastrin, histamine, and caffeine.[17] In response to maximal histamine stimulation, normal individuals secreted 24.0 mmoles H^+/hr compared to 42.3 mmoles for duodenal ulcer and 18.0 mmoles for gastric ulcer patients.[17] There was, however, considerable individual overlap between normals and both of the other groups.

Second, under normal conditions the gastric mucosa is more resistant to acid–pepsin digestion than is the duodenal mucosa. Patients with the extreme hypersecretion of gastrinoma develop duodenal or even jejunal ulcers six times more frequently than they develop gastric ulcers.[1] In general, the higher the secretory rate, the farther the ulcer is likely to be from the stomach.

Third, experimentally induced "stress" ulcers caused by various types of restraint, temperature modification, or forced exercise are almost exclusively gastric lesions.[18,19]

Finally, gastric and duodenal ulcer patients differ with regard to age, sex, social background, heredity, and a frequency of ABO blood group and secretor status.

These findings suggest that duodenal ulcers are due to hypersecretion of acid and pepsin, while gastric ulcers develop because of decreased mucosal resistance. However, current evidence suggests a basically similar etiology for duodenal and gastric ulcers—namely, too much acid and pepsin for the existing mucosal resistance.[20] It is clear that acid secretory rates cannot be used as a diagnostic tool (except in the case of gastrinoma) because of the overlap. There is also considerable evidence that many differences exist among patients with gastric ulcer and among patients with duodenal ulcer. However, it does seem obvious that in duodenal ulcer disease, increased acid–pepsin secretion usually plays a more obvious role than decreased mucosal resistance, while the opposite is true for gastric ulcer disease. Nevertheless, both elements of the equation play some part in the etiology of almost all peptic ulcers.[21]

3. Acid–Pepsin Secretion

Acid and pepsin secretion is readily quantified. As such, we know a good deal more about it than we do about mucosal resistance which is not only difficult to measure but hard to define. Increased secretion of acid and pepsin occurs in patients with duodenal ulcer disease, and, hence, most of our information pertaining to the etiology of hypersecretion comes from this patient group. It is important to realize that the following discussion does not refer to all patients with duodenal ulcer, nor does it refer to any particular type of patient. It does, however, represent an emerging picture of the pathophysiology of duodenal ulcer as it relates to acid and pepsin secretion.

3.1. Basal Acid Secretion

Basal acid secretion depends on the total secretory mass (parietal cell volume) and the stimuli which reach the parietal cells in the absence of a meal. The three physiologically significant stimulants of acid secretion are acetylcholine, histamine, and gastrin. The most obvious example of increased basal acid secretion occurs in patients with duodenal ulcer caused by high basal gastrin in the Zollinger–Ellison syndrome.[22] In this case, gastrin is continuously released from a tumor. The role of gastrin in basal acid secretion in other duodenal ulcer patients and in normal individuals is difficult to assess, for there is no specific, potent inhibitor of gastrin-stimulated acid secretion. Basal acid secretion in normal individuals also shows a distinctive circadian rhythm, peaking in the early morning hours. This rhythm is unrelated to serum gastrin values.[23]

Anticholinergics and histamine H_2-receptor antagonists inhibit basal acid secretion. These studies imply that there is continual release of small amounts of acetylcholine from neurons and histamine from histamine-containing paracrine cells in the mucosa. Increased release or decreased destruction of these secretagogues is a possible explanation for the increased basal secretory drive that occurs in some duodenal ulcer patients. There is disagreement on the amount of histamine found in the mucosa of patients with duodenal ulcer.[24,25] Recently, however, Peden et al.[26] measured histamine and histamine methyltransferase in 110 patients with duodenal ulcer and 62 control subjects. Ulcer patients had significantly lower levels of both histamine and the principal enzyme destroying it. These investigators noted, however, that the decreased histamine levels correlated better with the number of men in each group who were smokers than it did with the presence of duodenal ulcer. Thus, the only connection between histamine and increased basal secretion in ulcer patients is a tenuous one—namely, that ulcer patients apparently have a decreased ability to metabolize histamine.

3.2. Response to Secretagogues

Acid secretory responses to gastrin and histamine are greater in duodenal ulcer patients than controls.[27−29] The increased maximal acid output seen in this patient group is a function of increased parietal cell mass[30] (see below). Perhaps even more important, however, to the etiology of the disease is the fact that duodenal ulcer patients are hypersensitive to gastrin.[27,28] Isenberg et al.[27] studied the responses to graded doses of pentagastrin in 20 duodenal ulcer patients and 20 control subjects (Fig. 1). The mean dose of pentagastrin required for half-maximal acid secretion in the duodenal ulcer group was 92.1 ng/kg per hr compared to 246.8 ng in the control group. These figures were corrected for basal secretion and were normalized for the significant difference in maximal acid output in the two groups. The study demonstrated that non-duodenal ulcer subjects required 2.8 times more pentagastrin to produce the same amount of acid as patients with ulcers. In other words, much smaller

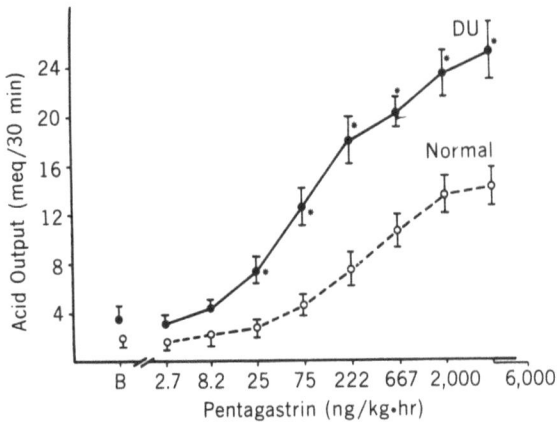

Fig. 1. Mean acid outputs to graded doses of pentagastrin in 20 patients with duodenal ulcer and 20 nonulcer controls. Bars represent S.E.M. Reproduced with permission from Isenberg et al.[27]

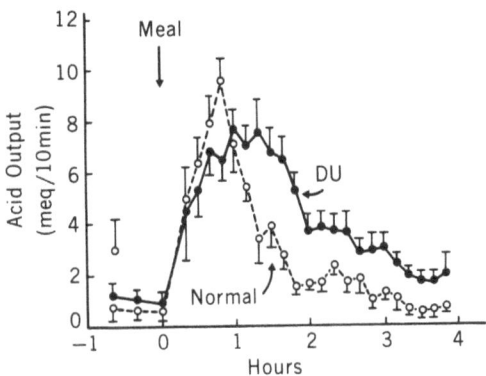

Fig. 3. Acid secretory responses to a meal in patients with duodenal ulcer and normal controls. DU patients secreted significantly more acid during the second, third, and fourth hours. Means and S.E.M. Reproduced with permission from Malagelada et al.[33]

amounts of exogenous gastrin are required for duodenal ulcer patients to secrete a fixed amount of acid. The same basic findings have been reported using the full gastrin-17 molecule.[28]

Lam et al.[28] have also demonstrated that duodenal ulcer patients are more sensitive to endogenously released gastrin than are normal control subjects. They studied acid secretion and serum gastrin levels in 25 patients with duodenal ulcer and 14 controls in response to graded doses of a peptone meal. At fixed serum gastrin levels, acid secretion was significantly higher to each dose of peptone in subjects with duodenal ulcer than it was in controls (Fig. 2).

3.3. Acid Secretory Response to a Meal

Acid secretion in response to a normal meal can be measured in man using the technique of intragastric titration. Although there is considerable overlap between individuals, duodenal ulcer patients secrete more acid and for a longer time than do normal individuals.[28,31,32] Malagelada et al.[33] found that most of the additional acid secretion in patients with duodenal ulcer occurred during the later stages of the response to a meal (Fig. 3). They employed a technique which allowed gastric pH

and emptying to vary in in a normal manner. In doing so, they demonstrated that the total amount of acid entering the duodenum was significantly increased during the third and fourth hours of the response to a meal in those individuals with duodenal ulcer. The actual load of acid present in the duodenum was significantly elevated 90 min after the start of a meal and remained so for the ensuing 3 hr. They interpreted their data as being consistent with postprandial regulatory defects which cause a prolonged gastric secretory response to food and an inappropriately high rate of delivery of acid from the stomach into the duodenum.[33]

3.4. Parietal Cell Mass

The most obvious explanation for increased acid secretory responses in duodenal ulcer patients is increased parietal cell mass. The normal stomach contains about 1 billion parietal cells, and this is approximately doubled in duodenal ulcer patients. Cox[34] examined over 200 gastric autopsy specimens and found an average of 1.8 billion parietal cells in stomachs from patients with duodenal ulcer and 0.8 billion in stomachs having gastric ulcers. Control stomachs ($N = 135$) had 1.0 billion parietal cells (Fig. 4). The correlation between the number of parietal cells and maximal acid output is well established both

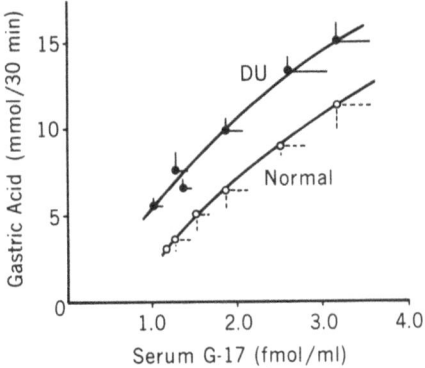

Fig. 2. Comparison between endogenous serum gastrin and corresponding rates of acid secretion in subjects with duodenal ulcer and normal controls. Horizontal and vertical lines represent S.E.M. Reproduced with permission from Lam et al.[28]

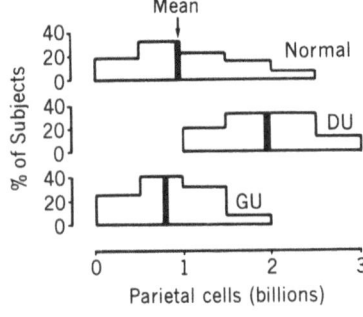

Fig. 4. Numbers of parietal cells present in the stomachs of normal individuals and subjects with duodenal or gastric ulcer expressed as percentage of population with a particular range. Data were obtained from autopsy specimens. Normal stomachs numbered 135. Redrawn with permission from Cox.[34]

in normal individuals[35,36] and in patients with peptic ulcer disease.[30]

The cause of increased secretory mass in duodenal ulcer patients is unknown and probably varies in different types of patients. There is strong evidence that it is hereditary in at least some groups.[30,37] The most obvious example of increased secretory mass occurs in patients with gastrinoma or Zollinger–Ellison syndrome.[38] The extreme gastric hyperplasia present in this disease is probably a direct effect of the trophic action of the hormone gastrin.[39] In fact, the reported increase in the size of the gastric mucosa of this group of patients was one of the factors which led to the discovery that gastrin stimulated mucosal growth throughout much of the gastrointestinal tract.[40] It is tempting to speculate that the increased parietal cell mass in some patients with nongastrinoma duodenal ulcer disease may also be caused by increased serum gastrin levels or an increased sensitivity to the trophic effects of gastrin. However, there are no data available on this subject and the only inference relates to elevated gastrin levels in some patients with duodenal ulcer.

3.5. Serum Gastrin Levels

Most studies report that fasting serum gastrin levels of patients with duodenal ulcer do not differ from those of normal subjects, but their postprandial serum gastrin responses are greater than normal (see review by Walsh[41]). Numerous groups, however, have reported series of patients with duodenal ulcer, increased fasting serum gastrin levels, and exaggerated serum gastrin response to meals.[42–44] While most studies indicate that serum gastrin is significantly elevated above normal in duodenal ulcer patients after a meal,[45,46] others have failed to find a difference.[32,33]

McGuigan and Trudeau[45] examined 22 patients with duodenal ulcer disease and 10 control subjects and found no differences in their mean fasting serum gastrin levels. Following the consumption of a 180 g meal of lean meat, serum gastrin rose within 15 min. The increase in the integrated gastrin response after 75 min in control subjects was 44% while gastrin increased 99% in the group with duodenal ulcer (Fig. 5). Thus, not only are patients with duodenal ulcer disease more sensitive to gastrin but according to most studies they also release more gastrin after eating.

Duodenal ulcer patients do not appear to have more antral

gastrin than normals, although the results of these studies are not clear. Malmstrom and Stadil[47] found no differences in antral gastrin between the two groups in one study. However, they later reported that duodenal ulcer patients had significantly less antral gastrin than normals.[48] Creutzfeldt et al.[49] reported significantly more antral gastrin in ulcer patients than in normals.

In normal individuals the most significant factor regulating gastrin release is the pH of the antral contents. In order for gastrin to be released, the gastric contents must be buffered above pH 3.0. As a meal is digested, its buffers saturated, and it begins to empty from the stomach, the pH falls, inhibiting further gastrin release and, hence, acid secretion. Walsh et al.[46] compared serum gastrin and gastric acid secretory responses at intragastric pHs 5.5 and 2.5 in duodenal ulcer patients and normal subjects. They used an amino acid meal which gave 60% of the acid secretory response to a steak meal. Ulcer patients had significantly higher rates of acid secretion and gastrin release at both pHs. More significant, however, was the finding that while serum gastrin was inhibited 80% at pH 2.5 in normal subjects, it was decreased only 50% in patients with duodenal ulcer. At the same time, acid secretion decreased 70% in normals but only 30% in the ulcer group (Fig. 6). The authors concluded that there was a defect in the inhibition of gastrin release by acid in patients with duodenal ulcer disease. This defect may account for the hypergastrinemia seen in ulcer patients following a meal. It could also explain the increased duration of the acid secretory response in these individuals.

Evidence that more gastrin is released in duodenal ulcer patients per unit of stimulation has recently been presented. Ou Tim et al.[50] found that bombesin released significantly more gastrin at all doses in ulcer patients compared to normals. Maximal gastrin levels were 609 pg/ml in the ulcer group compared to 153 pg/ml in the controls. Acid and pepsin secretion paralleled gastrin release. They also demonstrated a decreased inhibition of gastrin release at acidic luminal pHs.

As is the case with most endocrine evaluations, conclusions have been made regarding the role of gastrin in duodenal ulcer

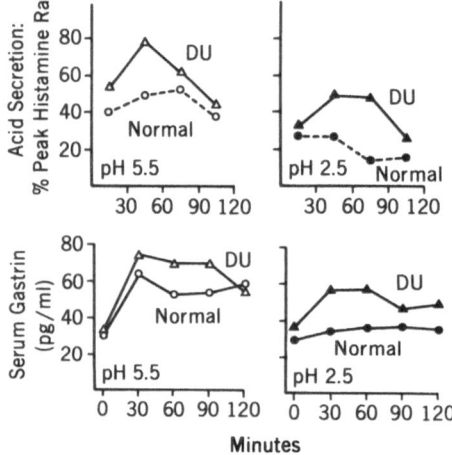

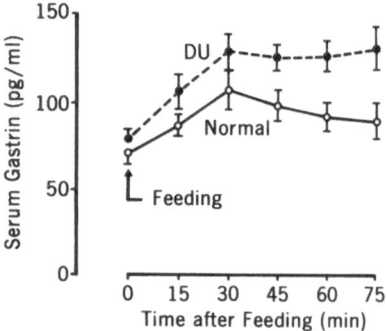

Fig. 5. Serum gastrin levels at various times after a meat meal in 22 patients with duodenal ulcer and 10 normal controls. Means and S.E.M. Gastrin levels were significantly higher in the DU group beginning at 45 min. Reproduced with permission from McGuigan and Trudeau.[45]

Fig. 6. Acid secretion (as percent maximum) in the top panels and serum gastrin levels in the bottom panels in response to amino acid meals held at pH 5.5 (left panels) or pH 2.5 (right panels). Values from patients with duodenal ulcer are compared to those obtained from normal controls. Reproduced with permission from Walsh et al.[46]

disease merely by measuring serum gastrin levels. We now know, however, that many physiological processes are not only regulated by changes in hormone concentration, but also by changes in the numbers or sensitivity of the receptors for that hormone. This is perfectly illustrated in newborn animals (including man) in the case of gastrin. Newborns have hypergastrinemia, but there is no acid secretory or trophic response to either exogenous or endogenous gastrin.[51] The lack of response is explained by the absence of receptors for the hormone. The receptors in the rat develop at the time of weaning along with the secretory and trophic response to gastrin. Corticosterone prematurely induces receptor development and the response to gastrin.[51] Gastrin itself can regulate the number of receptors present.[52] Chronic increases or decreases in serum gastrin lead to correspondingly parallel changes in receptor number. The up-regulation in response to chronically elevated serum gastrin, however, is preceded by a period of down-regulation as has been described for other peptide hormones.[53] Therefore, it is reasonable to assume that part of the hypersecretion present in duodenal ulcer patients and increased sensitivity to gastrin may be caused by changes at the level of the gastrin receptor. This area along with studies of histamine and cholinergic receptors on parietal cells should prove to be fruitful ones in forwarding our understanding of duodenal ulcer disease.

3.6. Pepsin Secretion

Most of this discussion has focused on acid secretion rather than pepsin secretion. As pointed out earlier, peptic activity is closely linked to the amount of acid present in the stomach. Investigation of pepsin in the etiology of duodenal ulcer disease has been complicated by the many isozymes of pepsin and by the lack of correlation between *in vitro* pepsin assays and the ulcerogenic potential of the protease activity present in gastric juice.[54] One study found that intragastric instillation of rabbit gastric juice into rats produced much more mucosal damage than an equal volume of cat gastric juice having the same concentration of acid and amount of peptic activity.[55] Another study found that intragastric perfusion of pentagastrin-stimulated rat gastric juice produced duodenal ulcers in rats, while the perfusion of a laboratory solution containing equal amounts of HCl and peptic activity had no effect.[56]

Because of the close relationship between acid secretion and pepsinogen secretion, pepsinogen activation, and pepsin activity, one would expect patients with duodenal ulcer disease to secrete more pepsinogen than normals. Samloff *et al.*[57] used a radioimmunoassay to measure serum group I pepsinogen in a variety of subjects. Three hundred healthy controls had a mean serum group I pepsinogen level of 110 ng/ml, while in 77 patients with duodenal ulcer it was 221 ng/ml (Fig. 7). Males had higher serum levels than females, but at all ages duodenal ulcer patients, including females, had significantly higher group I pepsinogen levels than the healthy male controls.

3.7. Summary

The studies summarized in this section present the beginnings of a picture involving changes in secretion in the pathophysiology of peptic ulcer, and more specifically duodenal ulcer. On the average, duodenal ulcer patients secrete more acid–pepsin than normal subjects in the basal state and after stimulation by a variety of agents. The mechanism behind the increased secretion is not the same in each case but may include:

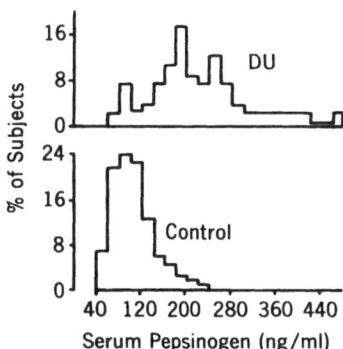

Fig. 7. Serum pepsinogen I levels in patients with duodenal ulcer and in normal control subjects. Reproduced with permission from Samloff *et al.*[57]

increased maximal secretory capacity due to increased secretory mass, increased sensitivity of target cells to secretagogues, increased levels of secretory stimulants, and decreased inhibition of acid secretion by acidic luminal contents.

Evidence exists and has been presented for each of these. However, it is important to realize that causality has not been established between these various factors. For example, decreased inhibition of gastrin release may lead to higher serum gastrin levels following a meal. This in turn could lead to increased numbers of parietal cells and increased acid secretion. There are, however, numerous individual cases which would not fit this scheme and, in fact, run counter to it. We have essentially no understanding of the regulation of parietal cell sensitivity at the receptor level or for that matter any knowledge of the receptor mechanism which regulates gastrin release from the G cell. Research into these areas should be extremely important to understanding the role of acid and pepsin secretion in the etiology of peptic ulcer.

At present, it is apparent that many physiological abnormalities in acid–pepsin secretion exist in patients with duodenal ulcer. Although a picture has begun to emerge, none of these abnormalities is present in all subjects with the disease. Thus, duodenal ulcer may best be viewed as a disease with many different causes.

4. Mucosal Resistance

Mucosal resistance is more difficult to measure than acid or pepsin secretion. In fact, the term itself refers to a variety of factors, some of which are poorly defined and others over which there is considerable disagreement as to their importance.

The concept of mucosal resistance and the notion that it is important have received a great deal of attention during the past 20 years. Efforts in this direction were primarily stimulated by the pioneering work of Davenport, begun in the mid 1960s, to define the so-called "gastric mucosal barrier." Actually, the fact that the normal gastric mucosa was highly impermeable to H+ was first emphasized by Teorell in 1939.[58] Code *et al.*[59] in 1955 coined the phrase "gastric mucosal barrier" to describe this relative impermeability. Davenport's first paper on this subject, which was done at the Mayo Clinic with Code, demonstrated that the disappearance of acid from the stimulated canine Heidenhain pouch after treatment with eugenol was not due to an effect on the secretory mechanism but to increased mucosal permeability which allowed the secreted acid to back-diffuse.[60] In other words, the gastric mucosal barrier had been broken.

In a large series of elegant experiments, Davenport went on to describe the functional nature of the barrier, the types of compounds capable of breaking the barrier, and the mechanisms by which they did so, and the consequences of this disruption. Despite this beginning, the physical and anatomical nature of the barrier is poorly understood.

If we think of the gastric mucosal barrier as what keeps the stomach from digesting itself, then we are actually dealing with mucosal resistance. Although one can determine that the gastric mucosal barrier has been broken by measuring the disappearance of H^+ from the lumen and the appearance of $Na^+, K^+,$ and fluid in the lumen,[61–64] it is still, except under experimental conditions, impossible to determine why the barrier has been weakened, or if weakened whether it has resulted in an ulcer. This observation suggested the possibility that gastric secretion in gastric ulcer patients was not abnormally low, but that secreted acid diffused through the mucosa before it could be measured. Davenport, responding to published accounts describing the high-volume, low-acid gastric juice of patients with gastric ulcer, wrote a comment with the title: "Is the apparent hyposecretion of acid by patients with gastric ulcer a consequence of a broken barrier to diffusion of hydrogen ions into the gastric mucosa?"[65] He has recently stated, "I don't know whether the answer to the question is yes or no, but the question did stir up the gastroenterological world for a while."[66]

The remainder of this chapter will examine in brief some of the more likely components or contributions to mucosal resistance. It is important to realize that none of these factors has been identified with certainty outside of the laboratory or in a specific instance of ulcer disease.

4.1. Cell Regeneration

The gastrointestinal mucosa has one of the most rapid rates of cellular turnover of any tissue in the body. The gastric mucosa replaces itself every 4–6 days and the duodenal mucosa 2–4 days.[67] If damaged cells are not replaced at the same rate as they are lost, mucosal resistance decreases. Therefore, factors which increase the rate of cell proliferation should give some protection against the development of ulcers. Because of the extremely rapid rate of cell turnover of the gastrointestinal mucosa, even the slightest change in the mitotic rate should have profound effects.

Hypophysectomy causes gastric mucosal atrophy and loss of cells.[68,69] These changes are largely prevented by growth hormone,[68] and Vanamee et al.[70] demonstrated that growth hormone partially, but significantly, protected rats from developing stress ulcers. In this experiment, growth hormone caused a dose-dependent decrease in the severity and number of ulcers developed in rats following a 24-hr period of restraint. The authors did not attempt to relate the effect of growth hormone to the rates of DNA synthesis or cell turnover, and the mechanism of its effect was not sought.

In a study similar in design to that of Vanamee et al.,[70] Takeuchi and Johnson[71] showed that gastrin was also capable of protecting against stress ulceration. In these studies, rats made hypogastrinemic by eating a liquid diet were much more susceptible to stress ulcer than animals fed solid food. Pentagastrin injection reduced the number and severity of ulcers in the liquid-fed rats. In all groups of rats, the ulcer index was

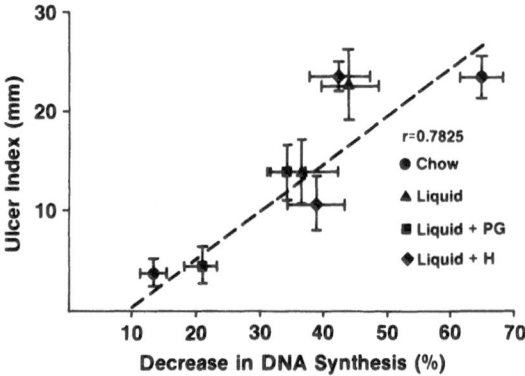

Fig. 8. Gastric ulcer index plotted against the percent decrease in DNA synthesis which occurred during the period of stress. Each point shows mean and S.E.M. for six observations. Two different time periods of stress are shown for each of the four groups. The correlation is $r = 0.782$, which is significant at $p < 0.01$. PG = rats injected with pentagastrin to stimulate DNA synthesis; H = rats injected with histamine as a control for acid secretion. Reproduced from Takeuchi and Johnson.[71]

directly related to the decrease in DNA synthesis seen during the period of stress (Fig. 8). Thus, we concluded that pentagastrin protected against ulcer formation because of its ability to stimulate mucosal growth.[71]

We also found that gastrin could protect against ulceration induced by aspirin.[72] Aspirin is commonly used to produce ulcers by topical application, and, although the process has been studied in depth, the exact mechanism of ulceration is unknown. However, it is likely that the mechanism differs significantly from that of stress-induced ulcers. Aspirin damage is accompanied by cell loss, for the DNA content of gastric washings increases after oral aspirin.[73,74] Croft[74] proposed that increased cell loss from the gastric mucosa, when associated with decreased cell production, disrupts the mucosal barrier. We found that decreased serum gastrin and the concomitant decrease in proliferation caused by feeding a liquid diet increased aspirin-induced damage. The damage could be partially prevented by increasing the mucosal cell proliferative rate with exogenous gastrin. The severity of lesions was directly proportional to the ratio of DNA loss to DNA synthesis.[72] Konturek et al.[75] have shown that epidermal growth factor also protects against aspirin-induced ulceration and suggested that protection was due to its ability to stimulate gastric mucosal DNA synthesis.[76] In the same study it was shown that prostaglandin also prevented aspirin-induced ulceration and the associated suppression of DNA synthesis. The antiulcer effects of both epidermal growth factor and prostaglandin occurred at doses insufficient to inhibit gastric acid secretion.

Thus, a number of agents are capable of preventing or reducing the severity of experimentally formed ulcers by virtue of their trophic activities. It should be pointed out that these effects occur over periods of several days of treatment and do not account for the cytoprotective effects of agents such as prostaglandin which can occur within minutes of administration.

4.2. Blood Flow

Among the many mysteries about ulcer remaining to be solved is why ulcers develop as focal defects rather than as diffuse mucosal damage. Except for those whose ulcers are caused by prolonged aspirin use, almost all patients with gastric ulcer have chronic atrophic gastritis. The ulcer always lies in an area involved by gastritis and tends to be at the uppermost (orad) border of the gastritis. In other words, the ulcer borders on healthy mucosa. Duodenal uclers as well as gastric ulcers develop in specific areas. A possible explanation is that the blood supply has been compromised in the area of ulcer development.

At least two mechanisms have been suggested to account for the role of blood flow in maintaining mucosal resistance. First, an adequate blood supply is necessary to deliver oxygen and nutrients to fuel aerobic metabolism. Second, tissue must be perfused sufficiently to remove metabolic wastes and H^+ from the mucosa. Using a rat hemorrhagic shock model, Menguy and Masters[77] showed that the gastric mucosa is extremely sensitive to decreased energy metabolism. Following hemorrhage, there was a 50–75% reduction in ATP levels coinciding with the development of mucosal lesions. In the same animals, liver tissue was much less severely affected and muscle not at all.[78] They found decreased glycogen levels in the mucosa and suggested that the vulnerability of gastric mucosal energy metabolism to ischemia resulted from a lack of stored glucose. They further proposed that the decrease in energy metabolism led to the rapid cellular necrosis during hypovolemic shock.[78] In other studies using an intramucosal pH probe, Kivilaakso et al.[79] studied the pH of the lamina propria in both dog and rabbit oxyntic gland mucosa exposed to acid during hemorrhagic shock. In the relatively permeable rabbit mucosa, the pH dropped quickly, coinciding with lesion formation. Ischemia decreased the pH of the canine mucosa more slowly, but when the ischemic tissue was exposed to 5 mM taurocholate, further disrupting the barrier, there was a rapid and drastic drop in pH which corresponded to severe lesion formation. The authors suggested that the critical determinant of ulceration during shock is not anoxia but an impaired capacity of the mucosa to remove or buffer the influx of acid.[79] In a similar set of experiments, Moody et al.[80] postulated that blood flow protects against barrier breakers such as aspirin by disposing of H^+ through buffers, dilution, or other processes which allow surface cells to withstand increased back-diffusion of acid.

In general, two types of experiments have been devised to demonstrate that mucosal blood flow plays an important role in the prevention of injury to the gastric mucosa. The first of these demonstrates that decreased mucosal blood flow is necessary for damage to occur. In the hypotensive rat, the gastric mucosa was ischemic and as little as 50 mmoles HCl produced lesions.[81] The normotensive rat with a healthy mucosa was resistant to 150 mmoles HCl. In other experiments, bile salt-induced damage to the stomach was exacerbated by reducing gastric mucosal blood flow.[82] Taurocholate alone increased mucosal blood flow and had only minor damaging effects. However, when combined with indomethacin to inhibit prostaglandin synthesis, blood flow was decreased and the number of lesions increased.[82]

The second presents evidence that increased mucosal blood flow prevents damage. Intraarterial infusion of isoproterenol significantly decreased lesions caused by topical aspirin or bile salts.[80,82] Prostaglandin has also been shown to protect against damage while at the same time increasing mucosal blood flow.[82]

The above observation indicates that prostaglandins may exert their protective effects by increasing mucosal blood flow. However, it should be pointed out that 16,16-dimethyl PGE_2 prevented ulceration in an amphibian gastric mucosal preparation lacking arterial perfusion.[83] Also, $PGF_{2\alpha}$, a vasoconstric-

tor, has antiulcer properties equipotent to PGE_2, which is a vasodilator.[84] Therefore, although considerable evidence suggests that mucosal perfusion influences the course of damage, one must ascertain whether changes in perfusion are coincidental or actually causative.

4.3. Mucus and Alkaline Secretion

The role of mucus in protecting the underlying gastrointestinal epithelium is unknown. The notion that mucus protects mucosa, and especially the gastric mucosa, is 36 years old.[85] Several mechanisms have been postulated to explain the protective action of mucus. These include its ability to buffer and neutralize acid, lubricant properties, antipeptic activity, and barrier function as an unstirred layer. Most firmly established is the ability of mucus to act as a lubricant and prevent actual physical damage. Recently, however, the discovery that the gastric mucosa secretes a significant amount of HCO_3^- has focused research on the ability of mucus to trap this secretion and establish a pH gradient between the mucosal cell surfaces and the luminal solution.

Heatley[85] hypothesized that mucus prevents mixing of the HCO_3^- with the excess HCl in the lumen allowing it to neutralize the small amount of acid diffusing through the gel to the mucosal cell surface. Flemström[86] has shown that the gastric mucosa secretes HCO_3^- in the absence of H^+ and that this amounts to 4–10% of maximal acid secretion. Williams and Turnberg[87] found that H^+ diffused through an unstirred layer of a control solution four times more rapidly than through an equivalent thickness of mucus. Pfeiffer[88] used linear diffusion and H^+ flux chambers to test the ability of various mucosubstances to retard the diffusion of H^+. Rabbit, whole, visible intestinal mucus, porcine gastric mucin, rat gastric neutral polysaccharide, and rat acid polysaccharide retarded H^+ diffusion in a dose-dependent manner. These findings suggested that mucus was responsible for allowing HCO_3^- to establish a pH gradient between the lumen and the epithelial cell surface.

Recently, Williams and Turnberg[89] used an antimony microelectrode to obtain evidence for a pH gradient across mucus adherent to rabbit gastric mucosa in vivo. Using an in vitro chamber with a slit lamp and a dissecting microscope to visualize the microelectrode, Takeuchi et al.[90] demonstrated a pH gradient from 3.0 in the lumen to 6.5 at the mucosal cell surface. The pH gradient was increased by increasing the amount of HCO_3^- on the serosal side and decreased by increasing the amount of acid on the luminal side. PGE_2 increased the thickness of the mucus layer from 441 to 637 μm. Using an in vivo preparation of rat duodenum, Flemström and Kivilaakso[91] found that acidification of the luminal solution from pH 7.6 to 5.0 doubled the amount of alkaline secretion; and if the pH were decreased to 2.0, alkaline secretion increased 600%. Even at luminal pH 2.0, the pH at the mucosal surface remained neutral. The authors concluded that alkaline secretion was sufficient to protect the duodenal mucosa from the amounts of acid encountered under normal conditions.

4.4. Cytoprotection

The term cytoprotection was used by Robert[84] to describe the ability of many prostaglandins to protect the gastric and intestinal mucosa from becoming inflamed and necrotic after being exposed to a variety of noxious agents. Cytoprotection was first described in the rat ileum with reference to the prevention of indomethacin-induced lesions by prostaglandin.[92]

Although the prostaglandins are potent inhibitors of gastric acid secretion, this does not appear to be the mechanism of cytoprotection. First, many prostaglandins are cytoprotective in doses as small as 1% of the threshold dose needed to inhibit gastric acid secretion.[93] Second, prostaglandins protect against a variety of damaging agents which are not acidic or do not use H^+ as the damaging agent. These include NaOH, hypertonic NaCl, alcohol, and boiling water.[93] Third, other antisecretory agents do not protect against damage by this same variety of agents.[84] Finally, prostaglandins can prevent damage induced by anti-inflammatory compounds in the nonacidic environment of the small intestine.[84]

Cytoprotection was originally described, and has most often been described, as the absence of visually evident lesions following treatment with prostaglandin in the presence of a damaging agent.[92,93] Little work has been done to histologically evaluate the protective effects of prostaglandin. Recently, Lacy and Ito[94] evaluated the morphological features of the effect of prostaglandin in preventing damage by absolute ethanol. They found that although orally administered 16,16-dimethyl PGE_2 prevented hemorrhagic necrotic lesions, it did not prevent the destruction of gastric mucosal cells. They suggested that the term *cytoprotection* was a misnomer, for cells were still destroyed in the presence of prostaglandin. They were unable to suggest a mechanism whereby prostaglandins prevent necrotic lesions. Thus, we really do not understand the structural correlates of the cytoprotection phenomenon.

The mechanism(s) underlying the protective action of prostaglandins has not been defined. Many have been suggested and are the subjects of a recent thorough review by Miller.[95] These are outside the scope of this chapter and include enhancement of mucosal blood flow, stimulation of mucus secretion, stimulation of HCO_3^- secretion, stimulation of DNA synthesis, stimulation of Na^+ and Cl^- transport, stabilization of tissue lysosomes, increased surface-active phospholipids, maintenance of mucosal sulfhydryl compounds, and dissolution of gastric mucosal folds.[95] Any mechanism which accounts for the protection afforded by prostaglandins must account for the following. First, essentially all prostaglandins have cytoprotective properties. Thus, enhancement of mucosal blood flow can be ruled out, for while prostaglandins A, E, and I are usually vasodilators, prostaglandins of the F type are constrictors. Second, the mechanism must account for the protection afforded to the wide variety of damaging agents. For example, it is difficult to see how the stimulation of alkaline secretion protects against damage by NaOH or boiling water. Finally, the protection afforded by prostaglandins appears to occur almost instantaneously following their administration. This is irrespective of whether they are administered orally or intravenously.[85] This time constraint rules out the stimulation of the synthesis of macromolecules (DNA, RNA, protein) as a general mechanism.

These characteristics of cytoprotection indicate that the mechanism may involve the enhancement of some general membrane characteristic. Along this line of thought, two concepts deserve mention. Recent studies by Szabo et al.[96] demonstrated that ethanol-induced damage is preceded by and associated with a significant decrease in mucosal sulfhydryl compounds. Pretreatment with $PGF_{2\beta}$ or sulfhydryl-containing drugs, such as cystamine, markedly reduced the formation of lesions. Agents such as N-ethylmaleimide, which inhibit sulfhydryl formation, caused damage and prevented the protection

induced by prostaglandin. Although numerous studies remain to be done, these findings suggest an interesting general mechanism at the membrane level.

Another mechanism operating at the cell membrane involves the surfactant properties of a variety of phospholipids. Lichtenberger et al.[97,98] have shown that endogenous and exogenous surface-active phospholipids can protect the gastric mucosa from damage by 0.6 N HCl. Previously thought to occur only in the lung, significant amounts of surfactant line the gastric mucosa.[99,100] Hills et al.[98] demonstrated that these compounds form a hydrophobic barrier over the mucosa, protecting it from damage by water-soluble substances. They have also shown that prostaglandins increased the concentration of intrinsic surfactants two- to sixfold.[97] Thus, endogenous prostaglandins may derive part of their protective function by mediating the amount of surfactant lining the mucosa.

These studies on the surfactant lining of the mucosa suggest yet another variable which may play a role in mucosal resistance. This concept is attractive because it involves general physicochemical events which might explain the rapidity and general nature of the cytoprotection offered by the prostaglandins. Much additional work is necessary to determine whether these compounds can protect against other types of damaging agents such as alcohol, aspirin and thermal injury.

However, if we are to understand the mechanisms of cytoprotection and its significance in the etiology of ulcer, the concept must be defined by histologists and morphologists using state-of-the-art techniques. Studies of protection by various agents which only quantify visible lesions tell little or nothing of what is occurring at the cellular level.

5. Conclusions

Peptic ulcer is best thought of as an imbalance between destructive factors acting on the mucosa and mucosal resistance. Both are probably important in every case of ulcer although the contribution of each certainly varies with different types of ulcer and in different individual cases of ulcer of the same type. Having said this, most evidence indicates that decreased mucosal resistance plays a greater role in gastric ulcer than it does in duodenal ulcer. On the other hand, increased acid and pepsin secretion is more important in the etiology of duodenal ulcer.

We are beginning to identify some of the factors which lead to increased acid secretion in duodenal ulcer. In general, these are readily defined and, therefore, measurable. Many have been found to correlate well with the presence of the disease in general. Although a pattern has emerged, however, so far it has been impossible to state the exact cause of any particular case of duodenal ulcer, excluding those produced by gastrinomas.

In order to further understand the mechanisms producing the factors which lead to increased secretion of acid, investigations will have to be carried out at the cellular, membrane, and organellar levels. Thus, for example, while patients with duodenal ulcer may show decreased inhibition of gastrin release in response to antral acidification, we have not begun to identify the defect responsible. Do the gastrin cells have increased numbers of receptors for protein products which release gastrin? Is their sensitivity increased? Are receptors which recognize changes in pH absent or insensitive? If somatostatin is responsible for the inhibition of gastrin release, are its levels sufficient? The questions go on and on.

The past 10 years has seen a great deal of research attention turned to what I have termed mucosal resistance as investigators have realized that acid and pepsin are not the only components of ulcer disease. Mucosal resistance has not been adequately defined and is, therefore, impossible to measure. However, we have made exciting progress in defining various components of resistance and measuring them. There are obviously many such components, with some still remaining to be identified. It is likely that a defect in any one of these could lead to an ulcer in a particular instance. Thus, except in extreme cases, it will probably be impossible to pinpoint specific alterations of mucosal resistance leading to a particular ulcer. Much of the excitement in this field has come from the concept of cytoprotection and the role the various prostaglandins may play in it. The term *cytoprotection*, however, has not been adequately defined histologically and means different things to different investigators (see Ref. 94). This will have to be rectified in order to properly interpret research results.

Investigations of mucosal resistance began with components easily measured and conceptualized such as blood flow and mucus secretion. These have been extended to include cell division and the concept of an alkaline unstirred layer. It is obvious that studies will soon encompass the cell membrane itself. Taken together, this work should provide a definition of mucosal resistance and perhaps its overall role in peptic ulcer disease.

References

1. Ellison, E. H., and S. D. Wilson. 1964. The Zollinger–Ellison syndrome: Re-appraisal and evaluation of 260 registered cases. *Ann. Surg.* 160:512–530.
2. Johnson, L. R. 1972. Pepsin secretion stimulated by topical hydrochloric and acetic acids. *Gastroenterology* 62:33–38.
3. Johnson, L. R. 1972. Regulation of pepsin secretion by topical acid in the stomach. *Am. J. Physiol.* 223:847–850.
4. Ivy, A. C., M. I. Grossman, and W. H. Bachrach. 1950. *Peptic Ulcer.* McGraw–Hill (Blakiston), New York.
5. Dragstedt, L. R. 1936. Acid ulcer. *Surg. Gynecol. Obstet.* 62:118–120.
6. Dameron, J. T., and O. H. Wangensteen. 1948. Effects produced by perfusion of jejunum with histamine and pilocarpine stimulated gastric juice. *Gastroenterology* 10:859–865.
7. Mann, F. C., and J. L. Bollman. 1932. Experimentally produced peptic ulcers; development and treatment. *J. Am. Med. Assoc.* 99:1576–1582.
8. Matzner, M. J., and C. Windwer. 1937. Pepsin versus hydrochloric acid in experimental production of gastric ulcer. *Am. J. Dig. Dis. Nutr.* 4:180–184.
9. Schiffrin, M. J., and A. A. Warren. 1942. Some factors concerned in the production of experimental ulceration of the GI tract in cats. *Am. J. Dig. Dis.* 9:205–209.
10. Perry, J. F., Jr., E. G. Yonchiro, P. M. Ya, H. D. Root, and O. H. Wangensteen. 1956. Digestive action of human gastric juice. *Proc. Soc. Exp. Biol. Med.* 92:237–240.
11. Goldberg, H. I., W. J. Dodds, S. Gee, C. Montgomery, and F. F. Zboralske. 1969. Role of acid and pepsin in acute experimental esophagitis. *Gastroenterology* 56:223–230.
12. Northrop, J. H. 1922. Mechanism of influence of acids and alkalies on the digestion of proteins by pepsin or trypsin. *J. Gen. Physiol.* 5:263–274.
13. Boyer, P. D., H. Lardy, and K. Myrbäck, eds. 1960. *The Enzymes,* 2nd ed., Volume 4. Academic Press, New York.
14. Bonnevie, O., H. Kallehauge, H. Wulff, and M. Wulff. 1971. Prognostic value of the augmented histamine test in ulcer disease and x-ray negative dyspepsia. *Scand. J. Gastroenterol.* 6:723–729.
15. Wormsley, K. G., and M. I. Grossman. 1965. Maximal histalog test in control subjects and patients with peptin ulcer. *Gut* 6:427–432.
16. Menguy, R. 1970. Pathophysiology of peptic ulcer. *Am. J. Surg.* 120:282–288.
17. Baron, J. H. 1978. Gastric secretion in duodenal ulcer disease. In: *Aspects of Peptic Ulceration.* J. E. Murphy, ed. Cambridge Medical Publications, Northampton.
18. Sander, L. D., A. M. Chander, and L. R. Johnson. 1975. Changes in liver and gastric mucosal hexosamine synthesis after restraint. *Gastroenterology* 68:285–293.
19. Takeuchi, K., S. Okabe, and K. Takagi. 1977. A new model of stress ulcer in rats with pylorus ligation and its pathogenesis. *Am. J. Dig. Dis.* 21:782–788.
20. Isenberg, J., C. T. Richardson, and J. S. Fordtran. 1978. Pathogenesis of peptic ulcer. In: *Gastrointestinal Disease.* M. H. Sleisenger and J. S. Fordtran, eds. Saunders, Philadelphia. pp. 792–806.
21. Rotter, J. I., and D. L. Rimoin. 1977. Peptic ulcer disease—A heterogeneous group of disorders? *Gastroenterology* 73:604–607.
22. Aoyagi, T., and W. H. J. Summerskill. 1966. Gastric secretion with ulcerogenic islet cell tumor: Importance of basal acid output. *Arch. Intern. Med.* 117:667–672.
23. Moore, J. G., and M. Wolfe. 1973. The relation of plasma gastrin to the circadian rhythm of gastric acid secretion in man. *Digestion* 9:97–105.
24. Troidl, H., W. Lorenz, H. Rohde, G. Hafner, and M. Ronzheimer. 1976. Histamine and peptic ulcer: A prospective study of mucosal histamine concentration in duodenal ulcer patients and in control subjects suffering from various gastrointestinal diseases. *Klin. Wochenschr.* 54:947–956.
25. Domschke, W., N. Subramanian, P. Mitznegg, H. W. Baenkler, S. Domschke, and E. Wansch. 1977. Gastric mucosal histamine in duodenal ulcer patients: Release by secretin. *Acta Hepato-Gastroenterol.* 24:444–446.
26. Peden, N. R., H. Callachan, D. M. Shepherd, and K. G. Wormsley. 1982. Gastric mucosal histamine and histamine methyltransferase in patients with duodenal ulcer. *Gut* 23:58–62.
27. Isenberg, J. I., M. I. Grossman, V. Maxwell, and J. H. Walsh. 1975. Increased sensitivity to stimulation of acid secretion by pentagastrin in duodenal ulcer. *J. Clin. Invest.* 55:330–337.
28. Lam, S. K., J. I. Isenberg, M. I. Grossman, W. H. Lane, and J. H. Walsh. 1980. Gastric acid secretion is abnormally sensitive to endogenous gastrin released after peptone test meals in duodenal ulcer patients. *J. Clin. Invest.* 65:555–562.
29. Kirkpatrick, J. R., J. H. Lawrie, A. P. M. Forrest, and H. Campbell. 1969. The short pentagastrin test in the investigation of gastric disease. *Gut* 10:760–762.
30. Cheng, F. C. Y., S. K. Lam, and G. B. Ong. 1977. Maximum acid output to graded doses of pentagastrin and its relation to parietal cell mass in Chinese patients with duodenal ulcer. *Gut* 18:827–832.
31. Fordtran, J. S., and J. H. Walsh. 1973. Gastric acid secretion rate and buffer content of the stomach after eating. *J. Clin. Invest.* 52:645–657.
32. Bodeman, G., A. Walan, and G. Lundquist. 1978. Food-stimulated acid secretion measured by intragastric titration with bicarbonate in patients with duodenal and gastric ulcer disease and in controls. *Scand. J. Gastroenterol.* 13:911–918.
33. Malagelada, J. R., G. F. Longstreth, T. B. Deering, W. H. J. Summerskill, and V.-L. W. Go. 1977. Gastric secretion and emptying after ordinary meals in duodenal ulcer. *Gastroenterology* 73:989–994.
34. Cox, A. M. 1952. Stomach size and its relation to chronic peptic ulcer. *Arch. Pathol.* 54:407–422.
35. Makhlouf, G. M., J. P. A. McManus, and W. I. Card. 1964. Dose–response curves for the effect of gastrin II on acid secretion in man. *Gut* 5:379–384.
36. Makhlouf, G. M., J. P. A. McManus, and W. I. Card. 1966.

Action of the pentapeptide (ICI 50, 123) on gastric secretion in man. *Gastroenterology* **51**:455–464.

37. Rotter, J. I., J. Q. Sones, I. M. Samloff, C. T. Richardson, J. M. Gursky, J. M. Walsh, and D. L. Rimoin. 1979. Duodenal ulcer disease associated with elevated serum pepsinogen I: An inherited autosomal dominant disorder. *N. Engl. J. Med.* **300**:63–66.

38. Ellison, E. H., and S. D. Wilson. 1967. Further observations on factors influencing the symptomatology manifest by patients with Zollinger–Ellison syndrome. In: *Gastric Secretion.* T. K. Shnitka, J. A. L. Gilbert, and R. C. Harrison, eds. Pergamon Press, Elmsford, N.Y. pp. 363–369.

39. Johnson, L. R. 1976. The trophic action of gastrointestinal hormones. *Gastroenterology* **70**:278–288.

40. Johnson, L. R., D. Aures, and L. Yuen. 1969. Pentagastrin-induced stimulation of protein synthesis in the gastrointestinal tract. *Am. J. Physiol.* **217**:251–254.

41. Walsh, J. H. 1978. Pathogenic role of the gastrins. In: *Gastrins and the Vagus.* J. F. Rehfeld and E. Amdoup, eds. Academic Press, New York. pp. 181–188.

42. DiMagno, E. P., and V.-L. W. Go. 1974. Malabsorption secondary to antral gastrin-cell hyperplasia. *Mayo Clin. Proc.* **49**: 727–730.

43. Lamers, C. B. M., C. M. Ruland, H. J. M. Joosten, H. C. M. Verkooyen, J. H. M. Van Tongeren, and J. E. Rehfeld. 1978. Hypergastrinemia of antral origin in duodenal ulcer. *Am. J. Dig. Dis.* **23**:998–1002.

44. Straus, E. 1978. Radioimmunoassay of gastrointestinal hormones. *Gastroenterology* **74**:141–152.

45. McGuigan, J. E., and W. L. Trudeau. 1973. Differences in rates of gastrin release in normal persons and patients with duodenal ulcer disease. *N. Engl. J. Med.* **288**:64–66.

46. Walsh, J. H., C. T. Richardson, and J. S. Fordtran. 1975. pH dependence on acid secretion and gastrin release in normal and ulcer subjects. *J. Clin. Invest.* **55**:462–468.

47. Malmstrom, J., and F. Stadil. 1975. Measurement of immunoreactive gastrin in gastric mucosa. *Scand. J. Gastroenterol.* **10**:433–439.

48. Malmstrom, J., F. Stadil, and J. F. Rehfeld. 1976. Gastrins in tissue. *Gastroenterology* **70**:697–703.

49. Creutzfeldt, W., R. Arnold, C. Creutzfeldt, and N. S. Track. 1976. Mucosal gastrin concentration, molecular forms of gastrin, number and ultrastructure of G-cells in patients with duodenal ulcer. *Gut* **17**:745–754.

50. Ou Tim, L. O., B. I. Hirschowitz, and E. Molina. 1983. Regulatory defects of gastric secretion in duodenal ulcer (DU) revealed by bombesin and G-17 dose response in DU and normals. *Gastroenterology* **84**:1267a.

51. Takeuchi, K., W. Peitsch, and L. R. Johnson. 1981. Mucosal gastrin receptor. V. Development in newborn rats. *Am. J. Physiol.* **240**:G163–G169.

52. Takeuchi, K., G. R. Speir, and L. R. Johnson. 1980. Mucosal gastrin receptor. III. Regulation by gastrin. *Am. J. Physiol.* **238**:G135–G140.

53. Speir, G. R., K. Takeuchi, W. Peitsch, and L. R. Johnson. 1982. Mucosal gastrin receptor. VII. Up and down regulation. *Am. J. Physiol.* **242**:G243–G249.

54. Samloff, I. M. 1981. Pepsins, peptic activity and peptic inhibitors. *J. Clin. Gastroenterol.* **3**(Suppl. 2):91–94.

55. Kamegashira, M., R. L. Goodale, Jr., J. W. Borner, U. Dhingra, and O. H. Wangensteen. 1977. Role of pepsin in species differences in erosive gastritis. *Arch. Surg.* **112**:193–197.

56. Joffe, S. N., N. B. Roberts, W. H. Taylor, and J. H. Baron. 1980. Exogenous and endogenous acid and pepsins in the pathogenesis of duodenal ulcer in the rat. *Dig. Dis. Sci.* **25**:837–841.

57. Samloff, I. M., W. M. Liebman, and N. M. Panitch. 1975. Serum group I pepsinogens by radioimmunoassay in control subjects and patients with peptic ulcer. *Gastroenterology* **69**:83–90.

58. Teorell, T. 1939. On the permeability of the stomach mucosa for acids and some other substances. *J. Gen. Physiol.* **23**:263–274.

59. Code, C. F., J. F. Scholer, A. L. Orvis, and J. A. Higgins. 1955.

Barrier offered by gastric mucosa to absorption of sodium. *Am. J. Physiol.* **183**:604a.

60. Davenport, H. W., H. A. Warner, and C. F. Code. 1964. Functional significance of the gastric mucosal barrier to sodium. *Gastroenterology* **47**:142–152.

61. Davenport, H. W. 1964. Gastric mucosal injury by fatty and acetylsalicylic acids. *Gastroenterology* **46**:245–253.

62. Davenport, H. W. 1965. Damage to the gastric mucosa: Effects of salicylates and stimulation. *Gastroenterology* **49**:189–196.

63. Davenport, H. W. 1965. Potassium fluxes across the resting and stimulated gastric mucosa: Injury by salicylic and acetic acids. *Gastroenterology* **49**:238–245.

64. Davenport, H. W. 1966. Fluid produced by the gastric mucosa during damage by acetic and salicylic acids. *Gastroenterology* **50**: 487–499.

65. Davenport, H. W. 1965. Is the apparent hyposecretion of acid by patients with gastric ulcer a consequence of a broken barrier to diffusion of hydrogen ions into the gastric mucosa? *Gut* **6**:513.

66. Davenport, H. W. 1982. The Gastric Mucosal Barrier—A Swan Song. Privately published, Ann Arbor, Mich.

67. MacDonald, W. C., J. S. Trier, and N. B. Everett. 1964. Cell proliferation and migration in the stomach, duodenum, and rectum of man: Radioautographic studies. *Gastroenterology* **46**:405–417.

68. Crean, G. P. 1963. The endocrine system and the stomach. *Vitam. Horm. (N.Y.)* **21**:215–280.

69. Jacobson, E. D., and T. J. Magnani. 1964. Some effects of hypophysectomy on gastrointestinal structure and function. *Gut* **5**: 472–479.

70. Vanamee, P., S. J. Winawer, P. Sherlock, M. Sonenberg, and M. Lipkin. 1970. Decreased incidence of restraint–stress induced gastric erosions in rats treated with bovine growth hormone. *Proc. Soc. Exp. Biol. Med.* **135**:259–262.

71. Takeuchi, K., and L. R. Johnson. 1979. Pentagastrin protects against stress ulceration in rats. *Gastroenterology* **76**:327–334.

72. Takeuchi, K., and L. R. Johnson. 1982. Effect of cell proliferation and loss on aspirin induced gastric damage in the rat. *Am. J. Physiol.* **243**:G463–G468.

73. Croft, D. N., D. J. Pollock, and N. F. Coghill. 1966. Cell loss from human gastric mucosa measured by the estimation of deoxyribonucleic acid (DNA) in gastric washings. *Gut* **7**:333–343.

74. Croft, D. N. 1977. Cell turnover and loss and the gastric mucosal barrier. *Am. J. Dig. Dis.* **22**:383–386.

75. Konturek, S. J., T. Radecki, T. Brozozowski, I. Piastucki, A. Dembinski, A. Dembinska-Kiec, A. Zmuda, R. Gruglewski, and H. Gregory. 1981. Gastric cytoprotection by epidermal growth factor. *Gastroenterology* **81**:838–843.

76. Johnson, L. R., and P. D. Guthrie. 1980. Stimulation of rat oxyntic gland mucosal growth by epidermal growth factor. *Am. J. Physiol.* **238**:G45–G49.

77. Menguy, R., and Y. F. Masters. 1974. Gastric mucosal energy metabolism and "stress ulceration." *Ann. Surg.* **180**:538–548.

78. Menguy, R., L. Desbaillets, and Y. F. Masters. 1974. Mechanism of stress ulcer: Influence of hypovolemic shock on energy metabolism in the gastric mucosa. *Gastroenterology* **66**:46–55.

79. Kivilaaksko, E., D. Fromm, and W. Silen. 1978. Relationship between ulceration and intramural pH of gastric mucosa during hemorrhagic shock. *Surgery* **84**:70–78.

80. Moody, F. G., J. McGreavy, C. Zalewski, L. Y. Cheung, and M. Simons. 1977. The cytoprotective effect of mucosal blood flow in experimental erosive gastritis. *Acta Physiol. Scand* (Spec. Suppl.) 35–43.

81. Mersereau, W. A., and E. J. Hinchey. 1973. Effect of gastric acidity on gastric ulceration induced by hemorrhage in the rat utilizing a gastric chamber technique. *Gastroenterology* **64**:1130–1135.

82. Whittle, B. J. R. 1977. Mechanisms underlying gastric mucosal damage induced by indomethacin and bile salts, and the actions of prostaglandins. *Br. J. Pharmacol.* **60**:455–460.

83. Barzilai, A., R. Schiessel, E. Kivilaakso, J. B. Matthews, L. A. Fleischer, G. Bartzokis, and W. Silen. 1980. Effect of 16,16

dimethyl prostaglandin E$_2$ on ulceration of isolated amphibian gastric mucosa. *Gastroenterology* **78**:1508–1512.

84. Robert, A. 1979. Cytoprotection by prostaglandins. *Gastroenterology* **77**:761–767.

85. Heatley, N. G. 1959. Mucosubstance as a barrier to diffusion. *Gastroenterology* **37**:313–317.

86. Flemström, G. 1977. Active alkalinization by amphibian gastric mucosa *in vitro*. *Am. J. Physiol.* **233**:E1–E12.

87. Williams, S. E., and L. A. Turnberg. 1980. Retardation of acid diffusion by pig gastric mucus: Potential role in mucosal protection. *Gastroenterology* **79**:299–304.

88. Pfeiffer, C. J. 1981. Experimental analysis of hydrogen ion diffusion in gastrointestinal mucus glycoprotein. *Am. J. Physiol.* **240**:G176–G182.

89. Williams, S. E., and L. A. Turnberg. 1981. Demonstration of a pH gradient across mucus adherent to rabbit gastric mucosa: Evidence for a "mucus–bicarbonate" barrier. *Gut* **22**:94–96.

90. Takeuchi, K., D. Magee, J. Critchlow, J. Matthews, and W. Silen. 1983. Studies of the pH gradient and thickness of frog gastric mucus gel. *Gastroenterology* **84**:331–340.

91. Flemström, G., and E. Kivilaakso. 1983. Demonstration of a pH gradient at the luminal surface of rat duodenum *in vivo* and its dependence on mucosal alkaline secretion. *Gastroenterology* **84**:787–794.

92. Robert, A. 1975. An intestinal disease produced experimentally by a prostaglandin deficiency. *Gastroenterology* **69**:1045–1047.

93. Robert, A., J. E. Nezamis, C. Lancaster, and A. J. Hanchar. 1979. Cytoprotection by prostaglandins in rats: Prevention of gastric necrosis produced by alcohol, HCl, NaOH, hypertonic NaCl, and thermal injury. *Gastroenterology* **77**:433–443.

94. Lacy, E. R., and S. Ito. 1982. Microscopic analysis of ethanol damage to rat gastric mucosa after treatment with a prostaglandin. *Gastroenterology* **83**:619–625.

95. Miller, T. A. 1983. Protective effects of prostaglandins against gastric mucosal damage: Current knowledge and proposed mechanisms. *Am. J. Physiol.*. **245**:G601–G623.

96. Szabo, S., J. S. Trier, and P. W. Frankel. 1981. Sulfhydryl compounds may mediate gastric cytoprotection. *Science* **214**:200–202.

97. Lichtenberger, L. M., L. A. Graziani, E. J. Dial, B. D. Butler, and B. A. Hills. 1983. Role of surface-active phospholipids in gastric cytoprotection. *Science* **219**:1327–1329.

98. Hills, B. A., B. D. Butler, and L. M. Lichtenberger. 1983. Gastric mucosal barrier: Hydrophobic lining to the lumen of the stomach. *Am. J. Physiol.* **244**:G561–G568.

99. Wassef, M. K., Y. N. Lin, and M. I. Horowitz. 1978. Phospholipid deacylating enzymes of rat stomach mucosa. *Biochim. Biophys. Acta* **528**:318–330.

100. Wassef, M. K., Y. N. Lin, and M. I. Horowitz. 1979. Molecular species of phosphatidylcholine from rat gastric mucosa. *Biochim. Biophys. Acta* **573**:222–226.

Malabsorption Syndromes

Henrik Westergaard and John M. Dietschy

1. Introduction

The term *malabsorption syndrome* is generally applied to disease processes that result in significant fecal losses of one or more of the three major caloric sources of the diet, i.e., fats, proteins, and carbohydrates.

Viral or bacterial enterocolitis may cause a brief period of malabsorption of electrolytes and water but are usually self-limited diseases and not included in the wide spectrum of more chronic diseases that cause the malabsorption syndrome. Fat absorption is a more complex process than protein or carbohydrate absorption and involves more and uniquely different steps in the overall absorptive process which renders fat absorption more vulnerable. Consequently, the most common feature of the malabsorption syndrome is increased fecal fat excretion, i.e., steatorrhea. For this reason, the terms *malabsorption syndrome* and *steatorrhea* are often used as synonyms. In many instances, however, patients with malabsorption syndrome also have increased loss of protein and carbohydrates in the feces but these losses are only infrequently documented clinically.[1]

A typical daily Western diet consists of 100–150 g of fat, 50–100 g of protein, and 300–400 g of carbohydrate as well as water, electrolytes, and vitamins. Normal intestinal absorption of dietary constituents is a very efficient process as evidenced by the fact that the daily stool weight is only 150–200 g of which more than 60% is water. The dry weight is only 20–30 g and consists mostly of bacteria with minimal amounts of fat, protein, and carbohydrate.[2]

The normal mechanisms of intestinal digestion and absorption are complex and depend on the functional integrity of the following four physiological systems: secretion of a mixture of digestive enzymes from the pancreas; maintenance of adequate concentrations of bile acids in the enterohepatic circulation; an adequate number of small intestinal mucosal cells to ensure a normal rate of absorption; and an adequate lymph and blood

supply to the intestinal villi to transport the absorbed dietary components to other organs. Bile acids are a unique requirement in fat absorption, and a compromised enterohepatic circulation does not interfere with protein or carbohydrate absorption. In contrast, disease processes that disrupt the normal function of the three other systems may cause malabsorption of fats, carbohydrates, and proteins. Therefore, the approach to the patient with a suspected malabsorption syndrome requires a complete understanding of the basic physiological mechanisms involved in intestinal digestion and absorption in order to define the specific defect with a limited number of tests.[3,4] This chapter reviews the normal mechanisms of fat, protein, and carbohydrate digestion and absorption, describes some of the clinical tests of digestive and absorptive functions, and examines the effect of specific diseases on these processes.

2. Lipid Digestion and Absorption

More than 90% of ingested lipid consists of triglycerides, i.e., three long-chain fatty acids esterified to the three hydroxyl groups of a glycerol molecule, and the remaining 5–10% is a mixture of phospholipids, sterols such as cholesterol, and fat-soluble vitamins. Lipid digestion begins in the stomach with two events. First, the lipids are mechanically broken down to smaller pieces by the churning action of the stomach which essentially functions as a mixing reservoir. As the lipids are insoluble in the water phase of the stomach contents, they are gradually converted to emulsion particles of a rather uniform size (1–2 μm) and will separate out as an oil phase on top of the water phase upon centrifugation of stomach contents. Second, an initial lipolysis commences in the stomach as the emulsion particles are attacked by lingual lipase, a recently discovered enzyme that is secreted from the von Ebner glands at the base of the tongue. The lingual lipase has a pH optimum of 3 to 5 and thus functions in the acidic milieu of the stomach and does not require cofactors or bile acids for activity. This enzyme only performs a limited lipolysis with release of mainly di- and monoglycerides from the emulsion particles. The quantitative importance of gastric lipolysis in adults has yet to be evaluated but lingual lipase plays a definite role in lipid digestion in the newborn.[5]

Henrik Westergaard and John M. Dietschy • Department of Internal Medicine, Southwestern Medical School, University of Texas Health Science Center, Dallas, Texas 75235.

The emulsion particles are gradually propelled into the duodenum by gastric peristalsis. The presence of fat in the duodenum causes release of two hormones, cholecystokinin and secretin, from the duodenal mucosa into the systemic circulation. The increase in cholecystokinin concentration results in gallbladder contraction, emptying of bile into the duodenum, and stimulation of pancreatic enzyme secretion. Secretin stimulates water and electrolyte secretion from the pancreatic acini and ducts. As a result, the emulsion particles in the duodenum are bathed in gallbladder bile with a high concentration of bile acids and pancreatic secretions rich in lipolytic enzymes. The most important lipolytic enzymes from the pancreas are lipase, cholesterol esterase, and various phospholipases. The pancreatic lipase attaches to the oil–water interface of the emulsion particles and requires a cofactor, called colipase, to be active. Colipase is also secreted from the pancreas and has been shown to complex with lipase in a 1 : 1 relationship in the presence of substrate. In the absence of colipase in *in vitro* systems, lipase is rapidly inactivated as it binds to the oil–water interface of the emulsion droplets. In the presence of colipase and substrate, the apparent association constant between lipase and colipase is increased by several orders of magnitude and it has been estimated that 70–90% of lipase and colipase will exist as a complex on the interface of the lipid droplets. The complex prevents substrate inactivation of lipase, and colipase serves to anchor the lipase to the substrate interface.[6] Colipase also has a binding site for bile acid micelles but whether this binding occurs during *in vivo* digestion remains to be proven. Lipase hydrolyzes the 1 and 3 ester bonds of the triglycerides, resulting in formation of fatty acids and monoglycerides as shown in Fig. 1 (step I). Cholesteryl esters and phospholipids in the emulsion particles are similarly hydrolyzed by cholesterol esterase and phospholipase A_2, respectively, and cholesterol, lysolecithin, and fatty acids are released to the aqueous environment. The

lipolytic products have only limited solubility in the aqueous phase of the intestinal contents as a result of the hydroxyl or carboxyl groups that can interact with water molecules.

Gallbladder bile has a high concentration of conjugated bile acids which are potent detergents and associate to form macromolecular complexes, called micelles, above a certain concentration. The micelles are roughly spherical with hydrophilic groups on the exterior surface and a lipophilic interior. The micelles have a high aqueous solubility and can solubilize the lipolytic products in the lipophilic interior by a process known as micellar solubilization (Fig. 1, step II). The secretion of bile acid micelles increases the solubility of the lipolytic products manyfold.[7] By this means, further lipolysis is not suppressed by product inhibition and the lipolytic products are rapidly removed from the emulsion particles. The fatty acids in the micelle interior are in rapid equilibrium with fatty acids in the aqueous phase and likewise monomeric bile acids in the aqueous phase are in equilibrium with bile acids in the micelles. The concentration of fatty acids and bile acids in the aqueous phase is determined by the partition coefficient of a particular fatty acid between bile acid micelles and water and the critical micellar concentration for a given bile acid, respectively. It has been estimated that the diffusion coefficient of mixed micelles is one order of magnitude below that of fatty acids in the water phase. The mass of fatty acids in the micellar phase, however, may be 10^2 to 10^3 higher than in the water phase, so that the mass of lipid moved per unit time is increased by 10- to 100-fold by micellar solubilization despite a slower rate of diffusion.

It was originally demonstrated that small intestinal contents during a meal could be separated into two phases by centrifugation: an oil phase and a micellar phase.[8] More recent studies, however, have shown that at least four different phases may coexist during lipid digestion.[9] These observations were made *in vitro* by watching lipid digestion on a microscope slide. Dur-

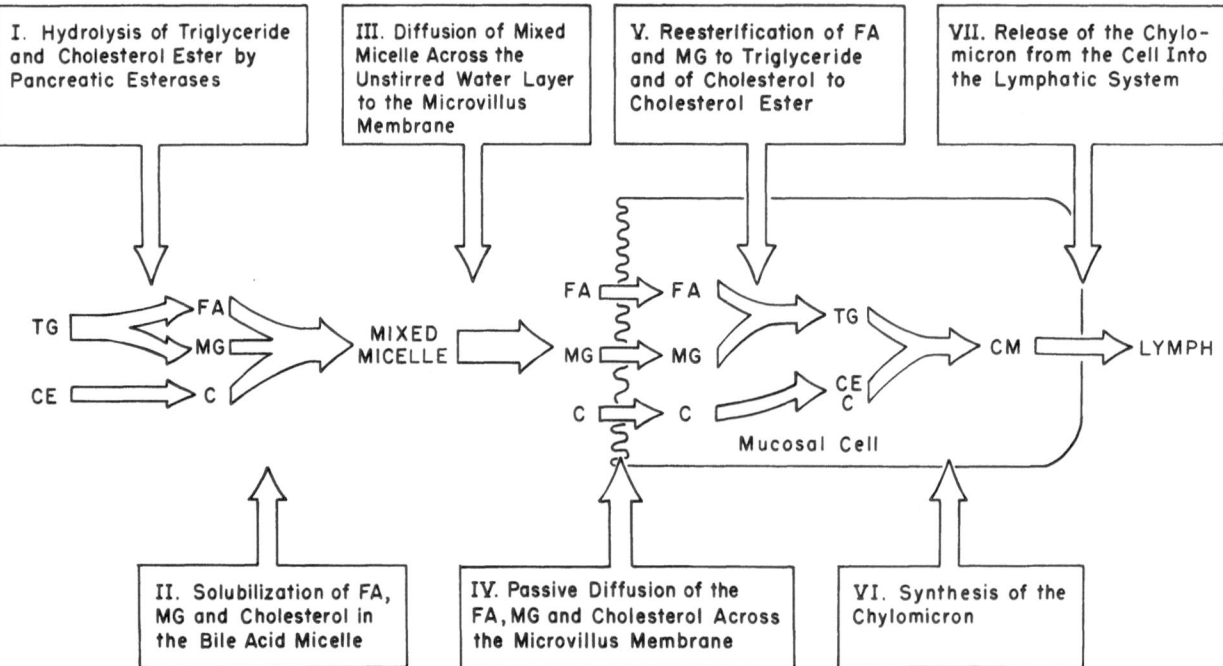

Fig. 1. The major steps involved in the digestion and absorption of dietary fat. TG, triglyceride; CE, cholesteryl ester; FA, long-chain fatty acids; MG, β-monoglyceride; C, unesterified cholesterol; CM, chylomicrons.

ing lipolysis the spherical oil droplets (triglycerides) gradually decreased in size and the first product formed was calcium soaps which are insoluble. The formation of calcium soaps was inhibited by monoglycerides and saturated micelles. The next phase observed under the microscope is called the viscous isotropic phase. The chemical composition of this phase is unknown but mixtures of protonated fatty acids and monoglycerides in water appear optically identical to the viscous isotropic phase. The geometric arrangement of the lipids in water is poorly understood, but it may be similar to the isotropic cubic phase of lipid–water systems.[10] Probes of nonpolar lipids in the oil droplet were quantitatively transferred into the viscous isotropic phase during lipolysis. So far, it has not been possible to isolate this phase. The isotropic viscous phase is recovered in the oil phase if saturated bile acid micelles are present and in the micellar phase in the presence of unsaturated micelles. The role of this phase during lipid digestion in vivo has not been established, but it may be of importance in the absorption of nonpolar lipids.[11] The last phase is the micellar phase consisting of mixed bile acid micelles. The primary function of this phase is to solubilize the lipolytic products as they are released from the oil droplets or the viscous isotropic phase and to transport these products to the sites of absorption at the microvillus membrane.

There are two barriers to lipid absorption in the small intestine—an unstirred layer of water overlying the villi of presumably varying dimensions, depending on peristaltic activity, and the lipid–protein moiety of the microvillus membrane.[12] The rate of movement of the mixed micelles across the unstirred water layer is proportional to the concentration gradient, the diffusion coefficient of the mixed micelle, and the absorptive surface area and inversely proportional to the thickness of the unstirred water layer. Micellar solubilization increases the concentration gradient of the lipolytic products across the unstirred water layer manyfold. The mixed micelle may thus be viewed as a carrier of a large mass of lipolytic products across this layer and by this means facilitates lipid absorption (Fig. 1, step III). The dependence of bile acid solubilization of the lipolytic products is related to the polarity of these compounds. Short- and medium-chain fatty acids are relatively polar compounds and thus do not require bile acid solubilization, but fatty acids become increasingly nonpolar as the chain length is increased and sterol molecules such as cholesterol are essentially insoluble in water. Long-chain fatty acids and cholesterol both require micellar solubilization to generate a high concentration gradient across the unstirred water layer to sustain a high rate of fat absorption. Accordingly, bile acid deficiency will have a profound effect on cholesterol absorption and a moderate effect on long-chain fatty acid absorption but medium- and short-chain fatty acid absorption will proceed unchanged. Thus, the physiological importance of micellar solubilization in intestinal fat absorption depends on the chemical characteristics of a particular class of lipid molecules.

At the microvillus membrane interface, the mixed micelles maintain a finite aqueous concentration of the lipolytic products as determined by the equilibrium constants, and it is assumed that actual uptake proceeds by diffusion of monomeric fatty acids, monoglycerides, and cholesterol from the aqueous phase into the lipid bilayer of the microvillus membrane and into the cytosol of the absorptive cells (Fig. 1, step IV). The rate of uptake is determined by the permeability coefficient of the particular lipid species and the aqueous concentration of that species at the interface. Once inside the cell, the fatty acids are rapidly bound to specific fatty acid-binding proteins[13] and

transported to the smooth endoplasmic reticulum (SER) where rapid reesterification takes place. These two processes serve to maintain a low fatty acid concentration inside the cell and thus sustain the concentration gradient across the microvillus membrane as fatty acids, monoglycerides, and cholesterol are constantly delivered to the aqueous environment at the interface by the mixed micelles. The net effect of these intraluminal processes, lipolysis and bile acid solubilization, is to generate a high concentration gradient between the intestinal lumen and the interior of the absorptive cells to facilitate a rapid rate of lipid uptake. This is illustrated by the fact that in normal man, lipid absorption is almost complete in the proximal jejunum.[14] Bile acid micelles are in equilibrium with monomeric bile acids which are also taken up by the enterocytes. The rate of uptake is low, however, since the conjugated bile acids have very low permeability coefficients in the proximal small intestine.[15] Consequently, a high concentration of bile acid micelles is maintained in the proximal small intestine and it is assumed that the micelles are reutilized in the solubilization process and shuttle back and forth across the unstirred water layer.

In the intestinal absorptive cells, the lipolytic products diffuse to the SER in the apical portion of the cells where reesterification takes place (Fig. 1, step V). The enzymes for these reesterification processes are localized in the SER which is corroborated morphologically by electron microscopy which shows accumulation of triglyceride droplets in the apical portion of the enterocytes in the early stage of lipid absorption.[16] Triglyceride resynthesis proceeds via both monoglyceride and α-glycerophosphate pathways but the monoglyceride pathway appears to be the most important and most efficient from an energetic point of view.[17] The newly resynthesized triglyceride droplets also acquire cholesteryl esters in their nonpolar interior and phospholipids on the surface, and are transported to the Golgi apparatus in the supranuclear portion of the cells where important apoproteins are incorporated into the surface of the particle and the triglyceride droplets are now converted to chylomicrons (Fig. 1, step VI). The three major apoproteins synthesized in intestinal epithelial cells are apo B (B-48), apo A-IV, and apo A-I of which apo B is crucial for actual excretion from the epithelial cells.[18] The chemical composition of chylomicrons is shown in Table I, and triglycerides are the major lipid components of these particles. The actual mechanism by which chylomicrons leave the intestinal cells is unknown but it has been shown that Golgi vesicles fuse with the basolateral membrane and release the chylomicrons to the central lacteals in the villus core (Fig. 1, step VII). The size of the chylomicrons (0.1–0.5 μm) precludes movement through the loose junctions between the endothelial cells of the capillaries but the spaces between the endothelial cells of the central lacteals are large enough to allow entry of the chylomicrons, which are then transported by the lymphatics to the systemic circulation. Thus, the lipolytic products of the digestive process require an intact lymphatic system for absorption, whereas the water-soluble products pass directly into the capillary network of the villus to reach the portal vein. In the systemic circulation, the chylomicrons gain additional apoproteins, including apo CII, which serves as an activator of lipoprotein lipase. This enzyme is located on the inner surface of capillaries in muscle, heart, and adipose tissue and its main function is to hydrolyze the triglycerides of the chylomicron particle.[19] Fatty acids are released and diffuse into the surrounding tissues. In muscle and heart, the fatty acids are oxidized for energy, whereas in adipose tissue, they may again be reesterified and stored as triglyceride. After a

Table I. Human Lymph Chylomicron Composition

Triglycerides	91%
Phospholipids	7.5%
Cholesterol	1.6%
Protein	1.3%

certain amount of triglycerides has been removed from the chylomicron particle, the particle is converted to a "remnant" which is rapidly cleared from the circulation by a specific transport system in the liver.[20] At this point, fat absorption is complete and most of the dietary triglycerides have been stored or utilized in peripheral tissues while most of the dietary cholesterol (and fat-soluble vitamins) have been taken up with the remnant in the liver.

3. Normal Enterohepatic Circulation of Bile Acids

As outlined in the previous section, normal fat digestion and absorption involve a complex series of events. The second step (Fig. 1) in this sequence is micellar solubilization of the lipolytic products which requires an adequate concentration of bile acid micelles in the proximal intestine during the digestive–absorptive process, which, in turn, requires an intact enterohepatic circulation of the bile acid pool. Any disease process that interrupts this recirculation may result in an insufficient intraluminal concentration of bile acid micelles for the solubilization process and a fat malabsorption syndrome may ensue.

Bile acids are synthesized from cholesterol in the liver and although they serve an important function in intestinal fat absorption, they can also be viewed as degradation products of cholesterol since bile acid loss in the feces is one important pathway for cholesterol excretion. During bile acid synthesis in

the hepatocytes, a three-carbon fragment is cleaved from the side chain of the cholesterol molecule and the carbon atom at the 24 position is oxidized to a carboxylic group. The A and B ring of the sterol nucleus is changed from a *trans* to a *cis* position and hydroxyl groups are added in the α position to carbon atom 7 (chenodeoxycholic acid) or 7 and 12 (cholic acid), which are the primary bile acids. Finally, these bile acids are conjugated to either glycine or taurine. These hydroxylation and conjugation processes lower the pK_a and increase the water solubility of the bile acids. The conjugated bile acids are excreted into the bile ducts by an active transport mechanism and stored in the gallbladder. The hepatic bile is converted to gallbladder bile by isosmotic salt and water absorption. Thus, gallbladder bile is transformed to a concentrated solution of conjugated bile acids.

In the distal small intestine and colon, the conjugated bile acids are exposed to intestinal bacteria that contain enzymes capable of deconjugating bile acids and of removing the hydroxyl groups in the C-7 position. Thus, the conjugated primary bile acids, glycocholic and taurocholic acid, are converted to deoxycholic acid, and glycochenodeoxycholic and taurochenodeoxycholic acid are converted to lithocholic acid. Deoxycholic and lithocholic acid are secondary bile acids and both are absorbed in the intestine, conjugated in the liver with glycine or taurine, and the conjugates of lithocholic acid are further sulfated before excretion into the bile. Thus, hepatic bile contains eight different bile acids in varying amounts—the glycine and taurine conjugates of cholic, chenodeoxycholic, deoxycholic, and lithocholic acid.[21]

These eight bile acids then constitute the bile acid pool which recirculates from liver to intestine and back to the liver via the portal vein in the enterohepatic circulation. The quantitative aspects of the dynamics of the enterohepatic circulation are shown in Fig. 2. The total bile acid pool size is only about 3 g. The daily bile acid synthesis from hepatic cholesterol is about 0.7 g which is balanced by a daily fecal loss of an equal amount. Bile acid synthesis is regulated by a negative feedback mecha-

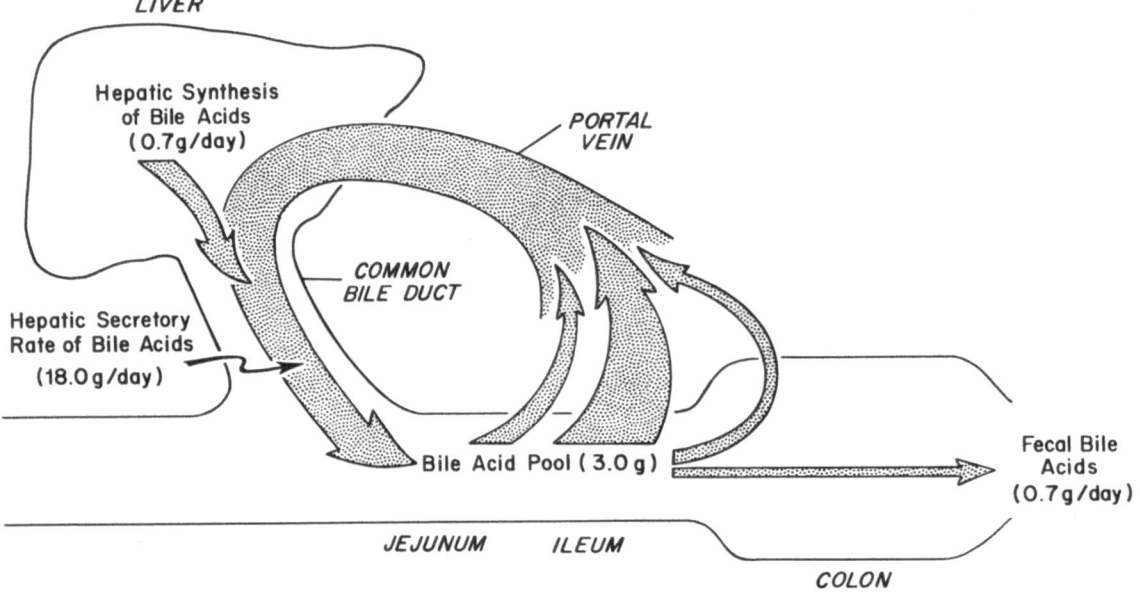

Fig. 2. Schematic illustration of the quantitative aspects of the enterohepatic circulation of the bile acid pool in normal man. The bile acid pool was assumed to recirculate an average of six times per day.

nism by the amount of bile acid returned to the liver. The bile acid pool recirculates from 6 to 10 times a day, depending on the number of meals, and, thus, the proximal small intestine receives from 18 to 30 g of bile acids daily which ensures an adequate number of bile acid micelles to facilitate fat absorption.[22]

The maintenance of the bile acid pool size depends on an efficient intestinal bile acid absorption. The rate of transport of the different bile acids depends on the chemical characteristics of the bile acid and the characteristics of the bile acid transport system in the intestinal epithelial cells. Bile acids can be absorbed at any level of the gastrointestinal tract by passive diffusion. The rate of this passive transport process is determined by the permeability coefficient of the microvillus membrane appropriate for each particular bile acid and the monomer concentration of the individual bile acids at the aqueous interface. The passive permeability coefficient varies inversely with the polarity of the different bile acids. The very polar bile acids such as glycocholic and taurocholic acid have very low passive permeability coefficients and are therefore only absorbed to a limited extent in the proximal small intestine. On the other hand, the unconjugated and dehydroxylated bile acids such as deoxycholic and lithocholic acid are less polar and have higher passive permeability coefficients, but these bile acids are not present in proximal small intestine under normal conditions. In general, the removal of one hydroxyl group from the bile acid molecule increases the passive permeability coefficient by a factor of 4, whereas removal of the glycine or taurine moiety increases the coefficient by 4 and 11, respectively.[15] In addition, bile acids are also absorbed by an active transport mechanism located in the ileum. The magnitude of this active transport again depends on the chemical structure of the individual bile acids. The maximal transport velocities vary directly with the number of hydroxyl groups on the bile acid molecule. Thus, the maximal transport capacity of a trihydroxy bile acid (cholic acid) is about eightfold higher than that of a dihydroxy bile acid and active transport of the monohydroxy bile acid is negligible.[23,24] Thus, the passive and active transport processes in the small intestine are complementary. The most polar bile acids with the lowest passive permeability coefficients also have the highest maximal transport velocities. Bile acid absorption operates very efficiently in man and less than 5% of the total circulating bile acid mass is lost into the feces per day.

The kinetics of the individual bile acids in the enterohepatic circulation vary markedly as the rate of absorption of a particular bile acid depends on its chemical structure. The kinetic parameters of the four major bile acids are summarized in Table II. Cholic and chenodeoxycholic acid, the two primary bile acids,

have the highest rates of daily synthesis, 312 and 212 mg, respectively, whereas the secondary bile acids, deoxycholic and lithocholic acid, both are synthesized at lower rates. The conjugated derivatives of cholic acid constitute the largest portion of the bile acid pool, approximately 1200 mg, with a fractional turnover rate of 0.26 per day. The pool sizes of the two dihydroxy bile acids, chenodeoxycholic and deoxycholic, are 1010 and 720 mg, respectively, with a fractional turnover rate of 0.21 and 0.26 per day, respectively.[22] Lastly, lithocholic acid has a very small pool size and a high turnover rate. This is probably due to the very low rate of intestinal absorption of this acid which is potentially toxic and is sulfated in the liver.

In summary, gallbladder bile normally contains a high concentration only of conjugated bile acids which are very polar and have low passive permeability coefficients. Consequently, when gallbladder emptying ensues at a meal, a relatively high bile acid concentration and, hence, micellar phase is maintained in the proximal small intestine to facilitate the rate of lipid absorption since only a small amount of bile acid is taken up by passive absorption. In the ileum, a certain fraction of the bile acid pool is deconjugated and dehydroxylated, increasing the passive permeability of these compounds. In addition, bile acids are absorbed by the active transport system in the ileum. The active and passive transport of bile acids in the ileum function very efficiently so that only a small amount of bile acids is spilled into the colon during each enterohepatic circulation. The bile acids are returned to the liver via the portal vein and the unconjugated bile acids are rapidly conjugated and reexcreted into the bile and small intestine.

4. Carbohydrate and Protein Digestion and Absorption

The normal steps in the digestion and absorption of carbohydrate and protein are outlined in Fig. 3, and may be compared with the major steps in lipid digestion and absorption as shown in Fig. 1. The major components of carbohydrate intake are disaccharides and starch, a glucose polymer, which is broken down to oligosaccharides by pancreatic amylase in the proximal jejunum. The disaccharides and oligosaccharides are very water soluble and they diffuse up to the microvillus membrane where they are further hydrolyzed by specific enzymes on the microvillus surface (maltase, lactase, and sucrase) to individual monosaccharides.[25]

Proteins are likewise broken down by pancreatic proteases (trypsin and chymotrypsin) to oligopeptides and amino acids, and diffuse up to the brush border membrane where oligopeptides are further hydrolyzed to dipeptides and amino acids.[25] The monosaccharides, dipeptides, and amino acids are transported into the absorptive cells by a Na^+-coupled, carrier-mediated process, which can transport these products against a high chemical gradient. The energy-requiring step is the active transport of Na^+ out of the cells via the Na^+, K^+-ATPase. The monosaccharides and amino acids leave the cells via a serosal carrier that does not require Na^+ coupling and diffuse into the capillary system of the villus and are then transported to the liver via the portal vein, and the absorptive process is then completed.

There are three major differences between the digestion and absorption of carbohydrates and proteins on the one hand and lipid digestion and absorption on the other. First, the nonpolar lipolytic products require solubilization by a micellar

Table II. Bile Acid Kinetics in Healthy Man[a]

Bile acid	Pool size (mg)	Fractional turnover rate (per day)	Synthesis rate (mg/day)
Cholic acid	1200	0.26	312
Deoxycholic acid	720	0.26	187
Chenodeoxycholic acid	1010	0.21	212
Lithocholic acid	80	1.00	80
Total	3010		791

[a]The kinetic data of bile acid metabolism in this table represent mean values obtained by different investigators.[22]

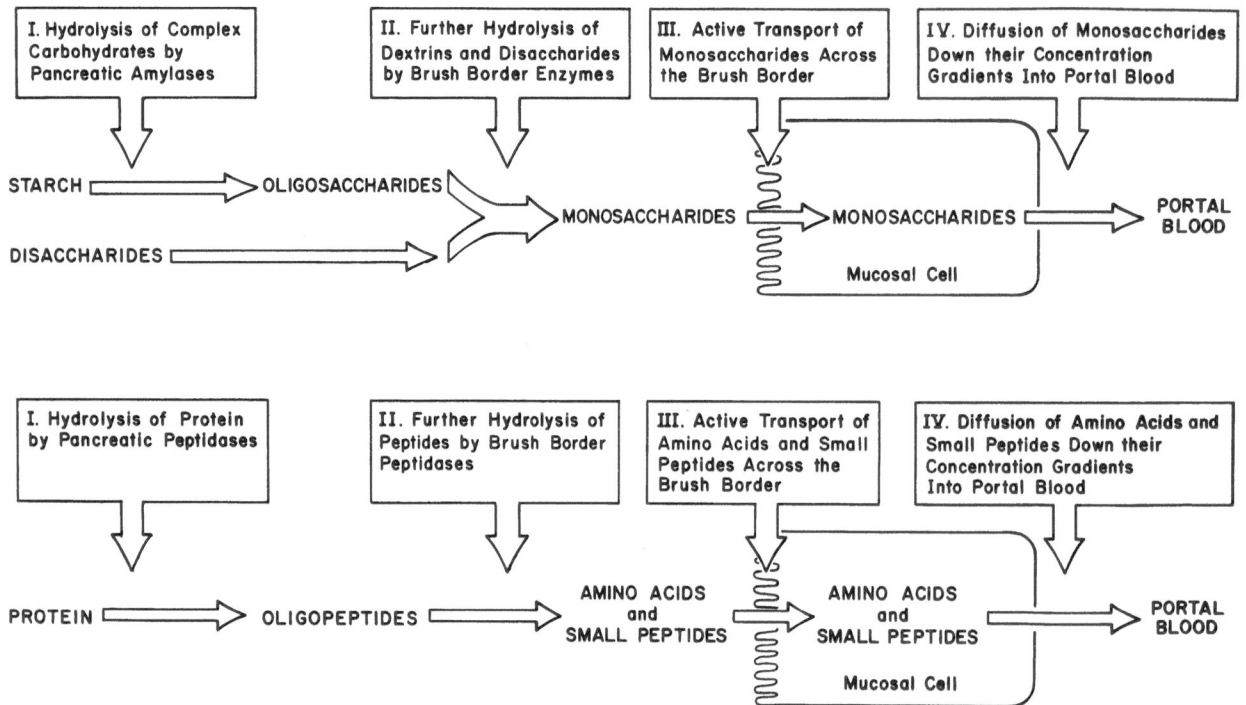

Fig. 3. The major steps involved in the digestion and absorption of dietary carbohydrate (upper panel) and dietary protein (lower panel).

phase to overcome the resistance of the unstirred water layer, whereas disaccharides and peptides are highly polar. Second, fatty acids and monoglycerides are reesterified and incorporated into chylomicrons prior to export from the absorptive cells, whereas monosaccharides and amino acids pass through the cells largely unchanged. Third, the chylomicrons are transported by the lymphatic system and the monosaccharides and amino acids by the portal vein.

Thus, it should be anticipated that disease processes that cause a decrease in pancreatic function or destroy a major proportion of the small intestinal mucosa will produce maldigestion or malabsorption of all three major constituents of the diet, whereas diseases that disturb the normal enterohepatic circulation of bile acids, block chylomicron synthesis, or interrupt normal lymphatic circulation will cause isolated maldigestion or malabsorption of fat, i.e., steatorrhea.

5. Tests of Intestinal Digestive and Absorptive Function

In the clinical approach to the patient with a suspected malabsorption syndrome, a variety of diagnostic tests may be utilized to identify the specific level at which the normal process of digestion and absorption is compromised. In specific diseases, certain features of the history or physical findings may provide important clues to the nature of the underlying disease. In addition, radiological studies of the gastrointestinal tract, definition of the type of anemia (microcytic, megaloblastic) present, and determination of a number of blood chemistry values may be helpful in directing the selection of further specific studies.

In general, a relatively limited number of diagnostic tests

that measure specific functions of the digestive and absorptive capacity will suffice to make a specific diagnosis. These tests are listed in Table III along with commonly encountered problems in the performance of the tests that may lead to erroneously high or low values.

The measurement of the daily fecal fat excretion is one of the most important tests in the diagnosis of a malabsorption syndrome. In normal man in the fasting state or on a diet devoid of lipids, there is a basal fecal fat excretion of 2–3 g/day (from desquamated cells and bacteria), which increases to 6–7 g/day on a diet with 100 g fat/day. Thus, normally a small amount of fat is excreted in the feces even on a fat-free diet. The important fact is that fat digestion and absorption are very efficient and more than 95% of ingested lipids are absorbed in the small intestine. Fecal fat determination requires that the patient ingests a diet with 70–100 g fat/day and the actual stool collection should not be initiated until the patient has been on this diet for a few days. The stools are collected quantitatively over 24-hr periods for 2 or 3 days and the fecal fat is measured by a modification of the van de Kamer method.[26] The normal upper limit is 7 g/day. If the patient excretes higher amounts of fecal fat, then a fat malabsorption syndrome is present, but this test does not identify the particular step(s) at which the normal fat digestion or absorption is interrupted. The magnitude of the steatorrhea may also be helpful in the differential diagnosis of fat malabsorption syndromes.[1] Certain diseases will produce mild defects in the absorptive process as outlined later in Table V. A valid determination of fecal fat excretion is critically dependent on an adequate fat intake and on quantitative stool collections. If fat intake is too low due to poor appetite or vomiting, or stools are not collected quantitatively, then fecal fat excretion may be falsely low or normal in a patient with a fat malabsorption syndrome.

Table III. Major Tests Used to Differentiate Types of Malabsorption

Test	Test useful in evaluation of	Factors which cause artifactually low values of the test	Factors which cause artifactually high values of the test
Quantitative fecal fat	Digestive and absorptive capacity for dietary fat	Inadequate fat intake Constipation Incomplete fecal collections	Ingestion of large quantities of castor oil or nut oils
Quantitative fecal nitrogen	Digestive and absorptive capacity for dietary protein	Inadequate protein intake Constipation Incomplete fecal collections	Excessive leakage of plasma proteins into the intestinal lumen (i.e., protein-losing enteropathy)
Duodenal intubation (Lundh test)	Exocrine pancreatic function Functional bile acid pool Bacterial cultures	Gastric acid hypersecretion	
D-xylose absorption test	Functional integrity of the jejunum Presence of massive bacterial overgrowth in the small bowel	Vomiting after administration Delayed gastric emptying Inadequate hydration Intrinsic renal disease Massive ascites Incomplete urine collection Old age	Contamination of urine with feces
Vitamin B_{12} absorption test	Functional integrity of the ileum Presence of massive bacterial overgrowth in the small bowel	Vomiting after administration Inadequate hydration Intrinsic renal disease Incomplete urine collections Presence of subtotal or total gastrectomy	Contamination of urine with feces

Determination of fecal nitrogen excretion provides a measurement of protein digestion and absorption. A normal adult excretes about 2.0–2.5 g of fecal nitrogen per day on a dietary intake of 80–100 g protein. Fecal nitrogen is derived from desquamated intestinal cells, intestinal secretions, ingested protein, and bacteria. If protein digestion or absorption is compromised by a specific disease process, fecal nitrogen excretion is increased (> 2.5 g/day). Excessive loss of plasma proteins into the gastrointestinal tract as seen in protein-losing gastroenteropathy may cause falsely elevated fecal nitrogen excretion. Falsely low values will result if protein intake is inadequate or stool collection is incomplete.

Duodenal intubation tests are useful for determination of several steps in the digestive process. This test is performed with a small tube introduced into the duodenum, and stimulation of gallbladder emptying and pancreatic secretion by either a liquid meal or i.v. administration of cholecystokinin. Duodenal aspirates are obtained over a given time period and can be analyzed for pancreatic enzymes (amylase, lipase, or trypsin) and bile acid concentrations. The aspirates can also be processed for bacterial cultures (aerobic and anaerobic) if bacterial overgrowth is suspected. The test will provide information on pancreatic function as the concentrations of the pancreatic enzymes are severely reduced in chronic pancreatic insufficiency.[27] Furthermore, the measurement of total bile acid concentration provides information on the functional size of the bile acid pool.[28] A decreased concentration may indicate disruption of the enterohepatic circulation. Gastric hypersecretion may cause lower concentrations of pancreatic enzymes and bile acids by dilution. Hypersecretion of hydrochloric acid may also result in lipase deficiency since this enzyme is denatured at low pH.

The D-xylose and vitamin B_{12} absorption tests are useful in the evaluation of the functional absorptive capacity of the jejunum and ileum, respectively. D-xylose is a pentose that is absorbed predominantly in the proximal small intestine and partially metabolized in the body. The remainder is excreted unchanged in the urine. The test is performed by giving an oral dose of 25 g of D-xylose in the fasting state followed by quantitative urine collection over the next 5 hr. In normal persons, the urinary excretion of D-xylose is 5 g or more in the 5-hr period. Patients with extensive jejunal mucosal disease may excrete only 1–2 g in 5 hr. As listed in Table III, there are a number of conditions associated with a falsely low urinary D-xylose excretion.

The Schilling test is a measure of vitamin B_{12} absorption. Vitamin B_{12} ingested with the food is bound to intrinsic factor (IF) in the stomach and proximal small intestine and this complex is absorbed by an active transport mechanism in the ileum. The Schilling test may thus be used to evaluate the absorptive capacity of the ileum. Conventionally, the Schilling test is performed by first giving a large nonradioactive dose of vitamin B_{12} parenterally to saturate the vitamin B_{12}-binding proteins in the body. Radiolabeled vitamin B_{12} is then given orally and urine is collected for 24 hr. In normal persons, more than 7% of the radiolabeled vitamin B_{12} is excreted in the urine in 24 hr. Patients with pernicious anemia will have low excretion ($< 7\%$) that is corrected with administration of IF whereas patients with ileal disease will continue to have low excretion. Again, a number of conditions may result in erroneously low values. Bacterial overgrowth of the small intestine may result in both low D-xylose and vitamin B_{12} excretion since the bacteria can use D-xylose as a metabolic substrate and also bind the vitamin B_{12}–IF complex and thus prevent absorption.

Recently, several noninvasive tests have been introduced to diagnose malabsorption of specific compounds since duodenal intubation is inconvenient for the patient and fecal fat collection is both inconvenient and time-consuming.

Most of these tests have been devised to facilitate the diagnosis of exocrine pancreatic insufficiency. One approach has been to administer radiolabeled triglyceride ([^{14}C]triolein) with a meal and to measure the amount excreted in the stools, appearing in the blood, or being expired as $^{14}CO_2$ in the breath over a given time period. A major problem with this type of test is that the specific activity of the radiolabeled fatty acids liberated from the [^{14}C]triolein during lipolysis is diluted manyfold, and leads to underestimation of the true rate of appearance in the blood or in the breath. So far, these tests have been found to be too insensitive and nonspecific to be used routinely in the diagnosis of pancreatic insufficiency.[29]

In addition to these functional tests of intestinal digestion and absorption, small bowel biopsy is also a very valuable tool in identifying specific diseases that affect the small bowl mucosa and cause characteristic histological changes. The biopsies are obtained by means of a small biopsy capsule that is passed through the mouth down into the proximal small intestine. The mucosa is suctioned into the capsule, cut off, and returned via a tube connected to the capsule. With this device, multiple biopsies may be obtained. The biopsies are examined microscopically for changes in mucosal morphology that may lead to a diagnosis of a specific small intestinal disease. The diseases where the small bowel biopsy is diagnostic or may contribute to the diagnosis are listed in Table IV with a brief description of the histological changes observed in each of these diseases.[30]

6. Diseases Affecting Normal Digestion or Absorption

As outlined in the previous sections, the digestion and absorption of lipids, carbohydrates, and proteins depend on a series of steps (Figs. 1 and 3) and one or more of these steps may be affected by a particular disease to cause a malabsorption syndrome. Consequently, the list of diseases that can cause malabsorption is very long. In this section, representative diseases that result in defects at specific steps of the normal absorptive sequence will be discussed. A more detailed discussion of the clinical aspects of many of these disorders and other causes of malabsorption may be found in recent reviews.[3,31]

6.1. Decreased Pancreatic Function

The digestion of lipids, carbohydrates, and proteins requires secretion of an adequate amount of pancreatic enzymes, and several diseases may affect the functional secretory capacity of the pancreas so that the secretory response is inadequate. One of the most common diseases in this group is recurrent pancreatitis that results in progressive destruction of the pancreatic parenchyma. Overt pancreatic insufficiency will not develop, however, until more than 90% of the gland is destroyed. Acute and chronic pancreatitis are often associated with alcoholism and are also encountered in patients with biliary tract disease (gallstones) or hypertriglyceridemia. Typically, these patients have a long history of recurrent bouts of severe abdominal pain, nausea, and vomiting followed by the development of a malabsorption syndrome. Occasionally, pancreatic insufficiency develops silently and the patient presents with complaints of weight loss and a change in bowel habits to bulky, greasy, malodorous stools suggestive of a malabsorption syndrome. Pancreatic carcinoma is another not infrequent cause of pancreatic insufficiency in the older age groups. The carcinoma is often localized in the head of the pancreas and causes obstruction of the main pancreatic duct or has diffusely destroyed most of the

Table IV. Summary of Representative Histological Findings in Small Bowel Biopsies in Specific Diseases Causing a Malabsorption Syndrome[a]

Diseases with specific diagnostic histological findings

A. *Abetalipoproteinemia:* Normal mucosal morphology and villus structure. The mucosal cells, however, are filled with lipid droplets in the fasting state.

B. *Amyloidosis:* Deposits of amyloid fibrils in the submucosa. Identification requires a special stain (Congo red) and viewing of the biopsies under polarized light (green birefringence).

C. *Collagenous sprue:* A possible variant of nontropical sprue with villus atrophy. The characteristic finding is a dense band of collagen beneath the mucosal cell layer.

D. *Eosinophilic gastroenteritis:* Variable villus structure from normal to flat mucosa with massive infiltration of eosinophils of the lamina propria.

E. *Immunodeficiency states:* Variable villus morphology from apparently normal to total villus atrophy. Mononuclear infiltration of the lamina propria but absence of plasma cells.

F. *Intestinal lymphangiectasia:* Swollen club-shaped villi with dilated central lacteals. Mucosal morphology is normal. Lipid-filled macrophages may be seen in the lamina propria.

G. *Intestinal lymphoma:* Broad, short villi to villus atrophy with elongated crypts. The lamina propria is very cellular due to massive infiltration with malignant lymphoid cells.

H. *Mastocytosis:* Normal villus morphology but heavy infiltration with mast cells in lamina propria, muscularis mucosa, and submucosa.

I. *Parasitic disease:* Infestation with intestinal parasites may cause a variable degree of villus atrophy. Specific parasites such as giardia lambia, coccidia, and cryptosporidia may be indentified adhering to the glycocalyx or invading the mucosa.

J. *Whipple's disease:* Short, blunted villi with widened villus core heavily infiltrated with PAS-positive macrophages. Mucosal structure relatively normal. By electron microscopy, many rod-shaped bacilli are seen in the villus core and within the macrophages.

Diseases with characteristic but nondiagnostic findings

K. *Dermatitis herpetiformis:* Varying degrees of villus atrophy and infiltration of lamina propria with mononuclear cells. The lesions are identical to those seen in the sprue syndromes.

L. *Nongranulomatous jejunitis:* Flattened villi and patchy jejunal ulcerations with mononuclear infiltrate in the lamina propria.

M. *Nontropical and tropical sprue:* Subtotal to total villus atrophy ("flat mucosa"), elongated crypts, and infiltration of the lamina propria with mononuclear cells. The changes are usually more severe in nontropical sprue.

N. *Scleroderma:* Normal villus morphology with fibrosis of the submucosa and muscle layer.

[a]This table, adapted from Wilson and Dietschy,[1] lists specific diseases of the small intestinal mucosa where the histological changes are either diagnostic (A to J) or compatible (K to N) with a specific disease.

normal pancreatic tissue. In both instances, the cancer results in an inadequate secretion of pancreatic enzymes. Cystic fibrosis is a relatively common congenital disorder that regularly leads to development of chronic pancreatic insufficiency during childhood. Finally, pancreatic insufficiency may also result from surgical resection of the pancreas or obstruction of the main duct due to a pseudocyst.

The common denominator to all these diseases is deficient pancreatic enzyme secretion and, hence, insufficient enzymatic breakdown of all three major caloric sources in the diet. Theoretically, the malabsorption syndrome produced by pancreatic insufficiency should manifest malabsorption of lipid and pro-

Table V. Representative Values of Malabsorption Tests in Specific Diseases Causing a Malabsorption Syndrome[a]

Probable site of defect	Examples of specific diseases	Fecal fat excretion (g/24 hr)	Fecal nitrogen excretion (g/24 hr)	D-Xylose absorption (g/5 hr)	Vitamin B_{12} absorption (%/24 hr)	Peroral intestinal biopsy
Representative normal values		<6	<2.0	>4.5	>7.0	
Decreased pancreatic function (I)	Chronic pancreatitis	37.0 ± 5.0	4.7 ± 0.6	6.1 ± 0.7	9.2 ± 2.0	Normal
	Pancreatic cancer	41.0 ± 7.0	6.0 ± 0.9	5.5 ± 0.6	8.4 ± 2.0	Normal
	Cystic fibrosis	25.0 ± 4.0	4.2 ± 0.6	6.5 ± 1.0	—	Normal
Insufficient functional bile acid intestinal lumen (II)	Obstructive jaundice	16.0 ± 2.0	1.2 ± 0.1	6.5 ± 0.9	12.0 ± 1.0	Normal
	Blind loop syndrome	17.0 ± 2.0	1.8 ± 0.2	3.0 ± 0.5	0.9 ± 0.3	Normal
	Ileal dysfunction syndrome	22.0 ± 5.0	2.4 ± 0.2	5.5 ± 0.7	2.1 ± 0.8	Normal
Increased unstirred water layer resistance (III)	Diseases affecting villous and small bowel motility	?	?	?	?	Normal
Diseases of the intestinal villi (IV, V)	Gluten enteropathy	28.0 ± 2.0	5.0 ± 1.2	2.0 ± 0.3	3.5 ± 2.0	Compatible
	Tropical sprue	16.0 ± 0.1	4.0 ± 2.0	2.2 ± 0.6	5.1 ± 1.3	Compatible
	Nongranulomatous jejunitis	27.0 ± 5.0	—	3.4 ± 1.1	2.0 ± 2.0	Compatible
	Whipple's disease	34.0 ± 5.0	3.8 ± 0.5	3.7 ± 0.4	12.8 ± 3.7	Diagnostic
	Amyloidosis	18.0 ± 2.0	3.0 ± 0.1	3.0 ± 0.3	—	Diagnostic
	Eosinophilic gastroenteritis	14.0 ± 2.0	—	2.3 ± 0.7	3.5 ± 2.1	Diagnostic
Inability to synthesize chylomicrons (VI)	Abetalipoproteinemia	18.0 ± 2.0	—	6.2 ± 1.3	19.0 ± 2.6	Diagnostic
Obstruction to intestinal lymphatics (VII)	Lymphoma, tumor of mesenteric nodes	10.0 ± 5.0	3.5 ± 2.0	4.8 ± 1.5	13.0 ± 6.2	Usually
		12.0 ± 7.0	3.2 ± 1.0	5.4 ± 2.0	—	Normal

[a]All data are expressed as mean values ± 1 S.E. and are derived from data presented Wilson and Dietschy.[1] The Roman numerals in the first column correspond to the probable site of defect in digestion or absorption as illustrated in Fig. 1.

tein, as measured by excretion in the stools, and a meal test should provoke only low concentrations of pancreatic enzymes in duodenal aspirates. In contrast, tests of intestinal absorptive function should be normal. These findings are commonly encountered in pancreatic insufficiency, irrespective of cause, as illustrated in the data in Table V which have been compiled from published series of patients with chronic pancreatitis, pancreatic cancer, and cystic fibrosis.[1] Generally, in severe cases of pancreatic insufficiency, fecal fat excretion is increased into the range of 25–50 g/day. Since proteins are also maldigested, there will be an associated increase in fecal nitrogen excretion (azotorrhea).

Duodenal aspirates during a test meal will show very low concentrations of pancreatic enzymes.[27] D-xylose and vitamin B_{12} absorption are most often normal as the functional integrity of the small bowel mucosa is not compromised by pancreatic dysfunction. In some cases, however, a decreased vitamin B_{12} absorption may be observed as liberation of vitamin B_{12} from certain binding factors other than IF depends on adequate proteolysis. Finally, biopsy of the small intestinal mucosa is normal. Patients with pancreatic insufficiency generally respond to oral replacement therapy with pancreatic enzymes as manifested by a decrease of steatorrhea and weight gain.

6.2. Decreased Micellar Solubilization of Lipolytic Products

The second step in fat absorption (Fig. 1) involves solubilization of the lipolytic products by bile acid micelles. This process, in turn, depends on the functional integrity of the enterohepatic circulation of bile acids. Many different diseases may interfere with one or more aspects of the enterohepatic circulation. These diseases can be conveniently divided into three groups of disorders according to the specific site of inter-

ruption of the enterohepatic circulation and each has a distinctly different pathophysiology.

The first group of disorders includes any severe, diffuse liver disease where there is decreased bile acid synthesis and any disease that blocks the outflow of hepatic or gallbladder bile such as intrahepatic cholestasis or extrahepatic obstruction due to common bile duct stones or cancer, strictures or cancer of the head of the pancreas. Since the micellar solubilization process is involved only in fat absorption and not in carbohydrate or protein absorption, it would be anticipated that duodenal aspirates would show low bile acid and normal pancreatic enzyme concentrations and that only fat malabsorption would result from these disorders. As shown in Table V, these predictions are verified in this group of patients, who have mild steatorrhea while other absorptive function tests are normal. Obviously, most of these patients have jaundice and the major clinical problem is the differential diagnosis of the cause of jaundice and not of steatorrhea.

The second group also consists of a diverse number of diseases with the common feature of bacterial overgrowth in the small intestine causing the so-called blind loop or stagnant loop syndrome. This syndrome may be seen with multiple jejunal diverticula, strictures of the small intestine, long afferent loops, and surgically created blind loops. In addition, diseases that interfere with small intestinal motility like diabetes and scleroderma will favor bacterial overgrowth as normal peristaltic activity that sweeps the small intestine between meals may be obliterated. The small intestine is colonized by colonic bacteria that include bacteroides, lactobacilli, clostridia, and coliforms. A number of these bacteria contain enzymes (deconjugase and 7-α-hydroxylase) which are capable of deconjugating and dehydroxylating the conjugated bile acids in the proximal small intestine. Both processes decrease the polarity of the bile acids and, hence, increase passive permeability so that the unconjugated bile acids are rapidly absorbed from the jejunum. This, in turn, may cause a progressive decrease in bile acid concentration with a functional deficiency of bile acid micelles. Furthermore, the bacteria may bind the vitamin B_{12}–IF complex and digest D-xylose so these two absorptive tests may also be low.

Typically, patients with the blind loop syndrome have modest steatorrhea, about 12 g/24 hr, and low vitamin B_{12} and D-xylose absorption as shown in Table V. The diagnosis of a blind loop syndrome may be suggested by X-ray studies of the small intestine and confirmed by jejunal aspiration that will show high bacterial colony counts, low total bile acid concentrations, and the presence of unconjugated bile acids.[28] Furthermore, these absorptive abnormalities should return to normal when the patients are treated with an appropriate antibiotic, usually tetracycline, that will eradicate the abnormal flora. Sometimes, surgical correction of discrete lesions (e.g., stricture) may obviate the need for antibiotic treatment.

The third group of disorders of the enterohepatic bile acid circulation is commonly referred to as the ileal dysfunction syndrome. This syndrome includes diseases of the ileal mucosa such as granulomatous ileitis (Crohn's disease), lymphoma, and tuberculosis that may all cause extensive destruction of the ileal mucosa. Surgical resections of the ileum due to disease processes such as Crohn's disease, gangrene, or irradiation damage will also result in loss of functional ileal mucosa area. Since bile acids normally are predominantly reabsorbed in the ileum, these disease processes result in decreased reabsorption and the bile acids are spilled into the colon. The amount of bile acids lost depends on the extent of the mucosal lesion. In the colon, small amounts of bile acid may be reabsorbed by passive permeation but, more importantly, these substances, especially the dihydroxy bile acids, induce isosmotic water and electrolyte secretion by the colonic mucosa. With limited ileal disease, the liver can usually compensate for the increased bile acid loss by increasing bile acid synthesis and thus maintain a normal pool size. With more extensive lesions, liver synthesis can no longer keep pace with the increased loss which results in a decreased rate of bile acid secretion in the bile, deficient micellar solubilization in the intestine and fat malabsorption. Patients with limited ileal disease may thus present with watery diarrhea but near-normal fecal fat excretion whereas patients with more extensive lesions have diarrhea and fat malabsorption.

In ileal dysfunction, the steatorrhea may be more severe than that seen in obstructive jaundice or the blind loop syndrome. In the two former groups, the primary defect is low bile acid concentration but the mucosal integrity of the small intestine is preserved. Thus, the rate of fat absorption in the jejunum per unit length of intestine is decreased. However, the ileum represents a reserve capacity for fat absorption so that this process can proceed throughout the small intestine, although at a reduced rate, so that steatorrhea is mild. In extensive ileal dysfunction, however, there is both bile acid deficiency and loss of mucosal surface area so that typically the steatorrhea is more severe (Table V). D-xylose absorption is normal, but vitamin B_{12} absorption may be reduced depending upon the extent of the ileal lesion.

6.3. Increased Unstirred Layer Resistance

The third major step in the absorptive process is diffusion of the mixed micelles, small peptides, and oligosaccharides from the luminal solution, where these digestive products are constantly mixed by the peristaltic activity of the intestine, across the unstirred water up to the microvillus membrane of the absorptive cells. Since the mixed micelles are much larger than the peptides and saccharides, they diffuse across the unstirred water layer at a slower rate. Thus, fat absorption is more sensitive to changes in the effective resistance of this layer. The effective thickness of this barrier has been found to be 100–300 μm under different experimental *in vitro* conditions[12] and around 600 μm during intestinal perfusions in humans.[32] Thus far, it has not been technically possible to measure the unstirred layer resistance under physiological conditions when digestion and absorption are occurring simultaneously. Physiologically, there are at least two mechanical processes in the small intestine that may contribute to a decrease in this resistance. First, there are forceful contractions of the circular and longitudinal muscle layers of the small intestine that generate the peristaltic activity where the luminal contents are moved back and forth in a vigorous mixing process. Second, the intestinal villi presumably have independent contractile movements. Both processes increase the stirring of the luminal contents and may thus reduce the dimensions of this diffusion barrier. Therefore, in healthy individuals with normal digestive functions, this barrier may offer less resistance to lipid uptake than these measurements would suggest.

Although specific diseases that change the resistance of this barrier have yet to be described, it might be anticipated that disorders of intestinal or villus motility would cause an increase in the functional thickness of the unstirred layer which in turn would affect the rate of uptake of lipids and other dietary constituents and ultimately result in a malabsorption syndrome. Decreased small bowel motility is sometimes observed in diseases such as diabetes mellitus, scleroderma, and some neurological disorders and steatorrhea is often a concomitant feature. It should be noted, however, that decreased intestinal motility is also a favorable condition for bacterial overgrowth, so that the fat malabsorption syndrome may have a mixed etiology. A precise definition of an increased diffusion barrier as a single cause of a fat malabsorption syndrome must await a method that allows measurement of this barrier under physiological conditions.

6.4. Diseases of the Intestinal Mucosa

Transport of the digestive products across the microvillus membrane into the cytosol of the absorptive cells is the fourth major step in the absorptive process. The digestive processes continue at the microvillus membrane where smaller peptides and saccharides are broken down to dipeptides, amino acids, and monosaccharides before they associate with specific carriers in the membrane for the translocation process. Diminished activity of a specific microvillus enzyme such as lactase may cause an isolated defect in lactose absorption. Other diseases, however, may affect the normal function of the mucosal cells either by extensive destruction of these cells, leading to a decreased absorptive surface area, or by interfering with the transport of the nutrients from the cells to the capillary network or central lacteals of the villi. A common feature of these diseases is malabsorption of nearly all of the normal constituents of the diet including water and electrolytes but the symptoms and clinical findings will vary according to the specific pathological lesions of the individual disease. Thus, it is not possible to generalize about the typical laboratory findings in this group of diseases. It should be stressed, however, that small bowel biopsy is of particular diagnostic value in this group of disorders as described in Tables IV and V.

Among the more common diseases in this group are the sprue syndromes, nontropical and tropical sprue. Nontropical sprue or gluten-sensitive enteropathy may develop in certain individuals where the gluten content of wheat, barley, and rye become toxic to the mucosal cells by an unknown mechanism. The normal villus structure disappears as a result of progressive destruction of these cells and the mucosa becomes flat with deep crypts and the submucosa is invaded by inflammatory cells. The characteristic lesions are most extensive in the jejunum but may extend to the ileum in severe cases. Tropical sprue, on the other hand, is presumed to be caused by unidentified intestinal bacteria that are found in tropical areas of the world and that affect both the ileum and the jejunum. The hallmark of both diseases is a loss of functional mucosal surface area with a resulting malabsorption syndrome. Fecal fat excretion is increased in both diseases (Table V) and the steatorrhea is usually more pronounced in nontropical sprue. D-xylose absorption is also decreased in both tropical and nontropical sprue. In gluten enteropathy, vitamin B_{12} absorption is commonly normal but may be abnormal depending on the extent of the mucosal lesion in the ileum, whereas it is usually low in tropical sprue due to more extensive ileal disease. A small bowel biopsy will show blunted villi or even villus atrophy with a flat mucosa (Table IV). These characteristic histological findings in combination with the results of the absorptive function tests are essentially diagnostic of a sprue syndrome. The differential diagnosis between nontropical and tropical sprue may at times be difficult but the history of previous exposure to a tropical environment is suggestive of tropical sprue. The treatment of the two diseases is completely different. In gluten enteropathy, it is mandatory that the patients adhere to a strict gluten-free diet for life which, in most cases, leads to rapid improvement, whereas patients with tropical sprue respond to a brief course of antibiotic treatment.

Another interesting but rare disorder in this group is Whipple's disease which is a systemic disease that invariably involves the small bowel. The pathological lesions in Whipple's disease are distinctly different from those of the sprue syndromes and mainly involve the lamina propria. The absorptive cells are normal but the villi are short and wide due to heavy infiltration of PAS-positive macrophages. With the higher resolution of electron microscopy, many rod-shaped bacteria are seen in the lamina propria and within the macrophages. The bacteria have so far eluded classification. Another common histological finding is dilated central lacteals which may be due to lymphatic obstruction in the villi or in the mesenteric lymph nodes by macrophage infiltration. Steatorrhea is the most prominent feature of the malabsorption syndrome in Whipple's disease, and the lymphatic obstruction is probably a contributory factor. D-xylose absorption is moderately decreased, whereas vitamin B_{12} absorption is normal (Table V). The disease was previously fatal, but after the presumed infectious nature was recognized, antibiotic treatment has usually been successful in eradicating the malabsorption syndrome and restoring the villus morphology toward normal.

Amyloidosis is another example of a systemic disease that frequently involves the gastrointestinal tract and may cause a malabsorption syndrome. The typical histological lesion is deposition of amyloid fibrils around small vessels in the lamina propria and submucosa, which can be observed when a small bowel biopsy is stained with Congo red and examined under polarized light. The presence of amyloid is revealed by a typical green birefringence of the amyloid deposits. Amyloid may also be deposited in the muscle layers or around the autonomic plexus in the bowel wall. With advanced disease, the amyloid may interfere with both the vascular perfusion and the motility of the small bowel. The pathophysiology of the malabsorption syndrome is poorly understood but may be due to decreased motility with intestinal stasis or to vascular insufficiency, both of which interfere with the normal function of the mucosal cells. Specific therapy of amyloidosis is not available and the disease progresses relentlessly.

There are several other rare diseases that affect the functional integrity of the small bowel mucosa such as eosinophilic gastroenteritis and nongranulomatous jejunitis as listed in Table V, and disorders such as intestinal lymphoma, immunodeficiency syndromes, and mastocytosis that may present with a malabsorption syndrome.

6.5. Absent Chylomicron Synthesis

The assembly of chylomicrons in the mucosal cells requires incorporation of apoproteins on the surface of the chylomicron particle prior to excretion from the cells. One very rare inherited disease is characterized by absent apo B protein synthesis and is called abetalipoproteinemia. Chylomicrons deficient in apo B are not excreted from the cells which become progressively filled with lipid particles even after a prolonged fast. Lipid uptake and reesterification presumably occur at a normal rate but export out of the cells is exceedingly low. Consequently, one of the unique features of this disease is isolated steatorrhea with normal pancreatic function and normal D-xylose and vitamin B_{12} absorption. The diagnosis is suggested by the finding of lipid-laden mucosal cells in a small bowel biopsy and is confirmed by the absence of apo B in plasma lipoproteins.

6.6. Lymphatic Obstruction

Finally, since the absorbed lipids are transported from the mucosal cells to the systemic circulation by the lymphatics, any disease process that obstructs lymph flow from the small intestine may cause fat malabsorption. These diseases include intestinal lymphangiectasia, a rare congenital disease with malformation of the lymphatic vessels in the intestine, and diseases involving the mesenteric lymph nodes such as lymphoma, metastatic cancer, and carcinoid. These patients usually have mild steatorrhea, normal pancreatic function, and normal D-xylose and vitamin B_{12} absorption. The diagnosis may be suspected when a small bowel biopsy shows dilated central lacteals and lipid accumulation in the mucosal cells. Lymphatic obstruction may also lead to rupture of the lymphatics and a chylomicron-rich lymph is spilled into the periotoneal cavity producing chylous ascites. These patients may obtain symptomatic benefit from a low-fat diet supplemented with medium-chain triglycerides which are absorbed via the portal vein.

7. Summary

Malabsorption syndromes include a variety of diseases in which the normal mechanisms of fat, protein, and carbohydrate digestion and/or absorption are disturbed. These three major caloric sources share several common steps in the overall absorptive process and diseases interfering with any of these may result in malabsorption of fats, carbohydrates, and proteins. Fat absorption, however, is the most complex process and therefore more prone to injury so that fat malabsorption is the most common feature of a malabsorption syndrome. The normal process of digestion and absorption depends on the functional integrity of pancreatic secretion and enterohepatic circulation of bile acids, the preservation of a normal mucosal cell population, and a sufficient vascular and lymphatic flow in the intestinal villi for removal of the absorbed substances.

Digestion and absorption: Fat, protein, and carbohydrate are digested by pancreatic enzymes to fatty acids and monoglycerides, smaller peptides, and oligosaccharides. The lipolytic products are insoluble in water and require micellar solubilization by bile acids whereas the products of protein and carbohydrate digestion are soluble in water. The mixed micelles, peptides, and oligosaccharides diffuse across the unstirred water layer up to the microvillus membrane, where the fatty acids and monoglycerides leave the micelles and are taken up by the mucosal cells by passive diffusion. The peptides and oligosaccharides are broken down further by specific enzymes in the microvillus membrane to dipeptides, amino acids, and monosaccharides which are taken up by an active carrier-mediated process. In the mucosal cells, the products of lipid digestion are reesterified, incorporated into chylomicrons, and excreted into the lymphatics, whereas amino acids and monosaccharides diffuse unchanged through the mucosal cells and are removed by the capillary network in the villus.

Enterohepatic circulation: The pool of 3 g of bile acids recirculates between liver and intestine from 6 to 10 times a day. Due to an efficient reabsorption in the ileum via active transport of conjugated bile acids and high passive permeability of deconjugated bile acids, only about 700 mg of bile acids is lost in the feces each day. The fecal loss is balanced by hepatic synthesis of an equal amount of bile acids to maintain a constant pool size.

Diagnostic tests in malabsorption syndromes: Fecal fat excretion is the most useful test to verify a malabsorption syndrome. A meal test with duodenal aspiration is helpful to document disturbances of the digestive phases (decreased pancreatic excretory function, decreased bile acid concentration, or bacterial overgrowth). The absorptive functions of the jejunum and ileum may be evaluated with the D-xylose and vitamin B_{12} absorption tests, respectively. Finally, small bowel biopsy is useful to define characteristic lesions of the jejunal mucosa.

Characteristic examples of malabsorption syndromes: (1) Decreased pancreatic function: Severe steatorrhea that is ameliorated by replacement therapy with pancreatic enzymes is typical. (2) Deficient micellar solubilization: A low intraluminal bile acid concentration may result from cholestasis, bacterial overgrowth of the small intestine, and ileal dysfunction. (3) Increased unstirred layer resistance: The thickness of the diffusion barrier overlying the villus tips may be increased by small bowel hypomotility but has not been documented in man. (4) Diseases of the intestinal mucosa: This group of diseases result in loss of mucosal surface area and typically malabsorption of all three caloric sources. One characteristic example is gluten enteropathy. Small bowel biopsy is of particular value in these diseases. (5) Abetalipoproteinemia: Inherited lack of apo B protein synthesis in the mucosal cells results in inability to excrete chylomicrons. (6) Lymphatic obstruction: These diseases are characterized by mild steatorrhea as the rate of chylomicron removal from the mucosal cells is decreased.

References

1. Wilson, F. A., and J. M. Dietschy. 1971. Differential diagnostic approach to clinical problems of malabsorption. *Gastroenterology* **61**:911–931.

2. *Geigy Scientific Tables.* 1981. Volume 8, pp. 151–158. Ciba-Geigy, Basel, Switzerland.

3. Sleisenger, M. H., and J. S. Fordtran. 1978. *Gastrointestinal Disease.* Saunders, Philadelphia. pp. 241–251, 272–297.

4. Johnston, J. M. 1967. Mechanism of fat absorption. In: *Handbook of Physiology,* Section 6. C. F. Code, section ed. Amercian Physiological Society, Washington, D.C. pp. 1353–1375.

5. Hamosh, M., J. W. Scanlon, D. Ganot, M. Likel, K. B. Scanlon, and P. Hamosh. 1981. Fat digestion in the newborn. *J. Clin. Invest.* **67**:838–846.

6. Borgstrom, B., C. Erlandson-Albertson, and T. Wieloch. 1979. Pancreatic colipase: Chemistry and physiology. *J. Lipid Res.* **20**:805–816.

7. Carey, M., and D. M. Small. 1970. The characteristics of mixed micellar solutions with particular reference to bile. *Am. J. Med.* **49**:590–608.

8. Hofmann, A. F., and B. Borgstrom. 1964. Physicochemical state of lipids in intestinal content during their digestion and absorption. *Fed. Proc.* **21**:43–50.

9. Patton, J. S., and M. C. Carey. 1979. Watching fat digestion. *Science* **204**:145–148.

10. Lusatti, V., A. Tardieu, T. Gulik-Kosywicki, E. Rivas, and F. Reiss-Hutton. 1968. Structure of the cubic phases of lipid–water systems. *Nature (London)* **220**:485–488.

11. Lundstrom, M., H. Ljusberg-Wahren, K. Larsson, and B. Borgstrom. 1981. Aqueous lipid phases of relevance to intestinal fat digestion and absorption. *Lipids* **16**:749–754.

12. Westergaard, H., and J. M. Dietschy. 1974. Delineation of the dimensions and permeability characteristics of the two major diffusion barriers to passive mucosal uptake in the rabbit intestine. *J. Clin. Invest.* **54**:718–732.

13. Ockner, R. K., and J. A. Manning. 1974. Fatty acid-binding protein in small intestine: Identification, isolation, and evidence for its role in cellular fatty acid transport. *J. Clin. Invest.* **54**:326–338.

14. Borgstrom, B., A. Dahlquist, G. Lundh, and J. Sjovall. 1957. Studies on intestinal digestion and absorption. *J. Clin. Invest.* **36**:1521–1536.

15. Wilson, F. A., and J. M. Dietschy. 1972. Characterization of bile acid absorption across the unstirred water layer and brush border of the rat jejunum. *J. Clin. Invest.* **51**:3015–3025.

16. Cardell, R. R., S. Badenhausen, and K. P. Porter. 1967. Intestinal triglyceride absorption in the rat: An electron microscopical study. *J. Cell Biol.* **34**:123–155.

17. Johnston, J. M. 1978. Esterification reactions in the intestinal mucosa and lipid absorption. In: *Disturbances in Lipid and Lipoprotein Metabolism.* J. M. Dietschy, A. M. Gotto, and J. A. Ontko, eds. American Physiological Society, Washington, D.C. pp. 57–69.

18. Green, P. H. R., and R. M. Glickman. 1981. Intestinal lipoprotein metabolism. *J. Lipid Res.* **22**:1153–1173.

19. Fielding, C. J., and P. E. Fielding, 1976. Chylomicron protein content and the rate of lipoprotein lipase activity. *J. Lipid Res.* **17**:419–423.

20. Sherrill, B., and J. M. Dietschy. 1978. Characterization of the sinusoidal transport process for uptake of chylomicrons by the liver. *J. Biol. Chem.* **253**:1859–1867.

21. Hofmann, A. F. 1977. The enterohepatic circulation of bile acids in man. In: *Clinics in Gastroenterology,* Volume 6. Saunders, Philadelphia. pp. 3–24.

22. Matern, S., and W. Gerok. 1979. Pathophysiology of the enterohepatic circulation of bile acids. *Rev. Physiol. Biochem. Pharmacol.* **85**:125–204.

23. Dietschy, J. M. 1968. Mechanisms for the intestinal absorption of bile acids. *J. Lipid Res.* **9**:297–309.

24. Schiff, E. R., N. C. Small, and J. M. Dietschy. 1972. Characterization of the kinetics of the passive and active transport mechanisms for bile acid absorption in the small intestine and colon of the rat. *J. Clin. Invest.* **51**:1351–1362.

25. Silk, D. B., and A. M. Dawson. 1979. Intestinal absorption of carbohydrate and protein in man. *Int. Rev. Physiol.* **19**:151–204.

26. Jover, G. 1968. Detection of malabsorption. In: *Malabsorption*

Syndromes. W. W. Shingleton and W. O. Dobbins, eds. Thomas, Springfield, Ill. pp. 44–53.

27. DiMagno, E. P., V.-L. W. Go, and H. J. Summerskill. 1973. Relations between pancreatic enzyme outputs and malabsorption in severe pancreatic insufficiency. *N. Engl. J. Med.* **288**:813–815.

28. Westergaard, H. 1977. Duodenal bile acid concentrations in fat malabsorption syndromes. *Scand. J. Gastroenterol.* **12**:115–122.

29. DiMagno, E. P. 1982. Diagnosis of chronic pancreatitis: Are noninvasive tests of exocrine pancreatic function sensitive and specific? *Gastroenterology* **83**:143–146.

30. Trier, J. S. 1977. Diagnostic usefulness of small intestinal biopsy. *Viewpoints Dig. Dis.* **9**(1).

31. Greenberger, N. J., and K. J. Isselbacher. 1980. Disorders of absorption. In: *Principles of Internal Medicine,* K. J. Isselbacher, R. D. Adams, E. Braunwald, R. G. Petersdorf, and J. D. Wilson, eds. McGraw-Hill, New York. pp. 1392–1409.

32. Read, N. W., D. C. Barbes, R. J. Levin, and C. D. Holdsworth. 1977. Unstirred layer and kinetics of electrogenic glucose absorption in the human jejunum in situ. *Gut* **18**:865–876.

CHAPTER 8

Pathophysiology of Calcium Absorptive Disorders

Stanley J. Birge and Louis V. Avioli

1. Introduction

Calcium is the most abundant cation and fifth most common inorganic element of the human body. It not only serves as the principal component of the skeleton, imparting to it the structural integrity essential to support the increasing body size of the individual during growth, but also plays a vital role in a variety of essential physiological and biochemical processes. Thus, maintenance of circulating Ca^{2+} within a narrow range is critical for the animal's survival. The homeostasis of Ca^{2+} in blood and extracellular fluid represents one of the most exquisite biological control systems in man and is achieved by the interaction of a variety of hormones on the skeleton, kidneys, and intestine. It is evident that the availability of dietary Ca^{2+} is an important determinant of Ca^{2+} homeostasis. In this chapter we shall review the normal physiological regulation of Ca^{2+} absorption in man and the alterations in intestinal Ca^{2+} absorption which characterize a variety of clinical disorders.

2. Regulation of Intestinal Calcium Absorption

The average diet consumed daily by humans contains approximately 400 to 1000 mg Ca^{2+}. It and related alkaline earth cations, unlike Na^+ and K^+, are only partially absorbed from the intestinal lumen. The net absorption of Ca^{2+} represents the summation of two unidirectional processes: the total transfer of Ca^{2+} from intestinal lumen to plasma; and the total transfer of Ca^{2+} from plasma to lumen, or endogenously secreted Ca^{2+}. Fecal Ca^{2+} consists not only of unabsorbed ingested Ca^{2+} but also that Ca^{2+} endogenously secreted into the gastrointestinal tract which is not absorbed. About 85% of the Ca^{2+} entering the intestine in this manner is secreted proximal to the site of absorp-

tion, presented to the absorptive sites, and assumed to be handled with the same efficiency as the Ca^{2+} of dietary origin. Total intestinal Ca^{2+} excretion in man averages 0.194 ± 0.073 g/day, whereas the secreted Ca^{2+} which is not absorbed, or endogenous fecal Ca^{2+}, averages 0.130 ± 0.047 g/cm per day.[1] Ca^{2+} excretion of the human jejunum is passive and approximates 0.008 mmole/hr per 20 cm.[2] Although endogenously secreted Ca^{2+} appears to play an insignificant role in maintaining circulating Ca^{2+} levels in man, it may contribute to the adaptive response of lower mammals when subjected to varying levels of dietary Ca^{2+}[3] Approximately 15% of the total intestinal Ca^{2+} secretion is assumed to be nonabsorbable, even under conditions when dietary Ca^{2+} is completely absorbed. These values are based on the assumption that dietary Ca^{2+} mixes homogeneously with endogenously secreted Ca^{2+}, that the absorptive efficiencies for both are identical, and that a portion of the endogenous Ca^{2+} is unabsorbed because it enters the gut distal to the absorptive site(s). Although in the past, phosphate has been alleged to interfere with Ca^{2+} absorption as the result of the formation of poorly soluble calcium phosphates, high-phosphate intakes have minimal effects on Ca^{2+} absorption.[4–6] The exception is undigestible organic phosphates (such as phytates or hexaphosphoinositol occurring in bran) which can form calcium salts. High concentrations of dietary Mg^{2+} can decrease Ca^{2+} absorption in the dog,[7] sheep,[8] and man[9] by presumably competing for the Ca^{2+} transport site. A variety of calcium salts have been used to supplement the diet including its lactate, chloride, gluconate, carbonate, or sulfate. There seems to be little difference in the intestinal utilization of Ca^{2+} from any of these salts.[10] However, there are a number of factors which when included in the diet can either influence the availability of Ca^{2+} to its site of transport or alter directly the transport process (Table I).

As an integral part of the Ca^{2+} homeostatic mechanism, the intestine has the ability to adapt to Ca^{2+} deprivation by increasing the absorption of dietary Ca^{2+}. This adaptation is achieved by the interplay of principally two hormones, para-

Stanley J. Birge and Louis V. Avioli • Department of Medicine, Washington University School of Medicine at The Jewish Hospital of St. Louis, St. Louis, Missouri 63110.

Table I. Dietary Factors Altering Intestinal Absorption of Ca²⁺

Increased absorption	Decreased absorption
Antibiotics	Oxalate
Penicillin	Unabsorbed organic phosphates
Neomycin	Tetracycline
Chloramphenicol	Unabsorbed fatty acids
Amino acids	
Tryptophan	
Lysine	
Arinine	
Lactose	

thyroid hormone (PTH) and 1,25(OH)$_2$D. These two hormones modulate the rates of Ca²⁺ transfer across not only the intestine but also bone and kidney (Fig. 1). Under normal physiological conditions, production and secretion of PTH is dependent upon minute changes in the circulating concentration of Ca²⁺. Thus, a reduction in dietary Ca²⁺ results in increased levels of PTH in the circulation. The increased intestinal absorption of Ca²⁺, however, cannot be attributed to a direct effect of the hormone on the intestine. Instead the hormone in conjunction with the reduced levels of Ca²⁺ stimulates the production of 1,25(OH)$_2$D from its precursor 25(OH)D.[11,12] The former sterol is approximately 500 times more potent than 25(OH)D in stimulating intestinal Ca²⁺ transport and is the effector hormone in the regulation of Ca²⁺ absorption. Estrogens also influence Ca²⁺ absorption, but again its effect is secondary to the induction of secondary hyperparathyroidism by decreasing the transfer of Ca²⁺ from bone. In addition, there is some evidence suggesting that estrogens may have a direct or indirect effect on

the renal production of 1,25(OH)$_2$D.[13,14] Thus, in the third trimester of pregnancy, the rapidly increasing maternal requirement for Ca²⁺ is met by increasing the efficiency of intestinal Ca²⁺ absorption through the increased production of 1,25(OH)$_2$D.[15] In this instance, it is of interest that the placenta is also capable of contributing to the expanded production of 1,25(OH)$_2$D.[16]

In man,[17–19] as in animals,[20,21] intestinal absorption of Ca²⁺ decreases with advancing age. Similarly, the adaptation to low dietary Ca²⁺ is also blunted in the elderly.[2] These changes can both be attributed to a progressive resistance to vitamin D which develops in the process of aging.[19,21] In addition, the renal production of 1,25(OH)$_2$D may also be impaired in elderly men[19] and experimental animals.[22,23] In contrast to young rats, PTH failed to stimulate the synthesis of the sterol in adult rats.[24] Consequently, the dietary intake of Ca²⁺ required to maintain positive Ca²⁺ balance progressively increases from 400 mg/day to 1500 mg/day by the age of 65.[25]

3. Sites of Calcium Absorption

The localization of the actual intestinal site(s) of Ca²⁺ absorption in man and animals has preoccupied a host of investigators for considerable time. It seems fairly well established that absorption occurs primarily in the upper portion of the small intestine of most species tested thus far (rat, guinea pig, rabbit, horse, mouse, and chick) and that the most efficient mechanism of absorption resides in the duodenum. In the golden hamster, however, the efficiency of Ca²⁺ absorption is greatest in the ileum.[26] *In vitro* experiments with everted intestinal loops,[27] as well as *in vivo* intestinal perfusion studies in the rat[28] and dog,[29] have demonstrated that the rate of Ca²⁺ absorption per unit segment (in terms of the amount of Ca²⁺ transferred per unit time) is maximal in the duodenum. However, when one also considers the transit time of Ca²⁺ through the intestinal tract, it becomes immediately obvious that more distal intestinal segments contain the major effective sites of Ca²⁺ absorption in both rat and dog. Observations by Birge *et al.*[30] and Wensel *et al.*[31] suggest that similar but more sensitive mechanisms of Ca²⁺ absorption apply in man. These studies demonstrated that, although the calculated rate of Ca²⁺ absorption by the human duodenum was three times that of the rest of the gut, the intestinal segment distal to the duodenum was more sensitive to factors controlling Ca²⁺ absorption under normal physiological conditions.[30] The studies also showed that flux from lumen to blood was independent of intraluminal Ca²⁺ concentration and was maximal in the more proximal intestinal segments.[31]

Recent studies have focused on the colon and cecum as sites for active Ca²⁺ transport. In the rat utilizing voltage-clamp techniques, both electroneutral and electrogenic-mediated Ca²⁺ fluxes in the mucosal-to-secrosal direction can be demonstrated which conceivably represent Na⁺/Ca²⁺ exchange and Ca²⁺-ATPase-mediated transport.[32] Both vitamin D and dietary Ca²⁺ restriction enhanced mucosal-to-secrosal fluxes without changing secretory fluxes.[33] Other investigators suggest, however, that the secretory fluxes may be vitamin D dependent and carrier mediated,[34] while still others demonstrate a decrease in secretion fluxes with Ca²⁺ deprivation.[35] It is evident from these studies that the large intestine must now also be considered as playing a potentially significant role in Ca²⁺ homeostasis, particularly during Ca²⁺ deprivation or as a consequence of the loss of ileal absorptive function.

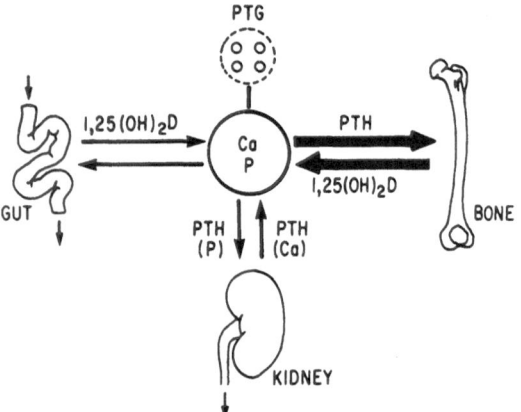

Fig. 1. Schematic representation of calcium and phosphorus homeostasis. The major determinant of plasma calcium and phosphorus is the exchange of mineral between bone and plasma. The rate of exchange and net flux is dependent upon both parathyroid hormone (PTH) and 1,25(OH)$_2$D. Calcium and phosphorus into plasma is accomplished by (1) increased production of PTH, PTH stimulating bone turnover through a primary effect on osteoblasts and a secondary (osteoblast-mediated) effect on osteoclasts, and (2) increased production of 1,25(OH)$_2$D by the kidney, 1,25(OH)$_2$D stimulating increased intestinal absorption of calcium and phosphorus and increased osteoclastic activity and the mobilization of calcium and phosphorus from bone.

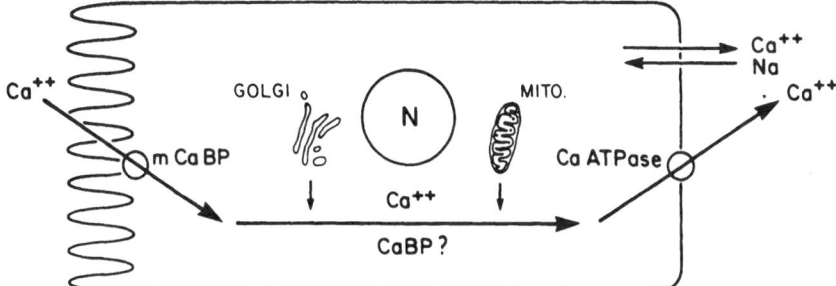

Fig. 2. Schematic representation of Ca^{2+} translocation across the intestinal epithelial cell.

4. Mechanism of Calcium Absorption

A number of mechanisms have been identified which mediate the intestinal transport of Ca^{2+}. These include simple passive ionic diffusion, facilitated diffusion, and active transport. The primary driving force for "passive ionic diffusion" is the electrochemical gradient across the intestinal mucosa with the rate of diffusion linearly related to the magnitude of this gradient. The kinetics of this movement are those of a first-order relationship. Therefore, unlike carrier-mediated facilitated diffusion or active transport, this mechanism is not capable of being "saturated" by the transported substance. A direct sensitive means of studying membrane transport has been established by Ussing.[36] This method, known as the short-circuit technique, permits continuous measurement of transepithelial influx in the absence of a membrane potential. An unambiguous way of documenting passive transport is to measure the bidirectional flux rates of an ion across the mucosal cell. If the ion is passively transported, the observed ratio of the mucosal-to-secrosal flux (J_{ms}) to the secrosal-to-mucosal flux (J_{sm}) will conform to the equation of Ussing[36]:

$$\frac{J_{ms}}{J_{sm}} = \frac{C_L}{C_P} \exp ZF \frac{\Psi_L - \Psi_P}{RT}$$

where C_L and C_P are the concentrations of the ion on the mucosal (lumen) and serosal (plasma) sides of the membranes, respectively, Z the ionic charge, F the Faraday number, R the gas constant, and T the absolute temperature; $\Psi_L - \Psi_P$ represents the potential difference across the membrane (or between lumen and plasma).

Although a precise definition of "active transport" is probably impossible, it may be simply defined as a carrier-mediated process, directly coupled to energy-yielding reactions within the cell, which effects a net transport of a substance against an electrochemical gradient. It is generally believed that the so-called carrier proteins are either loosely bound to membranes or confined to a surface compartment located between the cell membrane and the cell wall. Despite the general acceptance that carrier proteins are involved in transport, most if not all of the evidence supporting this contention is still indirect. Active transport yields a flux ratio different from that predictable from the Ussing equation and, because of its dependence on an energy-yielding reaction, is sensitive to inhibitors of cellular metabolism.

Finally, "facilitated diffusion" may be defined as a passive process with driving force that of the electrochemical gradient across the membrane. Since the transported ion is simply moving down its thermodynamic gradient, the mucosal flux ratio of facilitated diffusion is, like that of passive transport, identical to the theoretical flux ratio of Ussing. However, the kinetics of ion movement which obtain when a substance is transported by facilitated diffusion are not those of the first-order relationship described for simple passive diffusion but instead are consistent with those features presently considered characteristic of a carrier-mediated transport system.

Despite evidence to the contrary,[37,38] it seems well established that active transport contributes significantly to the intestinal absorption of Ca^{2+} in man and a variety of other animal species. The in vivo potential difference (PD) between the intestinal lumen and blood is positive with respect to the lumen and varies between 5 and 40 mV depending on the intestinal site and species.[26,39] Thus, in the absence of active transport, the blood/lumen flux ratio of a cation should be less than 1.0 under steady-state conditions. Of note in this regard is that Ca^{2+} transport ratios often drop to values below 1.0 in the presence of anaerobiosis or metabolic inhibitors of active transport.[40,41] Wensel et al., in experiments wherein the intact small intestine was perfused in vivo,[31] have demonstrated that, in man, Ca^{2+} is absorbed from the intestinal lumen against a concentration gradient. Using intestinal perfusion techniques with test solutions containing 1.0, 2.5, 5, and 10 mM Ca^{2+} (as calcium gluconate), Ireland and Fordtran concluded that jejunal Ca^{2+} transport in man was not only active and carrier mediated but also dependent on passive Ca^{2+} diffusion secondary to concentration gradients.[2] These investigators demonstrated that, with an estimated jejunal PD of 3 mV (lumen side negative), Ca^{2+} was absorbed against both electrical and concentration gradients when water flow was minimal. Active Ca^{2+} transport has also been observed in duodenal segments of young animals in the absence of concentration gradients and against a mean transmural potential of 5.2 mV.[40] Using in vitro gut sac procedures, a number of investigators have demonstrated that, in the duodenum of the rat, rabbit, and guinea pig, Ca^{2+} is transferred against a chemical gradient, that the "Ca^{2+} pump" can be saturated by increasing the concentration of Ca^{2+} in the precursor compartment and can be inhibited by hypothermia, anaerobiosis, and a variety of metabolic poisons. In the rat duodenum the apparent K_m value for the Ca^{2+} uptake process is of the order of 2.0 mM.[41] At low luminal Ca^{2+} concentrations Ca^{2+} absorption is mediated by this active process, but at high Ca^{2+} concentrations exchange diffusion and simple diffusion play an increasing important role in net Ca^{2+} absorption.[38,42] In the rat neonate, passive transport mechanisms prevail throughout the intestine since the vitamin D-dependent active transport process has not been induced.[43] The intestinal receptor protein for $1,25(OH)_2D$ does not appear until after the 14th postnatal day.[44]

A role of phosphate in mediating active Ca^{2+} transport has been suggested,[37] making the intestinal transport mechanism analogous to the phosphate-dependent Ca^{2+} transport in the kidney.[45] However, active transport of Ca^{2+} by the rat duodenum occurs in the absence of inorganic phosphate.[40,41,46,47] In the colon, stimulation of active transport of Ca^{2+} by vitamin D occurs in the absence of an effect of the vitamin on the active transport of inorganic phosphate, thus providing a clear dissociation between these two transport processes. The observed facilitation of Ca^{2+} transport by the addition of phosphate to the luminal compartment may be attributed to the restoration of depleted intracellular phosphate seen in the vitamin D-deficient experimental models used.[48,49]

In summary, the bulk of evidence accumulated to date supports the contention that in the duodenum, Ca^{2+} is transported by an active carrier-mediated energy-dependent process. Passive transport or facilitated diffusion or both would appear to be the primary regulating mechanism(s) for the absorption of Ca^{2+} in more distal intestinal segments. The higher pump efficiency of the more proximal intestinal segments may be due to intrinsic differences in the energy-generating systems or greater efficiencies of the intestinal cells in this location, a larger number of epithelial cells per unit length of intestine, and/or an increment in the amount of carrier essential to promote maximal active transport. The physiological significance of the duodenal transport pump appears to have been well established in that its transport capacity is adaptive and decreases with age, is greater in pregnant than in nonpregnant states, as well as during periods of low Ca^{2+} intake, and is stimulated by vitamin D and its biologically active metabolites. In general, the accumulated data are consistent with the fact that an active transport mechanism can increase the intestinal absorption of Ca^{2+} facultatively to meet the requirements of the organism. Despite repeated observations that the most rapid rate of of Ca^{2+} transport per unit length of intestine is in the small intestine just beyond the pylorus, the transit time through the duodenum is normally so short that, in reality, the largest fraction of the total Ca^{2+} absorbed is probably from intestinal sites distal to the duodenum where passive and/or facilitated diffusion predominate.

5. The Role of Vitamin D

Vitamin D is the singularly most important factor regulating the intestinal absorption of Ca^{2+} in man and animals. It had been well established that a 10- to 15-hr lag period obtained after vitamin D administration before stimulated intestinal Ca^{2+} transport was detectable.[50] We now know that vitamin D, itself biologically inactive, is subsequently converted to a variety of metabolites with varying degrees of effectiveness in stimulating Ca^{2+} absorption by the intestine. It appears that the lag phase between administration and biological effectiveness can be explained by: (1) the prerequisite rate-limiting hepatic conversion of vitamin D_3 to 25-hydroxycholecalciferol $[25(OH)D_3]$[51–53] (2) the subsequent transport of $25(OH)D_3$ in plasma to the kidney by a vitamin D-binding protein, immunologically identical to group-specific component (Gc) proteins[54–56]; and (3) the renal conversion of $25(OH)D_3$ to either $1,25(OH)_2D_3$,[57–59] a metabolite which is approximately 1000 times as active in stimulating intestinal transport as is the parent vitamin D_3 substance,[60–63] or to $24,25(OH)_2D_3$, a substance with a biological effectiveness on intestinal Ca^{2+} transport equal to that of $25(OH)D_3$.[64] $25,26(OH)_2D_3$ has also been discovered as a nor-

mal metabolite of vitamin D_3. Although its site of synthesis and physiological significance are still unknown,[65] it also is capable of stimulating intestinal Ca^{2+} transport.[65,66]

Although circulating levels of $25(OH)D_3$[67] and $24,25(OH)_2D_3$[68] are much higher than those reported for $1,25(OH)_2D_3$,[69] the latter is primarily responsible for the increments in Ca^{2+} absorption observed following vitamin D_3 feeding to rachitic animals. The exact sequence of subcellular molecular events which characterize the intestinal response to $1,25(OH)_2D_3$ (or other vitamin D_3 metabolites) continues to be an area of active investigation. Results of experiments with inhibitors of DNA-directed RNA synthesis are consistent with the interpretation that the stimulation of Ca^{2+} transport by $1,25(OH)_2D$ is at least in part mediated by *de novo* protein synthesis and therefore analogous to the mechanism of steroid hormone action. A cytoplasmic receptor highly specific for $1,25(OH)_2D_3$ has been demonstrated in rat intestine and partially purified from chick intestine with an association constant of 1×10^{-9} M[70,71] and a sedimentation coefficient of 3.5 S. $1,25(OH)_2D_3$ binding to this cytoplasmic receptor appears to represent an essential reaction prior to its subsequent binding to the nuclear chromatin of intestinal cells.[72,74] Additional studies on the nature of $1,25(OH)_2D_3$ nuclear binding have demonstrated an acidic protein with saturable binding sites, a dissociation constant of 2×10^{-9} M, and high specificity.[75] The sedimentation coefficient of 3.5 S observed for this protein is similar to that of the cytoplasmic protein which binds $1,25(OH)_2D_3$ *in vivo* after vitamin D_3 feeding. It is now postulated that $1,25(OH)_2D$ on entering the cell binds to the 3.5 S cytoplasmic receptor protein. Through a temperature-dependent process, the sterol–receptor complex is translocated to the nuclear chromatin.[73,74] A second protein has been identified in the cytosol of the intestine which also binds $1,25(OH)_2D$ and has a sedimentation coefficient of 5.6 S.[75] In contrast to the smaller 3.5 S binding protein, the affinity for $25(OH)D_3$ ($K_a \simeq 4 \times 10^{-10}$ M) exceeds that for $1,25(OH)_2D$.[76] The latter protein appears to be a conjugate of the serum vitamin D-binding protein and cytosolic actin[77,78] and can be identified in all nucleated tissues.[76] Its function and origin remain in doubt.[79] On the other hand, there is increasing evidence that the 3.5 S $1,25(OH)_2D$-binding protein does in fact mediate the biological expression of the cell's response to the sterol. The receptor protein has been identified in target tissues of $1,25(OH)_2D$ action[80–82] and clinical states of altered $1,25(OH)_2D$ action have been associated with a variety of defects or absence of the $1,25(OH)_2D$ receptor protein.[83]

Studies with isomeric forms of $25(OH)D_3$, $24,25(OH)_2D_3$, and $25,26(OH)_2D_3$ have demonstrated that structural modifications in the vitamin D_3 molecule do lead to varied differences in biological activity. In studies wherein the intestinal response to $1\alpha,25(OH)_2D_3$ was compared to structurally related analogs $1\alpha(OH)D_3$, $3B-1\alpha(OH)D_3$, $5,6$-*trans*-D_3, and dihydrotachysterol (DHT_3), structure–function relationships have been established.[84] It appears that a minimal planar blueprint is necessary for biological activity of vitamin D metabolites or structurally related synthetic analogs. The latter is represented by a molecule which incorporates a $1\alpha OH$ (or "pseudo" $1\alpha OH$), a side chain "R," on a seco-steroid backbone without a 3BOH or C-19 carbon atom.[84] A "seco-steroid" is a steroid with a ring which has undergone fission—the "B" ring in the case of vitamin D_3; a "pseudo" $1\alpha OH$ refers to compounds with carbon frameworks established by a 180° rotation about the 5,6 bond of vitamin D_3 so that the 3BOH group is in the same geometric position as the 1αOH of the natural 1α,25(OH)$_2$D$_3$ metabolite. Pseudo-1 hydroxyl-containing analogs include isotachysterol$_3$, 5,6-*trans*-D$_3$, DHT$_3$ and 25(OH)DHT$_3$. These geometric analogs of 1α,25(OH)$_2$D$_3$ are active in stimulating intestinal Ca^{2+} absorption and as such have been important therapeutic tools in treating patients with renal insufficiency and impaired endogenous 1α hydroxylation of vitamin D.

The molecular events resulting from the initiation of transcriptional processes which ultimately lead to an increase in Ca^{2+} transport across the intestinal epithelium have yet to be fully delineated. It has been demonstrated that vitamin D and its active metabolites stimulate the synthesis of a number of cell proteins. These include a Ca^{2+}-binding protein (CaBP),[85–88] alkaline phosphatase,[89–92] actin,[93] an intestinal membrane CaBP,[94] and a mitochondrial protein.[95] In addition, stimulation of cAMP,[96,97] alterations in membrane phospholipids,[98,102] and the phosphorylation of membrane proteins[103] have also been reported. These effects have all been alleged to contribute to the vitamin D-induced alterations in Ca^{2+} transport. In order to evaluate the potential role of these various molecular changes, it is necessary to appreciate which steps in the cellular translocation of Ca^{2+} are regulated by vitamin D.

Transport of Ca^{2+} across the intestinal epithelium can be considered to involve three steps: (1) transport across the brush border membrane, (2) transport across the cell cytosol to the basolateral membrane, and (3) transport across the basolateral membrane. The first step is along a downhill electrochemical gradient whereas the third step is against a steep electrochemical gradient and is considered to be the rate-limiting step (Fig. 1). The early studies of Schachter and others[104,105] demonstrated that Ca^{2+} uptake by the mucosal surface of the intestinal epithelial cell is not an active process but one of facilitated diffusion which is carrier mediated. Unlike the Na$^+$ dependency of sugar and amino acid transport, movement of Ca^{2+} across the brush border is independent of Na$^+$.[106,107] That this uptake of Ca^{2+} across the brush border is stimulated by vitamin D has been amply demonstrated by numerous investigators.[46,104,106,108,109] However, the mechanism of the vitamin's action at this site is still unknown. Some insight into this mechanism can be derived from the studies of Wong and Norman using the polyene antibiotic filipin which interacts with membrane lipid.[109] In the intestine, filipin enhances the mucosal uptake of Ca^{2+} in both vitamin D-deficient and -replete chicks without altering overall net mucosal–serosal Ca^{2+} flux. These studies suggest that alterations in the brush border lipid membrane structure can lead to a specific increase in the permeability of the membrane to Ca^{2+} and the accessibility of a latent transport system for Ca^{2+}. Rasmussen and co-workers[101,102] have demonstrated that 1,25(OH)$_2$D$_3$ can indeed alter phospholipid metabolism and composition of the brush border membrane. These changes in membrane lipid composition occur within a few hours and do not depend on new protein synthesis.[110] Similarly, the vitamin-induced increase in Ca^{2+} permeability of brush border membrane vesicles also occurs within hours of 1,25(OH)$_2$D$_3$ treatment and is not blocked by inhibitors of protein synthesis.[111,112] Vitamin D-dependent extrusion of Ca^{2+} across the basolateral membrane is not evident until 8–12 hr after treatment with the sterol and is blocked by inhibitors of both transcription and translation.[112] Thus, there are at least two phases in the cellular response to 1,25(OH)$_2$D$_3$. The first is an alteration in brush border membrane resulting in the facilitated diffusion of Ca^{2+} into the cell.

This phase is followed by the stimulation of active extrusion of Ca^{2+} across the basolateral membrane. The delayed response of the second phase is attributed to the requirement for the *de novo* synthesis of protein(s) associated with the active transport of Ca^{2+} out of the cell.

Because of the potential toxicity of the Ca^{2+} entering the cell, a number of mechanisms have been proposed to transport the Ca^{2+} through the cell cytosol to the Ca^{2+} pump at the basolateral membrane. One such mechanism involves the mitochondria. These organelles have an impressive ability to concentrate Ca^{2+} and indeed following vitamin D administration an increase in Ca^{2+} and phosphate granules can be demonstrated.[113] However, this increase in apparent uptake of Ca^{2+} is observed only during the initial phase of the cellular response to 1,25(OH)$_2$D and not during the second phase of active Ca^{2+} transport when intracellular levels of Ca^{2+} are restored presumably to normal.[114] Thus, it is unlikely that mitochondria participate in the transcellular movement of Ca^{2+} but serve instead as a sump in the prevention of Ca^{2+} toxicity resulting from the increased accumulation of cell Ca^{2+} during the initial phase of 1,25(OH)$_2$D$_3$ action. Another organelle which may serve as a shuttle for Ca^{2+} are Golgi-derived membrane vesicles. One of the earliest responses of the intestine to 1,25(OH)$_2$D is the increased accumulation of Ca^{2+} by these vesicles[115,116] occurring within 30 min of the vitamin's administration. Pretreatment with cycloheximide inhibits both Golgi Ca^{2+} uptake and intestinal Ca^{2+} transport.[117] On the other hand, their capacity to accumulate Ca^{2+} in response to vitamin D may reflect the presence of newly synthesized Ca^{2+} transport protein(s) destined for the cell membrane.

At the basolateral membrane, two mechanisms are believed to be involved in the efflux of Ca^{2+} against the electrochemical gradient. One mechanism involves a Na$^+$/Ca^{2+} exchange based on a number of studies demonstrating the requirement of the Ca^{2+} pump for Na$^+$[106,118] and the analogy to Na$^+$,Ca^{2+} exchange pumps in other plasma membrane systems.[119] In addition, a second Ca^{2+}-pumping mechanism has been identified in the enterocytes which is mediated by a Ca^{2+}-ATPase.[120,121,127] This enzyme is localized exclusively to the basolateral membrane and is activated by 10^{-7} M Ca^{2+}. In contrast, the Ca^{2+}-ATPase localized predominately along the brush border is activated by 10^{-3} M Ca^{2+} and cannot be distinguished from Mg^{2+}-dependent alkaline phosphatase. Like the Ca^{2+}-ATPase in membranes of red blood cells[123] or muscles,[124] the Ca^{2+}-ATPase of the enterocytes also formed a hydroxylamine-sensitive, phosphoenzyme intermediate which appears to be calmodulin regulated.[125,126] In addition, recent evidence suggests that the enzyme in the intestine is stimulated by vitamin D.[127]

Knowing the postulated sites for vitamin D action, it is still difficult to relate the known biochemical alterations induced by vitamin D in the enterocyte to the Ca^{2+} transport process. One intriguing observation is the rapid rise in cAMP which peaks between 30 and 90 min after the *in vitro* addition of 1,25(OH)$_2$D.[128] A second rise in cAMP occurs coincident with second phase in the stimulation of Ca^{2+} transport by the vitamin. These observations have been reviewed recently.[129] The initial delay in the cAMP response to 1,25(OH)$_2$D, however, is not typical of a cAMP-mediated hormone response. Considerable question remains as to whether the increase in cAMP is a primary event or secondary to other metabolic events such as an increase in intracellular phosphate and ATP induced by vitamin D.[130]

A number of CaBP have been identified in the intestinal epithelial cell membrane and cytosol. The first CaBP described has proven to be a cytosolic protein with a molecular weight of 9000–13,000 in mammalian intestine and ranging up to 25,000–28,000 in avian intestine.[131,132] The protein has been identified in multiple animal species[131,133–135] and a variety of tissues including kidney and pancreas.[136,137] In each instance and by definition, the protein is induced *de novo* by 1,25(OH)$_2$D.[130,138,139] Its role in Ca^{2+} transport is suggested by the high affinity of the protein for Ca^{2+} ($K_d \simeq 10^{-5}$ M),[140,141] the appearance of the protein parallels the increase in Ca^{2+} absorption of the second phase of the enterocyte response to 1,25(OH)$_2$D,[86,87,141,142] and the concentration of the protein in the various segments of the intestine correlates with the relative abilities of these segments to absorb Ca^{2+}.[143] On the other hand, CaBP cannot be detected at a time when the initial increment in Ca^{2+} transport is demonstrated,[144–146] and secondly, CaBP product can be blocked without inhibition of 1,25(OH)$_2$D-stimulated Ca^{2+} transport.[112] Although CaBP may not participate in membrane transport of Ca^{2+}, some role for the protein seems likely in the intestinal response to vitamin D. At present, it is postulated that the protein may serve as an intracellular carrier of Ca^{2+}, to facilitate the transfer of Ca^{2+} from mitochondria or Golgi to the Ca^{2+} pump of the basolateral membrane or as a buffer to protect the cell from the influx of Ca^{2+} induced by the vitamin. With regard to the latter hypothesis, Buckley and Bronner[119] have demonstrated that in the presence of 1,25(OH)$_2$D, the level of CaBP in the intestine was directly proportional to the dietary calcium content. In addition to the well-characterized cytosolic CaBP, two CaBP have been isolated from the brush border membrane with approximate molecular weights of 18,500 and 200,000, respectively.[92,147,148] The role of these proteins in the carrier-mediated entry of Ca^{2+} into the cell remains conjectural at this time.

Other proteins stimulated by vitamin D are brush border alkaline phosphatase[16,89] and Ca^{2+}-ATPase[149,150] although there is considerable doubt whether these two enzymes are not in fact the same enzyme.[16,151] Both enzyme activities are proportionally extracted from the membranes by butanol and inhibited equally by L-phenylalanine. Again, the Ca^{2+}-ATPase activity of the brush border which is stimulated by millimolar concentrations of Ca^{2+} is to be distinguished from the Ca^{2+} pump ATPase of the basolateral membrane stimulated by micromolar Ca^{2+} concentrations. The participation of either of these enzyme activities in either vitamin D-stimulated Ca^{2+} or phosphate transport is considered unlikely because L-phenylalanine does not inhibit intestinal Ca^{2+} or phosphate transport. Second, the changes in alkaline phosphatase activity do not correlate well with the changes in Ca^{2+} transport induced by vitamin D. On the other hand, this ectoenzyme may play a role in increasing the availability of dietary organic phosphate to intestinal absorption. Other proteins not identified are also synthesized in response to vitamin D and are a part of a generalized stimulation of RNA and protein synthesis.[112,146] This response of the intestinal epithelium reflects the generalized trophic effect of the vitamin characterized by an increase in DNA synthesis and mucosal proliferation.[152,153]

6. Clinical Disorders Associated with Alterations in Calcium Absorption

As noted in Table II, a variety of conditions can lead to an alteration in the absorption of Ca^{2+}. Regrettably, the techniques

Table II. Clinical States of Altered Ca^{2+} Absorption

Increased absorption
  Physiological states
    Pregnancy
    Lactation
  Pathological states
    Sarcoidosis and granulomatous disorders
    Primary hyperparathyroidism
    Diabetes
    Idiopathic hypercalciuric syndromes
    Phosphorus depletion
  Pharmacologically induced
    Milk-alkali syndrome
    Vitamin D intoxication
    Estrogens
    Nonabsorbable antacids
Malabsorption
  Physiological states
    Aging
  Pathological states
    Intrinsic bowel disease
    Hepatobiliary disease
    Renal disease
    Hyperthyroidism
    Hypoparathyroidism
  Pharmacologically induced
    Anticonvulsants (dilantin, phenobarbital)
    Thiazides
    Chemotherapeutic agents
    Glucocorticoids
    Antibiotics
  Dietary factors
    Vitamin D deficiency
    Mg^{2+} deficiency
    Ca^{2+} deficiency
    Iron deficiency
    High-Na^{+} diet

for measuring Ca^{2+} absorption remain cumbersome, time-consuming, and inaccurate. These various methods have been reviewed adequately elsewhere.[154,155] As a consequence of our inability to adequately assess Ca^{2+} absorption directly, one must rely for practical purposes on the clinical manifestations of increased or decreased availability of Ca^{2+}. Thus, it is appropriate to review in general terms the clinical and biochemical expression of these states of altered Ca^{2+} homeostasis. First with respect to states of intestinal hyperabsorption of Ca^{2+}, two basic mechanisms can be considered: physiologic and pathologic. Physiologic stimulation of Ca^{2+} absorption reflects the body's response to an increased requirement for Ca^{2+}. Under these conditions, urinary excretion of Ca^{2+} is either low or normal. On the other hand, pathologic states of increased Ca^{2+} absorption are in general associated with hypercalciuria (> 300 mg men, > 250 mg women, or > 4 mg/kg body wt),[157] the exception being early primary hyperparathyroidism. Urinary excretion of Ca^{2+} can be a very sensitive indicator of excessive assimilation of dietary Ca^{2+} with changes being evident before increments in the serum Ca^{2+} can be appreciated. Its usefulness as a diagnostic tool requires a knowledge of the dietary Ca^{2+} content during and preceding the urine collection since urinary excretion of Ca^{2+} is directly proportional to dietary Ca^{2+} intake in normal individuals.[156] Malabsorption of Ca^{2+} on the other hand is expressed clinically as osteopenia as a result of the

Table III. Etiology of Hypoparathyroidism and Parathyroid Hormone-Resistant States

---

Hypoparathyroidism
  Idiopathic
  Surgical
  Familial
  Hypomagnesium
Parathyroid hormone-resistant states
  Uremia
  Vitamin D deficiency (resistance)
  Hypomagnesium
  Pseudohypoparathyroidism

---

negative Ca^{2+} balance. Urinary excretion of Ca^{2+} has not proven to be a reliable estimate of impaired Ca^{2+} transport at the level of the intestine. This is in part due to the fact that the kidney's capacity to conserve Ca^{2+} is limited in contrast to its ability, for example, to conserve Na^+ or phosphate. Serum Ca^{2+} levels are maintained within the normal range by the various homeostatic mechanisms described previously; there again the serum Ca^{2+} poorly reflects the altered state of Ca^{2+} homeostasis. Only under selected circumstances, specifically states of hypoparathyroidism or states of PTH resistance, does this homeostatic mechanism fail, Ca^{2+} cannot be adequately mobilized from the skeleton, and the serum Ca^{2+} falls. These states of hypoparathyroidism and PTH resistance are listed in Table III. In the absence of these states, intestinal malabsorption of Ca^{2+} can result in the asymptomatic depletion of the skeleton of its mineral content. Regrettably, skeletal fracture may be the first and only manifestation of the malabsorption. The cellular response of bone depends on numerous factors which are poorly understood. As a simplification, malabsorption associated with vitamin D deficiency is characteristically associated with a picture of osteomalacia—a disease of bone formation and a failure to mineralize the newly formed bone matrix. Malabsorption of Ca^{2+} without a deficiency of vitamin D results in a skeletal response characterized by osteoporosis with or without evidence of secondary hyperparathyroidism.[158] Growing children appear to be an exception by developing classical rickets with pure Ca^{2+} deficiency.[159,160] In the following discussion, we will attempt to review briefly the mechanisms of the altered Ca^{2+} absorption in these clinical states included in Table II.

7. Increased Absorption of Calcium

Pregnancy and *lactation* are well-recognized stimulants to intestinal Ca^{2+} absorption[161,162] resulting in greater than a twofold increase in total Ca^{2+} absorbed. Despite this dramatic increase in response to the fetal and maternal Ca^{2+} requirements, the mechanism mediating this response is not understood. Heaney and Skillman[162] were able to demonstrate the increased absorption by the 20th week of pregnancy (the earliest time studied) at which time fetal Ca^{2+} storage is insignificant. One can postulate that the positive Ca^{2+} balance at this time will be utilized later in pregnancy for the fetus. The fascinating question is then: what factors induce this adaptation of the intestine to anticipate the needs of the mother later in her pregnancy? A rise in maternal PTH does not occur until the last

few weeks of pregnancy.[163] On the other hand, maternal serum concentrations of $1,25(OH)_2D$ rise progressively with pregnancy.[15,164–166] The origin of the apparent increased production of $1,25(OH)_2D$ has been attributed to the stimulation of the renal 1α-hydroxylase by prolactin or estrogens[13,166] or to production by the fetal–placental unit.[168–171] These studies have failed to appreciate the sex steroid hormone-induced rise in vitamin D-binding protein. Accordingly, estimates of free $1,25(OH)_2D$ do not increase in the maternal circulation until the last few weeks of pregnancy and again late in lactation.[172] In addition, subsequent studies have failed to demonstrate increased concentrations of $1,25(OH)_2D$ in women with prolactinomas and elevated levels of prolactin in their serum.[172] Nor did these same investigators observe an increase in the intestinal absorption of Ca^{2+}, thus making it unlikely that prolactin is stimulating directly intestinal absorption of Ca^{2+}. Similarly, the role of estrogens in the enhanced absorption of Ca^{2+} is disputed. *In vitro* studies fail to demonstrate a direct effect of the hormone on renal production of $1,25(OH)_2D$, suggesting that its effects may be mediated through enhanced secretion of PTH.[14] More recent data suggest that estrogen may induce an increase in the intestinal content of the $1,25(OH)_2D$ receptor protein.[173] Finally, it is of interest that intestinal absorption of Ca^{2+} is increased during pregnancy and lactation in the vitamin D-deficient animal.[174]

Sarcoidosis is a disease of unknown origin characterized by a granulomatous infiltrate of multiple tissues. During the active phase of the disease, increased absorption of Ca^{2+} from the intestine occurs which results in hypercalciuria and occasionally hypercalcemia.[175,176] These changes have been attributed to the finding of elevated levels of $1,25(OH)_2D$[177,178] which reflects an abnormal metabolism of the latter compound. Administration of a large dose of vitamin D suppressed $1,25(OH)_2D$ concentrations in normal patients but, in patients with both active and inactive sarcoidosis, increased the levels of $1,25(OH)_2D$.[179] Of great interest is the origin of the sterol since the excess production persists following bilateral nephrectomy.[180] More recently, production of $1,25(OH)_2D$ by a variety of different tissues has been reported.[181] One can speculate, therefore, that one of the cellular elements of the granulomatous reaction may be the origin of the excessive production of the sterol. This phenomenon may not be restricted to sarcoidosis since other granulomatous diseases have been associated with hypercalcemia, as for example, coccidiomycosis,[182] histoplasmosis,[183] berylliosis,[184] and tuberculosis.[185,186] There is some question whether these disease entities share with sarcoidosis an increased sensitivity to vitamin D.[187]

Diabetes mellitus has been recognized as being associated with osteopenia, usually subclinical,[188–190] which has initiated the search for the pathogenesis of this disorder. For the most part, these studies have utilized the streptozotocin-induced diabetic rat model and have revealed profound alterations in vitamin D metabolism and Ca^{2+} homeostasis. To avoid the confusion that exists in this area, it is necessary to make the distinction between the acute and the chronic (greater than 30 days) diabetic rat. The latter appears to be the more appropriate model for human diabetes mellitus. Biochemically, the chronic diabetic rat model displays an elevation in serum Ca^{2+} and depressed levels of PTH.[191] Serum concentrations of $1,25(OH)_2D$ are depressed while those of $24,25(OH)_2D$ are increased.[192,193] Paradoxically, intestinal absorption of Ca^{2+} is increased and is associated with a marked hypercalciuria.[194] This increased absorption of Ca^{2+} can be attributed to (1) increased dietary intake,[194] (2) hypertrophy of the intestine,[195,196] and (3) a fivefold increase in the intestinal mucosal receptors for 1,25(OH)$_2$D.[193] Thus, the modest reduction in the circulating concentration of the sterol is more than compensated for by the increase in its tissue receptor protein. In diabetic children, a similar trend in the relative concentrations of the vitamin D metabolites is evident[197] as well as increased intestinal absorption of Ca^{2+}.[198]

Idiopathic hypercalciuria is a common finding in patients with recurrent renal calculi and in family members of such patients. An additional 5% of the population is estimated to have idiopathic hypercalciuria in the absence of renal lithiasis.[199] The pathogenesis of these disorders are still in question and therefore the diagnosis must be one of exclusion. Under the heading of idiopathic hypercalciuria, two major subgroups have been identified on the basis of the response to a low-Ca^{2+} diet (< 400 mg/24 hr) or on the basis of the calcium/chromium ratio in a 2-hr fasting morning urine collection followed by a 100-mg oral Ca^{2+} load and two additional urine collections over the next 4 hr.[200,201] The basic distinction between these two groups is that in those patients labeled "absorptive" hypercalciurics, excessive urinary excretion of Ca^{2+} is corrected with the reduction in dietary Ca^{2+}, whereas in those patients designated "resorptive" hypercalciurics, the excessive excretion of Ca^{2+} persists after an overnight fast or on a low-Ca^{2+} diet. The latter group can be further subdivided on the basis of their response to a Ca^{2+} load. The majority of patients in this subgroup apparently have a primary decrease in the renal tubular reabsorption of Ca^{2+} and are therefore termed renal hypercalciurics.[202,203] Their serum Ca^{2+} tends to be low while their immunoreactive PTH levels are elevated as is the urinary excretion of cAMP and both suppress with Ca^{2+} loading[201,204] or thiazide therapy.[201,205] Hydrochlorothiazide increases the tubular reabsorption of Ca^{2+} by mechanisms which are not well understood and is therefore an effective form of therapy in all patients with idiopathic hypercalciuria. Two other groups have been included under resorptive hypercalciuria but perhaps inappropriately.[204] The first are patients with nonsuppressible hyperparathyroidism who in many instances have an underlying renal Ca^{2+} "leak" and presumably tertiary hyperparathyroidism. The second subgroup includes those patients who appear to have a primary renal phosphate leak; the mechanism of the increased intestinal absorption of Ca^{2+} and hypercalciuria in this patient population will be discussed below. Both the "absorptive" and the "resorptive" hypercalciuric syndromes have an increased intestinal absorption of Ca^{2+}.[201,206,207] In the latter group, the increased intestinal Ca^{2+} absorption can be explained on the basis of the secondary hyperparathyroidism and the resultant increase in the production of 1,25(OH)$_2$D. The bone biopsy lends additional credence to these physiological changes demonstrating increased osteoblastic and osteoclastic activity.[204] Indeed, thiazide therapy which corrects the renal Ca^{2+} loss, restores to normal the elevated serum concentrations of 1,25(OH)$_2$D[208] and normalizes the intestinal Ca^{2+} absorption.[206,208] In the absorptive hypercalciurias, the pathogenesis of the increased intestinal Ca^{2+} absorption is not as clearly understood. PTH levels are probably suppressed in this disorder which is reflected in the bone histology.[208] Both osteoclastic and osteoblastic activity are reduced with an increase in the fraction of inactive osteoid. Although thiazide therapy corrects the renal excretion of Ca^{2+}, it does not normalize the increased intestinal absorption.[206,208] This could be explained by the hypothesis that circulating levels of 1,25(OH)$_2$D are elevated in this disorder due to a primary phos-

phate leak and hypophosphatemia.[209,210] Unfortunately, there is not general agreement as to whether patients with absorptive hypercalciuria have reduced concentrations of phosphate in their serum.[211] Second, the role of phosphate depletion in the regulation of 1,25(OH)$_2$D synthesis can also be questioned.[212] Finally, Zerwekh and Pak did observe that there was no change in the slightly elevated levels of 1,25(OH)$_2$D in their absorptive hypercalciuric patients following thiazide treatment.[208] Perhaps another interpretation of the pathogenesis of this disorder is that the regulation of the renal 1α-hydroxylase by Ca^{2+} is impaired resulting in the inappropriate elevation of the dihydroxy metabolite of vitamin D relative to the Ca^{2+} demands of the individual.

Phosphate depletion results in a recognized clinical entity characterized by a myopathy, hypercalciuria, accelerated bone resorption, osteomalacia, and increased intestinal absorption of Ca^{2+}.[213–218] The pathophysiological processes involved in the various features of this disorder are complex and as yet the mechanism of the enhanced intestinal absorption of Ca^{2+} remains to be elucidated. The observed changes in the tubular reabsorption of Ca^{2+} and phosphate cannot be attributed to the reduction in circulating PTH. Thyroparathyroidectomy cannot abolish this adaptive response of the kidney.[219–221] With the discovery of the renal conversion of 25(OH)D to 1,25(OH)$_2$D, it was postulated that phosphate depletion resulted in the stimulation of 1,25(OH)$_2$D production. The increase in the sterol thus led to the increase in the intestinal absorption of Ca^{2+} and the rise in serum Ca^{2+}.[222,223] Indeed, increased production of 1,25(OH)$_2$D does occur as a result of phosphorus depletion.[224] However, it soon became apparent that the changes in intestinal Ca^{2+} absorption and serum Ca^{2+} did not require the renal metabolism of 25(OH)D and could be demonstrated in the vitamin D-deficient animal model.[225,226] Thus, the observed adaptations to phosphate depletion are mediated by factors which are independent of the thyroid, the parathyroids, and vitamin D metabolism. Consequently, it seems reasonable to postulate that a heretofore unrecognized factor, presumably humoral, can be implicated in the renal, intestinal, skeletal (and perhaps muscle) response to phosphorus depletion.[213,227]

Pharmacologically induced states of increased Ca^{2+} absorption can be of major clinical significance. Before being adequately appreciated, the milk-alkali syndrome was a significant cause of irreversible renal failure. The syndrome is characterized by the excessive ingestion of Ca^{2+} and alkali usually for the treatment of peptic symptoms resulting in alkalosis and uremia and in its chronic form hypercalcemia and calcinosis may be superimposed upon the renal impairment attributed to the alkalosis. The development of renal calcinosis leads to further reduction of glomerular filtration and increased retention of Ca^{2+}.[228–230] The increased intestinal absorption of Ca^{2+} is due only to the tremendous Ca^{2+} load presented to the intestine. Altering luminal pH over a wide range has no effect on Ca^{2+} transport, although normal gastric acidity is required to solubilize CaCO$_3$ preparations. The systemic alkalosis, however, may contribute to a reduction in the glomerular filtration rate and thus the retention of Ca^{2+} as would any underlying renal disease. However, alkalosis is usually associated with a rise rather than a fall in inulin and creatinine clearance.[231] The rising serum Ca^{2+} would reduce the glomerular filtration rate and thereby further exacerbate Ca^{2+} retention, hypercalcemia, and renal calcinosis. Vitamin D intoxication, although less common than previously reported, still presents a potentially serious problem in the use of vitamin D therapeutically.[232] With the advent of the more potent, faster acting analogs of vitamin D, $1\alpha(OH)D_3$ and $1\alpha,25(OH)_2D_3$, the incidence of vitamin D intoxication promises to increase. The mechanism of the increased intestinal absorption of Ca^{2+} is of interest in that it is mediated by $25(OH)D$ and its interaction with the $1,25(OH)_2D$ receptor in the intestine, $1,25(OH)_2D$ production being suppressed by the hypercalcemia.[233,234] Like vitamin D, $25(OH)D$ has a half-life of approximately 3 weeks,[235] being stored in adipose tissue and muscle.[236,237] Consequently, following the withdrawal of the vitamin D, 2–6 months or more may be required before hypercalcemia and the enhanced absorption of Ca^{2+} is restored to normal.[238] The influence of estrogens on Ca^{2+} absorption in man has been difficult to ascertain because the changes are not dramatic.[239,240] However, Ca^{2+} balance studies[241,242] and radioactive isotopic Ca^{2+} studies[243] suggest that estrogens induce a positive Ca^{2+} balance in women which is associated with a reduction in fecal Ca^{2+} and increased intestinal absorption of Ca^{2+}. This effect of estrogens on intestinal Ca^{2+} absorption is probably secondary to a reduction in bone resorption and secondary hyperparathyroidism.[242,244] A direct stimulation of intestinal Ca^{2+} absorption by estrogens is unlikely since the steroid probably inhibits the transport process.[245] Finally, the nonabsorbable antacids containing aluminum or magnesium hydroxide constitute a significant pharmacological stimulation of intestinal Ca^{2+} absorption. These agents act by blocking the intestinal absorption of phosphate leading to hypophosphatemia.[246] As discussed above, the hypophosphatemia leads to the stimulation of intestinal Ca^{2+} absorption by mechanisms which are poorly understood.

8. Malabsorption of Calcium

8.1. Physiological States

Perhaps the most common cause of decreased intestinal absorption of Ca^{2+} is the process of aging. Evidence has accumulated that in both the experimental animal[20,40,247] and man,[16,17] intestinal absorption of Ca^{2+} decreases progressively in the adult with advancing age. After the age of 65, absorption of Ca^{2+} is less than one-third that measured in the third and fourth decades of life. The decrease in absorption occurred prior to menopause in both males and females and therefore cannot be attributed entirely to the loss of sex steroids. A number of factors have been identified which contribute to this decline. First, the ability of the intestine to respond to the challenge of a low-Ca^{2+} diet is impaired in the elderly.[2] This suggests that in the elderly the adaptive response is impaired either at the level of the production of the adaptive hormones [PTH and $1,25(OH)_2D$] or at the level of the intestinal response to the hormones. There is now evidence to suggest that aging impairs the adaptive response to Ca^{2+} deprivation at both levels. In man[19] and in the laboratory animal model,[22,23] aging results in decreased synthesis of $1,25(OH)_2D$. Gallagher et al. [19] have demonstrated in osteoporotic women that not only are the levels of $1,25(OH)_2D$ reduced but there is no correlation between the levels of the sterol and the measured absorption of Ca^{2+}. These data can be interpreted by postulating a resistance of the intestine to the action of $1,25(OH)_2D$. Indeed, unpublished studies in our laboratory have demonstrated that the intestinal cytosolic receptor protein for $1,25(OH)_2D$ decreases in parallel with the reduction of Ca^{2+} transport associated with aging.

8.2. Pathological States Associated with Malabsorption of Calcium

One of the major causes of osteomalacia are diseases of the bowel or surgical resection of the intestine. It is obvious that simple reduction of the absorptive surface area as in Crohn's disease, nontropical sprue, celiac disease, and intestinal bypass surgery will lead to the malabsorption of Ca^{2+}. However, other factors have been identified in these conditions which also contribute to the malabsorption. Of these factors, loss of bile salts and/or steatorrhea are clearly important. As noted earlier, increased concentrations of unabsorbed fatty acids within the intestinal lumen bind Ca^{2+} and limit the accessibility of the Ca^{2+} to its transport site. In addition, the increase in luminal fat concentrations impairs the absorption of vitamin D. Thus, the steatorrhea associated with pancreatic insufficiency has been shown to contribute to the malabsorption of vitamin D and osteomalacia.[248,249] Like cholesterol, calciferol is a very nonpolar lipid with similar solubility characteristics. To permit absorption of these compounds by the intestinal mucosa, solubilization by micellar formation is essential. Thus, absorption of vitamin D and cholesterol is markedly impaired in experimental bile duct obstruction.[250] Biliary cirrhosis and other chronic cholestatic syndromes with impairment of bile salt secretion result in marked reduction of serum $25(OH)D$ levels and osteomalacia,[251–253] whereas pure cirrhosis with loss of parenchymal function leads to only modest reductions of serum $25(OH)D$ levels.[252] Although the liver is the primary site for the hydroxylation of vitamin D, the enzymatic reserve is apparently sufficient to sustain marked losses of hepatic parenchymal tissue. However, the hydroxylase may be inhibited by the high concentrations of bile salts anticipated in the liver of cholestatic disease.[254] Other investigators have shown no impairment in the hydroxylation of vitamin D by patients with severe forms of biliary cirrhosis.[255] A third factor to be considered is the loss of vitamin D by interruption of its enterohepatic circulation. The sterol as $25(OH)D$ has been demonstrated to be secreted with the bile into the duodenum and reabsorbed in the ileum.[256] The significance of this enterohepatic circulation remains controversial. Perhaps the most common gastrointestinal condition associated with Ca^{2+} malabsorption is gastric surgery and in particular subtotal gastrectomy and gastrojejunostomy.[257,258] An incidence of metabolic bone disease of up to 42% has been reported in patients with partial gastrectomy.[258] Vitamin D deficiency has been suggested to be one of the major causes of the bone disease of these patients due to its impaired absorption[259] or to poor intake.[260] In the absence of vitamin D deficiency, the mechanism of the deranged mineral metabolism remains to be elucidated. Factors such as inadequate Ca^{2+} intake, intestinal hurry, and inadequate acidification have been suggested.[261]

Two endocrine disorders are associated with significant malabsorption of Ca^{2+}. The most obvious is that of hypoparathyroidism.[30] The loss of circulating PTH and the rise in serum phosphate result in the decreased renal production of $1,25(OH)_2D$. The malabsorption of Ca^{2+} can therefore be attributed to the low levels of $1,25(OH)_2D$. Recent evidence suggests that PTH may also have a direct effect on intestinal Ca^{2+} transport[262] so that its absence may further contribute to the reduced intestinal absorption of Ca^{2+}. The second less obvious endocrine disorder leading to malabsorption of Ca^{2+} is that of hyperthyroidism.[263,264] Increased concentrations of thyroid hormone stimulate bone resorption[265,266] resulting in the increased serum Ca^{2+} and phosphorus observed in this disorder.[267,268] As a consequence of the hypercalcemia, PTH secretion is suppressed.[269,270] Hence, as noted above, the reduction in Ca^{2+} absorption observed in the hyperthyroid state[263,264] can be attributed to the altered metabolism of vitamin D. Indeed, serum concentrations of $1,25(OH)_2D$ are reduced in patients with active hyperthyroidism.[271]

Chronic renal failure is characterized by a marked impairment in the intestinal absorption of Ca^{2+}.[272-280] Patients with end-stage renal failure are also unable to adapt to alterations in dietary Ca^{2+} and, as a consequence, develop large fecal Ca^{2+} losses during periods of malnutrition of compromised Ca^{2+} intake, as may occur with low-protein dietary regimens.[275,276] Alterations in Ca^{2+} absorption usually are detectable when serum creatinine values are greater than 2.5 mg/100 ml,[275] although an inverse correlation between BUN level and Ca^{2+} absorption has been reported for patients with BUN levels between 25 and 150 mg/100 ml.[279] In patients with chronic renal disease, an intestinal defect in Ca^{2+} absorption has been demonstrated in the duodenum, jejunum, and ileum.[277,281] There are reports that distal non-vitamin D-dependent sites where Ca^{2+} absorption is primarily passive[8] are apparently functionally intact.[277] These findings are consistent with observations that uremic patients can absorb Ca^{2+} normally when dietary Ca^{2+} intake is augmented to 4 to 10 g/day[282,283] and others showing a direct relation between a positive Ca^{2+} balance (i.e., retained Ca^{2+}) and Ca^{2+} intake.[284] The intestinal malabsorption of Ca^{2+} observed in patients with end-stage renal disease is resistant to vitamin D therapy[275,286-291,311] [including $1\alpha(OH)D$ and $1,25(OH)_2D$], unaltered by chronic intermittent hemodialysis,[273,277,285] and ultimately unrelated to the skeletal pathology since the progression of the osteodystrophic bone lesions is not attended by further deterioration of the intestinal absorption of Ca^{2+}.[278,313]

The specific pathogenesis and exact nature of the intestinal Ca^{2+} transport defect in the chronic uremic state are still uncertain. There is mounting evidence obtained from experimental uremic models that renal insufficiency not only affects cellular Ca^{2+} transport parameters directly,[292,293,294,296] but also results in derangements in vitamin D metabolism, with defective production of $1,25(OH)_2D_3$.[297,299,300] Since $1,25(OH)_2D_3$ functions primarily to regulate intestinal absorption of Ca^{2+}, acquired alterations in its production by the diseased kidney obviously play a major role in initiating and perpetuating the intestinal malabsorption of Ca^{2+}. In this regard, it should be emphasized that Ca^{2+} absorption can be reduced even further when uremic patients are subjected to nephrectomy.[301] The specific intestinal transport abnormality resulting from defective $1,25(OH)_2D_3$ production in uremic man is still virtually unknown, since a vitamin D-activated intestinal ATPase[99,100] and mitochondrial activities,[302] as well as intestinal cellular proliferation[296,303] and CaBP synthesis[135,304] are all reportedly defective in the chronic uremic state. Moreover, reports of raised serum calcitonin levels in patients with chronic renal failure[305] and calcitonin inhibition of vitamin D-induced intestinal Ca^{2+} absorption[306] should also be considered.

Derangements in protein,[307] amino acid, and mitochondrial metabolism and ATPase activity in non-vitamin D-dependent tissues, well established in the chronic uremic state, could be attributed to one or more retained uremic toxins. The possibility exists therefore that uremic toxins *per se* may impose certain conditioned alterations in the established orderly sequence of intestinal cell multiplication and maturation and in the translocational and transcriptional processes normally regulating intestinal protein synthesis. Synthesized proteins could possibly be structurally or catalytically unable to effect the translocation of Ca^{2+} across the intestinal cell. That vitamin D responsiveness and metabolism are probably unaltered early in the course of uremia is reflected in reports of normal intestinal Ca^{2+} absorption in patients with serum creatinine concentrations lower than 2.5 mg/100 ml or filtration rates higher than 25-30 ml/min[295,308,309] and in others citing the cure of bone lesions in patients with early renal failure by treatment with vitamin D alone in doses similar to those used to cure nutritional rickets.[310] As the uremic state advances, each of the aforementioned non-vitamin D-related derangements in Ca^{2+} transport and intestinal absorption could obtain, albeit with varying degrees of severity. Increments in intestinal juice K^+ concentrations, well documented in uremic patients with secondary hyperaldosteronism, further compromise Ca^{2+} absorption, since K^+ interferes with Ca^{2+} transport.[296] With progressive decrease in functioning renal mass, the additional insult imposed by decreased $1,25(OH)_2D_3$ production obviously must contribute to the intestinal malabsorption of Ca^{2+} and observed "resistance" to vitamin D therapy. Regardless of the cause, the Ca^{2+} malabsorption attending the clinical course of patients with chronic renal failure, especially those with restricted dietary intake, adds another insult to Ca^{2+} homeostasis by providing an additional stimulus to parathyroid hyperplasia, elevated circulating PTH levels, and progressive skeletal deterioration.

Ca^{2+} malabsorption which is resistant to vitamin D therapy has also been observed in patients with either the nephrotic syndrome[297,298] or classical forms of renal tubular acidosis.[314-316] Since nitrogenous wastes and uremic toxins do not accumulate in either instance and functional renal mass is not severely impaired, the cause(s) of the malabsorptive defect in these "tubular" forms of renal disease is most probably related to either the persistent hypoalbuminemia (nephrotic syndrome) or systemic acidosis (renal tubular acidosis). The relationship between the net intestinal absorption of Ca^{2+} and hypotonicity of circulating fluids in man is still unknown, as is the effect of chronic acidosis on either the bioactivation of vitamin D_3 or the intestinal response to vitamin D_3 treatment. Since in animals Ca^{2+} absorption is dependent on circulating oncotic pressure and net lumen–plasma water movement,[54] one might reasonably postulate that clinical disorders characterized by chronic hypotonicity of circulating fluids might, in fact, lead to an increase in plasma–lumen water flux and a subsequent decrease in the intestinal absorption of Ca^{2+}. The malabsorption of Ca^{2+} attending the course of patients with renal tubular acidosis may also be multifactorial. It can, however, be completely corrected by alkali therapy alone.[317] However, although a systemic acidosis may directly impair the lumen–plasma transport of Ca^{2+}, the renal conversion of $25(OH)_2D_3$ to $1,25(OH)_2D_3$ is pH dependent and an acidic pH is inhibitory for the reaction.[317,318]

9. Drug-Induced Inhibition of Calcium Absorption

Anticonvulsant drugs have become recognized as a significant cause of disordered mineral homeostasis and osteomalacia with a reported incidence ranging from 4 to 40% in those receiving these drugs.[319,320] A number of factors have been identified which predispose the epileptic population to the development of osteopenia. Of these, the most prominent are the level of vitamin D intake and the exposure to sunlight.[319,321] Thus, the incidence of disordered mineral metabolism is greatest in institutionalized epileptic patients who are confined indoors.[322] Also of importance are the dose and duration of drug therapy and use of multiple anticonvulsant drugs.[323] Use of acetazolamide or a ketogenic diet to produce a systemic acidosis further compromises the already disordered state of mineral homeostasis by accelerating the loss of mineral from bone and inhibiting the conversion of 25(OH)D to 1,25(OH)$_2$D.[323] The mechanism of the anticonvulsant drug action leading to osteomalacia appears to be basically twofold. The first is an alteration of the metabolism of vitamin D and the second is a direct inhibition of ion transport across the intestinal epithelium and by bone cells. It is well recognized that the common anticonvulsants induce an increase in the hepatic P-450 mixed-function oxidase activity which results in the increased hydroxylation of vitamin D and 25(OH)D to biologically inactive more polar metabolites which are eliminated in the bile and urine.[324,325] Hence, there is a marked decrease in the serum half-life of vitamin D and 25(OH)D. In an apparent response to hypocalcemia and increased levels of PTH, plasma levels of 1,25(OH)$_2$D are elevated in individuals receiving anticonvulsants[326] which suggests that other factors must be invoked to account for the breakdown in Ca^{2+} homeostasis. There is indeed good evidence that anticonvulsant drugs, particularly phenytoin, have a direct inhibitory effect on the intestinal transport of Ca^{2+}.[327] Consequently, patients receiving combined drug therapy require above-normal circulating levels of vitamin D metabolites in order to maintain normal intestinal Ca^{2+} absorption.[328]

Glucocorticoid steroids represent a second common drug-induced alteration in mineral homeostasis which includes the malabsorption of Ca^{2+}[329] and the often profound loss of mineral from the skeleton. The pathophysiology of the events leading to steroid-induced osteoporosis are indeed complex. At the level of bone, the glucocorticoids have a number of actions which have been identified and reviewed elsewhere[323] resulting in a relative increase in osteoclastic activity over osteoblastic activity. The inhibition of intestinal Ca^{2+} absorption has been a matter of some controversy. Originally, steroids were considered to impair the metabolism of vitamin D to its metabolically active derivatives. However, more recent studies in experimental animals have failed to demonstrate that cortisone alters the conversion of vitamin D to 25(OH)D and the conversion of 25(OH)D to 1,25(OH)$_2$D.[330,331] On the other hand, the clearance of 1,25(OH)$_2$D may be accelerated by its metabolism to more polar biologically inactive metabolites[332] to account for the reduction in 1,25(OH)$_2$D levels observed in patients with systemic lupus erythematosis or glomerulonephritis treated chronically with steroids.[333] The extent of this reduction, however, would not appear to be sufficient to explain the degree of malabsorption of Ca^{2+}. In short-term studies (2 weeks) in normal volunteers, moderate doses of prednisone produced significant increases in the circulating levels of 1,25(OH)$_2$D.[334] These changes are consistent with the interpretation that glucocorticoids have a direct effect on the intestinal epithelium, inhibiting the transport of Ca^{2+} at this site and thereby inducing a compensatory increase in PTH secretion.

Chlorothiazide diuretic agents have a profound effect on the tubular reabsorption of Ca^{2+} resulting in hypocalciuria. This mechanism of the diuretic action appears to be a direct stimulation of the distal nephron Ca^{2+} reabsorption and independent of the PTH status of the experimental model.[335-337]

In addition, the thiazide diuretics inhibit the intestinal absorption of Ca^{2+} in patients with idiopathic hypercalciuria.[338,339] This inhibition of intestinal absorption of Ca^{2+} could be a direct effect of the diuretic on the intestine mediated through the prevention of secondary hyperparathyroidism through the action of the diuretic on the renal conservation of Ca^{2+}. In *in vitro* studies, thiazides have been shown to have no apparent effect on intestinal Ca^{2+} transport.[340] Furthermore, the diuretic had no effect on the increased intestinal absorption of Ca^{2+} induced in the rat by feeding a low-Ca^{2+} diet.[340] Thus, it would appear that the reduced intestinal absorption of Ca^{2+} observed in idiopathic hypercalciuria is the consequence of the suppression of PTH secretion and reduced synthesis of 1,25(OH)$_2$D$_3$. Indeed, patients with increased bone turnover in association with idiopathic hypercalciuria experience a reduction in the cellular activity of bone consistent with reduced levels of PTH.[338]

Cytotoxic chemotherapeutic agents can induce profound malabsorption of Ca^{2+} by destroying the intestinal epithelium which is uniquely sensitive to these agents because of the rapid turnover of the intestinal mucosal epithelium.

10. Nutritional Factors

A number of nutritional factors have been identified which are required for the maintenance of normal intestinal Ca^{2+} absorption. The most obvious of these is *vitamin D*. Although vitamin D can be obtained from a variety of dietary sources, the major sources are limited to milk and certain cereals which have been supplemented with ergosterol (vitamin D$_2$). Even in those areas in which food products are supplemented with vitamin D, it would appear that man's requirements for vitamin D are met primarily by the endogenous production of calciferol (vitamin D$_3$) in skin where 7-dihydrocholesterol is converted to calciferol under the influence of ultraviolet light.[341] Thus, the major circulating form of vitamin D is that of vitamin D$_3$.[342,343] This would suggest that the deficiency state would require both deprivation of sunlight as well as a dietary deficiency of the vitamin. In addition to the simple nutritional deficiency of vitamin D, one must also be cognizant of a growing list of vitamin D-resistant states which are also characterized by a malabsorption of Ca^{2+}. Examples include the malabsorption of renal insufficiency, aging, and a number of rare familial disorders as discussed previously.

Magnesium and iron deficiency probably constitutes yet another example of vitamin D resistance. The effects of Mg^{2+} deficiency on Ca^{2+} homeostasis are diverse and include an impaired secretion of PTH,[344,345] an inhibition of the skeletal response to PTH,[346] and an apparent resistance to the action of vitamin D on the intestine.[347] The influence of iron deficiency on Ca^{2+} homeostasis has not been adequately characterized. In the rat, chronic iron deficiency results in a significant decrease in intestinal Ca^{2+} absorption.[348] It has been the authors' experience that chronic iron deficiency in man can lead to severe osteomalacia and intestinal malabsorption of Ca^{2+} which is completely reversed by the replacement of the depleted iron stores.

Finally, dietary deficiency of Ca^{2+} must also be considered as a cause of reduced Ca^{2+} absorption. Considerable attention has been focused recently on the Ca^{2+} content of the American diet because of the growing appreciation of the relationship between Ca^{2+} intake and the subsequent development of osteoporosis later in life.[349] The results of these studies indicate that the majority of Caucasian women are consuming less than the recommended dietary allowances of Ca^{2+}. Second, these studies also revealed that the current recommended dietary allowances of Ca^{2+} are more a statement of philosophy than sound criteria based on relevant scientific data. Third, these criteria do not take into consideration the bioavailability of the dietary Ca^{2+}, the interactions between Ca^{2+} and other dietary components, or the nutritional status of the individual, all of which may profoundly impact on the individual's Ca^{2+} requirements. These considerations have been reviewed by Allen.[350] Clearly, further studies are needed to delineate the changing Ca^{2+} needs of the individual as he or she progresses through life. These studies will require a greater awareness of the physiologic role of Ca^{2+} in the development and maturation of the skeleton, in the pathogenesis of essential hypertension and eclampsia and as yet unrecognized pathological processes which are the consequence of dietary Ca^{2+} deficiency or altered intestinal absorption of Ca^{2+}.

References

1. Heaney, R. P., and T. G. Skilman. 1964. Secretion and excretion of calcium by the human gastrointestinal tract. *J. Lab. Clin. Med.* **64**:29–35.
2. Ireland, P., and J. S. Fordtran. 1973. Effect of dietary calcium and age on jejunal calcium absorption in humans studied by intestinal perfusion. *J. Clin. Invest.* **52**:2672–2681.
3. Anonymous 1975. Calcium transport in the ileum. *Nutr. Rev.* **33**:84–85.
4. Irving, J. R. 1964. Dynamics and function of phosphorus. In: *Mineral Metabolism*, Volume 2, Part A. C. L. Comar and F. Bronner, eds. Academic Press, New York. pp. 249–313.
5. Spencer, H., J. Menczel, I. Lewin, and J. Samachson. 1965. Effect of high intake on calcium and phosphorus metabolism in man. *J. Nutr.* **86**:125–132.
6. Cramer, C. F. 1968. Effect of Ca/P ratio and pH on calcium and phosphorus absorption from dog gut loops *in vivo*. *Can. J. Physiol. Pharmacol.* **46**:171–173.
7. Cramer, D. F., and J. Dueck. 1962. *In vivo* transport of calcium from healed Thiry-Vella fistulas in dogs. *Am. J. Physiol.* **202**:161–164.
8. Care, A. D., and A. T. Van't Klooster. 1965. *In vivo* transport of magnesium and other cations across the wall of the gastrointestinal tract of sheep. *J. Physiol. (London)* **177**:174–191.
9. Brannan, P. G., P. Vergne-Marini, C. Y. C. Pak, A. R. Hull, and J. S. Fordtran. 1976. Magnesium absorption in the human small intestine. *J. Clin. Invest.* **57**:1412–1418.
10. Patton, M. B., and T. S. Sutton. 1952. The utilization of calcium from lactate, gluconate, sulfate and carbonate salts by young college women. *J. Nutr.* **48**:443–452.
11. Treschel, U., J. P. Boryour, and H. Fleisch. 1979. Regulation of metabolism of 25-hydroxyvitamin D_3 in primary cultures of chick kidney cells. *J. Clin. Invest.* **64**:206–210.
12. Bar, A., S. Hurwitz, and A. Masz. 1980. The 25-hydroxycholecalciferol-1-hydroxylase activity of chick kidney cells: Direct effect of parathyroid hormone. *FEBS Lett.* **113**:328–330.
13. Baksi, S. N., and A. D. Kenney. 1978. Does estradiol stimulate *in vivo* production of 1,25-dihydroxyvitamin D_3 in the rat? *Life Sci.* **22**:787–791.
14. Henry, H. L. 1981. 25(OH)D_3 metabolism in kidney cell cultures: Lack of a direct effect of estradiol. *Am. J. Physiol.* **240**:E119–E124.
15. Kamar, R., W. R. Cohen, P. Silva, and F. H. Epstein. 1979. Elevated 1,25-dihydroxyvitamin D plasma levels in normal human pregnancy and lactation. *J. Clin. Invest.* **63**:342–344.
16. Tanaka, Y., B. Halloran, H. F. Schnoes, and H. F. DeLuca. 1979.

17. Bullamore, J. R., R. Wilkinson, J. C. Gallagher, B. E. C. Nordin, and D. H. Marshall. 1970. Effect of age on calcium absorption. *Lancet* **2**:535–537.
18. Alevizaki, C. C., D. C. Ikkos, and P. J. Singhelakis. 1973. Progressive decrease of true intestinal calcium absorption with age in normal man. *Nucl. Med.* **14**:760–762.
19. Gallagher, J. C., B. L. Riggs, J. Eisman, A. Hamstra, S. Arnaud, and H. F. DeLuca. 1979. Intestinal calcium absorption and serum vitamin D metabolites in normal subjects and osteoporotic patients. *J. Clin. Invest.* **64**:729–736.
20. Hansard, S. L., C. L. Comar, and G. D. Davis. 1954. Effects of age upon the physiological behavior of calcium in cattle. *Am. J. Physiol.* **177**:383–389.
21. Armbrecht, H. J., T. V. Zenser, and B. B. Davis. 1980. Effect of vitamin D metabolites on intestinal calcium absorption and calcium-binding proteins in young and adult rats. *Endocrinology* **106**:469–475.
22. Gray, R. W., and S. R. Gambert. 1982. Effect of age on plasma 1,25(OH)$_2$D vitamin D in the rat. *Age* **5**:54–56.
23. Armbrecht, H. J., T. V. Zenser, and B. B. Davis. 1980. Effect of age on the conversion of 25-hydroxyvitamin D_3 to 1,25-dihydroxyvitamin D_3 by kidney of rat. *J. Clin. Invest.* **66**:1118–1123.
24. Armbrecht, H. J., N. Wongsurawat, T. V. Zenser, and B. B. Davis. 1982. Differential effects of parathyroid hormone on the renal 1,25-dihydroxyvitamin D_3 and 24,25-dihydroxyvitamin D_3 production of young and adult rats. *Endocrinology* **11**:1339–1344.
25. Heaney, R. P., R. R. Recker, and P. D. Saville. 1978. Menopausal changes in Ca balance performance. *J. Lab. Clin. Med.* **92**:953–968.
26. Wilson, T. H. 1962. *Intestinal Absorption*. Saunders, Philadelphia.
27. Schachter, D. 1963. Vitamin D and the active transport of calcium by the small intestine. In: *The Transfer of Calcium and Strontium across Biological Membranes*. R. H. Wasserman, ed. Academic Press, New York. pp. 197–210.
28. Urban, E., and H. P. Schedl. 1969. Comparison of *in vivo* and *in vitro* effects of vitamin D in calcium transport in the rat. *Am. J. Physiol.* **217**:126–130.
29. Cramer, D. F. 1963. Qualitative studies on the absorption and excretion of calcium from Thiry-Vella intestinal loops in the dog. In: *The Transfer of Calcium and Strontium across Biological Membranes*. R. H. Wasserman, ed. Academic Press, New York. pp. 75–83.
30. Birge, S. J., W. A. Peck, M. Berman, and G. D. Whedon. 1969. Study of calcium absorption in man: A kinetic analysis and physiologic model. *J. Clin. Invest.* **48**:1705–1713.
31. Wensel, R. H., C. Rich, A. C. Brown, and W. Volwiler 1969. Absorption of calcium measured by incubation and perfusion of the intact human small intestine. *J. Clin. Invest.* **48**:1768–1775.
32. Nellans, H. N., and R. S. Goldsmith. 1981. Transepithelial calcium transport by rat cecum: High-efficiency absorptive site. *Am. J. Physiol.* **240**:G424–G431.
33. Favus, M. J., S. C. Kathpalia, F. L. Coe, and A. E. Mond. 1980. Effects of diet calcium and 1,25-dihydroxyvitamin D_3 on colon calcium active transport. *Am. J. Physiol.* **238**:G75–G78.
34. Lee, D. B. N., M. W. Walling, U. Gafter, V. Silis, and J. W. Coburn. 1980. Calcium and inorganic phosphate transport in rat colon. *J. Clin. Invest.* **65**:1326–1331.
35. Petith, M. M., and H. P. Schedl. 1976. Intestinal adaptation to dietary calcium restriction: *In vivo* cecal and colonic calcium transport in the rat. *Gastroenterology* **71**:1039–1042.
36. Ussing, H. H. 1949. The distinction by means of tracers between active transport and diffusion. *Acta Physiol. Scand.* **19**:43–46.
37. Helbock, H. J., J. G. Forte, and P. Saltman. 1966. The mechanism of calcium transport by rat intestine. *Biochim. Biophys. Acta* **126**:81–85.
38. Dumont, P. A., P. F. Curran, and A. K. Solomon. 1960. Calcium

In vitro production of 1,25-dihydroxyvitamin D_3 by rat placental tissue. *Proc. Natl. Acad. Sci. USA* **76**:5033.

and strontium in rat small intestine: Their fluxes and their effect on Na flux. *J. Gen. Physiol.* **43**:1119–1136.

39. Scott, D. 1965. Factors influencing the secretion and absorption of calcium and magnesium in the small intestine of the sheep. *Q. J. Exp. Physiol.* **50**:312–328.

40. Walling, M. W., and S. S. Rothman. 1969. Phosphate-independent, carrier-mediated transport of calcium by rat intestine. *Am. J. Physiol.* **217**:1144–1148.

41. Martin, D. L., and H. DeLuca. 1960. Calcium transport and the role of vitamin D. *Arch. Biochem. Biophys.* **134**:139–149.

42. Wasserman, R. H., and F. H. Kallfelz. 1962. Vitamin D_3 and the unidirectional calcium fluxes across the rachitic chick duodenum. *Am. J. Physiol.* **203**:221–229.

43. Halloran, B. P., and H. F. DeLuca. 1980. Calcium transport in small intestine during early development: Role of vitamin D. *Am. J. Physiol.* **239**:G473–G479.

44. Halloran, B. P., and H. F. DeLuca. 1981. Appearance of the intestinal cytosolic receptor for 1,25-dihydroxyvitamin D_3 during neonatal development in the rat. *J. Biol. Chem.* **256**:7338–7342.

45. Borle, A. B. 1970. Kinetic analysis of calcium movements in cell cultures. IV. Effects of phosphate and parathyroid hormone in kidney cells. *Endocrinology* **86**:1389–1393.

46. Adams, T. H., and A. W. Norman. 1970. Studies on the mechanism of action on calciferol. *J. Biol. Chem.* **245**:4421–4431.

47. Michalska, L., J. Wrobel, and R. Niemiro. 1972. Phosphate and vitamin D sensitive movement of calcium in rat distal ileum. *Acta Biochim. Pol.* **19**:333–339.

48. Brantbar, N., B. S. Levine, M. W. Wading, and J. W. Coburn. 1981. Intestinal absorption of calcium: Role of dietary phosphate and vitamin D. *Am. J. Physiol.* **241**:G49–G53.

49. Birge, S. J., and R. Miller. 1977. The role of phosphate in the action of vitamin D in the intestine. *J. Clin. Invest.* **60**:980–988.

50. DeLuca, H. F. 1969. Recent advances in the metabolism and function of vitamin D. *Fed. Proc.* **28**:1678–1689.

51. Horsting, M., and H. F. DeLuca. 1969. *In vitro* production of 25-hydroxycholecalciferol. *Biochem. Biophys. Res. Commun.* **36**:251–255.

52. Ponchon, G., A. L. Kennan, and H. F. DeLuca. 1969. Activation of vitamin D by the liver. *J. Clin. Invest.* **48**:2032–2037.

53. Blunt, J. W., Y. Tanaka, and H. F. DeLuca. 1968. The biological activity of 25-hydroxycholecalciferol, a metabolite of vitamin D_3. *Proc. Natl. Acad. Sci. USA* **61**:717–718.

54. Smith, J. E., and D. S. Goodman. 1971. The turnover and transport of vitamin D and of a polar metabolite with the properties of 25-hydroxycholecalciferol in human plasma. *J. Clin. Invest.* **50**:2159–2168.

55. Haddad, J. G., Jr., and K. J. Chyu. 1971. 25-Hydroxycholecalciferol binding globulin in human plasma. *Biochim. Biophys. Acta* **248**:471–478.

56. Daiger, S. P., M. S. Schanfield, and L. L. Cavilli-Storza. 1975. Group specific component (Gc) proteins bind vitamin D and 25OHD₃. *Proc. Natl. Acad. Sci. USA* **72**:2076–2080.

57. Gray, R. W., J. L. Omdahl, J. G. Ghazarian, and H. F. DeLuca. 1972. 25-Hydroxycholecalciferol-1-hydroxylase. *J. Biol. Chem.* **247**:7528–7532.

58. Ghazarian, J. G., Y. Tanaka, and H. F. DeLuca. 1975. The biochemistry of the chick kidney mitochrondrial 25-hydroxyvitamin D_3-1α-hydroxylase and its regulation. In: *Calcium Regulating Hormones.* R. V. Talmage, M. Owen, and J. A. Parson, eds. Excerpta Medica, Amsterdam. pp. 381–390.

59. Pedersen, J. I., J. G. Ghazarian, N. Orme-Johnson, and H. F. DeLuca. 1976. Isolation of chick renal mitochondria ferredoxin active in the 25-hydroxyvitamin D_3-1α-hydroxylase system. *J. Biol. Chem.* **251**:3933–3941.

60. Norman, A. W., J. F. Myrtle, R. J. Midget, and H. G. Nowicki 1971. 1,25-Dihydroxycholecalciferol: Identification of the proposed active form of vitamin D_3 in the intestine. *Science* **173**:51–54.

61. Holick, M. F., H. K. Schnoes, and H. F. DeLuca. 1971. Identification of 1,25-dihydroxycholecalciferol, a form of vitamin D_3

metabolically active in the intestine. *Proc. Natl. Acad. Sci. USA* **68**:803–804.

62. Lawson, D. E. M., P. W. Wilson, and E. Kodicek. 1969. New vitamin D metabolite localized in intestinal cell nuclei. *Nature (London)* **222**:171–172.

63. Myrtle, J. F., and A. W. Norman. 1970. Vitamin D: A cholecalciferol metabolite highly active in promoting intestinal calcium transport. *Science* **171**:79–82.

64. DeLuca, H. F. 1976. Recent advances in our understanding of the vitamin D endocrine system. *J. Lab. Clin. Med.* **87**:7–26.

65. Lam, H. Y., H. K. Schnoes, and H. F. DeLuca. 1975. Synthesis and biological activity of 25,26-dihydroxycholecalciferol. *Steroids* **25**:247–256.

66. Suda, T., H. F. DeLuca, H. K. Schnoes, *et al.* 1970. 25,26-Dihydroxycholecalciferol, a metabolite of vitamin D_3 with intestinal calcium transport activity. *Biochemistry* **9**:4776–4780.

67. Haddad, J. G., Jr., and J. C. Kyung. 1971. Competitive protein-binding radioassay for 25-hydroxycholecalciferol. *J. Clin. Endocrinol. Metab.* **33**:992–995.

68. Taylor, C. M., S. E. Hughes, and P. de Silva. 1976. Competitive protein binding assay for 24,25-dihydroxycholecalciferol. *Biochem. Biophys. Res. Commun.* **70**:1243–1249.

69. Brumbaugh, P. F., D. H. Haussler, R. Bressler, and M. R. Haussler. 1974. Radioreceptor assay for 1α,25-Dihydroxyvitamin D_3. *Science* **183**:1089–1091.

70. Tsai, H. C., and A. W. Norman. 1973. Studies on calciferol metabolism: Evidence for a cytoplasmic receptor for 1,25-dihydroxyvitamin D_3 in the intestinal mucosa. *J. Biol. Chem.* **248**:5967–5975.

71. Feldman, D., T. A. Melain, M. A. Hirst, T. L. Chen, and K. W. Colston. 1979. Characterization of a cytoplasmic receptor-like binder for 1α,25-dihydroxyvitamin D_3 in rat intestinal mucosa. *J. Biol. Chem.* **254**:10378–10384.

72. Tsai, H. C., and A. W. Norman. 1973. Studies on calciferol metabolism. I. Evidence for a cytoplasmic receptor for 1,25-dihydroxyvitamin D_3 in the intestinal mucosa. *J. Biol. Chem.* **248**:5967–5975.

73. Pike, J. W. 1981. Intestinal 1,25-dihydroxyvitamin D_3 receptors: Hormone-dependent uptake and saturability of nuclear components *in vitro*. *Life Sci.* **28**:957–963.

74. Walters, M. R., W. Hunziker, and A. W. Norman. 1980. Unoccupied 1,25-dihydroxyvitamin D_3 receptors: Nuclear/cytosol ratio depends on ionic strength. *J. Biol. Chem.* **255**:6799–6805.

75. Haddad, J. G., T. J. Hahn, and S. J. Birge. 1973. Vitamin D metabolites specific binding by rat intestinal cytosol. *Biochim. Biophys. Acta* **329**:93–97.

76. Haddad, J. G., and S. J. Birge. 1975. Widespread, specific binding of 25-hydroxycholecalciferol in rat tissues. *J. Biol. Chem.* **250**:299–303.

77. Haddad, J. G. 1982. Evidence that actin is the DBP binding component in human skeletal muscle. *Arch. Biochem. Biophys.* **213**:538–544.

78. Van Baelen, H., R. Bouillon, and P. DeMoor. 1980. Vitamin D-binding protein (Gc-globulin) binds actin. *J. Biol. Chem.* **255**:2270–2272.

79. Imawari, M., Y. Matsuzaki, K. Mitamura, and T. Osuga. 1982. Synthesis of serum and cytosol vitamin D-binding proteins of rat liver and kidney. *J. Biol. Chem.* **257**:8153–8157.

80. Colston, K., M. Hirst, and D. Feldman. 1980. Organ distribution of the cytoplasmic 1,25-dihydroxycholecalciferol receptor in various mouse tissues. *Endocrinology* **107**:1916–1922.

81. Kream, B. E., M. Jose, S. Yamada, and H. F. DeLuca. 1977. A specific high-affinity binding macromolecule for 1,25-dihydroxyvitamin D_3 in fetal bone. *Science* **197**:1086–1088.

82. Haddad, J. G. 1982. Evidence that actin is the DBP binding component in human skeletal muscle. *Arch. Biochem. Biophys.* **213**:538–544.

83. Eil, C., U. A. Liberman, J. F. Rosen, and S. J. Marx. 1981. Peripheral resistance to vitamin D: Deficient nuclear uptake of [3H] $1,25(OH)_2$ D_3 in skin fibroblasts. *Clin. Res.* **29**:538a.

84. Norman, A. W., D. A. Procsal, W. H. Okamura, and R. M. Wing. 1975. Structure–function studies of the interaction of the hormonally active form of vitamin D₃, 1,25-dihydroxyvitamin D₃, with the intestine. *J. Steroid Biochem.* **6**:461–467.

85. Wasserman, R. H., and A. N. Taylor. 1966. Vitamin D₃-induced calcium binding protein in chick intestinal mucosa. *Science* **152**:791–793.

86. Wasserman, R. H., and A. N. Taylor. 1968. Vitamin D-dependent calcium binding protein: Response to some physiological and nutritional variables. *J. Biol. Chem.* **243**:3987–3993.

87. MacGregor, R. R., J. W. Hamilton, and D. V. Cohn. 1970. The induction of calcium binding protein biosynthesis in intestine by vitamin D₃. *Biochim. Biophys. Acta* **222**:482–490.

88. Drescher, D., and H. F. DeLuca. 1971. Vitamin D stimulated calcium binding protein from rat intestinal mucosa: Purification and some properties. *Biochemistry* **10**:2302–2307.

89. Bachelet, M., A. Ulmann, and B. Lacour. 1979. Early stimulation of alkaline phosphatase activity in response to 1α,25-dihydroxycholecalciferol. *Biochem. Biophys. Res. Commun.* **89**:694–700.

90. Haussler, M. R., L. A. Nagode, and H. Rasmussen. 1970. Induction of intestinal brush border alkaline phosphatase by vitamin D identity with Ca-ATPase. *Nature (London)* **228**:1199–1201.

91. Holdsworth, E. S. 1970. The effect of vitamin D on enzyme activities in the mucosal cells of the chick small intestine. *J. Membr. Biol.* **3**:43–53.

92. Moriuchi, S., and H. F. DeLuca. 1976. The effect of vitamin D₃ metabolites on membrane proteins of chick duodenal brush borders. *Arch. Biochem. Biophys.* **174**:367–372.

93. Wilson, P. W., and D. E. M. Lawson. 1978. Incorporation of [³H] leucine into an actin-like protein in response to 1,25 dihydroxycholecalciferol in chick intestine brush borders. *Biochem. J.* **173**:627–631.

94. Kowarski, S., and D. Schachter. 1980. Intestinal membrane calcium-binding protein. *J. Biol. Chem.* **255**:10834–10840.

95. Hobden, A. N., M. Harding, and D. E. M. Lanson. 1980. 1,25-Dihydroxycholecalciferol stimulation of a mitochondrial protein in chick intestinal cells. *Nature (London)* **288**:718–720.

96. Corradino, R. A. 1975. Involvement of cyclic AMP in the vitamin D mediated intestinal calcium absorptive mechanism. In: *Calcium Regulating Hormones.* R. V. Talmage, M. Owen, and J. A. Parsons, eds. Excerpta Medica, Amsterdam. pp. 346–361.

97. Corradino, R. A. 1974. Embryonic chick intestine in organ culture: Interaction of adenylate cyclase system and vitamin D₃-mediated calcium absorptive mechanism. *Endocrinology* **94**:1607–1614.

98. Fontaine, O., T. Matsumoto, D. B. P. Goodman, and H. Rasmussen. 1981. Liponomic control of Ca transport: Relationship to mechanism of action of 1,25-dihydroxyvitamin D₃. *Proc. Natl. Acad. Sci. USA* **78**:1751–1754.

99. Hosoya, N., A. Fujimori, and T. Watanabe. 1964. The action of vitamin D on the incorporation of ³²P into phospholipids. *J. Biochem. (Tokyo)* **56**:613–615.

100. Thompson, O. W., and H. F. DeLuca. 1964. Vitamin D and phospholipid metabolism. *J. Biol. Chem.* **239**:984–989.

101. Goodman, D. B. P., M. R. Haussler, and H. Rasmussen. 1972. Vitamin D induced alteration of microvillus membrane lipid composition. *Biochem. Biophys. Res. Commun.* **46**:80–86.

102. Matsumoto, T., O. Fontaine, and H. Rasmussen. 1981. Effect of 1,25 dihydroxyvitamin D₃ on phospholipid metabolism in chick duodenal mucosal cell. *J. Biol. Chem.* **256**:3354–3360.

103. Wilson, P. W., and D. E. M. Larson. 1981. Vitamin D dependent phosphorylation of an intestinal protein. *Nature (London)* **289**:600–602.

104. Schachter, D., S. Kowarski, J. D. Finkelstein, R-I. W. Ma. 1966. Tissue concentration differences during active transport of calcium by intestine. *Am. J. Physiol.* **211**:1131–1136.

105. Urban, E., and H. P. Schedl. 1969. Comparison of *in vivo* and *in vitro* effects of vitamin D in calcium transport in the rat. *Am. J. Physiol.* **217**:126–130.

106. Martin, D. L., and H. F. DeLuca. 1969. Influence of sodium on

107. Bar, A., and S. Hurwitz. 1969. The accumulation of calcium in laying fowl intestine *in vitro*. *Biochim. Biophys. Acta* **183**:591–600.

108. Harrison, H. E., and H. C. Harrison. 1960. Transfer of ⁴⁵Ca across intestinal wall *in vitro* in relation to action of vitamin D and cortisol. *Am. J. Physiol.* **199**:265–271.

109. Wong, R. G., and A. W. Norman. 1975. Studies on the mechanism of action of calciferol. VIII. The effects of dietary vitamin D and the polyene antibiotic, filipin, *in vitro*, on the intestinal cellular uptake of calcium. *J. Biol. Chem.* **250**:2411–2419.

110. Max, E. E., D. B. P. Goodman, and H. Rasmussen. 1978. Purification and characterization of chick intestine brush border membrane. *Biochim. Biophys. Acta* **511**:224–231.

111. Rasmussen, H., O. Fontaine, E. E. Max, and D. B. P. Goodman. 1979. The effect of 1α-hydroxyvitamin D₃ administration on calcium transport in chick intestine brush border membrane vesicles. *J. Biol. Chem.* **254**:2993–2999.

112. Bikle, D. D., D. T. Zobock, R. Morrissey, and R. H. Herman. 1978. Independance of 1,25-dihydroxyvitamin D₃-mediated calcium transport from de novo RNA and protein synthesis. *J. Biol. Chem.* **253**:484–488.

113. Sampson, H. W., J. L. Matthews, J. H. Martin, and A. S. Kunin. 1970. An electron microscopic localization of calcium in the small intestine of normal, rachitic and vitamin D-treated rats. *Calcif. Tissue Res.* **5**:305–316.

114. Bikle, D. D., R. L. Morrissey, and D. T. Zolock. 1979. The mechanism of action of vitamin D in the intestine. *Am. J. Clin. Nutr.* **32**:2322–2338.

115. MacLaughlin, J. A., M. M. Weiser, and R. A. Freedman. 1980. Biphasic recovery of vitamin D-dependent Ca²⁺ uptake by rat intestinal Golgi membranes. *Gastroenterology* **78**:325–332.

116. Freedman, R. A., J. A. MacLaughlin, and M. M. Weiser. 1981. Properties of Ca²⁺ uptake and release by Golgi membrane vesicles from rat intestine. *Arch. Biochem. Biophys.* **206**:233–241.

117. Weiser, M. M., J. H. Bloor, and A. Dasmahapatia. 1982. Intestinal calcium absorption and vitamin D metabolism. *J. Clin. Gastroenterol.* **4**:75–86.

118. Birge, S. J., S. C. Switzer, and D. R. Leonard. 1974. Influence of sodium and parathyroid hormone on calcium release from intestinal mucosal cells. *J. Clin. Invest.* **54**:702–709.

119. Buckley, M., and Bronner, F. 1980. Calcium-binding protein biosynthesis in the rat: Regulation by calcium and 1,25-dihydroxyvitamin D₃. *Arch. Biochem. Biophys.* **202**:235–241.

120. Birge, S. J., and H. R. Gilbert. 1974. Identification of an intestinal sodium and calcium-dependent phosphatase stimulated by parathyroid hormone. *J. Clin. Invest.* **54**:710–717.

121. Ghijsen, W. E. J. M., M. D. DeJong, and C. H. Van Os. 1980. Dissociation between Ca²⁺-ATPase and alkaline phosphatase activities in plasma membranes of rat duodenum. *Biochim. Biophys. Acta* **599**:538–551.

122. Nellans, H. N., and J. E. Popovitch. 1981. Calmodulin-regulated ATP-driven calcium transport by basolateral membranes of rat small intestine. *J. Biol. Chem.* **256**:9932–9936.

123. Lehninger, A. L. 1974. Role of phosphate and other proton-donating anions in respiration-coupled transport of Ca²⁺ by mitochondria. *Proc. Natl. Acad. Sci. USA* **71**:1520–1524.

124. Rottenberg, H., and A. Scarpa. 1974. Calcium uptake and membrane potential in mitochondria. *Biochemistry* **13**:4811–4817.

125. Nellans, H. N., and J. E. Popovitch. 1981. Calmodulin-regulated ATP-driven calcium transport by basolateral membranes of rat small intestine. *J. Biol. Chem.* **256**:9932–9936.

126. DeJong, H. R., W. E. J. M. Ghijsen, and C. H. Van Os. 1981. Phosphorylated intermediates of Ca-ATPase and alkaline phosphatase in plasma membranes from rat duodenal epithelium. *Biochim. Biophys. Acta.* **647**:140–149.

127. Schiffl, H. and U. Binswager. 1980. Calcium ATPase and intestinal calcium transport in uremic rats. *Am. J. Physiol.* **238**:G424–G428.

calcium transport by the rat small intestine. *Am. J. Physiol.* **216**:1351–1359.

128. Corradino, R. A. 1977. Cyclic AMP regulation of 1,25(OH)$_2$D$_3$-mediated intestinal calcium absorption mechanism. In: *Vitamin D: Biochemical, Chemical, and Clinical Aspects Related to Calcium Metabolism.* A. W. Norman, K. Schaefer, J. W. Coburn, H. F. DeLuca, D. Fraser, H. G. Grigoleit, and D. V. Herrath. eds. de Gruyter, Berlin. pp. 231–240.

129. Peck, W. A., and S. Klahr. 1979. Cyclic nucleotides in bone and mineral metabolism. *Adv. Cyclic Nucleotide Res.* 11:89–130.

130. Birge, S. J., and R. A. Miller. 1977. The role of phosphate in the action of vitamin D on the intestine. *J. Clin. Invest.* 60:980–988.

131. Hitchman, A. J. W., and J. E. Harrison. 1972. Calcium binding proteins in the duodenal mucosa of the chick, rat, pig and human. *Can. J. Biochem.* 50:758–765.

132. Huang, W. Y., D. V. Cohn, J. W. Hamilton, C. Fullmer, and R. H. Wasserman. 1975. Calcium-binding protein of bovine intestine. *J. Biol. Chem.* 250:7647–7655.

133. Wasserman, R. H., and A. N. Taylor. 1970. Evidence for a vitamin D$_3$-induced calcium-binding protein in new world primates. *Proc. Soc. Exp. Biol. Med.* 136:25–28.

134. Taylor, A. N., R. H. Wasserman, and J. Jowsey. 1968. A vitamin D-dependent calcium binding protein in canine intestinal mucosa. *Fed. Proc.* 27:675.

135. Huang, W. Y., D. V. Cohn, J. W. Hamilton, C. Fullmer, and R. H. Wasserman. 1975. Calcium-binding protein of bovine intestine. *J. Biol. Chem.* 250:7647–7655.

136. Taylor, A. N., and R. H. Wasserman. 1972. Vitamin D-induced calcium-binding protein: Comparative aspects in kidney and intestine. *Am. J. Physiol.* 223:110–114.

137. Morrissey, R. L., T. J. Bucci, R. N. Empson, Jr., and E. G. Lufkin. 1975. Calcium-binding protein: Its cellular localization in jejunum, kidney and pancreas. *Proc. Soc. Exp. Biol. Med.* 149:56–60.

138. Zerwekh, J. E., T. J. Lindell, and M. R. Haussler. 1976. Increased intestinal chromatin template activity. *J. Biol. Chem.* 251:2388–2394.

139. Freund, T., and F. Bronner. 1975. Stimulation *in vitro* by 1,25-dihydroxyvitamin D$_3$ of intestinal cell calcium uptake and calcium-binding protein. *Science* 190:1300–1301.

140. Parker, T. F., P. Vergne-Marini, A. R. Hull, C. Y. C. Pak, and J. S. Fordtran. 1974. Jejunal absorption and secretion of calcium in patients with chronic renal disease on hemodialysis. *J. Clin. Invest.* 54:358–365.

141. Bronner, F., and T. Freund. 1975. Intestinal CaBP: A new quantitative index of vitamin D deficiency in the rat. *Am. J. Physiol.* 229:689–694.

142. Freund, T., and F. Bronner. 1975. Regulation of intestinal calcium-binding protein by calcium intake in the rat. *Am. J. Physiol.* 228:861–869.

143. Taylor, A. N., and R. H. Wasserman. 1967. Vitamin D-induced calcium binding protein: Partial purification, electrophoretic visualization and tissue distribution. *Arch. Biochem. Biophys.* 119:536–540.

144. Spencer, R., M. Charman, P. W. Wilson, and E. E. M. Lawson. 1980. The relationship between vitamin D-stimulated calcium transport and intestinal calcium-binding protein in the chicken. *Biochem. J.* 170:93.

145. Morrissey, R. L., R. N. Empson, D. T. Zolock, D. D. Bikle, and T. J. Bucci. 1978. Intestinal response to 1α,25-dihydroxycholecalciferol. II. A timed study of the intracellular localization of calcium-binding protein. *Biochim. Biophys. Acta* 538:34–41.

146. Morrissey, R. L., D. T. Zolock, D. D. Bikle, R. N. Empson, Jr., and T. J. Bucci. 1978. Intestinal response to 1α,25-dihydroxycholecalciferol. I. RNA polymerase, alkaline phosphatase, calcium and phosphorus uptake *in vitro* and *in vivo* calcium transport accumulation. *Biochim. Biophys. Acta* 538:23–33.

147. Miller, A., T. H. Ueng, and F. Bronner. 1979. Isolation of a vitamin D-dependent calcium-binding protein from brush borders of rat duodenal mucosa. *FEBS Lett.* 103:319–322.

148. Kowarski, S., and D. Schachter. 1980. Intestinal membrane calcium-binding protein: Vitamin D-dependent membrane compo-

149. nent of the intestinal calcium transport mechanism. *J. Biol. Chem.* 255:10834–10840.

149. Haussler, M. R., L. A. Nagode, and H. Rasmussen. 1970. Induction of intestinal brush border alkaline phosphatase by vitamin D and identity with Ca-ATPase. *Nature (London)* 228:1199–1201.

150. Melancon, M. J., Jr., and H. F. DeLuca. 1970. Vitamin D stimulation of calcium dependent adenosine triphosphatase in chick intestinal brush borders. *Biochemistry* 9:1658–1664.

151. Norman, A. W., A. K. Mircheff, T. H. Adams, and A. Spielvogel. 1970. Studies on the mechanism of action of calciferol. III. Vitamin D mediated increase of intestinal brush border alkaline phosphatase activity. *Biochim. Biophys. Acta* 215:348–359.

152. Birge, S. J., and D. H. Alpers. 1973. Stimulation of intestinal mucosal proliferation by vitamin D. *Gastroenterology* 64:977–982.

153. Spielvogel, A. M., R. D. Farley, and A. W. Norman. 1972. Studies on the mechanism of action of calciferol. V. Turnover time of chick intestine epithelial cells in relation to the intestinal action of vitamin D. *Exp. Cell Res.* 74:359.

154. Litwak, L. 1969. Tracer studies of intestinal calcium absorption in man. *Am. J. Clin. Nutr.* 22:771–785.

155. Wootton, R., and J. Reeve. 1980. The relative merits of various techniques for measuring radiocalcium absorption. *Clin. Sci.* 58:287–293.

156. Peacock, M., A. Hodgkinson, and B. E. C. Nordin. 1967. Importance of dietary calcium in the definition of hypercalciuria. *Br. Med. J.* 3:469–471.

157. Coe, F. L., and M. J. Favus. 1981. Hypercalciuric states. *Miner. Electrolyte Metab.* 5:183–200.

158. Nordin, B. E. C. 1960. Osteomalacia, osteoporosis and calcium deficiency. *Clin. Orthop. Relat. Res.* 17:235–258.

159. Mariel, P. J., J. M. Pettifor, F. P. Ross, and F. H. Glorieux. 1982. Histological osteomalacia due to dietary Ca deficiency in children. *N. Engl. J. Med.* 307:584–588.

160. Taylor, G. 1976. Osteomalacia and Ca deficiency. *Br. Med. J.* 1:960–965.

161. Kostial, K., N. Gruden, and A. Durakovic. 1969. Intestinal absorption of calcium-47 and strontium-85 in lactating rats. *Calcif. Tissue Res.* 4:13–19.

162. Heaney, R. P., and T. G. Skillman. 1971. Calcium metabolism in normal human pregnancy. *J. Clin. Endocrinol.* 33:661–669.

163. Pitkin, R. M. 1975. Calcium metabolism in pregnancy: A review. *Am. J. Obstet. Gynecol.* 12:724–737.

164. Lund, B., and A. Selnes. 1979. Plasma 1,25-dihydroxy vitamin D levels in pregnancy and lactation. *Acta Endocrinol. (Copenhagen)* 92:230–235.

165. Wieland, P., J. A. Fischer, U. Treschel, H. R. Roth, K. Vetter, H. Schneider, and Λ. Huch. 1980. Perinatal parathyroid hormone, vitamin D metabolites, and calcitonin in man. *Am. J. Physiol.* 239:E385–390.

166. Steichen, J. J., R. C. Tsang, T. L. Gratton, A. Hamstra, and H. F. DeLuca. 1978. Vitamin D homeostasis in the perinatal period. *N. Engl. J. Med.* 302:315–319.

167. MacIntyre, I. 1977. Comparative aspects of the biochemistry of the regulation of vitamin D metabolism. In: *Vitamin D: Biochemical, Chemical and Clinical Aspects Related to Calcium Metabolism.* A. W. Norman, K. Schaefer, J. W. Coburn, H. F. DeLuca, D. Fraser, H. G. Grigoleit, and D. V. Herrath, eds. de Gruyter, Berlin. pp. 155–164.

168. Tanaka, Y., B. Halloran, H. K. Schnoes, and H. F. DeLuca. 1979. *In vitro* production of 1,25(OH)$_2$D$_3$ by rat placental tissue. *Proc. Natl. Acad. Sci. USA* 76:5033–5035.

169. Gray, T. K., G. E. Lester, and R. J. Lorenc. 1979. Evidence for extrarenal 1α hydroxylation of 25-hydroxyvitamin D$_3$ in pregnancy. *Science* 204:1311.

170. Weisman, Y., A. Harrell, S. Edelstein, M. David, Z. Spirer, and A. Golander. 1979. 1α-25 dihydroxyvitamin D$_3$ and 24,25 dihydroxyvitamin D$_3$ in vitro synthesis by human decidua and placenta. *Nature (London)* 281:317.

171. Bouillon, R.. F. A. van Assche, H. van Baelen, W. Heyns, and P.

DeMoor. 1981. Influence of the vitamin D-binding protein on the serum concentration of 1,25-dihydroxyvitamin D_3. *J. Clin. Invest.* **67**:589–596.

172. Kumar, R., C. F. Abboud, and B. L. Riggs. 1980. The effect of elevated prolactin levels on plasma 1,25-dihydroxyvitamin D and intestinal absorption of calcium. *Mayo Clin. Proc.* **55**:51–53.

173. Miller, R. A., J. E. Russell, and S. J. Birge. 1983. Quantitation of the receptor proteins for 1,25(OH)$_2$D in the intestine of the aging rat. *Calcif. Tissue Int.* **35**:697a.

174. DeLuca, H. F. 1980. Some new concepts emanating from a study of the metabolism and function of vitamin D. *Nutr. Rev.* **38**:169–182.

175. Goldstein, R. A., H. L. Israel, K. L. Becker, and C. F. Moore. 1971. The infrequency of hypercalcemia in sarcoidosis. *Am. J. Med.* **51**:21–30.

176. Bell, N. H., J. R. Gill, and F. C. Bartter. 1964. On the abnormal calcium absorption in sarcoidosis: Evidence of increased sensitivity to vitamin D. *Am. J. Med.* **36**:500–513.

177. Koide, Y., N. Kugai, S. Kimura, T. Fujita, N. Yamashita, T. Hiranioto, J. Sukegawa, E. Ogata, and K. Yamashita. 1980. Increased 1,25-dihydroxycholecalciferol as a cause of abnormal calcium metabolism in sarcoidosis. *J. Clin. Endocrinol. Metab.* **52**:494–498.

178. Bell, N. H., P. H. Stern, E. Pantzer, T. K. Sinkna, and H. F. DeLuca. 1979. Evidence that increased circulating 1α,25-dihydroxyvitamin D is the probable cause for abnormal calcium metabolism in sarcoidosis. *J. Clin. Invest.* **64**:218–225.

179. Stern, P. H., J. DeOlazabal, and N. H. Bell. 1980. Evidence for abnormal regulation of circulating 1α,25-dihydroxyvitamin D in patients with sarcoidosis and normal calcium metabolism. *J. Clin. Invest.* **66**:852–855.

180. Barbour, G. L., J. W. Coburn, E. Slatopolsky, A. W. Norman, and R. L. Horst. 1981. Hypercalcemia in an omiphoric patient with sarcoidosis: Evidence for extrarenal generation of 1,25-dihydroxyvitamin D. *N. Engl. J. Med.* **305**:440–443.

181. Howard, G. A., R. T. Turner, D. J. Sherrard, and D. J. Baylink. 1981. Human bone cells in culture metabolize 25(OH)D$_3$ to 1,25(OH)$_2$D$_3$ and 24,25(OH)$_2$D$_3$. *J. Biol. Chem.* **256**:7738–7740.

182. Lee, J. C., A. Catanzaro, J. G. Parthemore, B. Roach, and L. J. Deftos. 1977. Hypercalcemia in disseminated coccidiomycosis. *N. Engl. J. Med.* **297**:431–433.

183. Walker, J. V., D. Baran, Y. N. Yakub, and R. B. Freeman. 1977. Histoplasmosis with hypercalcemia, renal failure, and papillary necrosis: Confusion with sarcoidosis. *J. Am. Med. Assoc.* **237**:1350–1352.

184. Stoeckle, J. D., H. L. Hardy, and A. L. Weber. 1969. Chronic beryllium disease: Long term follow-up of sixty cases and selective review of the literature. *Am. J. Med.* **461**:545–561.

185. Halverson, J. O., M. K. Mohler, and O. Bergheim. 1917. Calcium in the blood in tuberculosis. *J. Am. Med. Assoc.* **68**:1309–1311.

186. Abbasi, A. D., J. K. Chemplavil, S. Farah, B. F. Muller, and A. R. Arnstein. 1979. Hypercalcemia in active preliminary tuberculosis. *Ann. Intern. Med.* **90**:324–328.

187. Gwinup, G., G. Romdazzo, and A. Elias. 1981. The influence of vitamin D intake on serum calcium in tuberculosis. *Acta Endocrinol. (Copenhagen)* **97**:114–117.

188. Levin, M. E., V. C. Boisseau, and L. V. Avioli. 1976. Effects of diabetes mellitus on bone mass in juvenile and adult-onset diabetes. *N. Engl. J. Med.* **294**:241–245.

189. McNair, P., S. Madsbad, C. Christiansen, O. K. Faber, I. Transbol, and C. Binder. 1978. Osteopenia in insulin treated diabetes: Its relation to age at onset, sex and duration of disease. *Diabetologia* **15**:87–90.

190. Rosenbloom, A. L., D. C. Lezotte, F. T. Weber, J. Gudat, D. R. Heller, M. L. Weber, S. Klein, and B. Kennedy. 1977. Diminution of bone mass in childhood diabetes. *Diabetes* **26**:1052–1055.

191. Hough, S., L. V. Avioli, M. A. Bergfeld, *et al.* 1981. Correction of abnormal bone and mineral metabolism in chronic streptozotocin-induced diabetes mellitus in the rat by insulin therapy. *Endocrinology* **108**:2228–2234.

192. Schneider, I., H. P. Schedl, T. McCain, and M. R. Haussler. 1977. Experimental diabetes reduces circulatory 1,25-dihydroxyvitamin D in the rat. *Science* **196**:1452–1459.

193. Seino, Y., R. I. Siena, Y. M. Sonn, A. Jafari, S. J. Birge, and L. V. Avioli. 1983. The duodenal 1,25(OH)$_2$D$_3$ receptor in rats with experimentally-induced diabetes. *Endocrinology* in press.

194. Hough, S., J. E. Russell, S. L. Teitelbaum, and L. V. Avioli. 1982. Calcium homeostasis in chronic streptozotocin-induced, diabetes mellitus in the rat. *Am. J. Physiol.* **242**:E451–E456.

195. Caspary, W. F. 1973. Effect of insulin and experimental diabetes mellitus on the digestive–absorptive function of the small intestine. *Digestion* **9**:248–263.

196. Miller, D. L., W. Hanson, H. P. Schedl, and J. W. Osborne. 1977. Proliferation rate and transit time of mucosal cells in small intestine of the diabetic rat. *Gastroenterology* **73**:1326–1332.

197. Fraser, T. E., N. H. White, S. Hough, J. V. Santiago, B. R. McGee, G. Bryce, J. Mallon, and L. V. Avioli. 1981. Alterations in circulating vitamin D metabolites in the young insulin-dependent diabetic. *J. Clin. Endocrinol. Metab.* **53**:1154–1159.

198. Witt, M. F., N. H. White, J. V. Santiago, and L. V. Avioli. 1982. Increased calcium absorption in type I diabetic children. *Pediatr. Res.* **16**:1129a.

199. Moore, E., F. L. Coe, E. McMann, and M. Favus. 1978. Idiopathic hypercalciuria in children: Prevalence and metabolic characteristics. *J. Pediatr.* **92**:900–910.

200. Pak, C. Y. C., R. Kaplan, H. Bone, J. Townsend, and O. Waters. 1975. A simple test for the diagnosis of absorptive, resorptive and renal hypercalciurias. *N. Engl. J. Med.* **292**:497–500.

201. Broadus, A. E., M. Dominguez, and F. C. Bartter. 1978. Pathophysiological studies in idiopathic hypercalciuria: Use of an oral calcium tolerance test to characterize distinctive hypercalciuric subgroups. *J. Clin. Endocrinol. Metab.* **47**:751–760.

202. Pak, C. Y. C., M. Ohata, E. C. Lawrence, and W. Snyder. 1974. The hypercalciurias causes, parathyroid functions and diagnostic criteria. *J. Clin. Invest.* **54**:387–400.

203. Nordin, B. E. C., M. Peacock, and R. Wilkinson. 1972. Hypercalciuria and calcium stone disease. In: *Clinics in Endocrinology and Metabolism*, Volume 1. MacIntyre, ed. Saunders, Philadelphia. p. 169–183.

204. Bordier, P., A. Ryckewart, J. Gueris, and H. Rasmussen. 1977. On the pathogenesis of so-called idiopathic hypercalciuria. *Am. J. Med.* **63**:398–409.

205. Coe, F. L., J. M. Canterbury, J. J. Firpo, and E. Reiss. 1973. Evidence for secondary hyperparathyroidism in idiopathic hypercalciuria. *J. Clin. Invest.* **52**:134–142.

206. Barilla, D. E., R. Tolentino, R. A. Kaplan, and C. Y. C. Pak. 1978. Selective effect of thiazide on intestinal absorption of calcium in absorptive and renal hypercalciurias. *Metabolism* **27**:125–131.

207. Kaplan, R. A., M. R. Haussler, L. J. Deftos, H. Bone, and C. Y. C. Pak. 1977. The role of 1α,25-dihydroxyvitamin D in the mediation of intestinal hyperabsorption of calcium in primary hyperparathyroidism and absorptive hypercalciuria. *J. Clin. Invest.* **59**:756–760.

208. Zerwekh, J. E., and C. Y. C. Pak. 1980. Selective effects of thiazide therapy on serum 1α,25-dihydroxyvitamin D and intestinal Ca absorption in renal and absorptive hypercalciurias. *Metabolism* **29**:13–17.

209. Shen, F. H, D. J. Baylink, R. L. Nielson, D. J. Sherrard, J. L. Ivey, and M. R. Haussler. 1977. Increased serum 1,25-dihydroxyvitamin D in idiopathic hypercalciuria. *J. Lab. Clin. Med.* **90**:955–962.

210. Gray, R. W., D. R. Wilz, A. E. Caldas, and J. Lemann. 1977. The importance of phosphate in regulating plasma 1α,25-(OH)$_2$-vitamin D levels in humans: Studies in healthy subjects, in calcium stone-formers and in patients with primary hyperparathyroidism. *J. Clin. Endocrinol. Metab.* **45**:299–306.

211. Barilla, D. E., J. E. Zerwekh, and C. Y. C. Pak. 1979. A critical

evaluation of the role of phosphate in the pathogenesis of absorptive hypercalciuria. *Miner. Electrolyte Metab.* **2**:302–309.

212. Fukase, M., S. J. Birge, L. Rifas, L. V. Avioli, and L. R. Chase. 1982. Regulation of 25 hydroxyvitamin D₃ 1-hydroxylase in serum-free monolayer culture of mouse kidney. *Endocrinology* **110**:1073–1075.

213. Birge, S. J. 1978. Vitamin D, muscle and phosphate homeostasis. *Miner. Electrolyte Metab.* **1**:57–64.

214. Lotz, M., E. Zisman, and F. C. Bartter. 1968. Evidence for a phosphorus depletion syndrome in man. *N. Engl. J. Med.* **278**: 409–415.

215. Knochel, J. P. 1977. The pathophysiology and clinical characteristics of severe hypophosphatemia. *Arch. Intern. Med.* **137**: 203–220.

216. Baher, G. R., P. Ackrill, W. R. Cuttell, T. C. Stamp, and L. Watson. 1974. Iatrogenic osteomalacia and myopathy due to phosphate depletion. *Br. Med. J.* **3**:150–152.

217. Dominguez, J. H., R. W. Gray, and J. Lemann, Jr. 1976. Dietary phosphate deprivation in women and men: Effects on mineral and acid balance, parathyroid hormone and metabolism of 25-OH-vitamin D. *J. Clin. Endocrinol. Metab.* **43**:1056–1068.

218. Kreusser, W. J., K. Kurokawa, E. Aznar, and S. G. Massry. 1978. Phosphate depletion: Effect on renal inorganic phosphorus and adenine nucleotides, urinary phosphate and calcium, and calcium balance. *Miner. Electrolyte Metab.* **1**:30–42.

219. Steele, T. H., and H. F. DeLuca. 1976. Influence of dietary phosphorus on renal phosphate reabsorption in the parathyroidectomized rat. *J. Clin. Invest.* **57**:867–874.

220. Trohler, U., J. B. Bonyour, and H. Fleisch. 1976. Inorganic phosphate homeostasis: Renal adaptation to the dietary intake in intact and thyroparathyroidectomized rats. *J. Clin. Invest.* **57**: 264–273.

221. Coburn, J. W., and S. G. Massry. 1970. Changes in serum and urinary calcium during phosphate depletion: Studies on mechanism. *J. Clin. Invest.* **49**:1073–1087.

222. Tanaka, Y., H. Frank, and H. F. DeLuca. 1973. Intestinal calcium transport: Stimulation by low phosphorus diets. *Science* **181**:564–566.

223. Hughes, M. R., P. F. Brumbaugh, M. R. Haussler, J. E. Wergedal, and D. J. Baylink. 1975. Regulation of serum 1α,25-dihydroxyvitamin D₃ by calcium and phosphate in the rat. *Science* **190**:578–590.

224. Baxter, L. A., and H. F. DeLuca. 1976. Stimulation of 25-hydroxyvitamin D₃-1α-hydroxylase by phosphate depletion. *J. Biol. Chem.* **251**:3158–3161.

225. Brantbar, N., M. W. Walling, and J. W. Coburn. 1979. Interaction between vitamin D deficiency and phosphorus depletion in the rat. *J. Clin. Invest.* **63**:335–341.

226. Cramer, C. F., and J. McMillan. 1980. Phosphorus adaptation in rats in absence of vitamin D or parathyroid glands. *Am. J. Physiol.* **239**:G261–G265.

227. Massry, S. G. 1975. Effect of phosphate depletion on renal tubular transport. In: *Phosphate Metabolism, Kidney and Bone.* L. V. Avioli, P. Bordier, H. Fleisch, S. G. Massry, and E. Slasopolsky, eds. Imprieve Fournie, France. p. 25.

228. Burnett, C. H., R. R. Commons, F. Albright, and J. E. Howard. 1949. Hypercalcemia without hypercalciuria or hypophosphatemia, calcinosis, and renal insufficiency: A syndrome following prolonged intake of milk and alkali. *N. Engl. J. Med.* **240**:787–794.

229. McMillan, E. E., and R. B. Freeman. 1965. The milk alkali syndrome: A study of the acute disorder with comments on the development of the chronic condition. *Medicine (Baltimore)* **44**: 485–501.

230. Punsar, S., and T. Somer. 1963. The milk-alkali syndrome: A report of three illustrative cases and a review of the literature. *Acta Med. Scand.* **173**:435–449.

231. Van Goidsenhoven, G. M. T., Q. V. Gray, A. V. Price, and P. H. Sanderson. 1954. The effect of prolonged administration of large doses of sodium bicarbonate in man. *Clin. Sci.* **13**:383–384.

232. Peterson, C. R. 1980. Vitamin D poisoning: Survey of causes in 21 patients with hypercalcemia. *Lancet* **1**:1164–1165.

233. Brumbaugh, P. F., and M. R. Haussler. 1973. 1α,24-Dihydroxyvitamin D₃ receptor: Competitive binding of vitamin D analogs. *Life Sci.* **13**:1737–1746.

234. Hughs, M. R., D. J. Baylink, P. G. Jones, and M. R. Haussler. 1976. Radioligand receptor assay for 25-hydroxyvitamin D₂/D₃: Application to hypervitaminosis D. *J. Clin. Invest.* **58**:61–70.

235. Mawer, E. B., G. A. Lumb, and S. W. Stanbury. 1969. Long biological half-life of vitamin D₃ and its polar metabolites in human serum. *Nature (London)* **222**:482–483.

236. Lumb, G. A., E. B. Mawer, and S. W. Stanbury. 1971. The a-parent vitamin D resistance of chronic renal failure. *Am. J. Med.* **50**:421–441.

237. Rosenstreich, S. J., C. Rich, and W. Wolwiler. 1971. Deposition in and release of vitamin D₃ from body fat: Evidence for a storage site in the rat. *J. Clin. Invest.* **50**:679–687.

238. Shetty, K. R., K. Ajlouni, P. S. Rosenfeld, and T. C. Hagen. 1975. Protracted vitamin D intoxication. *Arch. Intern. Med.* **135**:986–988.

239. Young, M. M., C. Jasani, D. A. Smith, and B. E. C. Nordin. 1968. Some effects of ethinyl estradiol on calcium and phosphorus metabolism in osteoporosis. *Clin. Sci.* **34**:411–417.

240. Lafferty, F. W., G. E. Spencer, and O. H. Pearson. 1964. Effects of androgens, estrogens and high calcium intakes on bone formation and resorption in osteoporosis. *Am. J. Med.* **36**:514–528.

241. Albright, F., and E. C. Reifenstein, Jr. 1948. *The Parathyroid Glands and Metabolic Bone Disease.* Williams & Williams, Baltimore.

242. Recker, R. R., P. D. Saville, and R. P. Heaney. 1977. Effect of estrogens and calcium carbonate on bone loss in postmenopausal women. *Ann. Intern. Med.* **87**:649–655.

243. Caniggia, A., C. Gennari, V. Bianchi, and R. Guideri. 1963. Intestinal absorption of ⁴⁵Ca in senile osteoporosis. *Acta Med. Scand.* **173**:613–617.

244. Gallagher, J. C., B. L. Riggs, and H. F. DeLuca. 1980. Effect of estrogen on calcium absorption and serum vitamin D metabolites in postmenopausal osteoporosis. *J. Clin. Endocrinol. Metab.* **51**: 1359–1364.

245. Finkelstein, J. D., and D. Schachter. 1962. Active transport of calcium by intestine: The effects of hypophysectomy and growth hormone. *Am. J. Physiol.* **203**:873–880.

246. Shields, H. M. 1978. Rapid fall of serum phosphorus secondary to antacid therapy. *Gastroenterology* **75**:1137–1141.

247. Schachter, D., E. B. Dowdle, and H. Shenker. 1960. Active transport of calcium by the small intestine of the rat. *Am. J. Physiol.* **198**:263.

248. Thompson, G. R., B. Lewis, and C. C. Booth. 1966. Absorption of vitamin D₃-³H in control subjects and patients with intestinal malabsorption. *J. Clin. Invest.* **45**:94–102.

249. Scott, J., E. Elias, P. J. A. Moult, S. Barnes, and M. R. Wills. 1977. Rickets in adult cystic fibrosis with myopathy, pancreatic insufficiency and proximal renal tubular dysfunction. *Am. J. Med.* **63**:488–492.

250. Schachter, D., J. D. Finkelstein, and S. Kowarski. 1964. Metabolism of vitamin D. I. Preparation of radioactive vitamin D and its intestinal absorption in the rat. *J. Clin. Invest.* **43**:787–796.

251. Atkinson, M., B. E. C. Nordin, and S. Sherlock. 1956. Malabsorption and bone disease in prolonged obstructive jaundice. *Q. J. Med.* **25**:299–312.

252. Schoen, M. S., J. Lindenbaum, M. S. Roginsky, and P. R. Holt. 1978. Significance of serum level of 25-hydroxycholecalciferol in gastrointestinal disease. *Am. J. Dig. Dis.* **23**:137–142.

253. Compston, J. E., and R. P. H. Thompson. 1977. Intestinal absorption of 25-hydroxyvitamin D and osteomalacia in primary biliary cirrhosis. *Lancet* **2**:721–724.

254. Bolt, M., M. Sitrin, M. Favus, I. Rosenberg. 1977. Hepatic vitamin D 25-hydroxylase: Inhibition by bile duct ligation as bile salts *in vitro*. *Clin. Res.* **25**:606a.

255. Krawitt, E. L., M. J. Grundman and E. B. Mawer. 1977. Absorp-

tion, hydroxylation, and excretion of vitamin D_3 in primary biliary cirrhosis. *Lancet* 2:1246–1249.

256. Arnaud, S. B., R. S. Goldsmith, P. W. Lambert, *et al.* 1975. 25-Hydroxyvitamin D_3: Evidence of an enterohepatic circulation in man. *Proc. Soc. Exp. Biol. Med.* 149:570–572.

257. Sitrin, M., S. Meredith, and I. H. Rosenberg. 1978. Vitamin D deficiency and bone disease in gastrointestinal disorders. *Arch. Intern. Med.* 138:886–888.

258. Imawari, M., K. Kozawa, Y. Akanuma, S. Koizumi, H. Itakura, and K. Kosaka. 1980. Serum 25-hydroxyvitamin D and vitamin D-binding protein levels and mineral metabolism after partial and total gastrectomy. *Gastroenterology* 79:255–258.

259. Stamp, T. C. B. 1974. Intestinal absorption of 25-hydroxycholecalciferol. *Lancet* 2:121–123.

260. Thompson, G. R., B. Lewis, and C. C. Booth. 1966. Vitamin-D absorption after partial gastrectomy. *Lancet* 1:457–458.

261. Hillman, H. S. 1968. Postgastrectomy malnutrition. *Gut* 9:576–584.

262. Nemere, I., and C. M. Szego. 1981. Early actions of parathyroid hormone and 1,25-dihydroxycholecalciferol on isolated epithelial cells from rat intestine. I. Limited lysosomal enzyme release and calcium uptake. *Endocrinology* 108:1450–1462.

263. Cook, P. B., J. R. Nassim, and J. Collins. 1959. The effects of thyrotoxicosis upon the metabolism of calcium, phosphorus and nitrogen. *Q. J. Med.* 28:505–529.

264. Schafer, R. B., and D. H. Gregory. 1972. Calcium malabsorption in hyperthyroidism. *Gastroenterology* 63:235–239.

265. Krane, S. M., G. L. Brownell, J. B. Stanbury, and H. Corrigan. 1956. The effect of thyroid disease on calcium metabolism in man. *J. Clin. Invest.* 35:874–887.

266. Mundy, G. R., J. L. Shapiro, J. G. Bandelin, E. M. Canalis, and L. G. Raisz. 1976. Direct stimulation of bone resorption by thyroid hormones. *J. Clin. Invest.* 58:529–534.

267. Baxter, J. D., and P. K. Bondy. 1966. Hypercalcemia of thyrotoxicosis. *Ann. Intern. Med.* 65:429–442.

268. Burman, K. D., J. M. Monchik, J. M. Earll, and L. Wartofsky. 1976. Ionized and total serum calcium and parathyroid hormone in hyperthyroidism. *Ann. Intern. Med.* 84:668–671.

269. Bouillon, R., and P. DeMoor. 1974. Parathyroid function in patients with hyper- or hypothyroidism. *J. Clin. Endocrinol. Metab.* 38:999–1004.

270. Mosekilde, L., and M. S. Christensen. 1977. Decreased parathyroid function in hyperthyroidism: Interrelationships between serum parathyroid hormone, calcium–phosphorus metabolism and thyroid function. *Acta Endocrinol. (Copenhagen)* 84:566–575.

271. Bouillon, R., E. Muls, and P. DeMoor. 1980. Influence of thyroid function on the serum concentration of 1,25-dihydroxyvitamin D. *J. Clin. Endocrinol. Metab.* 51:793–797.

272. Kaye, M., and M. Silverman. 1965. Calcium metabolism in chronic renal failure. *J. Lab. Clin. Med.* 66:535–548.

273. Genuth, S. M., V. Vertes, and J. Leonards. 1969. Oral calcium absorption in patients with renal failure treated by chronic hemodialysis. *Metabolism* 18:124–131.

274. Liu, S. H., and H. I. Chu. 1943. Studies of calcium and phosphorus metabolism with special reference to the pathogenesis and effect of dihydrotachysterol (AT10) and iron. *Medicine (Baltimore)* 22:103–161.

275. Coburn, J. W., M. H. Koppel, A. S. Brickman, and S. G. Massry. 1973. Study of intestinal absorption of calcium in patients with renal failure. *Kidney Int.* 2:264–272.

276. Parker, T. F., P. Vergne-Marini, A. R. Hull, C. Y. C. Pak, and J. S. Fordtran. 1974. Jejunal absorption and secretion of calcium in patients with chronic renal disease on hemodialysis. *J. Clin. Invest.* 54:358–365.

277. Brickman, A. S., J. W. Coburn, P. H. Rowe, S. G. Massry, and A. W. Norman. 1974. Impaired calcium absorption in uremic man: Evidence for defective absorption in the proximal small intestine. *J. Lab. Clin. Med.* 84:791–801.

278. Ogg, C. S. 1968. The intestinal absorption of ^{47}Ca by patients in chronic renal failure. *Clin. Sci.* 34:467–471.

279. Recker, R. R., and P. D. Saville. 1971. Calcium absorption in renal failure: Its relation to blood urea nitrogen, dietary calcium intake, time on dialysis and other variables. *J. Lab. Clin. Med.* 78:380–388.

280. Charnard, J., J. Assailly, C. Bader, and J. L. Funck-Bretano. 1974. A rapid method for measurement of fractional intestinal absorption of calcium. *J. Nucl. Med.* 15:588–592.

281. Vergne-Marini, P., T. F. Parker, C. Y. C. Pak, A. R. Hull, H. F. DeLuca, and J. S. Fordtran. 1976. Jejunal and ileal calcium absorption in patients with chronic renal disease. *J. Clin. Invest.* 57:861–866.

282. Clarkson, E. M., S. J. McDonald, and H. E. deWardener. 1966. The effect of a high intake of calcium carbonate in normal subjects and patients with chronic renal failure. *Clin. Sci.* 30:425–438.

283. Clarkson, E. M., J. B. Eastwood, K. Koutsaimanis, and H. E. deWardener. 1973. Net intestinal absorption of calcium in patients with chronic renal failure. *Kidney Int.* 3:258–263.

284. Kopple, J. D., and J. W. Coburn. 1973. Metabolic studies of low calcium protein diets in uremia. II. Calcium, phosphorus, and magnesium. *Medicine (Baltimore)* 52:597–607.

285. Messener, R. P., H. T. Smith, F. L. Shapiro, and D. H. Gregory. 1969. The effect of hemodialysis, vitamin D and renal homotransplantation on the calcium malabsorption of chronic renal failure. *J. Lab. Clin. Med.* 74:472–481.

286. Rutherford, W. E., J. Blondin, K. Hruska, R. Kopelman, S. Klahr, and E. Slatopolsky. 1975. Effect of 25-hydroxycholecalciferol on calcium absorption in chronic renal disease. *Kidney Int.* 8:320–324.

287. Teitelbaum, S. L., J. M. Bone, P. M. Stein, J. J. Gilden, M. Bates, V. C. Boisseau, and L. V. Avioli. 1976. Calcifediol in chronic renal insufficiency. *J. Am. Med. Assoc.* 235:164–167.

288. Brickman, A. S., J. W. Coburn, and A. W. Norman. 1972. Action of 1,25-dihydroxycholecalciferol, a potent, kidney-produced metabolite of vitamin D, in uremic man. *N. Engl. J. Med.* 287:891–895.

289. Brickman, A. S., J. W. Coburn, G. R. Friedman, W. H. Okamura, S. G. Massry, and A. W. Norman. 1976. Comparison of effects of 1α-hydroxyvitamin D_3 and 1,25-dihydroxy-vitamin D_3 in man. *J. Clin. Invest.* 57:1540–1597.

290. Rutherford, W. E., K. Hruska, J. Blondin, M. Holick, H. F. DeLuca, S. Klahr, and E. Slatopolsky. 1975. The effect of 5,6-*trans* vitamin D_3 on calcium absorption in chronic renal disease. *J. Clin. Endocrinol. Metab.* 40:13–18.

291. Davie, M. W. J., T. M. Chalmers, J. O. Hunter, B. Pelc, and E. Kodicek. 1976. 1-Alphahydroxycholecalciferol in chronic renal failure. *Ann. Intern. Med.* 84:281–285.

292. Gonick, H. C., J. P. Drinkard, J. Hertoghe, and M. F. Rubini. 1974. A calcium kinetic approach to the problem of renal osteodystrophy. *Clin. Orthop. Relat. Res.* 100:315–336.

293. Ritz, E. 1970. Experimentelle untersuchugen zum intestinalen calcium transport bei uramie. *Z. Gesamte Exp. Med.* 152:313–324.

294. Kessner, D. M., and F. H. Epstein. 1965. Effect of renal insufficiency in gastrointestinal transport of calcium. *Am. J. Physiol.* 209:141–145.

295. Baerg, R. D., C. V. Kimberg, and E. Gershon. 1970. Effect of renal insufficiency on the active transport of calcium by the small intestine. *J. Clin. Invest.* 49:1288–1300.

296. McDermott, F. R., A. J. Galbraith, and M. K. Dalton. 1974. Effects of actue renal failure on ileal epithelial cell kinetics: Autoradiographic studies in the mouse. *Gastroenterology* 66:235–239.

297. Avioli, L. V., S. Birge, S. W. Lee, and E. Slatopolsky. 1968. The metabolic fate of vitamin D_3-^{3}H in chronic renal failure. *J. Clin. Invest.* 47:2239–2252.

298. Piel, C. F., B. S. Roof, and L. V. Avioli. 1973. Metabolism of tritiated 25-hydroxycholecalciferol in chronically uremic children before and after successful renal homotransplantation. *J. Clin. Endocrinol. Metab.* 37:944–948.

299. Gray, R. W., H. P. Weber, J. H. Dominguez, and J. Lemann, Jr. 1974. The metabolism of vitamin D_3 and 25-hydroxyvitamin D_3 in

normal and anephric humans. *J. Clin. Endocrinol. Metab.* **39**:1045–1056.

300. Mawer, E. B., C. M. Taylor, J. Backhouse, G. A. Lumb, and S. W. Stanbury. 1973. Failure of formation of 1,25-dihydroxycholecalciferol in chronic renal insufficiency. *Lancet* **1**:626–628.

301. Oettinger, C. W., R. Merrill, T. Blanton, and W. Briggs. 1974. Reduced calcium absorption after nephrectomy in uremic patients. *N. Engl. J. Med.* **291**:458–460.

302. Martin, D. L., M. J. Melancon, and H. F. DeLuca. 1969. Vitamin D-stimulated, calcium-dependent adenosine triphosphatase from brush borders of rat small intestine. *Biochem. Biophys. Res. Commun.* **35**:819–823.

303. Melancon, M. J., Jr., and H. F. DeLuca. 1970. Vitamin D stimulation of calcium dependent adenosine triphosphatase in chick intestinal brush borders. *Biochemistry* **9**:1658–1664.

304. Russell, J. E., and L. V. Avioli. 1974. The effect of chronic uremia on intestinal mitochondrial activy. *J. Lab. Clin. Med.* **84**:317–326.

305. Birge, S. J., and D. H. Alpers. 1973. Stimulations of intestinal mucosal proliferation by vitamin D. *Gastroenterology* **64**:977–982.

306. Piazolo, P., H. Hotz, K. Helmke, H. E. Franz, and M. Schleyer. 1975. Calcium-binding protein in the duodenal mucosa of uremic patients and normal subjects. *Kidney Int.* **8**:110–118.

307. Avioli, L. V., S. Scott, S. W. Lee, and H. F. DeLuca. 1969. Intestinal calcium absorption: Nature of defect in chronic renal disease. *Science* **166**:1154–1156.

308. Heyen, G., and P. Franchimont. 1974. Human calcitonin and serum-phosphate. *Lancet* **1**:627.

309. Olson, E. B., Jr., H. F. DeLuca, and J. T. Potts, Jr. 1972. Calcitonin inhibition of vitamin D-induced intestinal calcium absorption. *Endocrinology* **90**:151–157.

310. Shear, L. 1969. Internal redistribution of tissue protein synthesis in uremia. *J. Clin. Invest.* **48**:1252–1265.

311. Lumb, G. A., E. B. Mawer, and S. W. Stanbury. 1971. The apparent vitamin D resistance of chronic renal failure: A study of the physiology of vitamin D in man. *Am. J. Med.* **50**:421–441.

312. Schaefer, K., P. Schaefer, P. Koeppe, A. Opitz, and D. Hoffler. 1968. Uraemic osteopathy: The relationship between disturbances in intestinal calcium absorption and renal function. *Ger. Med. Month.* **13**:575–581.

313. Stanbury, S. W., G. A. Lumb, and E. B. Mawer. 1969. Osteodystrophy developing spontaneously in the course of chronic renal failure. *Arch. Intern. Med.* **124**:274–281.

314. Schachter, D. 1963. Vitamin D and the active transport of calcium by the small intestine. In: *The Transfer of Calcium and Strontium across Biological Membranes*, R. H. Wasserman, ed. Academic Press, New York. pp. 197–210.

315. Jones, J. H., D. K. Peters, D. B. Morgan, G. A. Coles, and N. P. Mallick. 1967. Observations on calcium metabolism in the nephrotic syndrome. *Q. J. Med.* **36**:301–320.

316. Emerson, K., Jr. and W. W. Beckman. 1945. Calcium metabolism in nephrosis. I. A description of an abnormality in calcium metabolism in children with nephrosis. *J. Clin. Invest.* **24**:564–572.

317. Albright, F., W. V. Consolazio, F. S. Coombs, H. W. Sulkowitch, and J. J. Talbott. 1940. Metabolic studies and therapy in a case of nephrocalcinosis with rickets and dwarfism. *Bull. Johns Hopkins Hosp.* **66**:7–33.

318. Albright, F., C. H. Burnett, W. Parsons, E. C. Reifenstein, Jr., and A. Roos. 1946. Osteomalacia and late rickets. *Medicine (Baltimore)* **25**:399–479.

319. Greenberg, A. J., H. McNamara, and W. W. McCrory. 1966. Metabolic balance studies in primary renal tubular acidosis: Effects of acidosis on external calcium and phosphorus balances. *J. Pediatr.* **69**:610–618.

320. Gray, R. W., J. L. Omdahl, J. G. Ghazarian, and H. F. DeLuca. 1972. 25-Hydroxycholecalciferol-1-hydroxylase. *J. Biol. Chem.* **247**:7528–7532.

321. Behar, J., and M. D. Kerstein. 1976. Intestinal calcium absorp-

tion: Differences in transport between duodenum and ileum. *Am. J. Physiol.* **230**:1255–1260.

322. Bikle, D., and H. Rasmussen. 1975. The ionic control of 1,25-hydroxyvitamin D_3 production in isolated chick renal tubules. *J. Clin. Invest.* **55**:292–298.

323. Rushton, M. L., H. G. Sammons, and B. H. B. Robinson. 1971. A study of calcium absorption using an automated fluorimetric assay procedure. *Clin. Chim. Acta* **35**:5–16.

324. Hahn, T. J., B. A. Hendin, C. R. Scharp, and J. G. Haddad, Jr. 1972. Effect of chronic anticonvulsant therapy on serum 25-hydroxycalciferol levels in adults. *N. Engl. J. Med.* **287**:900–904.

325. Rodbro, P., C. Christiansen, and M. Lund. 1974. Development of anticonvulsant osteomalacia in epileptic patients on phenytoin treatment. *Acta Neurol. Scand.* **50**:527–532.

326. Hahn, T. J., B. A. Hendin, C. R. Scharp, V. C. Boisseau, and J. G. Haddad, Jr. 1975. Serum 25-hydroxycalciferol levels and bone mass in children on chronic anticonvulsant therapy. *N. Engl. J. Med.* **292**:550–554.

327. Maclaren, N. K., and F. Lifshitz. 1973. Vitamin D dependency rickets in institutionalized, mentally retarded children on long-term anticonvulsant therapy. II. The response to 25-hydroxycholecalciferol and vitamin D. *Pediatr. Res.* **7**:914–923.

328. Hahn, T. J. 1980. Drug-induced disorders of vitamin D and mineral metabolism. *Clin. Endocrinol. Metab.* **9**:107–127.

329. Hahn, T. J., S. J. Birge, C. R. Scharp, and L. V. Avioli. 1972. Phenobarbital-induced alterations in vitamin D metabolism. *J. Clin. Invest.* **51**:741–748.

330. Kimberg, D. V., R. D. Bairg, and E. Gershon. 1971. Effect of cortisone treatment on the active transport of calcium by the small intestine. *J. Clin. Invest.* **50**:1309–1314.

331. Favus, M. J., D. V. Kimberg, and G. N. Millar. 1973. Effects of cortisone administration on the metabolism and localization of 25-hydroxycholecalciferol in the rat. *J. Clin. Invest.* **52**:1328–1331.

332. Carre, M., O. Ayigebede, L. Miravet, and H. Rasmussen. 1974. The effect of prednisolone on the metabolism and action of 25-hydroxy- and 1,25-dihydroxyvitamin D_3. *Proc. Natl. Acad. Sci. USA* **71**:2996–3000.

333. O'Regan, S., R. W. Chesney, A. Hamstra, J. A. Eisman, A. M. O'Gorman, and H. F. DeLuca. 1979. Reduced serum 1,25(OH)$_2$ vitamin D_3 levels in prednisone-treated adolescents with systemic lupus erythematosus. *Acta Paediatr. Scand.* **68**:109–111.

334. Hahn, T. J., D. T. Baran, and L. R. Halstead. 1979. Alteration in mineral and vitamin D metabolism produced by subacute prednisone administration in man. *Proceedings of the 61st Annual Meeting of the Endocrine Society.* p. 195.

335. Costanzo, L. S., and I. M. Weiner. 1974. On the hypocalciuric action of chlorothiazide. *J. Clin. Invest.* **54**:628–637.

336. Costanzo, L. S., and I. M. Weiner. 1976. Relationship between clearances of Ca and Na: Effect of distal diuretics and PTH. *Am. J. Physiol.* **230**:67–73.

337. Brickman, A. S., S. G. Massry, and J. W. Coburn. 1972. Changes in serum and urinary calcium during treatment with hydrochlorothiazide, studies on mechanisms. *J. Clin. Invest.* **51**:1877–1888.

338. Ehrig, U., J. E. Harrison, and D. R. Wilson. 1974. Effect of long-term thiazide therapy on intestinal calcium absorption in patients with recurrent renal calculi. *Metabolism* **23**:139–149.

339. Gursel, E. 1970. Effects of diuretics on renal and intestinal handling of calcium. *N.Y. State J. Med.* **1**:399–405.

340. Favus, M. J., F. I. Coe, S. C. Kathpalia, A. Porat, P. K. Sen, and L. M. Sherwood. 1982. Effects of chlorothiazide on 1,25-dihydroxyvitamin D_3, parathyroid hormone, and intestinal calcium absorption in the rat. *Am. J. Physiol.* **242**:G575–G581.

341. Holick, M. F., J. A. MacLaughlin, M. B. Clark, S. A. Holick, and J. T. Potts. 1980. Photosynthesis of previtamin D_3 in human skin and the physiologic consequences. *Science* **210**:203–205.

342. Haddad, J. G., and T. J. Hahn. 1973. Natural and synthetic sources of circulating 25-hydroxyvitamin D in man. *Nature (London)* **244**:515–517.

343. Lawson, D. E. M., A. A. Paul, A. E. Black, T. J. Cole, A. R.

Mandal, and M. Davie. 1979. Relative contributions of diet and sunlight to vitamin D state in the elderly. *Br. Med. J.* **2**:303–305.

344. Sherwood, L. M., and M. Abe. 1972. Magnesium ion, parathyroid hormone secretion and cyclic 3′5′-AMP. *Clin. Res.* **20**:756.

345. Muldowney, F. P., J. S. McKenna, L. H. Kyle, R. Freaney, and M. Swan. 1970. Parathormone-like effect of magnesium replenishment in steatorrhea. *N. Engl. J. Med.* **281**:61.

346. Freitag, J. J., K. J. Martin, M. B. Conrades, E. Bellorin-Font, S. Teitelbaum, S. Klahr, and E. Slatopolsky. 1979. Evidence for skeletal resistance to parathyroid hormone in magnesium deficiency. *J. Clin. Invest.* **64**:1238–1244.

347. Rosler, A., and D. Rabinowitz. 1973. Magnesium-induced reversal of vitamin-D resistance in hypoparathyroidism. *Lancet* **1**:803–805.

348. Mahoney, A. W., and D. G. Hendricks. 1975. Utilization of dietary calcium by iron-deficient rats. *Nutr. Metab.* **18**:6–15.

349. Heaney, R. P., J. C. Gallagher, C. C. Johnston, R. Neer, A. M. Parfitt, and G. D. Whedon. 1982. Calcium nutrition and bone health in the elderly. *Am. J. Clin. Nutr.* **36**:986–1013.

350. Allen, L. H. 1982. Calcium bioavailability and absorption: A review. *Am. J. Clin. Nutr.* **35**:783–808.

CHAPTER 9

Cystic Fibrosis

John A. Mangos

1. Introduction

Cystic fibrosis, an apparent inborn error of metabolism, is a common disease of children and young adults in countries with predominantly Caucasian populations. It was first described by Fanconi *et al.* in Switzerland following the recognition of the association of congenital cystic pancreatic fibrosis and bronchiectasis in 1936.[1] In this country, it was Andersen[2] who presented in 1938 the first complete description of the disease and introduced the term "cystic fibrosis of the pancreas." Although other names for the disease have been proposed in the past decades,[3,4] the shortened form of the original name for the disease, "cystic fibrosis," has become most common. Several detailed reviews have recently appeared in the literature.[3–6]

Cystic fibrosis is not rare. In this country, a patient with the disease is born in every 1500–2000 live births. Cystic fibrosis is recognized as the most common genetic disease in childhood; most of the affected individuals survive into adolescence and some into adulthood; it is the cause of virtually all cases of pancreatic enzyme deficiency and of the majority of cases of cirrhosis of the liver, portal hypertension, and malabsorption in the pediatric age group; it is also the most common nontuberculous pulmonary disease which causes death during childhood. Although cystic fibrosis is largely a disease of the Caucasian race, small but significant numbers of black and American Indian affected children have been observed; among Orientals it is very rare but not totally absent.[3,4]

2. Clinical Features

2.1. Respiratory

The onset of pulmonary involvement may occur at any time after birth but frequently appears before the age of 1 year. Coughing, wheezing, and stertorous breathing are the cardinal symptoms. Repeated bouts of pneumonitis resulting in bronchial

and peribronchial damage may ensue. Advancing pulmonary involvement is associated with increased chest anterior–posterior diameter and with clubbing of the fingers and toes. Decreased exercise tolerance, chronic productive cough, tachypnea, and chronic cor pulmonale may become increasingly severe. Death most frequently is the result of pneumonia, anoxia, and exhaustion after a long period of respiratory insufficiency. Nasal polyposis and sinusitis are frequently found in older patients.

2.2. Digestive

Intestinal obstruction (meconium ileus) is virtually pathognomonic of the disease and marks the appearance of cystic fibrosis at birth in 10–15% of patients. Obstruction of the small or large intestine from fecal-mucous plugs, intussusception, and vovulus occasionally occurs in older children and adults with the disease. Severe steatorrhea with large, foul-smelling stools, poor weight gain in spite of increased appetite, and abdominal distension are notable during the first year of life in untreated patients. Rectal prolapse occurs in approximately 20% of untreated patients in the first 3 years of life. Disaccharide intolerance and deficiency of intestinal lactase have been observed in some patients. Massive biliary cirrhosis may occur in 3–5% of patients, and at autopsy most patients are found to have to have focal areas of biliary cirrhosis. The incidence of diabetes mellitus in cystic fibrosis patients is higher than in the general population, and most patients have marked salivary gland enlargement.

2.3. Other Phenomena

Rates of linear growth, body weight, and osseous maturation are often depressed in patients with cystic fibrosis. Puberty is often delayed but the disease may show a slowing of its progress during and after adolescence. Heat exhaustion and shock may occur during profuse sweating. The danger is greatest in infants and in very ill patients with cystic fibrosis who cannot express their craving for salt. Sterility due to congenital obstruction or agenesis of the vas deferens is present in almost all male patients with cystic fibrosis. The nervous system is intact and

John A. Mangos • Department of Pediatrics, University of Texas Health Science Center, San Antonio, Texas 78284.

increased acuity of the senses of taste and smell has been reported in patients.

3. Prognosis

The prognosis of cystic fibrosis appears to have improved dramatically with treatment. In a recent national study, it was shown that the median age of death, which before treatment became available was 8 months, has been prolonged with the development of different methods of symptomatic therapy. With comprehensive treatment the median age of death in clinics where intensive prophylactic treatment is used increased, with median age at death up to 21 years being reported.[3,7]

4. Genetics

Cystic fibrosis is considered to be an autosomal recessive inherited disease expressed only in the homozygous state without X linkage.[3,4] Both sexes are affected with equal frequency at birth, although males may predominate later as a result of higher mortality in females. The disease has high incidence in sibs. The segregation frequency in the author's clinic population has been estimated to be $0.267 \pm$ S.D. 0.033 by a maximum likelihood method. This value is not significantly different from the value of 0.250 expected for a gene expressed only in the homozygous state (autosomal recessive). The gene frequency has been estimated to be 5% in this country. The cystic fibrosis gene is five times more common in the overall U.S. population than the gene for sickle-cell anemia, the next most frequent hereditary disease in this country which causes death in children.

5. Pathogenesis

Although cystic fibrosis has been known as a separate disease entity for over 45 years, its fundamental biochemical defect is still unknown. The pathogenesis of most clinical and pathological manifestations can be attributed to the two recognizable abnormalities in the function of exocrine glands: abnormal ionic composition of exocrine gland products, and abnormal physicochemical behavior of mucus in exocrine ducts and body cavities.

A striking increase of the concentrations of Na^+ and Cl^- and, to a lesser extent, of K^+ in the sweat of patients with cystic fibrosis was first described in 1953 by di Sant'Agnese et al.[7] This was the first indication of exocrine dysfunction in cystic fibrosis. The abnormality in the handling of these ions is present from birth; throughout life it is unrelated to the severity of disease or to the type of predominant organ involvement; it has led to the development of the "sweat test," the cardinal laboratory determination for the diagnosis of cystic fibrosis.[8]

Exocrine gland products containing mucus have an abnormal physicochemical behavior. They appear viscous, turbid, and precipitate and obstruct organ passages, thus giving rise to chronic obstructive pulmonary disease, pancreatic insufficiency, intestinal obstruction, hepatic cirrhosis, and other complications. Pathological changes, including atrophy, atresia, dilation, obstruction, tissue inflammation, and tissue destruction, follow.

Because of the abnormalities in the ionic composition of exocrine gland products, a generalized defect in the structure or function of all membranes of patients with cystic fibrosis was sought by investigators in the early years of research in this disease. Thus, accessible body fluids, organs, and tissues in which phenomena of ion transport across membranes occur have been and are being investigated for a generalized transport defect. Reviews of published information do not reveal general patterns of abnormal transmembrane transport in cystic fibrosis[3,6] Body fluids other than exocrine gland products have normal composition.[3,6] Major organs other than exocrine glands, such as the kidneys, brain, muscle, heart, appear to have normal function of ion handling unless secondarily affected by the complications of cystic fibrosis. Thus, the study of the transport abnormalities of cystic fibrosis has been limited to specialized tissues and organs.

In the studies of the pathogenesis of cystic fibrosis, investigators have focused their attention on two distinct areas: (1) the pathophysiology of cystic fibrosis and (2) the basic defect of cystic fibrosis.

5.1. Studies of the Pathophysiology of Cystic Fibrosis

The studies of the pathophysiology of exocrine glands in cystic fibrosis have progressed very slowly because of an apparent lack of complete knowledge of the normal function of these glands. As shown in Fig. 1, exocrine glands, the cardinal target organs of cystic fibrosis, have a common pattern of ductal architecture which serves their functional efficiency. In each gland one can recognize a *secretory segment* (acini + intercalated ducts in salivary glands, secretory coil in the human sweat gland, etc.) where the primary secretory fluid is produced, and an *excretory segment* which is composed of either a single duct (human sweat gland) or a system of ducts of increasing diameter which serves as a conduit delivering the final excretory product to the body surface or to a body cavity. In this chapter, *secretion* is the term used to describe processes of transfer of components of exocrine gland products from the blood and interstitium or from the intracellular space of the cells lining the ductal system of each gland into the duct lumen. *Reabsorption* is the process of transductal movements of components of the ductal fluid out of the duct system of the gland. *Excretion* refers to the final transfer of the formed exocrine gland product to the body surface or to a body cavity.

In order to interpret abnormal functional characteristics observed in patients with this disease, extensive studies of the "normal" function of various organs had to be conducted. This is best demonstrated by the efforts to elucidate the mechanisms involved in the development of the abnormality in the handling of Na^+ and Cl^- by the sweat gland of patients.

5.1.1. Eccrine Sweat Gland

In 1953, when di Sant'Agnese et al.[7] described the increased Na^+ and Cl^- concentrations in the sweat of children with cystic fibrosis, the pathophysiology of this abnormality could not be explained because of lack of convincing data about the physiology of the normal sweat gland. There was no direct information about the formation of sweat in the human sweat gland, and the existing hypotheses had many unclear and confusing points. The "abnormally" high concentrations of Na^+ and Cl^- in the sweat of patients could be due either to "abnormal" concentrations of these ions in the primary secretory fluid of the gland, to "abnormal" handling of ions by the duct, or to

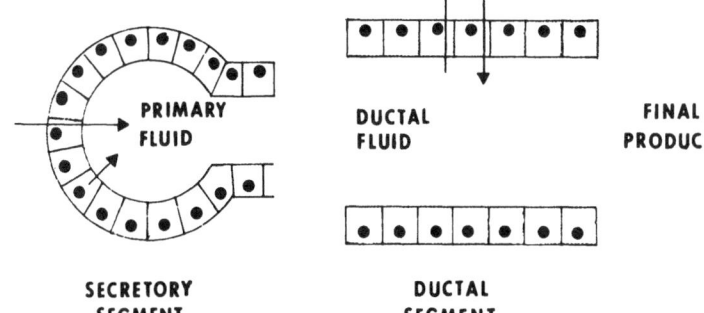

Fig. 1. Diagram showing the various segments of the duct system of an exocrine gland. Reprinted through the courtesy of Symposia Specialists, Inc., Division of Stratton Intercontinental Medical Book Corp., from *Fundamental Problems of Cystic Fibrosis and Related Diseases,* edited by J. A. Mangos and R. C. Talamo (1973).

"abnormal" permeability of the duct to water. As shown in Table I, it took 30 years of work in many laboratories around the world to reach a point of better, but still incomplete, understanding of the transport pathophysiology of the sweat gland in cystic fibrosis. A new, direct approach to the study of the individual sweat gland using micropuncture, microperfusion, and microanalysis had to be developed in order to answer some of the questions raised by di Sant'Agnese *et al.*

Today it is known that the human sweat gland secretes a fluid, which is either isotonic or slightly hypertonic to plasma, in the secretory coil of the gland.[11] This fluid is transformed into

hypotonic sweat through reabsorption of monovalent ions in excess of water in the duct of the gland,[11,13,15] which is relatively impermeable to water.[15] The fluid finally emerges as the hypotonic sweat on the surface of the skin. The tonicity and composition of the sweat of a given individual are affected by the age and sex of the subject, as well as by rare diseases other than cystic fibrosis.[4,10] In cystic fibrosis, the osmolality and monovalent ionic composition of the primary secretory fluid of the sweat gland are not different from those of subjects without the disease.[14] The increased concentration of Na^+ and Cl^- in the sweat of patients appeared to be due to decreased net transductal reabsorption and lack of diffusion of Na^+, as shown in the author's laboratory using *in vitro* microperfusion of single sweat ducts.[15,16] Recently, Quinton and associates presented evidence that the primary defect in cystic fibrosis may be in the transductal reabsorption of Cl^- rather than Na^+ in the sweat duct of patients.[17] The sweat ducts of patients were examined morphologically by light and electron microscopy, and they were found to be indistinguishable from those of normal subjects.[18] The sweat glands in cystic fibrosis are of normal size[19] and their density in the skin is identical to that of normals.[20] In an effort to explain this transport defect of the sweat gland in cystic fibrosis, the author proposed that the primary secretory fluid in the sweat glands of patients contains a humoral factor which inhibits the transductal reabsorption of Na^+ (or Cl^-) and causes the increased salinity of the sweat.[12] Using the rat parotid gland as an experimental tissue, it was demonstrated that retrograde perfusion of the duct system with sweat from patients with cystic fibrosis resulted in reduction of the transductal reabsorption by 45%.[12] Sweat from age- and sex-matched controls had no such inhibitory effect. This was considered an indication of the presence in the sweat of patients of a humorally transducible factor inhibiting the reabsorption of Na^+ in another exocrine gland. That the same factor can induce a similar defect in Na^+ reabsorption was later demonstrated by Kaiser *et al.* who perfused *in situ* individual sweat glands of normal subjects with sweat from patients with cystic fibrosis.[21] Subsequently, the author developed a system of *in vitro* perfusion of single sweat ducts, characterized the reabsorption of Na^+ and fluxes of water in the normal duct, and demonstrated both the abnormality in the sweat ducts of patients with cystic fibrosis and the transducibility of the defect to normal sweat ducts by perfusion with sweat from patients.[15,16] Although these studies have clarified to some degree the nature of the transport dysfuction of the sweat gland in cystic fibrosis, the biochemical nature, origin, and mode of action of this factor remain unclear pending its chemical characterization.

Table I. Historical Review of the Developments Leading to a Better Understanding of the Pathophysiology of the Sweat Gland in Cystic Fibrosis

Year	Topic	Investigators
1953	Detection of sweat gland electrolyte abnormality in cystic fibrosis	di Sant'Agnese *et al.*[7]
1956	Description of Na^+ excretion in sweat and hypotheses about function of human sweat gland	Schwartz and Thaysen[9]
1962	Description of effects of age, sex, and cystic fibrosis on Na^+ and Cl^- concentrations of human sweat	Lobeck and Huebner[10]
1965	Micropuncture of normal sweat gland and analysis of primary secretory fluid	Schulz *et al.*[11]
1967	Demonstration of a Na^+ transport inhibitory factor in the sweat of patients with cystic fibrosis	Mangos and McSherry[12]
1968	Study of single human sweat glands in health and in cystic fibrosis	Emrich *et al.*[13]
1969	Micropuncture of sweat glands of patients with cystic fibrosis and analysis of primary secretory fluid	Schultz[14]
1973	*In vitro* microperfusion of ducts from normal sweat glands and characterization of transductal fluxes of ions	Mangos[15]
1973	*In vitro* microperfusion of ducts from sweat glands of patients with cystic fibrosis and characterization of transductal fluxes of ions	Mangos[16]
1983	Demonstration that a defect in Cl^- reabsorption in the sweat duct may be responsible for the sweat electrolyte abnormality in cystic fibrosis	Quinton[17]

5.1.2. Pancreas

In the majority of patients with cystic fibrosis, there is significant insufficiency of pancreatic function due to pancreatic injury that usually begins during intrauterine life. It is the result of ductal obstruction with presumed autoactivation of pancreatic enzymes and destruction of the exocrine pancreas. This results in severe steatorrhea and azotorrhea which respond to administration of exogenous pancreatic enzymes. In 10–12% of the patients, stool fat and nitrogen excretion remain normal for many years and in some instances throughout life.[3–5] It is probably incorrect to assume that these patients have normal pancreatic function because malabsorption and maldigestion become evident only when the residual pancreatic function is less than 20% of what is expected from the intact pancreas. In these few patients with residual pancreatic function, the studies of Hadorn et al.[22] have shown that enzyme secretion was almost normal but fluid secretion in response to secretin was markedly decreased. The HCO_3^- concentration of the pancreatic juice and HCO_3^- output were low while the Cl^- concentration was high, even as the rate of flow of pancreatic juice increased. These investigators concluded that decreased responsiveness of the end-organ, pancreas, to the secretagogue hormone, secretin, may contribute to the obstruction of pancreatic ducts by thickened secretions during intrauterine life or later on. It is of interest that Crossley et al.[23] and Wilcken et al.[24] observed marked elevations of immunoreactive trypsin in the blood of nearly all newborns with cystic fibrosis. This observation would support the concept of ductal obstruction and spilling of concentrated enzymes from the duct system into the blood. Recently, Figarella et al. repeated the studies of Hadorn et al. and confirmed their findings.[25] The conclusions of both of these groups were that the patients with cystic fibrosis who have residual pancreatic function do not demonstrate a complete failure to secrete HCO_3^- in the pancreatic juice during stimulation with secretin but a less than normal responsiveness of the target organ (pancreas) to the hormonal stimulus. Studies of metabolic responses of other organs such as liver, rectal mucosa, and kidney to secretin[22] and observations during nasal application of secretin for 5 days in children with cystic fibrosis led Hadorn et al. to conclude that prolonged, sustained administration of secretin is beneficial to patients with cystic fibrosis and that an "imbalanced regulation of secretin as a consequence of an abnormality in the autonomic nervous system" may be the underlying abnormality of the pancreatic exocrine dysfunction in cystic fibrosis.[22]

5.1.3. Salivary Glands

Widespread abnormalities in the inorganic and organic composition of submaxillary and parotid salivas have been reported in patients with cystic fibrosis.[26] Gugler et al. suggested that alterations in the ionic composition of submaxillary saliva and other exocrine fluids, particularly in the concentration of Ca^{2+}, could result in a change in the physicochemical behavior of the glycoproteins present in these fluids.[27] Chernick and Barbero[28] believe that the high glycoprotein content of submaxillary saliva and other exocrine fluids may be due to increased synthesis and secretion of these macromolecules. Furthermore, they suggested that the secretory abnormality may be the result of an abnormal neurohumoral regulation of the involved glands.[28] Although small differences in the ionic composition of submaxillary, sublingual, and parotid salivas exist between patients with cystic fibrosis and normal controls,[26] the most pronounced and uniformly present differences in ionic composition are found in the fluid produced by the small submucosal salivary glands of the oral cavity.[29] Using the same methods as in the case of retrograde perfusion of the rat parotid with sweat from patients with cystic fibrosis, the author demonstrated that mixed saliva from patients inhibits the reabsorption of Na^+ in the same gland, thus suggesting the presence of a Na^+ transport inhibitory factor in the saliva.[30] It was shown that this inhibitory factor has the following characteristics: (1) it is not eliminated by dialysis of the saliva; (2) it is eliminated by exposure to glass and by repeated freeze–thawing; (3) its effect is not mediated by inhibition of the Na^+,K^+-sensitive ATPase; (4) its action is mimicked by positively charged polycations, such as polylysine, polyornithine, and protamine sulfate; and (5) its action is reversed by the addition of small amounts of heparin to the saliva. It was postulated that the Na^+ transport inhibitory factor of cystic fibrosis may be a positively charged macromolecule.[31]

These findings of the author were confirmed in 1974 by Taylor et al.,[32] who demonstrated that the saliva from submandibular, sublingual, and submucosal glands of patients contains the Na^+ transport inhibitory factor. No factor activity could be found in the parotid saliva by the same investigators. They suggested that this inhibitory effect is in some way related to the mucous content of the saliva from individual glands.

In a recent article describing the alterations of salivary gland structure and function in cystic fibrosis,[26] Martinez presented an exhaustive review of the literature which suggests that the only uniformly recorded difference in the ionic composition of saliva of patients with cystic fibrosis versus that of control subjects is the increased Ca^{2+} content of the saliva. This suggests that abnormalities in the handling of this ion in the acinar and ductal elements of salivary glands may be closely related to the basic biochemical, hereditary defect of cystic fibrosis.

5.1.4. Tracheobronchial Secretory System and Mucociliary Transport in the Lungs

The mucociliary transport system is one of the major pulmonary defense mechanisms; it removes from the airways inhaled particulate matter, including bacteria and natural debris. This system is composed of the cilia on the luminal surface of the epithelial cells and of the mucous layer which is considered to exist in two phases: a superficial layer of high viscosity (gel) and a periciliary layer of low viscosity (sol). The origin of the electrolytes and water comprising the two phases is not known. It is presumed that mucus originates from the submucosal bronchial glands and the surface goblet cells. The origin of the electrolytes and water could also be from the bronchial glands, but recent information suggests that the entire mucosa may be involved in transepithelial fluxes of electrolytes and water.[33]

In a recent review of the morphology of the airways in cystic fibrosis, Sturgess pointed out that no fundamental structural cellular defect has been detected among epithelial and secretory cells.[34] At birth, the bronchial mucosa of patients with cystic fibrosis is apparently normal and development may progress in a normal pattern during early childhood.[35] By quantitative analysis, the earliest change in the airways of patients has been shown to be a dilation of the acinar and ductal lumen of the submucosal glands. If this morphological alteration is associated with changes in the physicochemical behavior and/or composition of the mucus produced by these glands, this may be the change that leads to obstruction of the airways and impaction of secretory material in the glands and the lumen of the airways.

The pulmonary involvement in cystic fibrosis is basically due to two pathophysiological events: airway obstruction and infection. Airway obstruction would occur if one or more of the following factors was present: (1) hypersecretion of airway mucus; (2) abnormal physicochemical characteristics of the secreted mucus; (3) decreased rates of clearance of secretions from the airways. It has often been speculated that there is hypersecretion of airway mucus in cystic fibrosis. Evidence of such hypersecretion does not exist in the absence of infection in the lungs of infants with this disease.[35] Once chronic bacterial infection is established, there is no doubt that there is hypersecretion of mucus in the airways of patients with cystic fibrosis. Abnormal physicochemical characteristics of airway secretions are even more difficult to demonstrate. When sputum production is established in a patient, the respiratory mucus is already contaminated by bacteria, leukocytes, exudates, products of bacterial and leukocytic degradation, and other constituents. Even under those conditions, the available studies do not reveal specific alterations of viscosity characteristic of cystic fibrosis. For years it had been assumed that mucociliary clearance was impaired in the airways of patients with cystic fibrosis. Wood et al.[36] and Yeates et al.[37] reported impaired clearance in general, although there were several patients who had normal clearances. What the contribution of a primary or secondary failure of the mucociliary clearance is to the development of chronic obstructive airway disease is still under question. For example, patients with immotile cilia syndrome develop chronic obstructive lung disease but the rate of development is much slower than in cystic fibrosis. The role of infection in the development of chronic obstructive airway disease in cystic fibrosis is beyond any doubt although the conditions favoring colonization and infection of the lungs of patients with *Staphylococcus aureus* and *Pseudomonas aeruginosa* have so far escaped detection. Many centers for care of patients with cystic fibrosis report rates of airway infection with *Pseudomonas* up to 90%. Another factor that may contribute to the development of airway obstruction is the fact that more than 50% of patients have airway hyperreactivity resulting in reversible airway obstruction.[3] This factor, in combination with the other factors previously mentioned, probably contributes significantly to the development of airway obstruction and infection in patients with cystic fibrosis.

Are there any transport abnormalities in the airways which may favor the development of chronic obstructive airway disease in patients with cystic fibrosis? It has been demonstrated repeatedly that tracheobronchial secretions from patients with cystic fibrosis have significantly less water and more solids than sputum from other pathological conditions such as bronchiectasis.[38] Also, the Na^+ content has been reported to be decreased and the Ca^{2+} content to be increased. Recent studies by Boucher et al.[39] and Knowles et al.[40] have demonstrated increased reabsorption of Na^+ in the tracheal and bronchial epithelia of patients with cystic fibrosis. This was associated with higher than normal transepithelial electrical potential difference in the nose and airways of patients with cystic fibrosis. This increased potential returned to normal upon topical application of amiloride, suggesting that it was, most likely, due to increased reabsorption of Na^+. If this increased reabsorption of Na^+ is associated with reabsorption of water, this may be the underlying cause of the relative dehydration of the sputum of patients with cystic fibrosis. Reabsorption of fluid through the surface epithelium of the airways would, by necessity, abstract

fluid from the periciliary fluid which bathes the airway cilia. Such reduction in the size of the layer of periciliary fluid could result in collapse of the cilia and abnormalities in mucociliary clearance. This recent observation of transepithelial transport abnormalities in the airways of patients with cystic fibrosis could lead to a better understanding of the pathophysiology and pathogenesis of this disease.

Ca^{2+} is involved in the determination of the solubility of various macromolecules in solution. It is unclear what role the increased concentrations of Ca^{2+} in the sputum of patients with cystic fibrosis play in causing airway obstruction through mucus plugging. It is known that the abnormal electrophoretic patterns of proteins in the submandibular saliva of patients with cystic fibrosis can be corrected *in vitro* by the simple addition of a chelating agent to the saliva.[27] It is possible that similar interactions between Ca^{2+} and the macromolecules of the respiratory mucus may play an important role in the pathogenesis of airway obstruction in cystic fibrosis. Because of the absence of any data on the mechanisms of Ca^{2+} homeostasis in the airways, there is an obvious need for the characterization of such transepithelial transport processes for Ca^{2+} in mammalian airways and in those of controls and patients with cystic fibrosis.

5.1.5. Other Exocrine Glands

Inspissation of bile in the hepatic ductules results in microscopically demonstrable focal biliary cirrhosis in most patients with cystic fibrosis at autopsy.[3,4] A histochemical study of cystic fibrosis gallbladders reveals engorgement of epithelial cells with mucins and increased amounts of sulfomucins.[4] In the male genital system, there is atresia of the vas deferens, aspermia with rare exceptions, depressed concentrations of fructose, and elevated glucose concentrations in the seminal fluid.[4] The cervical mucus of women with cystic fibrosis appears desiccated and contains only one-fourth to one-third of the normal water content. Hair and nails of patients with cystic fibrosis have higher Na^+ and K^+ content than those of normal children.[3,4] An increase in sulfomucins and in the intensity of PAS staining has been detected by histochemical techniques.[3,4]

5.1.6. Other Cells and Tissues

Although cystic fibrosis is clearly an exocrinopathy, cells and tissues other than exocrine have been examined. It would be expected that an autosomal recessive genetic defect would be present in all cells of the patient's body. Since the demonstration by di Sant'Agnese et al.[7] of elevated Na^+ and Cl^- levels in the sweat of patients, many studies of the transport of electrolytes in erythrocytes have been reported.

5.1.6a. Erythrocytes. Because of the easy availability of these cells and their well-defined ion transport features, studies of transport of ions across the erythrocyte membrane have been conducted in cystic fibrosis. Despite conflicting reports, critical review of the available literature reveals no convincing differences between patients and controls in the transmembrane transport of Na^+, K^+, and Ca^{2+}, or in the levels of ATPases participating in the transport of ions.[3,4] Membranes of patients' red blood cells subjected to electrophoresis on polyacrylamide gels show an alteration in the electrophoretic pattern[3] which has been attributed to Ca^{2+} binding to certain membrane proteins.[3]

5.1.6b. Skin Fibroblasts. These cells have been cultured *in vitro* and have been used for many structural and functional studies.[6] No convincing differences in the composition of cell membranes between patients and controls have been detected. The numbers of ouabain-binding sites have been determined, and again no differences were detected. The fibroblasts from patients and heterozygotes seem to have increased amounts of metachromatically staining granules. Although earlier studies reported abnormal accumulation of glycosaminoglycans in fibroblasts from patients, more detailed studies, reported recently, failed to show any convincing differences.

5.1.6c. Leukocytes. Leukocytes cultured *in vitro* show metachromasia after 4 days in culture,[3,4] and they appear to produce the ciliary factor.

5.2. Studies in Search of the Basic Defect of Cystic Fibrosis

A discrete biochemical or structural defect is expected to be present in cystic fibrosis because of its autosomal recessive mode of inheritance. To date, no lesion has been described which could in some way provide a unifying hypothesis explaining all or most of the pathophysiological manifestations of this disease. Distinguishing between primary and secondary defects is impossible at this time, as is characterization of phenomena produced by the different therapeutic modalities used in this disease rather than by its basic defect. The number of scientific papers on this disease published to date exceeds 8000, of which half deal with the search for the basic defect. In a number of studies, the parents of patients with cystic fibrosis were used, reasoning that the basic defect should be demonstrable in heterozygotes. However, the unknown rate of mutations contributing to the high frequency of the gene, the often-raised question of paternity, and the possibility that the single dose of the abnormal gene could be masked have created many problems in these studies.

Most studies have begun as an effort to explain one or another of the pathophysiological manifestations of cystic fibrosis: the study of the Na$^+$ transport inhibitory action of sweat and saliva in an effort to explain the electrolyte abnormality in exocrine glands; the study of the ciliary inhibitor in an effort to explain the presumed defect in mucociliary transport leading to the production of obstructive lung disease.

5.2.1. Ciliary Factor(s)

In 1967, Spock *et al.*[41] first reported the existence of a factor in the serum of homozygotes and heterozygotes for the gene of cystic fibrosis. This factor appears to disrupt the normal ciliary beat pattern of rabbit tracheal explants. Similar findings, leading to cessation of ciliary motility, were observed using the oyster gill.[42] Attempts to reproduce these studies in other ciliated systems, including chick embryo trachea, gills of the freshwater mussel, human sperm, and ciliated protozoans, produced controversial results; some investigators were able to reproduce the ciliary inhibition whereas others could not.[3] A major problem in this area of research is that all methods currently used for the detection of the factor(s) are bioassays, in which most of the observed changes are measured subjectively and are difficult to interpret. The ciliary factor appears to be a basic protein with a low molecular weight of 4000–10,000, which can be labeled

with radioactive amino acids.[3,6] It is not known whether the factor is a normal product which accumulates because of the lack of normal degradation[3,4] or an abnormal protein, which is either produced in large quantities or cannot be degraded. Although it has not been proved, it is possible that the ciliary factor may alter the ciliary motility by affecting membrane transport in the ciliated cells. Media from cystic fibrosis fibroblast cultures have been reported to inhibit the Na$^+$,K$^+$-ATPase of red blood cells.[43] However, in cilia from rabbit trachea and oyster gill, no alterations in ATPase activity or ATP consumption were found in the presence of cystic fibrosis serum.[44] The ciliary factor has been considered to be a component of the complement system, but proof for this identity has not been presented. The ciliary factor is also present in the saliva and is reported to be closely associated with the salivary amylase.[45]

Recently, the mucociliary inhibitor was fractionated from the plasma of patients with cystic fibrosis in a pool containing protein with molecular weight between 6000 and 12,000.[46] Upon isoelectric focusing, the mucociliary inhibitor was demonstrated to be present in two fractions: one at pH 4.6–6.1 and one at pH 9.0. Subsequently, a similar protein fraction was isolated from the urine of patients with cystic fibrosis and was characterized by DEAE–cellulose chromatography, gel filtration, lectin-affinity chromatography, isoelectric focusing, laser light-scattering spectroscopy, and amino acid sequence analysis.[47] A similar glycopeptide was isolated in another laboratory from the serum of patients.[48] It is hoped that these molecules will soon be available for detailed studies of their metabolic activity on target cells and tissues.

5.2.2. Transport Inhibitory Factor(s)

The author's studies of the Na$^+$ transport inhibitory effect of sweat and saliva from patients with cystic fibrosis were described earlier. In addition, the author has shown that seminal fluid from patients with cystic fibrosis also shares this inhibitory activity.[4] These results may explain the abnormalities in the ionic composition of the sweat, saliva, and other exocrine products from patients with cystic fibrosis. Studies of the transepithelial transport of other substances in systems exposed to body fluids from patients and heterozygotes have been less conclusive. Plasma, saliva, and skin fibroblast culture media reportedly decrease ATPase activity in red cell membranes, but Na$^+$ influx and efflux are not altered by exposure to cystic fibrosis saliva.[3] Rat jejunal uptake of arbutin, a glucose analog, was reported to be inhibited during incubation with cystic fibrosis plasma.[49] Transport of [^{14}C]alanine was reduced by preincubation of rat jejunum with cystic fibrosis saliva but not serum[50]; other studies could not confirm this finding.[51] Sugar and amino acid transport in fibroblasts are not inhibited by cystic fibrosis saliva or dialyzed plasma.[52] Sera from patients with cystic fibrosis decreased short-circuit current and the short-circuit current response to glucose in rat jejunum,[53] suggesting interference with Na$^+$-dependent glucose transport in this system. The possibility that the transport inhibitory activity of these fluids involves inhibition of metabolic pathways,[54] or of energy production in target cells,[55] has been raised. No investigator has been able to separate heterozygotes from control populations on the basis of transport inhibition studies. Studies are currently under way to determine whether the isolated and partially characterized mucociliary inhibitor has any similarities or is identical to the transport inhibitory factor(s).

6. Animal Models

The inherent limitations of research on human subjects make the answers to the many questions regarding the genetic defect of cystic fibrosis difficult to obtain, and it has been necessary to seek ways of reproducing the abnormalities of this disease in animals. The search for an appropriate animal model has been hampered by the lack of a genetic equivalent to cystic fibrosis in lower species. Concentrating on the exocrine aspects of the disease, the author induced changes resembling some of those of cystic fibrosis in rats by the chronic administration of isoproterenol, which induces enlargement of the parotid and submaxillary glands.[56] Flow rates of saliva were decreased but salivary Na$^+$ concentrations were increased in the treated animals. Sera of these rats caused ciliary dyskinesia of the rabbit trachea *in vitro*. Isoproterenol also induces hypertrophy of bronchial submucosal glands and hyperplasia of goblet cells in the rat.[57] More recently, a different approach has been developed by Martinez *et al.*[58,59] who, by administering reserpine to rats, produced morphological and histochemical changes in various exocrine tissues and secretory abnormalities of the submaxillary gland resembling those of cystic fibrosis. Saliva from the treated animals has cilioinhibitory properties. Another method in the search for an animal model for cystic fibrosis is the screening for mutants of mice, which in the past has revealed mutants resembling other human diseases. A mutant mouse with recessively inherited exocrine pancreatic insufficiency has been discovered.[60]

7. Commentary

The preceding descriptions of the pathophysiology and research for the basic defect of cystic fibrosis demonstrate a variety of manifestations of membrane dysfunction in exocrine glands and other tissues and cells of patients: the presence in biological fluids of humoral factors affecting transmembrane transport of various ions and organic macromolecules; abnormalities in the secretion of macromolecules; suggestive changes in the activities of the enzyme system involved in biomembrane function and in systems of cellular energetics; abnormalities in complex biosystems, such as the ciliary system, in which membrane function is involved. What is missing in the studies of all these manifestations is the identification of the common denominator, the basic biochemical defect of cystic fibrosis. The basic defect of this autosomal recessive inherited disorder should be in the area of one protein or enzyme system but it has remained elusive in the 49 years the disease has been known. It is possible that the observed dysfunctions of membranes in various organs and systems of patients represent secondary, tertiary, or even further removed expressions of the basic defect of this disease. The various findings in the skin fibroblast cultures suggest that whatever the basic biochemical defect may be, it is operative in cells other than those in exocrine glands, and it can express itself *in vitro*. Yet in the clinical situation, cystic fibrosis expresses itself in exocrine glands or in systems involving exocrine glands of patients. What is the reason for this predilection of cystic fibrosis gene expression in exocrine glands? The control of secretion in exocrine glands is either neural via the two branches of the autonomic nervous system, or hormonal via the polypeptide secretory system (secretin, cholecystokinin, etc.). Most of the stimuli for secretion are exerted on the contraluminal

or lateral sides of the secretory cells, while secretion takes place on the luminal side. The highly differentiated secretory epithelial cells of exocrine glands as well as other cells possessing secretory function have elaborate systems of intracellular translation of stimuli into secretory events. The "stimulus–secretion coupling" through a series of biochemical steps is the most pronounced characteristic of secretory cells, and it would not be surprising if the biochemical error of cystic fibrosis was found to be localized in this system. In addition, recent information from various laboratories[61] suggests that in exocrine glands, a system of coupling of secretion–ductal modification of fluid exists which presents some analogies to the regulation of tubular transport by the magnitude of the glomerular filtration in the kidney. For example, the author has shown that the duct of the human eccrine sweat gland perfused *in vitro* would not transport Na$^+$ when perfused with saline, but transport would be activated if the perfusate contained natural sweat.[16] This indicates that the primary secretory fluid of this gland contains factors regulating transductal reabsorption of Na$^+$. The nature of these factors is still unclear and is under investigation.

The multiplicity of expression of exocrine gland dysfunction in cystic fibrosis suggests that the presumed common denominator of the disease must be a factor involved in one or more steps of the stimulus–secretion coupling system. This system, which extends from the various receptors to the actual sites of ion and secretory macromolecule translocation across membranes resulting in secretion, has been under study in recent years.[61] Generation of cyclic nucleotide synthesis in secretory cells, movements of regulatory ions, activation of enzymes through appropriate systems of kinases, and regulation of energy production and of protein synthesis are but a few of the areas in which research is being conducted. One of the major difficulties in the studies of these parameters of exocrine function has been the lack of appropriate cellular systems for *in vitro* studies. As pointed out by the author,[62] use of tissue slices or pieces for *in vitro* studies of exocrine gland cellular physiology is handicapped by many problems. In recent years, investigators have been successful in their efforts to dissociate exocrine glands using enzymatic and mechanical forces, and thus isolate more or less pure preparations of secretory cells for *in vitro* studies of secretion.[62–66] In the author's laboratory, the methods developed for the isolation of rat parotid acinar cells[62–64] have also been used for the successful isolation of morphologically and functionally intact human parotid acinar cells for *in vitro* studies of secretion.[67–70] The parotids were obtained from cadavers of kidney donors and from surgical specimens obtained at neck dissections (control cells), or from patients with cystic fibrosis immediately after death (CF cells). The studies of these cells revealed marked compositional and functional differences which indicate future directions of research for the identification of the common denominator of exocrine gland dysfunction in cystic fibrosis.

As shown in Table II, the cells from patients with cystic fibrosis had many easily identifiable differences from control cells. Even if these differences represented changes occurring in the cell cultures, they still demonstrated marked differences in the compositional and functional behavior of the cells from patients. The limitations of these experiments stem from the fact that they were performed during conditions of short-term cultures, less than 24 hr from the time the parotid tissue was obtained. Currently, methods for the long-term preservation of these cells, 15–21 days, under conditions of reaggregated cell

Table II. Differences in Cellular Composition and Fuction between Isolated Parotid Acinar Cells from Patients and Those from Control Subjects[a]

	Observed differences in cystic fibrosis
Cell sodium content	Increased
Cell potassium content	Decreased
Cell total calcium	Increased
Rates of transmembrane Na⁺ transport (efflux)	Decreased
Responses to cholinergic receptor stimulation	Decreased
Responses to β-adrenergic receptor stimulation	Increased
$^{45}C^{2+}$ efflux	Decreased

[a]The cells were studied *in vitro* under identical conditions of cell culture.[67–70]

Table III. The Intracellular Cystolic Concentration of Ionized Calcium in Isolated Parotid Acinar Cells from the Rat and Human Parotids[a]

	Rat parotid cells	Human parotid cells
Cytosolic $[Ca^{2+}]$ (μM)	0.73 ± S.D. 0.21*	1.12 ± S.D. 0.19*
Number of measurements	82	41

[a]The cells were isolated and embedded in agar before being placed in a superfusion chamber. Single cells were micropunctured with Ca^{2+}-selective microelectrodes containing the neutral ligand N,N'-di(ethoxycarbonyl)undecyl-N,N'-4,5-tetramethyl-3,6-dioxaoctane diacid diamide. The recorded transmembrane potentials were translated into Ca^{2+} activities from which the cytosolic Ca^{2+} concentrations were calculated. The human cells were obtained from five surgical specimens of parotid tissue. S.D., standard deviation from the mean. *Statistically significant ($p < 0.001$).

cultures on collagen films are being developed in the author's laboratory. It is hoped that extended cultures of isolated secretory cells from patients with cystic fibrosis will permit further exploration of the *in vitro* manifestations of this disease.

Calcium is present in all exocrine gland products and is considered one of the major regulatory agents in the secretory process of exocrine glands.[71] Exocrine gland products contain higher concentrations of total calcium than the same products of exocrine glands of control subjects.[5] These observations along with the findings from the author's laboratory suggest that an abnormality of the metabolism and actions of calcium may exist in the exocrine glands of patients with cystic fibrosis. However, the elevation in total cell calcium observed by the author is difficult to interpret. Calcium inside the cells exists in different forms and in various intracellular compartments: bound to cytosolic or structural proteins; sequestered in intracellular organelles such as zymogen granules, mitochondria, and microsomes; and as free ionized calcium in the cell cytosol.[71]

The ionized cytosolic calcium appears to be the functionally active portion of cellular calcium and is in dynamic equilibrium with the interstitial fluid across the cell membrane as well as with the calcium inside the various organelles and across their diffusion-limiting membranes.[71] The active, ionized portion of intracellular calcium is a small fraction of the total cell calcium and it is most important in the intracellular mediation of autonomic receptor activation, transmembrane fluxes of monovalent ions, and activity of many enzyme systems. These observations have led to the need for studies of the metabolism and fluxes of calcium in isolated cell systems in the hope of defining what the abnormality in calcium metabolism is in the cells of patients with cystic fibrosis. In recent years, calcium-sensitive microelectrodes have been developed for the measurements of ionized calcium activity inside various cells.[72] The author has developed a system of superfusion of a small agar droplet containing a few isolated parotid acinar cells. Impalements of single cells with calcium-sensitive electrodes (Fig. 2) has permitted measurements of intracellular ionized calcium concentrations in rat and human parotid acinar cells as shown in Table III. It is the firm belief of the author that utilization of similar techniques for measurements of intracellular activities of key ions may help focus the search for the basic defect of cystic fibrosis. Because of

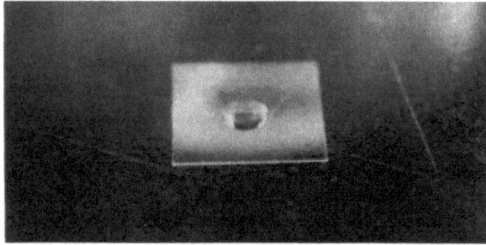

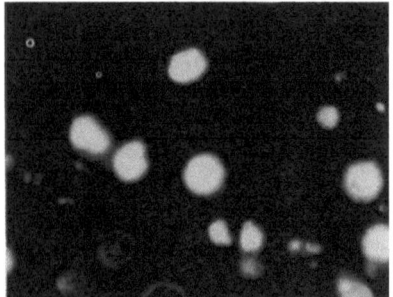
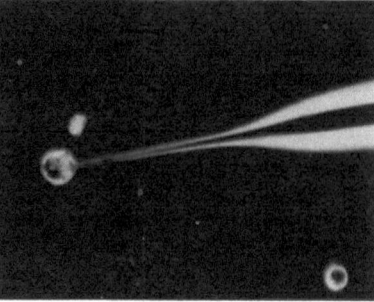

Fig. 2. Microphotographs demonstrating the method of direct measurements of Ca^{2+} concentration in the cytosol of parotid acinar cells. Upper: A coverslip with a drop of agar containing isolated parotid acinar cells. Lower left: A view of the cells in the agar droplet using high density of cells. Lower right: A Ca^{2+}-sensitive microelectrode with a beveled tip of approximately 1 μm has been inserted in a cell for direct measurements of the Ca^{2+} concentrations in the cytosol.

the pivotal role of calcium in the regulation of exocrine gland function, it may explain why the disease shows predilection for exocrine tissues and organs where Ca^{2+} ions are involved continuously in the regulation of the stimulus–secretion coupling and of other secretory mechanisms. This could then explain why no abnormalities, or minimal and ubiquitous abnormalities, are found in nonsecretory tissues and cells from patients. This hypothesis cannot explain why patients with cystic fibrosis have no generalized abnormalities of calcium metabolism,[3,4] unless specific factors act only intracellularly in secretory cells and, when released in body fluids, either become bound to larger molecules or inactivated. Further studies of membrane physiology of secretion and exocrine gland function appear to hold the key to the elucidation of the basic genetic defect and ultimate control of cystic fibrosis.

Research in cystic fibrosis has been characterized by controversies and often conflicting experimental results. The hallmark of the pathophysiology of the disease has been the increased salinity of the sweat which has been attributed to decreased reabsorption of Na^+ in the ducts of the sweat glands of patients. This has led to the search for and the demonstration of similar abnormalities in transmembrane Na^+ fluxes in other organs. In recent years, electrophysiological methodologies have been developed for the studies of ion fluxes in the mammalian airways.[39] When these studies were applied to the airways of patients with cystic fibrosis, increased bioelectric potentials with the lumen negative were detected and were interpreted as suggestive of increased reabsorption of Na^+ across the airway epithelium because of changes in Na^+ reabsorption upon exposure of the epithelium to amiloride.[40] This observation created a significant controversy because it was contrary to what had been believed to occur in this disease. More recently, it was observed that similar increases in the transductal bioelectric potential with the lumen negative were observed in the *in vitro* perfused ducts of sweat glands from patients with cystic fibrosis in the face of decreased reabsorption of NaCl.[17] These findings were interpreted to suggest decreased permeability of the ducts to Cl^- with secondary retention of Na^+ in the luminal fluid. In view of these data, it was suggested that it is possible that the observed bioelectric abnormalities in the airways of patients with cystic fibrosis may also be explained on the basis of decreased epithelial permeability to Cl^-. It is evident that the pathophysiological basis of this disease is far from being clarified and that even those issues that have been believed to be self-evident are open to questions in view of recent advances in our understanding of the physiology of biomembranes.

8. Summary

Cystic fibrosis, an apparent inborn error of metabolism, is a common disease of children and young adults of Caucasian origin. It has also been observed rarely among blacks and Orientals.

Cystic fibrosis is a generalized exocrinopathy which affects all exocrine glands of the body and causes abnormalities in the handling of ions and/or in the physicochemical behavior of mucous excretory products. Obstruction of ducts and passages with or without secondary infection causes respiratory, digestive, and other manifestations of this disease.

The inheritance of cystic fibrosis follows an autosomal recessive mode. The gene frequency in the U.S. population has been estimated to be 5%.

The eccrine sweat duct of patients with cystic fibrosis reabsorbs Na^+ and Cl^- in a defective way. The decreased reabsorption of these ions appears to be due to a nondialyzable Na^+ transport inhibitory factor which is present in the sweat. A similar factor is present in the saliva of patients. The serum of patients and heterozygotes contains a ciliotoxic factor.

The genetic defect is expressed in most tissues and cells of patients, but its *in vitro* pathogenetic effects are expressed mostly in exocrine glands.

Abnormalities of ion transport or of the stimulus–secretion coupling have been demonstrated in isolated parotid acinar cells from patients. It is speculated that a principal expression of the genetic defect is an abnormality of the transmembrane transport of Ca^{2+} in exocrine secretory cells. Such a defect may lead to a cascade of secondary changes in cell function resulting in the pathogenesis of this disease.

ACKNOWLEDGMENTS. The author's work has been supported by Grant AM-19200 from the Public Health Service and by grants from the Cystic Fibrosis Foundation and the University of Texas Health Science Center at San Antonio.

References

1. Fanconi, G., E. Uehlinger, and C. Kanuer. 1936. Das Coeliaksyndrom bei angeborener zystischer Pancreasfibromatose und Bronchiectasien. *Wien. Med. Wochenschr.* **86**:753–760.

2. Andersen, D. H. 1938. Cystic fibrosis of the pancreas and its relation to celiac disease. *Am. J. Dis. Child.* **56**:344–351.

3. Talamo, R. C., B. J. Rosenstein, and R. W. Berninger. 1983. Cystic fibrosis. In: *Metabolic Basis of Inherited Disease.* J. B. Stanbury, J. B. Wyngaarden, D. S. Fredrickson, J. L. Goldstein, and M. C. Brown, eds. McGraw-Hill, New York. pp. 1189–1197.

4. Lobeck, C. C. 1972. Cystic fibrosis. In: *Metabolic Basis of Inherited Disease.* J. B. Stanbury, J. B. Wyngaarden, and D. S. Fredrickson, eds. McGraw–Hill, New York. pp. 1605–1626.

5. Quinton, P. M., J. R. Martinez, and U. Hopfer, eds. 1982. *Fluid and Electrolyte Abnormalities in Exocrine Glands in Cystic Fibrosis.* San Francisco Press, San Francisco.

6. Davis, P. B., and P. A. di Sant'Agnese. 1980. A review: Cystic fibrosis at forty—Quo Vadis? *Pediatr. Res.* **14**:83–87.

7. di Sant'Agnese, P. A., R. C. Darling, G. A. Perera, and E. Shea. 1953. Abnormal electrolyte composition of sweat in cystic fibrosis of the pancreas. *Pediatrics* **12**:549–557.

8. Gibson, L. E., and R. E. Cooke. 1959. A test for concentration of electrolytes in cystic fibrosis of the pancreas utilizing pilocarpine by iontophoresis. *Pediatrics* **23**:545–552.

9. Schwartz, I. L., and J. H. Thaysen. 1956. Excretion of sodium and potassium in human sweat. *J. Clin. Invest.* **35**:114–123.

10. Lobeck, C. C., and D. Huebner. 1962. Effect of age, sex, and cystic fibrosis on the sodium and potassium content of human sweat. *Pediatrics* **30**:172–178.

11. Schulz, I., K. J. Ullrich, E. Fromter, H. Holzgrere, A. Frick, and J. Hegel. 1965. Micropunktion und elektrische Potentialmessung an Schweissdrusen des Menschen. *Pfluegers Arch.* **284**:360–372.

12. Mangos, J. A., and N. R. McSherry. 1967. Sodium transport: Inhibitory factor in sweat of patients with cystic fibrosis. *Science* **158**:135–136.

13. Emrich, J. M., E. Stoll, B. Friolet, J. P. Colombo, R. Richterich, and E. Rossi. 1968. Sweat composition in relation to rate of sweating in patients with cystic fibrosis of the pancreas. *Pediatr. Res.* **2**:464–476.

14. Schultz, I. J. 1969. Micropuncture studies of the sweat formation in cystic fibrosis patients. *J. Clin. Invest.* **48**:1470–1477.

15. Mangos, J. A. 1973. Microperfusion study of the sweat gland abnormality in cystic fibrosis. *Tex. Rep. Biol. Med.* **31**:651–663.

16. Mangos, J. A. 1973. Microperfusion study of the sweat gland abnormality in cystic fibrosis. *Tex. Rep. Biol. Med.* **31**:651–663.

17. Quinton, P. M. 1983. Chloride impermeability in cystic fibrosis. *Nature (London)* **301**:421–422.

18. Spicer, S. S., J. V. Briggman, and D. A. Baron. 1982. Morphological and cytochemical correlates of transport in sweat glands of normal and cystic fibrosis subjects. In: *Fluid and Electrolyte Abnormalities in Exocrine Glands in Cystic Fibrosis.* P. M. Quinton, J. R. Martinez, and U. Hopfer, eds. San Francisco Press, San Francisco. pp. 11–34.

19. Bartman, J., and B. H. Landing. 1966. Morphology of the sweat apparatus in cystic fibrosis. *Am. J. Clin. Pathol.* **45**:455–461.

20. Huebner, D., C. C. Lobeck, and N. R. McSherry. 1966. Density and secretory activity of sweat glands in patients with cystic fibrosis and in healthy controls. *Pediatrics* **38**:613–620.

21. Kaiser, D., E. Drack, and E. Rossi. 1970. Effect of cystic fibrosis sweat on sodium reabsorption by the normal sweat gland. *Lancet* **1**:1003–1004.

22. Hadorn, B., and A. A. Roscher. 1982. Exocrine pancreatic function in cystic fibrosis. In: *Fluid and Electrolyte Abnormalities in Exocrine Glands in Cystic Fibrosis.* P. M. Quinton, J. R. Martinez, and U. Hopfer, eds. San Francisco Press, San Francisco. pp. 182–192.

23. Crossley, J. R., R. B. Elliott, and P. A. Smith. 1979. Dried blood spot screening for cystic fibrosis in the newborn. *Lancet* **1**:472–474.

24. Wilcken, B., R. D. Brown, R. Urwin, and D. A. Brown. 1983. Cystic fibrosis screening by dried blood spot trypsin assay: Results in 75,000 newborn infants. *J. Pediatr.* **102**:383–387.

25. Figarella, C., J. F. Sauniere, L. Multigner, C. Galabert, M. Amourie, M. Filliat, H. Sarles, and L. P. Chazalette. 1981. Exocrine pancreatic function in cystic fibrosis patients without clinical pancreatic insufficiency with special reference to lactoferrin. *Proceedings of the 12th Annual Meeting of the European Working Group for Cystic Fibrosis,* Berne.

26. Martinez, J. R. 1982. Alterations in salivary gland structure and function in cystic fibrosis. In: *Fluid and Electrolyte Abnormalities in Exocrine Glands in Cystic Fibrosis.* P. M. Quinton, J. R. Martinez, and U. Hopfer, eds. San Francisco Press, San Francisco. pp. 125–142.

27. Gugler, E., J. C. Pallavicini, H. Swerdlow, and P. A. di Sant'Agnese. 1967. Role of calcium in submaxillary saliva of patients with cystic fibrosis. *J. Pediatr.* **71**:585–592.

28. Chernick, W. S., and G. J. Barbero. 1967. Reversal of submaxillary salivary alterations in cystic fibrosis by guanethidine. Fourth International Conference on Cystic Fibrosis of the Pancreas. *Mod. Probl. Paediatr.* **10**:125–132.

29. Wiesmann, U. N., T. F. Boat, and P. A. di Sant'Agnese. 1972. Flow rates and electrolytes in minor salivary gland saliva in normal subjects and patients with cystic fibrosis. *Lancet* **2**:510–512.

30. Mangos, J. A., N. R. McSherry, and P. J. Benke. 1967. A sodium transport inhibitory factor in the saliva of patients with cystic fibrosis of the pancreas. *Pediatr. Res.* **1**:436–442.

31. Mangos, J. A., and N. R. McSherry. 1968. Studies on the mechanism of inhibition of sodium transport in cystic fibrosis of the pancreas. *Pediatr. Res.* **2**:378–384.

32. Taylor, A., J. W. Mayo, T. F. Boat, and L. W. Matthews. 1974. Standardized assay for the sodium reabsorption inhibitory effect and studies of its salivary gland distribution in patients with cystic fibrosis. *Pediatr. Res.* **8**:861–865.

33. Nadel, J. A., B. Davis, and R. J. Phillips. 1979. Control of mucus secretion and ion transport in airways. *Am. Rev. Physiol.* **41**:369–391.

34. Sturgess, J. 1982. Morphological characteristics of the bronchial mucosa in cystic fibrosis. In: *Fluid and Electrolyte Abnormalities in Exocrine Glands in Cystic Fibrosis.* P. M. Quinton, J. R. Martinez, and U. Hopfer, eds. San Francisco Press, San Francisco. pp. 254–270.

35. Oppenheimer, E. 1981. Similarity of the tracheobronchial mucous glands and epithelium in infants with and without cystic fibrosis. *Hum. Pathol.* **12**:36–52.

36. Wood, R. E., A. Wanner, J. Hirsch, and P. M. Farrell. 1975. Tracheal mucociliary transport in patients with cystic fibrosis and its stimulation by terbutaline. *Am. Rev. Respir. Dis.* **111**:733–738.

37. Yeates, D. B., J. M. Sturgess, S. R. Kahn, H. Levison, and N. Aspin. 1976. Mucociliary transport in trachea of patients with cystic fibrosis. *Arch. Dis. Child.* **51**:28–34.

38. Wood, R. E., and G. J. Legris. 1982. The role of fluid and electrolyte transport abnormalities in the pulmonary pathophysiology of cystic fibrosis. In: *Fluid and Electrolyte Abnormalities in Exocrine Glands in Cystic Fibrosis.* P. M. Quinton, J. R. Martinez, and U. Hopfer, eds. San Francisco Press, San Francisco. pp. 289–298.

39. Boucher. R., C. J. Narvarte, C. Cotton, M. T. Stutts, M. R. Knowles, A. L. Finn, and J. T. Gatzy. 1982. Sodium absorption in mammalian airways. In: *Fluid and Electrolyte Abnormalities in Exocrine Glands in Cystic Fibrosis.* P. M. Quinton, J. R. Martinez, and U. Hopfer, eds. San Francisco Press, San Francisco. pp. 271–287.

40. Knowles, M. R., J. Gatzy, and R. Boucher. 1981. Increased bioelectric potential difference across respiratory epithelia in cystic fibrosis. *N. Engl. J. Med.* **305**:1483–1492.

41. Spock, A., H. M. C. Heich, and H. Cress. 1967. Abnormal serum factor in patients with cystic fibrosis of the pancreas. *Pediatr. Res.* **1**:173–177.

42. Bowman, R. H., L. H. Lockhart, and M. L. McCombs. 1969. Oyster ciliary inhibition by cystic fibrosis factor. *Science* **164**:325–326.

43. Schmoyer, I. R., and F. A. Baglia. 1974. Cystic fibrosis: Effect of media from cultured fibroblasts on ATPase activity. *Biochem. Biophys. Res. Commun.* **58**:1066–1070.

44. Farrell, P. M., G. N. Fox, and S. S. Spicer. 1976. Determination and characterization of ciliary ATPase in the presence of serum from cystic fibrosis patients. *Pediatr. Res.* **10**:127–133.

45. Doggett, R. G., and G. M. Harrison. 1973. Cystic fibrosis: *In vitro* reversal of the ciliostatic character of serum and salivary secretions by heparin. *Nature New Biol.* **243**:251–253.

46. Carson, S. D., and B. H. Bowman. 1982. Cystic fibrosis. I. Fractionation of the mucociliary inhibitor from plasma. *Pediatr. Res.* **16**:13–20.

47. McNeely, M. C., Y. C. Awasthi, D. R. Barrett, T. Iwasumi, L. Schneider, S. Srivastara, and B. H. Bowman. 1982. Cystic fibrosis. II. The urinary mucociliary inhibitor. *Pediatr. Res.* **16**:21–29.

48. Blitzer, M. G., and E. Shapiro. 1982. A purified serum glycopeptide from controls and cystic fibrosis patients. I. Comparison of their mucociliary activity on rabbit tracheal explants. *Pediatr. Res.* **16**:203–208.

49. Brown, G. A., A. Oshin, M. C. Goodchild, and C. M. Anderson. 1971. Inhibition of sugar transport by plasma from cystic fibrosis patients. *Lancet* **2**:639–640.

50. Morin, C. L., J. F. Desjeuz, and L. Authier. 1973. Effect of saliva and serum from patients with cystic fibrosis on intestinal uptake of amnio acids in rat. *Biomedicine* **19**:133–138.

51. Taussig, L. M., and J. D. Gardner. 1972. Effects of saliva and plasma from cystic fibrosis patients on membrane transport. *Lancet* **1**:1367–1369.

52. Benke, P. J., M. Erbstoeszer, and H. C. Pitot. 1972. Transport of labelled compounds in control and cystic fibrosis cells *in vitro*. *Lancet* **1**:182–184.

53. Araki, H., M. Field, and H. Shwachman. 1975. A new assay for cystic fibrosis factor: Effects of sera from patients with cystic fibrosis on the *in vitro* electrical properties of rat jejunum. *Pediatr. Res.* **9**:932–936.

54. Shapiro, B. L., S. M. Lee, and W. J. Warwick. 1970. The pentose phosphate pathway in cystic fibrosis erythrocytes. *Biochem. Biophys. Res. Commun.* **39**:816–819.

55. Bargman, G. J., J. E. Changus, M. Sukup, A. Vale, and H. Pitot. 1975. Effects of cystic fibrosis saliva on inorganic phosphorous esterification by rat liver mitochondria. *Pediatr. Res.* **9**:31.

56. Mangos, J. A., N. R. McSherry, P. J. Benke, and A. Spock. 1969. Studies on the pathogenesis of cystic fibrosis: The isoproterenol-treated rat as an experimental model. In: *Proceedings of the Fifth Cystic Fibrosis Conference*. D. Lawson, ed. Cystic Fibrosis Trust, London. pp. 25–34.

57. Sturgess, J., and L. Reid. 1973. The effect of isoprenaline and pilocarpine on (a) bronchial mucus-secreting tissue and (b) pancreas, salivary gland, heart, thymus, liver and spleen. *Br. J. Exp. Pathol.* **54**:388–402.

58. Martinez, J. R., E. Adelstein, D. Quissel, and G. J. Barbero. 1975. The chronically reserpinized rat as a possible model for cystic fibrosis. I. Submaxillary gland morphology and ultrastructure. *Pediatr. Res.* **9**:463–469.

59. Martinez, J. R., P. C. Adshead, D. Quissell, and G. J. Barbero. 1975. The chronically reserpinized rat as a possible model for cystic fibrosis. II. Composition and cilioinhibitory effects of submaxillary saliva. *Pediatr. Res.* **9**:470–476.

60. Pivetta, O. H., and E. L. Green. 1973. Exocrine pancreatic insufficiency: A new recessive mutation in mice. *J. Hered.* **64**:301–302.

61. Mangos, J. A., G. J. Bargman, J. R. Martinez, and O. M. Rennert. 1976. Physiology and pharmacology of secretion and cystic fibrosis. In: *Cystic Fibrosis: Projections into the Future*. J. A. Mangos and R. C. Talamo, eds. Stratton Intercontinental Medical Book Corp., New York. pp. 311–336.

62. Mangos, J. A., N. R. McSherry, and F. Butcher. 1975. Dispersed parotid acinar cells. I. Morphological and functional characterization. *Am. J. Physiol.* **229**:553–559.

63. Mangos, J. A., N. R. McSherry, and T. Barber. 1975. Dispersed parotid acinar cells. II. Characterization of adrenergic receptors. *Am. J. Physiol.* **229**:560–565.

64. Mangos, J. A., N. R. McSherry, and T. Barber. 1975. Dispersed rat parotid acinar cells. III. Characterization of cholinergic receptors. *Am. J. Physiol.* **229**:566–569.

65. Amsterdam, A., and J. D. Jamieson. 1972. Structural and functional characterization of isolated pancreatic exocrine cells. *Proc. Natl. Acad. Sci. USA* **69**:3028–3032.

66. Quissell, D. O., and R. S. Redman. 1979. Functional characteristics of dispersed rat submandibular cells. *Proc. Natl. Acad. Sci. USA* **76**:2789–2793.

67. Mangos, J. A. 1979. Morphological and functional characterization of isolated human parotid acinar cells *in vitro*. *J. Dent. Res.* **58**:2028–2035.

68. Mangos, J. A. 1980. Characterization of cholinergic and adrenergic receptors in isolated human parotid acinar cells. *J. Dent. Res.* **59**:156–163.

69. Mangos, J. A., and W. H. Donnelly. 1981. Isolated parotid acinar cells from patients with cystic fibrosis: Morphology and composition. *J. Dent. Res.* **60**:19–25.

70. Mangos, J. A. 1981. Isolated parotid acinar cells from patients with cystic fibrosis: Functional characterization. *J. Dent. Res.* **60**:797–804.

71. Rubin, R. P. 1982. *Calcium and Cellular Secretion*. Plenum Press, New York.

72. O'Doherty, J., J. F. Garcia-Diaz, and W. M. Armstrong. 1979. Sodium selective liquid ion-exchanger microelectrodes for intracellular measurements. *Science* **203**:1349–1351.

CHAPTER 10

Disorders of Glomerular Filtration

Roland C. Blantz and Juan C. Pelayo

1. Introduction

In the first edition of this volume, this chapter dealing with disorders of glomerular filtration described, by necessity, few specific mechanisms of altered glomerular ultrafiltration, because the stage of development of this field of investigation was in a rather early phase of data accumulation. Since that time, there has been a considerable acceleration in the rate of accumulation of knowledge in this area. Therefore, in this second edition, we will diminish our prior emphasis on methodology and discussion of the modeling predictions of the relative impacts of the four individual determinants of glomerular ultrafiltration: (1) nephron plasma flow (RPF); (2) systemic oncotic pressure (π_A); (3) the glomerular capillary hydrostatic pressure gradient (ΔP); and (4) the glomerular ultrafiltration coefficient (LpA). We refer the reader to the first edition of this text for this more methodologic and theoretical discussion. We will now focus upon what is currently known regarding the specific mechanisms contributing to the regulation of glomerular ultrafiltration in a variety of physiologic and pathophysiologic conditions. Because of the increasing complexity of the regulatory processes, the descriptions of these mechanisms will be classified by physiologic conditions rather than focusing specifically on the effects of changes in the respective determinants of glomerular ultrafiltration.

2. Some General Truths

The glomerulus is characterized by a high-resistance afferent arteriole, a somewhat lower-resistance efferent arteriole, and the glomerular capillary, a series of parallel conduits of large potential ultrafiltering surface area but very low net vascular resistance.[2] The glomerular capillary membrane is constituted by an inner, fenestrated endothelial cell layer, the glomerular basement membrane, and the epithelial cell layer on the outer aspect of the capillary. Bowman's space is a rather specialized interstitial space because of its low colloid osmotic pressure and lack of the "tight" pressure–volume relationship common to most other interstitial spaces adjacent to capillaries. The latter characteristics contribute to the high rates of ultrafiltration observed across the glomerular capillary into Bowman's space. The very high value for the glomerular ultrafiltration coefficient, a product of glomerular ultrafiltering surface area (A) and the hydraulic conductivity of the glomerular capillary membrane (Lp), is also a major factor leading to the high rate of glomerular ultrafiltration.

As a consequence of the physical characteristics of the glomerulus, the relatively high surface area membrane–hydraulic conductivity, and the fact that the ultrafiltrate is nearly protein free (~ 3 mg albumin/100 ml), the protein and red blood cells in glomerular blood flow are significantly concentrated along the length of the glomerular capillary from afferent to efferent arteriole. A factor which may significantly determine the high partition of protein, particularly albumin, across the glomerular capillary membrane, is the anionic characteristic or lower isoelectric point of albumin, the major oncotically active serum protein, and the electronegative surface charge of most of the structures which constitute the glomerular capillary membrane.[3] As previously noted, it is interesting from an evolutionary standpoint that mammals have developed a filtration membrane with a polyanionic character and a major oncotically active plasma protein with a negative surface charge.[1] Albumin would then seem particularly suitable in maintaining a high percentage of protein within the plasma volume, thereby favoring a high partition of extracellular fluid in the plasma relative to interstitial spaces.

The rise in protein concentration which occurs along the length of the glomerular capillary as a consequence of glomerular ultrafiltration is sufficiently large at the end of the glomerular capillary that the oncotic pressure generated by this protein concentration is essentially equal to the measured hydrostatic pressure difference between glomerular capillary and Bowman's space (ΔP) in the rat, a condition which has been described as filtration pressure equilibrium.[4–6] The directly

Roland C. Blantz and Juan C. Pelayo • Department of Medicine, University of California, San Diego, School of Medicine, La Jolla, California 92093, and Veterans Administration Medical Center, San Diego, California 92161.

measured ΔP in the hydropenic rat is approximately 35 mm Hg,[4,7] a value which is typical for the oncotic pressure at the efferent arteriole, as assessed either directly from the protein concentration in surface efferent peritubular capillary blood samples or predicted from values for systemic protein concentration and the filtration fraction typical for the hydropenic rat.

Studies in the dog have suggested that filtration pressure equilibrium may not be attained in this species, primarily as a consequence of lower values for LpA in relation to the normal values for RPF in this species.[8] However, this lack of filtration pressure equilibrium may not be universal within the species since Israelit et al., in earlier studies, demonstrated filtration pressure equilibrium in micropuncture studies in the dog.[9]

There is no direct evidence for or against the presence of filtration pressure equilibrium in the human. However, there is a definite trend for the filtration fraction to vary among species and this appears to correlate more or less inversely with the magnitude of systemic protein concentration and oncotic pressure within each species. Since the systemic oncotic pressure would influence the filtration fraction most strongly at filtration pressure equilibrium, this general inverse relation might suggest that filtration pressure equilibrium might obtain in more mammalian species in the unanesthetized state than some of the current literature would suggest

3. Some General Observations on the Mechanism of Change in GFR in the Physiologic Setting

Since the glomerulus is an ultrafilter driven by hydrostatic pressure, meaning that the positive force producing glomerular ultrafiltration is the difference in hydrostatic pressure between glomerular capillary and Bowman's space, it would seem reasonable that changes in this hydrostatic force should primarily influence changes in the rate of glomerular filtration. The older literature would suggest that changes in glomerular capillary hydrostatic pressure, via a balance in changes in afferent and efferent arteriolar resistances, and, to a lesser extent, alterations in tubular pressure, must mediate the physiologic regulation of GFR. The largest body of direct evidence pertinent to normal mechanisms mediating the regulation of glomerular ultrafiltration has been derived from studies in the Munich–Wistar rat, a strain with surface glomeruli which permit direct assessment of all the pertinent pressures and flows affecting ultrafiltration. Conclusions derived from these data are based upon the normal condition of filtration pressure equilibrium which has been observed for this strain. Based upon the large quantity of physiologic data derived from this experimental animal, one must conclude that the rate of nephron filtration is regulated primarily by variation in the forces that oppose glomerular ultrafiltration (variations in the profile of oncotic pressure along the capillary and the glomerular ultrafiltration coefficient) rather than alterations in the hydrostatic pressures.[10,11] The above conclusion requires some further explanation. First, studies have demonstrated that in spite of modest induced changes in glomerular capillary hydrostatic pressure (P_G), ΔP remains remarkably constant (due to a tendency for Bowman's space pressure to change in parallel with P_G). Changes in ΔP in the rat are rarely the major determinant regulating alterations in nephron filtration rate. Since the oncotic pressure rises rapidly along the length of the glomerular capillary in the normally hydrated rat, increases in the rate of nephron plasma flow will, of necessity, reduce the rate of rise in oncotic pressure and decrease the mean, integrated oncotic pressure opposing the glomerular ultrafiltration and

increase the consequent mean effective filtration pressure ($\overline{\Delta P} - \overline{\pi}$).

3.1. Effects of Nephron Plasma Flow

For the reasons described above, one would predict that changes in the rate of nephron plasma flow should be the dominant mode of altering nephron filtration rate. Data derived in studies in the Munich–Wistar rat in a variety of physiologic states support this prediction of a highly flow-dependent process of glomerular filtration.[10,11] The degree of flow dependence of GFR decreases significantly if filtration pressure equilibrium is not attained either spontaneously as has been suggested in certain species, or at very high rates of nephron plasma flow where filtration pressure equilibrium can no longer be maintained. In the Munich–Wistar rat, increases in nephron plasma flow to values 150–200% of the hydropenic nephron plasma flow produce a disequilibrium of the effective filtration pressure, such that ΔP significantly exceeds the oncotic pressure at the efferent end of the glomerular capillary.

3.2. π_A and ΔP

Changes in π_A and ΔP do affect the final nephron filtration rate and do so by altering the filtration fraction. In general, increases in ΔP and reductions in π_A (through dilution of serum protein concentration) will increase the filtration fraction and the converse is also the case. At filtration pressure equilibrium, increases in ΔP raise the value, π_E, to which oncotic pressure can rise, along the length of the capillary, and decreases in ΔP reduce this limiting value of π_E for any input systemic oncotic pressure, π_A. At any value for ΔP, reductions in π_A at filtration pressure equilibrium will result in increases in filtration fraction and single nephron GFR (SNGFR) since, at filtration pressure equilibrium, the limiting value to which this lower value for π_A can rise (π_E) is still defined by the ΔP.

3.3. Effects of LpA

While filtration equilibrium persists, changes in LpA will exert no effect upon SNGFR since there will be reciprocal changes in the mean effective filtration pressure ($\overline{EFP} = \overline{\Delta P} - \overline{\pi}$). The profile of effective filtration pressure along the length of the glomerular capillary will be altered because as LpA varies, the rate of ultrafiltration will change and, consequently, the rate of rise in oncotic pressure along the capillary length (Fig. 1).[12] Specific data derived from the Munich–Wistar rat demonstrate that if filtration equilibrium occurs, as in the hydropenic condition, large reductions (~ 50%) in LpA are required before SNGFR is decreased by this mechanism. The reduction in LpA must be sufficient to result in disequilibrium of the effective filtration pressure before the decrease in LpA can contribute to a reduction in SNGFR. If filtration pressure equilibrium occurs, there are significant constraints upon the degree to which alterations in LpA will be expressed as changes in SNGFR.[12]

4. Potential Interactions among the Determinants of Glomerular Ultrafiltration

In earlier studies on glomerular ultrafiltration during the past decade, the influences of the respective determinants of

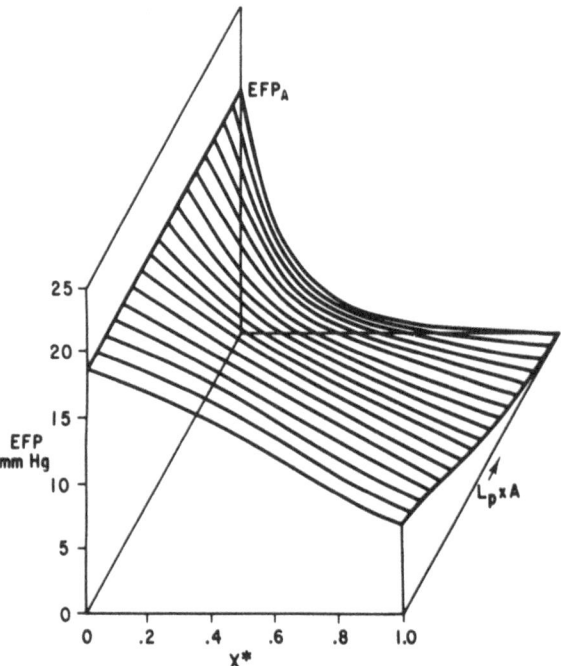

Fig. 1. A three-dimensional graph depicting the effect of changing values of the glomerular permeability coefficient (LpA) upon the profile of effective filtration pressure (EFP) along the glomerular capillary length (X*). At higher values of LpA, filtration pressure equilibrium results and, as long as equilibrium obtains, reductions in LpA are associated with reciprocal increases in the mean effective filtration pressure ($\overline{EFP}$). Only after EFP no longer achieves equilibrium ($\overline{EFP}_E > 0$) will reductions in LpA result in a reduced LpA · $\overline{EFP}$ product or filtration rate (SNGFR = $\overline{EFP}$ · L$_p$A).

glomerular ultrafiltration were assessed. As stated, studies in the Munich–Wistar rat provided the conclusions that nephron filtration rate is highly flow dependent and that major alterations in LpA can contribute to reductions in SNGFR. However, it has become evident that the determinants of glomerular ultrafiltration (RPF, ΔP, π_A, and LpA) are not wholly independent variables. Several studies have demonstrated a direct correlation of the absolute value for π_A and LpA.[5,12–15] More recent studies from our laboratory have shown that this direct relationship of π_A and LpA is not mediated by an indirect influence of π_A-induced alterations in ΔP, RPF, or systemic hematocrit, issues which were in question prior to this study.[15] The exact explanation for such a relationship has not been provided, but possibilities include: (1) concentration polarization of proteins within the glomerular capillary; and (2) an effect of glomerular oncotic pressure on endothelial cell volume and fenestra diameter. We have also demonstrated an inverse relation of ΔP and LpA. No specific mechanism for this association has been provided although it is tempting to speculate that ΔP might produce physical alterations in the glomerular endothelial surface. There is no significant association and interdependence of RPF and LpA, although one might predict that a higher rate of RPF might be associated with increases in functional glomerular capillary surface area for ultrafiltration.

From a teleologic standpoint, the direct relationship of π_A and LpA might limit the effect of decreased protein concentration upon the changes in GFR one would predict from an acute or chronic reduction in systemic oncotic pressure (e.g., after the infusion of large quantities of isotonic saline). Although no acute relation of changes in RPF and LpA have been documented, growth in the rat is associated with more or less parallel increases in RPF and LpA, probably on a purely structural basis.[16] The net effect of such a relationship in the Munich–Wistar rat is to maintain filtration pressure equilibrium in the hydropenic state in spite of progressive reductions in afferent and efferent arteriolar structures.

5. The Influence of Humoral and Hormonal Substances upon Glomerular Ultrafiltration

5.1. Initial Studies Employing Parenteral Infusions

Early studies examined the effects of a variety of hormonal and humoral substances by infusing the latter into the vasculature via the vein or the artery. Obviously, there were limitations inherent to the interpretation of data derived from such studies. First, the studies require that reasonable estimates of dose of the substances be made in advance such that physiologic concentrations will be achieved in both plasma and interstitium. Second, such studies require that the substances gain access to the respective effector or responding cell from the vasculature—which may not be the normal route whereby the biologically active substance gains access to its respective receptors. The early studies utilized such hormones and humoral substances as angiotensin II (AII),[17,18] adrenergic agonists,[17] prostaglandins,[19] histamine,[20] acetylcholine, and bradykinin.[19,21] Most of these agents do circulate in plasma for at least brief periods and exert an effect from the blood; however, most are also generated locally, often outside the vasculature and in specific concentrations that are difficult to assess accurately. Systemic infusions of AII and norepinephrine undoubtedly result in a much larger systemic or extrarenal effect relative to the local renal influence than would local release of these substances within the kidney. Therefore, these initial studies may not duplicate the physiologic reality, but results do demonstrate the physiologic potential of these agents. In addition, a totally independent effect is not proven and the quantitative importance of the physiologic effects cannot be accurately assessed.

5.2. Historical View

With the above limitations in mind, we will describe the results of these earlier studies for the purposes of a historical view as much as physiologic relevance. Descriptions will be limited to those studies in which single-nephron analyses were conducted and direct assessments of flows, pressures, individual segmental resistances, and LpA were obtained.

Parenteral infusion studies utilizing catecholamines, AII,[17,18] and antidiuretic hormone (ADH)[22] have provided evidence for the site of action of these agents [afferent (AR) versus efferent arteriolar resistance (ER)] and effects on LpA which have proven somewhat at odds with later studies utilizing different methods of evaluation such as antagonists of hormonal activity. As we will demonstrate, these differences may be directly related to the mode of administration or generation of the hormone and reflect differences in the expression of hormonal effects between substances circulating in the plasma and substances generated locally or at adrenergic neural junctions.

6. Angiotensin II

Myers et al.[17] were the first to evaluate the influences of exogenous AII on glomerular hemodynamics in the hydropenic Munich–Wistar rat. The infusion of pressor doses of synthetic AII increased ER to a larger extent than AR, the net effect of these alterations being increased ΔP (the consequence of constant P_t and elevated P_G) and decreased RPF. Attainment of filtration pressure equilibrium (EFP $\cong$ 0) in both control and experimental conditions precluded an assessment of potential effects of AII on the glomerular ultrafiltration coefficient (LpA or K_f). When the rise in renal perfusion pressure was prevented (by aortic constriction) in these studies, an increase in pre- and postglomerular vascular resistance was still observed; however, AR rose only by 25% while ER increased with AII infusion by more than 100%. Blantz et al.[18] examined the effects of both native and synthetic AII in plasma volume-expanded Munich–Wistar rats. Using pressor quantities of native AII, systemic blood pressure rose approximately 20 mm Hg, AR doubled, but ER increased to a somewhat greater degree. As a consequence of these alterations, P_G rose by about 12%, while RPF fell to about 50% of control values. Two important differences in conclusions were derived from these studies. First, the changes in AR were not merely the result of autoregulation of nephron blood flow in response to the increase in systemic blood pressure, because similar doses of synthetic AII ([Asn^1,Val^5]-AII) produced nearly identical changes in glomerular capillary hydrostatic pressure, nephron blood flow, AR, and ER, but had no effect on systemic blood pressure and peripheral resistances. However, there remains some question about the major site(s) of action of AII on renal vascular resistances. This study also demonstrated that AII decreased LpA, suggesting an effector cell responsive to AII other than vascular smooth muscle. Intrarenal infusion studies support the view that AII acts primarily, if not wholly, at the efferent arteriole and within the glomerular capillaries. Of interest, an assessment of the responses of isolated afferent and efferent renal microvessels to AII concluded that AII has a direct effect only on the efferent arteriole.[23] Employing micropuncture, Steiner and Blantz[24] have examined the efficacy of maximal doses of saralasin in reversing the renal hemodynamic effects of parenterally infused AII. In spite of the acute nature of this evaluation, not all the effects of parenteral AII infusion were neutralized. Saralasin infusion restored both the increases in ER and the decreases in LpA to the normal control values; however, the AII-induced increase in AR was not completely reversed. These studies suggest that the influence of saralasin may be selective and nonuniform, but no specific reason for the inability to restore AR to normal has been provided. On the other hand, the results of these investigations and the fact that autoregulation of nephron blood flow failed to fully explain the observed alteration in AR, in studies in which AII was infused systemically, may indicate that AII effects on AR are indirectly mediated by other vasoconstrictor systems (i.e., the adrenergic nervous system) when this agent is parenterally administered. Later studies by Baylis and Brenner[25] and Ichikawa et al.[26] confirmed the findings of Blantz et al.[18] in that AII, infused either intravenously or into the renal artery, produced a significant decrease in LpA or K_f.

However, the specific mechanism producing this AII effect on LpA was not immediately apparent in these initial studies. The studies of Hornych et al.,[27] Sraer et al.,[28,29] and Ausiello et al.[30] have suggested some insights into possible mechanisms. These earlier studies suggested that when AII was infused parenterally, the glomerulus, as evaluated by scanning electron microscopy, underwent significant changes which involved contraction, probably of the mesangial cell in the glomerular stalk, and narrowing of some of the capillary conduits on the surface of the glomerulus. Sraer et al.[28] and, later, others[31–34] have demonstrated AII-specific receptors within the glomerulus and specifically within the mesangial cell. Ausiello et al.[30] and Mahieu et al.[35] demonstated that cultured mesangial cells contract on exposure to AII and also ADH. In addition, specific responses to AII, i.e., contraction, have been identified in isolated whole glomeruli.[27] Thus, considerable experimental evidence has been accumulated to suggest that the LpA-lowering effect of AII may be the consequence of reductions in total glomerular capillary filtering surface area (A) perhaps by virtue of the action of AII to promote glomerular contraction. The mesangial cell may be the responsive contractile element because of its rich content of intracellular contractile filaments. These filaments have been shown to contain actin and myosin proteins. Of related interest, the contractile response of smooth muscle cells to AII is known to be dependent upon transmembrane Ca^{2+} transport.[36,37] Ichikawa et al.[26] conducted studies to determine whether the effects of AII on glomerular hemodynamics are sensitive to changes in Ca^{2+} transport. When verapamil or manganese, antagonists of Ca^{2+} entry into smooth muscle cells, is infused, the action of AII on glomerular hemodynamics is eliminated.[26] Specifically, all of the effects of AII upon LpA and glomerular vascular resistances were totally prevented or reversed. In a sense, this effect of blockers of Ca^{2+} entry may be nonspecific but supports the concept that the mesangial cell, an adapted smooth muscle cell, may be the critical cell mediating this reduction in LpA.

Figure 2 depicts graphically some of the reasonable alternative mechanisms whereby AII could exert this effect upon LpA. First, AII could decrease surface area for glomerular ultrafiltration via mesangial cell contraction and reduction in the radius of each of several capillary conduits. Although this seems the most straightforward explanation in keeping with all observations on this issue, if this were the sole mechanism, capillary vascular resistance should increase and the reduction in P_G along each capillary conduit should also be increased. Measurements of P_G during AII infusion revealed no increase in the heterogeneity of P_G, the range of values being similar to that observed in noninfused control rats (2–3 mm Hg).[18] One would predict that the range of P_G values along the glomerular capillary should have increased to approximately 18 mm Hg, if AII exerted its effect on LpA solely through a reduction in the capillary radius and surface area. Second, Zimmerhackl et al.[38] observed no changes in the diameter of glomerular capillaries by video scanning techniques after AII.

7. The Adrenergic Nervous System

Morphologic studies have conclusively demonstrated that the kidney is a highly innervated organ, with adrenergic fibers richly supplying renal vasculature, extraglomerular mesangium, and renal tubules[39,40]; also radioligand binding receptor studies have characterized both α- and β-adrenergic receptors in renal tissue membranes[41–43] and isolated glomeruli.[44,45] The physiologic basis for a role of the renal adrenergic nervous system in the regulation of glomerular ultrafiltration has been substantiated by a variety of experimental approaches, such as renal nerve stimulation, adrenergic agonist infusion, and renal dener-

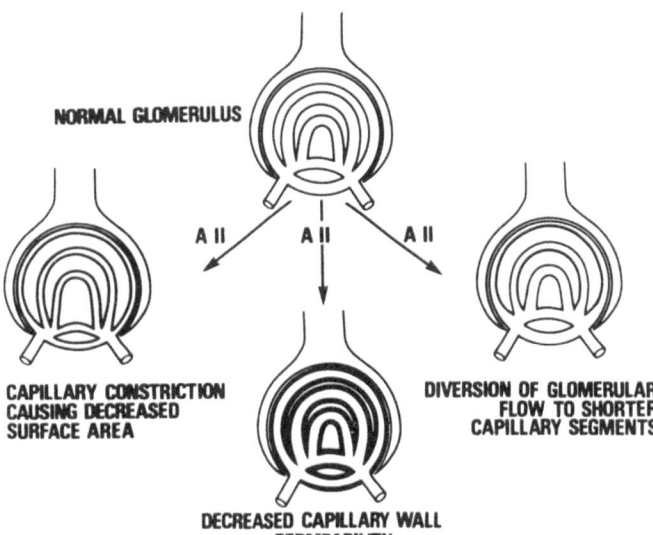

NORMAL GLOMERULUS

A II   A II   A II

CAPILLARY CONSTRICTION
CAUSING DECREASED
SURFACE AREA

DIVERSION OF GLOMERULAR
FLOW TO SHORTER
CAPILLARY SEGMENTS

DECREASED CAPILLARY WALL
PERMEABILITY

Fig. 2. Three potential mechanisms whereby angiotensin II (AII) could decrease the glomerular ultrafiltration coefficient (LpA). A schematized normal glomerulus is shown at the top of the figure. AII could decrease LpA (left to right) by: (1) capillary constriction causing decreased surface area; (2) decreased capillary wall hydraulic conductivity to water; or (3) by diversion of glomerular flow to shorter capillary segments (or a combination of these events). Reprinted with permission of Springer-Verlag.[22]

vation. Myers et al.[17] demonstrated that norepinephrine, a potent vasoconstrictor and a naturally occurring neurotransmitter, produces glomerular hemodynamic alterations similar to those observed with AII when administered parenterally in pressor doses to hydropenic Munich–Wistar rats. However, results differed from AII in that with norepinephrine all of the observed increase in AR could be accounted for by an autoregulatory response to the increase in systemic blood pressure. However, Andreucci et al.,[46] utilizing similar parenteral infusions of norepinephrine, concluded that the afferent arteriole was the major site of action for this catecholamine. The attainment of filtration pressure equilibrium in these studies prevented any conclusion regarding potential effects of norepinephrine on LpA or K_f.

The occurrence of multiple types of adrenergic receptors in the kidney suggests that the net response to increased adrenergic activity at the glomerulus may represent either α- or β-adrenergic receptor stimulation or a combination thereof, since norepinephrine binds to these different types of receptors.[47] In this regard, the use of specific adrenergic agonists and antagonists has identified, at the whole kidney level, opposing influences on renal hemodynamics for these receptors (reviewed in Ref. 48). Pelayo et al.[49] have carried out studies in plasma volume-expanded Munich–Wistar rats to further characterize the individual contributions of both α and β components of the adrenergic nervous system to the control of glomerular ultrafiltration, through potential alterations in LpA as well as the other determinants of GFR. Infusion of methoxamine, an α_1-adrenergic agonist, significantly increased mean arterial pressure by approximately 20%; however, SNGFR was maintained constant by an offsetting increment in AR and unchanged ER, events that led to near constancy of nephron plasma flow and glomerular capillary hydrostatic pressure. Of importance, methoxamine increased the LpA or K_f to values higher than control. Conversely, isoproterenol (a $\beta_{1,2}$-adrenergic agonist) caused a significant reduction in LpA, but unchanged RPF. The potentially negative effect of a reduced LpA on SNGFR was overcome by a proportional increase in ΔP. This augmentation in ΔP was primarily the consequence of a significant reduction in the Bowman's space hydrostatic pressure. It has long been recognized that renin re-

lease is partially under adrenergic control, and much is now known concerning the role of renal α- and β-adrenergic receptors in that regulation (reviewed in Ref. 50). While stimulation of the β-adrenergic receptors increases renin release both in vivo and in vitro, stimulation of the renal α-receptors, by most accounts, seems to inhibit renin release again in vivo and in vitro (reviewed in Refs. 51, 52). The effect of isoproterenol in decreasing LpA appears to require the presence of intrarenal AII activity, since either [Sar¹,Ala⁸]-AII or MK 421 (an angiotensin-converting enzyme inhibitor) infusion eliminated the effects of isoproterenol on LpA and ΔP.[49] The observation that AII inhibition in rats undergoing plasma volume expansion did not yield supranormal LpA values suggests that methoxamine may exert an independent effect in increasing LpA.[49] However, these studies did not permit conclusions as to the respective contributions to changes in surface area (A) and changes in hydraulic conductance (Lp) to the observed LpA alterations during administration of these adrenergic agonists.

The mechanisms by which direct electrical renal nerve stimulation affects glomerular ultrafiltration have been assessed by Hermansson et al.[53]; 2- to 5-Hz stimulation produces vasoconstriction at both afferent and efferent vessels, thereby decreasing nephron plasma flow and, according to a preliminary study by Kon et al.,[54] a reduction in the glomerular ultrafiltration coefficient. Pelayo and Blantz[55] have conducted a direct analysis of glomerular ultrafiltration and its determinants in euvolemic Munich–Wistar rats before and during 3-Hz stimulation. This study indicated that the decrement in SNGFR was the result of decreased ΔP and RPF, which were the consequence of increased afferent (↑↑ AR) and efferent (↑ ER) glomerular vascular resistances, since LpA remained unaffected. Of interest, in the same studies, the effects of renal nerve stimulation during AII inhibition, although qualitatively similar, were of much lower magnitude than those observed during stimulation alone. The aforementioned studies indicate that the renal adrenergic nervous system can regulate SNGFR. The balance of potential effects on RPF, ΔP, and LpA seems to depend, in major part, upon the intensity of renal nerve stimulation with vasoconstriction alone at lower frequencies and alterations in ΔP and LpA at higher frequencies. These physiologic effects of increased adrenergic activity appear to be the partial result of activation of intrarenal AII systems.

Pelayo *et al.* have also conducted micropuncture studies in order to characterize the functional consequences of acute elimination of the renal adrenergic nervous system.[56] Studies in the euvolemic Munich–Wistar rat revealed that acute elimination of renal innervation produces: (1) constancy of SNGFR and no alterations in any of the individual determinants of glomerular ultrafiltration and (2) a reduction in APR in the absence of any alterations in directly measured peritubular oncotic and hydrostatic pressure and renal interstitial pressure.[56] However, acute renal denervation in the hydropenic or somewhat volume-deficient rat yielded somewhat different results as a consequence of the presumably greater renal adrenergic activity in this condition.[35] Although SNGFR was not altered from control, innervated rats, P_G (hence ΔP) was increased and LpA was significantly decreased. The same investigation examined the effects of blockers of AII activity and converting enzyme inhibitors in the acutely denervated rat, because reductions in LpA (and increases in ΔP) have often been associated with increased intrarenal AII activity. The interpretation of these events is that after removal of adrenergic innervation in the hydropenic rat, intrarenal AII activity is increased, possibly via activation of the tubuloglomerular feedback system, which acts to partially compensate for the removal of adrenergic activity, but at the expense of a reduction in LpA. These studies do point out certain aspects of the potentially complex interrelationship between renal adrenergic influences and the renal effects of AII.

8. ADH, cAMP

A consensus of experimental evidence indicates that ADH exhibits somewhat divergent physiologic influences: (1) to increase osmotic water (hydraulic) permeability of the collecting duct and (in some species) a portion of the distal tubule as well,[57] and (2) to produce vasoconstriction of the systemic vasculature.[58] When the effect of ADH on glomerular hemodynamics was investigated in Munich–Wistar rats undergoing chronic water diuresis, a reduction in LpA or K_f and an increase in ΔP were observed.[22] The increment in ΔP associated with antidiuresis was due to consistent reductions in Bowman's space hydrostatic pressure rather than to increases in glomerular capillary hydrostatic pressure, the former a result of the fall in urine flow rate. These opposing influences ($\downarrow$ LpA and $\uparrow \Delta P$) effectively neutralized each other leading to near constancy of SNGFR and whole kidney GFR. Of importance is the observation that not only AII but ADH as well, were found to produce contraction of rat glomerular mesangial cells in tissue culture.[30,35] Based on the foregoing, it has been proposed that ADH may be a potentially important physiologic regulator of mesangial cell contraction and, therefore, glomerular capillary surface area and LpA or K_f. The sequence of steps that ADH initiates (that eventually increase the permeability to water in renal tubules) is one of many examples of the AMP-mediated action of hormones on receptor cells. However, the effects of ADH on cAMP generation observed in isolated glomeruli have been inconsistent,[59–61] but, according to one report, ADH did not induce an increase in mesangial cyclic nucleotide formation, at least at the concentration that stimulates contraction.[30] The aforementioned finding and the observation that the action of ADH on LpA or K_f is independent of intermediary role for AII

suggest that the effect of ADH on glomerular capillaries may not be via augmented intraglomerular synthesis of cAMP. Supporting this view is the fact that various vasoactive humoral substances capable of enhancing glomerular cAMP formation affect LpA or K_f through an intermediary action of AII.[62] An actual physiologic role for ADH in the control of glomerular ultrafiltration has not been demonstrated unambiguously.

Investigations have provided evidence that dibutyryl cAMP is capable of influencing the process of glomerular ultrafiltration, in large part, by reducing the glomerular ultrafiltration coefficient (LpA or K_f).[22] Administration of dibutyryl cAMP induced a significant decline in SNGFR. This finding was associated with increases in ΔP, ER, and total arteriolar resistance and decreases in RPF and LpA or K_f. The observed alterations in the determinants of SNGFR elicited by infusion of dibutyryl cAMP are typical of changes seen with increased AII activity. During infusion of saralasin, a competitive antagonist of AII, the effects of dibutyryl cAMP on glomerular hemodynamics were effectively reversed.[62] It was concluded, therefore, that the action of dibutyryl cAMP on the glomerular microcirculation may involve an intermediate effect of AII. However, it remains uncertain whether dibutyryl cAMP, administered exogeneously, mimics precisely the actions of endogenously generated cAMP on glomerular hemodynamics in response to local hormonal activation.

9. Prostaglandins, Thromboxanes, Kinins

Renal prostaglandins and thromboxanes are synthesized in both the cortex and the medulla[63,64] (reviewed in Refs. 65–67). The main sites of synthesis are: (1) glomeruli (rat glomerular mesangial and epithelial cells) and blood vessels in the cortex and (2) interstitial cells and collecting ducts in the medulla. Current evidence points to prostaglandins and thromboxanes as local vasoactive hormones capable of influencing: (1) renal hemodynamics,[65,69] (2) salt and water excretion,[68] and (3) stimulating renin release.[70] In most species, PGE_2, PGI_2, and PGD_2 are vasodilators, whereas endoperoxides and TXA_2 are vasoconstrictors.[69] Early studies had shown that infusions of potent vasodilators such as prostaglandins have little effect on whole kidney GFR or SNGFR, despite large measured increases in renal blood flow.[71–73] Subsequently, studies by Baylis *et al.*[19] defined the actions of mildly vasodepressor doses of PGE_1 on SNGFR and its determinants in normal hydropenic Munich–Wistar rats. PGE_1 infusion resulted in significant increases in RPF due to decreases in AR and ER, the reduction in AR being proportionately greater; however, SNGFR did not increase. Since ΔP or π_A remained essentially unchanged, the explanation for the constancy of SNGFR was that the increase in RPF was completely neutralized by a significant reduction in LpA or K_f. Schlondorff *et al.*[61] provided evidence for a direct stimulatory action of prostaglandins I_2, E_1, E_2, and A_2 on adenylate cyclase in isolated glomerular preparations of rat kidneys. This finding lends support to the possibility that prostaglandins could influence glomerular hemodynamics, perhaps via cAMP, through changes in smooth muscle tension and renin release.

Schor *et al.*[62] studied the role of the renin–angiotensin system in the vasoconstrictor effect induced by prostaglandins (specifically, PGI_2 and PGE_2) on the renal microcirculation of euvolemic Munich–Wistar rats. Administration of subvasodepressor doses of PGE_2 and PGI_2 to rats with suppressed endogenous prostaglandin generation elicited a significant reduction of total kidney GFR and SNGFR primarily due to reductions in RPF and LpA or K_f. The former was the consequence of increases in ER and total arteriolar vascular resistances, events that also led to augmentation of ΔP. It is noteworthy that these glomerular hemodynamic alterations are similar to those observed during increased AII activity. The concurrent administration of saralasin with either PGI_2 or PGE_2 essentially abolished the alterations in RPF, LpA, ER, RVR (renal vascular resistance), and ΔP. This inhibition of endogenous AII eliminated the action of PGE_2 or PGI_2 as a vasoconstrictor and converted its activity to that of an effective vasodilator ($\uparrow$ RPF, $\downarrow$ RVR).

Recently, Ichikawa et al.[74] examined the glomerular response to severe congestive heart failure evoked by ligation of the left coronary artery in male Munich–Wistar rats in a control situation and in the setting of endogenous prostaglandin inhibition (indomethacin treatment). Severe congestive heart failure in the rat was characterized by an increase in ER that led to an increase in P_G, so that SNGFR fell only slightly despite significant reductions in RPF and LpA or K_f. Indomethacin administration brought about a further increment in ER and decrements in SNGFR and RPF. Of importance, endogenous prostaglandin inhibition led to a further reduction in LpA or K_f. Similarly, in vitro studies have demonstrated that prostaglandins modulate AII-stimulated glomerular contraction.[75] In the aggregate, these and other studies indicate a significant protective role for renal prostaglandins to preserve glomerular hemodynamics in the face of vasoconstrictor influences. Nevertheless, the assessment of the potential contribution(s) of endogenous prostaglandins to the control of glomerular ultrafiltration in physiologic and pathophysiologic states, although extensively examined, remains to be fully delineated.

Although not as well studied, the renal kallikrein–kinin system has been implicated in the regulation of renal blood flow and sodium excretion. Bradykinin, infused to Munich–Wistar rats,[19] elicited no observable changes in SNGFR, π_A, or ΔP, but significantly increased RPF, an effect that was counteracted by a decrease in LpA or K_f. Whether this negative effect of bradykinin on LpA or K_f is AII mediated, as was the case for PGE_1 and PGE_2, is a matter of speculation at present.

10. Parathyroid Hormone (PTH), Ca^{2+}

The effects of PTH deprivation and PTH (synthetic bovine 1–34 fragment PTH) infusion on the determinants of glomerular ultrafiltration, in particular LpA or K_f, have been examined.[76] In acutely thyroparathyroidectomized Munich–Wistar rats, LpA or K_f (at filtration pressure equilibrium) was found to be significantly higher than normal values (at filtration pressure disequilibrium) in control animals.[76] Conversely, the infusion of synthetic PTH, in nonthyroparathyroidectomized and thyroparathyroidectomized rats, brought about a reduction in SNGFR.[76] This decrement in SNGFR was solely the consequence of a significant decline in LpA or K_f, since the other determinants of SNGFR, namely, RPF, π_A, and ΔP, remained essentially unchanged. Furthermore, studies by Humes et al.[77] characterized the effects of hypercalcemia on the determinants of glomerular ultrafiltration. Hypercalcemia in nonthyroparathyroidectomized rats was accompanied by significant declines in whole kidney GFR and SNGFR and these alterations were largely the result of a decrease ($\sim 60\%$) in LpA or K_f. It is of interest that these effects of acute hypercalcemia were largely

eliminated in rats that underwent acute thyroparathyroidectomy before the induction of a hypercalcemic state.

By contrast, infusion of a submaximally phosphaturic dose of PTH into hypercalcemic acutely thyroparathyroidectomized rats essentially reproduced the reduction in SNGFR, GFR, and LpA or K_f previously observed in nonthyroparathyroidectomized hypercalcemic rats.[77] Therefore, in the aggregate, these findings suggest that the decrease in glomerular ultrafiltration associated with hypercalcemia is due primarily to the decrement in LpA or K_f and that this effect on LpA or K_f is somewhat dependent upon the presence of PTH.

The glomeruli possess specific receptors for PTH and there is also evidence indicating that PTH induces glomerular cyclic nucleotide formation (cAMP as well as cGMP).[78,79]

As previously discussed, cAMP is one of the stimuli for renin release; thus, Schor et al.[62] investigated the possibility that PTH regulates LpA or K_f by virtue of its ability to promote cAMP formation through changes in local AII activity. PTH infusion to indomethacin-treated Munich–Wistar rats exerted a mild vasoconstrictor effect leading to decreases in RPF and LpA or K_f and increases in ΔP, and the net effect was slightly decreased SNGFR.[67] Because saralasin treatment reversed the alterations in RPF, LpA or K_f, ΔP and, therefore, in SNGFR, it was concluded that the PTH-induced alterations of glomerular hemodynamics require the action of AII.[62] In all likelihood, these findings suggest that PTH may play a physiologic role in the control of glomerular ultrafiltration. However, the extent to which PTH contributes to the regulation of glomerular ultrafiltration in altered physiologic states and disease processes (i.e., chronic renal failure) as well, remains to be determined.

11. Papaverine, Acetylcholine, Histamine, Methylpredinisolone

Deen et al.[80] were the first to demonstrate low values of LpA or K_f using the vasodilator papaverine in the plasma-loaded Munich–Wistar rat. Later studies by Baylis et al.[19] and Ichikawa and Brenner[20] extended our understanding of the mechanisms by which other vasodilators (i.e., acetylcholine and histamine, respectively) modify SNGFR and its determinants. Both vasodilators were noted to produce a significant reduction in LpΛ or K_f, thereby preventing measurable increases in SNGFR in spite of increases in RPF. Furthermore, infusion of diphenhydramine (an H_1-receptor antagonist), but not metiamide (an H_2-receptor antagonist) completely reversed the histamine-induced alterations in RPF and LpA or K_f.[20] Thus, available evidence supports the view that the failure of some potent vasodilators to positively influence SNGFR and total kidney GFR, despite large increments in RPF, is primarily due to their concomitant actions to reduce LpA or K_f.

It is generally agreed that chronic administration of pharmacological doses of glucocorticoid hormones causes significant increases in glomerular ultrafiltration at the single nephron and whole kidney level.[81,82] This increase in GFR persists even when the sodium balance is negative.[81,82] Baylis and Brenner[83] examined the mechanism(s) by which chronic treatment with methylprednisolone alters GFR. Their investigation conclusively demonstrated that the observed increases in SNGFR were flow dependent, since the other determinants of SNGFR remained unaffected. Because filtration pressure equilibrium persisted in rats chronically treated with methylprednisolone, it was not possible to determine any potential effect of this treatment on LpA or K_f. However, it is not clear whether this glucocorticoid exerts a direct or indirect (via an endogenous vasodilator) action on vascular smooth muscle.

12.   The Final Common Pathway Viewpoint

Over the last decade, the subject of mechanisms of action of a variety of hormones and humoral agents on glomerular ultrafiltration (*vide supra*) has been under extensive investigation. AII, ADH, PGI_2, PGE_2, and PGE_1, PTH, bradykinin, acetylcholine, histamine, and adrenergic agonists have been recognized as influential in regulating the process of glomerular ultrafiltration, primarily by an effect upon LpA or K_f. Many of these substances not only decrease LpA, but exert varied effects upon RPF, suggesting that changes in the major pre- and postcapillary vascular resistances do not cause the decrease in LpA. Many of the hormonal and humoral agents also stimulate the glomerular production of cAMP or cGMP, and dibutyryl cAMP decreases LpA. Furthermore, cAMP is a potent stimulus for renal renin release. These findings have raised the possibility that cyclic nucleotides act as the "final common pathway" through which most of these hormonal and humoral substances modulate LpA, via their ability to promote local intrarenal AII generation with AII acting as a direct effector. Recently, Schor et al.[62] have shown that the LpA-lowering effects of dibutyryl cAMP, PTH, PGI_2, and PGE_2 can be totally prevented by the infusion of saralasin, the AII receptor antagonist. These investigators concluded that all hormones tested, with the exception of ADH, act via the action of AII upon LpA. This conclusion is in keeping with studies on mesangial cells in culture in which only AII and ADH possess the capacity to cause contraction.[30,35]

In general, the data fit with AII being the "final common pathway" rather than cAMP. cAMP may act, then, indirectly on LpA to activate or liberate AII via a direct effect upon either the glomerulus and/or the macula densa cells.[84,88]

13.   Glomerular Ultrafiltration in Altered Physiologic States

13.1.   Responses to Acute Alterations in Systemic Blood Pressure

It has been well established that renal blood flow is remarkably well maintained in spite of wide variations in systemic blood pressure. Robertson et al. demonstrated that in the hydropenic Munich–Wistar rat (at filtration pressure equilibrium) the regulation of nephron filtration rate is closely linked to the autoregulation of renal blood flow.[89] During constriction of the aorta, SNGFR decreased more or less in proportion to reductions in RPF. A small contribution to the reduction in SNGFR was a documented decrease in ΔP as a consequence of a small decrease in P_G. SNGFR fell when the mean arterial pressure fell to levels of 70–80 mm Hg below the autoregulatory levels for renal blood flow. The ΔP and P_G were sustained during reductions in mean arterial pressure by a combination of afferent arteriolar dilation and a modest degree of efferent arteriolar constriction.

Deen et al. examined the effects of significant increases in mean arterial pressure upon SNGFR in plasma-expanded rats (5% body wt) utilizing bilateral carotid occlusion to produce an increase in blood pressure.[90] RPF increased significantly as a consequence, in spite of the fact that bilateral carotid occlusion

has been thought to increase adrenergic activity. Of interest, LpA did not change in spite of a large acute increase in RPF. Although autoregulation of RPF was reasonably complete in plasma-expanded rats in response to reductions in mean arterial pressure from 111 to 87 mm Hg, there was no observable autoregulation of RPF when mean arterial pressure was increased from 111 to 148 mm Hg.

In more recent studies, Ichikawa and Brenner reexamined glomerular responses to aortic constriction in euvolemic rats (surgical losses replaced with plasma).[91] Major reductions in mean arterial pressure produced significant reductions in LpA or K_f, increases in ER, decreases in AR, and relative maintenance of P_G and ΔP was achieved in normal and indomethacin-treated rats. Infusion of saralasin prevented both the decrease in K_f and the decrease in ER, the latter resulting in major reductions in P_G and ΔP, but no effect on the autoregulatory reduction in vascular resistance at the afferent arteriole. These results suggest that major reductions in mean arterial pressure produced by aortic constriction produce not only autoregulatory adjustments but increased AII activity which appears to: (1) reduce K_f and (2) increase ER and contribute to the preservation of P_G during reductions in arterial pressure.

Navar et al. have obtained results in the dog which differ from the above.[92] They observed little or no autoregulation of nephron filtration rate when micropuncture collections were obtained from the proximal tubule, but reasonably efficient autoregulation of SNGFR was observed when collections were obtained from the distal tubule. The results not only differ significantly from those obtained in the rat but also suggest that tubuloglomerular feedback mechanisms and maintenance of tubular flow to the distal tubule are required for full expression of autoregulation of nephron filtration rate.

13.2.   Alterations in Volume Status

Numerous micropuncture studies have investigated the glomerular responses to acute and chronic alterations in the status of the extracellular fluid volume. Extracellular volume expansion with saline (10% body wt i.v.) results in a proportional reduction in both AR and ER in the rat, such as to maintain P_G constant and similar to hydropenic values.[95] Saline expansion also decreased ΔP without a significant change in LpA and increased RPF, thereby producing a significant increase in SNGFR. When plasma volume is expanded with parenterally administered isoncotic plasma, there also occur major alterations in SNGFR and its determinants.[11,15,90] The balance of change in vascular resistance differs qualitatively in plasma volume expansion from that observed with saline in the rat in that AR is reduced to a greater extent than ER; the net effect is a tendency for P_G to increase significantly above values in both hydropenia and saline expansion. Because LpA remained unchanged, the increase in SNGFR during isoncotic plasma expansion was primarily a plasma flow-dependent effect. Blantz et al. [13] have examined the effect of hyperoncotic albumin expansion and the respective role of oncotic pressure, nephron plasma flow, and glomerular ultrafiltration coefficient upon the filtration process. In these studies, significant increases in SNGFR were associated with ↑ RPF—a consequence of ↓↓ AR and ↓ ER, ↑ ΔP, and ↑ LpA. Of importance is the study by Blantz[93] in which the effect of mannitol and the attendant increase in plasma volume upon glomerular ultrafiltration was analyzed in the hydropenic rat. SNGFR markedly rose after mannitol infusion; this increase was due to both increased RPF and decreased π_A. Since EFP was disequilibrated after mannitol, LpA could be calculated accurately and was significantly lower than the minimum estimate in hydropenia. Thus, data obtained in these two studies suggest that LpA varied directly with both large increases in π_A following hyperoncotic albumin infusion and significant decreases in π_A after mannitol infusion. Later studies by Tucker and Blantz[15] and Baylis et al.,[14] specifically designed to examine this relationship, also noted a significant direct correlation between π_A and LpA. At present, the specific mechanisms that link changes in π_A to alterations in LpA remain unknown.

In chronically NaCl-depleted Munich-Wistar rats, Steiner et al.[94] have found that nephron filtration rate is decreased when compared to normovolemic littermates maintained on a normal NaCl intake and this finding was the result of reductions in both RPF and LpA. In paired studies in NaCl-depleted rats, investigations have shown that infusion of saralasin completely restored RPF to the normovolemic values by eliminating the vasoconstriction, suggesting that these important vascular changes in chronic NaCl depletion are the result of increased AII generation and action on the afferent and efferent arterioles.[94] However, the reduction in LpA observed with chronic NaCl depletion was not reversed or normalized by saralasin infusion (at a dosage capable of normalizing AII-induced reductions in LpA). However, acute volume repletion with plasma expansion did completely normalize LpA.[94] Later studies from the same laboratory have indicated that long-term but not short-term converting enzyme inhibitor treatment restored LpA to normal values in chronic NaCl-depleted rats.[95] Of related interest, the studies by Schor et al.[96] have demonstrated that, in addition to AII, prostaglandins also seem to be involved in the adaptations of the renal microcirculation to chronic variations (i.e., reduction) in dietary NaCl intake. In contrast, in states with higher NaCl intake, maintenance of SNGFR and RPF was found to be less critically dependent on local prostaglandin synthesis.[96] Since the exact role of hormonal agents such as ADH and/or the adrenergic nervous system remain unknown in these physiologically altered states, it cannot be concluded that AII and prostaglandins are the only hormonal systems which participate in the regulatory control of glomerular function in the setting of NaCl depletion.

13.3. Altered Ureteral Pressure

The rate of glomerular filtration is relatively well maintained in spite of significant degrees of partial obstruction of the urinary collecting system and increases in tubular pressure.[6,97,98] Significantly compensating alterations occur in the determinants influencing SNGFR in modifying or ameliorating the effect of increased tubular pressure. Micropuncture studies have demonstrated that SNGFR and tubular fluid flow rate are reduced within about 30 min of ureteral obstruction in hydropenic and plasma-expanded rats, and in rats undergoing moderate mannitol diuresis.[99–101] Kallskog and Wolgast[102] found that AR decreased while RPF remained constant and P_G increased. However, Blantz et al.[99] demonstrated in the hydropenic rat that, when all segmental vascular resistances were quantitated, there was no change in either AR or ER and RPF remained constant when ureteral pressure was increased to 25 cm H_2O. The rise in P_G observed was entirely the result of increases in downstream hydrostatic pressure and vascular resistances beyond the peritubular capillaries. It should be noted that the reduction in SNGFR appeared to result from a decrease

in LpA. The filtration response to elevated ureteral pressure was modified by the prior state of volume expansion since, in plasma-expanded rats, SNGFR fell entirely as a result of a decrease in ΔP. More recently, Dal Canton et al.[103] have shown increases in RPF 1 hr after complete unilateral ureteral ligation. Conversely, the same group of investigators[104] and Arendshorst et al.[105] have shown that unilateral ureteral obstruction for 24 hr produces a marked increase in AR, accounting for a fall both in P_G and in RPF, an effect which was not reversed by ureteral release. Of note is the finding that the cause of the reduction in GFR is also quite dissimilar in bilateral ureteral obstruction, since the fall in SNGFR after bilateral ureteral obstruction was only associated with a marked reduction in ΔP (secondary to ↑ P_t and = P_G).[106] The reasons for these distinct results remained to be determined, and it is apparent that no universal generalizations can be applied to the effects of increased ureteral pressure upon glomerular hemodynamics.

With more chronic and sustained increases in ureteral pressure (i.e., rats with hereditary hydronephrosis), Humes et al.[107] have demonstrated that SNGFR values were maintained near normal because of an elevation in ΔP (due to ↑ P_G and = P_t), effects which neutralized a lower RPF and a possible decrease in LpA or K_f. These observations suggest that the increase in ER was greater than that in AR. Ichikawa and Brenner[108] have studied the effects of chronic mild partial ureteral obstruction (surgically created) on glomerular hemodynamics and the local intrarenal vasoconstrictor–vasodilator interactions in this setting as well. SNGFR and RPF in obstructed kidneys were essentially identical to values in unobstructed kidneys; however, the observed increase in P_G served to neutralize the markedly reduced LpA or K_f. Administration of indomethacin or meclofenamate to rats with obstructed kidneys caused arteriolar vascular resistances to increase and SNGFR and RPF to fall. Both angiotensin generation and prostaglandin–thromboxane synthesis are enhanced in obstructed kidneys. Therefore, there is reason to believe that the dynamic interplay of these two intrarenal hormone systems in the setting of ureteral obstruction may cause specific adjustments at the renal microcirculatory level to preserve glomerular ultrafiltration.

13.4. Increased Renal Venous Pressure

The effects of decreases in renal perfusion pressure produced by increasing renal venous pressure on glomerular hemodynamics are not well understood. Studies in hydropenic dogs have demonstrated that vasodilation and autoregulation of GFR occur following acute increments (10–35 mm Hg) in renal venous pressure.[109,110] Conversely, investigations on nondiuretic rats indicate that GFR is decreased by a 15–25 mm Hg increase in renal venous pressure.[102,111–113]. Kallskog and Wolgast[102] have examined the effects of increased renal venous pressure in the rat. Despite increases in renal venous pressures to about 15 mm Hg, RPF was not altered and AR and ER were not changed. P_G remained constant (as evaluated by stop-flow methods) but peritubular capillary hydrostatic pressure rose significantly. Tucker and Blantz[113] found strikingly similar results with partial renal venous occlusion in the rat. The major abnormality defined by both laboratories was a reduction in nephron filtration rate, which resulted primarily from the reduction in ΔP. However, Dilley et al.,[114] investigating the functional glomerular adaptations to increased renal venous pressure in the euvolemic Munich–Wistar rat, arrived at a markedly different conclusion. Increased venous pressure (4 to 22 mm Hg) elicited a reduction in SNGFR (determined without perturbation of the initial elevated intratubular pressure), primarily the consequence of decreased RPF and LpA. ΔP remained unaffected during the experimental period, since P_G and P_{BS} both increased by 7–8 mm Hg. The mechanisms whereby acute increases in renal venous pressure affect glomerular hemodynamics are less well defined. Preliminary reports suggest that the prostaglandin and renin–angiotensin systems and the renal nerves play a role in the acute adaptations of the renal microcirculation to partial renal venous compression.[115,116]

13.5. Pregnancy

There is abundant evidence that GFR is increased during uncomplicated human pregnancy and this appears to be true in rat pregnancy as well (reviewed in Refs. 117, 118). To date, studies in the pregnant and pseudopregnant euvolemic Munich–Wistar rat have demonstrated increases in GFR, RPF, and plasma volume at days 15–12,[119,120] day 9,[121] but not at day 6[122] of gestation. The mechanism(s) whereby GFR is increased in pregnancy has been evaluated by direct micropuncture techniques. Baylis's studies[119,121] have indicated that the increase in SNGFR that occurred in euvolemic 9- to 12-day pregnant and 9-day pseudopregnant rats was the result of the increases in single-nephron plasma flow, since the small but significant fall observed in ΔP would tend to reduce SNGFR. Filtration pressure equilibrium was observed in both control and experimental groups; thus, it was not possible to calculate a unique value for LpA or K_f. In this investigation, the fact that the glomerular hemodynamic alterations observed in pseudopregnant rats were markedly similar to those seen in pregnant rats was taken to indicate that the fetoplacental unit is not an obligatory requirement for the initiation or maintenance of the elevated GFR. The presence of the placenta, however, is required in order for GFR and RPF to persist elevated in the late stage of pregnancy in the rat.[123] These findings are somewhat in contrast with the work of Dal Canton et al.[120] who observed a reduction (~ 15%) in LpA or K_f by day 15 of gestation in the hydropenic as well as saline-expanded Munich–Wistar rat. Supranormal values for SNGFR persisted, however, because this fall in LpA or K_f was neutralized by a concomitant increment in ΔP (a consequence of = P_{BS} and ↑ P_G) and a concurrent increase in single-nephron plasma flow. The discrepancy in results may be accounted for by differences in: (1) the gestational age and (2) the status of the extracellular fluid volume of the pregnant rats under investigation.

The reasons for the increase in glomerular blood flow occurring during pregnancy are not fully elucidated; however, it seems likely to be the result of both renal vasodilation (perhaps prostaglandin mediated) and plasma volume expansion (by increasing the stroke volume component of cardiac output).

14. Influences of Tubuloglomerular Feedback System on the Process of Glomerular Filtration

14.1. General View

Regulation of renal NaCl and water excretion requires some coordination between the rate of glomerular ultrafiltration (the load function) and the rate of tubular reabsorption of NaCl and water (the transport function). The phenomenon of glomerular tubular balance contributes significantly to this coordination by exhibiting changes in tubular reabsorptive rate, more or less, in proportion to alterations in load. A complementary mechanism which also contributes to this coordination of load and transport rates is the tubuloglomerular feedback system.[124] Although all details regarding specific aspects of the afferent and efferent mechanisms of this feedback system are to be delineated, there remains little doubt that a relationship exists whereby changes in the rate of tubular reabsorption exert a major influence in the regulation of nephron filtration rate.[125–127] We will focus upon those examples in normal and altered physiologic states in which there is reasonable evidence that either activation or inhibition of tubuloglomerular feedback mechanisms contributes significantly to the final physiologic picture.

14.2. Activation of Tubuloglomerular Feedback Mechanisms

Many of the single-nephron microperfusion studies demonstrate that increases in late proximal tubular fluid flow rate produce reductions in filtration rate of the same nephron.[126–128] One would predict that physiologic conditions associated with reductions in absolute proximal tubular reabsorption which are not associated with parallel reductions in distal tubular transport should also result in decreased nephron and kidney filtration rate. Benzolamide, a carbonic anhydrase inhibitor which decreases absolute proximal tubular reabsorption, does result in reductions in nephron filtration rate, and evidence has been provided that this decrease in SNGFR is due to activation of tubuloglomerular feedback mechanisms.[129,130] These modest reductions in nephron filtration rate (~ 7 nl/min) were entirely the result of parallel increases in AR and ER and reductions in RPF.

Recent observations by Schnermann et al. suggest that tubuloglomerular feedback mechanisms are also activated by the infusion of hypertonic saline into the renal artery.[131] In this circumstance, late proximal flow was increased, in all probability, by a combination of reductions in proximal tubular reabsorption and transient increases in SNGFR (due to osmotic effects of hypertonic saline on renal vessels leading to increased renal blood flow). The net effect was a reduction in SNGFR via activation of tubuloglomerular feedback mechanisms. We have also provided evidence that tubuloglomerular feedback activation may contribute to the final nephron filtration rate following acute renal denervation in hydropenic rats.[132] Using similar techniques as utilized in the study with benzolamide, it was observed that if oil blocks were inserted into late proximal tubules prior to renal denervation, SNGFR was significantly higher in those nephrons than in undisturbed nephrons, suggesting that if no "signal" of increased flow into the loop of Henle was perceived, SNGFR was significantly higher. Therefore, several examples have been documented whereby activation of tubuloglomerular feedback mechanisms contributes significantly to the final physiologic result.

14.3. Inhibition of Tubuloglomerular Feedback

In a recent study from our laboratory, we examined the effects of modest degrees of acute hyperglycemia upon overall renal function in the Munich–Wistar rat.[133] At plasma glucose values of 430 mg/100 ml the nephron filtration rate was slightly increased in spite of major increases in both late proximal and early distal tubular flow rates. Since the predicted response to such elevations in tubular fluid flow rates should be a reduction in SNGFR, an inhibition of tubuloglomerular feedback activity seemed likely. Specific testing using microperfusion techniques and native tubular fluid (normoglycemic and hyperglycemic) revealed that SNGFR responses to elevated late proximal tubular flow rate were totally inhibited in hyperglycemic rats when hyperglycemic tubular fluid was utilized as the test solution.[134] It was concluded that approximately one-half of the inhibition of feedback in hyperglycemia was attributable to the presence of elevated glucose concentrations in tubular fluid beyond the proximal tubule, and the remainder of this effect due to extratubular effects of hyperglycemia (possibly elevated renal interstitial hydrostatic pressure). This latter conclusion with regard to potential mechanisms is compatible with the data of Persson *et al.* which suggested that reductions in peritubular capillary oncotic pressure were associated with partial inhibition of tubuloglomerular feedback activity, as tested by single-nephron microperfusion techniques.[135,136]

Application of diuretics which act within the loop of Henle, when added to tubular luminal fluid, results in inhibition of normal tubuloglomerular feedback.[137] We have recently examined the renal responses to large, systemic doses of furosemide in rats in which urine volume losses either were or were not replaced acutely.[138] As was the case with acute hyperglycemia, SNGFR did not decrease in spite of large increases in early distal tubular flow rate, suggesting suppression of feedback activity. However, it is of interest that furosemide did not produce vasodilation in the rat even when volume was carefully maintained by intravenous fluid replacement. This latter finding at least suggests that other systems contribute to the maintenance of arteriolar tone within the kidney aside from tubuloglomerular feedback mechanisms.

It is generally accepted that volume expansion, particularly with saline solutions, suppresses the activity of tubuloglomerular feedback mechanisms. Examination of all of the studies relating to this issue leads to the conclusion that inhibition of tubuloglomerular feedback activity is partial and not complete. It remains possible that a greater stimulus or microperfusion rate is required to decrease SNGFR after volume expansion because of the normally greater late proximal tubular flow rate in this condition. Studies by Moore *et al.* do provide convincing evidence for a major effect of acute alterations in volume status on the magnitude of feedback response to a given stimulus.[139] This alteration in feedback activity was not readily attributable to differences in the normal late proximal flow rate in the various conditions of volume status.

14.4. Information on the Mechanism whereby Nephron Filtration Rate Changes with Activation of Tubuloglomerular Feedback

Some debate continues over the specific efferent mechanisms whereby tubuloglomerular feedback activation produces a change in nephron filtration rate. Studies we have reported regarding the mechanism of SNGFR reduction after benzolamide revealed that this feedback-induced reduction was solely the result of decreases in RPF.[129] The stimulus to feedback activity was modest (an increase in late proximal flow rate of at most 7–8 nl/min), but all nephrons were simultaneously activated. There was no significant changes in either hydrostatic pressures or LpA and, therefore, these factors did not contribute to the

changes in SNGFR. Briggs and Wright developed techniques at the single-nephron level to examine the mechanisms whereby SNGFR decreased when late proximal microperfusion rate was increased to 40 nl/min[140] These authors concluded that the decrease in SNGFR was the consequence of reductions in both ΔP and RPF. Ichikawa utilized similar techniques, with the exception that glomerular capillary pressure was measured directly, and concluded that the decrease in SNGFR was the result of reductions in RPF and LpA.[141] The latter authors did not observe any changes in ΔP. The only firm conclusion one can make with confidence is that changes in RPF play a significant role in producing feedback-induced changes in SNGFR.

15. Pathophysiologic Conditions

15.1. Acute Renal Failure

The etiologies of clinical acute renal failure are multiple, and any attempt to design experimental models of acute renal failure in animals is fraught with the possibility that such attempts will not adequately duplicate or mimic the clinical disease. Experimental models can realistically, but somewhat arbitrarily, be divided into three categories: (1) ischemic models; (2) nephrotoxic models; and (3) models associated with an increased filtered load of biologic materials which can be toxic.[142] Several studies have utilized these three types of experimental models and data derived have expanded our knowledge of the specific mechanisms leading to reduction in GFR. The various considerations are as follows: (1) tubular obstruction; (2) transepithelial ''back-leak'' of solutes; (3) reductions in LpA; and (4) vasoconstriction leading to reductions in RPF and/or glomerular capillary hydrostatic pressure.

15.2. Ischemic Models

Experimental models have utilized either 60–90 min of unilateral renal artery clamping[143,144] or 60–90 min of norepinephrine infusion into one renal artery[145,146] to produce acute renal failure. These approaches provide total ischemia for significant time periods, a condition which many would argue is not a reasonable substitute for acute renal failure in the human, since total renal ischemia for any period is difficult to document in the clinical experience. Several studies have examined the mechanism of reduction in GFR after prolonged norepinephrine infusion. In an early study,[145] data suggested, indirectly, that a primary reduction in GFR occurred, which was likely the consequence of a decrease in LpA. This conclusion was supported by the finding of marked morphologic alterations in glomerular epithelial cells as evaluated by scanning electron microscopy. Tubular pressure was decreased below normal values. Later studies arrived at somewhat different findings.[146] Tubular pressure was significantly increased. No glomerular morphologic alterations were observed. Results suggested that the major cause of reductions in GFR was tubular obstruction and possibly a modest reduction in LpA.[146] In both studies, restoration of modest reductions in renal blood flow did not improve GFR.

Recent studies have revealed that acute renal failure resulting from 60–90 min of renal artery clamping was primarily the consequence of reductions in LpA.[144] Morphologic alterations were observed in both glomerular endothelial and epithelial cells by scanning electron microscopy after ischemia.

Tubular pressure was equal to control values and, therefore, no evidence for significant tubular obstruction was found. Significant differences of opinion persist as to the mechanisms of reductions in GFR after renal ischemia; however, it is clear from all recent studies that reductions in renal blood flow do not contribute to the persistent decrease in GFR. It is of interest that Myers *et al.* have examined the mechanism of acute renal failure after cardiovascular surgery in the human utilizing the clearances of inulin and neutral dextrans.[147] These studies concluded that reductions in GFR were due to a combination of: (1) tubular obstruction, (2) back-leak of solutes as large as inulin (~ 50% of filtered inulin), and (3) a primary reduction in glomerular ultrafiltration by some mechanism (reduced LpA or vasoconstriction). The degree to which this clinical form of renal failure is ischemic and to which degree nephrotoxic remains speculation.

15.3. Nephrotoxic Models

The administration of mercury and uranium salts was initially utilized in early models of experimental acute nephrotoxic renal failure. In early studies with mercuric chloride nephrotoxic renal failure, Bank *et al.* observed that lissamine green was no longer retained within the proximal tubular lumen, suggesting that mercury toxicity was associated with significant transepithelial "back-leak" of solutes.[148] A substantial percentage of the reduction in whole kidney GFR after uranyl nitrate administration has been shown to be the result of transepithelial back-leak of solutes.[52] Microinjection of small volumes containing [^{14}C]inulin and [^{3}H]mannitol into proximal tubules in control animals revealed essentially 100% recovery of these radiolabeled solutes in the urine of the microinjected kidney. However, in rats which have received uranyl nitrate, only about 50% of these microinjected substances were retained within the tubule and these radiolabeled substances were detected in the urine of the contralateral kidney, demonstrating that these materials "back-leaked" into the plasma across the damaged epithelia. There was considerable variation in the degree of "back-leak" observed among rats receiving uranyl nitrate which may be due to true variations in tubular damage or due to differences in the concentration gradient of these solutes between urine or tubular fluid and plasma, the driving force for inulin and mannitol "back-leak."[5]

The reduction in GFR after uranyl nitrate administration is also the consequence of primary alterations in ultrafiltration at the glomerulus. Studies in the rat and dog revealed that RPF was well maintained acutely after uranyl nitrate. However, LpA was consistently reduced acutely and constituted the sole mechanism whereby nephron filtration rate was decreased.[5] Stein *et al.* came to similar conclusions in studies conducted in the dog.[149] At the later time of evaluation after uranyl nitrate administration than examined by Blantz,[5] Stein *et al.*[149] found significant glomerular morphologic changes localized primarily to the epithelial cells which might have served as a basis for the decrease in LpA. More recently, Avasthi *et al.*[15] observed a reduction in the diameter and density of glomerular endothelial fenestrae by scanning electron microscopy, utilizing a protocol nearly identical for uranyl nitrate as previously used by Blantz.[5] Whether these various morphologic changes within the glomerulus are the basis for reduction in LpA remains an issue.

Aminoglycoside antibiotics such as gentamicin have proven a major cause of acute renal failure or, at least, transient

azotemia, since their introduction two decades ago. Baylis *et al.*[151] demonstrated that the reduction in nephron filtration rate was due to significant changes in K_f or LpA and modest decreases in RPF. Although these authors supplied no morphologic basis for the reduction in glomerular ultrafiltration, other studies have observed glomerular endothelial alterations[152] similar to those observed after uranyl nitrate[153] in a model of gentamicin nephrotoxicity. We had suggested in the first edition of this text[1] that the cationic nature of uranium and gentamicin might provide the basis for glomerular changes because of the anionic surface charge of many of the structures which comprise the glomerular capillary wall. An alternate mechanism has been provided by results from recent studies on gentamicin[154] and uranyl nitrate.[153] Both investigations suggested that maneuvers such as pretreatment with angiotensin-coverting enzyme inhibitor and, in the case of uranyl nitrate, plasma volume expansion resulted in restoration of normal values for LpA (or K_f) in spite of equivalent degrees of tubular damage[153,154] These more recent results suggest that the alterations in LpA in these two models are functional in origin and not the consequence of deposition of these cationic molecules into the glomerular membrane with secondary structural alterations.

15.4. Increased Filtered Load of Biologic Materials

There are less data on the specific mechanisms leading to reductions in GFR in this type of acute renal failure. In a model of uric acid nephropathy, Conger and Falk demonstrated that tubular obstruction and reductions in renal blood flow were the major reason for the reductions in GFR.[155] The increases in vascular resistance which led to reductions in RPF were distal to the efferent arteriole and may have been due to uric acid crystal deposition in tubules and renal interstitium. The mechanism of renal failure with increased filtered load of myeloma proteins, rhabdomyolysis, and various cationic materials may be similar to those described for uric acid nephropathy.[142]

In summary, the mechanisms of acute renal failure are heterogeneous and consist of glomerular and nonglomerular events. Reductions in LpA appear important to several models of acute renal failure. Although several studies have supplied a potential morphologic basis for the decrease in LpA, there are data suggesting a functional or reversible basis for alterations in LpA. The nonglomerular causes include: (1) tubular back-leak of solutes into the circulation across damaged epithelia; and (2) tubular obstruction, both mechanisms in keeping with original concepts that acute renal failure is primarily a disorder of the tubular epithelium.

16. Glomerular Immune Injury

Considerable information has been accumulated over the past several years regarding the specific mechanisms whereby immune injury to the glomerulus leads to reductions in GFR. Allison *et al.* demonstrated that several months after establishment of an anti-glomerular basement membrane antibody (GBMAb) and an immune-complex-induced glomerulonephritis, the reduction in nephron filtration rate was quite heterogeneous, varying from nearly zero to slightly above normal for age and weight.[156] Glomerular capillary pressures, as estimated indirectly from stop-flow methods, were also quite heterogeneous. Although not directly computed in this study, results do suggest that a reduction in LpA may have contributed to the lower values for SNGFR in these early experimental models of glomerular immune injury.

Maddox *et al.* later studied a mild form of nephrotoxic serum nephritis (40 μg of antibody) 2 weeks after the immune insult.[157] Although the nephron filtration rate was not reduced below control values, measurements of the determinants of glomerular ultrafiltration utilizing direct measurements of glomerular capillary hydrostatic pressure in Munich–Wistar rats revealed that glomerular ultrafiltration was decreased to values of about 0.03 nl/sec per mm Hg, values below 50% of the normal value. SNGFR was sustained by concomitant increases in ΔP and RPF. Later studies which examined the sieving characteristics of dextrans in conjunction with histologic analyses suggested that this decrease in K_f or LpA was due to reductions in the effective capillary surface area for ultrafiltration.[158]

Blantz and Wilson conducted a series of studies of the glomerular response to large doses of anti-glomerular basement membrane antibody and performed direct physiologic measurements within 60 min of the administration of 225–450 μg of antibody (50–100% antigenic saturation).[159,160] Nephron filtration rate decreased almost immediately to about 50% of control values in this proliferative model of glomerular immune injury. RPF fell significantly with large doses of anti-GBM Ab and ΔP rose, effects that in the aggregate were neutralizing. The major reason for the large decrease in SNGFR was a reduction in LpA from 0.075 to 0.013 nl/sec per kidney wt per mm Hg.[159] Later studies revealed that prior complement depletion of rats prevented essentially all of the antibody-induced vasoconstriction and most of the reduction in LpA.[160] The beneficial effect of complement depletion upon LpA appears to

have been mediated by the prevention of polymorphonuclear leukocyte (PMN) infiltration of the glomerular capillaries and attachment to the GBM. Electron micrographs revealed that in rats with intact complement systems, the PMN strips the endothelial cell from the underlying GBM and becomes attached to this membrane (see Fig. 3). One must conclude that the major reason for LpA reduction was attachment of PMNs within capillaries and consequent reduction in effective capillary surface area. PMNs not only occupy capillary surface area but partially obstruct capillary lumina and divert glomerular capillary plasma flow away from certain capillary conduits, resulting in net loss of ultrafiltering surface area. The complement system via the anaphylatoxins C5a and possibly C3a appears to attract the PMN to the GBM. This role of the PMN has been further elucidated by recent studies from our laboratory in leukocyte-depleted rats. Anti-GBM Ab does not decrease SNGFR or LpA in these rats.[161]

The mechanism whereby acute administration of anti-GBM Ab in high doses produces vasoconstriction is less well defined. No deposition of immunoglobulins and morphologic alterations were observed in either afferent or efferent arterioles which might provide a structural basis for increased vascular resistance. More recent studies have examined the mechanism of this complement-dependent vasoconstriction. Infusion of saralasin did not prevent vasoconstriction after anti-GBM Ab and, in fact, appeared to increase the intensity of immune-induced vasoconstriction.[162] However, the α-adrenergic antagonist phentolamine totally prevented vasoconstriction, suggesting some unexplained connection between complement components and α-adrenergic receptor activation.[162]

An additional factor which may have contributed to the reduction in LpA after the acute administration of anti-GBM Ab is the detachment of the endothelial cell from the underlying

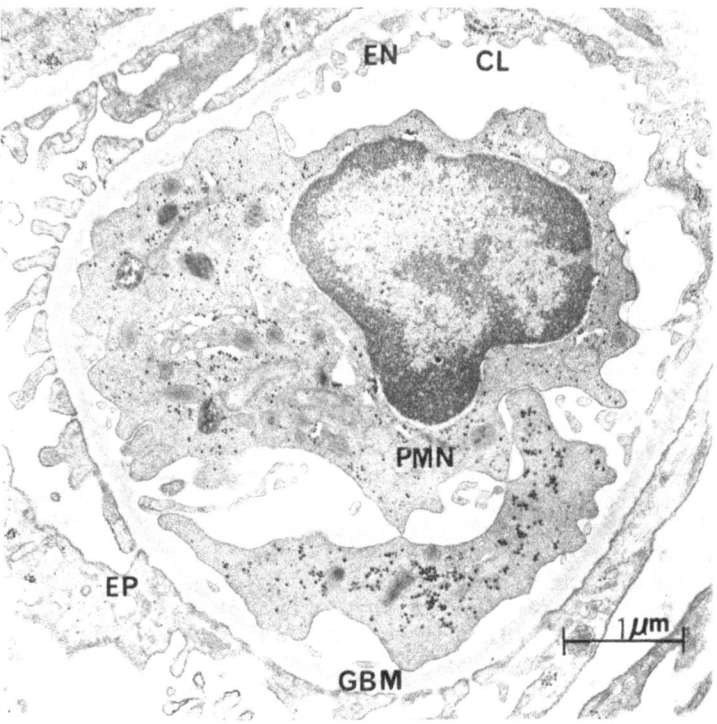

Fig. 3. An electron micrograph of the glomerular capillary from a rat taken 60 min after the infusion of rabbit anti-rat glomerular basement membrane antibody. The endothelial (EN) cytoplasm is being stripped from the glomerular basement membrane (GBM) by polymorphonuclear leukocytes (PMN) which have accumulated in the capillary lumen (CL). Some focal fusion of the epithelial (EP) cell foot processes is also apparent. 21,000×.

GBM after binding of antibody.[159,160] One can assume that this separation of endothelium from the GBM may create a new, less well-stirred compartment interposed between the endothelial cell and the underlying membrane. This less well-stirred compartment *in vivo* might alter hydraulic conductance (Lp) of the complex glomerular membrane by allowing concentration of larger proteins and elevated colloid osmotic pressure within this space due to the absence of "stirring" shear rates at the endothelial surface of the normal capillary. The specific mechanism whereby endothelial cell detachment from the GBM occurs is not known, but antibody fixation to the GBM may interfere with normal cell attachment or release of cationic molecules from the PMN may cause this structural alteration.

Since histamine release has been thought to be an integral part of the inflammatory response, histamine blockers have been utilized experimentally and clinically to modify the course of glomerulonephritis, and certain studies have suggested reductions in proteinuria with antihistamine treatments in experimental glomerulonephritis. We have recently examined the effects of large doses of H_1 and H_2 receptor blockers in the acute anti-GBM Ab model.[163] There were no beneficial effects observed upon either the immune-induced alterations in LpA, RPF, or nephron filtration rate during histamine blocker infusion.

The studies described above utilized anti-GBM Ab and nephrotoxic serum nephritis models and judgments as to mechanisms may not be universal to all forms of glomerular immune injury. Another model which has been examined is that of Heymann's nephritis or autologous immune complex nephropathy, a model of membranous nephropathy rather than the proliferative glomerulonephritis characteristic of acute anti-GBM Ab. In a study by Ichikawa *et al.*, rats immunized with renal tubule epithelial antigen ($F \times 1A$) were studied at 5 months after development of glomerulonephritis.[164] There was marked heterogeneity of glomerular immunoglobulin deposition and the degree of glomerular physiologic abnormalities among nephrons within a diseased kidney. The dispersion of values for SNGFR was primarily the consequence of a heterogeneity of both K_f and Q_A (RPF), nephron plasma flow among glomerular units (both of which varied directly with the resulting SNGFR). This heterogeneity of glomerular dysfunction was true only for the autologous model. If heterologous antiserum was injected into rats, a more homogeneous result was obtained, but again, reductions in K_f and Q_A were the mechanisms whereby SNGFR was decreased. As observed in the acute anti-GBM models. ΔP was significantly increased, a factor which partially ameliorated the effects of decreased K_f and Q_A.

Therefore, in the variety of experimental immune models examined, reductions in SNGFR are largely the consequence of reductions in LpA or K_f with reductions in RPF occurring in certain forms with greater immune insults. The ΔP is, in general, increased in most experimental models examined. Acute immune insults are characterized by homogeneous effects among nephron units but heterogeneity of glomerular dysfunction increases with duration of time after the initial insult for reasons that are not entirely clear.[165]

Another form of glomerular injury can be induced by the infusion of puromycin aminonucleoside. Systemic administration of this agent produces a reduction in SNGFR and abnormalities of glomerular permselectivity to macromolecules and albuminuria.[166] Recent studies have utilized infusion of puromycin into one renal artery in order to eliminate systemic or nonrenal contributions to the interpretation of renal mechanisms producing the decrease in nephron filtration rate.[167] The major reason for the decrease in SNGFR is a significant decrease in K_f. Morphologic changes in the glomerulus were confined to fusion of epithelial cell foot processes.

17. Glomerular Effects of Systemic Hypertension

The experimental model in which the glomerular effects of sustained systemic hypertension have been best studied is that of two-kidney Goldblatt hypertension. Schweitzer and Gertz examined the unclipped kidney of a two-kidney Goldblatt model with 4 weeks of rather severe hypertension (180 mm Hg mean arterial pressure).[168] There were increases in renal vascular resistances but glomerular capillary hydrostatic pressure was significantly elevated. LpA was modestly reduced in the unclipped kidney. Steiner *et al.* have also examined glomerular function in the unclipped kidney in a milder form of the two-kidney Goldblatt model (~ 20 mm Hg increases in mean arterial pressure).[169] The glomerular capillary hydrostatic pressure was also increased in this much milder form of hypertension. In addition, LpA was significantly decreased, which essentially neutralized the effect of increased ΔP. These studies raise the intriguing possibility that the elevation in glomerular capillary hydrostatic pressure may be causally related to the reduction in LpA in this form of systemic hypertension. No morphologic basis for the alteration in LpA was provided in either of these studies.

Specific information on other, NaCl-induced forms of hypertension is less readily available. Azar *et al.* have examined renal function in the later stages of NaCl-induced hypertension.[170] Unfortunately, this study represents a major nephron loss model, which confuses the interpretations of the resulting glomerular determinants. RPF and glomerular capillary hydrostatic pressure were both elevated. In retrospect, it would also appear that the LpA was also decreased. It is difficult to be certain if this latter reduction is the consequence of hypertension or related to the severe nephron loss observed (as will be discussed later). Definite sclerotic changes were observed in these glomeruli by histologic analysis.

18. Isolated Glomeruli *in Vitro*

Over the past decade, numerous micropuncture studies performed in a unique strain of Munich–Wistar rats have detailed the dynamics of glomerular ultrafiltration in physiologic as well as in altered physiologic states. More recently, at least two groups of investigators have reported the development of new *in vitro* techniques designed to further examine glomerular function.

Savin and Terreros[171] were the first to report an imaginative *in vitro* oncometric method for the determination of LpA or K_f in the isolated single glomerulus of a number of species. These *in vitro* estimates of LpA or K_f in the rat glomeruli are in close agreement with those obtained *in vivo*; however, in the dog, values for LpA or K_f are three to fivefold higher than the *in vivo* values.[171] The use of this method for determining LpA or K_f is substantiated by a number of interesting reports in which changes of LpA or K_f in various pathophysiologic states were assessed by this technique. In summary, these investigators demonstrated in rat glomeruli a reduction in LpA or K_f in uranyl nitrate-[172] and mercuric chloride-induced[172] acute renal failure and *in vivo* volume depletion.[173]

The ingenious and sophisticated application of advanced technology by Osgood et al.[174] has led to the development of a new technique for the evaluation of glomerular hemodynamics in which isolated glomeruli are perfused in vitro. In general, the methodology employed is analogous to the in vitro perfusion of isolated tubular segments. The validity of this new method is illustrated by preliminary reports by Osgood et al.[174,175] in which they studied the isolated perfused dog glomerulus in control conditions. The obtained value for LpA or K_f in the above study compares reasonably well with the in vivo value. A recent investigation by this group examined the effect of plasma colloid osmotic pressure on LpA or K_f; the observed effect of protein was the opposite of that reported in vivo.[175] This finding was taken to suggest that the in vivo effects may be mediated by indirect mechanisms.

19. Causes of Progressive Reduction in GFR

A reduction in the functional renal mass, whether the consequence of a disease process or surgical ablation, is associated with significant functional as well as morphologic changes in the surviving nephron units.[176–178] A compensatory functional and structural hypertrophy in the remaining nephrons partially offsets the loss of function that otherwise would have occurred.[176–178] The marked increase in SNGFR in the remaining nephrons observed 2 weeks after unilateral nephrectomy, is entirely due to greatly augmented glomerular pressures (ΔP) and flows (RPF), and closely correlated with the magnitude of renal mass that has been lost.[179] Studies by Hostetter et al.[180] have demonstrated at 1 week, following one and five-sixths nephrectomy, striking abnormalities of all three glomerular cell types associated with the aforementioned glomerular hemodynamic changes and proteinuria. Of importance is the observation that these glomerular morphologic alterations progress from early disruptions of epithelial and endothelial cell integrity and mesangial widening to an ultimate glomerular sclerosis and azotemia.[178,180] On the basis of these data, it has been suggested that compensatory increases in SNGFR—via increases in glomerular pressures and flows—may have maladaptive consequences by damaging remaining glomeruli and potentially contributing to the progressive destruction of the remaining glomeruli in renal diseases of diverse origin.[178,180]

In other studies, unilateral nephrectomy accelerates both the progression of glomerular sclerosis and the deterioration of renal function induced by a variety of noxious influences.[178] Further support for this concept has been provided by recent studies in the rat. Azar et al.[170] in the "postsalt" hypertension model and Dworkin et al.[181] in the "desoxycorticosterone acetate–salt" hypertension model have shown an association between glomerular pathology and increases in glomerular capillary pressures and flows. However, no firm conclusions are available to explain how such hemodynamic alterations lead to further glomerular destruction.

It has recently been suggested that high protein intake in the presence of renal injury contributes to the hyperfiltration of remaining glomeruli and, therefore, to their eventual destruction.[182] In this regard, a reduction in the dietary protein intake from 24 to 6% has been shown to significantly attenuate the glomerular hemodynamic changes and to lessen the glomerular morphologic alterations and proteinuria in rats with ablation of 90% of renal mass and with "desoxycorticosterone acetate–salt" hypertension.[180,181] Protein restriction has also been

shown to affect (retard) the rate of progression of nephrotoxic serum nephritis in rats[183,184] and the lupuslike nephropathy of the NZB/NZW mouse.[185] It should be noted that studies in Munich–Wistar rats, chronically fed diets of very low protein content, have demonstrated significant reductions ($\sim$ 35%) in GFR and SNGFR, the consequence of reduced single-nephron plasma flow and LpA.[186] In the remnant kidney model of chronic renal failure, long-term dietary restriction of phosphorus has been shown to retard functional deterioration and to reduce this histologic damage, but the mechanism for this protective effect has not been elucidated.[187] In addition to nutritional factors which may affect the rate of progressive reduction in GFR, the role of hormonal factors (i.e., PTH and thyroid hormone) has recently been assessed. A report by Tomford et al.[188] has indicated that thyroparathyroidectomy prevents the development of chronic renal failure associated with nephrotoxic serum nephritis in rats. Further studies by the same investigators[189] were designed to determine whether this protective effect is mediated through the removal of thyroid hormone or PTH. It was concluded that the protective effect afforded by thyroparathyroidectomy on the functional determination in the nephrotoxic serum nephritis model is mediated by the removal of thyroid hormone.[189] Whether the protective effect of thyroidectomy is hemodynamically mediated has not as yet been examined.

In the aggregate, these investigations support the hypothesis that hyperfiltration by acting upon remaining glomerular capillaries may be responsible for progression of renal failure after an initial insult has reduced functional nephron mass. Although the intrarenal mechanisms leading to hyperfiltration in the remaining glomeruli are a matter of speculation, an impaired autoregulatory mechanism (i.e., tubuloglomerular feedback) may be in part responsible.

20. Summary

Considerable information has accumulated over the past several years on specific mechanisms which underlie a variety of disorders of glomerular filtration. In a variety of altered physiologic states and pathophysiologic conditions, alterations in RPF and LpA remain the dominant mechanisms whereby changes in glomerular ultrafiltration occur, but alterations in hydrostatic forces also contribute a secondary or modulating influence. The specific reasons for alterations in LpA vary among the conditions discussed but include hormonal effects which are presumably functional in nature, specific damage or physical alterations to the glomerular capillary membrane, and, in certain instances, by mechanisms which remain to be elucidated.

References

1. Blantz, R. C. 1978. Disorders of glomerular filtration. In: *Physiology of Membrane Disorders*. T. E. Andreoli, J. F. Hoffman, and D. D. Fanestil, eds. Plenum Press, New York, pp. 967–985.
2. Blantz, R. C. 1980. Segmental renal vascular resistance: Single nephron. *Annu. Rev. Physiol.* **42**:573–588.
3. Chang, R. L. S., W. M. Deen, C. R. Robertson, and B. M. Brenner. 1975. Permselectivity of the glomerular capillary wall. III. Restricted transport of polyanions. *Kidney Int.* **8**:212–218.
4. Brenner, B. M., J. L. Troy, and T. M. Daugharty. 1971. The dynamics of glomerular ultrafiltration in the rat. *J. Clin. Invest.* **50**:1776–1780

5. Blantz, R. C. 1975. The mechanism of acute renal failure after uranyl nitrate. *J. Clin. Invest.* **55**:621–635.

6. Andreucci, V. E., J. Herrera-Acosta, F. C. Rector, Jr., and D. W. Seldin. 1971. Effective glomerular filtration pressure and single nephron filtration rate during hydropenia, elevated ureteral pressure, and acute volume expansion with isotonic saline. *J. Clin. Invest.* **50**:2230–2234.

7. Blantz, R. C., A. H. Israelit, F. C. Rector, Jr., and D. W. Seldin. 1972. The relation of distal tubular NaCl delivery and glomerular hydrostatic pressure. *Kidney Int.* **2**:22–32.

8. Ott, C. E., G. R. Marchand, J. Diaz-Buxo, and F. G. Knox. 1977. The determinants of glomerular filtration in the dog. *Am. J. Physiol.* **231**:235–239.

9. Israelit, A. H., F. C. Rector, Jr., and D. W. Seldin. 1971. Glomerular hydrostatic pressure (P_G) and effective filtration pressure (EFP) in the dog during hydropenia (H), acute volume expansion (VE), and aortic constriction (AC). *Clin. Res.* **19**:534A.

10. Tucker, B. J., and R. C. Blantz. 1977. An analysis of the determinants of nephron filtration rate. *Am. J. Physiol.* **232**:F477–F483.

11. Brenner, B. M., J. L. Troy, T. M. Daugharty, W. M. Deen, and C. R. Robertson. 1972. Dynamics of glomerular ultrafiltration in the rat II. Plasma flow dependence of GFR. *Am. J. Physiol.* **223**:1184–1190.

12. Blantz, R. C. 1977. Dynamics of glomerular ultrafiltration in the rat. *Fed. Proc.* **36**:2602–2608.

13. Blantz, R. C., F. C. Rector, Jr., and D. W. Seldin. 1974. The effect of hyperoncotic albumin expansion upon glomerular ultrafiltration in the rat. *Kidney Int.* **6**:209–221.

14. Baylis, G., I. Ichikawa, W. T. Willis, C. B. Wilson, and B. M. Brenner. 1977. Dynamics of glomerular ultrafiltration. IX. Effects of plasma protein concentration. *Am. J. Physiol.* **232**:F58–F71.

15. Tucker, B. J., and R. C. Blantz. 1981. Effects of glomerular filtration dynamics on the glomerular permeability coefficient. *Am. J. Physiol.* **240**:F245–F254.

16. Tucker, B. J., and R. C. Blantz. 1977. Factors determining superficial nephron filtration in the mature, growing rat. *Am. J. Physiol.* **232**:F97–F104.

17. Myers, B. D., W. Deen, and B. M. Brenner. 1975. Effects of norepinephrine and angiotensin II on the determinants of glomerular ultrafiltration and proximal tubule fluid reabsorption in the rat. *Circ. Res.* **37**:101–110.

18. Blantz, R. C., K. S. Konnen, and B. J. Tucker. 1976. Angiotensin II effects upon the glomerular microcirculation and ultrafiltration coefficient of the rat. *J. Clin. Invest.* **57**:419–434.

19. Baylis, C., W. M. Deen, B. D. Myers, and B. M. Brenner. 1977. Effects of some vasodilator drugs on transcapillary fluid exchange in the renal cortex. *Am. J. Physiol.* **230**:1148–1158.

20. Ichikawa, I., and B. M. Brenner. 1979. Mechanism of action of histamine and histamine antagonists on the glomerular microcirculation of the rat. *Circ. Res.* **45**:737–745.

21. Blantz, R. C. 1980. The glomerulus, passive filter or regulatory organ? *Klin. Wochenschr.* **58**:957–964.

22. Ichikawa, I., and B. M. Brenner. 1977. Evidence for glomerular actions of ADH and dibutyryl cyclic AMP in the rat. *Am. J. Physiol.* **233**:F102–F117.

23. Edwards, R. M. 1983. Response of isolated renal microvessels to intraluminal pressure, norepinephrine and angiotensin II. *Kidney Int.* **23**:243a.

24. Steiner, R. W., and R. C. Blantz. 1979. Acute reversal by saralasin of multiple intrarenal effects of angiotensin II. *Am. J. Physiol.* **237**:F386–F391.

25. Baylis, C., and B. M. Brenner. 1978. Modulation by prostaglandin synthesis inhibitors of the action of exogenous angiotensin II on glomerular ultrafiltration in the rat. *Circ. Res.* **43**:889–898.

26. Ichikawa, I., J. F. Miele, and B. M. Brenner. 1979. Reversal of renal cortical actions of angiotensin II by verapamil and manganese. *Kidney Int.* **16**:137–147.

27. Hornych, H., M. Beaufils, and G. Richet. 1972. The effect of exogenous angiotensin on superficial and deep glomeruli in the rat kidney. *Kidney Int.* **2**:336–343.

28. Sraer, J. D., J. Sraer, R. Ardaillou, and O. Mimoune. 1974. Evidence for renal glomerular receptors for angiotensin II. *Kidney Int.* **6**:241–246.

29. Sraer, J., L. Baud, J. P. Cosyns, P. Verroust, M. P. Nivez, and R. Ardaillou. 1977. High affinity binding of ^{125}I-angiotensin II to rat glomerular basement membranes. *J. Clin. Invest.* **59**:69–81.

30. Ausiello, D. A., J. I. Kreisberg, C. Roy, and M. J. Karnovsky. 1980. Contraction of cultured rat glomerular cells of apparent mesangial origin after stimulation with angiotensin II and arginine vasopressin. *J. Clin. Invest.* **65**:754–760.

31. Beaufils, M., J. Sraer, C. Leprevx, and R. Ardaillou. 1976. Angiotensin II binding to renal glomeruli from sodium-loaded and sodium-depleted rats. *Am. J. Physiol.* **230**:1187–1193.

32. Osborne, M. J., B. Droz, P. Meyer, and F. Morel. 1975. Angiotensin II: Renal localization in glomerular mesangial cells by autoradiography. *Kidney Int.* **8**:245–254.

33. Brown, G. P., J. G. Douglas, and J. Krontiris-Litowitz. 1980. Properties of angiotensin II receptors of isolated rat glomeruli: Factors influencing binding affinity and comparative binding of angiotensin analogs. *Endocrinology* **106**:1923–1929.

34. Barnes, L. D., M. N. Guyman, M. D. Lifschitz, and J. I. Kreisberg. 1981. Angiotensin II receptors in mesangial cells cultured from rat renal glomeruli. *Kidney Int.* **19**:163a.

35. Mahieu, P. R., J. B. Foidart, C. H. Dubois, C. A. Dechenne, and J. Deheneffe. 1980. Tissue culture of normal rat glomeruli: Contractile activity of the cultured mesangial cells. *Invest. Cell Pathol.* **3**:121–128.

36. Blanc, E., J. Sraer, J. D. Sraer, L. Baud, and R. Ardaillou. 1978. Ca^{++} and Mg^{++} dependence of angiotensin II binding to isolated rat glomeruli. *Biochem. Pharmacol.* **27**:517–524.

37. Freer, R. J. 1975. Calcium and angiotensin tachyphylaxis in rat uterine smooth muscle. *Am. J. Physiol.* **228**:1423–1430.

38. Zimmerhackl, B., N. Parekh, H. Kuecherer, and M. Steinhausen. 1982. Influence of angiotensin II on red cell velocity, red cell flux and capillary diameter of surface glomeruli of rats. *Pfluegers Arch.* **394**(Suppl.):R25a.

39. Muller, J., and L. Barajas. 1972. Electron microscopic and histochemical evidence for a tubular innervation in the renal cortex of the monkey. *J. Ultrastruct. Res.* **41**:533–549.

40. Barajas, L., and J. Muller. 1973. The innervation of the juxtaglomerular apparatus and surrounding tubules: A qualitative analysis by serial section electron microscopy. *J. Ultrastruct. Res.* **43**:107–132.

41. Gavendo, S., S. Kapuler, I. Serban, I. Iania, E. Ben-David, and H. Eliahou. 1980. Beta-1 adrenergic receptors in kidney tubular cell membrane in the rat. *Kidney Int.* **17**:764–770.

42. Snavely, M. D., E. Moustafa, H. J. Motulsky, L. C. Mahan, and P. A. Insel. 1982. Beta-adrenergic receptor subtypes in the rat renal cortex: Selective regulation of beta$_1$-adrenergic receptors by pheochromocytoma. *Circ. Res.* **51**:504–513.

43. Snavely, M. D., and P. A. Insel. 1982. Characterization of alpha-adrenergic receptor subtypes in the rat renal cortex: Differential regulation of alpha$_1$-and alpha$_2$-adrenergic receptors by guanylnucleotides and Na$^+$. *Mol. Pharmacol.* **22**:532–546.

44. Felder, R., J. C. Pelayo, A. Wargo, L. Schoelkopf, M. Cooke, P. Jose, and G. Eisner. 1980. Glomerular adrenergic receptors. *Physiologist* **23**:162a.

45. Felder, R., L. Schoelkopf, J. C. Pelayo, M. Blecher, P. Calcagno, G. Eisner, and P. Jose. 1981. Adrenergic and dopaminergic receptors in glomeruli and cortical tubules. *Kidney Int.* **19**:239a.

46. Andreucci, V. E., A. Dal Canton, A. Corradi, R. Stanziale, and L. Migone. 1976. Role of the efferent arteriole in glomerular hemodynamics of superficial nephrons. *Kidney Int.* **9**:475–480.

47. Hoffman, B. B., and R. J. Lefkowitz. 1980. Radioligand binding studies of adrenergic receptors: New insights into molecular and physiological regulation. *Annu. Rev. Pharmacol. Toxicol.* **20**:581–608.

48. Schrier, R. W. 1974. Effects of adrenergic nervous system and catecholamines on systemic and renal hemodynamics, sodium and water excretion, and renin secretion. *Kidney Int.* **6**:291–306.

49. Pelayo, J. C., B. J. Tucker, and R. C. Blantz. 1983. Differential

effects of methoxamine and isoproterenol on glomerular hemodynamics. *Kidney Int.* 23:247a.

50. DiBona, G. F. 1982. The functions of the renal nerves. *Rev. Physiol. Biochem. Pharmacol.* 94:76–181.

51. Keeton, T. K., and W, B. Campbell. 1980. The pharmacologic alteration of renin release. *Pharmacol. Rev.* 32:81–212.

52. Fray, J. C. S. 1980. Stimulus–secretion coupling of renin. *Circ. Res.* 47:485–493.

53. Hermansson, K., M. Larson, O. Kallskog, and M. Wolgast. 1981. Influence of renal nerve activity on arteriolar resistance, ultrafiltration dynamics and fluid reabsorption. *Pfluegers Arch.* 389:85–90.

54. Kon, V., J. L. Troy, and I. Ichikawa. 1982. Effector loci for vasomotor control by renal nerves. *Clin. Res.* 30:454A.

55. Pelayo, J. C., M. G. Ziegler, and R. C. Blantz. 1984. Angiotensin II in adrenergic-induced alterations in glomerular hemodynamics. *Am. J. Physiol.* 247:F799–F807.

56. Pelayo, J. C., M. G. Ziegler, P. A. Jose, and R. C. Blantz. 1983. Renal denervation in the rat: Analysis of glomerular and proximal tubular function. *Am. J. Physiol.* 244:F70–F77.

57. Jamison, R. L. 1981. Urine concentration and dilution. The roles of antidiuretic hormone and urea. In: *The Kidney*, Volume 1, 2nd ed. B. M. Brenner and F. C. Rector, Jr., eds. Saunders, Philadelphia. pp. 495–550.

58. Hays, R. M., and B. D. Levine. 1981. Pathophysiology of water metabolism. In: *The Kidney*, Volume 1, 2nd ed. B. M. Brenner and F. C. Rector, Jr., eds. Saunders, Philadelphia. pp. 777–840.

59. Imbert, M., D. Chabardes, and F. Morel. 1974. Hormone-sensitive adenylate cyclase in isolated rabbit glomeruli. *Mol. Cell. Endocrinol.* 1:295–304.

60. Sraer, J., R. Ardaillou, N. Loreau, and J. D. Sraer. 1974. Evidence for parathyroid hormone sensitive adenylate cyclase in rat glomeruli. *Mol. Cell. Endocrinol.* 1:285–294.

61. Schlondorff, D., P. Yoo, and B. E. Alpert. 1978. Stimulation of adenylate cyclase in isolated rat glomeruli by prostaglandins. *Am. J. Physiol.* 235:458–464.

62. Schor, N., I. Ichikawa, and B. M. Brenner. 1981. Mechanisms of action of various hormones and vasoactive substances on glomerular ultrafiltration in the rat. *Kidney Int.* 20:442–451.

63. Hassid, A., and M. J. Dunn. 1980. Microsomal prostaglandin biosynthesis of human kidney. *J. Biol. Chem.* 255:2472–2475.

64. Stollenwerk, P. A., M. Aikawa, and M. J. Dunn. 1981. Prostaglandin and thromboxane synthesis by rat glomerular epithelial cells. *Kidney Int.* 20:469–474.

65. McGiff, J. C., and P. Y.-K. Wong. 1979. Compartmentalization of prostaglandins and prostacyclin within the kidney: Implications for renal function. *Fed. Proc.* 38:89–93.

66. Dunn, M. J., and E. J. Zambraski. 1980. Renal effects of drugs that inhibit prostaglandin synthesis. *Kidney Int.* 18:609–622.

67. McGiff, J. C. 1981. Prostaglandins, prostacyclin, and thromboxanes. *Annu. Rev. Pharmacol. Toxicol.* 21:479–509.

68. Stokes, J. B. 1981. Integrated actions of renal medullary prostaglandins in the control of water excretion. *Am. J. Physiol.* 240:F471–F480.

69. Dunn, M. J., and L. L. Hood. 1977. Prostaglandins and the kidney. *Am. J. Physiol.* 233:F169–F184.

70. Terragno, N. A. 1981. Prostaglandins, antidiuretic hormone and renin angiotensin system. *Hypertension* 3(Suppl. II):II-65–II-70.

71. Baer, P. G., L. G. Navar, and A. C. Guyton. 1970. Renal autoregulation, filtration rate and electrolyte excretion during vasodilation. *Am. J. Physiol.* 219:619–625.

72. Johnson, H. H., J. P. Herzog, and D. P. Lavler. 1967. Effect of prostaglandin E_1 on renal hemodynamics, sodium and water excretion. *Am. J. Physiol.* 213:939–946.

73. Strandhoy, J. W., C. E. Ott, E. G. Schneider, L. R. Willis, N. P. Beck, B. B. Davis, and F. G. Knox. 1974. Effects of prostaglandin E_1 and E_2 on renal sodium reabsorption and Starling forces. *Am. J. Physiol.* 226:1015–1021.

74. Ichikawa, I., J. M. Pfeffer, M. A. Pfeffer, T. H. Hostetter, E. Braunwald, and B. M. Brenner. 1983. Glomerular response to severe congestive heart failure in the rat. *Kidney Int.* 23:245a.

75. Scharschmidt, L. A., M. J. Dunn, and A. Norris. 1983. Prostaglandins modulate angiotensin II stimulated glomerular contractility. *Kidney Int.* 23:284a.

76. Ichikawa, I., H. D. Humes, T. P. Dousa, and B. M. Brenner. 1978. Influence of parathyroid hormone on glomerular ultrafiltration in the rat. *Am. J. Physiol.* 234:F393–F401.

77. Humes, H. D., I. Ichikawa, J. L. Troy, and B. M. Brenner. 1978. Evidence for a parathyroid hormone-dependent influence of calcium on the glomerular ultrafiltration coefficient. *J. Clin. Invest.* 61:32–40.

78. Dousa, T. P., L. D. Barnes, S. H. Ong, and A. L. Steiner. 1977. Immunohistochemical localization of $3',5'$-cyclic AMP and $3',5'$-cyclic GMP in rat renal cortex: Effect of parathyroid hormone. *Proc. Natl. Acad. Sci. USA* 74:3569–3573.

79. Torres, V. E., T. E. Northrup, R. M. Edwards, S. V. Shah, and T. P. Dousa. 1978. Modulation of cyclic nucleotides in isolated rat glomeruli. *J. Clin. Invest.* 62:1334–1343.

80. Deen, W. M., C. R. Robertson, and B. M. Brenner. 1974. Glomerular ultrafiltration. *Fed. Proc.* 33:14–20.

81. Davis, J. O., and D. S. Howell. 1953. Comparative effect of ACTH, cortisone and DOCA on renal function, electrolyte excretion and water exchange in normal dogs. *Endocrinology* 52:245–255.

82. De Bermudez, L., and J. P. Hayslett. 1972. Effect of methylprednisolone on renal function and the zonal distribution of blood flow in the rat. *Circ. Res.* 33:44–52.

83. Baylis, C., and B. M. Brenner. 1978. Mechanism of the glucocorticoid-induced increase in glomerular filtration rate. *Am. J. Physiol.* 234:F166–F170.

84. Campbell, W. B., R. M. Graham, and E. K. Jackson. 1979. Role of renal prostaglandins in sympathetically mediated renin release in the rat. *J. Clin. Invest.* 64:448–456.

85. Hofbauer, K. G., A. Konrads, K. Schwartz, and U. Werner. 1978. Role of cyclic AMP in the regulation of renin release from the isolated perfused rat kidney. *Klin. Wochenschr.* 56(Suppl. 1):51–59.

86. Michelakis, A. M., J. Caudle, and G. W. Liddle. 1969. *In vitro* stimulation of renin production by epinephrine, norepinephrine and cyclic AMP. *Proc. Soc. Exp. Biol. Med.* 130:748–753.

87. Okahara, T., T. Abe, and K. Yamamoto. 1977. Effects of dibutyryl cyclic AMP and propranolol on renin secretion in dogs. *Proc. Soc. Exp. Biol. Med.* 159:213–218.

88. Wimer, N., D. S. Chokshi, and N. G. Walkenhorst. 1971. Effects of cyclic AMP, sympathomimetic amines and adrenergic receptors antagonist on renin secretion. *Circ. Res.* 29:239–248.

89. Robertson, C. R., W. M. Deen, J. L. Troy, and B. M. Brenner. 1972. Dynamics of glomerular ultrafiltration in the rat. III. Hemodynamics and autoregulation. *Am. J. Physiol.* 223:1191–1200.

90. Deen, W. M., C. R. Robertson, and B. M. Brenner. 1973. The dynamics of glomerular filtration coefficient. *J. Clin. Invest.* 52:1500–1508.

91. Ichikawa, I., and B. M. Brenner. 1980. Importance of efferent arteriolar vascular tone in regulation of proximal tubule fluid reabsorption and glomerulotubular balance in the rat. *J. Clin. Invest.* 65:1192–1201.

92. Navar, L. G., D. W. Ploth, and P. D. Bell. 1980. Distal tubular feedback control of renal hemodynamics and autoregulation. *Annu. Rev. Physiol.* 42:557–571.

93. Blantz, R. C. 1974. Effect of mannitol on glomerular ultrafiltration in the hydropenic rat. *J. Clin. Invest.* 54:1135–1143.

94. Steiner, R. W., B. J. Tucker, and R. C. Blantz. 1979. Glomerular hemodynamics in rats with chronic sodium depletion. *J. Clin. Invest.* 64:503–512.

95. Tucker, B. J., and R. C. Blantz. 1983. Mechanism of altered glomerular hemodynamics during chronic sodium depletion. *Am. J. Physiol.* 244:F11–F18.

96. Schor, N., I. Ichikawa, and B. M. Brenner. 1980. Glomerular adaptations to chronic dietary salt restriction or excess. *Am. J. Physiol.* 238:F428–F436.

97. Kiil, F., and K. Aukland. 1961. Renal concentration mechanism

and hemodynamics at increased ureteral pressure during osmotic and saline diuresis. *Scand. J. Clin. Invest.* **13**:276–287.

98. Suki, W. N., A. G. Guthrie, M. Martinez-Maldonado, and G. Eknoyan. 1971. Effects of ureteral pressure elevation on renal hemodynamics and urine concentration. *Am. J. Physiol.* **220**:38–43.

99. Blantz, R. C., K. Konnen, and B. J. Tucker. 1975. Glomerular filtration response to elevated ureteral pressure in both the hydropenic and plasma expanded rat. *Circ. Res.* **37**:819–829.

100. DiBona, G. F. 1971. Effect of mannitol diuresis and ureteral occlusion on distal tubular reabsorption. *Am. J. Physiol.* **221**:511–514.

101. Steinhausen, M. 1967. Measurement of tubular urine flows and tubular reabsorption under increased ureteral pressure. *Pfluegers Arch.* **298**:105–130.

102. Kallskog, O., and M. Wolgast. 1975. Effect of elevated interstitial pressure on the renal cortical hemodynamics. *Acta Physiol. Scand.* **95**:364–372.

103. Dal Canton, A., R. Stanziale, A. Corradi, V. E. Andreucci, and L. Migone. 1977. Effects of acute ureteral obstruction on glomerular hemodynamics in the rat kidney. *Kidney Int.* **12**:403–411.

104. Dal Canton, A., A. Corradi, R. Stanziale, G. Maruccio, and L. Migone. 1979. Effects of 24 hours unilateral ureteral obstruction on glomerular hemodynamics in rat kidney. *Kidney Int.* **15**:457–462.

105. Arendshorst, W. J., W. F. Finn, and C. W. Gottschalk. 1974. Nephron stop-flow pressure response to obstruction for 24 hours in the rat kidney. *J. Clin. Invest.* **53**:1497–1500.

106. Dal Canton, A., A. Corradi, R. Stanziale, G. Maruccio, and L. Migone. 1980. Glomerular hemodynamics before and after release of 24-hour bilateral ureteral obstruction. *Kidney Int.* **17**:491–496.

107. Humes, H. D., R. A. Dieppa, and B. M. Brenner. 1980. Glomerular hemodynamics in rats with hereditary hydronephrosis. *Invest. Urol.* **18**:46–51.

108. Ichikawa, I., and B. M. Brenner. 1979. Local intrarenal vasoconstrictor–vasodilator interactions in mild partial ureteral obstruction. *Am. J. Physiol.* **236**:F131–F140.

109. Kishimoto, T., M. Maekawa, Y. Abe, and K. Yamamoto. 1973. Intrarenal distribution of blood flow and renin release during renal venous pressure elevation. *Kidney Int.* **4**:259–266.

110. Miyazake, M., and J. McNay. 1971. Redistribution of renal cortical blood flow during ureteral occlusion and renal venous constriction. *Proc. Soc. Exp. Biol. Med.* **138**:454–461.

111. Lewy, J. E., and E. E. Windhager. 1968. Peritubular control of proximal tubular fluid reabsorption in the rat kidney. *Am. J. Physiol.* **214**:943–954.

112. Rodicio, J., J. Herrera-Acosta, J. C. Sellman, F. C. Rector, Jr., and D. W. Seldin. 1969. Studies on glomerulo-tubular balance during aortic constriction, ureteral obstruction and venous occlusion in hydropenic and saline-expanded rats. *Nephron* **6**:437–456.

113. Tucker, B. J., and R. C. Blantz. 1978. Determinants of proximal tubular reabsorption as mechanisms of glomerulotubular balance. *Am. J. Physiol.* **235**:F142–F150.

114. Dilley, J. R., A. Corradi, and W. J. Arendshorst. 1983. Glomerular ultrafiltration dynamics during increased renal venous pressure. *Am. J. Physiol.* **244**:F650–F658.

115. Corradi, A., and W. J. Arendshorst. 1982. Effect of increased renal venous pressure on renal hemodynamics in the rat. *Kidney Int.* **21**:241a.

116. Corradi, A., and W. J. Arendshorst. 1982. Influence of renal nerves and prostaglandins in renal hemodynamic responses to elevated renal venous pressure in the rat. *Fed. Proc.* **41**:1256a.

117. Davison, J. M., and W. Dunlop. 1980. Renal hemodynamics and tubular function in normal human pregnancy. *Kidney Int.* **18**:152–161.

118. Alexander, E. A., S. Churchill, and H. H. Bengele. 1980. Renal hemodynamics and volume homeostasis during pregnancy in the rat. *Kidney Int.* **18**:173–178.

119. Baylis, C. 1980. The mechanism of the increase in glomerular filtration rate in the 12-day pregnant rat. *J. Physiol. (London)* **305**:405–414.

120. Dal Canton, A., G. Conte, C. Sposito, G. Fuiano, R. Guasco, D. Russo, M. Sabbatini, F. Uccello, and V. E. Andreucci. 1982. Effects of pregnancy on glomerular hemodynamics: Micropuncture study in the rat. *Kidney Int.* **22**:608–612.

121. Baylis, C. 1982. Glomerular ultrafiltration in the pseudopregnant rat. *Am. J. Physiol.* **234**:F300–F305.

122. Baylis, C. 1979/80. The effect of early pregnancy on glomerular filtration rate and plasma volume in the rat. *Renal Physiol.* **2**:333–339.

123. Matthews, B. M., and D. W. Taylor. 1960. Effects of pregnancy on inulin and para-amino hippurate clearances in the anesthetized rat. *J. Physiol. (London)* **151**:385–389.

124. Blantz, R. C., and J. C. Pelayo. 1984. A functional role for the tubuloglomerular feedback mechanism. *Kidney Int.* **25**:739–746.

125. Thurau, K. 1964. Renal hemodynamics. *Am. J. Med.* **36**:698–719.

126. Schnermann, J., F. S. Wright, J. M. Davis, W. V. Stackelberg, and G. Grill. 1970. Regulation of superficial nephron filtration rate by tubuloglomerular feedback. *Pfluegers Arch.* **318**:147–175.

127. Blantz, R. C., and K. S. Konnen. 1977. Relation of distal tubule delivery rate and reabsorptive rate to nephron filtration. *Am. J. Physiol.* **233**:F315–F324.

128. Hierholzer, K., R. Muller-Suur, H.-U. Gutsche, M. Butz, and I. Lichtenstein. 1974. Filtration in surface glomeruli as regulated by flow rate through the loop of Henle. *Pfluegers Arch.* **352**:315–337.

129. Tucker, B. J., R. W. Steiner, L. C. Gushwa, and R. C. Blantz. 1978. Studies on the tubuloglomerular feedback system in the rat: The mechanism of reduction in filtration rate with benzolamide. *J. Clin. Invest.* **62**:993–1004.

130. Tucker, B. J., and R. C. Blantz. 1980. Studies on the mechanism of reduction in glomerular filtration rate after benzolamide. *Pfluegers Arch.* **388**:211–216.

131. Schnermann, J., J. Briggs, and F. S. Wright. 1981. Feedback mediated reduction of glomerular filtration rate during infusion of hypertonic saline. *Kidney Int.* **20**:462–468.

132. Pelayo, J. C. and R. C. Blantz. 1984. Analysis of renal denervation in the hydropenic rat: Interactions with angiotensin II. *Am. J. Physiol.* **246**:F87–F95.

133. Blantz, R. C., B. J. Tucker, L. Gushwa, and O. W. Peterson. 1983. Mechanism of diuresis following acute modest hyperglycemia in the rat. *Am. J. Physiol.* **244**:F185–F194.

134. Blantz, R. C., O. W. Peterson, L. C. Gushwa, and B. J. Tucker. 1982. Effect of modest hyperglycemia on tubuloglomerular feedback activity. *Kidney Int.* **22**:S206–S212.

135. Persson, A. E. G., R. Muller-Suur, and G. Selen. 1976. Peritubular capillary oncotic pressure as a modifier of tubuloglomerular feedback. *Kidney Int.* **19**:595–603.

136. Persson, A. E. G., R. Muller-Suur, and G. Selen. 1979. Capillary oncotic pressure as a modifier for tubuloglomerular feedback. *Am. J. Physiol.* **236**:F97–F102.

137. Wright, F. S., and J. Schnermann. 1974. Interference with feedback control of glomerular filtration rate by furosemide, triflocin and cyanide. *J. Clin. Invest.* **53**:1695–1708.

138. Tucker, B. J., and R. C. Blantz. 1984. Effect of furosemide administration on glomerular and tubular dynamics in the rat. *Kidney Int.* **26**:112–121.

139. Moore, L. C., S. Yarimazu, G. Schubert, P. C. Weber, and J. Schnermann. 1980. Dynamics of tubuloglomerular feedback adaptation to acute and chronic changes in body fluid volume. *Pfluegers Arch.* **387**:39–45.

140. Briggs, J. P., and F. S. Wright. 1979. Feedback control of glomerular filtration rate: Site of the effector mechanism. *Am. J. Physiol.* **236**:F40–F47.

141. Ichikawa, I. 1982. Direct analysis of the effector mechanism of the tubuloglomerular feedback system. *Am. J. Physiol.* **243**:F447–F455.

142. Blantz, R. C. 1985. Intrinsic renal failure: Acute, In: *The Kidney: Normal and Abnormal Function*. D. Seldin and G. Giebisch, eds. Raven Press, New York, in press.

143. Tanner, G. A., K. L. Sloan, and S. Sophason. 1973. Effects of renal artery occlusion on kidney function in the rat. *Kidney Int.* 4:377–389.

144. Williams, R. H., C. E. Thomas, L. G. Navar, and A. P. Evan. 1981. Hemodynamic and single nephron function during the maintenance phase of ischemic acute renal failure in the dog. *Kidney Int.* 19:503–515.

145. Cox, J. W., R. W. Baehler, H. Sharma, T. O'Dorisio, R. W. Osgood, J. H. Stein, and T. F. Ferris. 1974. Studies on the mechanism of oliguria in a model of acute renal failure. *J. Clin. Invest.* 53:1546–1558.

146. Conger, J. D., J. B. Robinette, and S. J. Guggenheim. 1981. Effect of acetylcholine on the early phase of reversible norepinephrine-induced acute renal failure. *Kidney Int.* 19:399–409.

147. Myers, B., F. Chui, M. Hilberman, and A. Michaels. 1979. Transtubular leakage of glomerular filtrate in human acute renal failure. *Am. J. Physiol.* 237:F319–F325.

148. Bank, N., B. F. Mutz, and H. Aynedian. 1967. The role of "leakage" of tubular fluid in anuria due to mercury poisoning. *J. Clin. Invest.* 46:695–704.

149. Stein, J. H., J. Gottschall, R. W. Osgood, and T. F. Ferris. 1975. Pathophysiology of nephrotoxic model of acute renal failure. *Kidney Int.* 8:27–41.

150. Avasthi, P. S., A. P. Evan, and D. Hay. 1980. Glomerular endothelial cells in uranyl nitrate-induced acute renal failure in rats. *J. Clin. Invest.* 65:121–127.

151. Baylis, C., H. G. Rennke, and B. M. Brenner. 1977. Mechanisms of the defects in glomerular ultrafiltration associated with gentamicin administration. *Kidney Int.* 12:344–352.

152. Avasthi, P. S., A. P. Evan, J. W. Huser, and F. C. Luft. 1981. Effect of gentamicin on glomerular ultrastructure. *J. Lab. Clin. Med.* 98:444–454.

153. Blantz, R. C., J. C. Pelayo, L. C. Gushwa, P. S. Avasthi, and A. P. Evan. 1983. Evidence of a functional basis for the glomerular microcirculatory and ultrastructural alterations during uranyl-nitrate-induced acute renal failure in the rat. *Kidney Int.* in press.

154. Schor, N., I. Ichikawa, H. G. Rennke, J. L. Troy, and B. M. Brenner. 1981. Pathophysiology of altered glomerular function in aminoglycoside treated rats. *Kidney Int.* 19:288–296.

155. Conger, J., and S. Falk. 1977. Intrarenal dynamics in the pathogenesis and prevention of acute urate nephropathy. *J. Clin. Invest.* 59:786–793.

156. Allison, M. E. M., C. B. Wilson, and C. W. Gottschalk. 1974. Pathophysiology of experimental glomerulonephritis in rats. *J. Clin. Invest.* 53:1402–1423.

157. Maddox, D. A., C. M. Bennett, W. M. Deen, R. J. Glassock, D. Knutson, T. M. Daugharty, and B. M. Brenner. 1975. Determinants of glomerular filtration in experimental glomerulonephritis in the rat. *J. Clin. Invest.* 55:305–318.

158. Bennett, C. M., R. J. Glassock, R. L. S. Chang, W. M. Deen, C. R. Robertson, and B. M. Brenner. 1976. Permselectivity of the glomerular capillary wall: Studies of experimental glomerulonephritis in the rat using dextran sulfate. *J. Clin. Invest.* 57:1287–1299.

159. Blantz, R. C., and C. B. Wilson. 1976. Acute effects of antiglomerular basement membrane antibody on the process of glomerular filtration in the rat. *J. Clin. Invest.* 58:899–911.

160. Blantz, R. C., B. J. Tucker, and C. B. Wilson. 1978. The acute effects of anti-glomerular basement membrane antibody on the process of glomerular filtration in the rat. II. Influence of dose and complement depletion. *J. Clin. Invest.* 61:910–921.

161. Tucker, B. J., C. B. Wilson, L. C. Gushwa, and R. C. Blantz. 1985. Effect of leukocyte (LC) depletion on glomerular dynamics during acute glomerular immune injury. *Kidney Int.* in press.

162. Blantz, R. C., B. J. Tucker, L. Gushwa, O. Peterson, and C. B. Wilson. 1981. Glomerular immune injury in the rat: The influence of AII and α-adrenergic inhibitors. *Kidney Int.* 20:452–461.

163. Wilson, C. B., L. C. Gushwa, O. W. Peterson, B. J. Tucker, and R. C. Blantz. 1981. Glomerular immune injury in the rat: The effect of antagonists of histamine activity. *Kidney Int.* 20:628–635.

164. Ichikawa, I., J. R. Hoyer, M. W. Seiler, and B. M. Brenner. 1982. Mechanisms of tubuloglomerular balance in the setting of heterogeneous glomerular injury. *J. Clin. Invest.* 69:185–198.

165. Blantz, R. C., T. M. Hostetter, and B. M. Brenner. 1979. Functional adaptations of the kidney to immunological injury. In: *Immunologic Mechanisms of Renal Disease.* Ċ. Wilson, B. Brenner, and J. Stein, eds. Churchill–Livingstone, Edinburgh. pp. 122–143.

166. Bohrer, M. D., C. Baylis, C. R. Robertson, and B. M. Brenner. 1977. Mechanism of the puromycin-induced defects in the transglomerular passage of water and macromolecules. *J. Clin. Invest.* 60:152–161.

167. Ichikawa, I., H. G. Rennke, J. R. Hoyer, K. F. Badr, N. Schor, J. L. Troy, C. P. Lechene, and B. M. Brenner. 1983. Role for intrarenal mechanisms in the impaired salt excretion of experimental nephrotic syndrome. *J. Clin. Invest.* 71:91–103.

168. Schweitzer, G., and K. H. Gertz. 1979. Changes of hemodynamics and glomerular ultrafiltration in renal hypertension of rats. *Kidney Int.* 15:134–143.

169. Steiner, R. W., B. J. Tucker, L. C. Gushwa, J. Gifford, C. B. Wilson, and R. C. Blantz. 1982. Glomerular hemodynamics in a moderate form of Goldblatt hypertension in the rat. *Hypertension* 4:51–57.

170. Azar, S., M. A. Johnson, B. Hertel, and L. Tobian. 1977. Single nephron pressures, flows and resistances in hypertensive kidneys with nephrosclerosis. *Kidney Int.* 11:28–36.

171. Savin, V. J., and D. Terreros. 1982. A study of filtration in single isolated mammalian glomeruli. *Kidney Int.* 20:188–197.

172. Cachia, R., V. J. Savin, R. V. Patak, and S. M. Ridge. 1981. Effect of mercury chloride and uranyl nitrate on ultrafiltration coefficient in isolated glomeruli. *Kidney Int.* 19:245a.

173. Ridge, S. M., R. V. Patak, and V. J. Savin. 1981. Effect of salt depletion and hemorrhage on ultrafiltration coefficient in isolated glomeruli. *Kidney Int.* 19:254a.

174. Osgood, R. W., M. Hanley, H. J. Reineck, C. B. Wilson, M. Venkatachalam, and J. H. Stein. 1981. Isolated perfusion of isolated dog glomeruli. *Kidney Int.* 19:251a.

175. Osgood, R. W., H. J. Reineck, and J. H. Stein. 1982. Methodologic considerations in the study of glomerular ultrafiltration. *Am. J. Physiol.* 242:F1–F7.

176. Bricker, N. S., and L. G. Fine. 1981. The renal response to progressive nephron loss. In: *The Kidney*, Volume 1, 2nd ed. B. M. Brenner and F. C. Rector, Jr., eds. Saunders, Philadelphia. pp. 1056–1096.

177. Hayslett, J. P. 1979. Functional adaptation to reduction in renal mass. *Physiol. Rev.* 59:137–164.

178. Hostetter, T. H., and B. M. Brenner. 1981. Glomerular adaptations to renal injury. In: *Contemporary Issues in Nephrology*, Volume 7. B. M. Brenner and J. H. Stein, eds. Churchill–Livingstone, Edinburgh. pp. 1–27.

179. Deen, W. M., D. A. Maddox, C. R. Robertson, and B. M. Brenner. 1974. Dynamics of glomerular ultrafiltration in the rat. VII. Response to reduced renal mass. *Am. J. Physiol.* 227:556–562.

180. Hostetter, T. H., J. L. Olson, H. G. Rennke, M. A. Venkatachalam, and B. M. Brenner. 1981. Hyperfiltration in remnant nephrons: A potentially adverse response to renal ablation. *Am. J. Physiol.* 241:F85–F93.

181. Dworkin, L. D., T. H. Hostetter, H. G. Rennke, and B. M. Brenner. 1982. Evidence for a hemodynamic basis for glomerular injury in hypertension. *Kidney Int.* 21:229a.

182. Brenner, B. M., T. W. Meyer, and T. H. Hostetter. 1982. Dietary protein intake and the progressive nature of kidney disease: The role of hemodynamically mediated glomerular injury in the pathogenesis of progressive glomerular sclerosis in aging, renal ablation, and intrinsic renal disease. *N. Engl. J. Med.* 307:632–659.

183. Farr, L. E., and J. E. Smadel. 1939. The effect of dietary protein on the course of nephrotoxic nephritis in rats. *J. Exp. Med.* 70:615–627.

184. Neugarten, J., H. Feiner, R. G. Schacht, and D. S. Baldwin. 1982. Ameliorative effect of dietary protein restriction on the course of nephrotoxic serum nephritis. *Clin. Res.* **30**:541A.

185. Friend, P. S., G. Fernandes, R. A. Good, A. F. Michael, and E. J. Yunis. 1978. Dietary restrictions early and late: Effects on the nephropathy of the NZB × NZW mouse. *Lab. Invest.* **38**:629–632.

186. Ichikawa, I., M. L. Purkerson, S. Klahr, J. L. Troy, M. Martinez-Maldonado, and B. M. Brenner. 1980. Mechanism of reduced glomerular filtration rate in chronic malnutrition. *J. Clin. Invest.* **65**:982–988.

187. Karlinsky, M. L., L. Haut, B. Buddington, N. A. Schrier, and A. C. Alfrey. 1980. Preservation of renal function in experimental glomerulonephritis. *Kidney Int.* **17**:293–302.

188. Tomford, C. R., M. L. Karlinsky, B. Buddington, and A. C. Alfrey. 1981. Effect of thyroparathyroidectomy and parathyroidectomy on renal function and the nephrotic syndrome in rat nephrotoxic serum nephritis. *J. Clin. Invest.* **68**:655–664.

189. Tomford, C. R., B. Buddington, and A. C. Alfrey. 1982. Thyroidectomy prevents chronic renal insufficiency in nephrotoxic serum nephritis in rats. *Kidney Int.* **21**:231a.

The Hypertonic and Hypotonic Syndromes

R. Michael Culpepper, Steven C. Hebert, and Thomas E. Andreoli

1. Introduction

Extracellular fluid (ECF) osmotic homeostasis is regulated by a neurorenal axis involving thirst, antidiuretic hormone (ADH), and ADH-responsive renal epithelia (see Chapter 38). Osmoregulatory disorders can best be understood through consideration of two of the cardinal physiologic processes involved in osmotic homeostasis: the water repletion reaction and the cell volume regulatory response. The former provides a frame of reference for understanding the pathogenesis of these disorders, while the latter provides a means for considering the changes in brain volume which can attend osmoregulatory disorders.

1.1. Regulation of ECF Osmolality

In normal individuals, the serum osmolality is virtually constant from day to day and the serum Na^+ concentration is a highly accurate index to body water osmolality. In fact, the reported ranges of normal values for serum Na^+ concentrations or osmolalities in healthy individuals depend on small differences in body water osmolality among individuals, rather than on variations in the solute to water ratio of body fluids in a given individual. A detailed analysis of the water repletion reaction will be presented in subsequent sections of this chapter, but Fig. 1 presents a brief analysis of the key elements involved in water homeostasis. The solid arrows indicate mechanisms activated by changes in effective ECF osmolality, and consequently by changes in cell volume; the dashed arrows indicate mechanisms activated by changes in effective circulating volume.

Figure 1 illustrates the sensor elements that adjust water balance. The osmoreceptors, both for ADH release and for thirst, respond to as little as a 2% increase in ECF osmolality produced by solutes which cause shrinkage of osmoreceptor cells, for example, by NaCl but not by urea. The osmoreceptors stimulate the release of ADH from storage sites in the posterior pituitary and stimulate water ingestion.

A second way of stimulating both ADH release and thirst involves volume-mediated stimuli which can operate independently of changes in plasma osmolality. When the effective circulating volume is reduced by approximately 10%, these volume-dependent mechanisms stimulate ADH release. Activation of extrarenal baroreceptors by blood volume depletion produces afferent signals, carried by cranial nerves IX and X, which result in nonosmotic ADH release and thirst. Volume contraction may also stimulate thirst by way of angiotensin II release.

The cardinal characteristics of the antidiuretic response depend primarily on the integrated activity of two regions of the nephron (see Chapter 38): the medullary thick ascending limb of Henle, or urinary diluting segment; and the collecting duct, or urinary concentrating segment. The medullary thick ascending limb absorbs a large fraction of the filtered load of NaCl, and the absorbed salt accounts for a significant fraction of the renal medullary interstitial hypertonicity. However, since the medullary thick limb is water impermeable, salt abstraction from this nephron segment accounts simultaneously: for the development of medullary hypertonicity, thus permitting maximal antidiuresis in the presence of ADH; and for the appearance of maximally dilute urine in early distal convoluted segments, thus permitting maximal water diuresis in the absence of ADH. When ADH is absent, the water permeability of collecting ducts is low and there is reduced absorption of tubular fluid, which escapes unchanged as hypotonic urine. During antidiuresis, ADH increases the water permeability of collecting ducts and there is osmotic equilibration of tubular fluid with the hypertonic medullary interstitium and, consequently, conservation of body water.

R. Michael Culpepper and Steven C. Hebert • Division of Nephrology, University of Texas Medical School, Houston, Texas 77225. Thomas E. Andreoli • Departments of Internal Medicine, and Physiology and Cell Biology, University of Texas Medical School, Houston, Texas 77225. Present address of S. C. H.: Department of Internal Medicine, Brigham and Women's Hospital, Boston, Massachusetts.

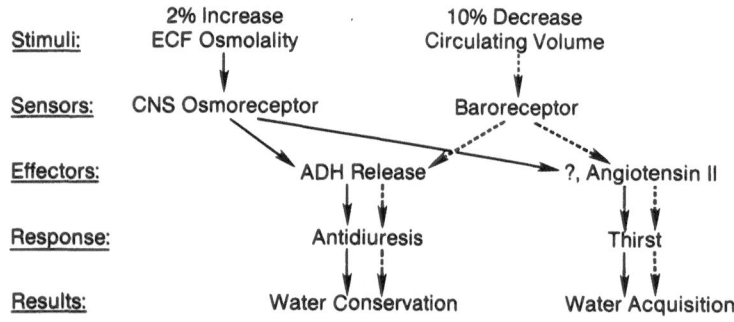

Fig. 1. A schematic illustration of the water repletion reaction. Solid arrows indicate osmotically stimulated pathways and dashed arrows indicate volume-stimulated pathways.

Certain features of the water repletion reaction are particularly noteworthy. First, the redundant mechanisms of thirst and the antidiuretic effect protect osmotic homeostasis. Second, the water repletion reaction operates at varying levels of sensitivity. Osmotic release of vasopressin and osmotic stimuli to thirst require only small (1–2%) changes in effective ECF osmolality, while nonosmotic stimuli to thirst and to ADH release occur only in association with rather large (8–10%) reductions in effective circulating volume.

1.2. Regulation of Cell Volume

Fluid exchange between extracellular and intracellular compartments normally functions to maintain constancy of cell volume and to maintain a negligible hydrostatic pressure gradient between cells and the ECF. Since cell membranes are freely permeable to water, these goals are achieved when the ECF osmolality is normal and when intracellular and extracellular osmolalities are identical. The water repletion reaction (Fig. 1) acts to maintain a normal ECF osmolality. The general mechanism for maintaining identical osmolalities between intracellular and extracellular fluids involves balancing the tendency for dissipative, ionic leak processes to reach equilibrium and the conservative action of the ubiquitous Na^+,K^+-ATPase to extrude Na^+ from cells while pumping K^+ into cells.

To be specific, both Na^+ leakage from the ECF into cells and K^+ leakage from cells into the ECF are counterbalanced exactly by active outward Na^+ transport coupled to active inward K^+ transport, both mediated via the Na^+,K^+ATPase. These active transport events maintain the intracellular cation (and therefore osmolar) content equal to that of ECF, making cells operationally impermeable to both Na^+ and K^+.

Thus, in normal individuals, the Na^+,K^+-ATPase maintains the equality of cation concentrations across cell membranes, while the water repletion reaction (Fig. 1) determines the water to solute ratio in body fluids. When the effective ECF osmolality is increased or decreased, additional processes, to be considered in subsequent sections of this chapter, are required to maintain the constancy of cell volume.

1.3. The Clinical Syndromes: Definitions

It is evident that either hypertonic or hypotonic deviations from osmotic homeostasis will depend on derangements in the operation of the water repletion reaction. The clinical manifestations of these disorders will be referable primarily to alterations in cell volume, particularly in the central nervous system, or to changes in effective circulating volume.

The *hypertonic syndromes* include those disorders in which the ratio of solutes to water in body fluids is increased. These disorders occur when water intake is less than the sum of renal plus extrarenal water losses; in the steady state, net water balance may be zero. Depending on the underlying pathophysiology, the hypertonic syndromes may be grouped into the following general categories:

1. Pituitary diabetes insipidus, in which there is absent or diminished production and secretion of ADH
2. Nephrogenic diabetes insipidus, either familial or acquired, in which collecting duct cells are partially or completely unresponsive to ADH
3. Solute diuresis, in which excessively high rates of solute delivery overwhelm quantitatively the ability of the loop of Henle to dissociate solute and water absorption
4. Renal concentrating disorders, in which there is impaired generation of a hypertonic medullary interstitium by renal countercurrent multiplication and exchange processes
5. Extrarenal water loss, in which losses of naturally hypotonic fluids, such as sweat or gastrointestinal fluids, are greatly in excess of normal and are not replaced with free water

The *hypotonic syndromes* include those disorders in which the ratio of solutes to water in body fluids is reduced, and the serum osmolality and serum Na^+ concentration are reduced in parallel. The hypotonic syndromes develop when water intake exceeds the sum of renal plus extrarenal water losses; in chronic hyponatremia, water intake and water output may be equal. Thus, body fluid hypotonicity may occur when there is a primary increase in water ingestion (primary polydipsia). However, the most common reason for clinically significant hyponatremia is a disturbance in water excretion because of an inability of the kidney to excrete a maximally dilute urine. This inability to dilute urine maximally may occur because of a reduction in the rate of filtrate delivery to the diluting segment, i.e., the thick ascending limb of Henle; because of sustained nonosmotic ADH release; or because of a combination of these factors.

2. Antidiuretic Hormone

Renal conservation or elimination of water as a part of the operation of osmoregulatory mechanisms is governed by the level of circulating ADH. This section will consider the biosynthetic origin of ADH within hypothalamic nuclei, the transport of ADH to storage sites within the posterior pituitary, the functional chemistry of ADH relating to activity and metabolism, and the stimuli for both osmotic and nonosmotic release of ADH from the posterior pituitary into the blood.

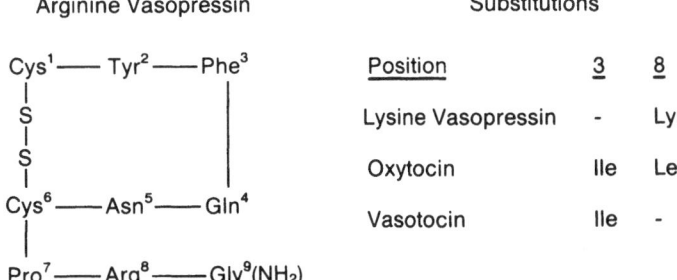

Fig. 2. Chemical structure of the major posterior pituitary hormones.

2.1. The Neurohypophysis

The neurohypophysis consists of the hypothalamic nuclei containing the cell bodies of the magnocellular neurons which synthesize vasopressin; the axonal processes of these neurons, which form the supraoptico-hypophyseal tract; and the termini of these neurons within the posterior lobe of the pituitary. The supraoptic nucleus (SON) and the paraventricular nucleus (PVN) are located in the hypothalamus juxtaposed to the third ventricle. Immunocytochemical staining has demonstrated the presence of cells containing vasopressin and oxytocin in both nuclei,[1,2] but current evidence indicates that the hormones are located in separate cells.[2,3] The unmyelinated axons of the magnocellular neurons terminate in the posterior lobe (pars nervosa) of the pituitary, comprising about 40% of the bulk of the gland. These axons terminate in sacculations which contain electronlucent vesicles and which abut on basement membranes near posterior pituitary capillaries. Preterminal dilations which contain Gomori-positive granules but no vesicles are seen removed from capillary contact.[4]

2.2. Hormone Biosynthesis, Transport, and Metabolism

The hormones elaborated by most mammalian neurohypophyses, oxytocin and arginine vasopressin (AVP), are nonapeptides of approximately 1100 daltons (Fig. 2). A sulfhydryl bond between the cysteine residues at positions 1 and 6 forms a single cystine moiety joining a ring of 20 atoms. AVP is the ADH in all mammals except the suborder Suina, in which lysine vasopressin occurs; the ADH among lower vertebrates is arginine vasotocin. Figure 2 also illustrates the amino acid substitutions which differentiate the three hormones.

2.2.1. Neurosecretory Granules

Scharrer and Scharrer[5] first identified neurosecretory cells by their content of Gomori-stainable granules alongside classical Nissl bodies, their long axonal processes, and their close association with an intricate capillary bed. Stainable neurosecretory granules (NSGs) have been demonstrated in cell bodies of the SON and PVN, along the length of the supraoptico-hypophyseal tract, and in nerve endings in the posterior pituitary.[6] Neurosecretory material accumulates at the proximal stump of a severed pituitary stalk,[7] and during dehydration NSGs disappear from the posterior pituitary in proportion to reductions in vasopressin content; repletion of posterior pituitary NSGs occurs with hydration.[7] These observations led Scharrer and Scharrer to conclude that the posterior pituitary is a storage depot for ADH produced in hypothalamic cells and transported via the axons as NSGs.[5] Axonal NSG transport rates of about 200 mm/day have been compared to the approximately 4 mm/day rate for axonal protoplasmic flow to infer facilitated axonal transport of NSGs.

Weinstein et al.[8] demonstrated the presence of vasopressin in NSGs isolated from dog posterior pituitary, showed the isolated particles to be morphologically identical to in situ NSGs and the isolated granules to contain biologically active vasopressin which could be released by treatments that disrupted the granule membrane. Studies of NSGs from bovine posterior pituitaries[9] have indicated that NSGs are packets of hormone and an associated binding protein, a neurophysin.

The neurophysins are saline-soluble, sulfur-rich proteins of 9000–10,000 daltons which form insoluble, ionic complexes when added to neurohypophyseal hormones.[10] In bovine, murine, and porcine species, two major (NpI and NpII) and one minor (Npc) neurophysins have been identified,[2,10] while in man, at least two major neurophysins exist.[10,11] The cosedimentation of oxytocin with bovine NpI and vasopressin with bovine NpII[10,12] in equimolar ratios demonstrated the unique association of one neuropeptide with one neurophysin: stimuli for vasopressin release lead to a preferential rise in serum of radioimmunoassayable NpII.[2,11]

2.2.2. Hormone Biosynthesis

Sachs et al.[13,14] have characterized the biosynthetic origins of the neurohypophyseal hormones and the neurophysins, the steps for which are schematized in Fig. 3. They demonstrated that vasopressin isolated from the hypothalamus after infusion of [35S]cysteine into the third ventricle of dog brain showed a two- to threefold greater incorporation of label than AVP isolated from the posterior pituitary; that labeled AVP failed to appear in the posterior pituitary when the pituitary stalk was sectioned; and that hypothalamo-median eminence slices incorporated[35S]cysteine into vasopressin in vitro, while pituitary stalk and posterior pituitary tissue was incapable of such synthesis.[13,15] The time course for appearance of labeled vasopressin after ventricular injection of [35S]cysteine showed rapid incorporation at the level of the SON, and appearance in the posterior pituitary after about 1.5 hr.

In vitro trypsin digestion of large, labeled proteins isolated from SON, and presumed to be Np–AVP complexes, yielded 10,000-dalton peptides identical to neurophysin, and 1000-dalton peptides which comigrated with AVP on gel separation and were bound by antibodies to AVP.[16,17] These results, together with autoradiographic evidence that 35S-labeled proteins are transported down the supraoptico-hypophyseal tract within

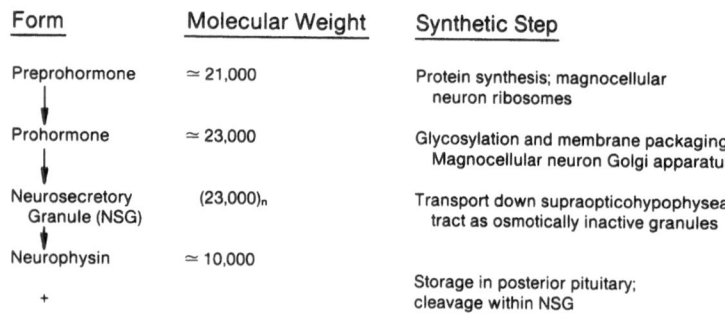

Form	Molecular Weight	Synthetic Step
Preprohormone	$\simeq 21,000$	Protein synthesis; magnocellular neuron ribosomes
Prohormone	$\simeq 23,000$	Glycosylation and membrane packaging; Magnocellular neuron Golgi apparatus
Neurosecretory Granule (NSG)	$(23,000)_n$	Transport down supraopticohypophyseal tract as osmotically inactive granules
Neurophysin	$\simeq 10,000$	
+		Storage in posterior pituitary; cleavage within NSG
Hormone	$\simeq 1,100$	

Fig. 3. Flow diagram for the pathway of posterior pituitary hormone biosynthesis.

NSGs[16] established the principle that intragranule processing (by an unknown "maturase" enzyme) of a prohormone, which is synthesized in ribosomes within magnocellular perikarya, occurs along the supraoptico-hypophyseal tract, and liberates vasopressin and oxytocin from the precursor molecule. The two prohormones, termed propressophysin and prooxyphysin,[9] have molecular weights of about 20,000.

2.2.3. Metabolism

AVP entering the circulation is distributed as unbound hormone in a volume approximating that of the extracellular space.[18] The finding of rapid AVP clearance in the perfused nonfiltering kidney in the dog[19] suggests specific, receptor-mediated clearance. With reference to Fig. 3, AVP may undergo proteolytic cleavage[20]: within the liver, at the 1,6 -S-S- disulfide bond; within the brain, at the 6,7 position; and in a variety of tissues, by hydrolysis of the peptide bond between the cysteine residue in position 1 and tyrosine in position 2. A peptidase has been isolated from renal plasma membranes which cleaves glycinamide and results in biological inactivation.[21]

Renal excretion of ADH is estimated to account for about one-fourth of total metabolic clearance.[18,19] In man, total clearance of ADH, representing both metabolic degradation and renal excretion, are in the range of 2–4 ml/min per kg body wt, yielding biological half-lives in the range of 30–40 min.[18]

2.3. Vasopressin: Structure–Function Relations

The structures of oxytocin and AVP (Fig. 2) differ at positions 3 and 8. These amino acid substitutions confer different biologic activities on the hormones. AVP has nearly comparable pressor and antidiuretic activities, both of which are 100 times greater than those of oxytocin, while the latter is about 20 times more uterotonic than AVP.[22]

The synthesis of analogs to the neurohypophyseal hormones has provided further understanding of the structural requirements for the various biologic activities. For instance, selectivity of antidiuretic over pressor activity has been achieved for analogs of AVP. Deamination of the 1-hemicystine residue to 1-deamino-arginine vasopressin (dAVP) increases antidiuretic activity fourfold but does not affect pressor activity.[23,24] Second, substitution of lipophilic amino acids for glutamine in position 4 [e.g., 4-valine-AVP (VAVP) or 4-threonine-AVP derivatives (dTAVP, dTDAVP)] creates further prolongation and specificity for antidiuretic activity.[25,26] Finally, the substitution of D-arginine in position 8 [D-arginine

vasopressin (DAVP)] reduces antidiuretic activity far less than pressor activity, so that the antidiuretic/pressor activity ratio of DAVP is 28.[23,24] These structural modifications of AVP have cumulative effects. Thus, in comparison to AVP: 1-deamino-[8-D-arginine]-vasopressin (dDAVP) has a longer duration of action, three times the antidiuretic effect, and reduced pressor activity[23]; 1-deamino,4-valine-[8-D-arginine]-vasopressin (dVDAVP) has a more prolonged duration of action, four times the antidiuretic potency, and undetectable pressor effects, making it among the most specific antidiuretic peptides reported.[23]

2.3.1. Antidiuretic Antagonists

Recently, the production of a series of peptides which contain a pentamethylene ring attached at position 1, O-ethyl- or O-methyl-tyrosine substitutions at position 2, and valine substitution for glutamine at position 4[27] has yielded a class of specific antidiuretic antagonists. These analogs competitively inhibit the antidiuretic response to exogenous and endogenous ADH and provoke a water diuresis in normally hydrated rats that is equal in intensity to that seen in vasopressin-deficient Brattleboro rats.[28] They have also been shown to inhibit competitively lysine vasopressin binding and adenylate cyclase activation in renal medullary membrane preparations.[29]

2.3.2. Conformation of Pituitary Hormones

Urry and Walter[30,31] have proposed that the conformations of oxytocin and lysine vasopressin include a β-turn involving the 20-membered ring and a second β-turn involving the COOH-terminal acyclic amino acid sequence. The aromatic side chains of tyrosine and phenylalanine in lysine vasopressin undergo a stacking interaction which prevents the tyrosine hydroxyl from folding over the ring and participating in hydrogen bonding as occurs in oxytocin. The molecule is pictured as having a hydrophobic surface composed of the exposed aromatic side chains and the hydrocarbon portions; and a hydrophilic cluster containing the carboxamides at positions 4, 5, and 9 plus the basic tail.[25] Substitution of D-tyrosine for L-tyrosine at position 2 in AVP, a maneuver which changes the topochemical orientation of the amino acid, results in no loss of adenylate cyclase activating activity, consistent with the lack of involvement of tyrosine in ring stability in AVP.[25] The importance of the Asn^5 CO has been demonstrated by the substitution of N^4,N^4-dimethylasparagine in lysine vasopressin, resulting in a peptide with only 3% of the activity of the native hormone.[32]

2.4. The Control of ADH Release

Release of ADH from the posterior pituitary in response to small changes in plasma osmolality is required in order to maintain plasma osmolality at a constant level. ADH may also be released when the plasma osmolality is less than normal via mechanisms activated by a decreased effective circulating volume. This section considers ADH release by osmotic and nonosmotic stimuli, the quantitative relations between these two variables, and the mechanism of ADH release from NSGs.

2.4.1. Osmotic Regulation of ADH Release

The relations between changes in plasma osmolality and ADH release were delineated by Verney[33] who postulated the presence of osmoreceptors, located in the distribution of the internal carotid arteries, which stimulated ADH release when plasma osmolality was raised by solutes to which osmoreceptors were impermeable. In his studies, increasing the osmolality of blood perfusing the internal carotid arteries by as little as 2% with hypertonic NaCl caused reductions in urine volume during a maximum water diuresis. Since the blood–brain barrier effectively limits passage of urea, intracarotid infusions of hypertonic urea solutions would dehydrate osmoreceptors were they located within the blood–brain barrier; thus, the absence of an antidiuretic response to urea provided support for an osmoreceptor locus outside the blood–brain barrier.[34] Discrete lesions of the organum vasculosum of the lamina terminalis (OVLT), a circumventricular organ of the anteroventral third ventricle which lies outside the blood–brain barrier, are associated with a deficient release of AVP to osmotic stimuli in the conscious dog.[35] Thus, the osmoreceptors may reside in this circumventricular organ, outside the blood–brain barrier.

It has been argued that specific sodium receptors in the brain modulate ADH release.[36] However, measurements of intraventricular (CSF) sodium concentrations during intracarotid infusions of hypertonic solutions of 1 M NaCl, 2 M sucrose, and 4.6 M urea each demonstrated an increase in CSF sodium concentrations, but antidiuresis occurred only with infusions of sucrose or hypertonic NaCl. This lack of an antidiuretic response to urea, which raised CSF sodium concentration significantly, argues in favor of primary osmoreceptor mediation of ADH release.

The relationship between plasma osmolality and plasma AVP concentrations determined by radioimmunoassay[37–39] is shown in Fig. 4. Plasma AVP levels are in the range of 0.5–1.5 pg/ml for plasma osmolalities below 280 mOsm/kg, while above a plasma osmolality of 280 mOsm/kg, plasma AVP rises in proportion to plasma osmolality according to the relation: plasma AVP = 0.38 (plasma osmolality − 280).[38,39] Thus, a plasma osmolality of 280 mOsm/kg may be considered the "osmotic threshold" for AVP release. In practical terms, a 1 mOsm/kg rise in plasma osmolality translates into an increase in plasma AVP levels of about 0.3 pg/ml and in urine osmolality of about 95 mOsm/liter.[40] This "gain" factor means that maximal urine concentrations are produced at a plasma osmolality of about 294 mOsm/liter and a plasma AVP level of about 5 pg/ml.[40]

2.4.2. Nonosmotic Regulation of ADH Release

There has accumulated a large body of evidence indicating that volume-mediated release of ADH may occur as a consequence of stimuli arising from "volume receptors," or bar-

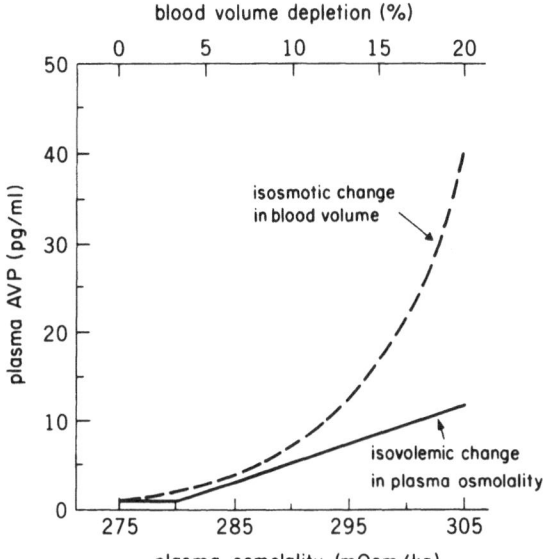

Fig. 4. Osmotic and nonosmotic control of plasma AVP. Adapted from Dunn *et al.*[38] and Robertson *et al.*[40]

oreceptors. Gauer and Henry[41] termed loci in the systemic venous circulation, the right side of the heart, and the left atrium the "low"-pressure baroreceptors, and loci within the systemic arterial system, "high"-pressure baroreceptors.

Positive pressure breathing and the upright position, which reduce pressure in the low-pressure regions of the vascular bed, produce antidiuresis. Negative pressure breathing, and its accompanying higher central venous pressures, produces a water diuresis which can be abolished by administering exogenous ADH,[42] thereby providing indirect evidence that the diuresis is mediated by suppression of ADH release. Balloon distension of the left atrium produces increases in urine volume[41] and, in electrophysiologic studies[43] of the neurohypophysis, this same maneuver inhibits electrical activity in cells of the SON. Section of the vagus nerve abolishes the inhibitory influence of atrial distension.

Hemorrhage sufficient to reduce arterial pressure in experimental animals has caused increases in a circulating antidiuretic substance, confirmed by radioimmunoassay to be vasopressin.[38,44] Stimulation of arterial baroreceptors by balloon distension of the carotid bifurcation or by an increase in systemic blood pressure inhibits electrical activity of supraoptic neurons, while local anesthesia of the carotid bifurcation abolishes the reflex.[43]

Release of ADH, demonstrated by bioassay, has been shown to occur in unanesthetized dogs with a 10–15% reduction in blood volume; this degree of hemorrhage left mean arterial pressure unchanged while central venous pressure fell.[45] It appears that nonosmotic ADH release is preferentially controlled by the low-pressure baroreceptor system.[46] In the rat, isosmotic volume contraction produced by intraperitoneal glycerol stimulated vasopressin release at a "threshold" of about 8% plasma volume contraction.[38] Acute plasma volume contraction in man, produced by hemofiltration, resulted in detectable rises in plasma vasopressin concentrations at as little as 3% plasma volume reduction.[47]

2.4.3. Quantitative Aspects of Osmotic and Nonosmotic Stimuli to ADH Release

It is evident that the regulation of plasma AVP levels depends on both osmotic and volume-mediated, nonosmotic stimuli. The relation between these two sets of stimuli and plasma AVP levels has been defined by the work of Dunn *et al.*[38] in the rat (Fig. 4). Osmotic stimuli produce linear increases in AVP release so long as blood volume is near normal. However, with blood volume depletion of greater than 10%, plasma AVP concentrations rise in a near-exponential fashion. Therefore, during severe volume contraction, volume-mediated ADH release overrides osmotically mediated ADH release. In addition, decreases in central venous pressures reduce the osmotic threshold and increase sensitivity for osmotic ADH release, while increases in central venous pressures elevate the threshold and dampen the sensitivity for osmotic ADH release.[38,48]

2.4.4. Chemical Mediators of ADH Release

ADH release may also be modulated by agents which have either systemic hemodynamic effects or CNS actions. Schrier *et al.*[49] have summarized the results of studies which provide evidence that α- and β-adrenergic agents mediate vasopressin release through nonosmotic, hemodynamic stimuli. In addition, adrenergic agents may stimulate release of ADH directly through a neurotransmitter function, but Sklar and Schrier[50] have concluded that it is currently impossible to accept a definitive role for adrenergic regulation of ADH release at the level of the hypothalamus.

In salt-depleted, unanesthetized dogs, both renin and angiotensin II increased plasma ADH levels estimated by bioassay,[51] and comparable rises in plasma AVP concentrations, measured by radioimmunoassay, obtained in man.[52] Other studies in man have demonstrated that the angiotensin II-converting enzyme inhibitor, captopril, produces a decrease in volume-mediated vasopressin release.[47]

While it has long been known that morphine induces antidiuresis, a number of studies have demonstrated that opiate antagonists cause a decrease in both basal and osmotically induced ADH release.[53] This finding plus the demonstration of enkephalins within the neurohypophysis[54] suggest a role for endogenous opiates in modulating ADH release.

2.5. Mechanism of Neurosecretion

Magnocellular neurons secreting vasopressin respond to osmotic stimulation by firing in a characteristic phasic pattern with bursts of 5–15 Hz separated by silent periods. Stimulation of vasopressin neurons by progressive dehydration causes a progressive recruitment of neurons into phasic activity,[55] suggesting a titration of hormone release relative to the magnitude of osmolar derangement. Although cells of the SON generate action potentials after application of hyperosmotic solutions *in vitro*,[56] it is believed that the magnocellular neurons are at least one synapse removed from the osmoreceptors.[43]

Current evidence, including freeze–fracture and electron microscopy studies, indicates that neurohypophyseal secretion occurs by Ca^{2+}-dependent exocytosis, a quantal process.[57] The exocytotic events involve fusion of NSG membranes with plasma membranes, granule opening at the site of fusion, and release of granule material, including ADH and neurophysin, into the extracellular space.[58] Micromolar concentrations of Ca^{2+} induced fusion of isolated secretory vesicles prepared from bovine neurohypophyses, and vasopressin release was demonstrated to occur concomitant to membrane fusion among vesicles.[59] Elimination of Ca^{2+} or application of Ca^{2+} channel blocking agents inhibits AVP release from isolated neurohypophyses, while Ca^{2+} ionophores increase AVP release.[58]

Thus, the release of vasopressin in response to either osmotic or nonosmotic stimuli ultimately depends upon the generation of Na^+-dependent, tetrodotoxin-sensitive action potentials in cells of either the SON or the PVN; the influx of extracellular Ca^{2+} into cells upon membrane depolarization; and the Ca^{2+}-dependent fusion of neurosecretory vesicles with the cell membrane of the axon terminal, releasing AVP into the circulation via an exocytotic process.

3. Thirst

Thirst is regulated by many of the same factors which determine ADH release so that water acquisition acts in concert with water conservation. The response of thirst to osmotic (hypertonic) stimuli is sufficiently powerful that significant hypertonicity does not develop, even in the total absence of ADH, in conscious individuals who have free access to water. The osmotic threshold for thirst stimulation in humans and other primates is reached with a 2–3% increase in plasma osmolality, a value only slightly higher than that which stimulates ADH release.[40,59] Thirst-mediated water intake also attends pronounced falls in effective ECF volume and, like volume-mediated ADH release, may continue despite significant diminution in body fluid osmolality.

3.1. Osmotic Regulation of Thirst

Injections of hypertonic NaCl into the medial hypothalamus have elicited excessive drinking in water-replete goats, suggesting osmoreceptor involvement in the regulation of water intake.[60] Ablation of the OVLT of sheep has reduced the water intake stimulated by intracarotid hypertonic NaCl infusion.[61] Thus, an osmoreceptor for thirst stimulation is located in or close to the OVLT, just as osmoregulation of ADH[35] seems to be controlled from this region of the brain.

Angiotensin II is a potent dipsogenic agent when injected directly into the third ventricle,[62] and intraventricular infusion of the angiotensin II inhibitor saralasin has slowed the onset of drinking in dehydrated animals. Angiotensin II may be important in the drinking response to a nonosmotic stimulus. Angiotensin II receptors have been demonstrated on the OVLT, and the dipsogenic effect of intraventricular angiotensin II is ablated when the OVLT is destroyed.[62]

3.2. Volume-Mediated Thirst

Underfilling of the low-pressure thoracic circulation elicits drinking[60] while crushing the left atrial appendage in sheep abolishes the drinking response to hypovolemia but leaves intact the response to hyperosmolality.[63] Peripherally generated angiotensin II may not participate in the hypovolemic thirst response; rather, the dipsogenic effects of angiotensin II may be mediated via a nerve-stimulated isorenin–angiotensin system within the brain.[60]

4. The Hypertonic Syndromes

ECF hyperosmolality is defined by an increased ratio of solute to water in the ECF. When caused by substances to which cell membranes are relatively impermeable (such as NaCl, mannitol, or, in the case of insulin deficiency, glucose), the ECF hypertonicity is directly associated with cellular dehydration and shrinkage. Either renal loss of water in excess of salt in the polyuric syndromes or excessive nonrenal loss of hypotonic body fluids leads to hypertonicity referable to hypernatremia and results in hypertonic volume depletion. This section considers a classification of the polyuric syndromes; the consequences of the hypernatremic state, referable mainly to changes in volume and osmotic composition of brain cells; and the differential diagnosis of the hypertonic syndromes.

4.1. Classification

With reference to Fig. 1, it is apparent that failure of water homeostasis may result either from inadequacy of ADH-dependent water conservation or from inadequacy of thirst-mediated water acquisition. Based on the underlying pathophysiology, the clinical circumstances which lead to hypernatremia may be grouped into the general categories outlined in Table I.

So long as free access to water is ensured, even a total deficiency of renal concentrating mechanisms does not ordinarily lead to hypertonicity. In fact, in those situations in which volume depletion develops, the nonosmotic stimulus to thirst and ADH release often causes these individuals to drink water in excess of solute and to become slightly hyponatremic. Most commonly, clinically significant hypertonic volume depletion is seen in the very young or very old, in whom either physical immaturity or debility prevents the translation of thirst into water-acquiring behavior. There also exists a small group of patients, those with "essential" hypernatremia, in whom the osmoregulatory centers are diseased, and in whom impaired osmotic stimulation of both ADH release and thirst is evident.

Pituitary diabetes insipidus is characterized by the failure of appropriate osmotic, volume, or chemical stimuli to evoke antidiuresis, and by a prompt response to exogenous vas-

Table I. Major Causes of Hypernatremia

Impaired thirst
   Coma
   Essential hypernatremia
Excessive water losses
   Renal
      Pituitary diabetes insipidus
      Nephrogenic diabetes insipidus
      Impaired medullary hypertonicity
   Extrarenal
      Sweating
      Gastrointestinal fluid loss
Solute diuresis
   Glucose
      Diabetic ketoacidosis
      Nonketotic hyperosmolar coma
   Other
      Mannitol administration
      Glycerol administration

opressin to diminish urine volume and raise urine osmolality. Congenital nephrogenic diabetes insipidus is identified by polyuria present since birth, generally in males, and by persistent unresponsiveness to exogenous ADH. Acquired nephrogenic diabetes insipidus is recognized by ADH unresponsiveness combined with a history of exposure to agents, such as lithium, demethylchlortetracycline, or methoxyflurane anesthesia, which antagonize the action of ADH on collecting ducts.

Disorders such as diabetes mellitus, which produces a solute diuresis, are characterized by an isotonic urine and by glycosuria. Therapeutic administration of mannitol or glycerol for reduction of intracranial pressure is, likewise, associated with production of an isotonic urine. In other words, hypertonic disorders associated with solute diuresis are characterized by high rates of solute excretion while the pituitary or nephrogenic diabetes insipidus syndromes are characterized by defects in water conservation without abnormalities in rates of solute excretion.

4.2. The Clinical Syndromes

4.2.1. "Essential" Hypernatremia

Chronic hypernatremia occurring in a setting of euvolemia, normal renal function, decreased thirst perception, and a normal renal response to exogenous vasopressin has been termed "essential hypernatremia."[64,65] These patients are hypodipsic and excrete an inappropriately dilute urine despite marked elevations of serum Na^+ concentrations and ECF osmolalities.

The osmotic threshold for ADH release is normal but the release of ADH at any level of plasma osmolality is markedly attenuated.[66] ADH release in response to effective baroreceptor stimulation via maneuvers that reduce either venous or arterial pressures is normal.[64–66] Given the association of a diminished sensation of thirst and a diminished release of AVP in response to osmotic stimulation, it is likely that essential hypernatremia represents a more or less specific defect of hypothalamic osmoreceptor function.

4.2.2. Pituitary Diabetes Insipidus

Pituitary diabetes insipidus is a polyuric syndrome that results from a lack of sufficient ADH to effect appropriate concentration of the urine for water conservation. The disease is identified by three primary findings: the persistence of an inappropriately dilute urine in the presence of strong osmotic or nonosmotic stimuli to ADH secretion; the absence of renal concentrating defects; and a rise in urine osmolality upon the administration of vasopressin.

Intracranial tumors and inflammatory processes of the basal meninges, such as syphilis and tuberculosis, accounted for the majority of cases of pituitary diabetes insipidus early in the century.[67] For the years 1972–1980, of 92 patients with pituitary diabetes insipidus reviewed by Moses and Notman,[68] 30% were idiopathic, 25% were related to intracranial tumors, 16% were secondary to head trauma, and 20% followed cranial surgery for tumor or hypophysectomy.

Experimental observations indicate that persistent polyuria develops only after an injury sufficiently high in the supraopticohypophyseal tract to cause bilateral neuron degeneration in the SON and PVN.[69] Preservation of as few as 15% of magno-

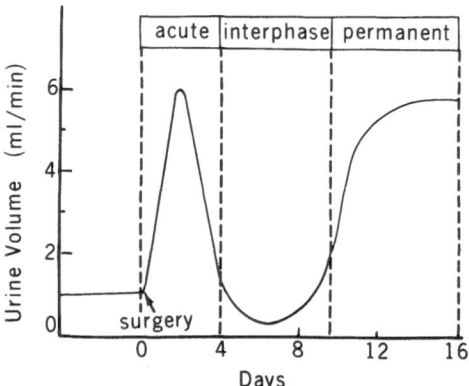

Fig. 5. The triphasic response of urinary volume following injury to the supraopticohypophyseal tract. Adapted from Fisher *et al.*[69]

cellular neurons has prevented polyuria entirely both in animals subjected to pituitary stalk transection[70] and in man with surgically induced hypothalamic trauma.[71] In short, while transient diabetes insipidus may accompany any injury to the neurohypophysis, permanent diabetes insipidus generally follows only neurohypophyseal damage high in the pituitary stalk.[72,73]

Injury to the neurohypophysis is often associated with a triphasic pattern of urinary excretion which results in permanent polyuria. Figure 5 depicts this response of an immediate rise in urine volume postinjury, accompanied by a fall in urine osmolality, which lasts 4–5 days; an intervening period of 5–7 days (the interphase) during which urine flow falls abruptly and urine osmolality rises; and a final phase consisting of a permanent, hyposthenuric polyuria. The initial phase appears related to injury-induced neuronal shock during which time no hormone release occurs.[72,73] During interphase, hormone apparently leaks from degenerating neurons, since the urinary excretion of water cannot be altered by maneuvers which render the plasma hypotonic and remove osmotic stimuli for ADH release[74] and since complete extirpation of the entire neurohypophyseal system prevents the appearance of an interphase.[73]

4.2.3. Nephrogenic Diabetes Insipidus

Nephrogenic diabetes insipidus is a polyuric disorder that is identified by the presence of normal rates of renal filtration and solute excretion, a persistently hypotonic urine, normal or high levels of either plasma AVP or urinary antidiuretic substance, and a failure of exogenous vasopressin to raise urine osmolality or reduce urine volume. Polyuric disorders due to failure of the renal tubule to respond to ADH may occur either as a relatively rare, sex-linked hereditary disorder or as a complication of drug therapy.

4.2.3a. Familial Disease. Familial nephrogenic diabetes insipidus is transmitted through a pattern consistent with sex-linked inheritance in that full expression of polyuria is seen only in males, female carriers have partial impairments in urinary concentrating ability, and male-to-male transmission is not known to occur.[75,76] The polyuria is evident from infancy and the clinical manifestations of the disorder in children are those of hypertonic volume depletion.

The cardinal abnormality in familial nephrogenic diabetes

insipidus is the failure of collecting ducts to increase their water permeability in response to ADH, resulting in the excretion of urine which is hypotonic to plasma. Robertson *et al.*,[37,39] using a specific radioimmunoassay for AVP, have shown that, in patients with nephrogenic diabetes insipidus, as in normal individuals, serum osmolalities greater than 280 mOsm/kg result in near-linear increments in serum AVP concentrations; while in pituitary diabetes insipidus, plasma AVP concentrations change negligibly or not at all in response to an osmotic challenge. Second, normal subjects and patients with pituitary diabetes insipidus exhibit a near-linear relationship between urine osmolality and plasma AVP concentrations, while patients with nephrogenic diabetes insipidus excrete a consistently hypotonic urine despite 15-fold variations in plasma AVP levels. These observations strongly support the hypothesis that familial nephrogenic diabetes insipidus is due to end-organ unresponsiveness to ADH.

Hereditary, vasopressin-unresponsive diabetes insipidus occurs in a genotypic strain of mice termed DI +/+ severe.[77] Stimulation of medullary adenylate cyclase activity by vasopressin is markedly reduced in these mice, although stimulation of cortical adenylate cyclase by parathyroid hormone is normal.[78] When compared to activity in normal mice, vasopressin-stimulated adenylate cyclase activity is modestly reduced in medullary collecting duct segments and severely reduced in medullary ascending limb segments.[79] However, the primary defect in these mice appears to be a failure of collecting tubules to raise intracellular levels of cAMP in response to vasopressin, perhaps due to an increased activity of collecting duct phosphodiesterase in these animals.[79] In addition, a defect in vasopressin-stimulated phosphorylation of a specific pair of proteins from the renal medulla of these mice has been identified.[80]

The relevancy of these studies to the pathogenesis of familial nephrogenic diabetes insipidus of humans is unclear. In some patients with familial nephrogenic diabetes insipidus, ADH does not increase rates of urinary cAMP excretion as it does in normal individuals,[81,82] while other patients with familial diabetes insipidus have an increase in urinary cAMP excretion in response to ADH comparable to that of normal persons.[83]

4.2.3b. Drug-Induced Nephrogenic Diabetes Insipidus. Demethylchlortetracycline causes a reversible vasopressin-resistant hyposthenuria which disappears shortly after discontinuing antibiotic therapy.[84] In toad urinary bladder, serosal demethylchlortetracycline inhibited the water permeability increase produced by either vasopressin or cAMP.[84] In human renal medulla, demethylchlortetracycline noncompetitively inhibited basal adenylate cyclase activity, ADH-stimulated adenylate cyclase activity, and cAMP-dependent protein kinase activity, but did not affect cyclic nucleotide phosphodiesterase activity.[85] These *in vitro* observations suggest that demethylchlortetracycline-induced nephrogenic diabetes insipidus may be due to inhibition of both cAMP accumulation and the action of cAMP on urinary membranes.

Methoxyflurane anesthesia may be complicated by a vasopressin-resistant polyuria and hyposthenuria that are related to the markedly increased serum concentration and urinary excretion of one of its metabolic products, inorganic fluoride.[84] Sodium fluoride causes a vasopressin-resistant polyuria in dogs,[86] and in the rat, inorganic fluoride seems to reduce collecting duct water permeability without affecting salt transport in the ascending limb.[87]

Finally, therapeutic serum lithium concentrations of 0.5–1.5 meq/liter produce reversible vasopressin-resistant diabetes insipidus in 12–30% of patients receiving lithium therapy. Urinary concentrating ability usually returns toward normal when lithium is discontinued.[84] Lithium specifically inhibited ADH-activated adenylate cyclase in mammalian renal medulla.[88] This inhibition is seen in both medullary thick ascending limbs and collecting ducts with acute introduction of lithium, but only in medullary collecting ducts in animals chronically treated with lithium.[89] Accordingly, it appears that the lithium-induced polyuria observed clinically may be the consequence of inhibition of vasopressin-stimulated cAMP formation in collecting ducts.

4.2.4. Osmotic Diuresis

Since the water permeability of the proximal nephron is sufficiently great that proximal tubular fluid remains isosmotic to plasma, the presence of a nonabsorbable solute (e.g., mannitol) in glomerular filtrate will result in the delivery of greater volumes of isotonic fluid to the loop of Henle. The same consequence occurs when the filtered load of an actively absorbed solute, such as glucose, exceeds the proximal tubular absorptive capacity for that solute, or when the filtered load of a partially absorbed solute, such as urea, is increased significantly. As a consequence, the capacity of the loop of Henle and collecting duct to modify the osmolality of urine is reduced. Thus, during progressive osmotic diuresis, urine osmolality falls in hydropenic individuals receiving ADH, while urine osmolality rises in normal subjects undergoing water diuresis[90]; in each case urine osmolality approaches serum osmolality.

4.2.5. Chronic Renal Disease

A virtually constant finding in chronic renal failure of diverse etiologies is the presence of refractory isosthenuria or persistent vasopressin-resistant hyposthenuria.[91,92] At a constant plasma urea concentration and a relatively normal salt intake, an individual with chronic renal disease and a glomerular filtration rate which is 20% of normal evidently excretes a greater fraction of the filtered solute load than occurs normally. Accordingly, it has been proposed that the isosthenuria of chronic renal disease is due in large part to the osmotic diuresis which occurs in remaining functioning nephrons, rather than to particular impairment of collecting duct function.[93]

In support of this view, Bricker et al.[93–95] have factored measurements of tubular function by the glomerular filtration rate in order to evaluate the function of residual nephrons. Their observations indicate that in experimental animal models of renal failure, the ratios of either positive or negative free water clearance to glomerular filtration rate are near normal. In contrast, other workers who have studied the pathogenesis of the concentrating defect in renal failure either in experimental animals[96] or in man[97,98] have argued that the concentrating ability of the diseased kidney is appreciably less than that which can be accounted for solely in terms of osmotic diuresis. Thus, additional factors besides the rate of solute excretion may be responsible for the relative polyuria of chronic renal failure.

Disturbances of the renal medullary countercurrent multiplication and exchange processes may also contribute to isosthenuria, particularly in diseases producing significant disruption of medullary architecture. In experimental pyelonephritis, a concentrating defect occurs early in the course of disease, without a significant reduction in glomerular filtration rate.[99] Gilbert et al.[100] have found a loss of the corticopapillary gradient for osmolality and urea accumulation and have concluded that a failure to recycle urea and accumulate interstitial solute contributes significantly to the concentrating defect in pyelonephritis. In sickle-cell disease, an early concentrating defect may occur in the absence of filtration rate reductions; it has been proposed[101,102] that the defect occurs because sickling of red blood cells within the renal medulla increases blood viscosity, reduces medullary blood flow, and disrupts the architecture of medullary vasa recta. Similar disturbances of medullary structure and/or function may account for the renal concentrating defect observed in medullary cystic disease[91] and in obstructive uropathy.[103]

4.2.6. Electrolyte Disorders

Both K^+ depletion and hypercalcemia cause a concentrating defect which is manifest primarily as a limitation in maximal urinary concentrating ability, rather than as persistent hyposthenuria. While both glomerular filtration rate and urinary diluting ability are near normal in hypokalemic polyuria, the Na^+ concentrations and osmolalities of the renal medulla and papilla are both decreased.[104] An increased urinary excretion of prostaglandin E may accompany hypokalemia, and the administration of indomethacin, an inhibitor of prostaglandin synthesis, has resulted in partial correction of the concentrating defect.[105] Thus, prostaglandin E inhibition of adenylate cyclase activation by ADH[106] may contribute to the pathogenesis of hypokalemic polyuria. K^+ depletion may result in polydipsia in some circumstances,[107] thus accentuating polyuria independently of the urinary concentrating defect.

The concentrating defect of hypercalcemic states is ordinarily accompanied by filtration rate reduction. Additionally, the reduction in medullary solute content observed in experimental hypercalcemic nephropathy,[108] and the inhibitory effect of Ca^{2+} on adenylate cyclase activation by vasopressin in hormone-sensitive epithelia[109] may contribute to the concentrating defect of hypercalcemia.

4.3. Hypertonic Encephalopathy

The cardinal clinical consequences of elevations in the effective ECF osmolality are hypertonic encephalopathy and volume contraction. Since most cell membranes are permeable to water, an increase in the effective ECF osmolality will inevitably lead to osmotic equilibration between cells and ECF, resulting in a parallel increase in intracellular osmolality. When ECF hypertonicity develops acutely, cells lose water and, in the CNS, the acute shrinkage in brain volume results in hypertonic encephalopathy. In chronic hypertonic states, CNS cells accumulate solutes and brain shrinkage is minimized as are CNS symptoms.[110] The relations among increases in effective ECF osmolality, changes in brain volume, and the occurrence of hypertonic encephalopathy depend on the magnitude of increase in ECF osmolality, the rapidity and duration of the increase, and the solute responsible for the increase in osmolality.

In rabbits made hypernatremic, early neurologic symptoms occur when the serum osmolality reaches 350 to 375 mOsm/kg water; nystagmus and ataxia occur at 375 to 400 mOsm/kg water; and coma, stupor, and death occur when serum osmolality is in the range of 400 to 435 mOsm/kg water.[111] Since experimental hypernatremic encephalopathy and death occur in the absence of any CNS pathologic changes other than brain shrinkage and a marked rise in brain NaCl content,[112] it is reasonable to conclude that the combination of hyperosmolality and cellular shrinkage is the major factor responsible for hypertonic encephalopathy.[110] This hypothesis coincides with clinical observations in different states of ECF hyperosmolality.

4.3.1. Cell Volume Adjustments to ECF Hypertonicity

The mechanisms by which CNS cells adjust their content of osmotically active solute and cell water during acute (1–2 hr) and chronic (2 hr–2 weeks) increases in ECF osmolality are shown for a number of common solutes in Table II. These data derive from experiments involving animals,[110,113,114] but similar changes are likely to occur in man during the development of hypertonic states. The term "idiogenic" osmoles refers to osmotically active solutes calculated as the difference between total cell osmolality and the sum of the osmolalities contributed by Na^+, K^+, and Cl^-.

During acute increases in osmolality by any of these endogenous or exogenous solutes, osmotic equilibrium between intracellular and extracellular water is achieved almost completely by cell water loss (Table II). In this case, the increase in cell osmolality is accounted for by increases in cell Na^+, K^+, and Cl^- concentrations with no contribution by idiogenic osmoles. The rapid change in brain cell volume appears to account for the severity of CNS symptoms and the high mortality referable to acute increases in effective ECF osmolality.

After the initial cell shrinkage due to acute elevations of ECF osmolality produced by endogenous solutes such as NaCl or glucose, brain cell volume returns toward normal (volume regulatory increase) by the generation of additional intracellular osmotically active solute, the "idiogenic" osmoles, and by uptake of electrolytes, principally KCl, from the ECF. The increase in intracellular solute causes water to reenter the cells from the ECF, thus restoring cell volume. Chronic elevation of ECF osmolality with exogenous solutes such as mannitol, glycerol, or sucrose does not result in production of "idiogenic" osmoles nor in an increase in cell volume.[110] This phenomenon provides a rationale for the use of these solutes to reduce brain volume during episodes of cerebral edema.

The extent to which brain cell volume regulation occurs by solute or electrolyte uptake, or by accumulation of organic idiogenic osmoles, is different for each of the endogenous solutes. About 50–60% of the increase in brain osmoles responsible for brain cell volume regulation during chronic hypernatremia is

due to amino acids.[110,115] The remaining 40–50% of cell volume regulation during hypernatremia results from the cellular accumulation of Na^+, K^+, and Cl^-. The transport mechanism mediating intracellular accumulation of the latter ions has not been explicitly defined, but may be similar to the coupled Na^+–K^+–$2Cl^-$ transport process responsible for hypertonic volume regulation in other cell types.[116] Dissipation of the organic osmoles induced by hypernatremia is not immediate upon return to the isotonic state but takes several hours to days.

During hyperglycemia, brain volume regulation is due not only to insulin-independent cellular uptake of glucose (20%), but also to electrolyte uptake and to accumulation of idiogenic osmoles. The latter, however, are not amino acids and their nature remains unknown. In contrast to the amino acids accumulated in hypernatremic states, the idiogenic osmoles that are accumulated during hyperglycemia dissipate rapidly with decreasing plasma glucose.[110] This difference may account for the observation that rapid reduction in serum Na^+ concentrations in diabetes insipidus patients with chronic hypernatremia may elicit seizures, while relatively rapid reductions in plasma glucose concentrations in nonketotic hyperglycemic coma generally improve CNS function.

5. The Hypotonic Syndromes

The hypotonic syndromes develop from disorders of the water repletion reaction (Fig. 1) in which free water is not excreted at a rate sufficient to maintain the serum Na^+ concentration, and consequently body fluid osmolality, at normal rather than subnormal concentrations. Such a circumstance might occur either because of a primary abnormality in the thirst mechanism, because of an inability of the kidney to excrete free water at a rate equal to water intake, or because of a combination of these events. This section will consider: a classification of the hypotonic syndromes; a clinical description of certain hypotonic syndromes; the mechanism for water intoxication and the adaptive processes which protect against water intoxication during hyponatremia; and the differential diagnosis and treatment of hyponatremia.

5.1. Classification

In accord with the water repletion reaction (Fig. 1), ECF hypoosmolality, representing dilution of body solutes with a total body excess of water, might occur as a result of excess water intake, decreased renal excretion of water, or a combina-

Table II. Brain Volume Adjustment during Hyperosmolality[a]

	Endogenous			Exogenous
	Na^+	Glucose	Urea	Mannitol, glycerol, sucrose
Acute (1–2 hr)				
Brain water	↓↓	↓↓	↓↓	↓↓
Electrolyte content	normal	normal	normal	normal
"Idiogenic" osmoles	absent	absent	absent	absent
Chronic (2 hr–2 weeks)				
Brain water	normal	normal	normal	↓↓
Electrolyte content	↑	↑	normal	normal
"Idiogenic" osmoles	↑↑↑	↑↑	↑	absent

[a]Original sources of data: Refs. 110, 114, 115.

Table III. The Hypotonic Syndrome

Mechanism	Disorders
Excessive water ingestion	Primary polydipsia
Decreased water excretion	
Decreased delivery to diluting segments (euvolemic)	Starvation Beer potomania

ADH excess (euvolemic or volume expanded)	SIADH
	  Drug-induced, trauma
	  ? K+ depletion

Mixed disorders (reduced effective circulating volume)	Volume-contracted: Addison's disease Volume-expanded: heart failure, cirrhosis

tion of these factors. Since in normal individuals, salt abstraction in the thick ascending limb of Henle results in the formation of a dilute distal tubular fluid having an osmolality of approximately 50 mOsm/kg water and a volume of about 18 liters (approximately 10% of the glomerular filtration rate),[117] the intake of water over the range of 5–15 liters daily will not lead to significant hyponatremia. The most common reason for clinically significant hyponatremia is a disturbance in the *rate* of water excretion and may occur even despite the kidney's ability to excrete a maximally dilute urine. This inability may occur because of a reduction in the rate of filtrate delivery to the diluting segment, because of sustained nonosmotic ADH release, or because of a combination of these factors. Table III presents a summary of the commonly encountered hyponatremic states based on this classification. While the primary derangement in the disorders is a defect in the rate of renal free water excretion, the development of hyponatremia also requires that free water intake exceed the rate of free water excretion; i.e., factors other than cellular shrinkage (an osmotic stimulus) stimulate thirst.

5.2. The Clinical Syndromes

5.2.1. Primary Polydipsia

Primary polydipsia is a rare disorder in which oral water intake exceeds 4–5 liters/day. The hyponatremia which may occur in primary polydipsia is generally slight, with serum Na+ concentrations averaging 135 meq/liter. However, a number of cases have been reported in which individuals consumed in excess of 20 liters of water daily and had serum Na+ concentrations less than 110 meq/liter in the face of maximal urinary dilution (< 75 mOsm/kg water).[118,119]

In one group of patients with primary polydipsia, ingestion of water over the range of 7–43 liters daily was associated with maximal urinary dilution of 37–95 mOsm/kg water, free water clearance of 12–36 liters daily, and serum Na+ concentrations less than 124 meq/liter.[120] When placed on water restriction, these patients excreted urine hypertonic to plasma, presumably as a result of osmotically stimulated ADH release, at plasma osmolalities of 242–272 mOsm/kg water, values considerably below the expected plasma osmolality threshold for ADH release of 280 mOsm/kg water. Thus, primary water drinkers who become profoundly hyponatremic may have a "reset osmostat" and may release ADH, thereby blunting urinary diluting capability, at a plasma osmolality which is distinctly below normal.

5.2.2. Reduced Salt Delivery to the Diluting Segment

While extreme reductions in effective circulating volume (e.g., ≅ 7–10% reduction in ECF volume) can stimulate the nonosmotic release of ADH and thereby reduce free water excretion, even minimal reductions in effective circulating volume may compromise partially the renal excretion of free water. This latter effect appears to be the result of reductions in the rate of salt, and therefore volume, delivery to the thick ascending limb of Henle's loop; this in turn reduces the volume of maximally dilute urine which can be excreted in the absence of ADH. In fact, this appears to be the primary pathogenic mechanism responsible for hyponatremia in euvolemic disorders such as beer potomania[121]; in volume-contracted states such as Addison's disease; and in edematous disorders such as congestive heart failure or cirrhosis.

An elegant analysis of hyponatremia in association with Na+ depletion in man was provided by McCance,[122] who showed that salt depletion induced by sweating regularly led to hyponatremia after a net loss of about 150–200 meq of Na+. Harrington[123] evaluated the contribution of vasopressin to this derangement by comparing the hyponatremic response to Na+ depletion in control versus homozygous Brattleboro rats, which have no ADH production. Since the hyponatremic response to Na+ depletion was remarkably similar in the two groups of animals, it appeared that water retention during Na+ depletion could occur without ADH. Berliner and Davidson[124] had shown that the urinary concentration which occurred in the absence of ADH resulted from a diminished rate of Na+ delivery to the diluting segment and Edwards et al.[125] showed that this effect occurred with relatively small changes in glomerular filtration rate. In this circumstance two factors serve to limit the excretion of a water load: the reduction in the volume of filtrate delivered to the thick ascending limb, because of increased proximal tubular avidity for Na+, reduces the volume of maximally dilute urine formed; and a reduced flow rate in the collecting ducts, consequent to reduced volume delivery from the thick ascending limb, allows partial equilibration of dilute luminal fluids with the hypertonic medullary interstitium, thus reducing the maximal dilution of the final urine.

5.2.2a. Beer Potomania: Euvolemic Hyponatremia.

The important effects of solute intake and salt delivery to the thick ascending limb on free water excretion are probably best exemplified by individuals with so-called *beer potomania*. These individuals develop profound hyponatremia in association with urine osmolalities in the range of 70 mOsm/kg water and no significant weight loss.[121,126] The maximal volume of free water that can be produced per day by such persons may be reduced from 15–18 liters to as low as 2–3 liters. Thus, the defect in free water excretion is not the degree to which urine is diluted, i.e., the osmolality of the urine, but rather in the rate of formation of dilute urine.

Patients with beer potomania derive a large part of their caloric intake from the ingestion of large volumes of solute-free beer. Therefore, this unusual dietary salt restriction increases the fractional rate of proximal Na+ absorption, diminishes the rate of salt delivery to diluting segments, and in turn limits the daily rate of formation of dilute urine. Another factor is related to the reduction in solutes ordinarily available for excretion, principally Na+ and urea derived from ingested protein. This point may be illustrated by a simple calculation. Since the minimum urinary osmolality ordinarily achieved in man is approximately 50 mOsm/kg water, the daily excretion of 15 liters of maximally dilute urine requires the excretion of 600–750 mOsm solute daily (15 liters × 50 mOsm/kg water). A reduction in daily solute excretion to as little as 300 mOsm, as is common in these individuals, would limit the maximal daily volume of free water excretion (and, therefore, free water ingestion) to 5–6 liters. Hyponatremia due to this mechanism may also occur during starvation, when solute intake may be dramatically reduced without parallel reductions in water intake.

5.2.3. Mixed Derangements

Hyponatremia also occurs commonly in states of true volume contraction and in certain edematous states, notably congestive heart failure and cirrhosis. The former disorders include patients in whom both ECF volume and total body water are reduced; the latter group includes those patients with deranged Starling forces, either local or systemic increases in venous pressure, which result in inadequate filling of the arterial tree. In both sets of disorders, two factors may contribute, individually or in unison, to a renal defect in water excretion: nonosmotic, volume-mediated ADH release; and reductions in the rate of Na^+ delivery to the diluting segment.

When effective circulating blood volume falls below a threshold value, plasma ADH levels rise sharply and can produce an antidiuretic effect even when the plasma osmolality is reduced below normal.[37,39] Volume-mediated, nonosmotic ADH release occurs primarily when circulatory dynamics are moderately to severely compromised and in that circumstance, volume-mediated stimuli override osmotically mediated ADH release, and hyponatremia ensues.

A second factor which is common to all of these disorders is the reduction in the maximal volume of free water produced each day due to a fall in the rate of filtrate delivery to diluting segments in the thick ascending limb.[122–125] The significance of volume contraction as a pathogenic factor in this type of hyponatremia can be gauged, as indicated above, by noting that hyponatremia occurs during volume contraction in Brattleboro rats with hypothalamic diabetes insipidus.[123,125] Finally, the development of hyponatremia in volume-contracted states requires that water ingestion continue in the face of hypotonicity. Volume-mediated, nonosmotic mechanisms appear to account for thirst in such circumstances.

5.2.4. Volume-Expanded States

Hyponatremia occurs in the advanced stages of disorders characterized by edema formation and a reduced effective circulating volume, particularly intractable heart failure and advanced hepatic cirrhosis with ascites. As noted above, reduced rates of salt delivery to diluting segments in these disorders clearly contribute to the impairment in water excretion. A limited number of radioimmunoassay measurements of plasma ADH levels in patients with heart failure or severe ascites also indicate that the plasma concentrations of this hormone are inappropriately high with respect to plasma osmolality.[127] Furthermore, since nonosmotic ADH release occurs only with significant reductions in blood volume (Fig. 4), it may be inferred that the occurrence of hyponatremia in congestive failure or cirrhosis indicates profound arterial underfilling. This observation correlates well with the ominous prognosis of hyponatremia in these disorders.

5.2.5. Volume-Contracted States

The hyponatremia occurring in association with volume-contracted states may be viewed, as indicated above, as a consequence of a reduction in Na^+ delivery rates to the loop of Henle. Thus, hyponatremia may commonly accompany prolonged administration of diuretics.[126,128] In this regard, two factors warrant particular consideration.

First, the most commonly used diuretics in clinical practice today include: "loop" diuretics such as furosemide and ethacrynic acid; and distal nephron diuretics, such as the thiazides. Because agents like furosemide inhibit salt absorption in the medullary thick limb of Henle,[129] they inhibit both urinary concentrating and diluting power. In contrast, agents such as thiazide diuretics inhibit salt absorption in cortical rather than medullary diluting segments. Consequently, thiazides inhibit urinary diluting power but do not appreciably limit urinary concentrating power.[130] Accordingly, the risk of diuretic-induced hyponatremia is greater with agents such as thiazides than with furosemide.

Second, Fichman et al.[131] have described a group of patients in whom hyponatremia associated with thiazide diuretics persisted even after correction of Na^+ depletion, but corrected with the repair of K^+ depletion. These authors postulated that K^+ depletion, per se, may have been responsible for sustained ADH release, since approximately half of their patients had hypertonic urines. It could also be argued that at least a portion of the hyponatremia may have been due to polydipsia which may accompany K^+ depletion.[107]

5.2.5a. Addison's Disease. In mineralocorticoid deficiency, the combination of ECF volume contraction, reduction in glomerular filtration rate, enhanced proximal tubular salt absorption, and volume-mediated, nonosmotic ADH release appear to be the major factors responsible for an inability to handle water loads.[126,128] In Brattleboro rats with hypothalamic diabetes insipidus, bilateral adrenalectomy resulted in hyponatremia which could be partially reversed by the concurrent administration of glucocorticoids and salt.[128] These results indicate clearly that, in hypoadrenalism, volume depletion can result in hyponatremia even in the absence of ADH. Glucocorticoids are also required for the complete correction of the defect in water excretion in Addison's disease,[128] presumably due to a glucocorticoid-mediated impairment of cardiac function which contributes to the reduction in effective circulating volume seen in Addison's disease.[132]

5.2.5b. Hypothyroidism. The cause for the hyponatremia occasionally seen in hypothyroid patients is not clear. It has been argued that hyponatremia in myxedema might occur because of sustained ADH release[133] or because of a "reset osmostat,"[126] i.e., normal modes for regulating plasma osmolality but at reduced plasma osmolalities. Alternatively, DeRubertis et al.[134] have inferred that salt delivery to the loop of Henle is reduced in hypothyroidism, and that this effect accounted for the defect in free water excretion. Regardless of the mechanism involved, appropriate treatment of myxedema is accompanied by the restoration of renal concentrating and diluting capacity.[135]

5.2.6. SIADH: ADH Excess

The second major group of disorders associated with hyponatremia includes those disturbances in which there is sustained release of ADH in the absence of either osmotic or nonosmotic stimuli. By definition, therefore, the diagnosis of such a disturbance requires that salt depletion be absent, so that there is no reduction in the rate of salt delivery to diluting segments; and that there is no reduction in effective circulating volume, so that volume-mediated, nonosmotic stimuli to ADH release are absent. In general, a primary excess of ADH occurs in two settings: the syndrome of inappropriate ADH release (SIADH); and as a consequence of drugs that enhance ADH release or ADH action.

5.2.6a. SIADH. Schwartz et al.[136] provided the first clear account of the occurrence of SIADH in a patient with bronchogenic carcinoma. Subsequently, SIADH has been observed in a variety of disorders, particularly pulmonary diseases, notably bronchogenic carcinoma, and cranial disorders.

It is now well established that, in most patients with SIADH, there is persistent production of ADH or an ADH-like peptide despite body fluid hypotonicity and an expanded effective circulating volume.[128,137] The largest single group of patients in whom this finding has been documented is that of Zerbe et al.[137] The most common derangement (37%) is wide fluctuations of ADH levels independent of osmotic or nonosmotic control. About 33% of the patients had an abnormally low osmotic threshold for ADH release; these patients were able to produce a maximally dilute urine if sufficiently hyponatremic. At higher plasma osmolalities, there existed a normal correlation between values of plasma ADH and plasma osmolality. In 16% of patients, there was sustained vasopressin release ("vasopressin leak") below serum osmolalities of 278 mOsm/kg water, and normal vasopressin release in response to osmotic stimuli. Approximately 14% of patients had no detectable abnormality of ADH levels, but a failure to dilute urine maximally. This type of SIADH therefore has a poorly understood pathogenesis.

As a result of the sustained release of ADH or ADH-like substances, patients who develop SIADH retain ingested water and become hyponatremic, and are modestly volume expanded, generally increasing body weight by 5–10%. The volume expansion results in reduced rates of proximal tubular Na^+ absorption, and consequently in a natriuresis, albeit at a total expansion of total body water. Leaf et al.[138] provided evidence that the increments in urinary salt excretion which accompanied exogenous ADH administration were referable to hormone-induced volume expansion due to water retention, rather than to a direct effect of ADH on renal tubular salt absorption. Administration of ADH, coupled with unrestricted fluid intake, initially resulted in hyponatremia, urinary concentration and antidiuresis, and a weight gain. After 3 days, body weight and the serum Na^+ concentration attained steady-state, volume-expanded values, and a natriuresis occurred. Fluid restriction corrected the hyponatremia, body weight declined, and urinary Na^+ excretion fell even in the face of continued ADH administration. This natriuresis observed with either hypotonic or isotonic volume expansion is believed to be due to a reduction in the fractional rate of proximal tubular salt absorption, i.e., to a resetting of glomerulotubular balance.

Since aldosterone secretion is stimulated by hyponatremia, it is possible that this mineralocorticoid may contribute to reducing renal Na^+ losses in volume-expanded hyponatremic patients with SIADH. There are also increased urinary losses of substances like uric acid, whose excretion rates vary directly with effective circulating volume and with rates of Na^+ excretion. Consequently, hypouricemia is common in SIADH. The

glomerular filtration rate is normal, as is adrenal and thyroid function.

5.2.6b. Pharmacologic Stimuli. Hyponatremia may occur as a consequence of drug therapy. As indicated above, diuretics may result in hyponatremia attendant to volume contraction or, less commonly, because of K^+ depletion.[131] Certain agents may either stimulate ADH release from the posterior pituitary or, as in the case of chlorpropamide,[139,140] potentiate the effects of ADH on renal tubules. Thus, each of these two classes of agents can result in an SIADH-like clinical syndrome, i.e., sustained hyponatremia in the absence of a reduction in effective circulating volume.

Finally, renal medullary prostaglandins, particularly of the E series, inhibit salt absorption in the medullary thick ascending limb and antagonize the effects of ADH on collecting tubules.[141,142] Furthermore, prostaglandins aid in the maintenance of glomerular blood flow during volume contraction.[143] Accordingly, it is reasonable to infer that aspirin or other nonsteroidal anti-inflammatory drugs might, by interfering with prostaglandin synthesis, lead to hyponatremia, particularly in volume-contracted patients. One recent report documented such an occurrence, with severe hyponatremia developing within 3 days of initiating ibuprofen therapy and disappearing with cessation of drug treatment.[144]

5.3. Water Intoxication

The syndrome of water intoxication varies in severity depending on two major factors: the degree of hyponatremia and the duration of hyponatremia. In acute hyponatremia with serum Na^+ concentrations of less than 120 meq/liter, the clinical syndrome is characterized by somnolence, seizures, and coma, with mortality rates as high as 50% having been reported.[145,146] The autopsy findings have been those of cerebral edema. In acute hyponatremia produced in experimental animals, the mortality rate is approximately 85%[146] and is associated with significant reductions in brain electrolyte concentrations and in brain osmolalities.

Chronic hyponatremia in man differs in two important respects from acute hyponatremia[145]: approximately half of the patients encountered with chronic hyponatremia are asymptomatic, even with serum Na^+ concentrations less than 125 meq/liter; and the fatality rate in chronic hyponatremia is nearly zero in asymptomatic patients and approximately 10–15% in symptomatic patients. In experimental animals, chronic hyponatremia results in a greater fall in brain Na^+, K^+, and Cl^-, for a given reduction in serum Na^+ concentrations, than in acute hyponatremia.[145] Consequently, in chronic hyponatremia, there is less of a rise in brain water content, for a given reduction in serum Na^+ concentration, than in acute hyponatremia, and a correspondingly lower mortality rate.[145,146]

These observations may be interpreted in terms of three general conclusions. First, the water permeability of the blood–brain barrier and of brain cells is sufficiently high that the brain approaches osmotic equilibrium with plasma quickly. Second, in acute hyponatremia, the approach to osmotic equilibrium between brain cells and the ECF primarily involves water gain; this water gain accounts both for the cerebral edema of acute hyponatremia and, in large part, for the high fatality rate of acute hyponatremia. Finally, for a given reduction in serum Na^+ concentration, brain electrolyte concentrations and brain water content are both lower in chronic, with respect to acute, hyponatremia. In other words, when chronic hyponatremia occurs, homeostatic mechanisms are activated which extrude solutes from cells, so that osmotic equilibration between the brain and the ECF occurs with smaller increases in brain volume. Consequently, the fatality rate of chronic hyponatremia is less than that of acute hyponatremia.

It is now recognized that a wide variety of cell types exhibit a volume regulatory decrease (VRD) in response to a hypotonic ECF environment. An initial period of cellular swelling occurs, in which the cell acts as an osmometer, and a subsequent cellular shrinkage accompanied by solute loss, principally electrolyte, follows. Moreover, it is probable that the VRD response involves the loss of intracellular electrolytes through the activation of membrane transport processes other than Na^+,K^+-ATPase, whose role in the VRD response is largely permissive. Furthermore, the nature of these solute efflux processes differs appreciably among different cell types.[147] In the present context, it is pertinent to note that, in brain, the VRD response involves principally a loss of intracellular KCl and NaCl.[114,115]

Finally, it should be emphasized that the occurrence of a VRD response in the setting of chronic hyponatremia has major therapeutic implications. It is well recognized that the rapid correction of chronic hyponatremia to normal serum Na^+ concentrations will, *per se*, lead to CNS disturbances, particularly in children. The reason for such an occurrence follows from a consideration of the VRD response. Thus, if brain electrolyte content is adaptively reduced in chronic hyponatremia, the rapid correction of serum Na^+ concentrations to normal levels will, in effect, result in an acute hypertonic encephalopathy. It is important to note that central pontine myelinolysis has now been documented to follow the rapid correction of hyponatremia both in experimental animals[148] and in man.[149]

6. Summary

This chapter has reviewed the pathophysiology of disorders which are manifest as derangements in the ratio of solute to water in body fluids, i.e., hypoosmolality or hyperosmolality. The pathologic consequences of these disorders are referable to changes in cell volume, especially in cells of the CNS. ECF hypoosmolality leads to cell swelling, and ECF hyperosmolality, when caused by solutes like NaCl to which cell membranes are impermeable, leads to cell shrinkage.

The water repletion reaction works to maintain a constant ECF osmolality. The dual stimuli of increased effective ECF osmolality, detected by CNS osmoreceptors, and decreased effective circulating vascular volume, detected by baroceptors located in both venous and arterial circulations, operate in synchrony to regulate the level of circulating ADH and the sensation of thirst. The latter elements work either to conserve body water via antidiuresis or to replenish body water via drinking. Conversely, dilution of body fluids, resulting in ECF hypoosmolality, normally serves to suppress thirst and water ingestion.

Regulation of cell volume is ordinarily accomplished by the cell membrane Na^+,K^+-ATPase, an energy-consuming pump which acts to maintain constant the intracellular cation, and thereby, osmolar, concentration. Acute changes in effective ECF osmolality are associated with rapid osmolar equilibration across cell membrane by water movement. When effective ECF osmolality is raised or lowered more slowly (over days), compensatory cell mechanisms operate to adjust cellular osmolar content and, thereby, minimize both water movement across cell membranes and changes in cell volume. In hyperosmolar states, osmotically active solute is added to cells; these solutes include NaCl, KCl, amino acids, and unidentified, "idiogenic" osmolar substances. In hypoosmolar states, cells adjust their osmotic content by dumping salts, principally KCl, from cell to ECF.

Significant ECF hyperosmolality is most often due to a failure of the water repletion action involving the abnormalities of ADH release or ADH action on the kidney; in any case, a concurrent failure of water ingestion need be present for ECF hyperosmolality to develop. A specific abnormality in osmoreceptor function has been identified in persons having "essential" hypernatremia. These individuals fail to release ADH or exhibit thirst in response to elevated ECF osmolality but respond normally to volume-mediated stimuli. Pituitary diabetes insipidus is a polyuric syndrome associated with absent or diminished secretion of ADH in response to any stimulus; nephrogenic diabetes insipidus, occurring either as a hereditary disorder or as a consequence of exposure to certain drugs, represents a partial or complete failure of collecting ducts to increase water permeability in response to ADH. Finally, abnormalities in renal concentrating ability may exist independent of the levels or action of ADH. In solute diuresis, excessively high rates of solute delivery to the loop of Henle overwhelm the kidney's ability to dissociate solute from water absorption and, in certain renal diseases, there is impaired generation of a hypertonic medullary interstitium by renal countercurrent multiplication processes.

Dilution of body fluids by primary ingestion of water rarely occurs due to the normal ability of the kidney to excrete free water over the large range of 10–15 liters daily. The daily volume of free water formed may be significantly reduced in states that decrease solute, and filtrate, delivery to the thick ascending limb of Henle. Rather, significant hyponatremia usually occurs when ADH release is stimulated by nonosmotic factors, e.g., volume-mediated, drug-stimulated, or autonomous release. Since volume-mediated stimulation of ADH release dominates osmotic stimulation, states of reduced effective circulating volume will continue to effect ADH release and antidiuresis even with significant dilution of body fluids. Release of ADH may also occur in response to certain chemical stimuli or be autonomous to any identifiable stimulus, as in SIADH.

The clinical consequences of these disorders depend upon the rapidity with which the ECF osmolar disturbance develops, the magnitude of the derangement, and, in the case of ECF hyperosmolality, the solute associated with the increase in osmolality. The clinical syndrome thus reflects the means by which CNS cells reach osmotic equilibrium with the ECF and the degree of cellular shrinkage or swelling involved in the cellular adjustment. The therapeutic approach to these syndromes must take into consideration the cellular adjustment mechanisms in order to avoid a reversal of the osmotic disequilibrium during correction of ECF osmolality.

References

1. Vandesande, F., and K. Dierickx. 1975. Identification of the vasopressin producing and of the oxytocin producing neurons in the hypothalamic magnocellular neurosecretory system of the rat. *Cell Tissue Res.* **164**:153–162.

2. Zimmerman, E. A., and A. G. Robinson. 1976. Hypothalamic neurons secreting vasopressin and neurophysin. *Kidney Int.* **10**:12–24.
3. Morris, J. F., H. W. Sokol, and H. Valtin. 1977. One neuron–one hormone? In: *Neurohypophysis.* A. M. Moses and L. Share, eds. Karger, Basel. pp. 58–66.
4. Cross, B. A., R. E. J. Dyball, R. G. Dyer, C. W. Jones, D. W. Lincoln, J. F. Morris, and B. T. Pickering. 1975. Endocrine neurons. *Recent Prog. Horm. Res.* **31**:243–294.
5. Scharrer, E., and B. Scharrer. 1954. Hormones produced by neurosecretory cells. *Recent Prog. Horm. Res.* **10**:183–240.
6. Zimmerman, E. A., and R. Defendi. 1977. Hypothalamic pathways containing oxytocin, vasopressin and associated neurophysins. In: *Neurohypophysis.* A. M. Moses and L. Share, eds. Karger, Basel. pp. 22–29.
7. Bargmann, W., and E. Scharrer. 1951. The site of origin of the hormones of the posterior pituitary. *Am. Sci.* **39**:255–259.
8. Weinstein, H. S., Malamed, and H. Sachs. 1961. Isolation of vasopressin-containing granules from the neurohypophysis of the dog. *Biochim. Biophys. Acta* **50**:386–389.
9. Russel, J. T., M. J. Brownstein, and H. Gainer. 1980. Biosynthesis of vasopressin, oxytocin and neurophysins: Isolation and characterization of two common precursors (propressophysin and prooxyphysin). *Endocrinology* **107**:1880–1891.
10. Pickering, B. T., and C. W. Jones. 1978. The neurophysins. In: *Hormonal Proteins and Peptides,* Volume 5. C. H. Li, ed. Academic Press, New York. pp. 103–158.
11. Robinson, A. G. 1975. Radioimmunoassay of neurophysin proteins: Utilization of specific neurophysin assays to demonstrate independent secretion of different neurophysins *in vivo. Ann. N.Y. Acad. Sci.* **248**:246–256.
12. Dean, C. R., D. B. Hope, and T. Kazic. 1968. The total hormone-binding capacity of the neurophysins and the oxytocin and vasopressin content of the posterior pituitary. *Br. J. Pharmacol.* **34**:193–194.
13. Sachs, H., P. Fawcett, Y. Takabatake, and R. Portanova. 1969. Biosynthesis and release of vasopressin and neurophysin. *Recent Prog. Horm. Res.* **25**:447–484.
14. Sachs, H., and Y. Takabatake. 1964. Evidence for a precursor in vasopressin biosynthesis. *Endocrinology* **75**:943–948.
15. Takabatake, Y., and H. Sachs. 1964. Vasopressin biosynthesis. III. *In vitro* studies. *Endocrinology* **75**:934–942.
16. Brownstein, M. J., J. T. Russell, and H. Gainer. 1980. Synthesis, transport and release of posterior pituitary hormones. *Science* **207**:373–378.
17. Russell, J. T., M. J. Brownstein, and H. Gainer. 1981. Time course of appearance and release of [35S] cysteine labeled neurophysins and peptides in the neurohypophysis. *Brain Res.* **205**:299–311.
18. Bauman, G., and J. F. Dingman. 1976. Distribution, blood transport, and degradation of antidiuretic hormone in man. *J. Clin. Invest* **57**:1109–1116.
19. Shade, R. E., and L. Share. 1977. Renal vasopressin clearance with reductions in renal blood flow in the dog. *Am. J. Physiol.* **232**:F341–F347.
20. Walter, R., and W. H. Simmons. 1977. Metabolism of neurohypophyseal hormones: Considerations from a molecular viewpoint. In: *Neurohypophysis.* A. M. Moses and L. Share, eds. Karger, Basel. pp. 167–188.
21. Nardacci, N. J., S. Mukhopadhyay, and B. J. Campbell. 1975. Partial purification and characterization of the antidiuretic hormone-activating enzyme from renal plasma membranes. *Biochim. Biophys. Acta* **377**:146–157.
22. du Vigneaud, V. 1969. Hormones of the mammalian posterior pituitary gland and their naturally occurring analogues. *Johns Hopkins Med. J.* **124**:53–65.
23. Sawyer, W. H., M. Acosta, L. Balaspiri, J. Judd, and M. Manning. 1974. Structural changes in the arginine vasopressin molecule that enhance antidiuretic activity and specificity. *Endocrinology* **94**:1106–1115.
24. Manning, M., and W. H. Sawyer. 1977. Structure–activity studies on oxytocin and vasopressin 1954–1976. in: *Neurohypophysis.* A. M. Moses and L. Share, eds. Karger, Basel. pp. 9–21.
25. Smith, C. W. 1981. Conformation–activity studies on oxytocin and vasopressin: Exploring the roles of the moieties within the hydrophilic cluster. In: *Neurohypophyseal Peptide Hormones and Other Biologically Active Peptides.* Elsevier, Amsterdam. pp. 23–35.
26. Manning, M., L. Balaspiri, J. Moehring, J. Haldar, and W. H. Sawyer. 1976. Synthesis and some pharmacological properties of deamino [4-threonine, 8-D-arginine] vasopressin and deamino [8-D-arginine] vasopressin, highly potent and specific antidiuretic peptides, and [8-D-arginine] vasopressin and deamino-arginine-vasopressin. *J. Med. Chem.* **19**:842–845.
27. Sawyer, W. H., and M. Manning. 1982. Effective antagonists of the antidiuretic action of vasopressin in rats. *Ann. N.Y. Acad. Sci.* **394**:464–472.
28. Sawyer, W. H., P. K. T. Pang, J. Seto, M. McEnroe, B. Lammek, and M. Manning. 1981. Vasopressin analogs that antagonize antidiuretic responses by rats to the antidiuretic hormone. *Science* **212**:49–51.
29. Stassen, F. L., R. W. Erickson, W. F. Huffman, J. Stefankiewicz, L. Sulat, and V. D. Wiebelhaus. 1982. Molecular mechanisms of novel antidiuretic antagonists: Analysis of the effects on vasopressin binding and adenylate cyclase activation in animal and human kidney. *J. Pharmacol. Exp. Ther.* **223**:50–54.
30. Urry, D. W., and R. Walter. 1971. Proposed conformation of oxytocin in solution. *Proc. Natl. Acad. Sci. USA* **68**:956–958.
31. Walter, R., J. D. Glickson, I. L. Schwartz, R. T. Havran, J. Meienhoffer, and D. W. Urry. 1972. Conformation of lysine vasopressin: A comparison with oxytocin. *Proc. Natl. Acad. Sci. USA* **69**:1920–1924.
32. Walter, R., G. L. Stahl, T. Caplaneris, P. Cordopatis, and D. Theodoropoulos. 1979. Active site studies of neurohypophyseal hormones: Synthesis and pharmacological properties of [5--(N4,N4-dimethyl/asparagine)] oxytocin. *J. Med. Chem.* **22**:890–893.
33. Verney, E. B. 1947. The antidiuretic hormone and the factors which determine its release. *Proc. R. Soc. London Ser. B* **135**:25–105.
34. Thrasher, T. N. 1982. Osmoreceptor mediation of thirst and vasopressin secretion in the dog. *Fed. Proc.* **41**:2528–2532.
35. Thrasher, T. N., L. C. Keil, and D. J. Ramsay. 1982. Lesions of the organum vasculosum of the lamina terminalis (OVLT) attenuate osmotically-induced drinking and vasopressin secretion in the dog. *Endocrinology* **110**:1837–1839.
36. Andersson, B., and K. Olsson. 1977. Evidence for periventricular sodium-sensitive receptors of importance in the regulation of ADH secretion. In: *Neurohypophysis.* A. M. Moses and L. Share, eds. Karger, Basel. pp. 118–127.
37. Robertson, G. L., E. A. Mahr, S. Athar, and T. Sinha. 1973. Development and clinical application of a new method for the radioimmunoassay of arginine vasopressin in human plasma. *J. Clin. Invest.* **52**:2340–2352.
38. Dunn, F. L., J. T. Brennan, A. E. Nelson, and G. L. Robertson. 1973. The role of blood osmolality and volume in regulating vasopressin secretion in the rat. *J. Clin. Invest.* **52**:3212–3219.
39. Robertson, G. L. 1974. Vasopressin in osmotic regulation in man. *Annu. Rev. Med.* **25**:315–322.
40. Robertson, G. L., R. L. Shelton, and S. Athar. 1976. The osmoregulation of vasopressin. *Kidney Int.* **10**:25–37.
41. Gauer, O. H., and J. P. Henry. 1963. Circulatory basis of fluid volume control. *Physiol. Rev.* **43**:423–481.
42. Murdaugh, H. V., H. O. Sieker, and F. Manfredi. 1959. Effect of altered intrathoracic pressure on renal hemodynamics, electrolyte excretion and water clearance. *J. Clin. Invest.* **38**:834–842.
43. Poulain, D. A., and J. B. Wakerley. 1982. Electrophysiology of hypothalamic magnocellular neurones secreting oxytocin and vasopressin. *Neuroscience* **7**:773–808.

44. Weinstein, H., R. M. Berne, and H. Sachs. 1960. Vasopressin in blood: Effect of hemorrhage. *Endocrinology* **66**:712–718.

45. Share, L. 1967. Vasopressin, its bioassay and the physiological control of its release. *Am. J. Med.* **42**:701–712.

46. Gupta, P. D., J. P. Henry, R. Sinclair, and R. von Baumgarten. 1966. Responses of atrial and aortic baroreceptors to nonhypotensive hemorrhage and to transfusion. *Am. J. Physiol.* **211**:1429–1437.

47. Caillens, H., W. Pruszczynski, A. Meyrier, K.-S. Ang, F. Rousselet, and R. Ardaillou. 1980. Relationship between change in volemia at constant osmolality and plasma antidiuretic hormone. *Miner. Electrolyte Metab.* **4**:161–171.

48. Quillen, E. W., and A. W. Cowley. 1983. Influence of volume changes on osmolality–vasopressin relationships in conscious dogs. *Am. J. Physiol.* **244**:H73–H79.

49. Schrier, R. W., T. Berl, R. J. Anderson, and K. M. McDonald. 1976. Non-osmotic regulation of renal water excretion. *Trans. Am. Clin. Climatol. Assoc.* **87**:161–169.

50. Sklar, A. H., and R. W. Schrier. 1983. Central nervous system mediators of vasopressin release. *Physiol. Rev.* **63**:1243–1280.

51. Brennan, L. A., J. P. Bonjour, and R. L. Malvin. 1971. ADH levels during salt depletion in dogs. *Eur. J. Clin. Invest.* **2**:43–46.

52. Uhlich, E., P. Weber, and U. Gröschel-Stewart. 1974. Angiotensin-stimulated vasopressin release in man; radioimmunologically determined plasma levels of vasopressin. *Acta Endocrinol. (Copenhagen) Suppl.* **184**:52.

53. Miller, M., and A. M. Moses. 1977. Clinical states due to alteration of ADH release and action. In: *Neurohypophysis.* A. M. Moses and L. Share, eds. Karger, Basel. pp. 153–166.

54. Martin, R., and K. H. Voigt. 1981. Enkephalins co-exist with oxytocin and vasopressin in nerve terminals of rat neurohypophysis. *Nature (London)* **289**:502–504.

55. Wakerly, J. B., D. A. Poulain, and D. Brown. 1978. Comparison of firing patterns in oxytocin- and vasopressin-releasing neurones during progressive dehydration. *Brain Res.* **148**:425–440.

56. Hatton, G. I., W. E. Armstrong, and W. A. Gregory. 1978. Spontaneous and osmotically-stimulated activity in slices of rat hypothalamus. *Brain Res. Bull.* **3**:497–508.

57. Theodosis, D. T., and J. J. Dreifuss. 1977. Ultrastructural evidence for exo-endocytosis in the neurohypophysis. In: *Neurohypophysis.* A. M. Moses and L. Share, eds. Karger, Basel. pp. 88–94.

58. Dreifuss, J. J. 1975. A review on neurosecretory granules: Their contents and mechanisms of release. *Ann. N.Y. Acad. Sci.* **248**:184–201.

59. Wood, R. J., E. T. Rolls, and B. J. Rolls. 1982. Physiological mechanisms for thirst in the nonhuman primate. *Am. J. Physiol.* **242**:R423–R428.

60. Andersson, B., and M. Rundgren. 1982. Thirst and its disorders. *Annu. Rev. Med.* **33**:231–239.

61. McKinley, M. J., D. A. Denton, L. G. Leksell, D. R. Mouw, B. A. Scoggins, M. H. Smith, R. S. Weisinger, and R. D. Wright. 1982. Osmoregulatory thirst in sheep is disrupted by ablation of the anterior wall of the optic recess. *Brain Res.* **236**:210–215.

62. Phillips, M. I., W. E. Hoffman, and S. L. Bealer. 1982. Dehydration and fluid balance: Central effects of angiotensin. *Fed. Proc.* **41**:2520–2527.

63. Zimmerman, M. B., E. H. Blaine, and E. M. Stricker. 1981. Water intake in hypovolemic sheep: Effects of crushing the left atrial appendage. *Science* **211**:489–491.

64. Mahoney, J. H., and A. D. Goodman. 1968. Hypernatremia due to hypodipsia and elevated threshold for vasopressin release. *N. Engl. J. Med.* **279**:1191–1196.

65. DeRubertis, F. R., M. F. Michelis, N. Beck, J. B. Field, and B. B. Davis. 1971. "Essential" hypernatremia due to ineffective osmotic and intact volume regulation of vasopressin secretion. *J. Clin. Invest.* **50**:97–111.

66. Halter, J. B., A. P. Goldbert, G. L. Robertson, and D. Porte. 1977. Selective osmoreceptor dysfunction in the syndrome of chronic hypernatremia. *J. Clin. Endocrinol. Metab.* **44**:609–616.

67. Fink, E. B. 1928. Diabetes insipidus. *Arch. Pathol. Lab. Med.* **6**:102–120.

68. Moses, A. M., and D. D. Notman. 1982. Diabetes insipidus and syndrome of inappropriate antidiuretic hormone secretion (SIADH). *Adv. Intern. Med.* **27**:73–100.

69. Fisher, C., W. R. Ingram, and S. W. Ranson. 1935. Relation of hypothalamico-hypophyseal system to diabetes insipidus. *Arch. Neurol. Psychiatry* **34**:124–163.

70. Heinbecker, P., and H. L. White. 1944. Hypothalamico-hypophysial system and its relation to water balance in the dog. *Am. J. Physiol.* **133**:582–593.

71. Rasmussen, A. T., and W. J. Gardner. 1940. Effects of hypophysial stalk resection on the hypophysis and hypothalamus of man. *Endocrinology* **27**:219–226.

72. Lipsett, M. B., J. P. MacLean, C. D. West, M. C. Li, and O. H. Pearson. 1956. An analysis of the polyuria induced by hypophysectomy in man. *J. Clin. Endocrinol. Metab.* **16**:183–195.

73. Weitzman, R. E., and C. R. Kleeman. 1979. The clinical physiology of water metabolism. Part II. Renal mechanisms for urinary concentration; diabetes insipidus. *West. J. Med.* **131**:486–515.

74. Mudd, R. H., H. W. Dodge, E. C. Clark, and R. V. Randall. 1957. Experimental diabetes insipidus: A study of the normal interphase. *Proc. Staff Meet. Mayo Clin.* **32**:94–108.

75. Williams, R. H., and C. Henry. 1945. Nephrogenic diabetes insipidus: Transmitted by females and appearing during infancy in males. *Ann. Intern. Med.* **27**:84–95.

76. Waring, A. J., L. Kajdi, and V. Tappan. 1945. A congenital defect of water metabolism. *Am. J. Dis. Child.* **69**:323–324.

77. Naik, D. V., and H. Valtin. 1969. Hereditary vasopressin-resistant urinary concentrating defects in mice. *Am. J. Physiol.* **217**:1183–1190.

78. Dousa, T. P., and H. Valtin. 1974. Cellular action of antidiuretic hormone in mice with inherited vasopressin-resistant urinary concentrating defects. *J. Clin. Invest.* **54**:753–762.

79. Jackson, B. A., R. M. Edwards, H. Valtin, and T. P. Dousa. 1980. Cellular action of vasopressin in medullary tubules of mice with hereditary nephrogenic diabetes insipidus. *J. Clin. Invest.* **66**:110–122.

80. Strewler, G. J., B. G. Fallon, and J. Orloff. 1981. Defective protein phosphorylation in renal medulla of vasopressin-resistant mice. *Biochem. Biophys. Res. Commun.* **103**:713–720.

81. Fichman, M. P., and G. Brooker. 1972. Deficient renal cyclic adenosine 3'-5' monophosphate production in nephrogenic diabetes insipidus. *J. Clin. Endocrinol. Metab.* **35**:35–47.

82. Bell, N. H., C. M. Clark, S. Avery, T. Sinha, C. W. Trygstad, and D. O. Allen. 1974. Demonstration of a defect in the formation of adenosine 3',5'-monophosphate in vasopressin-resistant diabetes insipidus. *Pediatr. Res.* **8**:223–230.

83. Uttley, W. S., B. Atkinson, A. Adams, and D. Shirling. 1978. Cyclic adenosine monophosphate excretion in urine of patients and carriers of congenital nephrogenic diabetes insipidus. *J. Inher. Metab. Dis.* **1**:75–77.

84. Singer, I., and J. N. Forrest. 1976. Drug-induced states of nephrogenic diabetes insipidus. *Kidney Int.* **10**:82–95.

85. Dousa, T. P., and D. M. Wilson. 1974. Effects of demethylchlortetracycline on cellular action of antidiuretic hormone *in vitro. Kidney Int.* **5**:279–284.

86. Frascino, J. D., J. O'Flaherty, C. Olmo, and S. Rivera. 1972. Effect of inorganic fluoride on the renal concentrating mechanism: Possible nephrotoxicity in man. *J. Lab. Clin. Med.* **79**:192–203.

87. Wallin, J. D., and R. A. Kaplan. 1977. Effect of sodium fluoride on concentrating and diluting ability in the rat. *Am. J. Physiol.* **232**:F335–F340.

88. Dousa, T. P. 1974. Interaction of lithium with vasopressin-sensitive cyclic AMP system of human renal medulla. *Endocrinology* **95**:1359–1366.

89. Jackson, B. A., R. M. Edwards, and T. P. Dousa. 1980. Lithium-induced polyuria: Effect of lithium on adenylate cyclase and adenosine 3'-5'-monophosphate phosphodiesterase in medullary ascending limb of Henle's loop and in medullary collecting tubules. *Endocrinology* **107**:1693–1698.

90. Gennari, F. J., and J. P. Kassirer. 1974. Osmotic diuresis. *N. Engl. J. Med.* **291**:714–720.

91. Holliday, M. A., T. J. Egan, and C. R. Morris. 1967. Pitressin-resistant hyposthenuria in chronic renal disease. *Am. J. Med.* **42**: 378–387.

92. Tannen, R. L., E. M. Regal, M. J. Dunn, and R. W. Schrier. 1969. Vasopressin-resistant hyposthenuria in advanced chronic renal disease. *N. Engl. J. Med.* **280**:1135–1141.

93. Bricker, N. S., R. R. Dewey, H. Lubowitz, J. Stokes, and T. Kirkensgaard. 1959. Observations on the concentrating and diluting mechanisms of the diseased kidney. *J. Clin. Invest.* **38**:516–523.

94. Bricker, N. S., S. Klahr, H. Lubowitz, and R. E. Rieselbach. 1965. Renal function in chronic renal disease. *Medicine (Baltimore)* **44**:263–288.

95. Lubowitz, H., M. L. Purkerson, and N. S. Bricker. 1966. Investigation of single nephrons in the chronically diseased (pyelonephritic) kidney of the rat using micropuncture techniques. *Nephron* **3**:73–83.

96. Bank, N., and H. S. Aynedjian. 1966. Individual nephron function in experimental bilateral pyelonephritis. II. Distal tubular sodium and water reabsorption and the concentrating defect. *J. Lab. Clin. Med.* **68**:728–739.

97. Mees, E. J. D. 1959. Role of osmotic diuresis in impairment of concentrating ability in renal disease. *Br. Med. J.* **1**:1156–1158.

98. Kleeman, C. R., D. A. Adams, and M. H. Maxwell. 1961. An evaluation of maximal water diuresis in chronic renal disease. 1. Normal solute intake. *J. Lab. Clin. Med.* **58**:169–184.

99. Gonick, H. C., G. Goldberg, M. E. Rubini, and L. B. Guze. 1965. Functional abnormalities in experimental pyelonephritis. 1. Studies of concentrating ability. *Nephron* **2**:193–206.

100. Gilbert, R. M., H. Weber, L. Turchin, L. G. Fine, J. J. Bourgoignie, and N. S. Bricker. 1976. A study of the intrarenal recycling of urea in the rat with chronic experimental pyelonephritis. *J.Clin. Invest.* **58**:1348–1357.

101. Perillie, P. E., and F. H. Epstein. 1963. Sickling phenomenon produced by hypertonic solutions: A possible explanation for the hyposthenuria of sicklemia. *J. Clin. Invest.* **42**:570–580.

102. van Eps, L. W. S., C. Pinedo-Veels, C. H. deVries, and J. de Koning. 1970. Nature of concentrating defect in sickle-cell nephropathy. *Lancet* **1**:450–452.

103. Bricker, N. S., E. I. Shwayri, J. B. Reardan, D. Kellog, J. P. Merrill, and J. H. Holmes. 1957. An abnormality in renal function resulting from urinary tract obstruction. *Am. J. Med.* **23**:554–564.

104. Manitius, A., H. Levitin, D. Beck, and F. H. Epstein. 1960. On the mechanisms of impairment of renal concentrating ability in potassium deficiency. *J. Clin. Invest.* **39**:684–692.

105. Galvez, O. G., B. W. Roberts, W. H. Bay, and T. F. Ferris. 1976. Studies on the mechanism of polyuria with hypokalemia. *Kidney Int.* **10**:583a.

106. Torikai, S., and K. Kurokawa. 1983. Effects of PGE_2 on vasopressin-dependent cell cAMP in isolated single segments. *Am. J. Physiol.* **245**:F58–F66.

107. Berl, T., S. L. Linas, G. A. Aisenbery, and R. J. Anderson. 1977. On the mechanism of polyuria in potassium depletion. *J. Clin. Invest.* **60**:620–625.

108. Manitius, A., H. Levitin, D. Beck, and F. H. Epstein. 1960. On the mechanism of impairment of renal concentrating ability in hypercalcemia. *J. Clin. Invest.* **39**:693–697.

109. Campbell, B. J., G. Woodward, and V. Broberg. 1972. Calcium-mediated interactions between the antidiuretic hormone and renal plasma membranes. *J. Biol. Chem.* **247**:6167–6175.

110. Arieff, A. I., R. Guisado, and V. C. Lazarowitz. 1977. The pathophysiology of hyperosmolar states. In: *Disturbances in Body Fluid Osmolality*. T. E. Andreoli, J. J. Grantham, and F. C. Rector, eds. American Physiological Society, Washington, D.C. pp. 227–250.

111. Dodge, P. R., J. F. Sotos, I. Gamstorp, D. DeVivo, M. Levy, and T. Rabe. 1962. Neurophysiologic disturbances in hypertonic dehydration. *Trans. Am. Neurol. Assoc.* **87**:33–36.

112. Sotos, J. F., P. R. Dodge, P. Meara, and N. B. Talbot. 1960. Studies in experimental hypertonicity: Pathogenesis of the clinical syndrome. biochemical abnormalities and cause of death. *Pediatrics* **26**:925–937.

113. Holliday, M. A., M. N. Kalayci, and J. Harrah. 1968. Factors that limit brain volume changes in response to acute and sustained hyper- and hyponatremia. *J. Clin. Invest.* **47**:1916–1928.

114. Arieff, A. I., and R. Guisado. 1976. Effects on the central nervous systems of hypernatremic and hyponatremic states. *Kidney Int.* **10**: 104–116.

115. Chan, P. H., and R. A. Fishman. 1979. Elevation of rat brain amino acids and idiogenic osmoles induced by hyperosmolality. *Brain Res.* **161**:293–301.

116. Cala, P. M. 1983. Volume regulation by red blood cells: Mechanism of ion transport. *Mol. Physiol.* **4**:33–52.

117. Gottschalk, C. W., and M. Mylle. 1959. Micropuncture study of the mammalian urinary concentrating mechanism: Evidence for the countercurrent hypothesis. *Am. J. Physiol.* **196**:927–936.

118. Langgård, H., and W. O. Smith. 1962. Self-induced water intoxication without predisposing illness. *N. Engl. J. Med.* **266**:378–381.

119. Rendell, M., D. McGrane, and M. Cuesta. 1978. Fatal compulsive water drinking. *J. Am. Med. Assoc.* **240**:2557–2559.

120. Hariprasad, M. K., R. P. Eisinger, I. M. Nadler, C. S. Padmanabhan, and B. D. Nidus. 1980. Hyponatremia in psychogenic polydipsia. *Arch. Intern. Med.* **140**:1639–1642.

121. Hilden, T., and T. L. Svendsen. 1975. Electrolyte disturbances in beer drinkers: A specific "hypo-osmolality syndrome." *Lancet* **2**: 245–246.

122. McCance, R. A. 1936. Experimental sodium chloride deficiency in man. *Proc. R. Soc. London Ser. B* **119**:245–268.

123. Harrington, A. R. 1972. Hyponatremia due to sodium depletion in the absence of vasopressin. *Am. J. Physiol.* **222**:768–774.

124. Berliner, R. W., and D. G. Davidson. 1957. Production of hypertonic urine in the absence of pituitary antidiuretic hormone. *J. Clin. Invest.* **36**:1416–1427.

125. Edwards, B. R., M. Gallai, and H. Valtin. 1980. Concentration of urine in the absence of ADH with minimal or no decrease in GFR. *Am. J. Physiol.* **239**:F84–F91.

126. Fanestil, D. D. 1977. Hyposmolar syndromes. In: *Disturbances in Body Fluid Osmolality*. American Physiological Society, Washington, D.C. pp. 267–284.

127. Szatalowicz, V. L., P. E. Arnold, C. Chaimovitz, D. Bichet, T. Bert, and R. W. Schrier. 1981. Radioimmunoassay of plasma arginine vasopressin in hyponatremic patients with congestive heart failure. *N. Engl. J. Med.* **305**:263–266.

128. Weitzman, R. E., and C. R. Kleeman. 1980. The clinical physiology of water metabolism. III. The water depletion (hyperosmolar) and water excess (hyposmolar) syndromes. *West. J. Med.* **132**:16–38.

129. Burg, M. B., and N. Green. 1973. Function of the thick ascending limb of Henle's loop. *Am. J. Physiol.* **224**:659–668.

130. Seldin, D. W., G. Eknoyan, W. N. Suki, and F. C. Rector. 1966. Localization of diuretic action from the pattern of water and electrolyte excretion. *Ann. N.Y. Acad. Sci.* **139**:328–343.

131. Fichman, M. P., H. Vorherr, C. R. Kleeman, and N. Telfer. 1971. Diuretic-induced hyponatremia. *Ann. Intern. Med.* **75**:853–863.

132. Schrier, R. W., and S. L. Linas. 1980. Mechanisms of the defect in water excretion in adrenal insufficiency. *Miner. Electrolyte Metab.* **4**:1–7.

133. Chinitz, A., and F. L. Turner. 1965. The association of primary hypothyroidism and inappropriate secretion of the antidiuretic hormone. *Arch. Intern. Med.* **116**:871–874.

134. DeRubertis, F. R., M. F. Michelis, M. E. Bloom, D. H. Mintz, J. B. Field, and B. B. Davis. 1971. Impaired water excretion in myxedema. *Am. J. Med.* **51**:41–53.

135. DiScala, V. A., and M. J. Kinney. 1971. Effects of myxedema on the renal diluting and concentrating mechanism. *Am. J. Med.* **50**: 325–335.

136. Schwartz, W. B., W. Bennett, S. Curelop, and F. C. Bartter. 1957. A syndrome of renal sodium loss and hyponatremia proba-

bly resulting from inappropriate secretion of antidiuretic hormone. *Am. J. Med.* **23**:529–542.

137. Zerbe, R. L., Stropes, and G. Robertson. 1980. Vasopressin function in the syndrome of inappropriate diuresis. *Annu. Rev. Med.* **31**:315–327.

138. Leaf, A., F. C. Bartter, R. F. Santos, and O. Wrong. 1953. Evidence in man that urinary electrolyte loss induced by polydipsia is a function of water retention. *J. Clin. Invest.* **32**:868–871.

139. Pokracki, F. J., A. G. Robinson, and S. M. Seif. 1981. Chlorpropamide effect: Measurement of neurophysin and vasopressin in humans and rats. *Metabolism* **30**:72–78.

140. Kusano, E., J. L. Braun-Werness, D. J. Vick, M. J. Keller, and T. P. Dousa. 1983. Chlorpropamide action on renal concentrating mechanism in rats with hypothalamic diabetes insipidus. *J. Clin. Invest.* **72**:1298–1313.

141. Culpepper, R. M., and T. E. Andreoli. 1983. Interactions among prostaglandin E_2, antidiuretic hormone, and cyclic adenosine monophosphate in modulating Cl absorption in single mouse medullary thick ascending limbs of Henle. *J. Clin. Invest.* **71**:1588–1601.

142. Grantham, J. J., and J. Orloff. 1968. Effect of prostaglandin E_1 on the permeability response of the isolated collecting tubule to vasopressin, adenosine 3′-5′-monophosphate and theophylline. *J. Clin. Invest.* **47**:1154–1161.

143. Clive, D. M., and J. S. Stoff. 1984. Renal syndromes associated with nonsteroidal antiinflammatory drugs. *N. Engl. J. Med.* **310**:563–572.

144. Blum, M., and A. Aviram. 1980. Ibuprofen induced hyponatremia. *Rheumatol. Rehabil.* **19**:258–259.

145. Arieff, A. I., F. Llach, and S. G. Massry. 1976. Neurological manifestations and morbidity of hyponatremia: Correlation with brain water and electrolytes. *Medicine (Baltimore)* **55**:121–129.

146. Pollock, A. S., and A. I. Arieff. 1980. Abnormalities of cell volume regulation and the functional consequences. *Am. J. Physiol.* **239**:F195–F205.

147. Grantham, J., and M. Linshaw. 1984. The metabolic response to hyponatremia. *Circ. Res.* in press.

148. Kleinschmidt-DeMasters, B. K., and M. D. Norenberg. 1981. Rapid correction of hyponatremia causes demyelination: Relation to central pontine myelinolysis. *Science* **211**:1068–1070.

149. Norenberg, M. D., and K. O. Leslie. 1982. Correction of hyponatremia and central pontine myelinolysis. *Am. J. Med.* **73**:882.

Disorders of Proton Secretion by the Kidney

Philip R. Steinmetz and Joseph Palmisano

1. Introduction

Rapid advances have been made in the understanding of the cellular mechanisms of proton transport in urinary epithelia. The polar epithelial cells of the kidney tubules as well as various urinary bladder preparations are made up of apical and basolateral cell membranes with very different properties. The site of H^+ secretion is at the apical membrane and for each H^+ secreted a HCO_3^- ion is generated and transported across the basolateral cell membrane. In the proximal renal tubule, H^+ secretion at the brush border occurs primarily via a Na^+/H^+ exchanger; in the tight epithelia of the turtle and toad bladder and, probably, the mammalian collecting tubules, the luminal membrane contains an electrogenic proton pump which appears to be a H^+-translocating ATPase. The efflux of HCO_3^- across the basolateral cell membranes of both leaky and tight epithelia appears to occur either via an anion exchanger or a conductive channel, but is not coupled directly to metabolic energy.

New techniques for the perfusion of individual nephron segments, for intracellular measurements, and for separation of apical and basolateral cell membranes have provided new insight into the specialization for ion transport in each nephron segment and begin to permit understanding of the functional organization of acid–base transport within the kidney.

Several of these advances in the physiology of urinary acidification have become applicable to the understanding of the clinical disorders of H^+ secretion by the kidney. The pathogenesis of the renal acidoses and the associated electrolyte disturbances can now be explored in terms of specific defects in membrane function.

2. Proton Transport across Urinary Epithelia

2.1. Electrogenic H^+ Transport in Tight Epithelia

2.1.1. H^+-Secreting Cell Population

The "tight" urinary epithelia derived from the urogenital sinus and the ureteric bud share several functional and morphologic characteristics. Thus, the renal collecting tubule of the mammalian kidney and certain model epithelia like the turtle and toad urinary bladders are capable of generating steep concentration gradients for Na^+ and H^+. Despite some species differences, these urinary epithelia show several similarities. They are made up of two major cell types. The granular and mitochondrion-rich cells of the bladders correspond to the principal and intercalated cells of the collecting tubule, respectively. Thus, the mitochondrion-rich cells and the intercalated cells both contain carbonic anhydrase (CA) and have prominent luminal microplicae or microvilli.[1-3] In Fig. 1, a scanning electron micrograph of the luminal surface of a sheet preparation of turtle bladder is shown. Most of the cells are granular cells. Four of the cells have characteristic microplicae and represent mitochondrion-rich cells. These cells contain abundant cytoplasmic CA. Considerable evidence is now available that the CA-rich cells are responsible for H^+ secretion. Husted et al.[3] showed that the luminal membrane area of CA cells is markedly reduced when the cells are made alkaline by blocking the efflux of HCO_3^- across the basolateral cell membrane with a disulfonic stilbene. This shrinkage of the luminal cell membrane area is shown in Fig. 2. The imposed cellular alkalinity by this and other maneuvers has no effect on the number of CA cells, but reduces the luminal membrane of the large-surface-area cells (Fig. 2, top) so that the surface areas of all CA cells become uniformly small (Fig. 2, bottom). In contrast, stimulation of H^+ secretion by CO_2 expands the luminal area of CA cells.[4,5] Morphometric studies[5] demonstrate an inverse relation between luminal membrane area and the volume percent of mem-

Philip R. Steinmetz and Joseph Palmisano • Division of Nephrology, Department of Medicine, University of Connecticut School of Medicine, Farmington, Connecticut 06032.

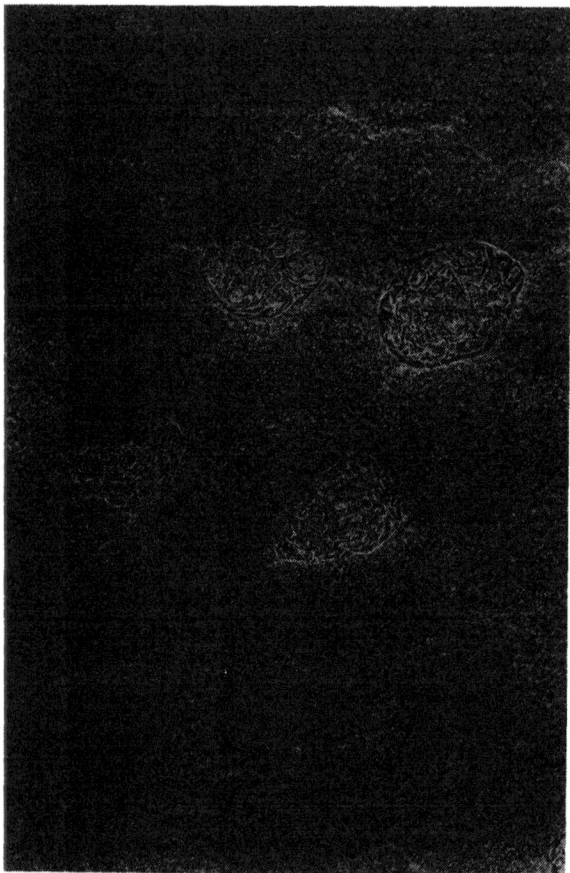

Fig. 1. Scanning electron micrograph of luminal surface of the turtle urinary bladder. The cells with prominent microridges (microplicae) are responsible for H⁺ secretion and are rich in mitochondria and carbonic anhydrase. The hexagonal cells with small microprojections are granular cells.

brane vesicles in the cytoplasm and suggest that these vesicles may be added or subtracted to the luminal membrane. Imposed acid–base changes affect the CA cells but not the granular cells.[5] Gluck et al.[6] showed by fluorescence techniques that the contents of the membrane vesicles are acid, consistent with the presence of H⁺ pumps in their membranes, and that membrane fusion occurs rapidly in response to increases in CO_2 tension. Studies of the morphology of the intercalated cells of the mammalian collecting duct also indicate that acid–base changes alter the extent of luminal membrane[7] and the volume percent of membrane vesicles.[8]

These various studies suggest rather strongly that the CA cells of the bladder preparations and the intercalated cells of the collecting tubule are specialized in the function of H⁺ secretion. For the turtle bladder, this evidence is further strengthened by the recent studies by Schwartz et al.[9] on separated epithelial cells. They showed that the CA cells, but not the granular cells, respond to acetazolamide with reduced O_2 consumption.[9]

2.1.2. Double Membrane Model for Urinary Acidification

The transport processes of urinary acidification are best analyzed according to the double membrane model of Koefoed-

Johnsen and Ussing.[10,11] From what is known about the electrochemical profile of the H⁺-secreting cells, the H⁺ pump is located at the luminal cell membrane as shown in Fig. 3, and the OH⁻ ions generated in series with the pump react with CO_2 to form HCO_3^-. This reaction is catalyzed by cytoplasmic CA. The HCO_3^- so formed moves down an electrochemical gradient across the basolateral cell membrane.[11,12] The precise nature of the exit step for HCO_3^- remains to be clarified. The basolateral cell membrane is rendered highly permeable by specific transport sites which can be blocked by the disulfonic stilbene, SITS. Since the transepithelial process of acidification is electrogenic, the efflux of HCO_3^- must be associated with the net transfer of a negative charge across the basolateral membrane. This charge transfer could be accomplished via a conductive HCO_3^- channel or, alternatively, via a more complex arrangement of an anion exchanger with a parallel conductive pathway that permits recycling of the counteranion. In Fig. 3, a HCO_3^- for Cl⁻ exchanger is depicted with a parallel Cl⁻ conductance. This model is favored because the HCO_3^- efflux depends on the presence of Cl⁻ in the serosal solution.[13] The arrangement shown is also consistent with the SITS sensitivity of the HCO_3^- exit.

Several of the characteristics of the acidification pathway have been clarified by studies of the isolated turtle bladder. In this preparation a number of simplifications can be made which permit closer examination of the transport processes at each cell membrane. When Na⁺ transport is abolished by serosal addition of ouabain or mucosal addition of amiloride, the spontaneous potential difference (PD) reverses from lumen negative to lumen positive.[12,14–16] Under many conditions the short-circuit current in ouabain-treated bladders is equivalent to the rate of H⁺ secretion (J_H) as measured directly by pH stat titration.[12,15,16] These and other studies[17,18] indicate that H⁺ transport is not coupled directly to the transport of Na⁺ or other electrolytes, but is closely coupled to metabolism. The rate of metabolism has been monitored as $^{14}CO_2$ evolution from glucose oxidation[18,19] and, under anaerobic conditions, as the rate of lactate production[20] while the active transport rate for H⁺ was altered experimentally. These studies on the sheet preparation of turtle bladder indicate that H⁺ translocation across the luminal membrane is directly coupled to the flow of metabolic energy and, hence, represents primary active transport. They also indicate that the luminal membranes of the H⁺-secreting cells have a low passive H⁺ permeability so that recycling of H⁺ across this membrane may be ignored. Older evidence obtained by pH stat titration studies[21] had already shown that the in vitro bladder had a low overall H⁺ permeability. These studies are summarized in Fig. 4. Net H⁺ transport (J_{net}) was measured during secretion against pH gradients. The rate of back diffusion ($-J_{pass}$) was negligible over the physiologic range of urine pH from 7.4 to 4.9. A rate of back-flow became detectable only when the luminal pH was lowered further below the physiologic range (shaded area). Because of the tightness of the bladder to H⁺, the measured net rates represented the active transport rates for H⁺. Of interest is the near-linear relation between J_H and ΔpH in this study. This relationship has been observed in many other studies in which J_H was measured as the short-circuit current in ouabain-treated bladders, but applies only over a limited luminal pH range of about 3 pH units. Consistent with the electrogenicity of the H⁺ pump, the transport rate is directly affected by the transepithelial PD. Since the luminal membrane has a low H⁺ permeability and a high resistance, most of this PD is thought to occur across the luminal membrane or the H⁺

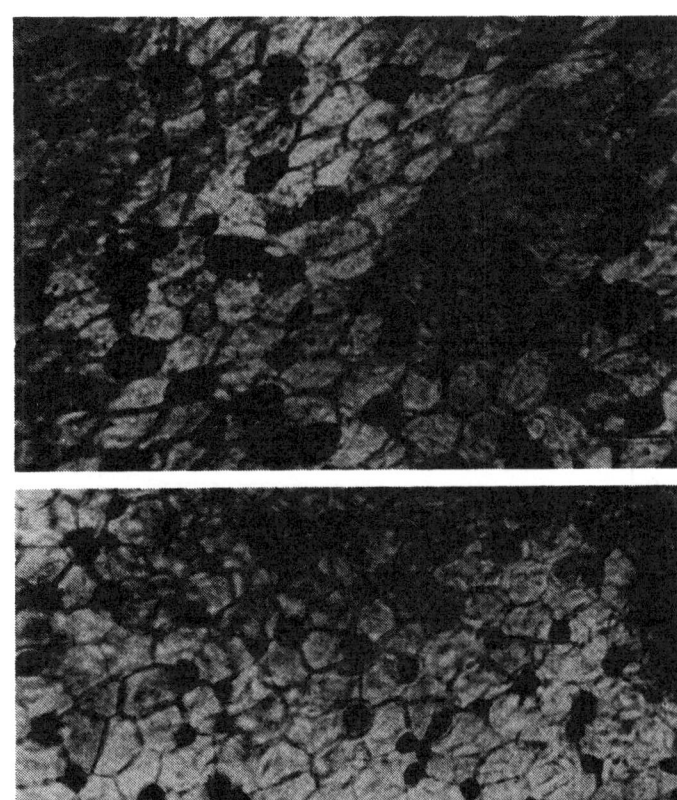

Fig. 2. View of luminal surface of paired hemibladders after histochemical localization of carbonic anhydrase (CA). In the control half (top), the CA-containing cells exhibit both large and small surface areas. In the half exposed to a disulfonic stilbene, the luminal areas of the large cells have been reduced so that all CA cells have small luminal membrane areas. Reproduced from Husted et al.[3] with permission.

pump complex *per se*. A lumen-negative PD as is normally generated by Na$^+$ transport clearly accelerates the H$^+$ transport rate in the turtle bladder[22] as well as in the toad bladder.[23] This effect represents indirect electrical coupling between Na$^+$ absorption and H$^+$ secretion. A similar indirect coupling occurs in the cortical collecting tubule[24] and possibly in other seg-

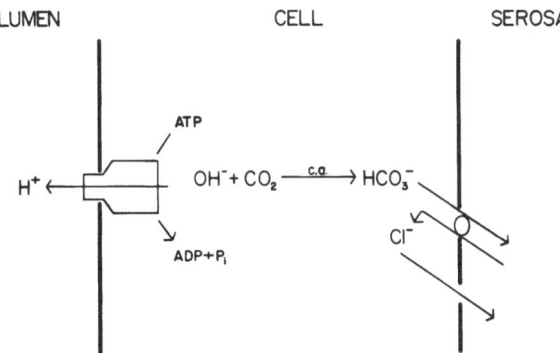

Fig. 3. Double membrane model for urinary acidification in a tight epithelium, such as the turtle bladder and the mammalian collecting duct (see text). Reproduced from Koeppen and Steinmetz[72] with permission.

ments of the distal mammalian nephron. Such electrical coupling may account for several of the clinical observations in which impaired Na$^+$ reabsorption is associated with reduced H$^+$ excretion. Examples are the distal acidifying defects seen in patients receiving amiloride or lithium, agents which inhibit Na$^+$ transport in the collecting tubules, and, thereby, reduce the lumen-negative voltage. With respect to the voltage dependence of H$^+$ transport, it is not surprising that a lumen-positive voltage ($\Delta\Psi$) reduces active H$^+$ secretion. It is of great interest that in the turtle bladder, J_H is inhibited in equivalent manner by a ΔpH and a $\Delta\Psi$ so that J_H is reduced to zero by either a pH gradient of 3 units or an opposing voltage of 180 mV.[25] The ΔpH and $\Delta\Psi$ are the major factors controlling the rate of H$^+$ translocation across the luminal cell membrane.

2.1.3. H$^+$-Translocating ATPase of Luminal Cell Membrane

Several lines of evidence indicate that the H$^+$ pump of the luminal membrane is a H$^+$-translocating ATPase. First of all, the isolated turtle bladder has a remarkable capacity to transport H$^+$ under strictly anaerobic conditions.[20,26] This anaerobic H$^+$ secretion continues for hours if glucose and CO$_2$ are made available to the epithelium. This characteristic by itself suggests that H$^+$ secretion is not driven by a redox pump. Anaerobic H$^+$

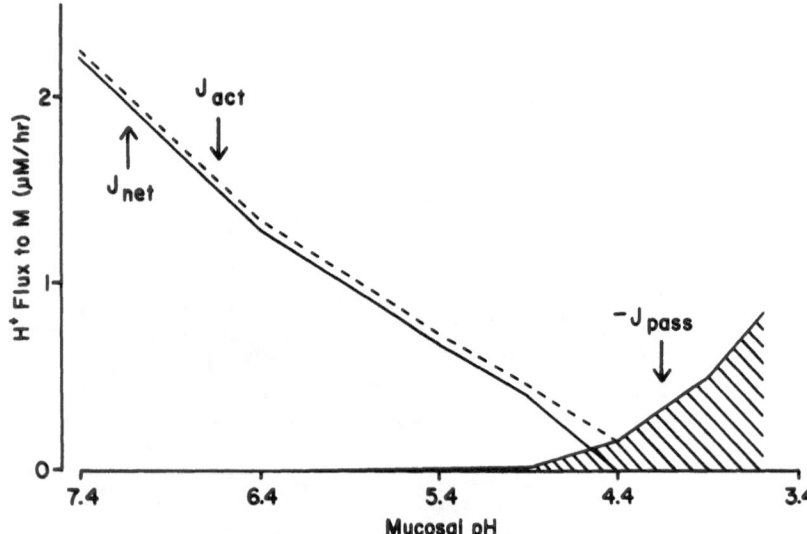

Fig. 4. Active H^+ transport against pH gradients in turtle urinary bladder. J_{net} represents the net rate of H^+ secretion measured by pH stat titration. The rate of back-diffusion is indicated by $-J_{pass}$ (shaded area). The active component of transport, J_{act}, is indicated by the dashed line. Reproduced from Steinmetz and Lawson[21] with permission.

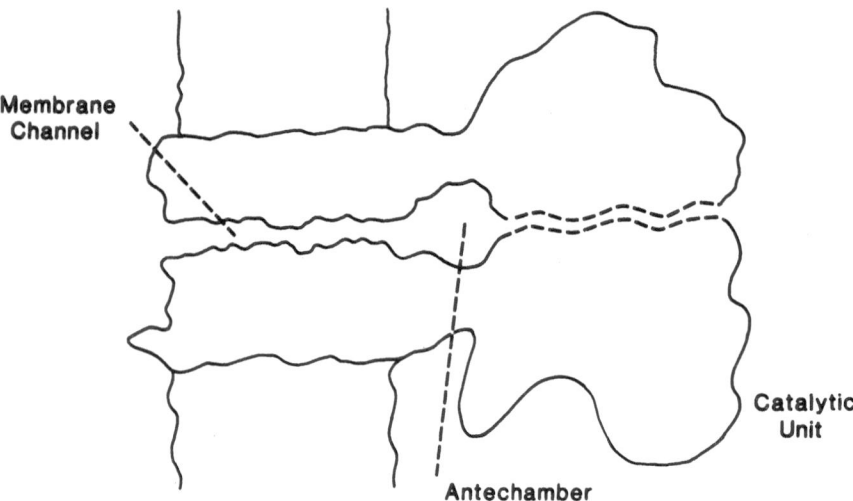

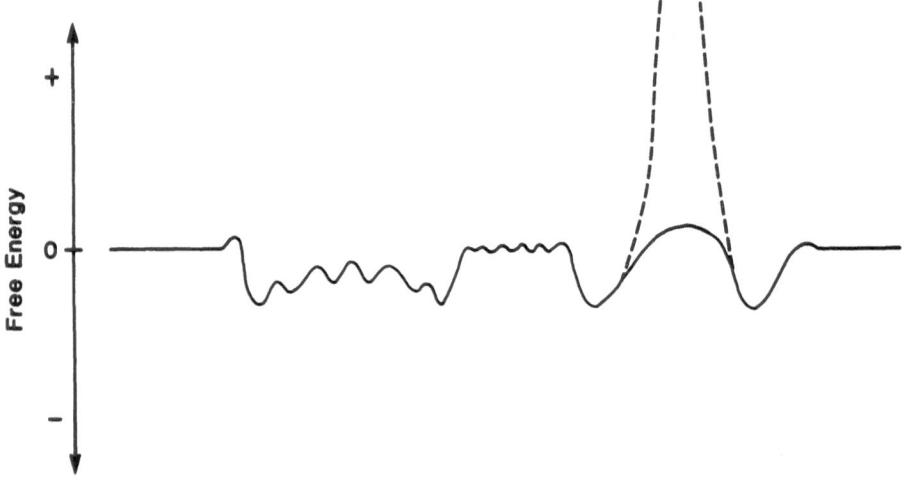

Fig. 5. Representation of H^+ pump complex. (Top) The major components are a membrane channel, an antechamber which serves as a buffer compartment, and the catalytic unit in which ATP hydrolysis is coupled to the translocation of H^+. (Bottom) Free energy profile for H^+ as it moves through the pump complex. Reproduced from Steinmetz and Andersen[11] with permission.

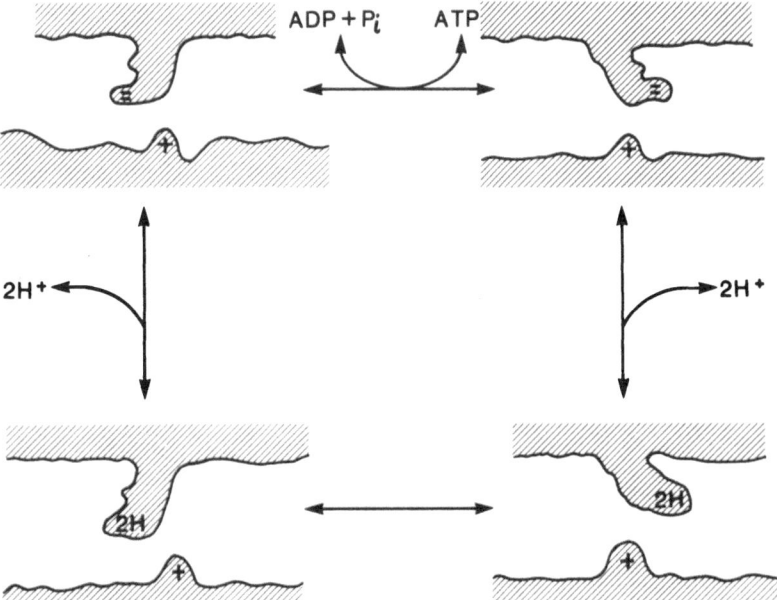

Fig. 6. Scheme for the active translocation reactions in the catalytic unit, adapted from Boyer.[30] The translocating groups denoted by two negative charges can face toward the right or left of the barrier for H^+ leak depicted by the positive charge. Protons can react with the translocating groups in either configuration to form $E''H_2$ and $E'H_2$, respectively; the neutral complex EH_2 can "translocate" freely across the barrier associated with the positive charge. The "translocation" of the unloaded E^{--} group, however, proceeds only if it is coupled to ATP hydrolysis (or synthesis). Reproduced from Steinmetz and Andersen[11] with permission.

secretion is inhibited by dicyclohexylcarbodiimide (DCCD) and Dio-9, but not by cyanide, ouabain, or oligomycin.[20] These studies suggest that the pump has several of the characteristics of a H^+-translocating ATPase. Dixon and Al-Awqati[27] carried out an interesting study in which they reversed the flow of H^+ through the pump by imposing a $\Delta\bar{\mu}_H$, a combination of ΔpH and $\Delta\Psi$, in excess of the reversal potential. They reported that the epithelial ATP levels were increased and concluded that the pump could be reversed to operate as an ATP-synthetase. Subsequent studies by Gluck et al.[28] and Youmans et al.[29] in membrane vesicles of turtle bladder provide further support for the existence of a H^+-ATPase in turtle bladder. The H^+-ATPase is not yet well defined. It resembles fungal plasma membrane ATPase in being electrogenic and resistant to ouabain and oligomycin. Although vanadate inhibits anaerobic H^+ secretion by the turtle bladder, it has not been proven that this inhibition occurs at the pump directly. Thus, a DCCD-sensitive fraction of H^+-ATPase in turtle bladder vesicles is vanadate resistant.[28] The ATPase of urinary acidification, therefore, may differ from the fungal H^+-ATPase which is vanadate sensitive and operates via a phosphorylated intermediate. It also differs from mitochondrial ATPase in being oligomycin resistant. At present, it appears to resemble most closely the H^+-ATPases of the membranes of lysosomes and chromaffin granules.[28]

Although the biochemical information on the H^+-ATPase remains incomplete, some relatively simple assumptions may be made about the steps involved in H^+ translocation by the H^+ pump complex. Figure 5 shows the major structural components of the pump complex based on information available from other H^+-ATPases. The active translocation reaction is thought to occur in the catalytic unit where ATP hydrolysis takes place. The translocated H^+ is buffered in an antechamber which is linked to a transmembrane channel which permits H^+ movement by an electrodiffusive mechanism. Since considerable experimental information is available for the turtle bladder on the J_H versus $\Delta\bar{\mu}_H$ relation, on the apparent force of the pump and the H^+/ATP stoichiometry, it has been useful to

attempt to model the active transport of H^+ through the pump complex. A possible scheme for the active translocation reactions in the catalytic unit is shown in Fig. 6. The translocation groups within the unit are denoted by two negative charges and can face toward either side of the barrier for H^+ leak (+). H^+ can react with the translocating groups in either configuration to form $E''H_2$ and $E'H_2$ (see Fig. 7). The neutral complex EH_2 can translocate freely across the barrier, whereas translocation of the unloaded group E^{--} is coupled to ATP hydrolysis. An overall reaction scheme for active H^+ transport is presented in Fig. 7. The kinetic transitions are indicated for the transmembrane channel (left) and the catalytic unit (right). The two components are coupled through variations in $[H^+]_a$, the H^+ concentration in the antechamber.

The rate of active H^+ transport, J_H, can be expressed as a function of the cellular and luminal H^+ concentrations and the transmembrane potential according to the model shown in Figs. 5–7 and on the basis of measured values in turtle bladder. Figure 8 shows a theoretical J_H versus $\Delta\Psi$ relationship based on relatively simple assumptions for the various rate constants. The theoretical relationship reveals a near-linear region over about 150 mV which resembles the experimentally observed J_H versus $\Delta\bar{\mu}$ relation. Such modeling also suggests that the true force of the pump (PMF) may be greater than the apparent PMF. A second suggestion is that when J_H is accelerated by a favorable $\Delta\bar{\mu}_H$, the linearity is lost. Recent experimental data indeed indicate saturation of J_H under such conditions. These considerations regarding the effects of $\Delta\Psi$ and ΔpH on the rate of H^+ secretion are pertinent for certain clinical disorders in which the formation of a pH gradient in the collecting duct is impaired.

2.1.4. Control of Active H^+ Transport

The control of the transport rate of H^+ by $\Delta\bar{\mu}_H$ is thought to reflect the intrinsic properties of the H^+ pump complex. For a given acid–base status of the epithelium, J_H can be varied virtually instantaneously between zero and maximal by the appro-

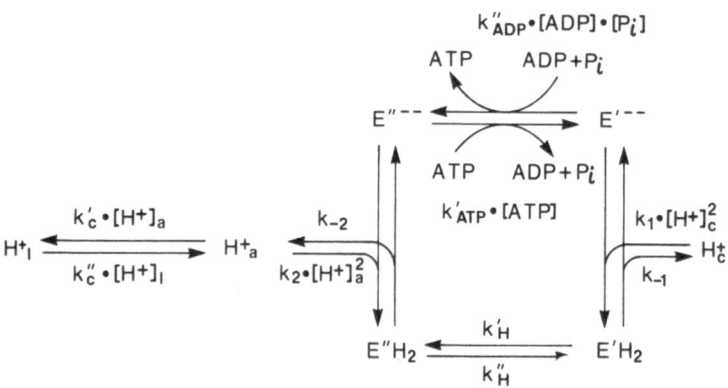

Fig. 7. Kinetic transitions in reaction scheme for active H^+ transport. The horizontal segment at left depicts H^+ translocation through the transmembrane channel, the square segment at right depicts the reactions in the catalytic unit. The two segments are intrinsically independent of each other, but coupled through the variations in $[H^+]_a$, the H^+ concentration in the antechamber. $[H^+]_l$ and $[H^+]_c$ are the luminal and cellular H^+ concentrations, respectively. [ATP], [ADP], and $[P_i]$ refer to concentrations in the cells. Reproduced from Steinmetz and Andersen[11] with permission.

priate ΔpH or $\Delta \Psi$. This control by $\Delta \bar{\mu}_H$ is preserved in bladders in which membrane fusion events are inhibited by colchicine. Hence, this control of J_H must occur across each H^+ pump complex that is in place at the luminal membrane.

Aside from this fundamental control mechanism for H^+ secretion, the rate of secretion can be altered markedly at constant $\Delta \bar{\mu}_H$. Thus, J_H is greatly stimulated by an increase in CO_2 tension and J_H is reduced when the cell is made alkaline. Also,

aldosterone will stimulate J_H in ouabain-treated bladders independently of its effect on Na^+ transport.

We referred earlier (Section 2.1.1) to the studies by Husted *et al.*,[3] Gluck *et al.*,[6] and Stetson and Steinmetz[5] to indicate that cellular acid–base changes can cause large changes in the luminal membrane area of CA cells. These studies demonstrated that CA cells are responsible for H^+ secretion. The most important conclusion of these studies is

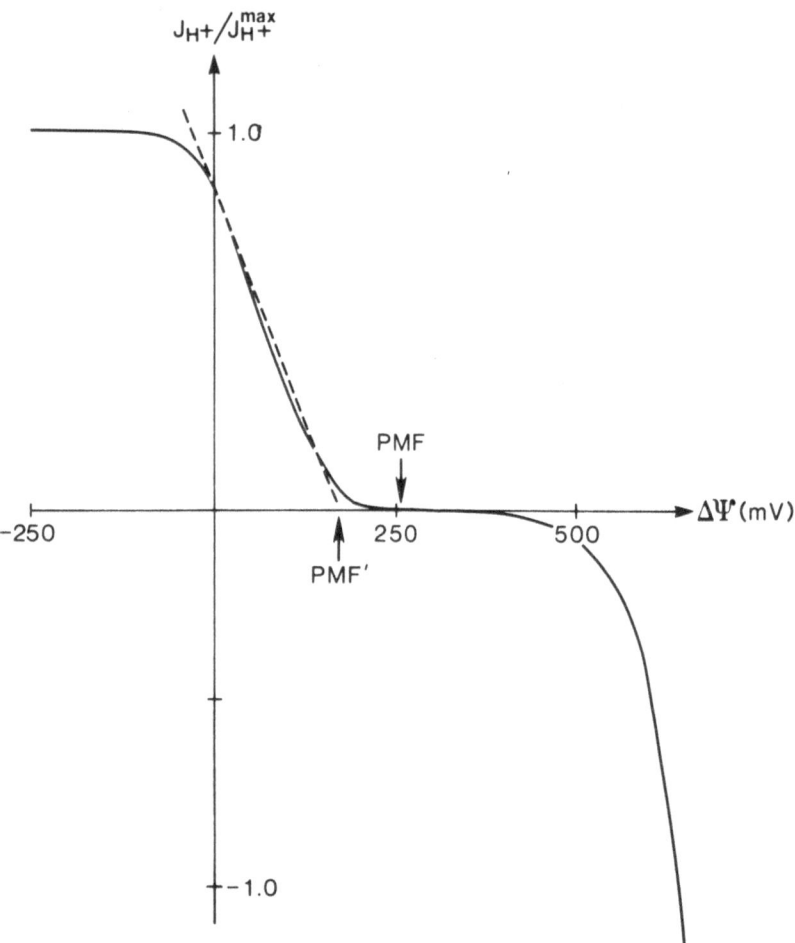

Fig. 8. Theoretical J_H versus $\Delta \Psi$ characteristic for the proposed H^+ pump. Abscissa: transmembrane potential. Ordinate: J_H/J_H^{max}. The relative values of the various rate constants are: $k_c = 1$, $k_{ATP} + k_{-1} = k_{-2} = 10^4$, $k_H' = k_H'' = 300$ (all in sec^{-1}). $\Delta G_{ATP} = 50$ kJ/mole, $k_1 = k_2 = 1$, and $[H^+]_l = [H^+]_c = 0.1$. $\Delta \Psi$ is assumed to be distributed equally across the transmembrane channel and the catalytic unit. The full curve is the theoretical J_H versus $\Delta \Psi$ characteristic; the arrows denote the position of the reversal potential of the pump; the true PMF = +259 mV at 25°C. The dashed line illustrates that the J_H versus $\Delta \Psi$ characteristic can be approximated by a straight line over an extended potential range. Note that the apparent PMF is only +173 mV or about two-thirds of the true PMF. Reproduced from Steinmetz and Andersen[11] with permission.

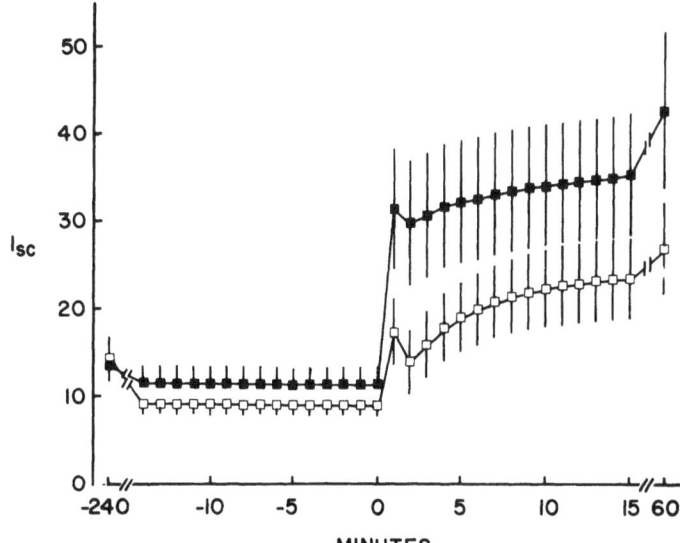

Fig. 9. Response of H^+ transport rate (I_{sc}) to rapid CO_2 addition in control (■) and colchicine-treated hemibladders (□). Colchicine (2×10^{-5} M) was added at $t = -240$ min and CO_2 was added at $t = 0$. Pretreatment with colchicine caused a marked reduction in the stimulation of I_{sc} by CO_2. Reproduced from Stetson and Steinmetz[5] with permission.

that the membrane vesicles in the cytoplasm contain H^+ pumps and constitute a reserve of pumps which can be added to the luminal surface in response to CO_2 stimulation. Conversely, removal of CO_2 or the serosal addition of HCO_3^- or SITS leads to endocytosis and shrinkage of the luminal area with a loss of H^+ pump sites.

How aldosterone stimulates H^+ secretion remains to be clarified. The stimulation may involve increased synthesis and subsequent insertion of H^+ pumps into the luminal membrane. Although the occurrence of pump addition and subtraction to and from the H^+-secreting apical membrane is now well established, the signals for exocytosis and endocytosis and the precise role of the cytoskeleton in the migration of vesicles remain to be explored. In general, inhibitors of cytoskeletal function reduce the stimulation of J_H by CO_2. Figure 9 shows how the CO_2 response is decreased in bladders pretreated with colchicine. The response is also decreased by 99% D_2O, cytochalasin B, and certain inhibitors of calmodulin.[5] In most studies to date, the acid–base signals for this kind of regulation of J_H have been very large. It is not known whether exocytosis and endocytosis are important for the fine regulation of J_H. Kinetic factors such as the supply of H^+ and the supply of metabolic energy may also influence the transport rate.

2.2. Na^+/H^+ Antiporter in Proximal Renal Tubule

Na^+/H^+ exchange has been invoked as a mechanism for H^+ secretion since the term was introduced by Homer Smith in the 1937 edition of his book, *The Physiology of the Kidney*.[31] It has long been known that stimulation of Na^+ reabsorption by the intact kidney is often associated with increased H^+ excretion[32,33] and it was customary in the 1950s and 1960s for renal physiologists to depict Na^+/H^+ exchange by a common transport system in the distal nephron.

In the previous section we discussed that H^+ secretion by tight urinary epithelia such as those of the distal nephron occurs by an electrogenic H^+ pump which is coupled to ATP hydrolysis rather than to Na^+ transport. Distal H^+ secretion therefore represents primary active transport rather than Na^+/H^+

exchange. True Na^+/H^+ exchange, however, is a major mechanism for H^+ secretion by the brush border of the proximal tubule. Evidence for direct coupling of the flows of Na^+ and H^+ in a common transporter has been obtained from studies of membrane vesicles of the brush border of proximal tubules. Brush border membranes can be isolated as a relatively pure preparation which spontaneously forms right-side-out membrane vesicles. Mürer *et al.*[34] first reported one-for-one coupling between Na^+ uptake and H^+ release in brush border vesicles of rat renal cortex. They were able to demonstrate acidification of the medium in which membrane vesicles were suspended following the sudden addition of Na^+. If the vesicles were preloaded with a buffer of low pH, the uptake of Na^+ into the vesicles was stimulated. In these and subsequent studies by Kinsella and Aronson,[35] a counterflow of either H^+ or Na^+ was induced by imposing a Na^+ or a H^+ gradient, respectively. The Na^+-for-H^+ exchange flows were not affected by maneuvers which change the membrane potential. The exchanger, therefore, operates in an electroneutral manner. Brush border membrane vesicles from rabbit renal cortex are similar to those obtained from the rat. The Na^+/H^+ antiport exchanges one Na^+ for one H^+ and the exchange is inhibited by Li^+, NH_4^+, and amiloride in high concentrations. In contrast, acetazolamide and furosemide appear to have no direct inhibitory effects on the exchange rate. Intravesicular pH changes can be monitored rapidly by means of fluorescence techniques with acridine orange. Kinetic studies by Warnock *et al.*[36] showed that the half-maximal H^+ flux occurred at an external [Na^+] of about 13 mM.

The role of Na^+/H^+ exchangers in the regulation of intracellular pH in nonpolar cells has been reviewed in Chapter 26 of this volume. The polar epithelial cells of various "leaky" epithelia such as the proximal renal tubule are equipped with a high density of these exchangers in their apical membranes. Under ordinary conditions the exchangers are driven by the Na^+ gradient across the apical cell membrane. The coupled transport of H^+ represents secondary active transport. The primary active transport step is located at the basolateral membrane where the Na^+ pump is driven by ATP hydrolysis and maintains low intracellular Na^+ concentrations. In the proximal tubule the Na^+

gradient serves as driving force not only for H^+ secretion, but also for several cotransport systems for such substrates as glucose, amino acids, and phosphate. It is of interest that in proximal tubular disorders the reabsorption of these substrates is often impaired simultaneously with HCO_3^- reabsorption.

2.3. HCO_3^-/Cl^- Exchange Transport

Anion-exchange transport systems have been described in a variety of animal cells. The HCO_3^-/Cl^- exchanger of the red cell is probably the most extensively studied anion exchanger.[37] In polar epithelial cells involved in H^+ secretion, the basolateral cell membrane contains transport sites for the efflux of HCO_3^- formed in series with the H^+ translocators at the luminal membrane. The precise nature of the HCO_3^- exit sites remains to be elucidated in "leaky" epithelia such as the proximal tubule as well as in the "tight" distal urinary epithelia. The efflux of HCO_3^- in both types of epithelia is inhibited by the disulfonic stilbenes[38–40] and by removal of ambient Cl^-.[13,41,42] These features suggest that the efflux of HCO_3^- is coupled to a counterflow of Cl^- by means of an anion exchanger. The efflux of HCO_3^-, however, is often associated with the transfer of one negative charge across the basolateral membrane. For example, in the tight urinary epithelia the overall process of urinary acidification is electrogenic. Studies in the turtle bladder suggest that the basolateral membrane contains a Cl^- conductance parallel to the HCO_3^-/Cl^- exchanger. As indicated in Fig. 3, Cl^- would recycle across the basolateral membrane, and, thereby, transfer the negative charge generated after H^+ extrusion. Alternatively, the HCO_3^-/Cl^- exchanger might operate in an electrogenic mode in which HCO_3^- efflux exceeds Cl^- influx.

Aside from the importance of HCO_3^- transport in the exit step of the acidification pathway, there is now evidence that several urinary epithelia may secrete HCO_3^-. In the turtle bladder, for example, HCO_3^- secretion can be demonstrated after inhibition of the Na^+ and the H^+ pumps.[43,44] Under ordinary conditions this secretion of HCO_3^- is coupled one-for-one to Cl^- absorption. A similar HCO_3^- secretory system has been described in the cortical collecting tubule.[45,46] The net excretion of acid, of course, would be reduced by any HCO_3^- secretion. Furthermore, the rate of HCO_3^- secretion would be influenced by the availability of Cl^- or other anions in the lumen for countertransport. Whether or not HCO_3^- secretion occurs in the human collecting duct remains to be determined.

3. Functional Organization of H^+ Secretion in the Kidney

3.1. HCO_3^- Reabsorption

The proximal tubule is responsible for the reabsorption of some 80–90% of the HCO_3^- filtered at the glomerulus. The rate of HCO_3^- reabsorption is faster than the rate of fluid reabsorption so that the luminal HCO_3^- concentrations are lowered to about one-third or one-half (depending on the animal species studied) of the concentration in the peritubular capillaries.[12,47] The gradients for H^+ and HCO_3^- established across the proximal tubular epithelium are much smaller than those attained across the epithelia of the collecting duct and turtle urinary bladder. In the proximal tubule the electrical gradients are even smaller than the concentration gradients.

Recent refinements of techniques for the measurement of pH, pCO_2, and $[HCO_3^-]$ in nanoliter volume of tubular fluid have enabled investigators to clarify the mechanism of HCO_3^- reabsorption by the proximal tubule and the role of CA. Before considering the handling of the different species of CO_2 in the renal cortex, we shall briefly summarize the transport processes across the proximal tubule cell. Figure 10 shows how H^+ ions are translocated across the luminal, or brush border membrane of the proximal tubule. The bulk of H^+ ions secreted into the lumen are translocated via the Na^+/H^+ exchanger described in Section 2.2. In addition, brush border membranes appear to contain a H^+-translocating ATPase which can be distinguished from mitochondrial ATPase.[48] This H^+-ATPase is thought to be responsible for the portion of proximal HCO_3^- reabsorption which is not dependent on the presence of Na^+.[49] Both transport systems secrete H^+ into the HCO_3^--containing glomerular filtrate. In the lumen of the proximal tubule, the following reaction takes place:

$$H^+ + HCO_3^- \rightleftharpoons H_2CO_3 \rightleftharpoons H_2O + CO_2$$

The first reaction is instantaneous. The second reaction, the dehydration of carbonic acid to CO_2, is catalyzed by CA which is available at the brush border membrane. In the absence of CA (i.e., after addition of CA inhibitors), the dehydration reaction is too slow to keep up with the secretion of H^+ into the tubular fluid.[50–54] Hence, the breakdown of H_2CO_3 lags and the *in vivo* tubular fluid pH is lower than it is at equilibrium, i.e., in a sample removed from the tubule at the same pCO_2. The dif-

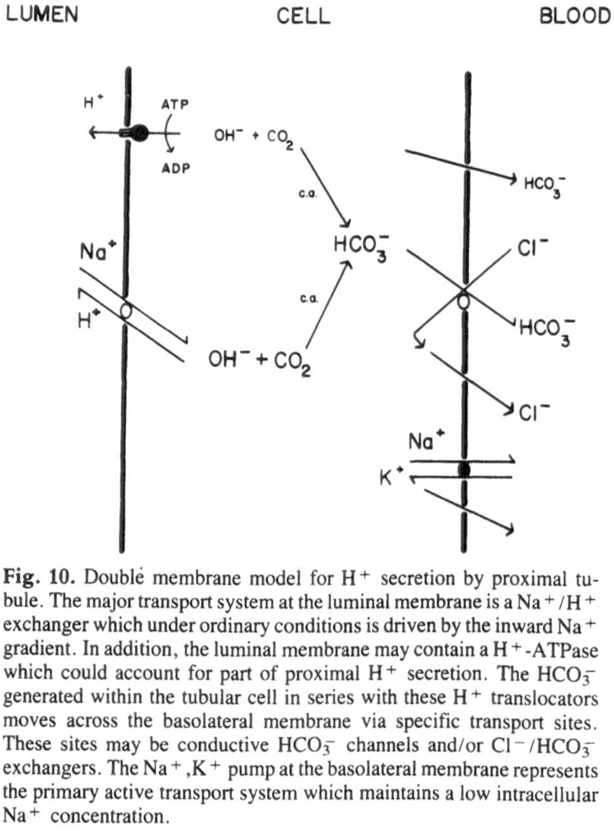

Fig. 10. Double membrane model for H^+ secretion by proximal tubule. The major transport system at the luminal membrane is a Na^+/H^+ exchanger which under ordinary conditions is driven by the inward Na^+ gradient. In addition, the luminal membrane may contain a H^+-ATPase which could account for part of proximal H^+ secretion. The HCO_3^- generated within the tubular cell in series with these H^+ translocators moves across the basolateral membrane via specific transport sites. These sites may be conductive HCO_3^- channels and/or Cl^-/HCO_3^- exchangers. The Na^+,K^+ pump at the basolateral membrane represents the primary active transport system which maintains a low intracellular Na^+ concentration.

ference between the *in vivo* pH and the equilibrium pH is referred to as the disequilibrium pH. The luminal CA plays an important role in dissipating the disequilibrium and, thereby, reducing the pH gradient against which H^+ ions are translocated from cell to lumen.[12,52]

The filtered HCO_3^- is dissipated to CO_2 and the CO_2 is highly diffusible across the proximal tubular and capillary membranes.[47] The disposition of this CO_2 is a fairly complex matter. Part of the CO_2 is used in the cytoplasm to re-form HCO_3^-. Hence, the HCO_3^- filtered is not the same as the HCO_3^- returned across the basolateral membrane to the peritubular capillaries. The pCO_2 of the proximal tubular fluid is about 25 mm higher than that of blood from the systemic circulation.[47,55] Across the structure within the renal cortex, however, pCO_2 gradients are either small or not detectable. Hence, the pCO_2 is high in all cortical convoluted tubules as well as in the peritubular capillaries.

Figure 10 shows how the HCO_3^- formed within the cells is transported across the basolateral cell membrane. At present, there is evidence for the existence of both a Cl^-/HCO_3^- exchanger[40–42] and a conductive HCO_3^- channel.[56] The structural characteristics of these efflux mechanisms and their relative importance in different animal species remain to be clarified.

Studies of single proximal tubule segments of the rabbit kidney have shown that the rates of HCO_3^- reabsorption are greater in the juxtamedullary nephrons than in the superficial nephrons.[57,58] In both populations of nephrons the reabsorptive rates decrease in the pars recta of the proximal tubule.[57,58]

The rate of HCO_3^- reabsorption is markedly reduced by CA inhibitors. This inhibition amounts to more than 80% in the superficial nephrons and to about 60% in the juxtamedullary nephrons.[59,60] The CA at the brush border membrane plays a critical role in proximal HCO_3^- reabsorption as judged from studies with dextran-bound CA inhibitors.[61] As shown in Fig. 10, CA is also abundantly present in the cytoplasm where it catalyzes the hydration reaction. A third site of action may be the exit step for HCO_3^- at the basolateral membrane.[56]

The regulation of HCO_3^- reabsorption by the proximal tubule involves acid–base control mechanisms as well as hemodynamic and hormonal factors. The reabsorption of HCO_3^- like that of Na^+ is influenced by glomerulotubular balance and by feedback control of GFR. This control derives from the fact that HCO_3^- like Cl^- is a major anion accompanying filtered Na^+. For example, expansion of the extracellular fluid volume reduces the threshold for HCO_3^- reabsorption and volume contraction increases fractional HCO_3^- reabsorption.[62,63] The reabsorption of HCO_3^- is stimulated by increases in the CO_2 tension and in general by peritubular acid–base changes which lower cell pH.[59,60,64,65] Reabsorption is also increased when the HCO_3^- concentration of the tubular fluid is high, a condition observed in chronic hypercapnia or metabolic alkalosis. High luminal HCO_3^- concentrations reduce the H^+ gradient across the brush border membrane and, thereby, facilitate H^+ translocation.[65,66] Chronic K^+ depletion also stimulates HCO_3^- reabsorption by a mechanism which remains to be explored. Severe hypophosphatemia inhibits HCO_3^- reabsorption and hypophosphatemia may be a factor in the impaired HCO_3^- reabsorption described in hyperparathyroidism.[67,68]

The HCO_3^- that escapes reabsorption by the convoluted and straight segments of the proximal tubules, perhaps 5–10% of the filtered HCO_3^-, is usually reabsorbed by the distal tubule

and the collecting duct system. In man, under ordinary conditions only a few milliequivalents of HCO_3^- is lost in the urine.

3.2. H^+ Secretion by Distal Tubule and Collecting Duct

With the exception of the thin and thick ascending limb of the loops of Henle, all segments of the distal nephron participate in urinary acidification. The distal convoluted tubule and the initial segments of the cortical collecting tubule have been studied extensively *in vivo* by micropuncture techniques.[50–52] The collecting duct system has been examined *in vitro* by perfusion of isolated cortical and medullary duct segments.[24,69–72] The papillary portion of the collecting duct has been approached with microcatheterization techniques.[73,74]

From these studies the understanding of distal urinary acidification has been greatly increased. It has become clear that all segments of the collecting tubule can contribute to H^+ secretion and can generate steep H^+ gradients under appropriate conditions.

The cortical collecting duct is quite versatile. It reabsorbs Na^+ and secretes K^+. The basolateral Na^+,K^+-ATPase serves as the primary active transport system for both Na^+ and K^+. The high capacity for K^+ secretion in this segment is determined by an unusually high K^+ permeability of the luminal membrane.[75,76] Mineralocorticoid hormone increases the rate of Na^+ entry across the luminal membrane and, thereby, accelerates the pump rate for both ions. The cortical collecting duct also secretes H^+. As in the turtle bladder, H^+ secretion is not directly coupled to Na^+ absorption. Inhibition of Na^+ absorption by ouabain or amiloride reverses the PD so that the lumen becomes positive[24,69] This lumen-positive PD is generated by H^+ secretion.[24] To complicate the analysis of acid–base transport in this segment, the rabbit cortical collecting duct also secretes HCO_3^- so that net acid–base transport is determined by the difference between the two transport processes.

The outer medullary segment of the collecting duct appears to be specialized in H^+ secretion.[70–72] The portion of this segment located in the inner stripe is capable of high rates of HCO_3^- reabsorption[70] and exhibits little net Na^+ transport *in vitro*.[71] Accordingly, it has a lumen-positive PD generated primarily by active H^+ secretion. Aldosterone has a direct stimulatory effect on H^+ secretion in this segment.[77] Such direct actions of aldosterone on H^+ secretion were first described in toad[78] and turtle bladder.[79]

Aldosterone also stimulates H^+ secretion indirectly in bladder preparations[78,79] as well as in cortical collecting ducts by increasing the PD generated by Na^+ transport. The lumen-negative voltage accelerates H^+ secretion by a mechanism of electric interaction.[11,23–25]

The urine pH decreases further along the papillary collecting duct consistent with H^+ secretion in this segment.[73,74]

3.3. Buffers and Net Acid Excretion

In general, the rate of H^+ secretion decreases as the pH gradient against which transport occurs increases. The net rate of H^+ transport is a function of the transport system and the permeability of the tubule segment.[12] In the proximal tubule, only modest pH gradients are achieved. In the distal segments of the nephron, however, the luminal membranes are relatively tight to H^+ and the force of the H^+ pumps permits the formation of pH gradients of about 3 units. As discussed above (Section 2.1), the rate of distal H^+ secretion is a function of the $\Delta\bar{\mu}_H$ across the transport pathway. Hence, the rate is much higher when the tubular fluid is alkaline than when it is acid. The presence of buffers in the tubular fluid permits high rates of H^+ secretion by minimizing the decrease in luminal pH. In the absence of buffers, the H^+ pumps of the distal nephron would rapidly lower the tubular fluid pH and, thereby, inhibit their own transport rate (Section 2.1).

In the intact kidney the H^+ secreted by the distal nephron react with three major buffer systems. First, the residual HCO_3^- is converted to H_2CO_3 and subsequently to CO_2 and H_2O. This process does not generate any new HCO_3^- for the body, it simply conserves the filtered HCO_3^-. Second, the filtered HPO_4^{2-} accepts secreted H^+ to form $H_2PO_4^-$. Third, the formation of NH_3 and NH_4^+ from glutamine and other precursors establishes a partial pressure for the highly diffusible NH_3 across the tubular structures. In sites where the tubular fluid is acid, NH_3 will be rapidly protonated to NH_4^+ which will remain trapped within the lumen. As NH_3 is converted to NH_4^+, additional NH_3 will diffuse into the tubules. Hence, the H^+-secreting tubules serve as a sink for NH_3 as long as the luminal fluid is acid. In quantitative terms, NH_3 is the most important H^+ acceptor for the H^+ secreted by the distal nephron.[12,47,80]

Net acid excretion by the kidney is usually defined in practical terms as the sum of ammonium excretion ($U_{NH_4}V$) and titratable acid excretion ($U_{TA}V$) minus any HCO_3^- excreted ($U_{HCO_3}V$), hence net acid excretion $= U_{NH_4}V + U_{TA}V - U_{HCO_3}V$. In man under ordinary conditions, net acid excretion ranges between 50 and 100 mmoles/24 hr. $U_{NH_4}V$ is the largest component and represents about two-thirds of the total. $U_{TA}V$ amounts to about one-third in subjects with normal renal function, whereas $U_{HCO_3}V$ is usually small or negligible in subjects on ordinary diets.

Net acid excretion represents only about 2% of total H^+ secretion by the kidney. In man, some 4500 mmoles of H^+ is secreted most of which is involved in the reclaiming of filtered HCO_3^-. The term "net acid excretion" is useful to assess the major components of renal H^+ secretion, and, therefore, to define the different renal tubular acidoses. Although the term defines the renal contribution to acid elimination, it does not provide much information about the biochemical steps and the metabolic processes by which new HCO_3^- ions are generated for the body. The renal physiologist is likely to analyze the transport steps across the tubular epithelium. Secreted H^+ ions react with the freely diffusible NH_3 to form NH_4^+ in the lumen. For each H^+ ion secreted, a HCO_3^- ion is added to the peritubular blood. The student of intermediary metabolism, on the other hand, may emphasize that it is the metabolism of the anionic amino acids, glutamate and aspartate, that leads to the formation of HCO_3^-.[81] With either analysis, the body gains one HCO_3^- ion for each NH_4^+ excreted.

4. The Renal Acidoses

The normal human kidney is capable of responding to large increases in endogenous acid production. The excretion of NH_4^+ is increased acutely when urine pH is lowered by administration of an acid load.[80] After chronic acid loading, $U_{NH_4}V$ is further increased some fivefold over control values. Titratable acid excretion may also be increased substantially by increasing the filtered load of PO_4^{3-} and other buffers.[80]

In patients with renal disease, acid–base balance is often maintained in a normal range so long as the GFR remains above about 25% of normal. As renal disease progresses, however, a metabolic acidosis develops. Net acid excretion becomes impaired and the plasma HCO_3^- concentration decreases until a new steady state is reached. As the filtered load of HCO_3^- decreases, the collecting tubule system receives less HCO_3^- from the proximal tubule so that more of distal H^+ secretion is used for the formation of titratable acid and NH_4^+. Hence, net acid excretion is again adjusted to the daily load of acid production. As the metabolic acidosis worsens, the patient often develops a slightly positive H^+ balance as excess H^+ ions are buffered by bone.[82,83]

There are many forms of renal dysfunction and renal parenchymal disease. Since H^+ is secreted by the renal tubules, one might expect greater impairment of acid excretion in primary tubular disease than in primary glomerular disease. Indeed, there are several forms of renal tubular disease in which the GFR remains close to normal and in which impaired acid excretion causes a hyperchloremic acidosis. More common, however, are the renal acidoses in which glomerular and tubular function are decreased together and in which the remaining nephron population is markedly reduced. In these renal acidoses, several anions such as SO_4^{2-} and PO_4^{3-} are retained and the anion gap (the difference between the serum Na^+ and the sum of Cl^- and HCO_3^-) is increased over normal. In the subsequent sections we review the consequences of (1) glomerular and tubulo-interstitial forms of renal disease associated with overall renal insufficiency, (2) renal hemodynamic alterations, and (3) a variety of disorders affecting specific segments of the renal tubules without much glomerular disease.

5. Reduced Nephron Population

Advanced renal parenchymal disease is the most common cause of renal acidosis. Because of renal and extrarenal compensatory mechanisms, metabolic acidosis is usually not encountered until the GFR falls below about 25% of normal.

In chronic renal failure, HCO_3^- reabsorption by the residual functioning nephrons is enhanced,[84–86] although this enhancement may be obscured by the opposing influence of the solute diuresis and volume overload which occur in chronic renal failure. Values for Tm_{HCO_3} per unit GFR in animals and humans with chronic renal insufficiency are enhanced when compared to values obtained in controls with an equivalent degree of volume expansion.[84,86] Cohn et al.[87] have shown in the remnant kidney model of chronic renal failure that brush border vesicles of dog proximal tubule have increased activity of the Na^+/H^+ exchanger. This adaptive change observed in vitro may be related to the enhanced HCO_3^- reabsorption in chronic renal failure. The specific influences of chronic acidosis, hyperphosphatemia, and secondary hyperparathyroidism upon the Na^+/H^+ exchanger and overall proximal HCO_3^- reabsorption remain to be defined.

In most instances of chronic renal disease, the rate of distal acidification remains intact when expressed as a function of the residual nephron population. Minimal urinary pH achieved during acidosis is often near 5; hence, the ability to generate and maintain a maximal blood to urine pH gradient is preserved. The excretion of titratable acid and NH_4^+ is normal or enhanced when factored for GFR in chronic renal failure.

In experimental and in human unilateral renal disease, excretion of titratable acid and NH_4^+ expressed per unit GFR is not significantly different in diseased and contralateral kidneys.[88,89] In animal models of unilateral renal disease, subsequent removal of the contralateral "normal" kidney results in increased excretion of titratable acid and NH_4^+.[90] Under these circumstances, excretion of titratable acid is increased out of proportion to the increased excretion of NH_4^+.

Titratable acid is composed largely of PO_4^{3-}. The increased excretion of titratable acid per unit GFR in chronic renal failure is due to both increased PO_4^{3-} filtration resulting from the hyperphosphatemia and decreased tubular PO_4^{3-} reabsorption resulting from secondary hyperparathyroidism.

Enhanced NH_4^+ excretion per nephron in chronic renal insufficiency appears to result from hyperplasia of the surviving renal tubules. MacLean and Hayslett[91] have demonstrated that in rats with a chronic reduction in GFR to about 20% of control and with normal systemic pH, the rate of NH_4^+ excretion per nephron was increased while the rate of NH_4^+ production per milligram of DNA, a measurement of cell number, was comparable to controls. The renal PO_4^{3-}-dependent glutaminase activity and concentration of ammoniagenic precursors were not different from controls. In this same study and in a study by Schoolwerth et al.,[92] rats with chronic reduction in GFR and systemic acidosis responded with enhanced NH_4^+ excretion per nephron and increased NH_4^+ production per milligram of DNA. Renal PO_4^{3-}-dependent glutaminase activity was increased while the concentrations of ammoniagenic precursors were decreased. Hence, it appears that in chronic renal disease under stable conditions, increased NH_4^+ production and excretion per nephron is achieved by tubular hyperplasia. A further enhancement of NH_4^+ production occurs in response to acidosis by induction of ammoniagenesis at the cellular level. Despite the capacity of the residual nephrons to increase the reabsorption of HCO_3^- and the excretion of NH_4^+ and titratable acid, metabolic acidosis ultimately develops as the number of surviving nephrons decreases. The metabolic acidosis typically is characterized by an increased anion gap as PO_4^{3-}, SO_4^{2-}, and other "unmeasured" anions are retained. The increased anion gap distinguishes the acidoses of the common forms of renal failure from the "renal tubular acidoses" in which GFR is preserved and the anion gap is normal. Hence, the acidoses of renal failure are usually not associated with hyperchloremia.

6. Renal Hypoperfusion

Underperfusion of the kidneys occurs in a wide range of clinical settings. Renal blood flow and GFR are reduced as a function of prerenal factors. Although this condition may lead to acute ischemic renal failure with tubular damage, it often remains a hemodynamic abnormality in which tubular function is preserved. Because of the reduced GFR and the high fractional reabsorption of fluid in the proximal tubule, the volume flow rates through the loops of Henle, the distal tubule, and the cortical and medullary collecting ducts are markedly reduced. The small quantities of urine formed contain little or no Na^+. Although this urine may contain high concentrations of K^+ and H^+, the actual excretion rates of these ions are reduced to the extent that hyperkalemia and metabolic acidosis may develop. The excretion of titratable acid and NH_4^+ is decreased because of a lack of buffer reaching the distal nephron and because of back-diffusion of NH_3. That the underperfused kidney is capable of excreting large amounts of acid and K^+ can be easily demonstrated if the prerenal disturbance is corrected. An example of

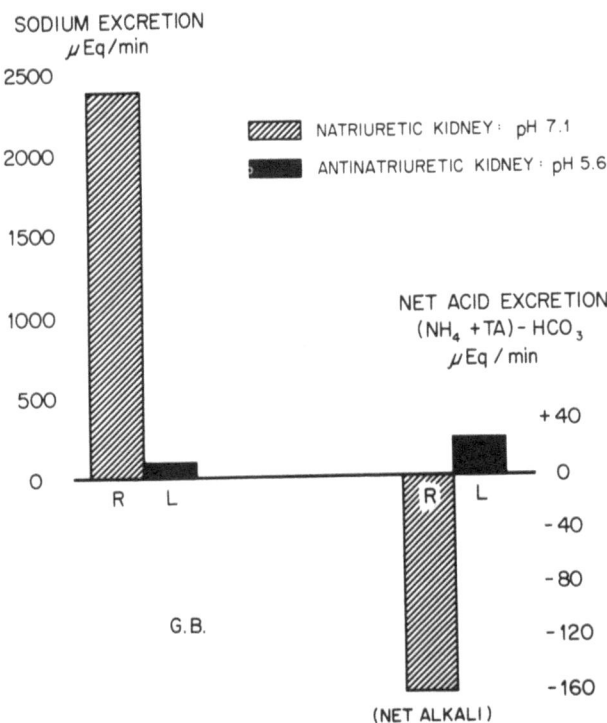

Fig. 11. Effect of sodium sulfate infusion on net acid excretion in a patient with stenosis of the left renal artery. Prior to sodium sulfate infusion, the left kidney conserved Na^+ to the extent that the urine flow rate was too low to be measured. During sodium sulfate infusion, the urine flow rate rose to 0.4 ml/min, and substantial quantities of net acid (black bar) were excreted by the kidney with renal artery stenosis. The contralateral kidney (shaded bars) underwent a massive Na^+ diuresis and excreted large quantities of net alkali. Reproduced from Steinmetz et al.[89] with permission.

the response of the underperfused kidney to increased volume flow to the distal nephron is shown in Fig. 11. A patient with severe stenosis of the left renal artery had a barely detectable urine flow rate in the affected kidney, the excretion rates of Na^+, K^+, and H^+ being negligible.[89] Following the intravenous infusion of sodium sulfate, the urine flow from the affected kidney became easily measurable at 0.4 ml/min, and Na^+ excretion rose to 77 µeq/min (black bar); the contralateral kidney (shaded bar) underwent a massive sodium sulfate diuresis. On the right side of the figure it can be seen that the affected kidney excreted net acid (black bar), whereas the contralateral kidney excreted large quantities of HCO_3^- and, therefore, net alkali (shaded bar). Although the kidney with renal artery stenosis continued to reabsorb a large fraction of the filtered HCO_3^- in the proximal tubule, the presence of Na_2SO_4 caused increased volume flow to the distal tubule and collecting duct and, thereby, increased not only the availability of buffer but also the electrical PD—lumen negative with respect to the peritubular capillary. Both factors in turn stimulated H^+ secretion. As discussed in Section 2.2, the active transport rate for H^+ is directly influenced by the electrochemical force across the active transport pathway. Although there is no direct, one-for-one, Na^+-for-H^+ exchange, the secretion of H^+ is indirectly coupled to Na^+ absorption via the PD. The unaffected kidney (shaded bar) was natriuretic from the beginning. It was able to respond normally to volume expansion by increasing Na^+ and HCO_3^- excretion, and the extent of the natriuresis was greatly increased by the sodium sulfate infusion.

This case illustrates the dependence of renal acid excretion on the availability of buffer and Na^+ in the lumen of the distal nephron and at the same time the importance of the state of Na^+ and HCO_3^- reabsorption in the proximal tubule. A $NaHCO_3^-$ diuresis originating in the proximal tubule can easily exceed the distal capacity for H^+ secretion. The reader is referred to the studies by Berliner et al.,[32] Schwartz et al.,[33] and others[93,94] for a variety of experimental observations indicating that increased Na^+ reabsorption may serve as a stimulus for the acidifying process.

7. Renal Tubular Acidosis

Renal tubular acidosis (RTA) is defined as a clinical syndrome of hyperchloremic acidosis caused by a renal tubular defect for net acid excretion and associated with normal or near-normal GFR. In contrast to the renal acidoses observed in patients with markedly reduced nephron populations, the disturbance in acid excretion can be attributed primarily to a disorder or dysfunction of the renal tubules or collecting ducts.

In many instances of RTA, the segment of the nephron responsible for the acidifying defect can be identified by relatively simple clinical determinations. Rodriquez-Soriano and Edelmann[95] distinguished proximal RTA caused by impaired HCO_3^- reabsorption in the proximal tubule from classic distal RTA which is characterized by an inability to establish adequate gradients for H^+ between blood and collecting duct urine. Morris et al.[96] developed a further classification in which classic distal RTA was named type 1 RTA, proximal RTA was named type 2, while type 3 was used to describe children with a com-

bination form of types 1 and 2. They designated a type 4 RTA in which impaired acid excretion is associated with reduced K^+ excretion and hyperkalemia.[96,97] Type 4 RTA is now a common form of RTA and occurs frequently as part of the clinical syndrome of hyporeninemic hypoaldosteronism. In subsequent years the term type 3 RTA has been abandoned, as it became clear that children with type 1 RTA may transiently waste HCO_3^- [98,99] and, hence, that type 3 RTA does not represent a separate entity. The term type 4 RTA, however, has persisted and the spectrum of type 4 RTA has expanded to include patients in whom the hyperkalemia is not mediated by hyporeninemic hypoaldosteronism but by tubular disease interfering directly with K^+ secretion.

For this chapter we classify renal tubular acidosis simply as proximal RTA, distal RTA, and RTA with hyperkalemia. Within each of these entities a variety of pathogenetic mechanisms may operate to cause the syndrome.

8. Proximal Renal Tubular Acidosis

8.1. Characteristics of Dysfunction

The proximal form of RTA is usually recognized on the basis of major features of proximal tubular dysfunction and on evidence that acidification by the collecting duct system is intact. Figure 12 shows that patients with proximal RTA can lower urine pH below 5.5 if their serum HCO_3^- is lowered enough to reduce the filtered load to the proximal tubule.[95] At serum HCO_3^- values below 14 mmoles/liter, distal acidification was shown to be intact. The dashed curve represents the mean response of healthy infants. As shown in Fig. 14, children with distal RTA are unable to lower urine pH even at very low serum HCO_3^- levels. The important characteristic of proximal RTA is that urinary HCO_3^- wasting is substantial at normal serum

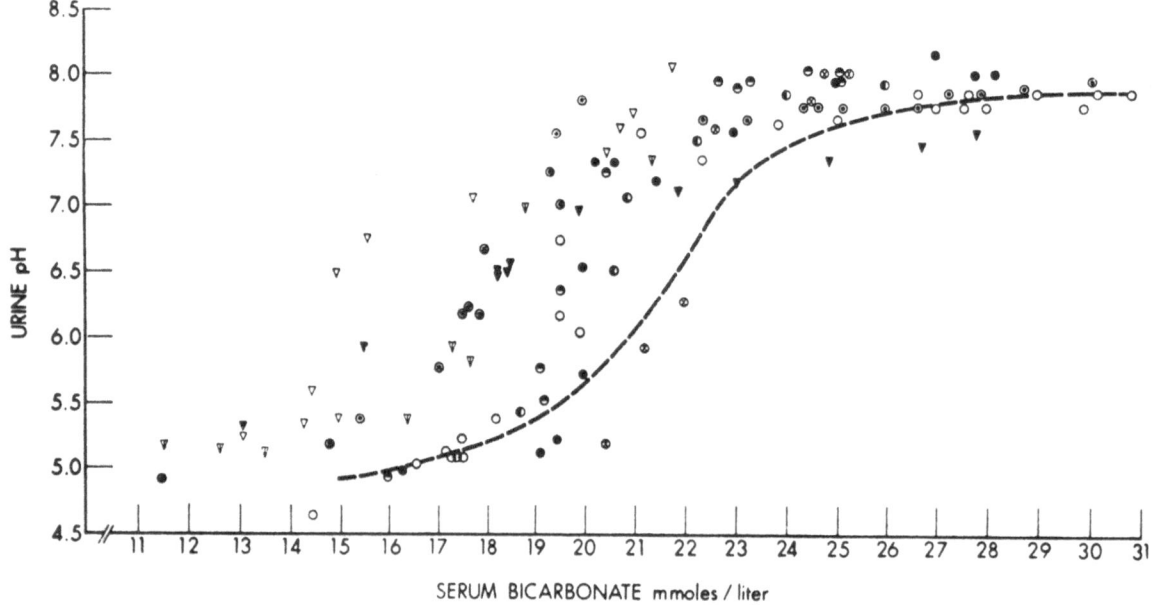

Fig. 12. Urine pH at different serum HCO_3^- levels in patients with proximal RTA. The dashed curve represents the mean response of healthy infants. Reproduced from Rodriguez-Soriano and Edelmann[95] with permission.

HCO_3^- levels, fractional excretion being at least 15%. In distal RTA, fractional excretion of HCO_3^- is usually less than 5% when acidosis is corrected. In most forms of proximal RTA, there is some evidence of impairment of other proximal tubular functions such as the reabsorption of glucose, PO_4^{3-}, and amino acids. The presence of the Fanconi syndrome in a patient with RTA points strongly to the proximal form of RTA.

8.2. Cellular Defects in HCO_3^- Reabsorption

The cellular mechanisms of HCO_3^- reabsorption across the two membranes of the proximal tubule cell have been reviewed above (Sections 2.2 and 3.1). The transepithelial processes involve many steps as shown in Fig. 10 each of which could be affected. Table I lists the possible defects that might account for impaired HCO_3^- reabsorption. Some information about the likelihood of the existence of these defects can be derived from certain diseases with known biochemical abnormalities which serve as experiments of nature. Thus, there is little doubt that impaired CA function can cause proximal RTA. CA inhibitors used chronically produce the syndrome in man. Furthermore, proximal RTA is a major feature of a recently described kindred with congenital absence of CA II, the major renal form of the enzyme.[100] The absence of CA II also appears to be responsible for the osteopetrosis observed in this kindred. Impaired CA function causes an isolated tubular acidosis without the dysfunctions commonly seen in the Fanconi syndrome. The frequent association of proximal RTA with impaired reabsorption of glucose, amino acids, and PO_4^{3-} is of great interest since these substrates traverse the brush border by means of cotransport systems with Na^+. Hence, their transport is driven by the Na^+ gradient across the apical membrane. The same driving force is responsible for the translocation of H^+ by the Na^+/H^+ exchanger in the luminal membrane. Since the Na^+ gradient depends primarily on the low intracellular Na^+ concentration maintained by outward Na^+ pumping at the basolateral membrane, increased intracellular Na^+ could explain all features of the tubular dysfunction. Impaired Na^+ transport as a result of energetic factors or a reduction in the number of Na^+ pumps at the basolateral membrane could account for a high intracellular Na^+ concentration. As discussed below, this possibility deserves consideration in certain diseases of metabolism such as hereditary fructose intolerance in which the energy-supplying reactions to the Na^+ pump are disturbed. Other possible mechanisms involve an increased Na^+ permeability of the proximal tubule cells and of course some abnormality in the H^+/Na^+ exchanger itself. Aside from these abnormalities of H^+ translocation across the apical cell membrane, HCO_3^- reabsorption could be affected at the exit step for HCO_3^- from cell to peritubular blood. In immunologically mediated tubular disease,

Table I. Cellular Defects in Bicarbonate Reabsorption

Impaired H^+/Na^+ antiport function
  Intrinsic defect
  Impaired Na^+ gradient across antiport
    Impaired Na^+ pumping across basolateral membrane
    Increased Na^+ permeability of tubular cell
Impaired carbonic anhydrase function
  Hydration reaction in cytoplasm (carbonic anhydrase II)
  Luminal membrane reaction
Impaired efflux of HCO_3^- across basolateral membrane

Table II. Clinical Forms of Proximal Renal Tubular Acidosis[a]

Selective proximal dysfunction
  Primary forms
    Infantile
    Adult
  Carbonic anhydrase (CA) deficiency
    Absence of CA II (osteopetrosis)
    CA inhibitors: acetazolamide, sulfonamide
Associated with other dysfunctions (Fanconi syndrome)
  Cystinosis
  Wilson's disease
  Lowe's syndrome
  Tyrosinosis
  Hereditary fructose intolerance
  Galactosemia
  Pyruvate carboxylase deficiency

  Hereditary elliptocytosis
  Phosphate depletion
  Hyperparathyroidism
  Vitamin D deficiency

  Dysproteinemia: multiple myeloma, amyloidosis, Sjögren's syndrome, nephrotic syndrome, monoclonal gammopathy

  Immunologically mediated tubular disease
  Renal transplantation
  Medullary cystic disease

  Drugs: outdated tetracycline, 6-mercaptopurine, streptozotocin
  Heavy metals: lead, mercury, cadmium, copper

[a]See text and Refs. 95, 96, 99–103.

the disease tends to be most active at the peritubular side where HCO_3^- efflux is thought to occur via specialized transmembrane proteins. The possibilities listed in Table I serve as a framework for investigating the precise pathogenesis of the different forms of proximal RTA.

8.3. Pathogenesis of Proximal RTA

Despite some interesting lines of evidence, the precise pathogenesis of many of the clinical forms of proximal RTA remains unknown. In Table II the major entities associated with proximal RTA are listed.[95,96,99–103] Selective proximal dysfunction for HCO_3^- reabsorption may occur as a sporadic or hereditary disorder in which the tubular defect appears to represent the primary abnormality. CA deficiency causes reduced proximal HCO_3^- reabsorption without associated Fanconi syndrome. Sly et al.[100] reported three siblings with the autosomal recessive syndrome of osteopetrosis with RTA and cerebral calcification. They demonstrated the virtual absence of CA II from the erythrocytes of these patients in the presence of normal CA I level. CA II (or CA-C) is the major cytosolic enzyme of the kidney and is distinct from the membrane-bound enzyme present in the brush border. It is present in the cytoplasm of proximal and distal tubules as well as in the intercalated cells of the collecting ducts.[1] The presence of CA II in the distal nephron may account for the distal component of the RTA in some of the patients with osteopetrosis.[100,104] Shapira et al.[105] described three members of a family with infantile RTA and nerve deafness and showed a genetic abnormality in red cell CA I (B). CA I, however, is not the major renal isoenzyme and Kendall and Tashian[106] showed that the virtually complete absence of CA I in man is not associated with RTA. Before reliable measurements of CA isoenzymes were available, investigators assessed the response of RTA patients to acetazolamide.[107] In these studies a reduced or absent response suggested that defective CA activity might be the cause of RTA. The availability of immunologic and biochemical techniques for the CA isoenzymes is likely to clarify the frequency of occurrence of CA deficiencies in RTA.

In many of the genetic disorders of metabolism, the proximal RTA is associated with the Fanconi syndrome and the impairment appears to be caused by an inadequate supply of metabolic energy to the transport systems. Since the primary active transport is accomplished by the Na^+,K^+-ATPase at the basolateral membrane, an impaired supply of ATP would increase cell $[Na^+]$ and, thereby, decrease the Na^+ gradient which drives the Na^+/H^+ exchanger and the cotransport systems for glucose, amino acids, and PO_4^{3-}.

Galactosemia and hereditary fructose intolerance are examples of metabolic disorders in which the biochemical defects have been defined.[108-111] In hereditary fructose intolerance, the tubular dysfunction has been studied extensively. Morris[109] demonstrated that fructose administration in patients with hereditary fructose intolerance induced a reversible defect with the characteristics of proximal RTA. Figure 13 shows a representative study on one of his patients. Fructose infusion caused a rise in urine pH and marked decreases in titratable acid and NH_4^+ excretion. The tubular dysfunction was reversible and also included impaired tubular reabsorption of amino acids and PO_4^{3-}. Since in this disorder the renal tubules are deficient in aldolase activity toward fructose-1-phosphate, the latter accumulates within the tubular cells when fructose is present. Since the enzymes involved in the phosphorylation and breakdown of fructose occur in the cortex of the normal kidney but not in the

medulla,[111] the medullary functions are little or not at all affected by fructose infusion in these patients. This observation is consistent with the evidence that the acidifying defect is limited to the cortical convoluted tubules.

PO_4^{3-} depletion, hyperparathyroidism, and vitamin D deficiency are grouped together since certain elements play a common role in the pathogenesis of proximal RTA. Hypophosphatemia *per se* has several effects on the net excretion rate of acid. It reduces proximal HCO_3^- reabsorption possibly by affecting intracellular PO_4^{3-} metabolism and the energetics of transport. Hypophosphatemia also reduces titratable acid excretion by reducing the filtered load of PO_4^{3-} available for buffering secreted H^+. Finally, it leads to a reduction in NH_3 production by the kidney. Hypophosphatemia is a major consequence of hyperparathyroidism as well as of vitamin D deficiency. It may mediate at least part of the proximal RTA in these conditions. Vitamin D deficiency is a major cause of proximal RTA, but in some systemic or primary renal diseases the conversion of 25-hydroxyvitamin D_3 to $1,25(OH)_2D_3$ may be impaired and the resulting deficiency of $1,25(OH)_2D_3$ may in turn cause a Fanconi syndrome with proximal RTA. Brewer *et al.*[112] demonstrated such an impaired conversion of D_3 in an experimental model of maleic acid-induced Fanconi syndrome. Injection of maleic acid[113] also reduces the renal cortical ATP levels and Na^+,K^+-ATPase activity[114,115] without effect on CA function.[116]

Disorders of protein metabolism are a rather common cause of the Fanconi syndrome. In multiple myeloma the reabsorption of light chains by the proximal tubule is thought to interfere with the transport processes of the epithelial cells. Not infrequently, the tubular damage precedes the other clinical manifestations of myeloma.[117] Amyloidosis may be associated with the Fanconi syndrome. This association is often related to the presence of Bence-Jones proteinuria or the nephrotic syndrome.[117-119] Sjögren's syndrome is usually associated with distal acidifying defects.[120] An associated proximal RTA, however, is sufficiently common to warrant listing in Table II.

Renal transplantation and immunologically mediated proximal tubular disease may be associated with the tubular dysfunction and RTA[121-124] Little is known about the pathogenesis of the tubular defect. On the basis of the peritubular location of the injury, it is attractive to speculate that one of the transport systems at the basolateral cell membrane would be affected. Drug toxicity[125] is likely to remain a cause of proximal RTA even though the offending drugs will change with medical practice. Thus, tubular dysfunction following outdated tetracycline is no longer being reported, whereas new forms of chemotherapy are carefully monitored to limit any proximal tubular dysfunction. Heavy metals have long been known to cause the Fanconi syndrome as well as varying degrees of acute renal failure.

9. Distal Renal Tubular Acidosis

9.1. Classic RTA

Classic distal RTA is characterized by an inability to lower urine pH appropriately in response to acidosis. It was the first form of RTA to be recognized.[82,126-129] Patients often present with nephrolithiasis, nephrocalcinosis, and hypercalciuria. Muscle weakness and hypokalemia are less common clinical presentations. Untreated children present with stunted growth. Although the entity contains a variety of hereditary and acquired

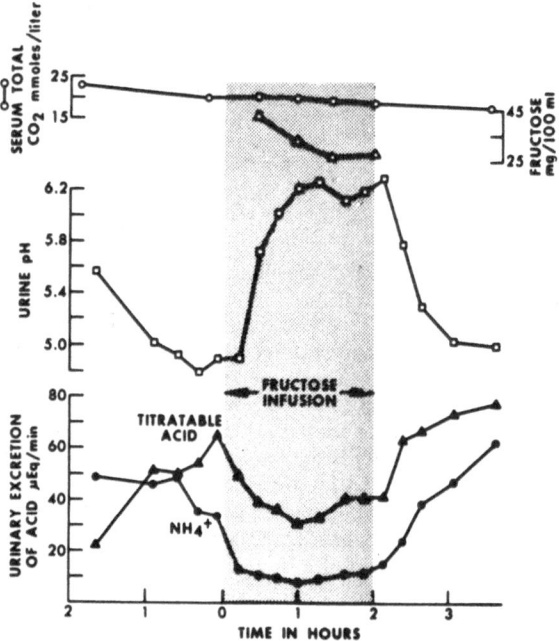

Fig. 13. Effect of intravenous infusion of fructose on renal acid excretion and urine pH in a patient with hereditary fructose intolerance. The fructose infusion was begun 6 hr after administration of ammonium chloride. Reproduced from Morris[109] with permission.

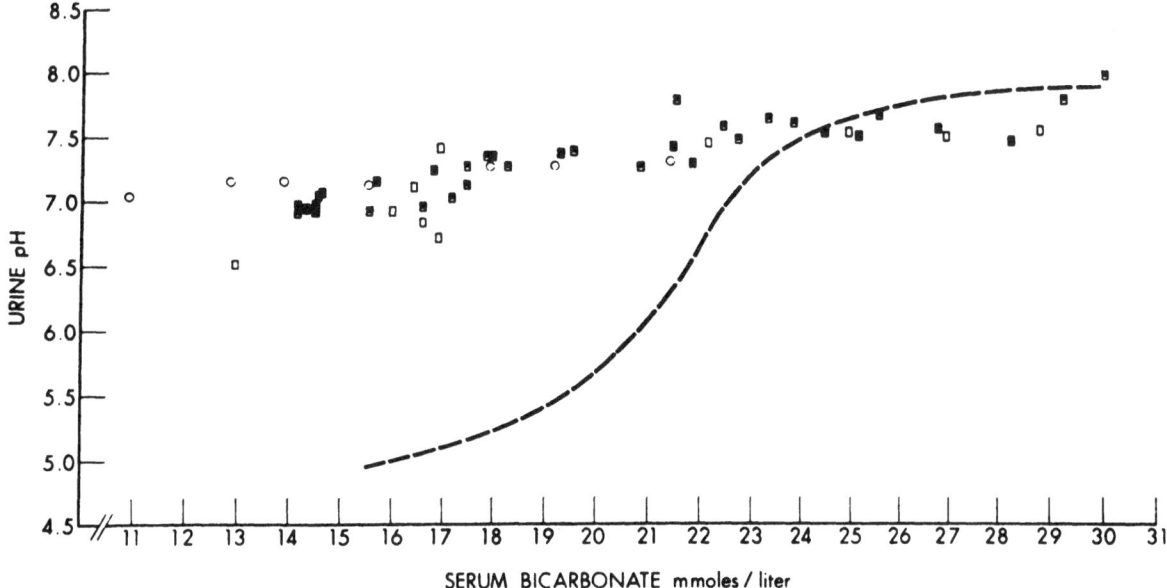

Fig. 14. Urine pH at different serum HCO_3^- levels in patients with distal RTA. The dashed curve represents the mean response of healthy infants. Reproduced from Rodriguez-Soriano and Edelmann[95] with permission.

forms of the syndrome, it has become better defined since proximal RTA and the hyperkalemic forms of RTA have been classified separately.[95–97]

In contrast to the results shown for proximal RTA (Fig. 12), patients with distal RTA are unable to lower urine pH even when the serum HCO_3^- levels are lowered below 12 mM, as shown for infants in Fig. 14. Several investigators have shown that in distal RTA, HCO_3^- reabsorption in the proximal tubule is normal.[95,96,128,129] In addition, several studies show that titratable acid excretion in distal RTA can be increased by PO_4^{3-} administration.[128] Total H^+ secretion by the proximal and distal nephron, therefore, has been considered to be near normal when urinary buffers are in ample supply.

The pathogenesis of distal RTA has been a subject of considerable study and debate. Seldin and Wilson[129] and others[96] attributed the defect to increased permeability of the distal nephron and excessive back-diffusion of secreted H^+. This explanation was consistent with the studies showing that total H^+ secretion was "normal" in the presence of ample buffer. Although increased leakiness of the distal nephron accounts for some forms of distal RTA, there is more and more evidence that in a majority of forms of distal RTA the defect is reduced active H^+ secretion. Halperin et al.[130] and Stinebaugh et al.[131] carried out a series of clinical studies in patients with RTA as well as experimental studies in dogs and rabbits[132] in which they established the importance of the urinary pCO_2 as an index of distal H^+ secretion. During HCO_3^- loading when the urine is alkaline the pCO_2 difference between urine and blood (U-B pCO_2) is a function of the rate of H^+ secretion by the distal tubule and collecting duct system. Ochwadt and Pitts[133] had shown that the pCO_2 of alkaline urines is often elevated and that this elevation results from the delayed dehydration of H_2CO_3 formed in distal nephron segments from filtered HCO_3^- and secreted H^+. Since CA is absent from the luminal membranes of the distal nephron, the dehydration of H_2CO_3 and, therefore, the formation of CO_2 is delayed. The urinary pCO_2 is elevated as a

result of this slow CO_2 generation and also as a function of slow CO_2 loss from alkaline urine in the lower urinary tract. Infusion of CA abolishes the U-B pCO_2 by accelerating the dehydration reaction[133] and also by increasing the rate of CO_2 diffusion in alkaline solutions.[134] Although the urinary pCO_2 is a complex function of the rate of distal H^+ secretion, the HCO_3^- concentration of the fluid in the collecting tubule system,[135,136] and the presence of nonbicarbonate buffers in the urine, the U-B pCO_2 determination can provide a valuable estimate of the magnitude of distal H^+ secretion.[131,132] The studies by Stinebaugh et al.[131] have demonstrated that impaired distal H^+ secretion is common in the pathogenesis of distal RTA. Figure 15 shows a comparison of U-B pCO_2 in 15 patients with distal RTA and 20 normal subjects. The results show that the U-B pCO_2 is significantly reduced in the patients with distal RTA at all concentrations of urinary HCO_3^-. Since patients with distal RTA frequently have associated defects in the renal concentrating mechanism, it was important to include controls with low urinary HCO_3^- concentrations.

The finding of impaired distal H^+ secretion by itself provides little information on the nature of the cellular defect of H^+ transport. Neither does it provide a unique interpretation for the inability of patients with distal RTA to lower their urine pH.

9.2. Cellular Defects in Distal RTA

As indicated above, the impaired secretion of H^+ by the distal nephron by itself does not explain the impaired gradient formation in distal RTA. H^+ secretion can be reduced by a variety of cellular mechanisms as shown in Table III. Most of these mechanisms do not involve the intrinsic properties of the H^+ pumps in the apical membrane of the H^+-secreting cell population. In acquired RTA, there is often extensive disease of the medullary structures with involvement of the corticomedullary regions which contain the segments of the collecting tubule system that contribute importantly to H^+ secretion. Reductions

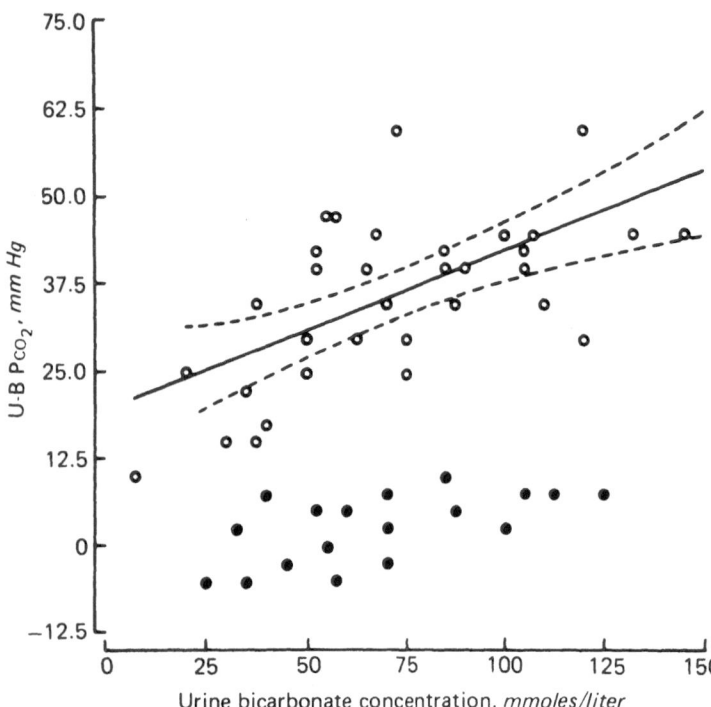

Fig. 15. Urine–blood pCO_2 gradient (U-B pCO_2) as a function of the urine HCO_3^- concentration in normal subjects (○) and patients with distal RTA(●).The U-B pCO_2 is lower at all concentrations of urine HCO_3^- in distal RTA patients. Reproduced from Stinebaugh[190] with permission.

in the H^+-secreting cell population or in the number of pumps in operation at the apical cell membrane could easily occur as a result of such disease. More physiologic reductions in H^+ secretion occur when Na^+ transport is inhibited by amiloride or lithium in collecting duct segments involved in H^+ secretion. Since H^+ secretion by tight urinary epithelia is electrogenic, it is influenced by the transepithelial PD (see Section 2.1). The lumen-negative potential generated by Na^+ transport normally accelerates H^+ secretion. Agents which interfere with Na^+ transport such as amiloride or lithium reduce the rate of H^+ secretion. The actual decrease in H^+ secretion, however, is not very large. Hence, the distal RTA seen in patients receiving amiloride or lithium is usually mild. Intrinsic pump defects constitute an interesting theoretical possibility for which no clinical evidence is available at this time.

Since the H^+ pump generates a OH^- ion for each H^+ translocated across the apical membrane of the epithelial cell

Table III. Cellular Defects in Collecting Tubule System in Distal Renal Tubular Acidosis

Reduced active H^+ secretion
  Reduced population of H^+-secreting cells
  Loss of H^+ pump sites in apical membrane
  Voltage-dependent reduction in H^+ secretion
  Intrinsic pump defects
Impaired disposition of OH^-
  Carbonic anhydrase deficiency
  Defect in exit step for HCO_3^-
Abnormal permeability
  Back-diffusion of H^+
  Diffusion of HCO_3^- into lumen
Increased HCO_3^- secretion

(Section 2.1 and Fig. 3), the disposition of OH^- is an importa. part of continued H^+ secretion. Within the cytoplasm the removal of OH^- is facilitated by CA which catalyzes the reaction of OH^- with CO_2. CA deficiency appears to be rare in man, but a few interesting instances have been reported. A CA II deficiency has been reported in a kindred with osteopetrosis and RTA.[100] The tubular defect had characteristics of both proximal and distal RTA. CA II is the cytoplasmic enzyme of the kidney. It is present in the proximal tubule as well as in the intercalated cells of the collecting duct system. The absence of this renal isoenzyme would be expected to slow the rate of H^+ secretion in the proximal as well as the distal nephron.

H^+ secretion could also be reduced by a defect in the exit step for HCO_3^-. As discussed in Section 2.1, the HCO_3^- formed in series with the H^+ pump moves from cytoplasm to peritubular space via specialized transport sites in the basolateral cell membrane. The peritubular region appears to be vulnerable to attack by immunologic and other disease processes.

An increased passive permeability of the distal nephron to H^+, OH^-, or HCO_3^- could explain the impaired gradient formation of distal RTA and such an explanation was indeed favored in the 1950s and 1960s by reviewers of the subject.[96,128,129] At present, the evidence suggests that increased passive H^+ permeability accounts for the pathogenesis of only a small group of patients with distal RTA. The acidifying defect induced by the polyene antibiotic amphotericin B is probably the best example of such a permeability defect. Patients receiving amphotericin B often develop distal RTA associated with urinary K^+ wasting.[135,137,138]

The defect is usually reversible upon discontinuation of amphotericin treatment. A comparable defect can be produced *in vitro* in turtle urinary bladder.

Steinmetz and Lawson[21,139] examined the nature of the defect in the isolated turtle bladder. They observed that the

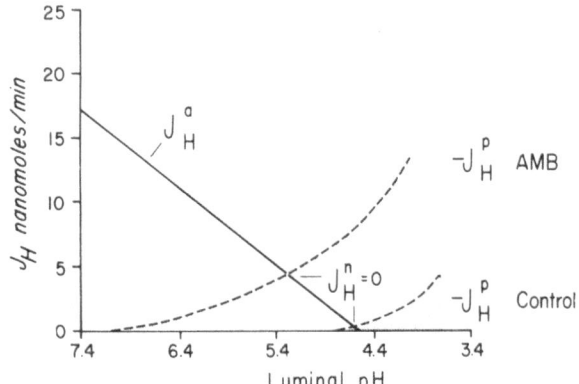

Fig. 16. Acidifying defect induced by amphotericin B (AMB). Back diffusion of H^+ ($-J_H^p$) was increased by luminal addition of AMB while active H^+ secretion (J_H^a) was unaffected. As a result the luminal pH at which net H^+ secretion (J_H^n) was zero was elevated.

luminal addition of amphotericin B had little or no effect on net H^+ secretion, J_H^n, in the absence of an electrochemical gradient across the epithelium. For this study, J_H^n was measured by pH stat titration in short-circuited bladders with both sides at pH 7.4. During secretion against an opposing H^+ gradient, however, J_H^n decreased and became zero at a luminal pH of about 5.4 instead of 4.4.[139] These results suggested that active H^+ secretion was unaffected by amphotericin B, but that back-diffusion of H^+ out of the mucosal (or luminal) compartment was increased. This interpretation was confirmed by demonstrating a direct increase in the rate of loss of H^+ from the mucosal compartment upon amphotericin addition at mucosal pH values below 4.4.[21] The effect of amphotericin on the active and passive H^+ flows in this urinary epithelium is shown in Fig. 16, which has the same format as Fig. 4 (see Section 2.1). Under control conditions, J_H^n is equal to J_H^a and back-diffusion of H^+, $-J_H^p$, is negligible over the physiologic range of acidification. Since H^+ secretion was not affected by amphotericin B in the absence of gradient J_H^a was the same. The back-flux is indicated by $-J_H^p$. The minimum pH that could be generated (pH of zero net H^+ secretion) was about 4.4 for the control observations and about 5.4 after exposure to amphotericin B. These studies, which were carried out in the absence of exogenous HCO_3^-, revealed that the acidifying defect induced by amphotericin is caused by increased passive permeability to H^+. The increased permeability to H^+ was associated with a large increase in K^+ permeability and smaller increases in Na^+ and Cl^- permeabilities.[139]

As indicated in Table III, a gradient defect in the distal nephron could be caused by a leak flow of H^+ out of the lumen or by a leak flow of HCO_3^- into the lumen. It is of interest that amphotericin B has little effect on the flow of HCO_3^- from serosa to mucosa in turtle bladder when the serosal solution contains 20 mM HCO_3^-.[140] The permeability defect induced by amphotericin, therefore, is selective for H^+ rather than for HCO_3^-. Since the permeability for K^+ was increased more than 20-fold, whereas that for Cl^- was only inconsistently enhanced, these studies indicate that the permeability increase is selective for cations, especially H^+ and K^+.

These studies are of interest from both a clinical and a theoretical point of view. The early renal manifestations of amphotericin toxicity are impaired acidification and urinary K^+ wasting. A single permeability defect of the luminal membrane

can account for both features. Increased quantities of H^+ and K^+ would move passively down their respective electrochemical gradients across a leaky luminal cell membrane. The loss of K^+ from and the entry of H^+ into the epithelial cells would accelerate the injury and, thereby, contribute to the development of further renal damage. If this sequence of events plays a role in the pathogenesis of the renal insufficiency, preventive measures, such as the early administration of HCO_3^- and K^+, may reduce tubular damage.[141]

The second aspect of these *in vitro* studies deals with the apparent cation selectivity of the amphotericin channels. Studies by Finkelstein and Holz,[142] Andreoli,[143] and Marty and Finkelstein[144] on the mode of action of amphotericin B in lipid bilayer membranes and in biological membranes have clarified how amphotericin B may create pores in membranes. The amphotericin molecule is about 24 Å long and resembles a phospholipid in having a long hydrophobic chain and a short hydrophilic end. The molecule spans about one-half of the lipid bilayer. In the model of Finkelstein and Holz,[142] amphotericin B (or nystatin) creates a channel made up of about eight amphotericin molecules with their chains of hydroxyl groups facing the center of the channel to form an aqueous pore surrounded by cholesterol molecules. To increase the permeability of thin lipid bilayers, amphotericin B must be added to both sides of the bilayer so that half-pores are formed from both sides and link up to form a complete channel. In artificial lipid membranes, the increased permeability is anion selective, whereas in biological membranes, the increase is cation selective. An explanation for this difference in the selectivity has been advanced by Marty and Finkelstein.[144] In the lipid bilayers having amphotericin present on both sides, two half-pores are converted into a complete pore by being held together in the middle of the bilayer through hydrogen bonds between the hydroxyl groups located at the hydrophobic ends of the polyene molecules. This channel arrangement would convey anion selectivity to the artificial membranes. In contrast, in biological membranes, only the outside layer of each cell membrane is exposed to amphotericin B, and a less stable array of amphotericin molecules is formed which traverses only the outer half of the membrane and makes contact with existing structures in the inner half. The characteristics of the half-pore *per se* with its lining of hydroxyl groups would favor the transfer of cations. Although this model remains to be defined further, we may conclude that the defect induced by amphotericin B in the luminal membrane of urinary epithelia is cation selective and that the principal leak pathway is transcellular rather than paracellular.

In terms of the experimental defect in urinary acidification induced in the intact animal, Roscoe *et al.*[145] have shown that amphotericin-treated rats retain the ability to elevate the urinary pCO_2 during HCO_3^- loading. This result suggests that H^+ secretion in the distal nephron is preserved in the absence of concentration gradients and that impaired gradient formation in this model is caused by increased back-diffusion of H^+ as it is in the defect induced *in vitro*.

Several tight urinary epithelia have been shown to be capable of HCO_3^- secretion into the urinary compartment. In the turtle bladder the secretion of HCO_3^- is coupled to Cl^- absorption under ordinary conditions.[43,44] HCO_3^- secretion is not limited to the reptilian urinary tract, as recent studies by McKinney and Burg[45] and Boyer and Burg[46] indicate that the cortical collecting duct of the rabbit can secrete HCO_3^- and that this process requires luminal Cl^-.[46] These studies have two major implications for distal urinary acidification in man. First, the secretion of acid by the distal nephron may represent a net rate, the difference between H$^+$ secretion and HCO$_3^-$ secretion. Second, the rate of HCO$_3^-$ secretion may be greatly influenced by the prevailing anion in the tubular fluid. Thus, replacement of luminal Cl$^-$ by SO$_4^{2-}$ would markedly decrease HCO$_3^-$ secretion. It is well known that induction of a sodium sulfate diuresis may lower the urine pH in patients with distal RTA.[146,147] This phenomenon might be explained as a result of reduced HCO$_3^-$ secretion by inhibition of the exchange secretion of HCO$_3^-$ for Cl$^-$ or alternatively it might be accounted for by a voltage effect on H$^+$ secretion. As a poorly reabsorbable anion, SO$_4^{2-}$ would increase the lumen-negative potential in the Na$^+$-transporting segments of the distal nephron and, thereby, would accelerate H$^+$ secretion. At present, it is not known whether either or both of these mechanisms operate in man.

Certain diuretics, especially mersalyl and ethacrynic acid, are also capable of lowering urine pH in distal RTA.[146] As in the case of sodium sulfate infusion, the lowering may either be voltage mediated or depend on inhibition of HCO$_3^-$ secretion.

9.3. Clinical Forms of Distal RTA

Primary distal RTA occurs in sporadic as well as familial forms. In affected families, different degrees of distal RTA occur with various combinations of hypercalciuria, nephrocalcinosis, and nephrolithiasis.[148,149] Buckalew et al.[148] have shown that in some families the disorders of calcium metabolism may antedate the RTA which suggests that some familial RTA may be secondary to nephrocalcinosis.

The recent described genetic disorder of absent CA II (or C) is of great interest since this isoenzyme is the cytoplasmic CA of both proximal and distal segments of the nephron. The patients with this deficiency of CA have an RTA with proximal HCO$_3^-$ wasting as well as impaired distal H$^+$ secretion.[100,104] The RTA is associated with osteopetrosis[100] presumably as a consequence of impaired CA function in osteoclasts. Distal RTA has also been described in a family with a CA-B deficiency of the red cells.[105] The significance of this observation remains to be clarified since the B isoenzyme does not appear to be of much importance for renal function.[106]

Disorders of calcium metabolism are commonly associated with distal RTA. Hypercalciuria with stone formation and calcium deposition in the renal parenchyma are common consequences of distal RTA.[98,99,101,148–150] These complications may further impair distal acidification. In a number of conditions listed in Table IV, the deposition of calcium in the renal medulla and in the cortical regions adjacent to the collecting ducts appears to play a primary role in the pathogenesis of distal RTA.[127–129,148–150] Thus, sustained hypercalcemia and hypercalciuria can cause impairment of urinary acidification. The distal RTA observed in vitamin D intoxication,[151] hyperparathyroidism,[82,101] and hyperthyroidism[152] are mediated by such a mechanism.

Among the acquired forms of distal RTA, a group of immunologic disorders are most common. Systemic lupus erythematosis, Sjögren's syndrome, and certain forms of chronic active liver disease associated with hyperglobulinemia all may become complicated by distal RTA.[120–122] Aside from the hyperglobulinemia, there is often an interstitial nephritis with or without involvement of the tubular basement membrane.[122]

Various hereditary diseases affect distal urinary acidification by mechanisms which are often secondary. In sickle-cell disease the renal papillae are the sites of ischemia and severe

Table IV. Clinical Forms of Distal Renal Tubular Acidosis[a]

Primary forms
  Infantile
  Adult
Carbonic anhydrase (CA) deficiency
  Absence of CA-C (osteopetrosis)
  Reduced CA-B
Nephrocalcinosis
  Hyperparathyroidism
  Hypervitaminosis D
  Hypercalciuria
  Hyperthyroidism
  Medullary sponge kidney
Immunologic disorders
  Hyperglobulinemias
  Systemic lupus erythematosis
  Sjögren's syndrome
  Primary biliary cirrhosis
  Chronic active hepatitis
  Tubular basement membrane disease
  Interstitial nephritis
  Renal transplanation
Hereditary systemic diseases
  Sickle-cell disease
  Hereditary elliptocytosis
  Fabry's syndrome
  Ehlers Danlos syndrome
Drug induced
  Amphotericin B
  Toluene
  Lithium
  Amiloride
Miscellaneous
  Obstructive nephropathy
  Pyelonephritis
  Hepatic cirrhosis

[a]See text and Refs. 95–105, 120–122, 128–131, 137–139, 148–158.

interstitial nephritis. Abnormal deposition of glycosphingolipids in the kidneys of patients with Fabry's disease may cause impaired acidification.[153] Diseases of connective tissue such as the Ehlers Danlos syndrome are occasionally associated with distal RTA.[95,96,99,101–103]

As discussed above, amphotericin B can induce an acute reversible form of distal RTA associated with marked K$^+$ wasting.[137] Toluene sniffing also may cause a transient acidifying defect with K$^+$ depletion.[154] These toxic defects are best attributed to acute increases in the H$^+$ and K$^+$ permeabilities of segments of the collecting duct which are normally almost impermeable to these ions.[135]

The reductions in distal H$^+$ secretion observed in patients receiving amiloride or lithium are probably mediated by a decreased voltage.[147,155] These drugs inhibit Na$^+$ transport by the collecting duct and, thereby, abolish the lumen-negative PD which normally facilitates H$^+$ secretion. In general, the acidifying defects so induced are mild and lead to incomplete forms of distal RTA.

In pyelonephritis and in obstructive nephropathy, a classic distal RTA may occur.[156] In these cases, some overall impairment of renal function is often present and the hyperchloremic acidosis may be associated with hyperkalemia rather than hypokalemia. The absence of hypoaldosteronism and the inability to lower urine pH below 5.5 justify classifying these cases as classic distal RTA.

Hepatic cirrhosis is associated with distal RTA.[157,158] The RTA is usually characterized by an inability to lower urine pH and by hypokalemia. Interestingly, correction of the K^+ depletion may restore the capacity to lower urine pH. This restoration of the ability to establish a pH gradient is associated with reduced NH_4^+ excretion. There is a reciprocal relation between NH_4^+ excretion and serum K^+ which probably accounts for this phenomenon.[159–161] Hypokalemia stimulates NH_3 production and thereby the availability of buffer. When normokalemia is restored, the availability of buffer is reduced and the urine pH is lowered despite the rather low rates of distal H^+ secretion.

10. Impaired Excretion of Net Acid and Potassium (Normal pH Gradient Formation)

10.1. Pathogenesis and Clinical Occurrence

Lathem[162] and Carroll and Farber[163] described a hyperchloremic acidosis associated with hyperkalemia in patients with pyelonephritis and only moderately impaired renal function. Twelve of the thirteen patients of Carroll and Farber lowered their urine pH below 5.5. Hence, these patients differ from those with classic RTA in their ability to generate a normal pH gradient across the collecting duct. Also, their acidosis is associated with hyperkalemia rather than hypokalemia.

Several interrelated mechanisms must be considered in the pathogenesis of this form of hyperchloremic acidosis. Aldosterone is known to stimulate K^+ secretion in the cortical collecting duct[75,76] and H^+ secretion in both cortical and medullary segments of the collecting duct.[24,71,77] Primary mineralocorticoid hormone deficiency results in impaired renal excretion of H^+ and K^+. Renin deficiency resulting from interstitial renal disease without significant impairment of glomerular function may produce a secondary hypoaldosteronism[164–166] with associated decreases in renal excretion of H^+ and K^+.

Hyperkalemic RTA may also be caused by direct injury to the K^+-secreting tubule[156] or by a primary tubular defect which renders this segment unresponsive to mineralocorticoid hormone.[167] In all forms of hyperkalemic RTA, net acid excretion is impaired as a result of decreased NH_4^+ excretion. This decreased NH_4^+ excretion is the result of both the primary renal disease and hyperkalemia.[160,168]

10.2. Hypoaldosteronism

Hypoaldosteronism may occur as part of a combined deficiency of glucocorticoid and mineralocorticoid hormones as in Addison's disease or it may represent a selective deficiency of aldosterone (Table V). Impaired renin secretion in certain forms of renal disease is a major cause of the latter. The syndrome of hyporeninemic hypoaldosteronism was first described by Hudson et al.[164] and subsequently defined by Gerstein et al.,[165] Vagnucci,[166] Schambelan et al.,[169] and DeFronzo.[170] The majority of the patients with selective hypoaldosteronism and hyperkalemic RTA have a renin deficiency. Table V lists the more common forms of renal disease which appear to affect the renin-producing cells of the juxtaglomerular apparatus. A common characteristic of these diseases is the extensive involvement

Table V. Clinical Disorders Causing Hyperchloremic Acidosis with Hyperkalemia[a]

Associated with primary adrenal insufficiency
    Addison's disease
    Bilateral adrenalectomy
    Isolated aldosterone deficiency
    Congenital enzymatic defects
Associated with secondary hypoaldosteronism
    Diabetic nephropathy
    Chronic interstitial nephritis
    Lead nephropathy
    Obstructive nephropathy
    Lupus nephritis
    Prostaglandin inhibition
    Converting enzyme inhibition
Associated with tubular defect for K^+ secretion
    Obstructive nephropathy
    Sickle-cell renal disease
    Chronic interstitial nephritis
    Renal transplantation
    Tubular resistance to mineralocorticoid
    Drug induced:
        Amiloride, spironolactone, triamterene
Hyperkalemia from K^+ redistribution
    Certain acidemias
    Insulin deficiency
    β-Adrenergic blockade

[a]See text and Refs. 162–171.

of the renal interstitium. Hyporeninemia may be worsened by drugs inhibiting prostaglandin synthesis.[171] Similarly, hypoaldosteronism may be worsened by angiotensin-converting enzyme inhibitors.

The mechanism of impaired H^+ secretion in hypoaldosteronism has become better understood as a result of studies of urinary acidification in vitro. In the isolated toad[78] and turtle urinary bladders,[79] aldosterone stimulates the rate of H^+ secretion by two mechanisms. The first mechanism involves direct stimulation of H^+ secretion independently of Na^+ transport. H^+ secretion is stimulated by aldosterone in short-circuited bladder preparations in which the PD is nullified or in ouabain-treated tissues in which Na^+ transport is abolished. Spironolactone, an antagonist of the aldosterone stimulation of Na^+ transport, fails to inhibit the stimulation of H^+ secretion.[172] The precise mechanism for the direct stimulation of H^+ secretion remains to be defined. It probably involves the addition of H^+ pumps to the luminal surface of the H^+-secreting cells either from a reserve of pumps in cytoplasmic membrane vesicles or from synthesis of new pumps or from both. The apparent protonmotive force of the pump is not increased by aldosterone, a result consistent with the clinical observation that patients with hypoaldosteronism maintain a normal ability to lower urine pH.

Recent studies by Stone et al.[77] demonstrate that aldosterone also stimulates urinary acidification directly in isolated perfused segments of the medullary collecting duct harvested from adrenalectomized rabbits. This stimulation was observed in the absence of sodium.

The second mechanism of stimulation of H^+ secretion is mediated by the increased lumen-negative PD which results from the stimulation of Na^+ reabsorption by aldosterone. Chronic administration of mineralocorticoid hormone causes large increases in the rate of Na^+ reabsorption by the initial collecting tubule and the cortical collecting duct.[75,76,173] Associated with this increase are increased PD, increased rate of K^+ secretion, and increased basolateral membrane area of the principal cells of the cortical collecting tubule.[174] Na^+,K^+-ATPase activity at the basolateral cell membrane is increased beyond the increase in membrane area,[175] presumably in response to increased Na^+ entry into the cells at the luminal membrane and increased work load.

The luminal membrane of these cells has an unusually high K^+ permeability compared to other segments of the nephron.[75,76] As a result, the K^+ taken up by the basolateral Na^+,K^+ pumps into the cells will be secreted across the luminal membrane rather than recycled across the basolateral membrane. In any event, the stimulation of K^+ secretion by mineralocorticoid hormone is considered to be closely coupled to increased Na^+ transport. The increased PD across the luminal membrane facilitates K^+ secretion from cell interior to lumen.

In hypoaldosteronism the aldosterone-sensitive transport systems of the cortical collecting tubule are inactive so that both Na^+ absorption and K^+ secretion are markedly reduced. Since almost all K^+ eliminated by the kidneys is secreted at this site, renal K^+ excretion is jeopardized and hyperkalemia results.

10.3. Primary Tubular Defects

Although renin deficiency and resultant hypoaldosteronism are common in the pathogenesis of hyperkalemic RTA, there are several forms of renal disease that may lead directly to impaired excretion of K^+ and H^+ without mediation by the renin–aldosterone system. Patients with predominantly interstitial nephritis, obstructive nephropathy, and sickle-cell renal disease may develop the syndrome despite normal renin and aldosterone levels (see Table V). In this group of patients, the hyperkalemia and acidosis fail to respond to treatment with mineralocorticoid hormone.[167,169,176] Presumably, the renal disease has affected the collecting tubules directly and impaired their major transport functions. Such direct inhibition of transport is illustrated by the action of the major collecting duct diuretics. Amiloride is known to inhibit the Na^+ entry step across the luminal membrane of tight urinary epithelia. When amiloride is added to the luminal perfusate in the cortical collecting duct, both Na^+ reabsorption and K^+ secretion are virtually abolished. K^+ secretion is inhibited because it depends on active K^+ uptake by the Na^+,K^+-ATPase at the basolateral cell membrane and this transport ATPase in turn requires Na^+ delivery across the luminal membrane.[177] Amiloride abolishes Na^+ entry into the cell and thereby the lumen-negative PD resulting from Na^+ transport; as a result, K^+ movement from cell interior to tubular lumen is markedly reduced. Thus, although amiloride acts primarily on the luminal Na^+ channels, it indirectly inhibits K^+ transport at both cell membranes. As a diuretic, spironolactone gives similar urinary excretion patterns as amiloride. It acts, however, by a different mechanism, namely by displacing aldosterone from its receptor sites in the principal cells of the collecting duct. Triamterene is another K^+-sparing diuretic acting at the same nephron segment by a less specific mechanism. All three diuretics may cause hyperkalemia and may impair acid excretion by a voltage-dependent mechanism (Section 9.2).

In a few patients, a combination of hypertension, hyperkalemia, and metabolic acidosis has been observed with hyporeninemia but without any evidence of renal disease. Gordon et al.[178] reported this combination in a 10-year-old girl with growth retardation and postulated a tubular dysfunction proximal to the K^+ secretory segment. The dysfunction would cause excessive Na^+ and Cl^- reabsorption, volume expansion, and secondary hyporeninemia with hypoaldosteronism. Sodium restriction corrected the volume excess and, thereby, reversed the syndrome.[178] Cases of a similar syndrome have been reported.[167,179,180] In these cases the syndrome was again not characterized by salt wasting and in addition failed to respond to treatment with mineralocorticoid hormone. The transport abnormality in these hormone-resistant cases remains to be elucidated. One possible explanation would be that the K^+-secreting segments have an abnormally high Cl^- permeability.[167] Shunting of Cl^- would reduce the PD and thus the rate of K^+ secretion.

Since hyperkalemia per se stimulates aldosterone secretion, it is at times difficult to assess the extent to which hyporeninemia has reduced aldosterone production. Similarly, the hyporeninemia may be caused by volume expansion rather than by disease of the renin-secreting cells of the juxtaglomerular apparatus. These factors should be considered if one tries to distinguish between a primary tubular defect and a dysfunction mediated by hyporeninemic hypoaldosteronism.

10.4. Impaired NH_4^+ Excretion

Several factors contribute to impaired net acid excretion in hyperkalemic RTA. As discussed above, the rate of H^+ secretion may be reduced as a result of hypoaldosteronism or as a result of direct involvement of the collecting duct in renal disease. The rate of H^+ secretion for a given population of H^+ pumps may be further decreased by a lack of buffer in the tubular fluid. The transport rate of each H^+ pump (Section 2.1) is greatly influenced by the pH gradient against which secretion occurs. In the absence of H^+ acceptors, the tubular fluid is rapidly acidified and the H^+ pump inhibits its own transport rate by the pH gradient it generates. In hyperkalemic RTA, renal NH_3 production is often reduced.[168] This production is inversely related to serum K^+ concentration, so that NH_3 is in short supply so long as the hyperkalemia persists.[160,168] In patients with renal insufficiency, reduced nephron mass may limit NH_3 production[90–92] or the transfer of NH_3 from cortical structures to the acidifying segments of the collecting ducts where NH_3 serves as a H^+ acceptor. Hence, impaired NH_4^+ excretion plays a role in the pathogenesis of hyperkalemic hyperchloremic acidosis as well as in the development of the ordinary renal acidosis of glomerular insufficiency. The availability of NH_3 to the collecting tubules and the size of the functioning H^+ pump population are factors which tend to influence the urine pH in opposite directions. When NH_3 production is reduced as in hypoaldosteronism with hyperkalemia, the pH gradients generated in the collecting ducts are usually normal. In contrast, patients with classic RTA are often hypokalemic and excrete substantial quantities of NH_3 which contribute to the inappropriately high pH of their urine.

10.5. Clinical Implications

The clinical presentation of the hyperkalemic forms of hyperchloremic acidosis is sufficiently different from that of proximal and distal RTA to warrant separate discussion. The major elements in the pathogenesis are the impaired excretion of K^+ and net acid. In most of the patients a primary or secondary aldosterone deficiency accounts for the tubular dysfunction. In a small but interesting group of patients, the primary disorder resides in the collecting duct system. A hereditary or acquired tubular defect impairs K^+ secretion and renders the collecting duct unresponsive to mineralocorticoid hormone.

The syndrome has become increasingly common and is usually discovered as a result of routine serum electrolyte determinations revealing hyperkalemia and hyperchloremia. Although one of the first reported cases with hypoaldosteronism[164] presented with Stokes–Adams attacks, such a presentation is relatively rare. Often, the hyperkalemia is relatively modest but is aggravated from time to time by dietary factors or by factors which cause K^+ redistribution from the cellular to the extracellular compartment, such as insulin deficiency in diabetics (see Table V). In contrast to patients with classic RTA, patients with hyperkalemic forms of RTA usually do not present with nephrolithiasis or nephrocalcinosis; they are able to lower urine pH appropriately; and they rarely waste large quantities of HCO_3^- into the urine as occurs in proximal RTA.

The major clinical goal in this group of patients is to clarify the pathophysiology of the syndrome. After having assessed the possible role of a shift of K^+ from the cellular to the extracellular fluid compartment, the mechanism of impaired excretion of K^+ must be evaluated. The most common pathogenesis is renin deficiency. If renin and aldosterone secretion rates are inappropriately low, the underlying renal disease is evaluated. Is there an interstitial nephritis or a preventable form of renal disease such as urinary obstruction or lead intoxication (Table V)? If secondary hypoaldosteronism is not caused by volume expansion, but reflects disease of the renin-producing cells of the juxtaglomerular apparatus, the hyperkalemia and the acidosis may be corrected by mineralocorticoid hormone administration. A positive response to mineralocorticoid hormone clarifies the pathogenesis of the syndrome and permits correction of the abnormalities. Fludrohydrocortisone should be administered with care to avoid fluid retention and heart failure.

A small group of patients will be refractory to treatment, with fludrohydrocortisone. In these mineralocorticoid-resistant cases, the collecting duct system may be involved directly in congenital or acquired disease. In such instances the syndrome may be improved by the use of one of the thiazide or loop diuretics which increase K^+ excretion. Even in mineralocorticoid-responsive cases, fluid retention may dictate combined treatment with fludrohydrocortisone and a thiazide diuretic. Excessive K^+ intake and K^+ redistribution should be addressed separately before a long-term program of management is undertaken.

11. Clinical Aspects of Renal Tubular Acidosis and Associated Disorders of Electrolyte Transport

11.1. Clinical Presentation

The hyperkalemic forms of RTA with impaired K^+ excretion have been reviewed above (Section 10). In this section we discuss the clinical aspects of the RTA which are associated with K^+ wasting rather than K^+ retention. The main forms of RTA, i.e., classic distal and proximal RTA, are characterized by varying degrees of hypokalemia and K^+ wasting.

The clinical manifestations of patients with RTA reflect the associated disturbances in calcium and phosphorus metabolism

and the extent of potassium depletion. If RTA is secondary to a systemic disease or to a toxic exposure, the manifestations of the primary disease may be in the foreground.

Calcium infarction of the kidneys on postmortem examination was described by Lightwood[181] in 1935 in six infants without obvious underlying renal disease. The calcium deposits occurred primarily in the renal medulla and the corticomedullary junctions. Microscopically, the deposits were not marked in the region of the collecting tubules. All six infants had a clinical record of a failure to thrive with anorexia, constipation, wasting, and muscle weakness. In subsequent reports by Butler *et al.*,[182] Albright *et al.*,[82] and Lightwood *et al.*,[127] the occurrence of hyperchloremia, acidosis, and dehydration were described and vitamin D intoxication and hyperparathyroidism were excluded as causes for the nephrocalcinosis. Hartmann[126] in 1939 may have been the first to recognize the importance of the impaired urinary acidification in a four-month-old child with metabolic acidosis treated successfully with sodium lactate.

In 1946 Albright *et al.*[82] described hyperchloremic acidosis resulting from tubular-insufficiency-without-glomerular insufficiency in adults. The disorder often began in childhood and persisted into adult life. It occurred in sporadic as well as in familial forms and may be first discovered in the adult.[82,127] Rickets and osteomalacia were common complications leading to bone pain and pseudofractures, especially in patients with long histories of hypercalciuria and hyperphosphaturia. Other early reports of RTA emphasized the frequent occurrence of kidney stones, thirst, and polyuria[127,183] and the presence of nephrocalcinosis on radiologic examination.

Since the pathophysiology of RTA has become better defined during the past 15 years, it has become feasible to distinguish between the disturbances of Ca^{2+} metabolism in proximal and distal RTA. Radiographic abnormalities of the skeleton are nowadays rarely seen in patients with classic distal RTA.[150] They are common, however, in proximal RTA. Children with proximal RTA often have rickets, whereas adults have osteopenia.[150] Nephrocalcinosis and nephrolithiasis on the other hand are common in distal RTA and virtually absent in proximal RTA.

In terms of the distinction between distal and proximal RTA, McSherry and Morris[98] and Morris and Sebastian[99] have emphasized that infants and children with classic distal RTA often spill increased amounts of HCO_3^- in the urine during periods of rapid growth. HCO_3^- wasting, however, is transient and these children do not develop permanent proximal tubular defects. McSherry and Morris[98] have shown that the growth retardation of children with distal RTA can be corrected by the administration of HCO_3^- in quantities adequate to restore the serum HCO_3^- to normal.

Hypokalemia and consequent K^+ depletion is a major complication of both proximal RTA and distal RTA of the classic type.[184,185] Patients with K^+ depletion may present with thirst and polyuria or with fatigue and muscle weakness. Occasionally, such a patient is admitted to a neurologic service because of severe paralysis.

11.2. K^+ Wasting

Renal K^+ wasting is a common complication of both proximal and distal RTA.[184,185]

In proximal RTA the urinary HCO_3^- losses are often large and associated with high rates of K^+ secretion. The presence of secondary hyperaldosteronism may accelerate this loss of K^+

into the urine. Since patients with proximal RTA usually excrete at least 15% of the filtered load of HCO_3^- at normal serum HCO_3^-, they require large quantities of alkaline salts. To control K^+ depletion, it is advisable to provide a substantial portion of the administered alkali in the form of potassium salts such as potassium gluconate, citrate, or acetate.[135] These patients often require from 5 to 10 meq alkali/kg body wt, about half of which might be given as the potassium salt.

In distal RTA the mechanism of K^+ wasting is different. HCO_3^- losses in the urine are usually modest. It is thought that K^+ secretion is increased as a function of decreased H^+ secretion by a mechanism of electrical interaction. Na^+ reabsorption renders the lumen negative with respect to the capillary in the distal tubule and certain segments of the collecting duct. Active H^+ secretion, on the other hand, would render the lumen positive. Under the usual circumstances, the net effect of H^+ secretion would be a reduction of the PD of Na^+ transport. When H^+ secretion is impaired, the full PD generated by Na^+ reabsorption is available as an electrical driving force for K^+ secretion. Gill et al.[186] showed that administration of $NaHCO_3$ or Na_2HPO_4 instead of NaCl to patients with distal RTA corrected not only the systemic acidosis but also the hypokalemia in several of their studies. In addition to this reciprocal behavior between the excretion of H^+ and K^+ on the basis of electrical coupling, K^+ excretion is often increased by secondary hyperaldosteronism. Sebastian et al.[185] observed two patients with distal RTA in whom K^+ wasting continued despite severe hypokalemia and correction of acidosis. In the acute phase of certain drug-induced forms of tubular disease, i.e., distal RTA caused by amphotericin B[137,138] or toluene and its derivatives,[135,154] the extent of K^+ depletion appears to be considerably more severe than the extent of metabolic acidosis. These observations suggest that K^+ wasting might be caused directly by the tubular injury. As discussed in Section 10.2, the defect in urinary acidification induced in vitro by amphotericin B is characterized by a marked increase in the permeability of the luminal membrane to H^+ and K^+. Since the transport of K^+ from cell interior to lumen is thought to occur passively down an electrochemical gradient, such an increased K^+ permeability could greatly increase K^+ secretion for a given driving force.

Aside from the severe acute forms of K^+ depletion which require vigorous replacement of K^+, many cases of chronic distal RTA behave like the patients studied by Gill et al.[186] in that hypokalemia is rather easily corrected. These patients usually require alkali in small quantities designed to match endogenous production of nonvolatile acid, about 1 or 2 meq/kg body wt. Hypokalemia may be corrected by giving sodium citrate or sodium bicarbonate alone. If hypokalemia persists, part or all of the alkali should be given as a potassium salt.[135]

11.3. Disturbances in Calcium and Phosphorus Metabolism

Although both proximal and distal RTA are associated with abnormal calcium metabolism, there are enough differences between the two forms of RTA to warrant separate discussion.

In patients with proximal RTA and the Fanconi syndrome, rickets and osteomalacia are common. Excessive phosphaturia and hypophosphatemia play a major role in the demineralization of bone. Vitamin D deficiency and hypocalcemia further contribute to the development of the bone disease. Hypercalciuria is also frequently present. Among the factors that contribute to the increased excretion of Ca^{2+} are the dissolution of

bone caused by acidosis[83] and Na^+ wasting. Since the reabsorption of Ca^{2+} and the reabsorption of Na^+ in the proximal tubule are rather closely correlated, the impaired HCO_3^- reabsorption causes increased excretion of both cations. It is of interest that, despite the hypercalciuria, nephrolithiasis and nephrocalcinosis are much less common in proximal RTA than in distal RTA. The reason for this difference is only partially understood. The fact that patients with the Fanconi syndrome elaborate an acid urine during acidosis and usually have a high urine flow when their acidosis is being treated may account for the rarity of these complications. The presence of amino acids and organic acids in the urine of patients with proximal RTA may also have a protective effect. The osteomalacia of patients with proximal RTA is usually not corrected by the administration of alkaline salts alone. Large doses of vitamin D and oral supplements of inorganic phosphate are often required to improve the bone disease. Thiazides may be used to reduce urinary Ca^{2+} excretion.[101,187,188]

In distal RTA, the excessive Ca^{2+} excretion is thought to be secondary to metabolic acidosis and dissolution of bone.[82,83] Nephrocalcinosis and nephrolithiasis are very common complications of the disease and have been attributed to the alkalinity of the urine and to the low rate of citrate excretion.[189] PO_4^{3-} excretion in distal RTA is increased when the metabolic acidosis is untreated. In contrast to the hyperphosphaturia of the Fanconi syndrome, which originates in the proximal tubule and is an essential feature of the syndrome, the hyperphosphaturia of distal RTA is thought to be a secondary phenomenon. Secondary hyperparathyroidism and metabolic acidosis appear to mediate the PO_4^{3-} wastage.

Although patients with distal RTA were described to have rickets and osteomalacia with bone pain and pseudofractures[82] in the 1940s and 1950s,[127] the incidence of these complications has decreased over the last 15 years. At least in radiographic studies, skeletal abnormalities have become rare in classic distal RTA.[150] Earlier detection of RTA and more vigorous treatment with alkali may have contributed to this decrease in bone involvement. McSherry and Morris[98] have shown that children with stunted growth will return to a normal growth curve upon adequate alkali treatment of the acidosis. When the plasma HCO_3^- is restored to normal, the quantities of Ca^{2+} excreted in the urine and, thereby, the hazards of Ca^{2+} deposition and nephrolithiasis are reduced.[99] Early treatment of distal RTA in children is likely to cause healing of the osteomalacia and to prevent the retardation of growth as well as the nephrocalcinosis. The plasma levels of HCO_3^-, K^+, calcium, and phosphorus should be monitored; and if treatment is adequate, hypercalciuria should no longer be present.

In some families, distal RTA may occur as a complication rather than a cause of hypercalciuria.[148,149] In such families, the primary effort should be directed toward the correction of hypercalciuria and the prevention of damage to the collecting tubules.

12. Summary

Urinary acidification along the nephron is brought about by the secretion of H^+ by transport systems located at the luminal membrane of the renal epithelial cells. In the proximal tubule the secretion of H^+ across the brush border occurs primarily via a Na^+/H^+ exchanger which is driven by the inward Na^+ gradient. In the distal nephron, especially in the cortical and medullary segments of the collecting ducts, H^+ secretion is thought to be driven by an electrogenic H^+ pump. H^+ secretion by "tight" urinary epithelia such as the collecting duct is coupled directly to ATP hydrolysis and, therefore, represents primary active transport. In contrast, the Na^+/H^+ exchanger of the proximal tubule is driven by concentration gradients generated by the basolateral Na^+ pump.

In series with the H^+ transport systems of the luminal membranes of both proximal and distal nephron are transport systems in the basolateral cell membrane designed to facilitate the efflux of HCO_3^- from cell to peritubular space. In general, for each H^+ secreted into the lumen, a HCO_3^- ion moves into the peritubular capillary.

H^+ secretion by the proximal tubule (including its straight portion) is responsible for the reabsorption of about 90% of the filtered load of HCO_3^-. The rate of H^+ secretion by the proximal tubule is regulated by the CO_2 tension, the state of the extracellular volume, and the Na^+ and H^+ gradients across the exchanger. H^+ secretion by the distal nephron is responsible for reabsorption of the remainder of the filtered HCO_3^- and for the formation of net acid for excretion. The H^+ transport rate is regulated by the electrochemical potential gradient for H^+ across the H^+ pump complex and by the number of H^+ pumps available at the luminal surface. Increased CO_2 tension and probably increased mineralocorticoid activity increase H^+ secretion by fusion to the luminal membrane of preexisting membrane vesicles containing H^+ pumps. Net acid excretion is determined not only by the size of the H^+ pump population, but also by the availability of H^+ acceptors such as NH_3 and PO_4^{3-} in the lumen of the distal nephron.

RTA is defined as a clinical syndrome of hyperchloremic acidosis associated with near-normal GFR and a renal tubular defect for the net excretion of acid. The disturbance in net acid excretion may be caused by impaired HCO_3^- reabsorption in the proximal tubule, by defects in the formation of transepithelial acid–base gradients in the cortical and medullary collecting ducts, by certain secretory defects that are associated with a lack of buffer (NH_3 or titratable acid), or by a combination of these. Impaired HCO_3^- reabsorption is characteristic of proximal RTA and is usually associated with aminoaciduria, glucosuria, and hyperphosphaturia. At normal plasma HCO_3^- levels, large quantities of HCO_3^- are lost in the urine. At low plasma HCO_3^- levels, the rate of impaired proximal H^+ secretion more closely matches the rate at which HCO_3^- is presented to the proximal tubule; as a result, HCO_3^- delivery to the distal nephron decreases and the distal nephron becomes capable of elaborating a maximally acid urine.

Defects in the formation of transepithelial acid–base gradients are characteristic of distal RTA. The urine pH is inappropriately high, even at low plasma HCO_3^- levels, but proximal HCO_3^- reabsorption is normal. Hyperchloremic acidosis may also result from impaired formation of NH_3 or titratable acid in the distal nephron without impaired gradient formation.

Cellular defects in HCO_3^- reabsorption by the proximal tubule are most often caused by an impairment of H^+ secretion via the Na^+/H^+ exchanger and are usually associated with other transport defects. The occurrence of proximal RTA in a variety of metabolic disorders suggests that the defect may lie in the supply of metabolic energy to the transport systems of the tubular cell. Defects in CA function or in the exit step for HCO_3^- from the tubule cell may play a role in certain forms of proximal RTA caused by toxic agents or by immunologically mediated peritubular disease.

Distal RTA may result from either impaired formation of transepithelial acid–base gradients or reduced availability of

buffer. The ability to lower urinary pH depends not only on the tightness of the collecting duct epithelium but also on the rate of active H^+ transport. An extensive loss of pump sites is a common cause of impaired distal acidification. When NH_3 production or the availability of PO_4^{3-} buffer is reduced, hyperchloremic acidosis may develop despite an ability to lower urine pH normally.

Electrolyte disorders associated with RTA are K^+ wasting and increased Ca^{2+} and phosphorus excretion. They are responsible for the clinical manifestations of RTA. In proximal RTA, urinary K^+ losses are often substantial and are likely to continue during treatment with alkaline salts. In distal RTA, smaller quantities of alkali and K^+ are required, except for certain acute forms in which K^+ wasting appears to be caused by the tubular defect itself. Another increasingly common form of RTA is associated with hyperkalemia. In this group of patients, impaired K^+ excretion is mediated either by renin deficiency and hypoaldosteronism or by a primary tubular disturbance of K^+ secretion.

References

1. Dobyan, D. C., and R. E. Bulger. 1982. Renal carbonic anhydrase. *Am. J. Physiol.* **243**:F311–F324.
2. Rosen, S. 1972. Localization of carbonic anhydrase activity in the vertebrate nephron. *Histochem. J.* **4**:35–48.
3. Husted, R. F., A. L. Mueller, R. G. Kessel, and P. R. Steinmetz. 1981. Surface characteristics of carbonic-anhydrase-rich cells in turtle urinary bladder. *Kidney Int.* **19**:491–502.
4. Cohen, L. H., R. F. Husted, A. Mueller, and P. R. Steinmetz. 1980. Surface characteristics of H^+ secreting cells in turtle bladder as a function of pH. *Clin. Res.* **28**:773a.
5. Stetson, D. L., and P. R. Steinmetz. 1983. Role of membrane fusion in CO_2 stimulation of proton secretion by turtle bladder. *Am. J. Physiol.* **245**:C113–C120.
6. Gluck, S., C. Cannon, and Q. Al-Awqati. 1982. Exocytosis regulates urinary acidification in turtle bladder by rapid insertion of H^+ pumps into the luminal membrane. *Proc. Natl. Acad. Sci. USA* **79**:4327–4331.
7. Hagège, J. 1972. Morphologie et histophysiologie des cellules intercalaires du tube urinaire des vertéles tetrapodes. *Année. Biol.* **11**:106–143.
8. Madsen, K. M., and C. C. Tisher. 1983. Celluler response to acute respiratory acidosis in rat medullary collecting duct. *Am. J. Physiol.* **245**:F670–F679.
9. Schwartz, J. H., D. Bethencourt, and S. Rosen. 1982. Specialized function of carbonic anhydrase-rich and granular cells of turtle bladder. *Am. J. Physiol.* **242**:F627–F633.
10. Koefoed-Johnsen, V., and H. H. Ussing. 1958. The nature of the frog skin potential. *Acta Physiol. Scand.* **42**:298–308.
11. Steinmetz, P. R., and O. S. Andersen. 1982. Electrogenic proton transport in epithelial membranes. *J. Membr. Biol.* **65**:155–174.
12. Steinmetz, P. R. 1974. Cellular mechanisms of urinary acidification. *Physiol. Rev.* **54**:890–956.
13. Fischer, J. L., R. F. Husted, and P. R. Steinmetz. 1983. Chloride dependence of the HCO_3 exit step in urinary acidification by the turtle bladder. *Am. J. Physiol.* **245**:F564–F568.
14. Solinger, R. E., C. F. Gonzalez, Y. E. Shamoo, H. R. Wyssbrod, and W. A. Brodsky. 1968. Effect of ouabain on ion transport mechanism in the isolated turtle bladder. *Am. J. Physiol.* **215**:249–260.
15. Schwartz, J. H. 1976. H^+ current response to CO_2 and carbonic anhydrase inhibition in turtle bladder. *Am. J. Physiol.* **231**:565–572.
16. Husted, R. F., and P. R. Steinmetz. 1979. The effects of amiloride and ouabain on urinary acidification by turtle bladder. *J. Pharmacol. Exp. Ther.* **210**:264–268.
17. Steinmetz, P. R., R. S. Omachi, and H. S. Frazier. 1967. Inde-

pendence of hydrogen ion secretion and transport of other electrolytes in turtle bladder. *J. Clin. Invest.* **46**:1541–1548.

18. Beauwens, R., and Q. Al-Awqati. 1976. Active H$^+$ transport in the turtle urinary bladder: Coupling of transport to glucose oxidation. *J. Gen. Physiol.* **68**:421–439.

19. Norby, L. H., and J. H. Schwartz. 1978. Relationship between the rate of H$^+$ transport and pathways of glucose metabolism by turtle urinary bladder. *J. Clin. Invest.* **62**:532–538.

20. Steinmetz, P. R., R. F. Husted, A. Mueller, and R. Beauwens. 1981. Coupling between H$^+$ transport and anaerobic glycolysis in turtle urinary bladder: Effects of inhibitors of H$^+$ ATPase. *J. Membr. Biol.* **59**:27–34.

21. Steinmetz, P. R., and L. R. Lawson. 1971. Effect of luminal pH on ion permeability and flows of Na$^+$ and H$^+$ in turtle bladder. *Am. J. Physiol.* **220**:1573–1580.

22. Steinmetz, P. R. 1967. Characteristics of hydrogen ion transport in urinary bladder of water turtle. *J. Clin. Invest.* **46**:1531–1540.

23. Ziegler, T. W., D. D. Fanestil, and J. H. Ludens. 1976. Influence of transepithelial potential difference on acidification in the toad urinary bladder. *Kidney Int.* **10**:279–286.

24. Koeppen, B. M., and S. I. Helman. 1982. Acidification of luminal fluid by the rabbit cortical collecting tubule perfused *in vitro*. *Am. J. Physiol.* **242**:F521–F531.

25. Al-Awqati, Q., A. Mueller, and P. R. Steinmetz. 1977. Transport of H$^+$ against electrochemical gradients in turtle urinary bladder. *Am. J. Physiol.* **233**:F502–F508.

26. Schwartz, J. H., and P. R. Steinmetz. 1977. Metabolic energy and PCO$_2$ as determinants of H$^+$ secretion by turtle urinary bladder. *Am. J. Physiol.* **233**:F145–F149.

27. Dixon, T. E., and Q. Al-Awqati. 1979. Urinary acidification in turtle bladder is due to a reversible proton-translocating ATPase. *Proc. Natl. Acad. Sci. USA* **76**:3135–3138.

28. Gluck, S., S. Kelly, and Q. Al-Awqati. 1982. The proton translocating ATPase responsible for urinary acidification. *J. Biol. Chem.* **257**:9230–9233.

29. Youmans, S. J., H. J. Worman, and W. A. Brodsky. 1983. ATPase activity and ATP-dependent proton translocation in plasma membrane vesicles of turtle bladder epithelial cells. *Biochim. Biophys. Acta* **730**:173–177.

30. Boyer, P. D. 1975. A model for conformational coupling of membrane potential and proton translocation to ATP synthesis and to active transport. *FEBS Lett.* **58**:1–6.

31. Smith, H. W. 1937. *The Physiology of the Kidney*. Oxford University Press, London. pp. 176–181.

32. Berliner, R. W., T. J. Kennedy, Jr., and J. Orloff. 1954. Factors affecting the transport of potassium and hydrogen ions by the renal tubules. *Arch. Int. Pharmacodyn. Ther.* **97**:299–312.

33. Schwartz, W. B., R. L. Jenson, and A. S. Relman. 1955. Acidification of the urine and increased ammonium excretion without change in acid–base equilibrium: Sodium reabsorption as a stimulus to the acidification process. *J. Clin. Invest.* **34**:673–680.

34. Mürer, H., U. Hopfer, and R. Kinne. 1976. Sodium/proton antiport in brush border membrane vesicles isolated from rat small intestine and kidney. *Biochem. J.* **154**:597–604.

35. Kinsella, J. L., and P. S. Aronson, 1981. Amiloride inhibition of the Na$^+$–H$^+$ exchanger in renal microvillus membrane vesicles. *Am. J. Physiol.* **241**:F374–F379.

36. Warnock, D. G., W. W. Reenstra, and V. J. Yee. 1982. Na$^+$/H$^+$ antiporter of brush border vesicles: Studies with acridine orange uptake. *Am. J. Physiol.* **242**:F733–F739.

37. Rothstein, A., Z. I. Cabantchik, and P. Knauf. 1976. Mechanism of anion transport in red blood cells: Role of membrane proteins. *Fed. Proc.* **35**:3–10.

38. Ehrenspeck, G., and W. A. Brodsky. 1976. Effects of 4-acetamido-4′-isothiocyano-2,2′-disulfonic stilbene on ion transport in turtle bladders. *Biochim. Biophys. Acta* **419**:555–558.

39. Cohen, L. H., A. Mueller, and P. R. Steinmetz. 1978. Inhibition of the bicarbonate exit step in urinary acidification by a disulfonic stilbene. *J. Clin. Invest.* **61**:981–986.

40. Ullrich, K. J., G. Capasso, G. Rumrich, F. Papavassiliou, and S. Kloss. 1977. Coupling between proximal tubular transport processes: Studies with ouabain, SITS, and HCO$_3^-$-free solutions. *Pfluegers Arch.* **368**:245–252.

41. Kleinman, J. G., R. A. Ware, and J. H. Schwartz. 1981. Anion transport regulates intracellular pH in renal cortical tissue. *Biochim. Biophys. Acta* **648**:87–92.

42. Boron, W. F., and E. L. Boulpaep. 1983. Intracellular pH regulation in the renal proximal tubule of the salamander: Basolateral HCO$_3^-$ transport. *J. Gen. Physiol.* **85**:53–94.

43. Leslie, B. R., J. H. Schwartz, and P. R. Steinmetz. 1973. Coupling between Cl$^-$ absorption and HCO$_3^-$ secretion in turtle urinary bladder. *Am. J. Physiol.* **225**:610–617.

44. Oliver, J. A., S. Himmelstein, and P. R. Steinmetz. 1975. Energy dependence of urinary bicarbonate secretion in turtle bladder. *J. Clin. Invest.* **55**:1003–1008.

45. McKinney, T. D., and M. B. Burg. 1978. Bicarbonate secretion by rabbit cortical collecting tubules *in vitro*. *J. Clin. Invest.* **61**:1421–1427.

46. Boyer, J., and M. B. Burg. 1981. Bicarbonate secretion by isolated perfused rabbit cortical collecting ducts. *Kidney Int.* **19**:223a.

47. Warnock, D. G., and F. C. Rector, Jr. 1981. Renal acidification mechanisms. In: *The Kidney*, 2nd ed. B. M. Brenner and F. C. Rector, Jr., eds. Saunders, Philadelphia. pp. 440–494.

48. Kinne-Saffran, E., R. Beauwens, and R. Kinne. 1982. An ATP-driven proton pump in brush border membranes from rat renal cortex. *J. Membr. Biol.* **64**:67–76.

49. Chan, Y. L., and G. Giebisch. 1981. Relationship between sodium and bicarbonate transport in the rat proximal convoluted tubule. *Am. J. Physiol.* **240**:F222–F230.

50. Rector, R. C., Jr., N. W. Carter, and D. W. Seldin. 1965. The mechanism of bicarbonate reabsorption in the proximal and distal tubules of the kidney. *J. Clin. Invest.* **44**:278–290.

51. Vieira, F. L., and G. Malnic. 1968. Hydrogen ion secretion by rat renal cortical tubules as studied by an antimony microelectrode. *Am. J. Physiol.* **214**:710–718.

52. Malnic, G., and P. R. Steinmetz. 1976. Transport processes in urinary acidification. *Kidney Int.* **9**:172–188.

53. Maren, T. H. 1967. Carbonic anhydrase: Chemistry, physiology, and inhibition. *Physiol. Rev.* **47**:595–781.

54. Maren, T. H. 1974. Chemistry of the renal reabsorption of bicarbonate. *Can. J. Physiol. Pharmacol.* **52**:1041–1050.

55. DuBose, T. D., L. R. Pucacco, D. W. Seldin, N. W. Carter, and J. P. Kokko. 1979. Microelectrode determination of pH and PCO$_2$ in rat proximal tubule after benzolamide: Evidence for hydrogen ion secretion. *Kidney Int.* **15**:624–629.

56. Burckhart, B. C., and E. Fromter. 1980. Bicarbonate transport across the peritubular membrane of rat kidney proximal tubule. In: *Hydrogen Ion Transport in Epithelia*. I. Schulz, G. Sachs, J. G. Forte, and K. J. Ullrich, Elsevier/North-Holland, Amsterdam. pp. 277–285.

57. Berry, C. A. 1982. Heterogeneity of tubular transport processes in the nephron. *Annu. Rev. Physiol.* **44**:181–202.

58. Jacobson, H. R. 1981. Functional segmentation of the mammalian nephron. *Am. J. Physiol.* **241**:F203–F218.

59. Jacobson, H. R. 1981. Effects of CO$_2$ and acetazolamide on bicarbonate and fluid transport in rabbit proximal tubules. *Am. J. Physiol.* **240**:F54–F62.

60. Berry, C. A., and D. G. Warnock. 1982. Acidification in the *in vitro* perfused tubule. *Kidney Int.* **22**:507–518.

61. Lucci, M. S., J. P. Tinker, I. M. Weiner, and T. D. DuBose, Jr. 1983. Function of proximal tubule carbonic anhydrase defined by selective inhibition. *Am. J. Physiol.* **245**:F443–F449.

62. Purkerson, M. L., H. Lubowitz, R. W. White, and N. S. Bricker. 1969. On the influence of extracellular fluid volume expansion on bicarbonate reabsorption in the rat. *J. Clin. Invest.* **48**:1754–1760.

63. Kurtzman, N. A. 1970. Regulation of renal bicarbonate reabsorption by extracellular volume. *J. Clin. Invest.* **49**:586–595.

64. Portwood, R. M., D. W. Seldin, F. C. Rector, Jr., and R. Cade. 1959. The relation of urinary CO$_2$ tension to bicarbonate excretion. *J. Clin. Invest.* **38**:770–776.

65. Sasaki, S., C. A. Berry, and F. C. Rector, Jr. 1982. Effect of

luminal and peritubular HCO_3^- concentrations and PCO_2 on HCO_3^- reabsorption in rabbit proximal convoluted tubules perfused *in vitro. J. Clin. Invest.* **70**:639–649.

66. Alpern, R. J., M. G. Cogan, and F. C. Rector, Jr. 1982. Effect of luminal bicarbonate concentrations on proximal acidification in the rat. *Am. J. Physiol.* **243**:F53–F59.

67. Morris, R. C., Jr., E. McSherry, and A. Sebastian. 1971. Modulation of experimental renal dysfunction of hereditary fructose intolerance by circulating parathyroid hormone. *Proc. Natl. Acad. Sci. USA* **68**:132–135.

68. Muldowney, F. P., J. F. Donohoe, D. V. Carroll, D. Powell, and R. Freaney. 1972. Parathyroid acidosis in uraemia. *Q. J. Med.* **41**:321–342.

69. Stoner, L. C., M. B. Burg, and J. Orloff. 1974. Ion transport in cortical collecting tubule: Effects of amiloride. *Am. J. Physiol.* **227**:453–459.

70. Lombard, W. E., J. P. Kokko, and H. R. Jacobson. 1983. Bicarbonate transport in cortical and outer medullary collecting tubules. *Am. J. Physiol.* **224**:F289–F296.

71. Stokes, J. B. 1982. Ion transport by the cortical and outer medullary collecting tubule. *Kidney Int.* **22**:473–484.

72. Koeppen, B. M., and P. R. Steinmetz. 1983. Basic mechanisms of urinary acidification. *Med. Clin. North Am.* **67**:753–770.

73. Ullrich, K. J., and F. Papavassiliou. 1981. Bicarbonate reabsorption in the papillary collecting duct of rats. *Pfluegers Arch.* **389**:271–275.

74. Graber, M. L., H. H. Bengele, J. H. Schwartz, and E. A. Alexander. 1981. pH and pCO_2 profiles of the rat inner medullary collecting duct. *Am. J. Physiol.* **241**:F659–F668.

75. O'Neil, R. G. 1981. Potassium secretion by the cortical collecting tubule. *Fed. Proc.* **40**:2403–2407.

76. Koeppen, B. M., B. A. Biagi, and G. H. Giebisch. 1983. Intracellular microelectrode characterization of the rabbit cortical collecting duct. *Am. J. Physiol.* **244**:F35–F47.

77. Stone, D. K., D. W. Seldin, J. P. Kokko, and H. R. Jacobson. 1983. Mineralocorticoid modulation of rabbit medullary collecting duct acidification: A sodium-independent effect. *J. Clin. Invest.* **72**:77–83.

78. Ludens, J. H., and D. D. Fanestil. 1974. Aldosterone stimulation of acidification of urine by isolated urinary bladder of the Colombian toad. *Am. J. Physiol.* **226**:1321–1326.

79. Al-Awqati, Q., L. H. Norby, A. Mueller, and P. R. Steinmetz. 1976. Characteristics of stimulation of H^+ transport by aldosterone in turtle urinary bladder. *J. Clin. Invest.* **58**:351–358.

80. Pitts, R. 1948. Renal excretion of acid. *Fed. Proc.* **7**:418–426.

81. Halperin, M. L., and R. L. Jungas. 1983. The metabolic production and renal disposal of hydrogen ions: An examination of the biochemical process. *Kidney Int.* **24**:709–713.

82. Albright, F., C. H. Burnett, W. Parson, E. C. Reifenstein, Jr., and A. Roos. 1946. Osteomalacia and late rickets. *Medicine (Baltimore)* **25**:399–479.

83. Lemann, J., Jr., J. R. Litzow, and E. J. Lennon. 1966. The effects of chronic acid loads in normal man: Further evidence for the participation of bone mineral in the defense against chronic metabolic acidosis. *J. Clin. Invest.* **45**:1608–1614.

84. Arruda, J. A. L., T. Carrasquillo, A. Cubria, D. R. Rademacher, and N. A. Kurtzman. 1976. Bicarbonate reabsorption in chronic renal failure. *Kidney Int.* **9**:481–488.

85. Schmidt, R. W., N. S. Bricker, and G. Gavellas. 1976. Bicarbonate reabsorption in the dog with experimental renal disease. *Kidney Int.* **10**:287–294.

86. Arruda, J. A. L., L. Nascimento, G. Arevalo, R. L. Baranowski, A. Cubria, C. Carrasquillo, C. Westenfelder, and N. A. Kurtzman. 1978. Bicarbonate reabsorption in chronic renal failure: Studies in man and the rat. *Pfluegers Arch.* **376**:193.

87. Cohn, D. E., K. A. Hruska, S. Klahr, and M. R. Hammerman. 1982. Increased Na^+–H^+ exchange in brush border vesicles from dogs with renal failure. *Am. J. Physiol.* **243**:293–299.

88. Morrin. P. A. F., N. S. Bricker, S. W. Kime, Jr., and C. Klein. 1962. Observations on the acidifying capacity of the experimentally diseased kidney in the dog. *J. Clin. Invest.* **41**:1297–1302.

89. Steinmetz, P. R., R. P. Eisinger, and J. Lowenstein. 1965. The excretion of acid in unilateral renal disease in man. *J. Clin. Invest.* **44**:582–591.

90. Dorhout-Mees, E. J., M. Machado, E. Slatopolsky, S. Klahr, and N. S. Bricker. 1966. The functional adaptation of the diseased kidney. III. Ammonium excretion. *J. Clin. Invest.* **45**:289–296.

91. MacLean, A. J., and J. P. Hayslett. 1975. Adaptive change in ammonia excretion in renal insufficiency. *Kidney Int.* **7**:397–404.

92. Schoolwerth, A. C., R. C. Sandler, P. M. Hoffman, and S. Klahr. 1975. Effects of nephron reduction and dietary protein content on ammoniagenesis in the rat. *Kidney Int.* **7**:397–404.

93. Bank, N., and W. B. Schwartz. 1960. The influence of anion penetrating ability on urinary acidification and the excretion of titratable acid. *J. Clin. Invest.* **39**:1516–1525.

94. Steinmetz, P. R. 1968. Excretion of acid by the kidney—Functional organization and cellular aspects of acidification. *N. Engl. J. Med.* **278**:1102–1109.

95. Rodriguez-Soriano, J., and C. M. Edelmann. 1969. Renal tubular acidosis. *Annu. Rev. Med.* **20**:363–382.

96. Morris, R. C., Jr., A. Sebastian, and E. McSherry, 1972. Renal acidosis. *Kidney Int.* **1**:322–340.

97. Morris, R. C., Jr. 1981. Renal tubular acidosis. *N. Engl. J. Med.* **304**:418–420.

98. McSherry, E., and R. C. Morris, Jr. 1978. Attainment and maintenance of normal stature with alkali therapy in infants and children with classic renal tubular acidosis. *J. Clin. Invest.* **61**:509.

99. Morris, R. C., Jr., and A. Sebastian. 1980. Disorders of the renal tubule that cause disorders of fluid, acid–base, and electrolyte metabolism. In: *Clinical Disorders of Fluid and Electrolyte Metabolism.* M. H. Maxwell and C. R. Kleeman, eds. McGraw-Hill, New York. pp. 883–946.

100. Sly, W. S., D. Hewett-Emmett, M. P. Whyte, Y.-S. L. Yu, and R. E. Tashian. 1983. Carbonic anhydrase II deficiency identified as the primary defect in the autosomal recessive syndrome of osteopetrosis with renal tubular acidosis and cerebral calcification. *Proc. Natl. Acad. Sci. USA* **80**:2752–2756.

101. Narins, R. G., and M. Goldberg. 1977. Renal tubular acidosis: Pathophysiology, diagnosis, and treatment. *Dis. Mon.* **23**:1–66.

102. Rector, F. C., Jr., and M. G. Cogan. 1978. The renal acidoses. *Hosp. Pract.* **15**:99–111.

103. Gennari, F. J., and J. J. Cohen. 1978. Renal tubular acidosis. *Annu. Rev. Med.* **29**:521–541.

104. Vainsel, M., P. Fondu, S. Cadranel, C. L. Rocmans, and W. Geptz. 1972. Osteopetrosis associated with proximal and distal tubular acidosis. *Acta Paediatr. Scand.* **61**:429–434.

105. Shapira, E., Y. Ben-Yoseph, F. G. Eyal, and A. Russell. 1974. Enzymatically inactive red cell carbonic anhydrase B in a family with renal tubular acidosis. *J. Clin. Invest.* **53**:59–63.

106. Kendall, A. G., and R. E. Tashian. 1977. Erythrocyte carbonic anhydrase I: Inherited deficiency in humans. *Science* **197**:471–472.

107. Donckerwolcke, R. A., G. J. Van Stekelenburg, and H. A. Tiddens. 1970. A case of bicarbonate-losing renal tubular acidosis with defective carbo-anhydrase activity. *Arch. Dis. Child.* **45**:769–773.

108. Komrower, G. M., V. Schwarz, A. Holzel, and L. Goldberg. 1956. A clinical and biochemical study of the galactosaemia: A possible explanation of the nature of the biochemical lesion. *Arch. Dis. Child.* **31**:254–264.

109. Morris, R. C., Jr. 1968. An experimental renal acidification defect in patients with hereditary fructose intolerance. I. Its resemblance to renal tubular acidosis. *J. Clin. Invest.* **47**:1389–1398.

110. Morris, R. C., Jr. 1968. An experimental renal acidification defect in patients with hereditary fructose intolerance. II. Its distinction from classic renal tubular acidosis; its resemblance to the renal acidification defect associated with the Fanconi syndrome of children with cystinosis. *J. Clin. Invest.* **47**:1648–1663.

111. Kranhold, J. R., D. Loh, and R. C. Morris, Jr. 1969. Renal fructose-metabolizing enzymes: Significance in hereditary fructose intolerance. *Science* **165**:402–403.

112. Brewer, E. D., H. C. Tsai, K. S. Szeto, and R. C. Morris, Jr.

1977. Maleic acid-induced impaired conversion of 25(OH)D$_3$ to 1,25-(OH)$_2$D$_3$: Implications for Fanconi's syndrome. *Kidney Int.* **12**:244–252.

113. Harrison, H. E., and H. C. Harrison. 1954. Experimental production of renal glucosuria, phosphaturia and aminoaciduria by injection of maleic acid. *Science* **120**:606–608.

114. Kramer, H. J., and H. C. Gonick. 1970. Experimental Fanconi syndrome. I. Effect of maleic acid on renal cortical Na-K-ATPase activity and ATP level. *J. Lab. Clin. Med.* **76**:799–808.

115. Rosen, V. J., H. J. Kramer, and H. C. Gonick. 1973. Experimental Fanconi syndrome. II. Effect of maleic acid on renal tubular ultrastructure. *Lab. Invest.* **28**:446–455.

116. Gougoux, A., G. Lemieux, and N. Lavoie. 1976. Maleate induced bicarbonaturia in the dog: A carbonic anhydrase-independent effect. *Am. J. Physiol.* **231**:1010–1017.

117. Maldonado, J. E., J. H. Velosa, R. A. Kyle, R. D. Wagoner, K. E. Holley, and R. M. Salassa. 1975. Fanconi syndrome in adults: A manifestation of a latent form of myeloma. *Am. J. Med.* **58**:354–364.

118. Sebastian, A., E. McSherry, I. Ueki, and R. C. Morris, Jr. 1968. Renal amyloidosis, nephrotic syndrome, and impaired renal tubular reabsorption of bicarbonate. *Ann. Intern. Med.* **69**:541–548.

119. Walker, B. R., F. Alexander, and P. J. Tannenbaum. 1971. Fanconi syndrome with renal tubular acidosis and light chain proteinuria. *Nephron* **8**:103–107.

120. Shioji, R., T. Furuyama, S. Onodera, H. Saito, H. Ito, and Y. Sasaki. 1970. Sjögren's syndrome and renal tubular acidosis. *Am. J. Med.* **48**:456–463.

121. Bergstein, J., and N. Litman. 1975. Interstitial nephritis with antitubular basement-membrane antibody. *N. Engl. J. Med.* **292**:875–878.

122. McCluskey, R. T. 1975. Anti-tubular basement-membrane (TBM) nephritis. *N. Engl. J. Med.* **292**:914–915.

123. Henderson, L. W., K. D. Nolph, J. B. Puschett, and M. Goldberg. 1968. Proximal tubular malfunction as a mechanism for diuresis after renal homotransplantation. *N. Engl. J. Med.* **278**:467–473.

124. Wilson, D. R., and A. A. Siddiqui. 1973. Renal tubular acidosis after kidney transplantation: Natural history and significance. *Ann. Intern. Med.* **79**:352–361.

125. Schein, P. S., M. J. O. O'Connell, J. Blom, S. Hubbard, L. T. Magrath, P. Bergevin, P. H. Wiernik, J. L. Ziegler, and V. T. DeVita. 1974. Clinical antitumor activity and toxicity of streptozotocin (NSC-85998). *Cancer* **34**:993–1000.

126. Hartmann, A. F. 1939. Clinical studies in acidosis and alkalosis. *Ann. Intern. Med.* **13**:940–956.

127. Lightwood, R., W. W. Payne, and J. A. Black. 1953. Infantile renal acidosis. *Pediatrics* **12**:628–644.

128. Reynolds, T. B. 1958. Observations on the pathogenesis of renal tubular acidosis. *Am. J. Med.* **25**:503–515.

129. Seldin, D. W., and J. D. Wilson. 1972. Renal tubular acidosis. In: *The Metabolic Basis of Inherited Disease.* J. B. Stanbury, J. B. Wyngaarden, and D. S. Fredrickson, eds. McGraw–Hill, New York. pp. 1548–1566.

130. Halperin, M. L., M. B. Goldstein, A. Haig, M. D. Johnson, and B. J. Stinebaugh. 1974. Studies on the pathogenesis of type 1 (distal) renal tubular acidosis as revealed by urinary PCO$_2$ tensions. *J. Clin. Invest.* **53**:669–677.

131. Stinebaugh, B. J., F. X. Schloeder, S. C. Tam, M. B. Goldstein, and M. L. Halperin. 1981. Pathogenesis of distal renal tubular acidosis. [Editorial Review]. *Kidney Int.* **19**:1–7.

132. Stinebaugh, B. J., R. Esquenari, F. X. Schloeder, W. N. Suki, M. B. Goldstein, and M. L. Halperin. 1980. Control of the urine–blood PCO$_2$ gradient in alkaline urine. *Kidney Int.* **17**:31–39.

133. Ochwadt, B. K., and R. F. Pitts. 1956. Effects of intravenous infusion of carbonic anhydrase on carbon dioxide tension of alkaline urine. *Am. J. Physiol.* **185**:426–429.

134. Gutknecht, J., M. A. Bisson, and D. C. Tosteson. 1977. Diffusion of carbon dioxide through lipid bilayer membranes: Effects

of carbonic anhydrase, bicarbonate, and unstirred layers. *J. Gen. Physiol.* **69**:779–794.

135. Steinmetz, P. R., Q. Al-Awqati, and W. J. Lawton. 1976. Nephrology rounds, University of Iowa Hospitals: Renal tubular acidosis. *Am. J. Med. Sci.* **271**:40–54.

136. Arruda, J. A. L., L. Nascimento, P. K. Mehta, D. R. Rademacher, J. T. Sehy, C. Westenfelder, and N. A. Kurtzman. 1977. The critical importance of urinary concentrating ability in the generation of urinary carbon dioxide tension. *J. Clin. Invest.* **60**:922–935.

137. McCurdy, D. K., M. Frederic, and J. R. Elkinton. 1968. Renal tubular acidosis due to amphotericin B. *N. Engl. J. Med.* **278**:124–131.

138. Burgess, J. L., and R. Birchall. 1972. Nephrotoxicity of amphotericin B, with emphasis on changes in tubular function. *Am. J. Med.* **53**:77–84.

139. Steinmetz, P. R., and L. R. Lawson. 1970. Defect in urinary acidification induced *in vitro* by amphotericin B. *J. Clin. Invest.* **49**:596–601.

140. Finn, J. T., L. H. Cohen, and P. R. Steinmetz. 1977. Acidifying defect induced by amphotericin B: Comparison of HCO$_3^-$ and H$^+$ permeabilities. *Kidney Int.* **11**:261–266.

141. Gouge, T. H., and V. T. Andriole. 1971. An experimental model of amphotericin B nephrotoxicity with renal tubular acidosis. *J. Lab. Clin. Med.* **78**:713–724.

142. Finkelstein, A., and R. Holz. 1973. Aqueous pores created in thin lipid membranes by the polyene antibiotics nystatin and amphotericin B. In: *Membranes, Lipid Bilayers, and Antibiotics,* Volume 2. G. Eisenman, ed. Dekker, New York. pp. 377–408.

143. Andreoli, T. E. 1974. The structure and function of amphotericin B–cholesterol pores in lipid bilayer membranes. *Ann. N.Y. Acad. Sci.* **235**:448–458.

144. Marty, A., and A. Finkelstein. 1975. Pores formed in lipid bilayer membranes by nystatin: Differences in its one-sided and two-sided action. *J. Gen. Physiol.* **65**:515–526.

145. Roscoe, J. M., M. B. Goldstein, M. L. Halperin, F. X. Schloeder, and B. J. Stinebaugh. 1977. Effect of amphotericin B on urine acidification in rats: Implications for the pathogenesis of distal renal tubular acidosis. *J. Lab. Clin. Med.* **89**:463–470.

146. Gyory, A. Z., and D. G. Edwards. 1971. Effect of mersalyl, ethacrynic acid and sodium sulphate infusion or urinary acidification in hereditary renal tubular acidosis. *Med. J. Aust.* **2**:940–945.

147. Arruda, J. A. L., and N. A. Kurtzman. 1980. Mechanisms and classification of deranged distal urinary acidification. *Am. J. Physiol.* **239**:F515–F523.

148. Buckalew, V. M., M. L. Purvis, M. G. Shulman, C. N. Herndon, and D. Rudman. 1974. Hereditary renal tubular acidosis. *Medicine (Baltimore)* **53**:229–254.

149. Feest, T. G., and O. M. Wrong. 1975. Inherited defects in distal tubular acidification. *Ann. Intern. Med.* **82**:584–585.

150. Brenner, R. J., D. B. Spring, A. Sebastian, E. M. McSherry, H. K. Genant, A. J. Palubinskas, and R. C. Morris, Jr. 1982. Incidence of radiographically evident bone disease, nephrocalcinosis, and nephrolithiasis in various types of renal tubular acidosis. *N. Engl. J. Med.* **307**:217–221.

151. Ferris, T., M. Kashgarian, H. Levitin, I. Brandt, and F. H. Epstein. 1961. Renal tubular acidosis and renal potassium wasting acquired as a result of hypercalcemic nephropathy. *N. Engl. J. Med.* **265**:924–928.

152. Huth, E. J., R. L. Mayock, and R. M. Kerr. 1959. Hyperthyroidism associated with renal tubular acidosis. *Am. J. Med.* **26**:818–826.

153. Yeoh, S. A. 1967. Fabry's disease with renal tubular acidosis. *Singapore Med. J.* **8**:275.

154. Taher, S. M., R. J. Anderson, R. McCartney, M. M. Popovtzer, and R. W. Schrier. 1974. Renal tubular acidosis associated with toluene "sniffing." *N. Engl. J. Med.* **290**:765–768.

155. Perez, G. O., J. R. Oster, and C. A. Vaamonde. 1975. Incomplete

syndrome of renal tubular acidosis induced by lithium carbonate. *J. Lab. Clin. Med.* **86**:386–394.

156. Batlle, D. C., J. A. L. Arruda, and N. A. Kurtzman. 1981. Hyperkalemic distal renal tubular acidosis associated with obstructive uropathy. *N. Engl. J. Med.* **304**:373–379.

157. Shear, L., H. L. Bonkowsky, and G. J. Gabuzda. 1969. Renal tubular acidosis in cirrhosis: A determinant of susceptibility to recurrent hepatic precoma. *N. Engl. J. Med.* **280**:1–3.

158. Oster, J. R., J. L. Hotchkiss, M. Carbon, and C. A. Vaamonde. 1975. Abnormal renal acidification in alcoholic liver disease. *J. Lab. Clin. Med.* **85**:987–1000.

159. Tannen, R. L. 1970. The effect of uncomplicated potassium depletion on urine acidification. *J. Clin. Invest.* **49**:813–827.

160. Tannen, R. L., E. Wedell, and R. Moore. 1973. Renal adaptation to a high potassium intake: The role of hydrogen ion. *J. Clin. Invest.* **52**:2089–2101.

161. Kamm, D. E., and G. L. Strope. 1973. Glutamine and glutamate metabolism in renal cortex from potassium-depleted rats. *Am. J. Physiol.* **224**:1241–1248.

162. Lathem, W. 1958. Hyperchloremic acidosis in chronic pyelonephritis. *N. Engl. J. Med.* **258**:1031–1036.

163. Carroll, H. J., and S. J. Farber. 1964. Hyperkalemia and hyperchloremic acidosis in chronic pyelonephritis. *Metabolism* **13**:808–817.

164. Hudson, J. B., A. V. Chobanian, and A. S. Relman. 1957. Hypoaldosteronism: A clinical study of a patient with an isolated adrenal mineralocorticoid deficiency, resulting in hyperkalemia and Stokes–Adams attacks. *N. Engl. J. Med.* **257**:529–536.

165. Gerstein, A. R., C. R. Kleeman, E. M. Gold, S. S. Franklin, M. H. Maxwell, H. C. Gonick, M. L. Feffer, and T. I. Steinman. 1968. Aldosterone deficiency in chronic renal failure. *Nephron* **5**:90–105.

166. Vagnucci, A. H. 1970. Selective aldosterone deficiency in chronic pyelonephritis. *Nephron.* **7**:524–537.

167. Schambelan, M., A. Sebastian, and F. C. Rector, Jr. 1981. Mineralocorticoid-resistant renal hyperkalemia without salt wasting (type II pseudohypoaldosteronism): Role of increased renal chloride reabsorption. *Kidney Int.* **19**:716–727.

168. Szylman, P., O. S. Better, C. Chaimowitz, and A. Rosler. 1976. Role of hyperkalemia in the metabolic acidosis of isolated hypoaldosteronism. *N. Engl. J. Med.* **294**:361–365.

169. Schambelan, M., A. Sebastian, and E. G. Biglieri. 1980. Prevalence, pathogenesis, and functional significance of aldosterone deficiency in hyperkalemic patients with chronic renal insufficiency. *Kidney Int.* **17**:89–101.

170. DeFronzo, R. A. 1980. Hyperkalemia and hyporeninemic hypoaldosteronism. *Kidney Int.* **17**:118–134.

171. Tan, S. Y., R. Shapiro, R. Franco, H. Stockard, and P. J. Mulrow. 1979. Indomethacin-induced prostaglandin inhibition with hyperkalemia. *Ann. Intern. Med.* **90**:783–785.

172. Mueller, A., and P. R. Steinmetz. 1978. Spironolactone: An aldosterone agonist in the stimulation of H^+ secretion by turtle urinary bladder. *J. Clin. Invest.* **61**:1666–1670.

173. O'Neil, R. G., and S. I. Helman. 1977. Transport characteristics of renal collecting tubules: Influences of DOCA and diet. *Am. J. Physiol.* **233**:F544–F558.

174. Wade, J. B., R. G. O'Neil, J. L. Pryor, and E. L. Boulpaep. 1979.

Modulation of cell membrane area in renal collecting tubules by corticosteroid hormones. *J. Cell Biol.* **81**:439–445.

175. O'Neil, R. G., and W. P. Dubinsky. 1983. Na-dependent mineralocorticoid regulation of cortical collecting duct (CCD) Na-K-ATPase. *Fed. Proc.* **42**:475.

176. Cogan, M. G., and A. I. Arieff. 1978. Sodium wasting, acidosis, and hyperkalemia induced by methicillin interstitial nephritis: Evidence for selective distal tubular dysfunction. *Am. J. Med.* **64**:500–507.

177. Husted, R. F., and P. R. Steinmetz. 1982. Mechanisms of K^+ transport in isolated turtle urinary bladder: Induction of active K^+ secretion in a K^+ absorbing epithelium. *J. Clin. Invest.* **70**:832–834.

178. Gordon, R. D., R. A. Geddes, C. G. K. Pawsey, and W. O'Halloran. 1970. Hypertension and severe hyperkalemia associated with suppression of renin and aldosterone and completely reversed by dietary sodium restriction. *Australas. Ann. Med.* **4**:287–294.

179. Arnold, J. E., and J. K. Healy. 1969. Hyperkalemia, hypertension, and systemic acidosis without renal failure associated with a tubular defect in potassium excretion. *Am. J. Med.* **47**:461–472.

180. Spitzer, A., C. M. Edelmann, Jr., L. D. Goldberg, and P. H. Henneman. 1973. Short stature, hyperkalemia and acidosis: A defect in renal transport of potassium. *Kidney Int.* **3**:251–257.

181. Lightwood, R. 1935. Calcium infarction of the kidneys in infants. British Paediatric Association. *Arch. Dis. Child* **10**:205–206.

182. Butler, A. M., J. F. Wilson, and S. J. Farber. 1936. Dehydration and acidosis with calcification at renal tubules. *Pediatrics* **8**:489–499.

183. Baines, G. H., J. A. Barclay, and W. T. Cooke. 1945. Nephrocalcinosis associated with hyperchloraemia and low plasma bicarbonate. *Q. J. Med.* **14**(N.S.):113.

184. Sebastian, A., E. McSherry, and R. C. Morris, Jr. 1971. On the mechanism of renal potassium wasting in renal tubular acidosis associated with the Fanconi syndrome (type 2 RTA). *J. Clin. Invest.* **50**:231–243.

185. Sebastian, A., E. McSherry, and R. C. Morris, Jr. 1971. Renal potassium wasting in renal tubular acidosis (RTA): Its occurrence in types 1 and 2 RTA despite sustained correction of systemic acidosis. *J. Clin. Invest.* **50**:667–678.

186. Gill, J. R., Jr. N. H. Bell, and F. C. Bartter. 1967. Impaired conservation of sodium and potassium in renal tubular acidosis and its correction by buffer anions. *Clin. Sci.* **33**:577–592.

187. Rampini, S., A. Fanconi, R. Illig, and A. Prader. 1968. Effect of hydrochlorothiazide on proximal renal tubular acidosis in a patient with idiopathic "de Toni-Debré-Fanconi syndrome." *Helv. Paediatr. Acta* **23**:13–21.

188. Donckerwolcke, R. A., G. J. Van Stekelenburg, and H. A. Tiddens. 1970. Therapy of bicarbonate-losing renal tubular acidosis. *Arch. Dis. Child.* **45**:774–779.

189. Morrissey, J. F., M. Ochoa, Jr., W. D. Lotspeich, and C. Waterhouse. 1963. Citrate excretion in renal tubular acidosis. *Ann. Intern. Med.* **58**:159–166.

190. Stinebaugh, B. J. 1978. Distal renal tubular acidosis: A review of current concepts of pathogenesis. *Proc. VII Int. Congr. Nephrol., Montreal, 1978.* In: Karger, Basel. pp. 337–344.

CHAPTER 13

Cystinosis and the Fanconi Syndrome

Jerry A. Schneider and Joseph D. Schulman

1. Introduction

Cystinosis is a recessively inherited metabolic disorder characterized biochemically by an elevated intracellular content of free (nonprotein) cystine which is compartmentalized within lysosomes. This results in crystal deposition in the cornea, conjunctiva, bone marrow, lymph nodes, other internal organs, and leukocytes. The metabolic error in this condition was recently found to be a defect in the carrier-mediated transport of cystine from the lysosome to the cytosol.[1-3]

Cystinosis is a common cause of the Fanconi syndrome in children, although the essential elements of the Fanconi syndrome appear in a variety of other hereditary and acquired disorders (Table I). Several recent reviews of cystinosis cover aspects of cystinosis beyond the scope of this chapter.[4-8]

2. Historical Resume

Recognition of the various clinical and pathological features of cystinosis and the Fanconi syndrome was fragmented in time and confounded by chance association with other disorders. (Further details and exact references may be found elsewhere[4].) This confusion began with the first recognition of the disease by Abderhalden in 1903 when he described cystine crystals in the liver and spleen at autopsy of a 21-month-old infant who died of "inanition." Two additional sibs had died previously of a similar illness. Since the urine of the child's father, paternal grandfather, and two sibs contained excessive quantities of cystine, he called the condition a "familial cystine diathesis." This led to an early and unwarranted view that cystinosis was a more severe expression of cystinuria.

Lignac's report in 1924 of cystine deposits in each of three infants with rickets, dwarfism, renal disease, and wasting pro-

vided some delineation of the clinical expression of cystinosis. The finding of ureteral cystine stones in one of these children further compounded the confusion with cystinuria. Subsequent reports of other cases of cystine storage without evidence of cystinuria by Russell and Barrie in 1936 and Beumer in 1937 further delineated the multisystem syndrome of "cystine storage disease." The distinction between cystinosis and cystinuria, first suspected on clinical grounds, was clearly established with the subsequent demonstration by Dent and Harris that patients with cystinuria excreted excessive quantities of other dibasic amino acids (lysine, arginine, and ornithine) in addition to cystine.

Additional aspects of the clinical expression of this disease emerged from the association of rickets and stunted growth in a child with glycosuria and albuminuria reported by Fanconi in 1931. Vitamin D-resistant rickets with spontaneous fractures was described in 1933 by deToni in a dwarfed child who also showed a low serum phosphate concentration, acidosis, proteinuria, and glucosuria. A report of a similar child by Debre in 1934 led Fanconi in 1936 to propose a syndrome of "nephrotic-glycosuric dwarfism and hypophosphatemic rickets."

The possibility that the Fanconi syndrome and cystinosis were but two aspects of the same entity, proposed by Beumer and Wepler in 1937, was supported by the demonstration of cystine crystals in the tissues at necropsy of one of Fanconi's original patients. A detailed study by Bickel in 1952 in a larger series of patients provided further evidence of the association of cystinosis with the Fanconi syndrome and progressive glomerular damage.

3. Clinical and Pathological Features

The clinical expression of cystinosis ranges widely from one family to another. The most basic distinction is between patients who have nephropathic as opposed to those who have benign cystinosis. The latter have no renal abnormality. Patients with nephropathic cystinosis are further classified based on the age at which first symptoms are noted. Patients with the most widely recognized form of nephropathic cystinosis become

Jerry A. Schneider • Metabolic Diseases Division, Department of Pediatrics, University of California at San Diego, La Jolla, California 92093. Joseph D. Schulman • Genetics and IVF Institute, Department of Obstetrics and Gynecology, Fairfax Hospital, Fairfax, Virginia 22031.

Table I. Causes of the Renal Fanconi Syndrome

Heritable disorders
  Cystinosis
    Idiopathic[133,134]
    Lowe's syndrome (oculo-cerebro-renal syndrome)[135]
    Tyrosinemia, type I[128,129]
    Familial nephrosis[136]
    Galactosemia[137]
    Glycogen storage disease[138]
    Hereditary fructose intolerance[139]
    Wilson's disease[117–119,140]
Other disorders
  Human kidney transplantation[141–143]
  Myeloma[130,132,144]
  Amyloidosis[130,145]
  Sjögren's syndrome[146,147]
  Nephrotic syndrome[131]
  Pancreatic carcinoma[133]
  Vitamin D deficiency[148,149]
Exogenous toxins
  Heavy metals (especially lead, mercury; including organic mercurials)[111,113–116,150]
  Nonmetals (Lysol, maleic acid, methyl-3-chromone, outdated tetracycline, streptozotocin, vitamin D)[120,124–127]

symptomatic near 1 year of age and are said to have infantile, or simply nephropathic, cystinosis. Patients who become symptomatic at later ages are said to have late-onset cystinosis. All forms of cystinosis exhibit autosomal recessive inheritance. Within a given family the type of cystinosis is consistent.

3.1. Nephropathic Cystinosis

The major clinical symptomatology of nephropathic cystinosis can, for the most part, be related to the progressive impairment of initially tubular, then glomerular function to produce an unremitting course.

3.1.1. Early Stage

3.1.1a. Clinical Course. Children with nephropathic cystinosis appear normal at birth and during the first 6 months or so of life. The first overt signs of the disease are usually produced by the renal tubular defect in water reabsorption. The resulting polyuria and polydipsia make affected children especially vulnerable to dehydration, and this leads to recurrent fever as one of the most common presenting symptoms. In addition, by 1 year of age they usually show growth retardation, rickets, acidosis, and other chemical evidence of renal tubular abnormalities, such as increased renal excretion of glucose, amino acids, phosphate, and potassium. There usually is some evidence of glomerular damage as well and the subsequent course is determined by the rate at which glomerular insufficiency progresses. Less overt clinical and biochemical evidence of the disease has been found at a much earlier age by careful examination of children known to be at risk because of an older affected sib.[9,10] Some patients show recurrent episodes of acute prostration, weakness, and cardiovascular collapse which can lead to early and sudden death. This has been observed during intravenous infusion of glucose but can also occur without known precipitating cause. This disturbance is associated with gross changes in serum electrolytes and most of

the symptomatology is thought to be related to profound hypokalemia. Three French-Canadian children from the province of Quebec have presented with clinical features of Bartter's syndrome and were subsequently found to have nephropathic cystinosis.[11–13]

Failure to thrive is one of the more prominent features of the disease. Affected children show growth failure within the first year of life and almost always remain below the third percentile in both height and weight throughout life. The serum growth hormone concentration is normal,[14] but in the later stage of their disease many patients become hypothyroid and require thyroxin replacement therapy.[15–18] Although mental development appears normal, it has recently become apparent that some patients show enlarged lateral ventricles on CAT scans.[19,20] It is not yet clear whether this is a sporadic occurrence in cystinosis or whether it is a characteristic of the disease in certain families.

Rickets develops at an early age in most patients despite intake of ordinary quantities of vitamin D (400 U/day). Frontal bossing, genu valgum, thickening of the wrists and ankles, rachitic rosary, and Harrison's groove are frequently seen. X-ray examination of the long bones shows the broadened and frayed epiphyses characteristic of rickets. The glomerular damage progresses in a sporadic but unremitting manner. Careful study of affected patients shows that glomerular dysfunction may remain constant for many months or even years, followed by periods of fairly rapid deterioration that seem to be unrelated to any environmental or biochemical factor yet detected.

Cystinotic children show a number of clinical features not obviously related to the renal abnormalities. The majority have blond hair and a fair complexion, often having substantially lighter pigmentation than their parents. However, several non-Caucasian children who appear to have typical nephropathic cystinosis have now been identified.[21,22] In the patients with fair complexion, exposure to sunlight produces skin tanning with far less tendency to sunburn than would be expected from the lack of skin pigment. Cystinotic patients tolerate infections well and show no unusual susceptibility to intercurrent illness.

Severe photophobia develops in most affected children within the first few years of life. Characteristic changes in the eye provide some of the most consistent features of the disease, and a detailed ophthalmological examination often establishes the clinical diagnosis. The first ophthalmic manifestations were described by Bürki in 1941.[23] Slit-lamp examination discloses homogeneously dispersed tinsellike refractile opacities in the cornea and the conjunctiva. Crystalline cystine has been identified by X-ray diffraction in the conjunctival but not in the corneal deposits.[24] The crystalline opacities in the periphery of the cornea occupy the entire corneal thickness. Early in the course of the disease, only the anterior one-third to two-thirds is affected in the central region of the cornea but with time the entire thickness is involved. These iridescent crystalline particles are so typical as to be virtually diagnostic of all forms of cystinosis; they appear before the full clinical manifestations of nephropathic cystinosis are expressed, but are not present at birth.[9,10] However, other disorders are known in which corneal deposits are found.[25]

In addition to the corneal and conjunctival changes, a peripheral retinopathy is present[26] which is characteristic only of the nephropathic forms of cystinosis. The pathological change consists of a generalized pigment disturbance that often assumes a depigmented patchy pattern with superimposed pigment clumps of irregular distribution, varying in size from about 1/10th disk diameter to a fine pepper like stippling. These changes affect the temporal side more extensively than the nasal and are marked in the peripheral regions of the retina, whereas the central regions are generally devoid of abnormal pigmentation during the first years of life. This retinopathy has permitted a diagnosis of cystinosis to be made in a child as young as 5 weeks of age, even though only a rare crystal was present in the iris at that age.[10] Evidently the retinal changes precede those in the cornea.

3.1.1b. Laboratory Findings.

The development of clinical symptoms of the Fanconi syndrome is paralleled by the appearance of laboratory evidence of renal tubular dysfunction. This includes glycosuria, with excretion of from traces to 5 g glucose/100 ml urine,[8,9] organic aciduria, and aminoaciduria. Excessive quantities of multiple amino acids appear in the urine and generally account for about 80% of the organic acids. Cystine excretion is generally increased in the same proportion as other amino acids. An increased phosphate excretion is usually found and a decreased intestinal absorption of phosphate has also been reported.[8] Proteinuria is often present and consists primarily of the so-called tubular protein associated with the Fanconi syndrome. This includes over 50 times the normal excretion of the light chain of γ-globulin.[27] The significance of this characteristic finding remains to be determined. In spite of systemic acidosis, the urine tends to be alkaline and contains increased amounts of NH_4^+. In addition, microscopic examination reveals many granular casts and occasional erythrocytes as glomerular damage progresses.

The blood sugar concentration is normal. Metabolic acidosis with diminished plasma CO_2 reflects the renal loss of HCO_3^-. Hypophosphatemia is related to the appearance of rickets, the activity of which is reflected in an increased serum alkaline phosphatase activity. An increased blood pyruvate concentration has been found in some but not all patients and presumably results from an effect of the cystine on SH-dependent enzymes.[28,29] Severe hypokalemia may be present. BUN and serum creatinine are usually in the normal range early in the disease even though creatinine and urea clearance are substantially diminished. There is often a significant degree of anemia before there is substantial renal failure. The erythrocyte sedimentation rate is usually elevated. The exact reason for this finding is not known but is related to a plasma rather than a red blood cell phenomenon.[30]

3.1.2. Late Symptoms and Findings

The clinical course of the disease is one of unremitting progression of glomerular damage, leading eventually to end-stage renal failure before puberty. The time course of the progression of glomerular damage is uneven, with periods of apparent stability of renal function interspersed with periods of rapid decline. As in other forms of renal dysfunction, children are able to maintain relatively normal degrees of activity until the renal insufficiency becomes advanced, so that only in the late stages are the children seriously incapacitated.

In some patients, advanced renal disease with elevated BUN and creatinine is found as early as 2 years of age, but this degree of uremia does not usually occur until 7 to 10 years of age. With end-stage glomerular damage, the aminoaciduria is often masked by the severely diminished GFR, so that the total amino acid excretion is in the normal range. Other aspects of the Fanconi syndrome may also be obscured in some patients by advanced glomerular damage.[29] In the same manner, the

hypophosphatemia and hypokalemia are corrected as renal deterioration progresses and are eventually replaced by hyperphosphatemia, hyperkalemia, and a secondary hypocalcemia. The acidosis may then become even more marked. Correction of the acidosis by administration of alkali salts can then result in tetany. Growth failure, progressive renal dysfunction, and acidosis may also ameliorate the clinical symptoms of rickets. Polyuria and polydipsia also diminish as glomerular insufficiency supervenes. The clinical symptomatology then becomes primarily that associated with severe renal failure and uremia. Weakness and lethargy with edema and congestive cardiac failure result from salt retention and from the profound anemia of renal failure.

The children remain dwarfed throughout life. It is not clear whether this is entirely related to renal failure. Some children have exhibited moderate degrees of "catch-up" growth following renal transplantation. As mentioned, these children have a normal serum growth hormone concentration, but often become hypothyroid during the course of their disease. This is presumed to be related to the deposition of cystine crystals in the thyroid gland.[14–17] An abnormal pituitary resistance to thyroid hormone negative feedback also appears to be present in many euthyroid patients.[18]

Now that cystinotic patients are being kept alive with renal transplants for many years after the development of renal failure, the possibility of functional impairment of other organs exists. As these children grow older, both the number of conjunctival and corneal crystals and the degree of retinopathy increase.[7,31] The retinopathy, which originally is only apparent at the equator of the eye, extends toward the posterior pole. In one study of 10 cystinotic children, aged 11 to 19 years, who had received renal transplants 3 to 9 years previously, none had abnormalities of visual acuity which were of subjective significance.[31] In another study, however, of 16 cystinotic post-transplant patients, 2 had marked impairment of vision and another was practically blind.[7] Another potential problem is suggested by the recent report of a 19-year-old cystinotic patient who had a very successful renal transplant, but who developed a right hemiparesis and drowned accidentally. Postmortem examination revealed bilateral necrosis and calcification of basal ganglia structures.[32] In addition, the authors have become aware of as yet unreported instances of portal hypertension in association with cystinosis.

3.2. Benign Cystinosis

What appears to be a completely benign variant of cystinosis was first described by Cogan and associates in 1957, and since then many additional cases have been reported.[33–38] The primary clinical distinction between the benign and the nephropathic variants of the disease is failure of the former patients to show either retinopathy or renal dysfunction. Nevertheless, they show crystalline deposits in the cornea, bone marrow, and leukocytes, but this causes no disability. These patients live to adult life, and their cystinosis is usually identified only incidentally during a routine ophthalmological examination with a slit lamp. This benign variant of the disease appears to represent one end of the spectrum of clinical expression. These patients tend to show a somewhat lower concentration of intracellular cystine[35,39] than do patients with nephropathic cystinosis. Kidney biopsies from two patients demonstrated no cystine crystals,[33,38] and the cystine content of the one biopsy which was measured was not elevated.[38] Thus, it appears that the renal function is normal in patients with the benign form of cystinosis because the kidney does not accumulate cystine.

3.3. Intermediate Forms

More recently, a third form of cystinosis has been described, called either later-onset, intermediate, or adolescent cystinosis.[39–49] These patients differ from those with benign cystinosis in that they do have renal dysfunction. They differ from the typical patients with nephropathic cystinosis because their disease is not apparent until an older age, they frequently do not have the complete Fanconi syndrome, and their glomerular insufficiency progresses more slowly. The age at which the first symptoms are noted in these patients has ranged from 18 months to 25 years.[50] When more than one individual in a sibship has this condition,[50] the age of onset and symptomatology are similar. These facts suggest that intermediate cystinosis may represent a range in the spectrum of different types of cystinosis rather than one simply defined disease. It is possible that these patients are "double heterozygotes" for the infantile nephropathic and benign forms of cystinosis.

All of the patients have shown crystalline material in the cornea and conjunctiva visible by slit-lamp examination, and have had typical cystine crystals in bone marrow aspirates. Likewise, they have all shown abnormally increased concentrations of cystine in leukocytes and cultured skin fibroblasts although actual values have tended to be lower than generally seen in patients with nephropathic cystinosis.[37,39,42,46,47] The presence of photophobia and retinopathy, however, has been variable. Unfortunately, most case reports do not comment on skin pigmentation, but a normal skin pigmentation has been reported in one child,[39] and a dark pigmentation in another patient.[45] Growth data on these patients are incomplete. It appears that some patients have near-normal growth, whereas others have delayed growth, but not usually to the extreme degree seen in infantile nephropathic cystinosis.

3.4. Genetics

Both the nephropathic and the benign forms of cystinosis show autosomal recessive patterns of inheritance. Preliminary evidence indicates that the same is true for the intermediate forms of cystinosis. The frequency of the disease in the general population is difficult to estimate. In a recent French study, the minimum incidence of cystinosis in the province of Brittany was calculated as 1 per 25,909, but for the rest of France was calculated at 1 per 326,440.[51] Genetic heterogeneity is suggested by the differences in clinical expression of the disease in different families.

3.5. Pathology of Cystinotic Tissues

For a detailed discussion of the pathology on this condition, the reader is referred to Spear[52,53] and to a recent review.[8] The specific and characteristic lesion in cystinosis is the deposition of cystine crystals in numerous organs, sparing the muscle and brain parenchyma, and heavily involving the reticuloendothelial cells of the bone marrow, liver, spleen, and lymphatic system. The solubility of cystine in acid solutions and formalin commonly used as fixatives undoubtedly contributes to the wide variation in the amount of crystal formation observed in histological preparations. Even in tissues fixed in absolute alcohol or frozen, instances are known in which essentially all crystals

were lost during the final exposure to the aqueous solutions used in staining. The cystine crystals are, therefore, best observed in their natural state in frozen or alcohol-fixed tissues before staining, using cross-polarizing filters and phase microscopy. Treatment with organic solvents will remove birefringent lipid materials. The cystine crystals show several different forms including, most frequently, clusters of brick-shaped birefringent crystals, and less commonly, nonbirefringent hexagonal forms.

3.6. Intracellular Location of Cystine

Cystinotic cells contain large amounts of free cystine in spite of normal activities of the soluble enzyme systems which reduce and further metabolize this amino acid.[54] It appeared that one explanation for this paradox might be the intracellular compartmentalization of cystine in a location where it is not available to these enzymes. To test this hypothesis, lysates of cystinotic leukocytes were separated into nuclear, granular (acid phosphate-rich), and soluble fractions, and each fraction was assayed for cystine. Most of the total free cystine was found in the granular fraction.[55] Similar results were obtained with lysates of cultured fibroblasts[56] and in both leukocytes and fibroblasts from patients with benign cystinosis.[35]

Compartmentalization of cystine appears to be a primary manifestation of cystinosis, rather than a consequence of prior crystallization of cystine within the cytoplasm. This is shown by several lines of evidence:

1. Cultured cystinotic fibroblasts examined by light or electron microscopy have never shown crystalline deposits. The same is true for cultured cystinotic lymphoblasts, and they too have a greatly increased intracellular cystine content.[57]
2. Only rare cystine crystals are seen in uncultured cystinotic leukocytes,[58,59] and crystals have not been observed in cystine-rich cystinotic leukocytes prepared from whole blood after using hypotonic saline to disrupt erythrocytes.[55]
3. Cystine compartmentalization occurs in the leukocytes and cultured fibroblasts of cystinotic heterozygotes (healthy carriers) in whose cells or tissues cystine crystals have never been observed.[60]

Several morphological and biochemical approaches have been employed in the effort to establish the specific type of subcellular organelle involved in the cystinotic process. With the exception of a single unconfirmed report in 1968,[61] all studies have concluded that cystine appears to be stored within the lysosomes. The major evidence for lysosomes being the site of cystine accumulation is as follows:

1. Electron microscopic examination of cystinotic lymph node by Patrick and Lake[62] revealed crystalline bodies within membrane-limited organelles having the appearance of lysosomes, and containing the lysosomal marker enzyme acid phosphatase. Lysosomal location of cystine crystals or noncrystalline amorphous inclusions has also been suggested by others on the basis of electron micrographic study of cystinotic conjunctiva,[38,63–65] cornea,[66] kidney,[53] rectal lamina propria, and lymphocytes.[58] Crystalline outlines have been described within lysosomes of Kupffer cells, but not hepatocytes, in a cystinotic child and fetus.[67,68] In one study, cystinotic conjunctival cells were incubated in vitro with ferritin, an electron-dense protein which upon phagocytosis enters lysosomes. The ferritin was then found within the same membrane-limited organelles as the cystine crystals.[65]

2. Cystine and lysosomal marker enzymes were found in the same fractions of sucrose density gradients of the sonicates of washed cystinotic leukocytes.[69]

3. Cystinotic fibroblasts cultured *in vitro* accumulate large amounts of penicillamine-cysteine disulfide (a cystine analog) within lysosomelike vacuoles when this compound is added in substantial quantity to the tissue culture medium, while normal fibroblasts under similar conditions do not demonstrate accumulation of this cystinelike compound.[70,71]

4. Highly purified human lysosome from cystinotic homozygotes and heterozygotes contain abnormally elevated concentrations of cystine.[72]

These studies, while seeming to prove the involvement of lysosomes in the cystinotic process, do not definitively exclude the possibility of other organelles being involved to some extent. It seems reasonable, however, to consider cystinosis as an example of a lysosomal storage disease, differing notably from other such disorders in the low molecular weight of the stored metabolite (M_r 240). Experiments on human[70,71] and mouse[73] cells suggest that the lysosomal membrane is relatively impermeable to nonmetabolized amino acids and peptides with a molecular weight greater than approximately 200; thus, normal lysosomes presumably have a mechanism permitting the efflux or metabolism of cystine. As described below, methods have recently been developed to study lysosomal cystine efflux, and this transport system was found to be defective in cystinotic cells.[1-3]

4. Chemistry and Metabolism of Cystine

It is beyond the scope of this chapter to give a detailed discussion of cystine chemistry and metabolism. Detailed reviews of these topics have recently been published.[4,8] Cystine is the least soluble of the amino acids. Studies of cystinosis have been hampered by the difficulties encountered in accurately measuring this amino acid in physiological samples. This difficulty has now been circumvented by the development of a rapid isotope-dilution assay which utilizes a specific cystine-binding protein isolated from *Escherichia coli*.[74]

The primary means of cysteine synthesis in mammals is by the conversion of methionine to cysteine which involves transsulfuration via S-adenosyl-L-methionine, S-adenosyl-L-homocysteine, L-homocysteine, and L-cystathionine. In normal man, methionine is an essential amino acid, although cystine, which is not, can replace 80–90% of the human requirement for methionine.[75] Children can grow normally on a diet containing adequate methionine but no cystine.[76] However, cystine is an essential amino acid for human diploid fibroblasts in culture,[77-79] for certain patients deficient in enzymes for methionine–cystine conversion,[80] and probably for human fetuses and neonates who normally have a lack of cystathionase.[81] Cystine degradation first involves the reduction of cystine to cysteine. Cysteine is then degraded to inorganic sulfate and taurine.[82] Cysteine has additional metabolic fates summarized elsewhere.[4,8,83] Of particular interest are its conversion to the tripeptide glutathione, its oxidation to cystine, and its incorporation into protein. All of the enzymes catalyzing these reactions exhibit normal activity in cystinotic tissues and cells.[8]

4.1. Cystine Metabolism in Cystinosis

The metabolic error in cystinosis was recently found to be a defect in the normal transport of cystine from the lysosome to the cytosol.[1-3] This finding was the culmination of many years of work by numerous laboratories. The following pages will briefly review some of the studies which led to the elucidation of the metabolic defect in this disease.

4.1.1. Intracellular Concentration of Free Cystine

Since histological examination of cystinotic tissues shows cystine crystals primarily in reticuloendothelial cells, the simplest explanation is that the crystals are first formed extracellularly and are then phagocytized. On the other hand, Barr concluded 34 years ago, on histological grounds, that the crystals form inside these cells and are not trapped by phagocytosis.[84] Two biochemical findings proved that Barr was correct. First, the fasting plasma concentration of free nonprotein cystine was normal in cystinotic patients.[29] Second, an increased intracellular concentration of free cystine was found in crystal-free cells from these patients (see below).

4.1.1a. Leukocyte Content of Cystine. Mixed populations of peripheral leukocytes from patients with cystinosis were found to contain much more free cystine than normal cells, and thus they provided the first meaningful biochemical measurement of cystine accumulation in this disease.[55] Cells containing crystals apparently did not withstand the brief exposure to hypotonic solutions used to lyse erythrocytes during the preparation procedure and were never observed in the final leukocyte preparations. Subsequent studies of such leukocyte preparations by electron microscopy have revealed no intracellular crystals.[30,58] The mean free cystine (i.e., nonprotein cystine) content of white cells from nine children with infantile nephropathic cystinosis was, nevertheless, 80 times normal, and from three patients with benign cystinosis, 30 times normal.[35,55] Patients with intermediate (late-onset) cystinosis tend to have free cystine content of leukocytes greater than found in patients with benign cystinosis, but less than found in patients with infantile nephropathic cystinosis.[37,39,41] However, many exceptions have been noted.[85] Subsequent studies have demonstrated that cystine stores are primarily in the polymorphonuclear leukocytes and monocytes rather than in lymphocytes.[86]

Although cyst(e)ine is known to occur primarily as cysteine within cells,[87-89] an elevation of cysteine (measured as its N-ethylmaleimide adduct) is not found in cystinotic cells.[55] This may be because only the disulfide form is trapped within cystinotic lysosomes, or because the intracellular cysteine may leach out of the leukocytes during preparation. The content of another reduced sulfhydryl compound of higher molecular weight, glutathione, is normal in leukocytes from patients with cystinosis.[90]

4.1.1b. Cystine Content of Cultured Cells.
Fibroblasts cultured *in vitro* from skin biopsies of patients with cystinosis and maintained in culture through several generations also contain greatly increased amounts of free cystine.[56] Fibroblasts from patients with all types of cystinosis grow normally and appear normal in culture. Cystine crystals have not been seen in these fibroblasts by either phase or electron microscopy.[30] Yet fibroblasts from patients with infantile

nephropathic cystinosis contain over 100 times the normal content of free cystine, and from patients with benign cystinosis, about 50 times the normal free cystine content.[35,56] Fibroblasts cultured from patients with the intermediate type of cystinosis tend to have a free cystine content greater than found in patients with benign cystinosis, but less than found in patients with infantile nephropathic cystinosis.[37,39,41] However, many exceptions occur.[91]

Continuous lymphoblast cultures (transformed by Epstein–Barr virus) have been established from patients with cystinosis. These cells contain 15 to 20 times more free cystine than continuous lymphoblast cultures from normal individuals.[57] Cystinotic skin fibroblasts transformed with SV40 virus contain 30 to 50 times more free cystine per milligram protein than do SV40-transformed cells from normal control individuals.[92]

4.2. Detection of Heterozygotes

The mean free cystine content of peripheral leukocytes from parents (obligate heterozygotes) of cystinotic children is five to six times greater than the mean normal value.[55] This provided the first biochemical identification of the heterozygote in this disorder. Although there is occasional overlap between control and heterozygote values, the difference between these groups is statistically significant ($p < 0.001$). Early data from fibroblast cultures indicated that they might also be used for heterozygote detection.[56] However, the values from fibroblast measurements have proven to be very erratic and cannot be recommended for this purpose.[85] Steinherz *et al.* have recently proposed that heterozygotes can be detected more accurately by measuring the half-time of cystine efflux from leukocytes which are preloaded by exposure to cystine dimethyl ester.[93] This method showed a good distinction between normal, heterozygote, and cystinotic leukocytes. It has the disadvantage, however, of requiring fresh blood from the individual being studied. Although simply measuring the cystine content of cells may be less exact, it allows individuals to be studied who live great distances from the specialized laboratory.[94]

4.3. Source of the Cystine in Cystinotic Cells

Until recently, the exact source of the cystine which accumulates in cystinotic cells was not known. The finding that the cystine content of cystinotic cells is somewhat proportional to the cystine content of the medium suggested that some of the accumulated cystine came from extracellular cyst(e)ine (cystine and/or cysteine).[91,95] However, Oshima *et al.*[96] found that after cystinotic fibroblasts are depleted of free cystine with cysteamine, 30 to 50% of the cystine reaccumulated in 24 hr in the absence of extracellular cystine. The possibility that the cystine in cystinotic fibroblasts is derived from methionine, cystathionine, or serine was ruled out by experiments with radiolabeled precursors.[91,97] The availability of methods to deplete cystinotic cells of their increased free cystine with either dithiothreitol[98] or cysteamine[99] made the further study of this problem possible. Thoene *et al.*[97] were able to show that after depletion of cystine from cystinotic fibroblasts with cysteamine, the cystine which reaccumulates in a cystine-free medium comes from the intracellular degradation of protein. More recently, Thoene and Lemons[100] have shown that extracellular protein can also serve as a source of the cystine which accumulates in cystinotic cells. These find-

ings have been misinterpreted to imply that protein degradation is abnormal in cystinotic cells.[101] This is not the case. Protein degradation simply provides a major source of the free cystine, which accumulates in the lysosomes of cystinotic cells.

4.4. The Metabolic Defect in Cystinosis

For over 20 years, investigators attempted to find an abnormal activity of an enzyme involved in cystine degradation.[4,8] None was ever found. All the enzymes known to use cystine or cysteine as substrates are nonlysosomal and all those studied were active in properly prepared cystinotic tissues. No lysosomal enzyme which utilizes cystine or cysteine as a substrate was ever confirmed. These findings led to the suggestion that cystinosis might be caused by a defect of cystine transport.[102,103] Exploratory transport studies were extremely difficult to interpret for several reasons:

1. The cystinotic cells which were studied usually contained a large load of cystine, whereas the control cells did not.
2. The cystine load in cystinotic cells is compartmentalized within lysosomes.
3. Cystine taken up by the cell is rapidly reduced to cysteine and then incorporated into either glutathione or protein.[96] Thus, investigators were forced to study the transport of a compound into a cell which was preloaded with that compound, which had a specific major pool of the compound, and which was rapidly metabolizing that compound.

It seemed reasonable to suppose that if cystinosis was caused by a defect in cyst(e)ine transport, it would be at the level of the lysosomal membrane rather than the plasma membrane. This would explain why cystine accumulates within cystinotic lysosomes. Studies of impure[4] and pure[72] cystinotic lysosomes suggested that other amino acids do not accumulate abnormally at this site. Methods for determining amino acid efflux from isolated lysosomes were inadequate,[104] but indirect evidence had been obtained from studies on whole cells that cystinotic lysosomes are similar to normal in their permeability to nonmetabolizable small molecules but not to cystinelike disulfides.[70,71] After Reeves described the use of amino acid methyl esters for measuring efflux of amino acids from isolated lysosomes,[105] Steinherz *et al.* applied this method to compare the efflux of cystine and other amino acids in normal and cystinotic leukocyte lysosomes.[106] Their results indicated that cystinotic lysosomes have normal efflux of leucine and thus do not have decreased permeability to L-amino acids in general. Cystine efflux was much slower than leucine efflux in both normals and cystinotics. The extremely slow normal rate of cystine efflux confounded interpretation of the comparisons with cystinotics. In contrast, when *intact* cystinotic leukocytes were loaded with cystine by exposure to cystine dimethyl ester, distended lysosomes were observed electron microscopically.[107] The clearance of cystine from such cells was far slower in cystinotic than in normal hyperloaded leukocytes, and heterozygotes had, on average, intermediate values.[107]

Utilizing cultured skin fibroblasts, Jonas *et al.* increased the cystine content of normal and heterozygous cells six- to sevenfold by incubation for 24 hr in the presence of 30 mM cysteine-glutathione mixed disulfide.[108] The cystine was located in the lysosomes of these cells and when placed in cystine-free medium both cell types rapidly lost their lysosomal cystine ($t_{1/2} = 20$ min), whereas the cystine content of cystinotic cells (whether or not they were first incubated in the mixed disulfide)

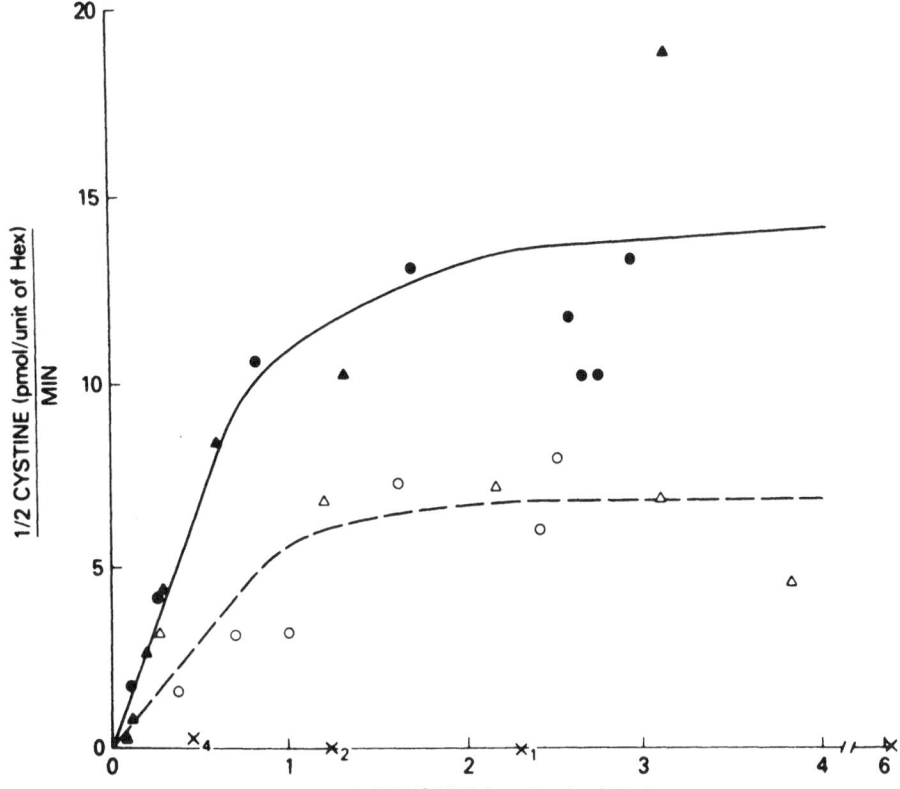

Fig. 1. Velocity of cystine egress from granular fractions isolated from leukocytes from four patients with cystinosis (X), two heterozygotes (○, △), and two normal control individuals (●, ▲), The abscissa shows the initial loading of the granular fractions expressed as cystine per unit hexosaminidase activity; the ordinate indicates initial velocity (30 min) expressed as cystine per minute per unit hexosaminidase activity.[2]

remained stable. In contrast to the findings in intact cells, cystine was not lost from isolated, cystine-loaded lysosomes from normal, heterozygous, or cystinotic fibroblasts.[108]

In contrast to the findings in isolated, cystine-loaded lysosomes from cultured fibroblasts,[108] Gahl et al. found that such lysosomes from normal peripheral leukocytes showed definite cystine efflux.[1,2] Furthermore, cystinotic lysosomes had a markedly retarded rate of cystine efflux, whereas heterozygous lysosomal cystine efflux rates were intermediate. When the rate of cystine efflux from leukocyte lysosomes was plotted against the initial cystine content of the loaded lysosome, heterozygotes were found to have a maximum velocity of cystine efflux approximately half of normal (Fig. 1).[2]

The initial difficulty in demonstrating cystine efflux from lysosomes obtained from cultured cells was found to be due to the requirement for ATP for this process.[3] In normal leukocyte lysosomes, the rate of cystine efflux was stimulated 2.5-fold by 1.25 M ATP.[2] In the cultured cell lysosomes (from Epstein–Barr virus-transformed lymphoblasts), cystine efflux from normal lysosomes required ATP, but was unresponsive to ATP in lysosomes from individuals with cystinosis (Fig. 2).[3] Lysosomes (from cultured lymphoblasts) appear to contain an ATPase which can be distinguished from mitochondrial and Na+, K+-ATPase on the basis of responses to various inhibitors.[3,109] This ATPase activity is necessary for the acidification of lysosomes in the presence of ATP and is present in cystinotic as well as normal lysosomes.[109] The pattern of response of this ATPase to various inhibitors is identical in cystinotic and normal lysosomes (Table I). Furthermore, inhibitors of this ATPase also inhibit cystine efflux from normal

lysosomes.[3,109] The proton translocator, carbonylcyanide-m-chlorophenylhydrazone (CCCP), does not inhibit ATPase activity, but does inhibit lysosomal cystine efflux. These data suggest that cystinosis is due to deficient activity of a carrier-mediated cystine transport system which is saturable, and is influenced by the pH across the lysosomal membrane.

5. Other Causes of the Fanconi Syndrome

The multiple causes of the renal Fanconi syndrome (Table I) have encouraged investigators to seek a unified concept to explain the tubular dysfunction which occurs.[110–112] Many of the agents which cause the renal Fanconi syndrome are known inhibitors of sulfhydryl-requiring enzymes. This is true of the heavy metals cadmium,[113] lead,[114–116] uranium,[115] mercury,[115] and copper in Wilson's disease,[117–119] maleic acid,[120] and cystine.[121–123] Stave and Schlaak produced a generalized aminoaciduria and glucosuria in rabbits by feeding them large amounts of cystine.[121,122] and Schwartz-Tiene et al. produced the full triad of aminoaciduria, glucosuria, and phosphaturia in rats fed large amounts of cystine.[123]

Renal tubular dysfunction has also resulted from the organic compounds methyl-3-chromone,[124] anhydro-4-epitetracycline,[125] Lysol (cresol),[126] and streptozotocin.[127] In tyrosinema, p-hydroxyphenyllactic acid and p-hydroxyphenylpyruvic acid accumulate[128,129] and are the likely causes of tubular damage. These compounds could each exert a toxic effect on the renal tubules by interaction with sulfhydryl-requiring enzymes, but other mechanisms may well apply.

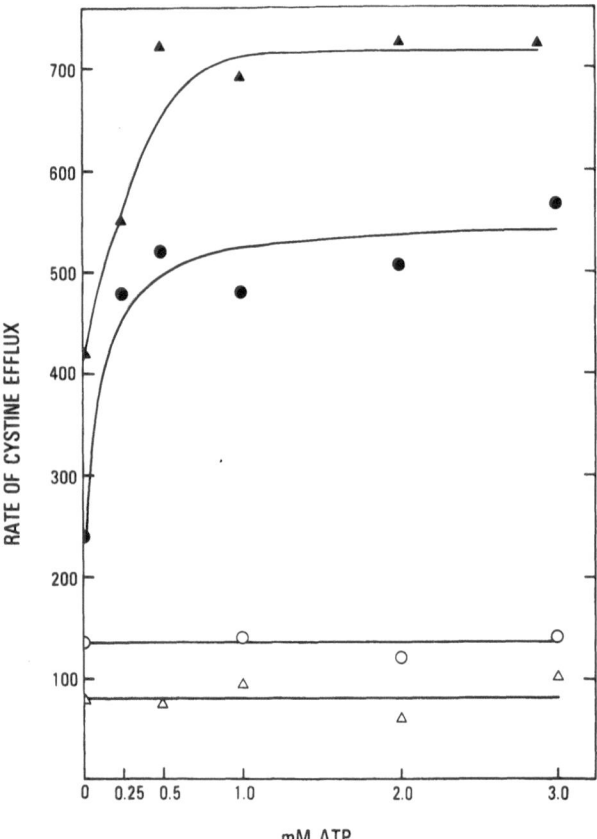

Fig. 2. Effect of ATP on the loss of cystine from granular fractions. Granular fractions were prepared from normal (●,▲) and cystinotic (○,△) cells that had been exposed to 1.0 and 0.03 mM cystine dimethyl ester, respectively. Duplicate granular fractions were incubated for 20 min at 37°C with different amounts of ATP/MgCl₂. The rate of cystine efflux was defined as the change in lysosomal cystine (nmoles/per unit of hexosaminidase activity) per minute × 100.[3]

In galactosemia and hereditary fructose intolerance, the toxic products presumably are galactose-1-phosphate and fructose-1-phosphate, respectively, but the process by which these products cause tubular damage is not understood. In Lowe's syndrome, idiopathic Fanconi syndrome, and glycogen storage disease, the toxic product is unknown. The events leading to renal tubular·damage in nonheritable disorders associated with the Fanconi syndrome are also poorly understood. Proteins filtered in excess through the glomerulus may lead to tubular damage in disorders such as nephrosis or myeloma,[130] but Shioji *et al.*[131] have reported a reversal of tubular dysfunction during continued proteinuria in a case of nephrosis. The Fanconi syndrome may also be the presenting symptom of multiple myeloma in the adult.[130,132]

6. Treatment

6.1. Symptomatic Treatment

The symptomatic treatment of the earlier stages of nephropathic cystinosis consists of providing an adequate fluid intake, correcting the metabolic acidosis and potassium deficit,

and healing the rickets. In most patients this is readily accomplished and results in a marked improvement in the child's personality, eating, activity, and "general well-being." As the disease progresses, treatment for renal insufficiency becomes the major concern. Patients with late-onset cystinosis often require treatment similar to that given patients with nephropathic cystinosis. Patients with benign cystinosis require no therapy.

The acidosis and hypokalemia are corrected by the use of a potassium-containing alkalinizing mixture such as a mixture of sodium and potassium citrate. In most patients, the rachitic changes are corrected with doses of 10,000 to 15,000 U/day of vitamin D (ergocalciferol, vitamin D₂). Some patients are more easily managed with equivalent doses of either dihydrotachysterol or 1,25-dihydroxycholecalciferol.

Cystinotic children may have extreme polyuria secondary to their renal tubular damage. Free access to water in large amounts is essential during both day and night. Vigorous intravenous fluid therapy is often necessary for management of viral or other illnesses causing vomiting or diarrhea. When receiving proper treatment, cystinotic children usually tolerate the typical infectious diseases of childhood without difficulty. It is important to monitor thyroid function in these children since many eventually become hypothyroid.

6.2. Specific Therapy

Attempts at specific therapy for nephropathic cystinosis have been difficult to evaluate for three reasons. First, no specific, objective parameter has been available for evaluating the therapy. Second, symptomatic therapy often causes temporary improvement, including improved glomerular function. Third, the natural course of the disease is not uniform, and some patients may go several years without a decrease in glomerular function in the absence of any specific therapy.

Ideally, a successful therapeutic regimen would rid these cystinotic children of their cystine stores. The amounts of free cystine in critical tissues should provide the best objective measurement of cystine accumulation and be helpful in evaluating therapy. Treatment might be effective without actually decreasing cystine storage, if it protected the actual site of cystine toxicity in the renal tubular and glomerular cells.

Evaluation of any form of therapy requires assessment of renal function over a prolonged period of time while the patient is symptomatically stable. Three approaches have been suggested for the therapy of this disease: specific drugs, diet low in methionine and cystine, and renal transplantation.

6.3. Specific Drugs

One of the first drugs suggested for the specific treatment of cystinosis was penicillamine (β,β-dimethylcysteine). Although several investigators thought it was helpful to cystinotic patients, two careful studies demonstrated no improvement with this drug.[29,151] A more recent study has shown that penicillamine (at 1.0 mM) does not deplete cystinotic fibroblasts of free cystine.[99] Another approach to treatment was suggested by the observation of Goldman *et al.*[98] that dithiothreitol was effective in lowering intracellular cystine in fibroblasts cultured from a patient with cystinosis.[152] This form of therapy must be evaluated more fully before it can be advocated for general use.

Another thiol compound has shown promising results *in vitro* and *in vivo*. Cysteamine at concentrations as low as 0.1 mM can remove, within 1 hr, all the free cystine from cultured cystinotic fibroblasts.[99] When tested for 1 hr at 1.0 mM, pH 7.4, 37°C, cysteamine removed nearly three times as much free cystine from cystinotic fibroblasts as did dithiothreitol.[99] This drug has now been used clinically in a number of patients with cystinosis.[99,153–158] The drug is available (for experimental use) as cysteamine-HCl. It is given orally in four divided doses for a total of about 60 mg (expressed as cysteamine free-base) per killigram body weight per day. The dose is monitored in each patient individually by maintaining the leukocyte free cystine content in the heterozygous range.[157] In patients who are slowly brought to full dose levels over a 3-month period, no toxicity has been observed.[158] The major drawback to this drug is its extremely objectionable smell and taste. Thoene and Lemons[156] have shown that a derivative of this drug, phosphocysteamine, is equally effective in depleting cystinotic cells of cystine, both in vitro and in vivo. Phosphocysteamine is odorless and tasteless and may have less toxicity than cysteamine.[156] It almost certainly is converted to cysteamine which is the presumed active agent. Unfortunately, it is available in only small amounts at present. If cysteamine proves effective for cystinosis, it is likely that phosphocysteamine will become more widely available for use.

In addition to the thiol-containing compounds, another compound with reducing properties, ascorbic acid, showed therapeutic potential in studies in vitro.[159] On the basis of this finding, a double-blind study was done to evaluate the use of high-dose ascorbic acid in cystinotic patients. Thirty-two patients received ascorbic acid (200 mg/kg body wt per day) and 32 received placebo. The study was terminated after about 2 years because there was no evidence that vitamin C was beneficial and accumulating evidence that it might be harmful. The serum creatinine concentration increased 0.53 mg/dl per year in patients receiving ascorbic acid and 0.24 mg/dl per year in patients receiving placebo ($p = 0.08$).[160]

6.4. Dietary Therapy

Based on the information known about cystinosis in the 1960s, it seemed reasonable to treat these children with a diet low in sulfur-containing amino acids. Such diets were derived either from natural proteins which were low in cystine and methionine or from mixtures of free amino acids. Investigators attcmptcd to give amounts of mcthioninc barcly adcquatc to maintain positive nitrogen balance, with the hope that this would decrease the cystine storage. The subsequent report[97] that protein turnover is a major source of the cystine storage in this disease would lead one to predict the futility of this approach. In fact, Seegmiller et al.[161] found that dietary therapy did not reduce intracellular cystine content and Bickel et al.[162] analyzed data from 21 patients so treated and found no evidence of clinical benefit.

6.5. Renal Transplantation

Since the primary abnormality leading to renal failure in cystinosis appears to be a genetically determined defect of the intracellular environment, cells without this genetic defect transplanted into a cystinotic patient should not accumulate cystine. Therefore, in theory, a cystinotic child should be as good a candidate for renal transplantation as any other child with chronic renal disease. In fact, reports from many centers indicate that cystinotic patients do as well with renal transplants as do other transplanted children.[7,163–166]

Cystinotic patients remain short after renal transplanation and their posttransplant growth rate is similar to that of other pediatric renal allograft recipients.[165] Anemia and renal osteodystrophy improve. Hepatosplenomegaly may or may not decrease after successful transplantation, but hepatic function does not deteriorate. Corneal crystal deposits and retinopathy tend to progress after transplantation. In the United States experience, visual function is altered minimally if at all,[31,165] and photophobia may worsen or improve.[31] The French experience with 16 cases over 10 years of age was worse with regard to visual function. One patient was totally blind, two had severe and six had mild impairment of vision.[7]

Several cystinotics are alive approximately a decade after renal transplantation. Most have done well. A recent report of extensive necrosis of the basal ganglia in a 19-year-old cystinotic patient who had received a renal transplant 9 years previously raises the possibility that CNS involvement may occur in older patients.[32] Puberty has occurred in the mid or late teens in several patients with whom the author (J.A.S.) is personally familiar. The transplanted kidneys do not develop the series of functional changes typical of cystinosis; they may, however, demonstrate reaccumulations of cystine which electron microscopic studies suggest are largely and perhaps entirely confined to interstitial and mesangial cells, presumably of host origin.[163,165–167] This phenomenon occurs in both heterozygous and cadaver donor kidneys. Despite theoretical concerns that heterozygote donor kidneys might be more prone than cadaver kidneys to cystine reaccumulation,[166] there is no evidence that, in practice, this is of real significance, while it is clearly important to have the best possible match between donor and recipient.

At present, sufficient favorable results from transplantation have accumulated to justify offering this procedure to affected patients in the terminal stages of nephropathic cystinosis, and to their families, with careful explanation of the prospects for success or failure.

7. Summary

Cystinosis is a recessively inherited metabolic disorder characterized biochemically by an abnormally high intracellular content of free cystine which results in cystine crystal deposition in the conjunctiva, bone marrow, lymph nodes, leukocytes, and internal organs. The metabolic error in this condition was recently found to be a defect in the carrier-mediated transport of cystine from the lysosome to the cytosol. There are at least three forms of this disease. The infantile nephropathic form has been the most thoroughly studied. Children with this form of cystinosis present in the first year of life with the renal tubular defects characteristic of the Fanconi syndrome and also have progressive renal glomerular damage which leads to end-stage kidney failure, usually before 10 years of age. Other patients have a completely benign form of cystinosis and are only discovered when an ophthalmological examination is done for another reason and reveals characteristic crystalline opacities in the cornea and conjunctiva. Finally, some patients have an intermediate type of cystinosis characterized by later onset of renal disease and renal failure than occurs in the infantile nephropathic variety.

A characteristic lesion in the peripheral retina, a patchy depigmentation, is usually found in the infantile nephropathic form, but never in the benign form. It has been noted in some, but not all patients with the intermediate form of cystinosis. Retinal lesions were detectable in a 5-week-old infant with nephropathic cystinosis as well as in an affected aborted fetus.

The renal defects in nephropathic cystinosis are probably caused by the large intracellular accumulation of free cystine. Cystine is known to inhibit many sulfhydryl-requiring enzymes, but the cystinotic cell has some protection because this amino acid is compartmentalized in these cells, and thus separated from other cellular enzymes. The intracellular site of the cystine storage is the lysosome.

The symptomatic management of patients with nephropathic cystinosis is usually not difficult but so far attempts at specific therapy have been unsuccessful. Renal transplantation has been lifesaving and is receiving more widespread acceptance. Cysteamine (mercaptoethylamine) depletes cystinotic cells of cystine and is now undergoing clinical trial.

Antenatal diagnosis is an established procedure for this condition. Affected fetuses have 20 to 100 times the usual amount of free cystine in most internal organs.

References

1. Gahl, W. A., F. Tietze, N. Bashan, R. Steinherz, and J. D. Schulman. 1982. Defective cystine exodus from isolated lysosome-rich fractions of cystinotic leukocytes. *J. Biol. Chem.* **257**:9570–9575.
2. Gahl, W. A., N. Bashan, F. Tietze, I. Bernardini, and J. D. Schulman. 1982. Cystine transport is defective in isolated leukocyte lysosomes from patients with cystinosis. *Science* **217**:1263–1265.
3. Jonas, A. J., M. Smith, and J. A. Schneider. 1982. ATP-dependent lysosomal cystine efflux is defective in cystinosis. *J. Biol. Chem.* **257**:13185–13188.
4. Schulman, J. D., ed. 1973. *Cystinosis.* DHEW Publication No. (NIH) 72-249. U.S. Government Printing Office, Washington, D.C.
5. Seegmiller, J. E. 1973. Cystinosis. In: *Lysosomes and Storage Diseases.* H. G. Hers and F. Van Hoof, eds. Academic Press, New York. pp. 485–518.
6. Schulman, J. D., S. H. Mudd, J. A. Schneider, S. P. Spielberg, L. Boxer, J. Oliver, L. Corash, and M. Shietz. 1980. Genetic disorders of glutathione and sulfur amino-acid metabolism: New biochemical insights and therapeutic approaches. *Ann. Intern. Med.* **93**:330–346.
7. Broyer, M., M. Guillot, M. C. Gubler, and R. Habib. 1981. Infantile cystinosis: A reappraisal of early and late symptoms. In: *Advances in Nephrology.* J. Hamburger, J. Crosnier, J.-P. Grunfeld, and M. H. Maxwell, eds. Year Book Medical Publishers, Chicago. pp. 137–166.
8. Schneider, J. A., and J. D. Schulman. 1982. Cystinosis and the Fanconi syndrome. in: *The Metabolic Basis of Inherited Disease.* J. B. Stanbury, J. B. Wyngaarden, D. S. Fredrickson, J. L. Goldstein, and M. S. Brown, eds. McGraw–Hill, New York. pp. 1844–1866.
9. Bickel, H. 1955. Die Entwicklung der biochemischen läsion bei der Lignac-Faconischen Krankheit. *Helv. Paediatr. Acta* **10**:259–268.
10. Schneider, J. A., V. Wong, and J. E. Seegmiller. 1969. The early diagnosis of cystinosis. *J. Pediatr.* **74**:114–116.
11. Lebel, M., J. H. Grose, E. Delage, and G. Crepin. 1977. Syndrome de Bartter associé a une cystinose: Association des Médicine de la Langue Francaise du Canada. *Congres Annuel.*
12. Lemire, J., B. S. Kaplan, and C. R. Scriver. 1978. Presentation of cystinosis as Bartter's syndrome and conversion to Fanconi syndrome on indomethacin treatment. *Pediatr. Res.* **12**:544.
13. O'Regan, S., J.-G. Mongeau, and P. Robitaille. 1980. A patient with cystinosis presenting with the features of Barrter syndrome. *Acta Paediatr. Belg.* **33**:51–52.
14. Lucky, A. W., P. M. Howley, K. Megylesi, S. P. Spielberg, and J. D. Schulman. 1977. Endocrine studies in cystinosis: Compensated primary hypothyroidism. *J. Pediatr.* **91**:204–210.
15. Chan, A. M., M. J. G. Lynch, J. D. Bailey, C. Ezrin, and D. Fraser. 1970. Hypothyroidism in cystinosis. *Am. J. Med.* **48**:678–692.
16. Burke, J. R., M. M. El-Bishti, M. N. Maisey, and C. Chantler. 1978. Hypothyroidism in children with cystinosis. *Arch. Dis. Child.* **53**:947–951.
17. Czernichow, P., G. Lenoir, M. P. Roy, R. Rappaport, and M. Broyer. 1978. Atteintes thyroidiennes au cours de la cystinose. *Arch. Fr. Pediatr.* **35**:930–938.
18. Bercu, B. B., S. Orloff, and J. D. Schulman. 1980. Partial pituitary resistance to thyroid hormone in cystinosis. *J. Clin. Endocrinol. Metab.* **51**:1262–1268.
19. Ehrich, J. H. H., L. Stoeppler, G. Offner, and J. Brodehl. 1979. Evidence for cerebral involvement in nephropathic cystinosis. *Neuropaediatrie* **10**:128–137.
20. Ehrich, J. H. H., G. Wolff, L. Stoeppler, R. Heyer, G. Offner, and J. Brodehl. 1979. Psychosozial-intellektuelle entwicklung bei kindern mit infantiler zystinose und hirnatrophie. *Klin. Paediatr.* **191**:483–492.
21. Jonas, A. J., and J. A. Schneider. 1982. Cystinosis in a black child. *J. Pediatr.* **100**:934–935.
22. Jonas, A. J., J. D. Schulman, R. Matalon, A. Velazquez, E. D. Brewer, H. Chen, M. G. Boyer, E. Brandwein, G. S. Arbus, C. R. Morris, and J. A. Schneider. 1982. Cystinosis in noncaucasian children. *Johns Hopkins Med. J.* **151**:117–121.
23. Bürki, V. E. 1941. Ueber die cystinkrankheit im klienkindesalter unter besonderer berucksichtigung des augenbefundes. *Ophthalmologica* **101**:257–272.
24. Frazier, P. D., and V. G. Wong. 1968. Cystinosis: Histologic and crystallographic examination of crystals in eye tissues. *Arch. Ophthalmol.* **80**:87–91.
25. Wong, V. G. 1973. The eye and cystinosis. In: *Cystinosis.* J. D. Schulman, ed. DHEW Publication No. (NIH) 72-249. U.S. Government Printing Office, Washington, D.C. pp. 23–35.
26. Wong, V. G., P. S. Lietman, and J. E. Seegmiller. 1967. Alterations of pigment epithelium in cystinosis. *Arch. Ophthalmol.* **77**:361–369.
27. Waldman, T. A., R. P. Mogielnicki, and W. Strober. 1973. The proteinuria of cystinosis: Its pattern and pathogenesis. In: *Cystinosis.* J. D. Schulman, ed. DHEW Publication No. (NIH) 72-249. U.S. Government Printing Office, Washington, D.C. pp. 55–66.
28. Clayton, B. E., and A. D. Patrick. 1961. Use of dimercaprol or penicillamine in the treatment of cystinosis. *Lancet* **2**:909–910.
29. Crawhall, J. C., P. S. Lietman, J. A. Schneider, and J.E. Seegmiller. 1968. Cystinosis: Plasma cystine and cysteine concentrations and the effect of D-penicillamine and dietary treatment. *Am. J. Med.* **44**:330–339.
30. Schulman, J. D. Unpublished observations.
31. Yamamoto, G. K., J. D. Schulman, J. A. Schneider, and V. G. Wong. 1979. Long-term ocular changes in cystinosis: Observations in renal transplant recipients. *J. Pediatr. Ophthalmol. Strobis.* **16**:21–25.
32. Levine, S., and G. Paparo. 1982. Brain lesions in a case of cystinosis. *Acta Neuropathol.* **57**:217–220.
33. Lietman, P. S., P. D. Frazier, V. G. Wong, D. Shotton, and J. E. Seegmiller. 1966. Adult cystinosis—A benign disorder. *Am. J. Med.* **40**:511–517.
34. Brubaker, R. F., V. G. Wong, J. D. Schulman, J. E. Seegmiller, and T. Kuwabara. 1970. Benign cystinosis: The clinical, biochemical and morphologic findings in a family with two affected siblings. *Am. J. Med.* **49**:546–550.
35. Schneider, J. A., V. Wong, K. H. Bradley, and J. E. Seegmiller. 1968. Biochemical comparisons of the adult and childhood forms of cystinosis. *N. Engl. J. Med.* **279**:1253–1257.

36. Kraus, E., and P. Lutz. 1971. Ocular cystine deposits in an adult. *Arch. Ophthalmol.* **85**:690–694.

37. Kroll, W. A., and K.-H. Lichte. 1973. Cystinosis: A review of the different forms and of recent advances. *Humangenetik* **20**:75–87.

38. Dodd, M. G., S. M. Pusin, and W. R. Green. 1978. Adult cystinosis: A case report. *Arch. Ophthalmol.* **96**:1054–1057.

39. Goldman, H., C. R. Scriver, K. Aaron, E. Delvin, and Z. Canlos. 1971. Adolescent cystinosis: Comparisons with infantile and adult forms. *Pediatrics* **47**:979–988.

40. Aaron, K., H. Goldman, and C. R. Scriver. 1971. Cystinosis; new observation: 1. Adolescent (type III) form. 2. Correction of phenotypes in vitro with dithiothreitol. In: *Inherited Disorders of Sulphur Metabolism.* N. A. J. Carson and D. N. Raine, eds. Churchill/Livingstone, Edinburgh. pp. 150–161.

41. Hooft, C., D. Carton, F. de Schrijver, M. H. Delbeke, W. Samijn, and J. Kint. 1971. Juvenile cystinosis in two siblings. In: *Inherited Disorders of Sulphur Metabolism.* N. A. J. Carson and D. N. Raine, eds. Churchill/Livingstone, Edinburgh. pp. 141–149.

42. Spear, G. S., R. J. Slusser, J. D. Schulman, and F. Alexander. 1971. Polykaryocytosis in the visceral glomerular epithelium in cystinosis with description of an unusual clinical variant. *Johns Hopkins Med. J.* **129**:83–99.

43. Pittman, G., S. Deodhar, J. D. Schulman, and J. E. Lando. 1971. Nephropathic cystinosis in a young adult—Report of a case. *Lab. Invest.* **24**:442–443.

44. Francois, J., M. Hanssens, R. Coppieters, and L. Evens. 1972. Cystinosis: A clinical and histopathologic study. *Am. J. Ophthalmol.* **73**:643–650.

45. Adams, B. K., and P. M. Naidoo. 1978. Cystinosis with Fanconi syndrome: A case report. *S. Afr. Med. Tydskr.* **53**:719–720.

46. Blanc-Brunat, N., F. Berthoux, S. Colon, and G. Janin. 1978. Cystinose a revelation tardive chez deux freres. *Arch. Fr. Pediatr.* **35**:486–503.

47. Weber, H.-P., E. Harms, and G. Knopfle. 1979. Adoleszenten-zystinose literatürübersicht mit eigener beobachtung. *Klin. Paediatr.* **191**:8–19.

48. Volcker, H. E., E. Harms, G. O. H. Naumann, and G. Weiss. 1979. Cystinosis in two non-infantile patients diagnosed by the ophthalmologist. *Metab. Pediatr. Ophthalmol.* **3**:161–165.

49. Manz, F., E. Harms, P. Lutz, R. Waldherr, and K. Scharer. 1982. Adolescent cystinosis: Renal function and morphology. *Eur. J. Pediatr.* **138**:354–357.

50. Schneider. J. A. 1983. Case report in preparation.

51. Bois, E., P. Feingold, P. Frenay, and M. L. Briard. 1976. Infantile cystinosis in France: Genetics, incidence, geographic distribution. *J. Med. Genet.* **13**:434–438.

52. Spear, G. S. 1973. The pathology of the kidney. In: *Cystinosis.* J. D. Schulman, ed. DHEW Publication No. (NIH) 72-249. U.S. Government Printing Office, Washington, D.C. pp. 37–53.

53. Spear, G. S. 1974. Pathology of the kidney in cystinosis. In: *Pathology Annual.* S. C. Sommers, ed. Appleton–Century–Crofts, New York. pp. 81–92.

54. Patrick, A. D. 1962. The degradative metabolism of L-cysteine and L-cystine *in vitro* by liver in cystinosis. *Biochem. J.* **83**:248–256.

55. Schneider, J. A., K. Bradley, and J. E. Seegmiller. 1967. Increased cystine in leukocytes from individuals homozygous and heterozygous for cystinosis. *Science* **157**:1321–1322.

56. Schneider, J. A., F. M. Rosenbloom, K. H. Bradley, and J. E. Seegmiller. 1967. Increased free-cystine content of fibroblasts cultured from patients with cystinosis. *Biochem. Biophys. Res. Commun.* **29**:527–531.

57. Schulman, J. D., K. H. Bradley, I. K. Berezesky, P. M. Grimley, W. E. Dodson, and M. S. Al-Aish. 1971. Biochemical, morphologic, and cytogenetic studies from individuals and from patients with cystinosis. *Pediatr. Res.* **5**:501–510.

58. Hummeler, K., B. A. Zajac, M. Genel, P. G. Holtzapple, and S. Segal. 1970. Human cystinosis: Intracellular deposition of cystine. *Science* **168**:859–860.

59. Korn, D. 1960. Demonstration of cystine crystals in peripheral white blood cells in a patient with cystinosis. *N. Engl. J. Med.* **262**:545–548.

60. Schulman, J. D., J. A. Schneider, K. Bradley, and J. E. Seegmiller. 1970. Heterozygote studies in cystinosis. *Clin. Chim. Acta* **29**:73–76.

61. Morecki, R., L. Paunier, and J. E. Hamilton. 1968. Intestinal mucosa in cystinosis: A fine structure study. *Arch. Pathol.* **86**:297–307.

62. Patrick, A. D., and B. D. Lake. 1968. Cystinosis: Electron microscopic evidence of lysosomal storage of cystine in lymph node. *J. Clin. Pathol.* **21**:571–575.

63. Wong, V. G., T. Kewabara, R. Brubaker, W. Olson, J. D. Schulman, and J. E. Seegmiller. 1970. Intralysosomal cystine crystals in cystinosis. *Invest. Ophthalmol.* **9**:83–88.

64. Verougstraete, C., J. Libert, and D. Toussaint. 1978. Cystinose de l'adolescence-etude clinique et histopathologique. *Bull. Soc. Belge Ophthalmol.* **180**:9–20.

65. Schulman, J. D., V. Wong, W. H. Olson, and J. E. Seegmiller. 1970. Lysosomal site of crystalline deposits in cystinosis as shown by ferritin uptake. *Arch. Pathol.* **90**:259–264.

66. Kenyon, E. R., and J. A. Sensenbrenner. 1974. Electron microscopy of cornea and conjunctiva in childhood cystinosis. *Am. J. Ophthalmol.* **78**:68–76.

67. Scotto, J. M., and H. G. Stralin. 1977. Ultrastructure of the liver in a case of childhood cystinosis. *Virchows Arch. A Pathol. Anat. Histol.* **377**:43–48.

68. Haynes, M.D., R. F. Carter, A. C. Pollard, and W. F. Carey. 1980. Light and electron microscopy of infantile and fetal tissues in cystinosis. *Micron* **11**:443–444.

69. Schulman, J. D., K. H. Bradley, and J. E. Seegmiller. 1969. Cystine: Compartmentalization within lysosomes in cystinotic leukocytes. *Science* **166**:1152–1154.

70. Schulman, J. D., and K. H. Bradley. 1970. Cystinosis: Selective induction of vacuolation in fibroblasts by L-cysteine-D-penicillamine disulfide. *Science* **169**:595–597.

71. Schulman, J. D., and K. H. Bradley. 1970. Metabolism of amino acids, peptides and disulfides in the lysosomes of fibroblasts cultured from normal individuals and those with cystinosis. *J. Exp. Med.* **132**:1090–1104.

72. Harms, E., and J. A. Schneider. 1979. The lysosomal localization of free-cystine in normal and cystinotic cells. *Clin. Res.* **27**:457A.

73. Ehrenreich, B. A., and Z. A. Cohn. 1969. The fate of peptides pinocytosed by macrophages *in vitro. J. Exp. Med.* **129**:227–245.

74. Oshima, R. G., R. C. Willis, C. F. Furlong, and J. A. Schneider. 1974. Binding assays for amino acids: The utilization of a cystine binding protein from *Escherichia coli* for the determination of acid-soluble cystine in small physiological samples. *J. Biol. Chem.* **249**:6033–6039.

75. Rose, W. C., and R. L. Wixom. 1955. The amino acid requirements of man. XIII. The sparing effect of cystine on the methionine requirement. *J. Biol. Chem.* **216**:763–773.

76. McKean, C.M. 1970. Growth of phenylketonuric children on chemically defined diets. *Lancet* **1**:148–149.

77. Eagle, H., K. A. Piez, and V. I. Oyama. 1961. The biosynthesis of cystine in human cell cultures. *J. Biol. Chem.* **236**:1425–1428.

78. Eagle, H., C. Washington, and S. M. Friedman. 1966. The synthesis of homocystine, cystathionine, and cystine by cultured diploid and heteroploid human cells. *Proc. Natl. Acad. Sci. USA* **56**:156–163.

79. Jacoby, L. B., and J. W. Littlefield. 1971. Mutant human fibroblast clones able to grow on homocysteine instead of cyst(e)ine. *Exp. Cell Res.* **69**:447–449.

80. Brenton, D. P., D. C. Cusworth, D. E. Dent, and E. E. Jones. 1966. Homocystinuria: Clinical and dietary studies. *Q. J. Med.* **35**:325–346.

81. Sturman, J. A., G. Gaull, and N. C. R. Raiha. 1970. Absence of cystathionase in human fetal liver: Is cystine essential? *Science* **169**:74–76.

82. Jacobsen, J. D., and L. H. Smith, Jr. 1968. Biochemistry and

physiology of taurine and taurine derivatives. *Physiol. Rev.* **48**:424–511.

83. Meister, A. 1965. *Biochemistry of the Amino Acids*, 2nd ed. Academic Press, New York.

84. Barr, H. S. 1951. Pathologie des Aminosauren-Diabetes. *Monatsschr. Kinderheilkd.* **99**:35.

85. Schneider, J. A. Unpublished observations.

86. Schulman, J. D., V. G. Wong, T. Kuwabara, K. H. Bradley, and J. E. Seegmiller. 1970. Intracellular cystine content of leukocyte populations in cystinosis. *Arch. Intern. Med.* **125**:660–664.

87. Crawhall, J. C., and S. Segal. 1966. The intracellular cysteine/cystine ratio in kidney cortex. *Biochem. J.* **99**:19c–20c.

88. Crawhall, J. C., and S. Segal. 1967. The intracellular ratio of cysteine and cystine in various tissue. *Biochem. J.* **105**:891–896.

89. Rosenberg, L. E., J. C. Crawhall, and S. Segal. 1967. Intestinal transport of cystine and cysteine in man: Evidence for separate mechanisms. *J. Clin. Invest.* **46**:30–34.

90. Schulman, J. D., J. A. Schneider, K. Bradley, and J. E. Seegmiller. 1971. Cystine, cysteine, and glutathione metabolism in normal and cystinotic fibroblasts *in vitro* and in cultured amniotic fluid cells. *Clin. Chim. Acta* **35**:383–388.

91. Crawhall, J. C., R. G. Oshima, and J. A. Schneider. 1977. Factors controlling the nonprotein cystine content of cystinotic fibroblasts. *Pediatr. Res.* **11**:41–45.

92. Oshima, R. G., O. L. Pellett, J. A. Robb, and J. A. Schneider. 1977. Transformation of human cystinotic fibroblasts by SV40: Characteristics of transformed cells with limited and unlimited growth potential. *J. Cell. Physiol.* **93**:129–136.

93. Steinherz, R., F. Tietze, T. Triche, A. Modesti, W. A. Gahl, and J. D. Schulman. 1982. Heterozygote detection in cystinosis, using leukocytes exposed to cystine dimethyl ester. *N. Engl. J. Med.* **306**:1468–1470.

94. Bowman, H., and J. A. Schneider. 1981. Prenatal diagnosis of nephropathic cystinosis: Pregnancy at risk ascertained through heterozygote diagnosis of parents. *Acta Paediatr Scand.* **70**:389–393.

95. Schulman, J. D., and K. H. Bradley. 1971. Cystinosis: Therapeutic implications of *in vitro* studies of cultured fibroblasts. *J. Pediatr.* **78**:833–836.

96. Oshima, R. G., W. J. Rhead, J. G. Thoene, and J. A. Schneider. 1976. Cystine metabolism in human fibroblasts: Comparison of normal, cystinotic and gamma-glutamylcysteine synthetase-deficient cells. *J. Biol. Chem.* **251**:4287–4293.

97. Thoene, J. G., D. G. Ritchie, R. G. Oshima, and J. A. Schneider. 1977. Cystinotic fibroblasts accumulate cystine from intracellular protein degradation. *Proc. Natl. Acad. Sci. USA* **74**:4505–4507.

98. Goldman, H., C. R. Scriver, K. Aaron, and L. Pinsky. 1970. Use of dithiothreitol to correct cystine storage in cultured cystinotic fibroblasts. *Lancet* **1**:811–812.

99. Thoene, J. G., R. G. Oshima, J. C. Crawhall, D. L. Olson, and J. A. Schneider. 1976. Cystinosis: Intracellular cystine depletion by aminothiols *in vitro* and *in vivo*. *J. Clin. Invest.* **58**:180–189.

100. Thoene, J. G., and R. Lemons. 1980. Modulation of the intracellular cystine content of cystinotic fibroblasts by extracellular albumin. *Pediatr. Res.* **14**:785–787.

101. States, B., J. Lee, and S. Segal. 1981. Uptake of cystine by cystine-depleted fibroblasts from patients with cystinosis. *Biochem. Biophys. Res. Commun.* **98**:290–296.

102. Segal, S. 1965. Disorders of amino acid transport. *Ann. Intern. Med.* **62**:847–851.

103. Schneider, J. A., K. Bradley, and J. E. Seegmiller. 1968. Transport and intracellular fate of cysteine-35S in leukocytes from normal subjects and patients with cystinosis. *Pediatr. Res.* **2**:441–450.

104. Reijngoud, D. J., and J. M. Tager. 1977. The permeability properties of the lysosomal membrane. *Biochim. Biophys. Acta* **472**:419–449.

105. Reeves, J. P. 1979. Accumulation of amino acids by lysosomes incubated with amino acid methyl esters. *J. Biol. Chem.* **254**:8914–8921.

106. Steinherz, R., F. Tietze, D. Raiford, W. A. Gahl, and J. D. Schulman. 1982. Patterns of amino acid efflux from isolated normal and cystinotic human leukocyte lysosomes. *J. Biol. Chem.* **257**:6041–6049.

107. Steinherz, R., F. Tietze, W. A. Gahl, T. J. Triche, H. Chiang, A. Modesti, and J. D. Schulman. 1982. Cystine accumulation and clearance by normal and cystinotic leukocytes exposed to cystine dimethyl ester. *Proc. Natl. Acad. Sci. USA* **79**:4446–4450.

108. Jonas, A. J., A. A. Greene, M. L. Smith, and J. A. Schneider. 1982. Cystine accumulation and loss in normal, heterozygous and cystinotic fibroblasts. *Proc. Natl. Acad. Sci. USA* **79**:4442–4445.

109. Jonas, A. J., M. L. Smith, W. S. Allison, P. K. Laikind, A. A. Greene, and J. A. Schneider. 1983. Protein translocating ATPase and lysosomal cystine transport. *J. Biol. Chem.* **258**:11727–11730.

110. Harrison, H. E. 1958. The Fanconi syndrome. *J. Chronic Dis.* **7**:346–355.

111. Schneider, J. A., and J. E. Seegmiller. 1972. Cystinosis and the Fanconi syndrome. In: *The Metabolic Basis of Inherited Disease*, 3rd ed. J. B. Stanbury, J. B. Wyngaarden, and D. S. Fredrickson, eds. McGraw–Hill, New York. pp. 1581–1604.

112. Morris, R. C., Jr., R. R. McInnes, C. J. Epstein, A. Sebastian, and C. R. Scriver. 1976. Genetic and metabolic injury of the kidney. In: *The Kidney*. B. M. Brenner and F. C. Rector, eds. Saunders, Philadelphia. pp. 1193–1256.

113. Nicaud, P., A. Lafitte, A. Gros, and J.-P. Guatier. 1942. Les lésions osseuses de l'intoxication chronique par le cadmium. Aspects radiologique à type de syndrome de milkman. Efficacité du traitement calcique et vitaminique (Vitamin D). *Bull. Mem. Soc. Med. Hop. (Paris)* **18**:204.

114. Wilson, V. K., M. L. Thomson, and C. E. Dent. 1953. Aminoaciduria in lead poisoning: A case in childhood. *Lancet* **2**:66–68.

115. Clarkson, T. W., and J. E. Kench. 1956. Urinary excretion of amino acids by men absorbing heavy metals. *Biochem. J.* **62**:361–372.

116. Chisolm, J. J., Jr. 1962. Aminoaciduria as a manifestation of renal tubular injury in lead intoxication and a comparison with patterns of aminoaciduria seen in other diseases. *J. Pediatr.* **60**:1–17.

117. Uzman, L. L., and D. Denny-Brown. 1948. Amino-aciduria in hepatolenticular degeneration (Wilson's disease). *Am. J. Med. Sci.* **215**:599–611.

118. Cooper, A. M., R. D. Eckhardt, W. W. Faloon, and C. S. Davidson. 1950. Investigation of the aminoacidura in Wilson's disease (hepatolenticular degeneration): Demonstration of a defect in renal function. *J. Clin. Invest.* **29**:265–278.

119. Bearn, A. G., T. F. Yu, and A. B. Gutman. 1957. Renal function in Wilson's disease. *J. Clin. Invest.* **36**:1107–1114.

120. Harrison, H. E., and H. C. Harrison. 1954. Experimental production of renal glycosuria, phosphaturia and aminoaciduria by injection of maleic acid. *Science* **120**:606–608.

121. Stave, U., and E. Schlaak. 1956. Aminosäuren-Verfütterung und tubulusschaden. I. Mitteilung. Aminoacidurie nach cystine-verfütterung bei kaninchen. *Z. Kinderheilkd.* **78**:261–274.

122. Stave, U. 1956. Aminosäuren-Verfütterung und tubulusschaden. II. Mitteilung. Die tubuläre funktion nach cystine-verfütterung bei kaninchen. *Z. Kinderheilkd.* **78**:275–282.

123. Schwartz-Tiene. E., P. Careddu, and N. Cabassa. 1957. Alterazioni funzionali e anatomiche renali nella intossicazione sperimentale da cistina. *Minerva Pediatr.* **9**:231–238.

124. Otten, J., and H. L. Vis. 1968. Acute reversible renal tubular dysfunction following intoxication with methyl-3-chromone. *J. Pediatr.* **73**:422–425.

125. Benitz, K.-F., and H. F. Diermeier. 1964. Renal toxicity of tetracycline degradation products. *Proc. Soc. Exp. Biol. Med.* **115**:930–935.

126. Spencer, A. G., and G. T. Franglen. 1952. Gross aminoaciduria following a Lysol burn. *Lancet* **1**:190–192.

127. Sadoff, L. 1970. Nephrotoxicity of streptozotocin. *Cancer Chemother. Rep.* **54**:457–459.

128. Gentz, J., R. Jagenburg, and R. Zetterstrom. 1965. Tyrosinemia: An inborn error of tyrosine metabolism with cirrhosis of the liver and multiple renal tubular defects (de Toni–Debré–Fanconi syndrome). *J. Pediatr.* **66**:670–696.

129. Buist, N. R. M., N. G. Kennaway, and J. H. Fellman. 1974. Disorders of tyrosine metabolism. In: *Heritable Disorders of Amino Acid Metabolism.* W. L. Nuhan, ed. Wiley, New York. pp. 160–176.

130. Maldonado, J. E., J. A. Velosa, R. A. Kyle, R. D. Wagoner, K. E. Holley, and R. M. Salassa. 1975. Fanconi syndrome in adults: A manifestation of a latent form of myeloma. *Am. J. Med.* **58**:354–364.

131. Shioji, R., Y. Sasaki, H. Saito, and T. Furuyama. 1974. Reversible tubular dysfunction associated with chronic renal failure in an adult patient with the nephrotic syndrome. *Clin. Nephrol.* **2**:76–80.

132. Smithline, N., J. P. Kassirer, and J. J. Cohen. 1976. Light-chain nephropathy: Tubular dysfunction and light-chain proteinuria. *N. Engl. J. Med.* **294**:71–74.

133. Hunt, D. D., G. Stearns, J. B. McKinely, E. Froning, P. Hicks, and M. Bonfiglio. 1966. Long-term study of family with Fanconi syndrome without cystinosis (deToni–Debré–Fanconi syndrome). *Am. J. Med.* **40**:492–510.

134. Smith, R., R. H. Lindenbaum, and R. J. Walton. 1976. Hypophosphataemic osteomalacia and Fanconi syndrome of adult onset with dominant inheritance. *Q. J. Med.* **179**:387–400.

135. Lowe, C. U., M. Terrey, and E. A. MacLachlan. 1952. Organicaciduria, decreased renal ammonia production, hydrophthalmos, and mental retardation; clinical entity. *Am. J. Dis. Child.* **83**:164–184.

136. Burke, E. C., K. E. Holley, and G. B. Stickler. 1973. Familial nephrotic syndrome with nephrocalcinosis and tubular dysfunction. *J. Pediatr.* **82**:202–206.

137. Segal, S. 1983. Disorders of galactose metabolism. In: *The Metabolic Basis of Inherited Disease.* J. B. Stanbury, J. B. Wyngaarden, D. S. Fredrickson, D. L. Goldstein, and M. S. Brown, eds. McGraw–Hill, New York. pp. 167–191.

138. Garty, R., and E. Tabachnik. 1974. The Fanconi syndrome associated with hepatic glycogenosis and abnormal metabolism of galactose. *J. Pediatr.* **85**:821–833.

139. Morris, R. C., Jr. 1968. An experimental renal acidification defect in patients with hereditary fructose intolerance. II. Its distinction from classic renal acidification defect associated with the Fanconi syndrome of children with cystinosis. *J. Clin. Invest.* **47**:1648–1663.

140. Bearn, A. G. 1972. Wilson's disease. In: *The Metabolic Basis of Inherited Disease.* J. B. Stanbury, J. B. Wyngaarden, and D. S. Fredrickson, eds. McGraw–Hill, New York. pp. 1033–1084.

141. Massry, S. G., H. G. Preuss, J. F. Maher, and G. E. Schreiner. 1967. Renal tubular acidosis after cadaver kidney homotransplantation. *Am. J. Med.* **42**:284–292.

142. Wilson, D. R., and A. A. Siddiqui. 1973. Renal tubular acidosis after kidney transplantation: Natural history and significance. *Ann. Intern. Med.* **79**:352–361.

143. Vertuno, L. L., H. G. Preuss, W. P. Argy, Jr., and G. E. Schreiner. 1974. Fanconi syndrome following homotransplantation. *Arch. Intern. Med.* **133**:302–305.

144. Harrison, J. F., and J. D. Blainey. 1967. Adult Fanconi syndrome with monoclonal abnormality of immunoglobulin light chains. *J. Clin. Pathol.* **20**:42–48.

145. Sebastian, A., E. McSherry, I. Veki, and R. C. Morris, Jr. 1968. Renal amyloidosis, nephrotic syndrome and impaired renal tubular reabsorption of bicarbonate. *Ann. Intern. Med.* **69**:541–548.

146. Kamm, D. E., and M. S. Fischer. 1972. Proximal renal tubular acidosis and the Fanconi syndrome in a patient with hypergammaglobulinemia. *Nephron.* **9**:208–219.

147. Ghozlan, R., B. Amor, and F. Belbarre. 1974. Syndrome de Sjogren avec acidose tubulaire; amino-acidurie et hyperglobulinemie. *Nouv. Presse Med.* **3**:1711–1713.

148. Chesney, R. W., and H. E. Harrison. 1975. Fanconi syndrome following bowel surgery and hepatitis reversed by 25-hydroxycholecalciferol. *J. Pediatr.* **86**:857–861.

149. Guiganar, J. P., and A. Torrado. 1973. Proximal tubular acidosis in vitamin D deficient rickets. *Acta Paediatr. Scand.* **62**:543–546.

150. Rützler, L. 1973. Passageres tubuläres syndrom durch vergiftung mit einer organischen quecksilber-verbindung (Glyceromerfen) bei einem 1-1/2 jährigen mächen. *Schweiz. Med. Wochenschr.* **103**:678–681.

151. Hambreaus, L., and O. Broberger. 1967. Penicillamine treatment of cystinosis. *Acta Paediatr. Scand.* **56**:243–248.

152. Dapape-Brigger, D., H. Goldman, C. R. Scriver, E. Delvin, and O. Mamer. 1977. The *in vivo* use of dithiothreitol in cystinosis. *Pediatr. Res.* **11**:124–131.

153. Roy, L. P., and A. C. Pollard. 1978. Cysteamine therapy for cystinosis. *Lancet* **2**:729–730.

154. Girardin. E. P., M. J. DeWolfe, and J. F. S. Crocker. 1979. Treatment of cystinosis with cysteamine. *J. Pediatr.* **94**:838–840.

155. Yudkoff, M., J. W. Foreman, and S. Segal. 1981. Effects of cysteamine therapy in nephropathic cystinosis. *N. Engl. J. Med.* **304**:141–145.

156. Thoene, J. G., and R. Lemons. 1980. Cystine depletion of cystinotic tissues by phosphocysteamine (WR638). *J. Pediatr.* **96**:1043–1044.

157. Schneider, J. A., J. D. Schulman, and J. G. Thoene. 1981. Cysteamine therapy in cystinosis. *N. Engl. J. Med.* **304**:1172.

158. Corden, B. J., J. D. Schulman, J. A. Schneider, and J. G. Thoene. 1981. Adverse reactions to oral cysteamine use in nephropathic cystinosis. *Dev. Pharmacol. Ther.* **3**:25–30.

159. Kroll, W. A., and J. A. Schneider. 1974. Decrease in free-cystine content of cultured cystinotic fibroblasts by ascorbic acid. *Science* **186**:1040–1042.

160. Schneider, J. A., J. J. Schlesselman, S. A. Mendoza, S. Orloff, J. G. Thoene, W. A. Kroll, A. D. Godfrey, and J. D. Schulman. 1979. Ineffectiveness of ascorbic acid therapy in nephropathic cystinosis. *N. Engl. J. Med.* **300**:756–759.

161. Seegmiller, J. E., T. Friedman, H. E. Harrison, V. Wong, and J. A. Schneider. 1968. Cystinosis. *Ann. Intern. Med.* **68**:883–905.

162. Bickel, H., P. Lutz, and H. Schmidt. 1973. The treatment of cystinosis with diet or drug. In: *Cystinosis.* J. D. Schulman, ed. DHEW Publication No. (NIH) 72-249. U.S. Government Printing Office, Washington, D.C. pp. 199–223.

163. Goodman, S. I., K. M. Hambidge, C. P. Mahoney, and G. E. Striker. 1973. Renal homotransplantation in the treatment of cystinosis. In: *Cystinosis.* J. D. Schulman. ed. DHEW Publication No. (NIH) 72-249. U.S. Government Printing Office, Washington, D.C. pp. 225–232.

164. Advisory Committee to the Renal Transplant Registry. 1975. Renal transplantation in congenital and metabolic diseases. A report from the ASC/NIH Renal Transplant Registry. *J. Am. Med. Assoc.* **232**:148–153.

165. Malakzadeh, M. H., H. B. Neustein, J. A. Schneider, A. J. Pennisi, R. B. Ettenger, C. H. Vittenbogaart, M. D. Kogut, and R. N. Fine. 1977. Cadaver renal transplantation in children with cystinosis. *Am. J. Med.* **63**:525–533.

166. West, J. C., S. I. Goodman, G. P. Schroter, P. A. Bloustein, K. M. Hambidge, and R. Weil. 1977. Pediatric kidney transplantation for cystinosis. *J. Pediatr. Surg.* **12**:651–655.

167. Mahoney, C. P., G. E. Striker, R. O. Hickman, G. B. Manning, and T. L. Marchiord. 1970. Renal transplantation for childhood cystinosis. *N. Engl. J. Med.* **283**:397–402.

Renal Tubular Defects in Phosphate and Amino Acid Transport

Ralph A. DeFronzo and Samuel O. Thier

1. Introduction

Although not commonly encountered in clinical practice, disorders of phosphate and amino acid transport are of major pathophysiological importance. Studies of these disorders have provided important insights into normal transport physiology and have served as a model of the way basic science principles can be used to explain clinical disease, and how clinical disease can provide new physiological principles. In addition, these disorders have provided much information on the genetics of transport, which forms the basis for sound genetic counseling.

In this chapter we will update recent advances in our knowledge of the disorders of phosphate and amino acid transport. Emphasis will be placed on the mechanisms responsible for the disease, the clinical manifestations of the pathophysiology, the patterns of inheritance, and therapy.

2. Defects in Phosphate Transport Processes

Defects in renal phosphate transport, uncommon in clinical medicine, have been encountered in five settings; (1) vitamin D-resistant rickets; (2) the Fanconi syndrome; (3) pseudohypoparathyroidism; (4) in association with certain mesenchymal tumors; and (5) in association with various endocrinopathies. In order to understand these clinical disorders of phosphate transport better, a brief review of the renal handling of phosphate is warranted.

2.1. Renal Handling of Phosphate

Classical clearance techniques in dog and in man have shown that at physiologic plasma phosphate levels, the phosphate clearance is substantially less than the glomerular filtration rate (GFR), with the fractional excretion of phosphate in the 10–15% range.[1] This indicates that the predominant renal mechanisms regulating phosphate excretion are filtration and net tubular reabsorption. The reabsorptive mechanism for phosphate illustrates the characteristics of an active transport system with limited capacity, similar to that for glucose. Thus, as the plasma phosphate concentration is progressively elevated, one can demonstrate a minimal threshold, a splay, and a maximal load at which the tubular reabsorptive system is completely saturated.[1] The Tm phenomenon is characteristic of both proximal and distal nephron phosphate transport systems.[1,2] It is important to note that the Tm for phosphate varies directly with the GFR.[1] Although phosphate secretion has been documented in lower forms,[3-7] such an effect was not demonstrable in the dog,[8] despite numerous maneuvers known to augment phosphate excretion. Similar studies in man have also failed to demonstrate phosphate secretion. Micropuncture studies in rat[9-11] and dog[12-15] have, likewise, failed to demonstrate *net* tubular secretion. These studies, however, do not exclude definitively a renal mechanism for phosphate secretion nor do they exclude a significant component of passive back-flux of phosphate from tubular cell into tubular lumen, as has been suggested by microperfusion studies.[16,17]

Micropuncture studies in Munich–Wistar rats have demonstrated that phosphate is completely ultrafiltered, the ratio of glomerular filtrate phosphate concentration in Bowman's space to plasma ultrafilterable phosphate being 1.0.[18,19] In the proximal tubule, 60–70% of the filtered phosphate load is reabsorbed by that portion of the pars convoluta accessible to micropuncture[9,13,20] (Fig. 1). However, phosphate transport in the proximal tubule is not homogeneous. Within the first 20% of the proximal tubule, some 30–40% of the filtered phosphate load is reabsorbed; in contrast, only 15–20% of filtered sodium and water is absorbed.[11,15,21] Consequently, $(TP/P)_{PO_4}$ decreases significantly in the early portion of the proximal tubule. Thereafter, tubular phosphate concentration remains constant, indi-

Ralph A. DeFronzo and Samuel O. Thier • Department of Medicine, Yale University School of Medicine, New Haven, Connecticut 06510.

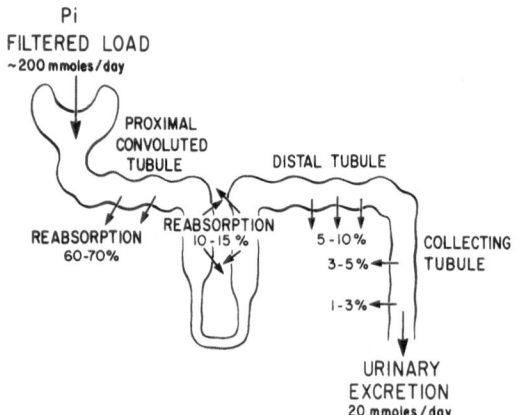

Pi
FILTERED LOAD
~200 mmoles/day

PROXIMAL
CONVOLUTED
TUBULE

DISTAL TUBULE

REABSORPTION
60-70%

REABSORPTION
10-15%

5-10%

3-5%

1-3%

COLLECTING
TUBULE

URINARY
EXCRETION
20 mmoles/day

Fig. 1. Sites of phosphate transport in the mammalian nephron. In man the plasma P_i is approximately 1.14 μmoles/ml and the GFR is about 120 ml/min. Thus, the normal human kidney filters on the order of 200 mmoles of P_i daily, of which about 10% or 20 mmoles is excreted in the urine.

cating parallel reabsorption of phosphate, sodium, and water. Comparison of late proximal and early distal micropuncture studies indicates that approximately 10–15% of the filtered phosphate is reabsorbed by this middle segment of the nephron[10,11,22,23] (Fig. 1). The decrease in fractional phosphate delivery between late proximal convoluted and early distal tubule is due to two factors: (1) continued phosphate reabsorption by the pars recta[11,21,24]; there appears to be no significant phosphate reabsorption by the thin descending and ascending or the thick ascending loops of Henle,[25] and (2) nephron heterogeneity. Several recent studies have demonstrated that in juxtamedullary nephrons, phosphate reabsorption by the proximal convoluted tubule and pars recta is significantly greater than in superficial nephrons.[24,26–29] Since 5–20% of the filtered phosphate load is normally excreted in the final urine, up to 10–20% of the filtered load may be reabsorbed in the late distal tubule and collecting duct.[9,21–24]

Studies employing free-flow micropuncture[21,30–32] and isolated cortical collecting tubules[33,34] have demonstrated that as much as 10–15% and 5% of the filtered phosphate load is reabsorbed by these distal nephron segments, respectively. There does not appear to be any significant phosphate transport in the medullary collecting duct.[35] The distal phosphate reabsorptive system is quite sensitive to parathyroid hormone (PTH), such that in its absence, up to 30–40% of the filtered phosphate may be reabsorbed in the late distal nephron.[22,36,37] In the presence of normal levels of PTH, this distal phosphate transport system plays a less prominent role, and its contribution, if any, in clinical disorders of phosphate transport remains to be delineated.

Changes in both GFR[37,38] and plasma phosphate concentration[39] have been shown to play an important role in phosphate excretion. Thus, both a decrease in GFR and hypophosphatemia enhance proximal tubular phosphate reabsorption, and vice versa. However, the two most important regulators of phosphate excretion are the plasma parathormone (PTH) concentration and dietary phosphate intake.[1,2,37,40,41] Increases in circulating PTH result in the rapid onset of phosphaturia without changes in GFR.[15,42,43] That this effect is direct is evidenced by a unilateral phosphaturia following direct

renal artery injection.[43] Micropuncture studies have shown that administration of PTH to intact animals results in a marked decrease in proximal tubular reabsorption of phosphate.[12–15] Since the distal phosphate reabsorptive system is inhibited in the presence of PTH, most of the incremental phosphate load delivered out of the proximal tubule appears in the final urine. The inhibitory effect of PTH on proximal phosphate reabsorption appears to be mediated by at least three distinct mechanisms: first, a direct effect involving cAMP as the mediator[20,22,44–46]; second, a PTH-induced reduction in proximal tubular sodium and fluid reabsorption[13–15,47,48]; and third, by alkalinization of proximal tubular fluid[49] with shift of the HPO_4^{2-}/HPO_4^{-} buffer equilibrium toward the less transportable HPO_4^{2-}.[50] Recently, Puschett *et al.* have questioned several of the time-honored concepts concerning the action of PTH.[51] As pointed out by these authors, most studies that have examined the effect of PTH on renal tubular phosphate transport have employed pharmacologic doses of the hormone. When a small, physiologic dose of the hormone was employed,[51] a phosphaturia ensued but there was no alteration in proximal tubular phosphate reabsorption. Additionally, there was no change in urinary or tissue cAMP. These results, if substantiated, indicate that physiologic changes in PTH primarily modulate phosphate transport at sites beyond the convoluted proximal tubule and that this phosphaturic effect is independent of the cAMP system. In contrast to the diminished proximal tubular phosphate reabsorption and resultant phosphaturia following PTH administration, parathyroidectomy has been shown to enhance proximal phosphate reabsorption markedly, with up to 70–85% of the filtered phosphate load being reabsorbed in the pars convoluta.[37]

The second most important regulator of renal phosphate transport is the dietary phosphate intake.[40,41,52,53] Following phosphate restriction there is a marked increase in overall renal reabsorptive capacity. With 24 hr of phosphate restriction there is a marked reduction in phosphate excretion and within 3 days the urine is virtually free of phosphate.[54,55] Conversely, a high-phosphate diet inhibits the ability of the kidney to reabsorb phosphate.[40,41,56] These changes occur independently of the plasma phosphate concentration and appear to involve an alteration in the intrinsic capacity of both proximal and distal phosphate transport systems.[40,41,52,53,56] They occur in the presence[53,56,57] or absence[40,41] of PTH. Furthermore, phosphate restriction totally blocks the phosphaturic effect of PTH.[58,59] These changes in phosphate transport are not related to alterations in sodium reabsorption and can be demonstrated *in vitro* using renal brush border membrane vesicles.[60–62] It is likely that effects of dietary phosphate are mediated via changes in the cellular phosphate concentration.

Vitamin D has also been shown to be antiphosphaturic with cholecalciferol, 25-hydroxycholecalciferol, and 1,25-dihydroxycholecalciferol all enhancing net renal phosphate reabsorption.[72,74] Micropuncture studies in the rat have localized this effect to the proximal tubule.[75] However, these observations have not been reproduced by others and the interaction of vitamin D with PTH and the physiologic importance of this effect of vitamin D on renal phosphate transport remains controversial.[76]

2.2. Cellular Mechanism of Phosphate Transport

Using the renal tubular brush border membrane vesicle preparation,[71,77–80] our understanding of the cellular transport of phosphate has been greatly enhanced (Fig. 2). Phosphate

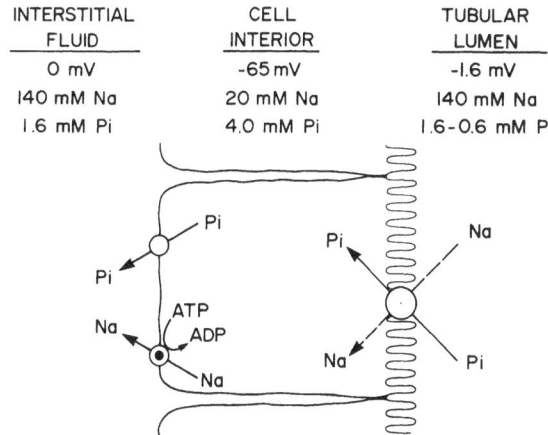

INTERSTITIAL FLUID	CELL INTERIOR	TUBULAR LUMEN
0 mV	-65 mV	-1.6 mV
140 mM Na	20 mM Na	140 mM Na
1.6 mM Pi	4.0 mM Pi	1.6-0.6 mM Pi

Fig. 2. Schematic representation of proximal tubular P_i reabsorption. Na^+-P_i are cotransported by the brush border membrane and phosphate exits passively at the contraluminal membrane. The energy for phosphate transport is provided by the Na^+ concentration gradient. Redrawn from Dousa and Kempson.[82]

transport across the proximal tubule appears to involve two discrete steps. The first involves the uptake of phosphate across the luminal basement membrane, occurs against an electrochemical gradient, is electroneutral, and occurs in cotransport with Na^+.[71,77-80] Phosphate transport across the luminal membrane is secondarily active and is fueled by the active extrusion of Na^+ by the basolateral membrane (Fig. 2). This active Na^+ pump is intimately associated with the enzyme Na^+,K^+-ATPase. The transport of phosphate from tubular lumen to cell interior to interstitial fluid is purely passive and proceeds down its electrochemical gradient. Phosphate uptake at the luminal basement membrane is the rate-limiting factor for phosphate transport.[71,77-80] That phosphate and Na^+ transport are intimately associated is demonstrated by the observation that Na^+ increases phosphate uptake by the luminal basement membrane in the absence of any transmembrane Na^+ gradient.[77-79]

Additional evidence for the important linkage between Na^+ and phosphate transport stems from studies examining the interaction between phosphate, glucose, and alanine. Thus, both glucose and alanine have been shown to inhibit phosphate transport in brush border vesicles. This represents an indirect effect in which the cotransport of Na^+ with glucose (or alanine) across the luminal membrane decreases the electrochemical Na^+ gradient necessary to energize the uphill transport of phosphate.[81]

The precise cellular mechanism(s) by which specific regulatory factors, such as PTH and dietary phosphate intake. alter phosphate transport by the luminal brush border membrane is not known. Three major hypotheses have recently been reviewed by Dousa and Kempson.[82] Of these the "phosphorylation" theory has received the most attention. Briefly, this hypothesis states that adenylate cyclase, located in the basolateral membrane, is stimulated by PTH and this leads to an increase in cellular cAMP. cAMP then activates a protein kinase located in the cytoplasm as well as in the luminal basement membrane. Activation of the protein kinase leads to a phosphorylation of specific protein(s) within the luminal basement membrane and this in turn regulates the rate of phosphate uptake.[71,82] Evidence also has been presented that an increase in cytosolic Ca^{2+} (as well as cAMP) may serve as the second messenger.[83]

The second hypothesis suggests that alkaline phosphatase is an important regulator of cellular phosphate transport. Evidence supporting this contention includes the following: alkaline phosphatase activity parallels the rate of luminal phosphate transport; alkaline phosphate inhibitors decrease phosphate reabsorption; and alkaline phosphatase activity is highest in the early proximal tubule, where phosphate transport is also highest.[82] However, agents (cycloheximide, actinomycin D, levamisol, EDTA) which inhibit alkaline phosphatase activity, do not lead to any decrease in phosphate transport.[82]

The most recent hypothesis put forth to explain the cellular regulation of phosphate transport involves changes in the rate of gluconeogenesis and subsequent alteration in the ratio of NAD^+ to NADH.[82] Specifically, PTH has been shown to stimulate gluconeogenesis, a process mediated by changes in both cellular Ca^{2+} and cAMP. The resultant stimulation of gluconeogenesis causes an increase in cytosolic NAD^+, which in turn inhibits Na^+-phosphate cotransport across the luminal basement membrane.[82] It should be noted that these three hypotheses are not mutually exclusive, nor do they preclude other, as yet unidentified regulatory processes for cellular phosphate transport.

2.3. Vitamin D-Resistant Rickets

Vitamin D-resistant rickets (VDRR) is a unique clinical disorder characterized by familial inheritance, usually X-linked dominant; rickets or osteomalacia; hypophosphatemia; decreased renal tubular reabsorption of phosphate; normal serum Ca^{2+} with normal to low urine Ca^{2+} excretion; diminished gastrointestinal absorption of both Ca^{2+} and phosphate; and resistance to large doses of vitamin D.

Albright et al.[84] originally proposed that the primary defect in VDRR was defective gastrointestinal absorption of Ca^{2+} with a then predictable sequence of events. This defect in Ca^{2+} absorption could be primary or secondary to an abnormality of vitamin D absorption or metabolism. As a result of the decrease in Ca^{2+} absorption, the serum Ca^{2+} would fall, and secondary hyperparathyroidism would ensue. The subsequent increase in circulating PTH would inhibit proximal tubular phosphate reabsorption, with resultant phosphaturia and a decrease in serum phosphate concentration. The fall in phosphate concentration would then lead to defective mineralization of bone (rickets or osteomalacia).

Although defective gastrointestinal absorption of Ca^{2+} has been repeatedly documented in patients with VDRR,[85-92] the original hypothesis of Albright et al. is no longer tenable. First, the serum Ca^{2+} concentration has uniformly been found to be normal.[87-102] Second, measurements of plasma PTH concentration have revealed either normal[89,91,96,100,103-108] or slightly elevated[94-98] levels. Third, although it is now clear that patients with VDRR have an abnormality in vitamin D metabolism,[105,107,109,110] plasma levels of $1,25-(OH)_2D$ are normal or only slightly decreased in untreated patients.[105-108,111-114] The implications of this abnormality in vitamin D metabolism will be discussed in a later section. The most cogent argument against a primary pathogenetic role for deranged vitamin D metabolism is the observation that both 25-hydroxycholecalciferol and 1,25-dihydroxycholecalciferol[90-92,104] have proven unsuccessful in correcting the hypophosphatemia and excessive phosphaturia despite markedly enhanced gastrointestinal Ca^{2+} absorption, positive Ca^{2+} balance, and even the development of hypercalcemia in some patients. Puschett *et al.* have shown that both 25-hydroxycholecalciferol and 1,25-dihydroxycholecalciferol are antiphosphaturic.[73,74] Although it has been postulated that the phosphaturic effect of PTH may be magnified in the absence of vitamin D, this has not been proven experimentally. Furthermore, if such a mechanism were responsible for the phosphaturia in patients with VDRR, one would have expected a fall in urinary phosphate excretion following treatment with both 25-hydroxycholecalciferol and 1,25-dihydroxycholecalciferol. As stated earlier, this does not occur. Another possibility that has not been considered is that, even though circulating levels of the active vitamin D metabolites may be normal, there may still be an abnormality in metabolism with the accumulation of nonpolar metabolites[109] which in themselves may be phosphaturic.

At present, the major controversy concerning the pathogenesis of VDRR revolves around the role of PTH versus a primary renal defect in renal phosphate handling. Early investigators[96-99,101,121,122] had shown that an acute calcium infusion, 15–30 mg/kg, resulted in a prompt increase in tubular reabsorption of phosphate (TRP). Since such an acute rise in serum Ca^{2+} concentration is known to be associated with a fall in plasma PTH level, these early investigators postulated that the excessive phosphate excretion and resultant hypophosphatemia in subjects with VDRR was secondary to hyperparathyroidism. With prolonged calcium infusions (> 6 days), the phosphate clearance has been variously reported either to return to the basal high levels[121] or to remain diminished.[122] The interpretation of these results, however, is obscured by the observation that Ca^{2+} itself may have a direct effect to enhance TRP.[123] In fact, Glorieux and Scriver[96] have so interpreted their data. Following the acute infusion of calcium in five subjects with VDRR, they observed a 20% increase in TRP. However, they were unable to correlate a fall in plasma PTH levels with the rise in TRP. Consequently, they concluded that the increased TRP following calcium infusion was secondary to a direct effect of Ca^{2+} to enhancing phosphate reabsorption, and not to a change in circulating PTH concentration.

If hyperparathyroidism were responsible for the phosphaturia, then one would expect to find plasma PTH levels to be inappropriately elevated; this is not the case. Roof *et al.* found normal circulating PTH levels in 17 children with VDRR[103] and Arnaud *et al.* similarly reported normal levels in an additional 13 subjects.[100] Puschett *et al.* followed two patients with VDRR for up to 2 years and reported multiple PTH determinations to be within normal limits.[91] In two additional patients, Brickman *et al.* also documented normal PTH levels with a fall following treatment with 1,25-dihydroxycholecalciferol and the induction of positive calcium balance, yet TRP remained unchanged.[89] Numerous recent reports also have documented normal circulating PTH levels in large numbers of patients with VDRR.[105,108] Only three groups have reported elevated plasma levels of PTH. Reitz and Weinstein reported five patients with VDRR and elevated PTH levels, but in three of these five subjects there was considerable overlap with the normal population.[94] Lewy *et al.* also reported five children with elevated PTH levels, but they emphasized that the TRP was disproportionately reduced in comparison to the mild elevation in PTH concentrations.[98] Hahn *et al.* have reported mildly to moderately elevated levels of circulating PTH in 14 untreated patients from five kindreds with VDRR.[124] Unfortunately, data concerning phosphate excretion in patients with primary hyperparathyroidism and similar elevations of serum PTH are not given for comparison. Furthermore, calcium infusion resulted in

suppression of serum PTH concentration and urinary cAMP excretion, yet TRP remained inappropriately low. Thus, the majority of evidence does not support hyperparathyroidism as the etiological mechanism of the phosphaturia in subjects with VDRR.

Although elevated circulating levels of PTH do not seem to be responsible for the decreased TRP in VDRR, the aforementioned studies do not exclude the possibility of increased sensitivity to normal levels of PTH. Glorieux and Scriver found a blunted phosphaturia following PTH infusion, but a normal increase in urinary cAMP and a normal decrease in urinary Ca^{2+} excretion.[96] They interpreted these results to indicate the loss of a specific PTH-modulated phosphate transport mechanism within the renal tubule cell. Recently, Lyles and Drezner[107] have been unable to reproduce the blunted phosphaturia noted by Glorieux and Scriver.[96] Following PTH infusion they found a similar decline in TmP/GFR in control and VDRR subjects. Even if correct, one could argue that the results of Glorieux and Scriver are due to increased sensitivity to endogenous PTH such that the infusion of exogenous PTH had little further effect on decreasing TRP. Riggs *et al.*[125] attempted to examine this problem by observing the change in TRP following parathyroidectomy in a patient with VDRR: following parathyroidectomy, the serum phosphate concentration returned to normal in association with an increase in TRP to 94%. Unfortunately, the diagnosis of VDRR was not clearly established in this patient since there was no family history, the disease was of adult onset, and the patient had renal glucosuria which is not normally associated with VDRR. In addition, there was no determination of TRP pre- or postoperatively, and postop the serum phosphate concentration failed to increase despite hypocalcemic tetany. Only after many years of vitamin D therapy was the TRP shown to be normal. In a study by Kleerekopper *et al.*,[126] partial parathyroidectomy failed to correct the renal phosphate wasting, and a recently reported patient with idiopathic hypoparathyroidism and VDRR had persistent phosphaturia even after the hypocalcemia was corrected.[127] These observations mitigate against PTH as a primary pathogenetic factor in the excessive phosphaturia. To further examine the role of PTH, Short *et al.* induced a state of functional hypoparathyroidism with a calcium infusion and then infused exogenous PTH.[128] With this technique, they were able to demonstrate a marked phosphaturia in patients with VDRR when compared to control subjects similarly studied. Indeed, these data suggest that there is increased sensitivity to PTH in VDRR, but they have yet to be substantiated by other investigators.

The opposing view, namely that the excessive phosphaturia in patients with VDRR is due to a primary defect in renal tubular phosphate reabsorption, has been championed by Glorieux and Scriver.[96,100] They interpreted the findings of normal circulating PTH levels and a blunted phosphaturia to exogenous PTH to be most consistent with the loss of a PTH-modulated renal tubular phosphate transport system. As pointed out earlier, however, such data are compatible with increased sensitivity to normal levels of PTH. These authors also demonstrated that the TRP was lowest in male hemizygotes, as would be predicted by the X-linked dominant pattern of inheritance and the Lyon hypothesis with inactivation of one of the X chromosomes in the female. Although they have interpreted these data to be most consistent with a primary renal tubular defect in phosphate transport, one could argue that in females there is an intermediate sensitivity to PTH. Other support for a primary defect in renal phosphate reabsorption comes from the work of Short *et al.*,[129] who found that specimens of jejunal mucosa obtained by intestinal biopsy in normal subjects demonstrated saturation kinetics of the phosphate transport systems. In five patients with VDRR, they found that the high-affinity transport system was absent. Since defects in intestinal transport may occur in association with renal tubular defects (see ensuing discussion on aminoaciduria), one could extrapolate that the same gastrointestinal defect in phosphate transport is also present in the kidney.

As mentioned previously, recent studies have suggested that patients with VDRR have an abnormality in vitamin D metabolism. When $1,25\text{-}(OH)_2D$ levels have been measured, they have generally been normal or in the low normal range.[105−108,111−114] However, since these individuals are hypophosphatemic and since hypophosphatemia is known to be an important stimulator of $25(OH)D\text{-}1\text{-}hydroxylase$ activity in the kidney,[130] one would have expected to observe increased levels of $1,25(OH)_2D$. Further evidence for a defect in vitamin D metabolism has been provided by studies which have attempted to augment $1,25(OH)_2D$ synthesis. Insogna et al.[105] found that, following 3 days of phosphate depletion, normal subjects demonstrated a 30% rise in circulating $1,25(OH)_2D$ levels and a similar rise in TmP/GFR. In contrast, in patients with VDRR, plasma $1,25(OH)_2D$ levels failed to increase and TmP/GFR did not change. Similarly, Lyles and Drezner[107] found that the rise in $1,25(OH)_2D$ in response to PTH infusion was fourfold greater in normal compared to VDRR patients. This disparate response occurred despite an equivalent increase in urinary cAMP and a similar decrease in TmP/GFR in the two groups.

The link between the defect in vitamin D metabolism and the renal phosphate wasting in patients with VDRR is not known at present. Nonetheless, it is likely that these two abnormalities are related. If the defect in renal phosphate reabsorption were related to decreased phosphate transport at the brush border membrane (Fig. 2), one would expect low intracellular renal tubular phosphate levels. This would be expected to increase, not decrease, $1,25(OH)_2D$ synthesis. In contrast, if there were a defect in phosphate exit from the cell at the basolateral membrane (Fig. 2), one would expect a normal to high intracellular phosphate concentration. This, in turn, would be expected to inhibit $1,25(OH)_2D$ synthesis and would decrease phosphate transport across the brush border membrane. The finding of normal intracellular total phosphorus concentrations in the *Hyp* mouse model of VDRR is consistent with this latter hypothesis.[131] Nonetheless, one would still expect phosphate depletion to stimulate $1,25(OH)_2D$ synthesis and such was not observed by Insogna et al.[105] It is possible that the 3-day period of phosphate depletion employed by these investigators was too short to cause a decrease in renal tubular phosphate concentration. Alternatively, it is possible that the defects in phosphate reabsorption and vitamin D metabolism are unrelated or are linked via some as yet unrecognized step.

The occurrence of a strain of mice (*Hyp/Y*) that have familial X-linked renal phosphate wasting, hypophosphatemia, and osteomalacia with normal serum Ca^{2+} levels[132,133] has provided an animal model that appears analogous to VDRR in man. Micropuncture studies[134] have demonstrated decreased proximal tubular phosphate reabsorption, a defect which is not corrected by parathyroidectomy. Furthermore, following a low-phosphorus diet the rise in TmP/GFR in *Hyp* mice is markedly attenuated compared to controls.[135] Phosphate uptake and efflux by renal cortical slices are normal,[133,136] indicating a normal flux of phosphate across the basolateral membrane in the *Hyp* mouse. In contrast, studies of brush border membrane vesicles prepared from *Hyp* mouse kidneys reveal a defect in the Na^+-dependent component of phosphate transport.[136,137] Although phosphate depletion *in vivo* enhanced the transport of phosphate by these membrane vesicles, it did not restore it to normal.[138,139] If the renal tubular defect in VDRR in man is similar to that in the *Hyp* mouse and resides in the brush border membrane, it is difficult to explain the abnormality in vitamin D metabolism. Nonetheless, Meyer et al. have shown that a similar defect in vitamin D metabolism is present in the *Hyp* mouse.[140] Baseline $1,25(OH)_2D$ levels are normal but are inappropriately low when compared to the ambient plasma phosphate concentration. Furthermore, phosphate depletion caused a fall, rather than an increase, in $1,25(OH)_2D$. More recently, direct measurements of $25(OH)D\text{-}1\text{-}hydroxylase$ activity in mouse kidney have documented low levels that fail to increase in response to phosphate depletion.[141]

In summary, the majority of evidence would suggest that the primary defect in VDRR involves a renal tubular defect in phosphate reabsorption by the proximal tubule. A second defect in Vitamin D metabolism also exists. This is characterized by a relative, rather than absolute, deficiency of $1,25(OH)_2D$. What role this abnormality plays in the pathogenesis of VDRR and how it relates to the defect in renal tubular phosphate transport are not clear at present. The failure of parathyroidectomy to reverse the renal phosphate wasting suggests that neither excessive PTH levels nor increased sensitivity to normal levels of PTH is responsible for the tubular defect in phosphate reabsorption.

2.4. Vitamin D-Dependent Rickets

Rickets resulting from insufficient amounts of the active metabolite(s) of vitamin D has been described in two clinical situations: dietary deficiency and vitamin D-dependent rickets (VDDR). The latter entity is characterized by hypocalcemia, hypophosphatemia, and severe rachitic bone lesions, and may be treated with large doses of vitamin D.[142] The response to vitamin D is in contrast to VDRR. Two distinct forms of VDDR have been delineated and these have been designated type I and type II. The clinical and laboratory features of both disorders are similar, but they can be distinguished by measuring the circulating levels of $1,25(OH)_2D$. The hormone level is reduced in type I and elevated in type II.

In patients with type I VDDR, serum $25(OH)D$ levels are normal in untreated patients and elevated in patients receiving vitamin D therapy.[143−145] Circulating $1,25(OH)_2D$ levels are depressed in both treated and untreated patients.[143,145,146] These observations suggest that deficiency of renal $25(OH)D\text{-}1\text{-}hydroxylase$ is responsible for the pathogenesis of this disorder. In keeping with this hypothesis, large doses of vitamin D or $25(OH)D$ are required to effect a therapeutic response whereas small physiologic doses of $1,25(OH)_2D$ heal the rachitic lesions and correct the biochemical abnormalities.[144,145,147−149]

In contrast to type I VDDR, patients with type II are characterized by elevated levels of $1,25(OH)_2D$[149−155] and are refractory to treatment with all forms of vitamin D, including $1,25(OH)_2D$.[149−155] Recent studies have employed cultured fibroblasts to examine the nature of the end-organ resistance in patients with type II VDDR. Fibroblasts cultured from skin of normal individuals demonstrate high-affinity uptake of $1,25(OH)_2D$ and the presence of vitamin D receptors similar to those in typical target tissues for vitamin D action.[156] In patients with type II VDDR, nuclear uptake of $[^3H]\text{-}1,25(OH)_2D$

is defective.[157–160] Both receptor and postreceptor abnormalities appear to contribute to the defect in 1,25(OH)$_2$D action.[157–160]

Common to VDDR, as well as dietary deficiency of vitamin D, is the presence of excessive phosphaturia and generalized aminoaciduria.[161–165] Both of these urinary abnormalities are related to the development of secondary hyperparathyroidism. Fraser, Scriver, and associates[164,166,167] demonstrated a characteristic evolution of vitamin D deficiency in infancy in which parathyroid hypersecretion appeared to correlate with the development of aminoaciduria and phosphaturia. During the earliest stage of vitamin D deficiency, the serum calcium was low, serum phosphorus was normal, and aminoaciduria and phosphaturia were absent. Plasma PTH levels were normal during this stage. This was followed by high circulating levels of PTH, a return of the serum calcium to normal, decreased TRP, and the onset of aminoaciduria. This stage was associated with the presence of high circulating levels of PTH. Following calcium infusion, TRP rose and the aminoaciduria disappeared. These authors interpreted this result to mean that secondary hyperparathyroidism was responsible for the renal leak of phosphate and amino acids. These observations have subsequently been confirmed and extended in a vitamin D-deficient rat model.[168–170] Prolonged exposure to hypocalcemia, with or without vitamin D deficiency, reproduces the aminoaciduria and phosphate wasting, both of which can be corrected by restoring the serum calcium concentration to normal.

Furthermore, a strong correlation was found between abnormal tubular function and parathyroid gland activity. Extirpation of the hyperplastic parathyroids returned the TRP to normal, and aminoaciduria disappeared despite persistent hypocalcemia. Thus, the excessive phosphaturia as well as aminoaciduria in patients with dietary deficiency and hereditary VDDR appears to be the result of secondary hyperparathyroidism, and can be cured with restoration of normal serum calcium concentrations with either calcium infusions or vitamin D administration. The effect of PTH on renal phosphate handling has previously been discussed, and *in vivo* and *in vitro* studies have demonstrated a similar inhibitory effect of PTH on amino acid transport.[171,172]

2.5. Pseudohypoparathyroidism

Pseudohypoparathyroidism is a heterogeneous disorder characterized by hypocalcemia and hyperphosphatemia, elevated PTH levels, and the variable presence of somatic abnormalities (short stature, round face, brachydactyly, impaired mental function) which have collectively been termed Albright's hereditary osteodystrophy.[173,174] The syndrome was first described in 1942 by Albright *et al.*[175] in three patients with apparent hypoparathyroidism but who failed to respond to parathyroid extract. Subsequently, numerous investigators demonstrated that the serum PTH levels are elevated in these patients[176–183] and Chase *et al.*[177] documented that PTH infusion failed to cause a rise in urinary cAMP or phosphate excretion. Later, Bell *et al.*[184] showed that dibutyryl cAMP administration could circumvent the intracellular block in PTH action and produce a phosphaturia. Because the biochemical features of the disease resemble those of PTH lack, the disorder has been referred to as pseudohypoparathyroidism, type I, to distinguish it from a subtype in which resistance to the phosphaturic effect of PTH is associated with a normal increase in urinary cAMP.[173,174] This latter group of patients usually does

not manifest any of the somatic abnormalities described above and has been termed pseudohypoparathyroidism, type II.

The intrarenal site of PTH resistance in pseudohypoparathyroidism is the proximal tubule, where the bulk of the filtered phosphate is reabsorbed. The inability of PTH to enhance cAMP excretion in patients with pseudohypoparathyroidism could be due to a defect in PTH binding to its receptor, a defect in the coupling mechanism linking the PTH–receptor complex with the postreceptor events, a defect in cAMP generation, an accelerated rate of cAMP degradation, or an inability of cAMP to exit from the cell. The last two possibilities seem unlikely. First, theophylline, a phosphodiesterase inhibitor, has no effect on the response to PTH.[173,185] Second, if an abnormality in cAMP exit from the cell was present, the intracellular cAMP concentration would be expected to be normal or high and the ability of dibutyryl cAMP to bypass the defect would not be explained. Two groups of investigators, using renal cortical plasma membranes from single patients with pseudohypoparathyroidism type I, found a normal stimulation of adenylate cyclase by PTH when the plasma membranes were incubated with saturating concentrations of ATP.[186,187] These observations indicate that, at least in the two patients studied, the PTH receptor and the cAMP catalytic moiety must be functionally intact. At subsaturating concentrations of ATP, however, the generation of adenylate cyclase was impaired but could be restored to normal with the addition of GTP. Recent studies have helped to clarify these findings.[174,188] As reviewed by Ross and Gilman,[188] the enzyme adenylate cyclase is comprised of three separate components: (1) hormone receptors which bind PTH; (2) a catalytic unit which regulates the conversion of ATP to cAMP; and (3) N protein, which is responsible for the coupling between receptor and catalytic units. N protein binds guanine nucleotides, possesses GTPase activity, and stimulates cAMP synthesis by guanine nucleotides. On the basis of this information, Drezner and Burch[187] have interpreted their results (i.e., impaired adenylate cyclase generation at subsaturating ATP concentrations and correction with GTP) to indicate a defect in N protein function.

The demonstration of N protein in erythrocyte membranes has allowed further clarification of the defect in patients with pseudohypoparathyroidism.[174,189–191] In approximately 50–60% of afflicted individuals, N protein activity was found to be diminished and these patients have been defined as having pseudohypoparathyroidism type Ia. Over 95% of this group have the somatic features of Albright's hereditary osteodystrophy. In contrast, the remaining 40–50% of patients have normal N protein activity and have been classified as pseudohypoparathyroidism type Ib.

Farfel and Bourne have assessed N protein activity in platelets,[192] cultured skin fibroblasts,[193] and transformed lymphoblasts.[194] All type Ia patients, who demonstrated a decrease in erythrocyte N protein activity, manifested a similar defect in the other three cell lines that were examined; platelets, fibroblasts, and lymphoblasts. The same cell types, when examined in pseudohypoparathyroidism type Ib patients showed quantitatively normal N protein activity. The persistence of the defect in cells propagated for many generations *in vitro* suggests that a genetic, rather than environmental, abnormality is responsible for the PTH resistance in type Ia patients.

The precise biochemical defect in patients with pseudohypoparathyroidism type Ib remains undefined at present. These individuals are resistant to the effects of PTH and uncommonly manifest any skeletal abnormalities.[173,174] In the patients reported by Farfel and Bourne,[193,194] adenylate cyclase activity and the accumulation of intracellular cAMP in response to a variety of hormones was found to be normal in cultured skin fibroblasts and transformed lymphocytes. In a preliminary report, Fischer[195] reported a biologically inactive PTH in several patients with pseudohypoparathyroidism type Ib and postulated that the end-organ resistance was due to receptor occupancy by the inert hormone. It is likely that type Ib will turn out to be a heterogeneous disorder in which multiple defects (including biologically inert PTH, abnormal PTH receptor, defective adenylate cyclase catalytic unit, abnormal coupling between receptor and catalytic unit, etc.) are responsible for the resistance to PTH.

Although considerable progress has been made in defining the renal tubular defect in PTH action, the metabolic defect(s) responsible for the hypocalcemia remains less well characterized. The increased plasma phosphate concentration, by complexing with Ca^{2+}, undoubtedly contributes to the hypocalcemia but other factors are likely to be of more importance. Because of the failure of PTH to cause a rise in the serum Ca^{2+} concentration,[180,184,196–199] it has been assumed that bone is unresponsive to the action of PTH. However, bone biopsies in individuals with pseudohypoparathyroidism have usually demonstrated osteitis fibrosis cystica and severe demineralization, consistent with excessive PTH bone remodeling.[180,181,200–204] Therefore, if bone tissue in patients with pseudohypoparathyroidism is unresponsive to PTH, the defect must be a partial one. A more likely explanation for the failure of PTH to elicit a hypercalcemic response is a low serum $1,25(OH)_2D$ level which fails to increase following PTH administration.[180,199,205–209] This explanation is supported by the observation that pharmacologic amounts of vitamin D[180,204,210,211] or physiologic amounts of $1,25(OH)_2D$[206,212–214] restore the normal calcemic response to PTH. It should be pointed out that, to the extent that $1,25(OH)_2D$ is deficient, impaired gastrointestinal absorption of Ca^{4+} will also contribute to the hypocalcemia observed in the postabsorptive state. Little is known about vitamin D metabolism in pseudohypoparathyroidism type II, so the contribution of $1,25(OH)_2D$ deficiency in the genesis of the bone resistance and hypocalcemia remains unknown.

2.6. Tumor-Induced Hypophosphatemic Osteomalacia

Several authors have reported severe adult-onset hypophosphatemic osteomalacia associated with benign mesenchymal tumors, usually of bone.[216–230] The syndrome has also been reported in association with prostatic carcinoma.[231] Characteristically, the serum Ca^{2+} concentration has been normal with a low serum phosphate concentration and markedly reduced tubular reabsorption of phsophate. In the seven patients in whom it was measured,[216,223,224,231] serum PTH concentration was normal. Following removal of the tumor, restoration of the serum phosphate to normal with healing of the osteomalacia has been reported.[216,218,219,221,223,224,227,228,230]

Several studies have given some insight into the cause of the renal phosphate wasting and osteomalacia. The finding of normal PTH levels in patients with this syndrome[216,223,224,231] excludes excessive secretion of this hormone as the etiological factor in the phosphaturia. Furthermore, one would not have expected increased PTH *per se* to lead to osteomalacia. Three groups have documented low $1,25(OH)_2D$ and normal $25(OH)D$ levels in patients with tumor-related hypophosphatemic os-

teomalaica.[223,230,231] Following surgical removal of the tumor in two patients,[223,230] $1,25(OH)_2D$ levels returned to normal and the osteomalacia was cured. In one patient, administration of physiologic amounts of 1,25-dihydroxycholecalciferol resulted in healing of the bone histology.[223] From these observations, two important conclusions can be drawn: (1) the tumor elaborates a substance that inhibits renal 1-hydroxylase activity, thereby impairing the conversion of $25(OH)D$ to $1,25(OH)_2D$, and (2) deficiency of $1,25(OH)_2D$ plays a causal role in the development of osteomalacia. Although the presence of hypophosphatemia undoubtedly contributes to osteomalacia, it is now clear that impaired vitamin D metabolism also plays a major role.

The above results raise questions concerning the pathogenesis of the renal phosphate wasting and hypophosphatemia in patients with the tumor-related osteomalacia syndrome. It was originally thought that the tumor produced a phosphaturic substance and this may still turn out to be the case. However, the ability of 1,25-dihydroxycholecalciferol to normalize the renal tubular maximum for phosphate reabsorption and to restore the plasma phosphate concentration to normal in the patient reported by Drezner and Feinglos[223] suggests that the phosphaturia may, at least in part, be secondary to $1,25(OH)_2D$ deficiency. Consistent with this, some evidence has been provided that in thyroparathyroidectomized animals, $1,25(OH)_2D$ may be antiphosphaturic.[73,232,233] Further studies will be needed to clarify whether it is $1,25(OH)_2D$ lack or a tumor-elaborated phosphaturic principle which is responsible for the hypophosphatemia.

Recent studies in patients with cancer-associated hypercalcemia[234–236] may have particular relevance to the above controversy. Although bone biopsies have not been performed in these patients and osteomalacia has not been recognized clinically, many of the biochemical abnormalities resemble those observed in the syndrome of oncogenic hypophosphatemic osteomalacia. Since the types of tumors associated with hypercalcemia are malignant and more aggressive in nature, it may be that patients do not survive long enough to develop the full-blown osteomalacia syndrome. In a series of 50 consecutive patients with cancer-associated hypercalcemia, Stewart[234,235] found that nephrogenous cAMP (an index of PTH-like action on the renal tubule) was elevated in 41. These 41 individuals had a decreased tubular maximum phosphate reabsorptive capacity and hypophosphatemia. Serum PTH concentrations, measured by four different antisera, revealed normal PTH levels. Circulating $1,25(OH)_2D$ concentrations were reduced. Thus, with the exception of hypercalcemia, these patients closely resemble those with tumor-induced hypophosphatemic osteomalacia. More recently, Stewart et al. have attempted to characterize further the phosphaturic substance in tumor extracts from patients with humoral hypercalcemia of malignancy.[236] In vitro, the tumor extract is capable of stimulating adenylate cyclase and glucose-6-phosphate dehydrogenase activity in a dose-dependent fashion, just as does PTH. Both adenylate cyclase and cytochemical bioactivity are the result of specific binding to renal PTH receptors since PTH analogs competitively inhibit the activity of the extract. However, preincubation with a PTH antiserum does not result in any loss of activity in the tumor extract. The molecular weight of the active principle in the tumor extract is much greater than that of PTH. Thus, although the substance elaborated by the tumor shares many of the biological effects of PTH, it is chemically and immunologically quite distinct. Similar studies in patients with the hypophosphatemic osteomalacia syndrome might reveal that the biologically active principle is similar to that reported in patients with humorally mediated hypercalcemia.

2.7. Endocrine Disorders

2.7.1. Hyperparathyroidism

The usual biochemical abnormalities of primary hyperparathyroidism are hypercalcemia, hypophosphatemia, and increased renal clearance of phosphate. A low serum phosphate concentration is found in approximately 50–60% of subjects with hyperparathyroidism, while a decrease in TRP is somewhat more frequent, occurring in about 60–70%.[237] As discussed previously, elevation of the plasma PTH concentration, acting through cAMP, inhibits proximal tubular phosphate reabsorption.[13,15,36,42–46] Since the phosphate reabsorptive capacity of the distal nephron is limited, increased urinary phosphate excretion results and is a useful laboratory determination in the diagnosis of hyperparathyroidism. Similarly, in states of secondary hyperparathyroidism (vitamin D deficiency, gastrointestinal disease), hypophosphatemia and diminished TRP are common. In the secondary hyperparathyroidism associated with renal insufficiency, the serum phosphate concentration is high because the decrease in GFR outweighs the effect of elevated PTH on the renal tubule. The fractional excretion of phosphate is high but the absolute excretion is low.

2.7.2. Acromegaly

Increased plasma phosphate concentration in association with an elevation in the maximal tubular phosphate transport capacity (Tm) has been reported in patients with growth hormone excess. Both acute and chronic administration of growth hormone have been shown to diminish phosphate excretion and increase the Tm for phosphate in man[238–242] without change in GFR or plasma phosphate concentration. A similar effect has been observed in dogs[243] and rats.[244] Following long-term growth hormone administration to healthy individuals, to patients with hypopituitarism, or to dogs, an increase in TmP/GFR has been documented.[241,245] These results indicate that the changes in renal tubular phosphate handling are independent of changes in renal hemodynamics and GFR. The effect of growth hormone to enhance TmP/GFR in acromegalic subjects is reversible by resection of the pituitary tumor.[240] Since the effect of growth hormone to increase TmP can be documented in parathyroidectomized animals, the presence of PTH is not necessary. At present, no micropuncture studies have been performed to localize the tubular site of action of growth hormone. However, Hammerman et al. have shown that growth hormone administration to dogs increases the Na^+-dependent phosphate flux from proximal tubular brush border membrane vesicles.[246]

2.7.3. Thyrocalcitonin

Administration of porcine, salmon, and synthetic thyrocalcitonin to man causes a phosphaturia, as well as a natriuresis.[247–252] The phosphaturic effect is independent of changes in PTH since it can be demonstrated in hypoparathyroidectomized man[251–253] and animals.[254,255] Since the phosphaturia occurs without any change in GFR or decline in plasma phosphate concentration, it most likely represents a di-

rect tubular action of the hormone. This is consistent with the demonstration of high-affinity calcitonin receptors on renal tubule cells.[256,257] Although calcitonin-stimulated adenylate cyclase has been demonstrated in renal tissue,[256–258] it is not clear that the phosphaturic action of the hormone is related to an increase in nephrogenous cAMP.[257,259–261] Utilizing re-collection micropuncture techniques in the rat, Gekle has located the phosphaturic effect of calcitonin to the proximal tubule.[262] Although there is substantial evidence that acute elevations in the plasma calcitonin level will cause a phosphaturia, it remains uncertain whether the hormone plays a significant role in the normal regulation of phosphate homeostasis. Furthermore, in patients with medullary carcinoma of the thyroid and chronically elevated calcitonin levels, the plasma phosphate concentration has been found to be normal.[263]

2.7.4. Ephinephrine

Epinephrine has been shown to consistently decrease urinary phosphate excretion in a variety of species including man,[264] dog,[265,266] and rat.[267] Since isoproterenol mimics the effect of epinephrine on renal phosphate transport,[267] its hypophosphaturic effect is presumably mediated via the β-adrenergic receptor. Although catecholamine receptors have been demonstrated in renal tissue[268] and epinephrine has been shown to stimulate renal cAMP formation,[269] it is not clear that epinephrine exerts a direct inhibitory effect on renal phosphate excretion. When epinephrine is infused in man[270,271] or animals,[272] the plasma phosphate concentration falls and the decline in renal phosphate excretion may be secondary to the hypophosphatemia. The precise contribution of epinephrine to acute and chronic phosphate homeostasis remains undefined at present.

2.7.5. Insulin

When hyperinsulinemia is created by glucose administration, a significant decline in urinary phosphate excretion results.[273–278] However, in all of these studies, significant glucosuria was present. Since glucose is known to inhibit phosphate transport by the renal tubule, it is possible that an antiphosphaturic effect of insulin may have been masked. Several studies[274,279–281] have documented a decrease in phosphate excretion following insulin administration, but these results are difficult to interpret since a concomitant fall in plasma phosphate concentration also occurred.

Extensive studies examining the effect of insulin on renal phosphate transport have been performed by DeFronzo et al.[282] During the time control study (NaCl infusion), whole kidney fractional phosphate excretion rose fourfold in association with a small rise in plasma phosphate concentration. During conditions of hyperglycemic hyperinsulinemia, fractional phosphate excretion fell despite a slight increase in plasma phosphate concentration. When hyperinsulinemia was created while maintaining plasma glucose constant at basal levels, fractional phosphate excretion fell markedly by 71%, even though plasma phosphate concentration remained unchanged. In all instances, the change in whole kidney fractional phosphate excretion was paralleled by a similar change in the fractional delivery of phosphate out of the proximal tubule. These results clearly indicate that insulin exerts a powerful effect to enhance proximal tubular phosphate reabsorption. Recently, Roy and Seely have reported similar findings employing clearance techniques in rats.[283] Using the isolated perfused kidney, Nizet *et al.*[284] have shown that the antiphosphaturic effect of insulin is a direct one, unrelated to changes in the plasma phosphate concentration. This effect of insulin on renal phosphate transport has important physiologic implications for the diabetic individual. In the presence of insulin deficiency, there is a shift of phosphate from intracellular to extracellular compartment and this results in an increase in the filtered phosphate load. In the presence of a glucose osmotic diuresis and insulin lack, a marked phosphaturia ensues.

2.7.6. Adrenal Hormones

The acute administration of glucocorticoids causes a rise in renal phosphate excretion,[285–287] which is unrelated to changes in the plasma phosphate concentration or GFR and occurs in the absence of PTH.[288,289] A similar phosphaturic effect has been documented after long-term glucocorticoid administration.[288,290] In contrast to the phosphaturic response to glucocorticoids, no increase in urinary phosphate excretion occurs in response to mineralocorticoids.[288,291]

3. Aminoacidurias

Aminoaciduria is the excretion of greater than normal quantities of amino acid in the urine, and may occur as the result of disordered metabolism, defective transport, or a combination of the two. This discussion will focus on the diseases of transport which, although accounting for a small percentage of the total of aminoacidurias, have proved to be extremely powerful experiments of nature in defining the physiology and genetics of transport systems.

Amino acids are derived from dietary protein intake and from *de novo* synthesis. Protein in the diet is digested in the intestinal lumen to the more absorbable oligopeptides and free amino acids. The oligopeptides, upon entering the intestinal epithelial cells, are further metabolized to free amino acids, which enter the circulation and are transported into cells throughout the body to be stored in free amino acid pools, metabolized, or incorporated into proteins. At the same time, free amino acids synthesized in the cells or produced by catabolism of cellular proteins may efflux into the circulation. Thus, the plasma amino acid concentrations are the result of a dynamic state and represent the balance between ingestion, cellular uptake, and cellular release. Under ordinary circumstances, the kidney contributes little to variations in plasma amino acid concentrations, since only a small fraction of the filtered load is excreted.

Although the handling of amino acids by the kidney has now been studied for over 75 years,[292] it is only in the past decade that the complexity of renal amino acid transport has become clearer.[293] In general, amino acids are filtered and net reabsorption of 95–99% of the filtered load[294] occurs largely in the proximal convoluted tubule.[295] Though the vast majority of amino acid reabsorption occurs in the proximal convoluted tubule, the highest luminal to cellular gradient is achieved in the pars recta.[296–298] A higher intracellular amino acid concentration with a lower transepithelial transport rate implies a reduced basolateral membrane efflux, and that has been observed.[296] The ability of the pars recta to generate very high intracellular

concentrations may permit reabsorption of the low concentrations of filtered amino acid that escaped reabsorption in the proximal convolution.

Apparent Tm values for certain amino acids (glycine, lysine, arginine) have been demonstrated, and clearance studies have suggested stereospecific transport systems which recognize neutral, dibasic, dicarboxylic amino acids and the imino acid–glycine group.[299–310] Studies utilizing kidney slices and isolated tubule cells demonstrated that amino acids were accumulated against electrochemical gradients by processes which were stereospecific, saturable, oxygen dependent, and Na+ dependent.[311–313] More recent autoradiographic studies in slices,[314–316] in isolated perfused tubules,[317] and studies *in vivo* suggest that a significant percentage of renal tubule cell amino acid accumulation occurs from the antiluminal surface by transport mechanisms which may differ from those in the brush border membrane.[318–321] Studies using membrane vesicles from brush border and basolateral membranes have expanded our knowledge of these processes.[322,323] Amino acid uptake across the brush border is coupled to electrogenic Na+ transport for all but the dibasic amino acids which undergo Na+-independent electrogenic transport. Na+-independent transport of proline occurs via a nonsaturable process made up of simple diffusion and of facilitated diffusion.[324–326] Competitive inhibition between groups of amino acids is demonstrable at the brush border and, again, with the exception of the dibasic amino acids, is dependent on the presence of Na+. In general, independent transport of neutral, dibasic, the imino acid–glycine group, and acidic amino acids can be demonstrated, but there is sharing of transport systems, particularly by glycine.

Transport at the basolateral membrane suggested by studies in kidney slices has been unequivocally demonstrated in isolated perfused tubules,[317] and again, has amino acid group specificity.[327] The Na+ requirements for basolateral membrane transport are less clearly defined. There appears to be less Na+ dependence than in the brush border membrane, and in some studies none has been evident.

It seems likely that some of the accumulated amino acids exit from the cell into the lumen only to be reabsorbed. If, however, the reabsorptive mechanism is defective, secretion may become evident.

In summary, amino acids are filtered and reabsorbed by highly specific mechanisms in the brush border membrane. Cellular accumulation also occurs from basolateral membranes with a portion of the accumulated amino acid possibly entering the lumen only to be reabsorbed along with the filtered load. When reabsorption is defective, accumulation from the basolateral surface and subsequent efflux into the tubular lumen or paracellular back-leak will produce a picture of amino acid secretion.

Aminoacidurias represent such a broad spectrum of disorders that it is necessary to dissect them on the basis of physiological and genetic considerations to allow a coherent discussion. In broadest terms, amino acids are presented to cells which have specific mechanisms in their membranes for transport to an intracellular site, where they may enter the free amino acid pool, be incorporated into proteins, or be further metabolized via enzymatically catalyzed steps. The accumulated amino acid or its metabolites may efflux from the cell either by diffusion or via a specific membrane mechanism. The general schema applies to all cells, but may be modified further when applied to polar epithelial cells, such as renal tubular or intestinal epithelium. In these epithelial cells, the brush border at the luminal

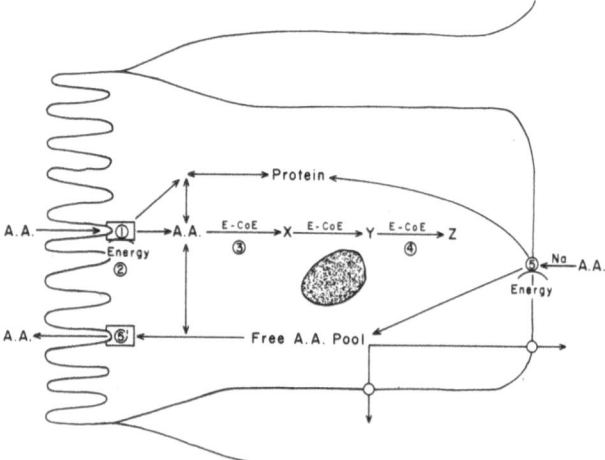

Fig. 3. Schematic drawing of a renal tubule cell illustrating potential mechanisms of disordered amino acid transport and metabolism. Mechanism 1, defective stereospecific membrane receptor in brush border, cotransport of Na$^+$ usually involved; mechanism 2, competition for energy (Na$^+$ cotransport); mechanism 3, metabolic block leading to accumulation, intracellularly, of transported amino acid; mechanism 4, metabolic block leading to accumulation of metabolites of transported amino acid (the metabolite may be an amino acid); mechanism 5, defective stereospecific exit step at basolateral membrane; mechanism 6,6', basolateral membrane entry step and brush border membrane exit step, potential pathway for secretion of amino acids.

cell surface may have transport mechanisms which differ from those in the basolateral cell membranes, and which are modified for transepithelial transport (reabsorption, absorption). In addition, differing efflux systems at the luminal versus basolateral surface can facilitate directing amino acid movement.

With this admittedly simplified schema (Fig. 3), one can anticipate the types of disorders classified as aminoacidurias and the mechanisms of amino acid interaction. First, the amino acid entering in cotransport with Na$^+$ may encounter a defective stereospecific transport mechanism (defect 1). Alternatively, the electrochemical driving force for Na$^+$ may be reduced, impairing active amino acid reabsorption (defect 2). In the former case, the defect may be specific for a group of amino acids (e.g., all neutral amino acids), or for a member of a specific group (e.g., methionine alone of all neutral amino acids). In the latter case, there will be an increased aminoaciduria, but it will lack specificity. Second, if an accumulated amino acid cannot be metabolized further, either because of defective enzyme–substrate or enzyme–coenzyme interaction (defect 3), the amino acid would efflux from the cell. If nonrenal cells were involved, the amino acid might accumulate in the blood and be filtered at higher than reabsorbable loads, leading to an aminoaciduria which would be classified as an overflow or non-transport defect aminoaciduria. Third, if this last sequence of events occurred several steps into the metabolic pathway (defect 4), metabolites would accumulate, efflux from cells, and be presented to the kidney for excretion. These metabolites might be substances not ordinarily presented to the kidney and for which no reabsorbing system exists. Such substances would be cleared at the same rate as inulin, but the failure to reabsorb these materials would not represent a primary tubular defect. If a metabolic defect were confined to the kidney, accumulation and efflux of amino acids could give the appearance of a transport defect or even of secre-

tion. Fourth, amino acids exiting through stereospecific pathways in the basolateral membrane may compete for efflux thereby accumulating in the cell and leaking back into the lumen producing the picture of a renal aminoaciduria (defect 5). Finally, there may be secretion (defect 6,6').

This schema also permits a classification of potential amino acid interactions. There may be competition for stereospecific transport, competition for Na$^+$ cotransport, competition for intracellular metabolism, competition for basolateral efflux, and competition for basolateral influx.

The remainder of the discussion will focus only on those disorders which present as transport abnormalities. In discussing these disorders, a logical though unproved assumption is made. It is assumed that genes are coded for proteins and that a genetic defect affecting the synthesis of proteins involved in amino acid transport will produce specific transport abnormalities. Thus, one may anticipate autosomal recessive disorders affecting (1) neutral amino acid transport; (2) dibasic amino acid transport; (3) imino acids and glycine; (4) dicarboxylic amino acids; and (5) β-amino acids or specific members of any of the groups (Table I). If a genetic defect leads to a metabolic abnormality that results in renal cell toxicity, the anticipated transport defect would be generalized. Thus, in galactosemia or hereditary fructose intolerance, the accumulation of galactose-1-phosphate or fructose-1-phosphate results in a generalized aminoaciduria. There may be other disorders, more likely inherited as autosomal dominant or X-linked traits, which will produce nonspecific defects in transport that cannot be readily tied to a specific protein abnormality. There may also be transport defects that appear to be inherited as dominant traits when in reality they represent the heterozygous expression of an autosomal recessive trait. Finally, there may be acquired exogenous toxins, such as heavy metals, or endogenous toxins, such as Bence–Jones proteins, which also result in tubular injury and in nonspecific transport abnormalities.

3.1. Hartnup Disease

Hartnup disease is a disorder characterized by a pellagra-like skin rash, cerebellar ataxia, variable psychiatric illness, and an inconsistent mental retardation.[328] The mental retardation has been noted in a minority of patients and may represent a biased selection, since screening for aminoacidurias was frequently carried out in patients who were mentally retarded. The renal transport defect in Hartnup disease was documented to be a persistent aminoaciduria affecting the monoaminomonocarboxylic amino acids, but sparing dibasic, dicarboxylic, and imino acids as well as glycine. A pathophysiologic connection between the renal aminoaciduria and the clinical syndrome was not readily apparent. Observations indicating that bacterial breakdown products of tryptophan, i.e., indican and indols, were excreted in increased concentrations in stool and urine of patients with Hartnup disease led to the consideration that a transport defect comparable to that located in the renal epithelium was present in the intestine.[329] If this were the case, the two major epithelial cells modified for transepithelial transport of amino acids would be assumed to be under the same genetic control for specific transport systems.

Milne *et al.* noted that after tryptophan feeding, increased amounts of free tryptophan appeared in the stool of Hartnup patients but not in comparably fed control individuals.[330] There was a smaller rise in plasma tryptophan concentration after oral ingestion in Hartnup patients versus controls. The

Table I. Amino Acid Transport Disorders

Amino acid group and individually affected group members	Disorder of transport
Monoaminomonocarboxylic amino acids	Hartnup disease
Methionine	Methionine malabsorption
Tryptophan	Blue diaper syndrome
Histidine	Histidinuria
Diamino amino acids	Cystinuria
Cystine	Isolated cystinuria
Lysine, arginine, ornithine	Lysinuric protein intolerance
Lysine	Isolated lysine malabsorption[a]
Imino acids and glycine	Iminoglycinuria
Proline, hydroxyproline	None described
Glycine	? Glycinuria
Dicarboxylic amino acids	Dicarboxylic aminoaciduria
β-Amino acids	None described

[a]Ref. 416.

Hartnup patients also had increased quantities of indican and indols in their urine and stool and, when bacteria were suppressed with neomycin, there was a reduction in indols and indicans and a far greater rise in tryptophan excretion in the stool. Scriver, employing a two-dimensional chromatography of urine and stool, noted a marked increase in free neutral amino acids in both urine and feces of patients with Hartnup disease.[331] Finally, Shih et al. were able to demonstrate defective tryptophan and methionine transport in biopsy specimens of intestinal mucosa.[332]

That the defect was specific for free amino acids was demonstrated by experiments proving that oligopeptide absorption of tryptophan-containing substances was normal in patients with Hartnup disease.[333,334] The demonstration of defective intestinal transport of neutral amino acids went a lot further toward explaining the clinical picture of Hartnup disease. Though it is estimated that approximately 60 mg of tryptophan is required to produce 1 mg of nicotinamide, half of the usual daily requirement for nicotinamide is derived from dietary tryptophan.[335,336] If there were defective absorption of tryptophan, particularly if there were periods of dietary lack of increased metabolic demand, a picture of niacin deficiency characterized by pellagra-like skin rash and neurological deficit might emerge. There has also been a suggestion that indol formation may inhibit nicotinamide synthesis, which might further reduce tryptophan absorption, leading to further indol production, etc.[328] This cycle of events could add to the problems of nicotinamide deficiency in the patient with Hartnup disease. In any event, it is clear that individuals with Hartnup disease may achieve significant improvement in both their neurological problems and their skin rash by the administration of nicotinamide.[337]

Thus, Hartnup disease provides the information that the renal tubule and intestinal epithelial cells are under the same genetic control. Demonstration of a defective transport system in either tissue should suggest the discovery of a comparable lesion in the other tissue. In the case of Hartnup disease, it was the presence of the intestinal transport defect which explained the clinical picture and allowed a rational approach to therapy. It was also the discovery of Hartnup disease which indicated that a single genetic abnormality could modify the control of the transport of all amino acids in the neutral group. It is worth noting that although the entire group of neutral amino acids is affected by

the defect in Hartnup disease, there are specific disorders for specific substrates within the neutral amino acid group. Therefore, methionine malabsorption with an associated defect in the kidney,[338] blue diaper syndrome (malabsorption of tryptophan),[339] and histidinuria with defective renal and intestinal transport of histidine[340] have been described without abnormalities for transport of the remainder of the neutral amino acids. These latter observations suggest that there are group-specific transport systems and more substrate-specific transport systems within the group, either of which may malfunction.

3.2. Cystinuria

Cystinuria, a transport defect involving cystine, lysine, arginine, and ornithine, is inherited as an autosomal recessive trait.[341] The disorder would be nothing more than a physiological curiosity were it not for the fact that cystine is the least soluble naturally occurring amino acid, and precipitates in the urinary tract forming calculi which can cause discomfort, obstruction, and potentially renal insufficiency. The disorder has been recognized since the early 19th century and was one of the classic inborn errors described by Garrod.[342] The first major clue to the abnormality in cystinuria was provided by Dent et al.,[343,344] who demonstrated that there was increased renal clearance of cystine and the dibasic amino acids at a time when the plasma levels of these substances were normal or low. Thus, with a normal or reduced filtered load, the renal tubule was incapable of reabsorbing the affected amino acids. Dent et al. postulated that there was a single stereospecific transport mechanism which recognized the unique structure of the dibasic amino acids and cystine, and that this mechanism was altered in cystinuria.

This postulate was challenged by several observations. First, in clearance studies, cystine clearance was noted to exceed inulin clearance, suggesting secretion.[345] Second, in animal studies, Rosenberg et al. were able to demonstrate that the dibasic amino acids lysine, arginine, and ornithine were competitive inhibitors, one of the other, in a kidney slice system, but that cystine neither competed with nor was inhibited by the dibasic amino acids.[346] These animal studies were extended by Fox et al.[347] to human kidney tissue, in which the observation that there was defective cellular accumulation of the dibasic amino acids and competition between the dibasic amino acids was consistent with Dent's hypothesis. There was, however, no defect in cystine accumulation by kidney tissue from cystinuric subjects, and cystine did not interact with the dibasic amino acids.

This potential discrepancy between a reasonable hypothesis and a series of observations inconsistent with the hypothesis has been partially resolved by the following observations: Wedeen and Thier noted that cellular accumulation of amino acids in kidney slices occurred in straight segments of the proximal tubule, rather than the pars convoluta, where the majority of amino acid reabsorption was thought to occur.[314] Wedeen and Weiner extended these studies, noting that whereas in kidney slices p-aminohippurate entered cells and was pumped into the lumen of proximal tubules, inulin did not penetrate the lumen of the proximal tubule.[315,316] These observations suggested that the lumina of proximal tubules in kidney slices were frequently sealed from communication with bathing medium. In this same system the nonmetabolizable amino acid, α-aminoisobutyric acid, was taken up by proximal tubule cells but was never apparent in the lumen. Wedeen and Weiner suggested that these observations indicated that much of cellular accumulation in kidney slices occurred from the periluminal surface.

Subsequent studies by Ausiello et al.[318] and Greth et al.[319] demonstrated that, following the injection of lysine or cystine into the tail vein of rats, uptake of these amino acids into proximal tubule cells could be demonstrated. When one ureter was ligated before tail vein injection, interrupting GFR, the cellular accumulation continued at only a minimally reduced rate on the ligated versus the unligated side. Thus, in vivo, much of cellular accumulation could be sustained from the nonluminal cell surface. In simultaneous clearance and uptake studies, the clearance of lysine could be markedly increased by arginine and the clearance of cystine markedly increased by lysine, whereas cellular accumulation of these amino acids actually rose; when luminal uptake of amino acid was inhibited, cellular accumulation might not only be sustained but enhanced. These observations seem most consistent with the hypothesis that there is an inherited disorder in the brush border of the renal tubule cell which affects the absorption of cystine and the dibasic amino acids; this mechanism is defective in cystinuria. Studies of rat tubule fragments and brush border membranes support and expand this hypothesis.[348–350] Uptake of cystine by isolated cortical tubules, which presumably expose more brush border to the incubation medium, occurs via high (0.055 mmole) and low (0.012 mmole) K_m systems. Lysine competes for the low, but not the high K_m system, and cystine inhibits lysine uptake. In brush border membrane vesicles, several cystine transport systems are demonstrable with the dibasic amino acids again sharing a low K_m system with cystine. Heteroexchange between cystine and lysine also occurs via the low K_m system. In contrast to kidney slices (see below) and tubule fragments where cystine is converted to cysteine in the cell, cystine remains unreduced in membrane vesicles, and becomes bound to the inside of the membrane.

Thus, a low-K_m, high-affinity brush border membrane transport system shared by cystine and the dibasic amino acids exists along with a high-K_m cystine transport mechanism that is unshared. There is also evidence for a second dibasic amino acid transport system of low affinity that is unaffected in cystinuria. Cystinuria appears to be a defect in the shared low-K_m high-affinity system. In addition, the reduction of cystine to cysteine in intact cells maintains a favorable gradient for cystine entry and avoids binding of cystine, thus facilitating transcellular movement.

There are, in the basal and lateral cell membranes, separate uptake systems for the dibasic amino acids and cystine which serve nutritive cell function and are comparable to uptake systems in other tissues in the body. In fact, studies of other tissues, such as leukocytes, indicate that there is no interaction between cystine and dibasic amino acids in that tissue.[351] The accumulation of amino acids from the basal and lateral cell surfaces leads to increased cellular levels of those amino acids and the potential for leakage of the amino acid into the lumen. If there is a defect in the brush border, then the luminal leakage of a substance such as cystine will be added to the already unreabsorbed filtrate and produce a picture of cystine clearances in excess of inulin clearance. What remains to be explained is the reason for an increase in cellular accumulation when lysine is added to cystine. This may be explained by the observations of Segal and his colleagues: first, that cystine exists intracellularly primarily as cysteine and reduced glutathione[352,353]; and second, that cysteine competes with the dibasic amino acids for efflux.[354] Thus, if one administered lysine simultaneously with cystine, the accumulation of lysine intracellularly by a method independent of cystine uptake would lead to increased levels of lysine and cystine and then competition for efflux. These processes could occur at the same time that there was interference with luminal reabsorption or luminal delivery.

The suggested renal defect in cystinuria would then be an abnormality in the brush border specific for that cell surface. If this hypothesis were correct, one would predict from the experience in Hartnup disease, that there would be a defect in the intestinal brush border as well. Studies of intestinal transport carried out by Milne et al. utilizing the same reasoning as applied to tryptophan in Hartnup disease, demonstrated defective absorption of the dibasic amino acids.[355,356] Studies by Thier et al.[357,358] and McCarthy et al.[359] on intestinal mucosa from cystinuric and control subjects demonstrated that in cystinuria there was defective transport of both cystine and the dibasic amino acids into intestinal epithelial cells. Furthermore, competition between cystine and the dibasic amino acids could be shown in the intestinal mucosa. This ability to demonstrate the defect in the intestine in vitro, while the defect could not be demonstrated in the kidney in vitro, may be explained in part by the observations of Alpers and Thier[360] that the intestinal cellular accumulation and incorporation of amino acids into protein occur predominantly from the brush border surface, with very little occurring from the serosal surface. This is in contrast to the marked polar differences seen in renal tubule cells. In addition, brush border membrane vesicles from intestine have only one transport system for cystine, and that is shared with the dibasic amino acids.

Studies of the intestinal transport defect in cystinuria led to some surprising observations on the genetics of the disorder. Harris et al.[361,362] had described the genetics of cystinuria as a homozygous individual being the product either of the mating of "truly" recessive parents (individuals with no abnormalities documented in their urine), or the product of parents who were "incompletely" recessive (individuals with increased excretions of cystine and dibasic amino acids in their urine). In the "truly" recessive family, siblings of the affected individual would have no urinary abnormalities, whereas in the "incompletely" recessive form, siblings would have increased excretions of the affected amino acids.

Table II. Genetics of Cystinuria[a]

Harris classification	Rosenberg classification

Completely recessive
    Family: No apparent abnormality
    Homozygote: Cystinuria, dibasic aminoaciduria, stone formation
	Type I
	    Family: No apparent abnormality
	    Homozygote: Cystinuria, dibasic aminoaciduria, stone formation
	        In vitro: Absent intestinal transport of cystine and dibasic amino acids
	        In vivo: Failure to increase plasma cystine after oral load

Incompletely recessive
    Family: Increased excretion of cystine and dibasic amino acids (less than homozygote)
    Homozygote: Cystinuria, dibasic aminoaciduria, stone formation. Indistinguishable from homozygote of completely recessive family
	Type II
	    Family: Increased excretion of cystine and dibasic amino acids (less than homozygote)
	    Homozygote: Cystinuria, dibasic aminoaciduria, stone formation
	        In vitro: Transport of cystine but not dibasic amino acids in intestine
	        In vivo: Failure to increase plasma cystine after oral load
	Type III
	    Family: Increased excretion of cystine and dibasic amino acids (less than homozygote)
	    Homozygote: Cystinuria, dibasic aminoaciduria, stone formation
	        In vitro: Transport of cystine and dibasic amino acids in intestine
	        In vivo: Normal increase in plasma cystine after oral load

[a] Reprinted from De Fronza, R. A., and Thiers, S. O. 1981. Inherited disorders of renal tubular function. In: *The Kidney*, 2nd ed. B. M. Brenner and F. C. Rector, Jr., eds. W. B. Saunders, Philadelphia, p. 1835.

In this system of Harris *et al.*, there was no way of discerning the type of inheritance from studying the homozygous individual directly. During studies of intestinal transport, Rosenberg *et al.*[363] observed that although most cystinurics had no intestinal transport of cystine and the dibasic amino acids, there were subpopulations of cystinuric subjects in which partial or even nearly complete transport of cystine and dibasic amino acids occurred. These investigators were able to dissect cystinuria into three genetic varieties, shown in Table II, in which the genetic type could be discerned from studies of the homozygote directly. The presentation of at least three genetic patterns raised the question of whether there was a single genetic mechanism with multiple potential defects, or whether there were several steps in the amino acid transport process with each step capable of presenting as a different genetic form of cystinuria. Fortuitous matings of individuals with completely and incompletely recessive cystinuria produced offspring with full-blown homozygous cystinuria, indicating that the genetic variance represented abnormalities in a single allele.[364,365]

Cystinuria, as contrasted to Hartnup disease, is a transport disorder in which the renal defect produces the clinical symptomatology. Treatment of cystinuria with high fluid and alkalinization to reduce the precipitation of cystine is generally successful. However, there are individuals who respond inadequately to that program who may be controlled with the use of the drug D-penicillamine.[366] Since D-penicillamine has numerous toxic manifestations, it should be used only when specifically indicated, and only under the supervision of a physician experienced in its administration. Mercaptoproprionylglycine (MPG) has been reported to be as effective as D-penicillamine, but less toxic. Initial studies in the United States, however, indicate that MPG has a very high incidence of side effects in patients previously sensitive to D-penicillamine.[367–374] Glutamine has been observed to reduce cystine excretion, but there have been conflicting reports of its efficacy.[375–378]

Cystinuria demonstrates, then, another group-specific transport system under the same genetic control in kidney and intestine. The renal lesion shows the presence of secretion of amino acids and is explicable most easily by viewing the renal tubule cell as a polar system, with different transport capabilities at the basal and brush border membranes. The intestinal epithelial cell does not appear to have the same functional polar characteristics, but does demonstrate the inherited defect in the brush border extremely well. In addition, the intestinal cell is a much more sensitive genetic marker than is the renal cell. As in Hartnup disease, there are disorders affecting subgroups of the cystine–dibasic amino acid category. Thus, there are disorders of dibasic aminoaciduria without cystinuria,[379,380] and cystinuria without dibasic aminoaciduria.[381] It is possible that these abnormalities represent genetic defects in brush border substrate-specific transport mechanisms. It is also important, however, that these defects be analyzed with the recognition that the dibasic amino acids and cystine are also transported independently at the basolateral membrane of the renal tubule cell. Thus, a defect in the basolateral membrane could produce either selective cystinuria or dibasic aminoaciduria by cellular accumulation of amino acid leading to inhibition of further uptake from the lumen or to back-diffusion.

3.3. Lysinuric Protein Intolerance

Lysinuric protein intolerance is a disorder of dibasic amino acid transport associated with hyperammonemia after protein loading. The disease, inherited as an autosomal recessive trait, may be expressed as asymptomatic dibasic aminoaciduria in the heterozygous state.[380] The asymptomatic aminoaciduria is then transmitted as an apparent dominant trait. Clinically, homozygous patients develop vomiting and diarrhea with high protein intake, and may be of short stature. Hepatosplenomegaly, neutropenia, and thrombocytopenia have all been reported. The hyperammonemia is thought to be related to insufficient dibasic amino acids to maintain full urea cycle function.[382] Administration of arginine or ornithine can reverse the hyperammonemia.

The disease lysinuric protein intolerance is expressed in both kidney and gut. An isolated defect in dibasic amino acid transport could easily be explained in the kidney because of the independent transport systems for cystine and dibasic amino acids at luminal and periluminal surfaces. In the gut, however, a defect in the brush border should involve cystine as well as the dibasic amino acids. In fact, the defect in lysinuric protein intolerance appears to reside in the basolateral membrane and to be a defect in efflux.[383,384] This would, if initial observations are borne out, be the first aminoaciduria defined as due to a basolateral transport defect.

3.4. Iminoglycinuria

In a classic set of experiments, Scriver et al.[310] reported that individuals with hyperprolinemia excreted not only increased quantities of proline but also hydroxyproline and glycine in their urine. These authors postulated that proline, hydroxyproline, and glycine shared a common transport mechanism in the kidney, and that when proline was filtered at greater than normal loads, the proline competed with hydroxyproline and glycine for reabsorption. The renal transport systems shared by the imino acids and glycine were subsequently shown to be defective in an autosomal recessive disorder of transport. Tada et al.[385] reported two patients with increased clearance of imino acids and glycine at normal plasma concentrations of these amino acids. Similar observations were made by Scriver and Wilson.[386] These latter investigators noted that in homozygous iminoglycinuria, the vast majority of filtered proline and hydroxyproline was cleared, but less than half of filtered glycine was cleared. Of particular importance was the observation that in the homozygous individual, infusion of proline or hydroxyproline did not increase glycine clearance any further. Furthermore, heterozygous individuals had glycinuria without iminoaciduria. These observations led Scriver and his co-workers to postulate that the disease iminoglycinuria was one affecting a high-capacity, low-affinity system shared by all three amino acids. The persistence of the majority of glycine reabsorption and a portion of proline and hydroxyproline reabsorption suggested secondary transport systems perhaps of higher affinity but of lower capacity. Studies in kidney slices and isolated renal tubules, which demonstrated three separate glycine transport systems and at least two separate systems for proline and hydroxyproline, supported this analysis.[387,388]

Thus, the transport disorder iminoglycinuria was predicted from studies of a disorder of imino acid metabolism. Although clinically of no significance, the disorder does define clearly a separate group-specific transport system for imino acids and glycine. This disorder, more than any other disease of amino acid transport, helped identify the presence of secondary, more specific transport mechanisms for individual members of amino acid groups. The observation that heterozygotes for iminoglycinuria may have glycinuria alone provides a potential explanation for the observed disorder hyperglycinuria, which appears to be inherited as an autosomal dominant trait.[389] It is possible that the hyperglycinuria is in fact heterozygous iminoglycinuria, transmitted as an apparent dominant trait. It is of course also possible that the hyperglycinuria represents a defect in the specific imino acid-independent glycine transport system, or a defect in the imino acid–glycine transport system affecting the affinity rather than the capacity of the transport system. A form of hyperglycinuria explicable on this latter basis has been reported in at least one family.[390] Finally, as predicted, a defect

in amino acid and glycine transport has been demonstrated in the intestinal epithelium of patients with this disorder.[391,392] Of special note is the fact that the intestinal defect is not demonstrable in all patients, and reasoning from the observations in cystinuria one might question whether there is a potential defect by which iminoglycinuria can be produced as a secretory leak process rather than as a brush border reabsorptive disorder.

3.5. Dicarboxylic Aminoaciduria

There have now been at least two recognized cases of dicarboxylic aminoaciduria.[393,394] Their importance goes beyond the small number of cases in that the disorder's occurrence was predicted on the basis of the classification of aminoacidurias developed through the study of genetic disorders.[395] In the cases reported thus far, the renal clearance of dicarboxylic amino acids exceeded creatinine clearance. Studies of the parents were possible in one case, and revealed a normal urinary amino acid excretion suggesting recessive inheritance. Defective intestinal absorption of glutamate was also demonstrable in one patient. There was mental retardation with hyperprolinemia and hypoglycemia noted in one, but not in the other case.

3.6. β-Aminoaciduria

Although there has not been a defect reported in which the transport of amino acids by the β-amino acid transport system is defective, we would predict that such a disorder will occur. The presence of a β-transport system has been defined by observations in the disorder hyperbeta-alaninemia.[396] In this disorder, characterized by CNS abnormalities, the presence of increased plasma levels and increased urinary excretion of β-alanine is accompanied by increased urinary excretion of β-aminoisobutyric acid and taurine. In a manner analogous to the definition of the imino acid–glycine system by the disease state of hyperprolinemia, there is again the definition of a transport system based on the saturation of the reabsorptive system secondary to a metabolic defect.

3.7. Generalized Aminoacidurias

Abnormal transport of the entire class of amino acids occurs with inherited disorders associated with accumulation of nephrotoxic materials, such as galactose-1-phosphate (galactosemia), fructose-1-phosphate (fructose intolerance), or copper (Wilson's disease). Acquired endogenous toxins such as light chains in multiple myeloma or more poorly defined substances in malignancy also produce generalized aminoacidurias. Nutritional disorders, tubular injury, or for that matter any structural or functional impairment of the proximal tubule, may produce generalized aminoaciduria and the broader disorder—the Fanconi syndrome.

3.8. Fanconi Syndrome

The Fanconi syndrome is a generalized disorder of renal tubular transport which results in phosphaturia, as well as aminoaciduria, glucosuria, and less constantly excessive urinary losses of bicarbonate, uric acid, potassium, and calcium. In a given individual, any one or all of these defects in tubular transport may be present. Clinically, it is manifested by osteomalacia or rickets, growth failure, chronic acidosis, and symptoms referable to hypokalemia. In children, the Fanconi syndrome is most commonly associated with cystinosis, whereas in adults, the most common causes are multiple myeloma, Wilson's disease, and renal toxins (outdated tetracycline, heavy metals, Lysol, vitamin D intoxication). Other causes include Lowe's syndrome, glycogen storage disease, fructose intolerance, tyrosinosis, galactosemia, amyloidosis, and malignancy. Many cases are idiopathic and occasionally the disease may be familial.

Classically, the Fanconi syndrome has been attributed to a generalized defect in proximal tubular reabsorption,[397] since amino acids, glucose, and phosphate are known to be primarily, if not exclusively, reabsorbed in this segment. This concept was strengthened by the finding of morphological alterations confined to the proximal tubule in both adults and children.[398,399] Many of the disease entities associated with the Fanconi syndrome are associated with extensive intracellular deposition of amino acids, sugars, and low-molecular-weight proteins,[400-407] or the ingestion of nephrotoxins.[408,409] These have been shown to result in disruption of the endoplasmic reticulum and mitochondrial network, structural changes which could lead to defects in renal tubular transport processes. Similar ultrastructural changes have also been reported with the maleic acid model of the Fanconi syndrome.[410,411]

The generalized nature of the transport defect (i.e., multiple amino acid groups, glucose, phosphate, uric acid, bicarbonate) in the Fanconi syndrome argues against a specific defective carrier transport system. A more likely explanation for the generalized transport defect is the inhibition of a common energy source at the level of the Krebs cycle or inhibition of transport enzymes including Na^+,K^+-ATPase[412] or an intrinsic defect of the membrane structure, that allows increased efflux of the transported substrate. Lastly, Na^+ has been shown to be important in the transport of many of the substances that are excreted in excessive quantities in the Fanconi syndrome. Thus, an abnormality in the handling of this ion by the brush border membrane could lead to a more generalized transport defect.

Bergeron et al.,[413] using stop-flow and single-nephron microperfusion techniques, have shown that the phosphaturia, as well as aminoaciduria and glucosuria, following the injection of sodium maleate into rats, is the result of efflux throughout both proximal and distal tubules, and is associated with major modifications in the relationship between mitochondria and perimitochondrial membranes of both proximal and distal tubules. The brush border structure of the proximal tubule was well preserved. Measurement of net influx of phosphate, amino acids, and glucose in the proximal tubule was normal. Similar findings have been reported in vitro employing renal cortical slices incubated with maleic acid.[414] These data, if they can be extrapolated to other causes of the Fanconi syndrome, suggest that the transport defect may be an efflux phenomenon and not related to defective proximal tubular reabsorption. The increased efflux from the early part of the proximal tubule can be corrected further downstream, since the reabsorptive capacity of the entire proximal tubule is very high. However, since the transport capacity of the distal nephron for phosphate, amino acids, and glucose is small, efflux of molecules at this site cannot be corrected and enhanced excretion occurs.

Bovee et al.[415] have described a dog Fanconi model which closely resembles the human model. Histological examination of renal tissues obtained from animals with normal renal function was unremarkable. Studies examining amino acid and glucose uptake by renal cortical slices were most consistent with a defect in influx, efflux being normal. These results are in contrast to those observed in the maleic acid model of the Fanconi syndrome. Clearly, further studies employing other models will be needed to clarify whether the Fanconi syndrome is a disorder of efflux versus influx and what role the distal nephron plays in the disturbed transport processes.

4. Summary

Clinical disorders of amino acid and phosphate transport are discussed in terms of normal transport physiology and the insight that these disorders have provided for the development of new physiological principles.

The pathogenesis of the major clinical disorders of phosphate transport (hyperparathyroidism, VDRR, VDDR, pseudohypoparathyroidism) is reviewed. VDRR appears to be the result of a primary renal tubular defect in phosphate reabsorption. Whether enhanced tubular sensitivity to PTH contributes to the renal phosphate wasting remains conjectural. The phosphaturia in conditions associated with VDDR appears to be related to excessive PTH secretion. Pseudohypoparathyroidism can be subdivided into two types. In type I, circulating PTH levels are high but the proximal tubule cannot respond by making cAMP. The impairment in renal phosphate excretion can be overcome by administering dibutyryl cAMP. Pseudohypoparathyroidism type II is also characterized by a blunted response to PTH but these individuals demonstrate a normal nephrogenous cAMP response.

A general schema for classifying disorders of amino acid transport is developed and potential sites of defective transport are identified. The currently recognized renal aminoacidurias are discussed within this framework and the predictive value of the classification is illustrated. Specific emphasis is given to cystinuria as a model for the physiologic and genetic principles learned from studying the disorders of amino acid transport. Specifically, the study of cystinuria illustrates (1) the presence of comparable genetically controlled transport mechanisms for amino acids in gut and renal tubular epithelial cells; (2) the intestinal mucosa as a sensitive genetic marker; (3) the presence of bidirectional amino acid transport by the renal tubule cell; and (4) the dissociation between cellular uptake and transepithelial transport of amino acids in the renal tubule cells.

References

1. Bijvoet, O. L. M. 1980. Indices for the measurement of renal handling of phosphate. In: Renal Handling of Phosphate. S. G. Massry and H. Fleisch, eds. Plenum Press, New York. pp. 1–37.
2. Agus, Z. S., S. Goldfarb, and A. Wasserstein. 1981. Disorders of calcium and phosphate balance. In: The Kidney, 2nd ed. B. M. Brenner and F. C. Rector, Jr., eds. W. B. Saunders, Philadelphia. pp. 940–1022.
3. Marshall, E. K., and A. L. Grafflin. 1933. Excretion of inorganic phosphate by the glomerular kidney. Proc. Soc. Exp. Biol. Med. 31:44–46.
4. Wolbach, R. A. 1970. Phlorizin and renal phosphate secretion in the spiny dog fish—Squalus acanthias. Am. J. Physiol. 219:886–888.
5. Walker, A. M., and C. C. Hudson. 1937. The role of the tubule in the excretion of inorganic phosphates by the amphibian kidney. Am. J. Physiol. 118:167–173.
6. Clark, N. B., and W. H. Dantzler. 1972. Renal tubular transport of calcium and phosphate in snakes: Role of parathyroid hormone. Am. J. Physiol. 223:1455–1468.

7. Levinsky, N. G., and D. G. Davidson. 1957. Renal action of parathyroid extract in the chicken. *Am. J. Physiol.* **191**:530–536.

8. Handler, J. S. 1962. A study of renal phosphate excretion in the dog. *Am. J. Physiol.* **202**:787–790.

9. Strickler, J. C., D. D. Thompson, R. M. Klose, and G. Giebisch. 1964. Micropuncture study of inorganic phosphate excretion in the rat. *J. Clin. Invest.* **43**:1596–1607.

10. Le Grimellec, C., N. Roinel, and F. Morel. 1974. Simultaneous Mg, Ca, P, K, and Cl analysis in rat tubular fluid. IV. During acute plasma phosphate loading. *Pfluegers Arch.* **346**:189–204.

11. Amiel, C., H. Kuntziger, and G. Richet. 1970. Micropuncture study of handling of phosphate by proximal and distal nephron in normal and parathyroidectomized rat: Evidence for distal reabsorption. *Arch. Ges. Physiol.* **317**:93–109.

12. Goldfarb, S., L. H. Beck, Z. S. Agus, and M. Goldberg. 1974. Dissociation between the sites of action of PTH and saline on renal phosphate reabsorption. *Clin. Res.* **22**:528A.

13. Agus, Z. S., J. B. Puschett, D. Senesky, and M. Goldberg. 1971. Mode of action of cyclic adenosine 3',5'-monophosphate on renal tubular phosphate reabsorption in the dog. *J. Clin. Invest.* **50**:617–626.

14. Wan, S. F. 1974. Micropuncture studies of phosphate transport in the proximal tubule of the dog. *J. Clin. Invest.* **53**:143–153.

15. Agus, Z. S., L. B. Gardner, L. H. Beck, and M. Goldberg. 1973. Effects of parathyroid hormone on renal tubular reabsorption of calcium, sodium and phosphate. *Am. J. Physiol.* **224**:1143–1148.

16. Lambert, P. P., R. Vanderveiken, J. P. DeKoster, R. J. Kahn, and M. DeMyttenaere. 1964. Study of PO₄ excretion by the stop-flow technique: Effects of PTH. *Nephron.* **1**:103–117.

17. Davis, B. B., L. H. Kedes, and J. V. Field. 1966. Demonstration of distal tubule flux of phosphate using modified stop-flow analysis. *Metabolism* **15**:482–491.

18. Le Grimellec, C., P. Poujeol, and C. de Rouffignac. 1975. ³H inulin and electrolyte concentrations in Bowman's capsule in rat kidney. *Pfluegers Arch.* **354**:117–131.

19. Harris, C. A., P. G. Baer, E. Chirito, and J. H. Dirks. 1974. Composition of mammalian glomerular filtrate. *Am. J. Physiol.* **227**:972–976.

20. Staum, B. B., R. J. Hamburger, and M. Goldberg. 1972. Tracer microinjection study of renal tubular phosphate reabsorption in the rat. *J. Clin. Invest.* **51**:2271–2276.

21. Amiel, C. 1980. Sites of renal tubular reabsorption of phosphate. In: *Renal Handling of Phosphate.* S. G. Massry and H. Fleisch, eds. Plenum Press, New York. pp. 39–57.

22. Chiu, P. J. S., Z. S. Agus, and M. Goldberg. 1974. Effect of parathyroidectomy on renal phosphate transport in the rat. *Clin. Res.* **23**:520A.

23. Murayama, Y., F. Morel, and C. Le Grimellec. 1972. Phosphate, calcium and magnesium transfers in proximal tubules and loops of Henle, as measured by single nephron microperfusion experiments in the rat. *Pfluegers Arch.* **333**:1–16.

24. Haramati, A., F. G. Knox. 1981. Is phosphate reabsorbed by the distal nephron? *Miner. Electrolyte Metab.* **6**:165–173.

25. Jamison, R. L., and J. F. Arrascue. 1980. Calcium and phosphate reabsorption by the loop of Henle. *Miner. Electrolyte Metab.* **4**:97–105.

26. de Rouffignac, C., F. Morel, and N. Roinel. 1976. Micropuncture study of water and electrolyte movement along the loop of Henle in *Psammomys* with special references to magnesium, calcium and phosphorus. *Pfluegers Arch.* **344**:309–326.

27. Knox, F., H. Osswald, G. Marchand, W. Spielman, J. Haas, T. Berndt, and S. Youngberg. 1977. Phosphate transport along the nephron. *Am. J. Physiol.* **233**:F261–F268.

28. Goldfarb, S. 1980. Juxtamedullary and superficial nephron phosphate reabsorption in the cat. *Am. J. Physiol.* **239**:F336–F342.

29. Haas, J., T. Berndt, and F. Knox. 1978. Nephron heterogeneity of phosphate reabsorption. *Am. J. Physiol.* **233**:F287–F300.

30. Pastoriza-Munoz, E., R. Colindres, W. Lassiter, and C. Lechene. 1978. Effect of parathyroid hormone on phosphate reabsorption in rat distal convolution. *Am. J. Physiol.* **235**:F321–F330.

31. Harris, C., M. Burnatowska, J. Seely, R. Sutton, G. Quamme, and J. Dirks. 1979. Effects of parathyroid hormone on electrolyte transport in the hamster nephron. *Am. J. Physiol.* **236**:F342–F348.

32. Poujeol, P., B. Corman, C. Touvay, and C. de Rouffignac. 1977. Phosphate reabsorption in rat nephron terminal segments: Intrarenal heterogeneity and strain differences. *Pfluegers Arch.* **371**:39–44.

33. Peraino, R., and W. Suki. 1980. Phosphate transport by isolated rabbit cortical collecting tubule. *Am. J. Physiol.* **238**:F358–F362.

34. Shareghi, G. R., and Z. S. Agus. 1982. Magnesium transport in the cortical thick ascending limb of Henle's loop of the rabbit. *J. Clin. Invest.* **69**:759–769.

35. Haas, J., T. Berndt, and F. Knox. 1980. Failure to detect papillary collecting duct phosphate reabsorption in thyroparathyroidectomized rats. *Miner. Electrolyte Metab.* **3**:146–150.

36. Beck, L. H., and M. Goldberg. 1974. Mechanism of the blunted phosphaturia in saline-loaded thyroparathyroidectomized dogs. *Kidney Int.* **6**:18–23.

37. Mudge, G. H., W. O. Berndt, and H. Valtin. 1973. Tubular transport of urea, glucose, phosphate, uric acid, sulfate and thiosulfate. In: *Handbook of Physiology* Section 8. J. Orloff and R. W. Berliner, eds. Williams & Wilkins, Baltimore, pp. 587–652.

38. Puschett, J. B., Z. S. Agus, D. Senesky, and M. Goldberg. 1972. Effects of saline loading and aortic obstruction on proximal phosphate transport. *Am. J. Physiol.* **223**:851–857.

39. Thompson, D. D., and H. H. Hiatt. 1957. Effects of phosphate loading and depletion on the renal excretion and reabsorption of inorganic phosphate. *J. Clin. Invest.* **36**:566–572.

40. Steele, T. H., and H. F. DeLuca. 1976. Influence of dietary phosphorus on renal phosphate reabsorption in the parathyroidectomized rat. *J. Clin. Invest.* **57**:867–874.

41. Trohler, H., J. P. Bonjour, and H. Fleisch. 1976. Inorganic phosphate homeostasis: Renal adaptation to the dietary intake in intact and thyroparathyroidectomized rats. *J. Clin. Invest.* **57**:264–273.

42. Samiy, A. H., P. F. Hirsch, and A. G. Ramsay. 1965. Localization of phosphaturic effect of parathyroid hormone in the nephron of the dog. *Am. J. Physiol.* **208**:73–77.

43. Pullman, T. N., A. R. Lavender, I. Aho, and H. Rasmussen. 1960. Direct renal action of a purified parathyroid extract. *Endocrinology* **67**:570–582.

44. Bell, N. H., S. Avery, T. Sinha, C. M. Clark, Jr., D. O. Allen, and C. Johnston, Jr. 1972. Effects of dibutyryl cyclic adenosine 3'-5'-monophosphate and parathyroid extract on calcium and phosphorus metabolism in hypoparathyroidism and pseudohypoparathyroidism. *J. Clin. Invest.* **51**:816–823.

45. Hamburger, R. J., N. L. Lawson, and V. W. Dennis. 1974. Effects of cyclic adenosine nucleotides on fluid absorption by different segments of proximal tubule. *Am. J. Physiol.* **227**:396–401.

46. Kinziger, H., C. Amiel, N. Roinel, and F. Morel. 1974. Effects of parathyroidectomy and cyclic AMP on renal transport of phosphate, calcium and magnesium. *Am. J. Physiol.* **227**:905–911.

47. Gekle, D. 1971. Der Einfluss von Parathormon auf die Nierenfunktion. I. Tierexperimentelle Untersuchungen. *Arch. Ges. Physiol.* **323**:96–120.

48. Nordin, B. E. C. 1960. The effect of intravenous parathyroid extract on urinary pH, bicarbonate and electrolyte excretion. *Clin. Sci.* **19**:311–319.

49. Puschett, J. B., and M. Goldberg. 1969. The relationship between the renal handling of phosphate and bicarbonate in man. *J. Lab. Clin. Med.* **73**:956–969.

50. Bank, N., H. S. Aneydjian, and S. W. Weinstein. 1974. A microperfusion study of phosphate reabsorption by the rat proximal renal tubule: Effect of parathyroid hormone. *J. Clin. Invest.* **54**:1040–1048.

51. Puschett, J. B. 1982. Are all of the renal tubular actions of parathyroid hormone mediated by the adenylate cyclase system? *Miner. Electrolyte Metab.* **7**:281–284.

52. Brazy, P. C., J. W. McKeown, R. H. Harris, and V. W. Dennis. 1980. Comparative effects of dietary phosphate, unilateral nephrectomy, and parathyroid hormone on phosphate transport by the rabbit proximal tubule. *Kidney Int.* **17**:788–800.

53. Haramati, A., J. A. Haas, and F. G. Knox. 1983. Adaptation of deep and superficial nephrons to changes in dietary phosphate intake. *Am. J. Physiol.* **244**:F265–F269.

54. Dominguez, J. H., R. W. Gray, and J. Lemann, Jr. 1976. Dietary phosphate deprivation in women and men: Effects on mineral and acid balances, parathyroid hormone and the metabolism of 25-OH vitamin D. *J. Clin. Endocrinol. Metab.* **43**:1056–1068.

55. Kreusser, W. J., K. Kurokawa, E. Aznar, and S. G. Massry. 1978. Phosphate depletion: Effect on renal inorganic phosphorus and adenine nucleotides, urinary phosphate and calcium, and calcium balances. *Miner. Electrolyte Metab.* **1**:30–42.

56. Muhlbauer, R. C., J. P. Bonjour, and H. Fleisch. 1977. Tubular localization of adaptation to dietary phosphate in rats. *Am. J. Physiol.* **233**:F342–F348.

57. Chambers, E. L., Jr., G. S. Gordan, L. Goldman, and E. C. Reifenstein, Jr. 1956. Tests for hyperparathyroidism: Tubular reabsorption of phosphate, phosphate deprivation, and calcium infusion. *J. Clin. Endocrinol. Metab.* **16**:1507–1521.

58. Steele, T. H. 1976. Renal resistance to parathyroid hormone during phosphorus deprivation. *J. Clin. Invest.* **58**:1461–1464.

59. Steele, T. H. 1977. Renal response to phosphorus deprivation: Effect of the parathyroids and bicarbonate. *Kidney Int.* **11**:327–334.

60. Barrett, P. Q., J. M. Gertner, and H. Rasmussen. 1980. Effect of dietary phosphate on transport properties of pig renal microvillus vesicles. *Am. J. Physiol.* **239**:F352–F359.

61. Kempson, S. A., and T. P. Dousa. 1979. Phosphate transport across renal cortical brush border membrane vesicles from rats stabilized on a normal, high or low phosphate diet. *Life Sci.* **24**:881–888.

62. Stoll, R., R. Kinne, H. Murer, H. Fleisch, and J. P. Bonjour. 1979. Phosphate transport by rat renal brush border membrane vesicles: Influence of dietary phosphate, thyroparathyroidectomy, and 1,25-dihydroxyvitamin D_3. *Pfluegers Arch.* **380**:47–52.

72. Harrison, H. E., and H. C. Harrison. 1941. The renal excretion of inorganic phosphate in relation to the action of vitamin D and parathyroid hormone. *J. Clin. Invest.* **20**:47–55.

73. Puschett, J. B., P. C. Fernandez, I. T. Boyle, R. W. Gray, J. L. Omdahl, and H. F. DeLuca. 1972. The acute renal tubular effects of 1,25-dihydroxycholecalciferol. *Proc. Soc. Exp. Biol. Med.* **141**:379–384.

74. Puschett, J. B., J. Moranz, and W. S. Kurnick. 1972. Evidence for a direct action of cholecalciferol and 25-hydroxycholecalciferol on the renal transport of phosphate, sodium and calcium. *J. Clin. Invest.* **51**:373–385.

75. Gekle, D., J. Stroder, and D. Rostock. 1971. The effect of vitamin D on renal inorganic phosphate reabsorption in normal rats, parathyroidectomized rats and rats with rickets. *Pediatr. Res.* **5**:40–52.

76. Avioli, L. 1980. Effects of vitamin D and its metabolites on renal handling of phosphate. In: *Renal Handling of Phosphate.* S. G. Massry and H. Fleisch, eds. Plenum Press, New York. pp. 197–208.

77. Kinne, R., W. Berner, N. Hoffmann, and H. Murer. 1976. Phosphate transport by isolated renal and intestinal plasma membranes. *Adv. Exp. Med. Biol.* **81**:265–277.

78. Murer, H., H. Stern, G. Burchkhardt, C. Storelli, and R. Kinne. 1980. Sodium-dependent transport of inorganic phosphate across the renal brush border membrane. *Adv. Exp. Med. Biol.* **128**:11–23.

79. Cheng, L., and B. Sacktor. 1981. Sodium gradient-dependent phosphate transport in renal brush border membrane vesicles. *J. Biol. Chem.* **256**:1556–1564.

80. Aronson, P. S. 1981. Identifying secondary active solute transport in epithelia. *Am. J. Physiol.* **240**:F1–F11.

81. Barrett, P. Q., and P. S. Aronson. 1982. Glucose and alanine inhibition of phosphate transport in renal microvillus membrane vesicles. *Am. J. Physiol.* **242**:F126–F131.

82. Dousa, T. P., and S. A. Kempson. 1982. Regulation of renal brush border membrane transport of phosphate. *Miner. Electrolyte Metab.* **7**:113–121.

83. Rasmussen, H., D. B. P. Goodman, N. Friedman, E. Allen, and K. Kurokawa. 1976. Ionic control of metabolism. In: *Handbook of Physiology*, Section 7, Volume 7. G. D. Aurbach, ed. American Physiological Society, Washington, D.C. pp. 225–264.

84. Albright, F., A. M. Butler, and E. Bloomberg. 1937. Resistance to vitamin D therapy. *Am. J. Dis. Child.* **54**:529–547.

85. Saville, P. D., J. R. Nassim, F. H. Stevenson, L. Mulligan, and M. Carey. 1955. The effect of A.T. 10 on calcium and phosphorus metabolism in resistant rickets. *Clin. Sci.* **14**:489–499.

86. Stickler, G. B. 1963. External calcium and phosphorus metabolism in resistant rickets. *J. Pediatr.* **63**:942–948.

87. Russell, R. G. G., R. Smith, C. Preston, R. J. Walton, C. G. Woods, R. G. Henderson, and A. W. Norman. 1975. The effect of 1,25-dihydroxycholecalciferol on renal tubular reabsorption of phosphate, intestinal absorption of calcium and bone histology in hypophosphatemic renal tubular rickets. *Clin. Sci. Mol. Med.* **48**:177–186.

88. Williams, T. F., and R. W. Winters. 1972. Familial (hereditary) vitamin D-resistant rickets with hypophosphatemia. In: *Metabolic Basis of Inherited Disease.* J. B. Stanbury, J. B. Wyngaarden, and D. W. Fredrickson, eds. McGraw-Hill, New York. pp. 1465–1485.

89. Brickman. A. S., J. W. Coburn, K. Kurokawa, J. E. Bethune, H. E. Harrison, and A. W. Norman. 1973. Actions of 1,25-dihydroxycholecalciferol in patients with hypophosphatemic vitamin D-resistant rickets. *N. Engl. J. Med.* **289**:495–498.

90. Earp, H. S., R. L. Ney, H. J. Gitelman, R. Richman, and H. F. DeLuca. 1970. Effects of 25-hydroxycholecalciferol in patients with familial hypophosphatemia and vitamin D-resistant rickets. *N. Engl. J. Med.* **283**:627–630.

91. Puschett, J. B., M. Genel, A. Rastegar, C. Anast, H. DeLuca, and A. Friedman. 1975. Long term therapy of vitamin D-resistant rickets with 25-hydroxycholecalciferol. *Clin. Pharmacol. Ther.* **17**:202–211.

92. Cohanim, M., H. F. DeLuca, and E. R. Yendt. 1972. Effect of prolonged treatment with 25-hydroxycholecalciferol in hypophosphatemic (vitamin D refractory) rickets and osteomalacia. *Johns Hopkins Med. J.* **131**:118–132.

93. Stickler, G. B., J. W. Beabout, and B. L. Riggs. 1970. Vitamin D-resistant rickets: Clinical experience with 41 typical familial hypophosphatemic patients and 2 atypical nonfamilial cases. *Mayo Clin. Proc.* **45**:197–218.

94. Reitz, R. E., and R. L. Weinstein. 1973. Parathyroid hormone secretion in familial vitamin D-resistant rickets. *N. Engl. J. Med.* **289**:941–945.

95. Glorieux, F. H., M. F. Holick, C. R. Scriver, and H. F. DeLuca. 1973. X-linked hypophosphataemic rickets: Inadequate therapeutic response to 1,25-dihydrocholecalciferol. *Lancet* **2**:287–289.

96. Glorieux, F., and C. R. Scriver. 1972. Loss of a parathyroid hormone sensitive component of phosphate transport in X-linked hypophosphatemia. *Science* **175**:997–999.

97. Barbour, B. H., S. J. Kronfield, and A. M. Pawlicki. 1966. Studies on the mechanism of phosphorus excretion in vitamin D-resistant rickets. *Nephron.* **3**:40–58.

98. Lewy, J. E., E. C. Carbana, H. A. Repetto, J. M. Canterbury, and E. Reiss. 1972. Serum parathyroid hormone in hypophosphatemic vitamin D-resistant rickets. *J. Pediatr.* **81**:294–300.

99. Field, M. H., and E. Reiss. 1960. Vitamin D-resistant rickets: The effect of calcium infusion on phosphate reabsorption. *J. Clin. Invest.* **30**:1807–1812.

100. Arnaud, C., F. Glorieux, and C. Scriver. 1971. Serum parathyroid hormone in X-linked hypophosphatemia. *Science* **173**:845–847.

101. Falls, W. F., N. W. Carter, F. C. Rector, and D. W. Seldin. 1968.

Familial vitamin D-resistant rickets: Study of six cases with evaluation of the pathogenetic role of secondary hyperparathyroidism. *Ann. Intern Med.* **68**:553–560.

102. Balsan, S., and M. Garabedian. 1972. 25-Hydroxycholecalciferol: A comparative study in deficiency rickets and different types of resistant rickets. *J. Clin. Invest.* **51**:749–759.

103. Roof, B. S., C. F. Piel, and G. S. Gordan. 1972. Nature of defect for familial vitamin D-resistant rickets (VDRR) based on parathormone (PTH) assay. *Clin. Res.* **20**:624.

104. Puschett, J. B., A. Rastegar, M. Genel, C. Anast, and H. F. DeLuca. 1974. Effects of 25-hydroxycholecalciferol on urinary electrolyte excretion in hypophosphatemic rickets. *Lancet* **2**:920–926.

105. Insogna, K. L., A. E. Broadus, and J. M. Gertner. 1983. Impaired phosphorus conservation and 1,25 dihydroxyvitamin D generation during phosphorus deprivation in familial hypophosphatemic rickets. *J. Clin. Invest.* **71**:1562–1569.

106. Chesney, R. W., R. B. Mazess, P. Rose, A. J. Hamstra, H. DeLuca, and A. L. Breed. 1983. Long-term influence of calcitriol and supplemental phosphate in X-linked hypophosphatemic rickets. *Pediatrics* **71**:559–567.

107. Lyles, K. W., and M. K. Drezner. 1982. Parathyroid hormone effects on serum 1,25-dihydroxyvitamin D levels in patients with X-linked hypophosphatemic rickets: Evidence for abnormal 25-hydroxyvitamin D-1-hydroxylase activity. *J. Clin. Endocrinol. Metab.* **54**:638–644.

108. Chesney, R. W., R. B. Mazess, P. Rose, A. Hamstra, and H. F. DeLuca. 1980. Supranormal 25-hydroxyvitamin D and subnormal 1,25-dihydroxyvitamin D. *Am. J. Dis. Child.* **134**:140–143.

109. Avioli, L. V., T. F. Williams, J. Lund, and H. F. DeLuca. 1967. Metabolism of D_3-^{3}H in vitamin D-resistant rickets and familial hypophosphatemia. *J. Clin. Invest.* **46**:1907–1915.

110. DeLuca, H. F., J. Lund, A. Rosenbloom, and C. C. Lobeck. 1967. Metabolism of tritiated vitamin D_3 in familial vitamin D-resistant rickets with hypophosphatemia. *J. Pediatr.* **70**:828–832.

111. Lyles, K. W., A. G. Clark, and M. K. Drezner. 1982. Serum 1,25-dihydroxyvitamin D levels in subjects with X-linked hypophosphatemic rickets and osteomalaica. *Calcif. Tissue Int.* **34**:125–130.

112. Lyles, K. W., and M. K. Drezner. 1982. Parathyroid hormone effects on serum 1,25-dihydroxyvitamin D levels in patients with X-linked hypophosphatemic rickets: Evidence for abnormal 25-hydroxyvitamin D-1-hydroxylase activity. *J. Clin. Endocrinol. Metab.* **54**:638–644.

113. Scriver, C. R., T. M. Reade, H. F. DeLuca, and A. J. Hamstra. 1978. Serum 1,25-dihydroxyvitamin D levels in normal subjects and in patients with hereditary rickets or bone disease. *N. Engl. J. Med.* **299**:976–980.

114. Drezner, M. K., and M. R. Haussler. 1979. Serum 1,25-dihydroxyvitamin D in bone disease. *N. Engl. J. Med.* **300**:435–436.

121. Cassimos, C., C. Tsenghi, S. Michael, A. Liaromati, and K. Metaxotou. 1963. Continuous intravenous infusion of calcium in vitamin D deficient rickets. *Pediatrics* **32**:272–279.

122. Fraser, D., J. M. Leeming, E. A. Cerwenka, and K. Kenyeres. 1959. Studies of the pathogenesis of the high clearances of phosphate in hypophosphatemic vitamin D refractory rickets of the same type. *Am. J. Dis. Child.* **98**:586–587.

123. Lavender, A. R., and T. N. Pullman. 1963. Changes in inorganic phosphate excretion induced by renal arterial infusion of calcium. *Am. J. Physiol.* **205**:1025–1032.

124. Hahn, T. J., C. R. Scharp, L. R. Halstead, J. G. Haddad, D. M. Karl, and L. V. Avioli. 1975. Parathyroid hormone status and renal responsiveness in familial hypophosphatemic rickets. *J. Clin. Endocrinol. Metab.* **41**:926–937.

125. Riggs, L., R. G. Sprague, J. Jowsey, and F. T. Maher. 1969. Adult onset vitamin D-resistant hypophosphatemic osteomalacia: Effect of total parathyroidectomy. *N. Engl. J. Med.* **281**:762–766.

126. Kleerekopper, M., R. Coffey, T. Greco, S. Nichols, N. Cooke, W. Murphy, and L. Avioli. 1977. Hypercalcemic hyper-

127. Lyles, K. W., C. R. McNamara, and M. K. Drezner. 1981. Coincidence of hypoparathyroidism and X-linked hypophosphatemic rickets: An experiment of nature illuminating the physiological factors which regulate the renal tubular defect. *Am. Soc. Bone Miner. Res.* **3**:22a.

128. Short, E., A. Sebastian, M. Spencer, and R. C. Morris. 1974. Hyperresponsiveness to the phosphaturic effect of parathyroid hormone in X-linked hypophosphatemic vitamin D-resistant rickets. *J. Clin. Invest.* **53**:75a.

129. Short, E. M., H. J. Binder, and L. E. Rosenberg. 1973. Familial hypophosphatemic rickets: Defective transport of inorganic phosphate by intestinal mucosa. *Science* **179**:700–702.

130. Haussler, M. R., M. Hughes, D. Baylink, E. Littledike, D. Cork, and M. Pitt. 1977. Influence of phosphate depletion on the biosynthesis and circulating level of 1,25 dihydroxyvitamin D. *Adv. Exp. Med. Biol.* **81**:233–250.

131. Tenehouse, H. S., C. R. Scriver, R. R. McInnes, and F. H. Glorieux. 1978. Renal handling of phosphate *in vivo* and *in vitro* by the X-linked hypophosphatemic male mouse: Evidence for a defect in the brush border membrane. *Kidney Int.* **14**:236–244.

132. Glorieux, F. H., C. R. Scriver, E. M. Eicher, J. L. Southland, and R. Travers. 1974. X-linked hypophosphatemia in HYP/Y mouse: A model of the human disease. *Pediatr. Res.* **8**:389.

133. Eicher, E. M., J. L. Southard, C. R. Scriver, and F. H. Glorieux. 1976. Mouse model for human familial hypophosphatemic (vitamin D resistant) rickets. *Proc. Natl. Acad. Sci. USA* **73**:4667–4671.

134. Cowgill, L. D., S. Goldfarb, K. Lau, E. Slatopolsky, and Z. Agus. 1979. Evidence for an intrinsic renal defect in mice with genetic hypophosphatemic rickets. *J. Clin. Invest.* **63**:1203–1210.

135. Muhlbauer, R., J. P. Bonjour, and H. Fleisch. 1981. Tubular handling of phosphate (Pi) and Ca in X-linked hypophosphatemic mice: Defective adaptation to Pi restriction associated with a renal Ca leak. In: *Hormonal Control of Calcium Metabolism.* D. V. Cohn, R. V. Talmage, and J. L. Matthews, eds. Excerpta Medica, Amsterdam. p. 435.

136. Tenenhouse, H. S., C. R. Scriver, R. R. McInnes, and F. H. Glorieux. 1978. Renal handling of phosphate *in vivo* by the X-linked hypophosphatemic male mouse: Evidence for a defect in the brush border membrane. *Kidney Int.* **14**:236–244.

137. Tenenhouse, H. S., and C. R. Scriver. 1978. The defect in transcellular transport of phosphate in the nephron is located in brush-border membranes in X-linked hypophosphatemia (*Hyp* mouse model). *Can. J. Biochem.* **56**:640–645.

138. Tenenhouse, H. S., and C. R. Scriver. 1979. Renal brush border membrane adaptation to phosphorus deprivation in the *Hyp/Y* mouse. *Nature (London)* **281**:225–227.

139. Tenenhouse, H. S., and C. R. Scriver. 1979. Renal adaptation to phosphate deprivation in the *Hyp* mouse with X-linked hypophosphatemia. *Can. J. Biochem.* **57**:938–946.

140. Meyer, R. A., R. W. Gray, and M. H. Meyer. 1980. Abnormal vitamin D metabolism in the X-linked hypophosphatemic mouse. *Endocrinology* **107**:1577–1581.

141. Lobaugh, B., and M. K. Drezner. 1983. Abnormal regulation of renal 25-hydroxy vitamin D-1-α-hydroxylase activity in the X-linked hypophosphatemic mouse. *J. Clin. Invest.* **71**:400–403.

142. Rasmussen, H., and C. Anast. 1983. Familial hypophosphatemic rickets and vitamin D-dependent rickets. In: *The Metabolic Basis of Inherited Disease,* 5th ed. J. B. Stanbury, J. B. Wyngaarden, D. S. Fredrickson, J. L. Goldstein, and M. S. Brown, eds. McGraw-Hill, New York. pp. 1743–1773.

143. Scriver, C. R., T. M. Reade, H. F. DeLuca, and A. J. Hamstra. 1978. Serum 1,25-dihydroxyvitamin D levels in normal subjects and in patients with hereditary rickets or bone disease. *N. Engl. J. Med.* **299**:976–979.

144. Reade, T. M., C. R. Scriver, F. H. Glorieux, B. Nogrady, E.

Delvin, R. Poirier, F. Holick, and H. F. DeLuca. 1975. Response to crystalline 1α-hydroxyvitamin D₃ in vitamin D-dependency. *Pediatr. Res.* **9**:593–599.

145. Balsan, S., M. Garabedian, M. Lieberherr, J. Gueris, and A. Ulmann. 1979. Serum 1,25-dihydroxyvitamin D concentrations in two different types of pseudo-deficiency rickets. In: *Vitamin D: Basic Research and Its Clinical Application.* A. W. Norman, K. Schaefer, D. V. Herrath, H.-G. Grigoleit, J. W. Colburn, H. F. DeLuca, E. B. Mawer, and T. Suda, eds. de Gruyter, Berlin. p. 1143.

146. DeLuca, H. F. 1978. Vitamin D metabolism and function. *Arch. Intern. Med.* **138**:836–847.

147. Prader, A., H. P. Kind, and H. F. DeLuca. 1976. Pseudovitamin D deficiency (vitamin D dependency). In: *Inborn Errors of Calcium and Bone Metabolism.* H. Bickel and J. Stern, eds. University Park Press, Baltimore. p. 115.

148. Fraser, D., S. W. Koch, H. P. Kind, M. F. Holick, Y. Tanaka, and H. F. DeLuca. 1973. Pathogenesis of hereditary vitamin D-dependent rickets: An inborn error of vitamin D metabolism involving defective conversion of 25-hydroxycholecalciferol to 1,25-dihydroxyvitamin D. *N. Engl. J. Med.* **289**:817–822.

149. Balsan, S., M. Garabedian, R. Sorginiard, M. F. Holick, and H. F. DeLuca. 1975. 1,25-Dihydroxyvitamin D₃ and 1α-hydroxyvitamin D₃ in children: Biologic and therapeutic effects in nutritional rickets and different types of vitamin D resistance. *Pediatr. Res.* **9**:586–593.

150. Brooks, M. H., N. H. Bell, L. Love, P. H. Stern, F. Orfei, S. F. Queener, A. J. Hamstra, and H. F. DeLuca. 1978. Vitamin D-dependent rickets type II: Resistance of target organs to 1,25-dihydroxyvitamin D. *N. Engl. J. Med.* **298**:996–999.

151. Marx, S. J., A. M. Speigel, E. M. Brown, D. G. Gardner, R. W. Downs, Jr., M. Attie, A. J. Hamstra, and H. F. DeLuca. 1978. A familial syndrome of decrease in sensitivity to 1,25-dihydroxyvitamin D. *J. Clin. Endocrinol. Metab.* **47**:1303–1310.

152. Zerwekh, J. E., K. Glass, J. Jowsey, and Y. C. Charles. 1979. A unique form of osteomalacia associated with end organ refractoriness to 1,25-dihydroxyvitamin D and apparent defective synthesis of 25-hydroxyvitamin D. *J. Clin. Endocrinol. Metab.* **49**:171–175.

153. Rosen, J. F., A. R. Fleischman, L. Finberg, A. J. Hamstra, and H. F. DeLuca. 1979. Rickets with alopecia: An inborn error of vitamin D metabolism. *J. Pediatr.* **94**:729–735.

154. Liberman, U. A., R. Samuel, A. Halabe, R. Kauli, S. Edelstein, Y. Weisman, S. E. Papapoulos, T. L. Clemens, L. J. Fraher, and J. L. H. O'Riordan. 1980. End-organ resistance to 1,25-dihydroxycholecalciferol. *Lancet* **1**:504–507.

155. Tsuchiya, Y., N. Matsuo, H. Cho, M. Kumagai, A. Yasaka, T. Suda, H. Orimo, and M. Shiraki. 1980. An unusual form of vitamin D-dependent rickets in a child: Alopecia and marked end-organ hyposensitivity to biologically active vitamin D. *J. Clin. Endocrinol. Metab.* **51**:685–690.

156. Eil, C., and S. J. Marx. 1981. Nuclear uptake of [³H]1,25-dihydroxycholecalciferol in dispersed fibroblasts cultured from normal human skin. *Proc. Natl. Acad. Sci. USA* **78**:2562–2563.

157. Eil, C., V. A. Liberman, J. F. Rosen, and S. J. Marx. 1981. A cellular defect in hereditary vitamin D dependent rickets type II: Defective nuclear uptake of 1,25 dihydroxyvitamin D in cultured skin fibroblasts. *N. Engl. J. Med.* **304**:1588–1591.

158. Feldman, D. T. Chen, C. Cone, M. Hirst, S. Shani, A. Benderli, and Z. Hochberg. 1982. Vitamin D resistant rickets with alopecia: Cultured skin fibroblasts exhibit defective cytoplasmic receptors and unresponsiveness to 1,25(OH)₂D₃. *J. Clin. Endocrinol. Metab.* **55**:1020–1022.

159. Clemens, T. L., J. S. Adams, B. A. Gilcrest, Y. Tsuchiya, N. Matsuo, T. Suda, and M. F. Holick. 1981. Receptors for 1,25-dihydroxyvitamin D₃ in cultured human skin keratinocytes and fibroblasts: A model for the study of the mode of action of 1,25-dihydroxyvitamin D₃. *Fifth Workshop on Vitamin D.* p. 45.

160. Griffin, J. E., J. S. Chandler, M. R. Haussler, and J. E. Zerwekh. 1982. Receptor-positive resistance to 1,25-dihydroxyvitamin D:

A new cause of osteomalacia associated with impaired induction of 24-hydroxylase in fibroblasts. *Clin. Res.* **30**:524A.

161. Fraser, D., S. W. Koch, P. Kind, M. F. Holick, Y. Tanaka, and H. DeLuca. 1973. Pathogenesis of hereditary vitamin D-dependent rickets. *N. Engl. J. Med.* **289**:817–822.

162. Arnaud, C., R. Maijer, T. Reade, C. R. Scriver, and D. T. Whelan. 1970. Vitamin D dependency: An inherited postnatal syndrome with secondary hyperparathyroidism. *Pediatrics* **46**: 871–880.

163. Hamilton, R., J. Harrison, D. Fraser, I. Radde, R. Monecki, and L. Paunier. 1970. The small intestine in vitamin D-dependent rickets. *Pediatrics* **45**:364–373.

164. Fraser, D., S. W. Kooh, and C. R. Scriver. 1967. Hyperparathyroidism as the cause of hyperaminoaciduria and phosphaturia in human vitamin D deficiency. *Pediatr. Res.* **1**:425–435.

165. Suster, P., and J. W. Paala. 1970. Pseudovitamin D-deficiency rickets. *J. Pediatr.* **76**:937.

166. Arnaud, C., F. Glorieux, and C. R. Scriver. 1972. Serum parathyroid hormone levels in acquired vitamin D deficiency of infancy. *Pediatrics* **49**:837–840.

167. Scriver, C. R. 1974. Rickets and the pathogenesis of impaired tubular transport of phosphate and other solutes. *Am. J. Med.* **57**:43–49.

168. Grose, J. H., and C. R. Scriver. 1968. Parathyroid-dependent phosphaturia and aminoaciduria in the vitamin D-deficient rat. *Am. J. Physiol.* **214**:370–377.

169. Engstrom, G. W., H. F. DeLuca, J. W. Cramer, and H. Steenbock. 1961. Vitamin D and urinary amino acid excretion in the rat. *Proc. Soc. Exp. Biol. Med.* **108**:105–108.

170. Chisholm, J. J., and H. E. Harrison. 1962. Aminoaciduria in vitamin D deficiency states in premature infants and older infants with rickets. *J. Pediatr.* **60**:206–219.

171. Cusworth, C. D., C. E. Dent, and C. R. Scriver. 1972. Primary hyperparathyroidism and hyperaminoaciduria. *Clin. Chim. Acta* **41**:355–361.

172. Weiss, I. W., K. Morgan, and J. M. Phang. 1972. Cyclic adenosine monophosphate-stimulated transport of amino acids in kidney cortex. *J. Biol. Chem.* **247**:760–764.

173. Drezner, M. K., and F. A. Neelon. 1983. Pseudohypoparathyroidism. In: *The Metabolic Basis of Inherited Disease,* 5th ed. J. B. Stanbury, J. B. Wyngaarden, D. S. Fredrickson, J. L. Goldstein, and M. S. Brown, eds. McGraw–Hill, New York. pp. 1508–1527.

174. Farfel, Z., and H. R. Bourne. 1982. Pseudohypoparathyroidism: Mutation affecting adenylate cyclase. *Miner. Electrolyte Metab.* **8**:227–236.

175. Albright, E., C. H. Burnett, P. H. Smith, and W. Parson. 1942. Pseudo-hypoparathyroidism, an example of Seabright–Bantam syndrome: Report of three cases. *Endocrinology* **30**:922–932.

176. Tashjian, A. H., Jr., A. G. Frantz, and J. B. Lee. 1966. Pseudohypoparathyroidism: Assays of parathyroid hormone and thyrocalcitonin. *Proc. Natl. Acad. Sci. USA* **56**:1138–1142.

177. Chase, L. R., G. L. Melson, and G. D. Aurbach. 1969. Pseudohypoparathyroidism: Defective excretion of 3′,5′-AMP in response to parathyroid hormone. *J. Clin. Invest.* **48**:1832–1844.

178. Lee, J. B., A. H. Tashjian, Jr., J. M. Streeto, and G. A. Frantz. 1968. Familial pseudohypoparathyroidism. *N. Engl. J. Med.* **279**: 1179–1184.

179. Werder, E. A., J. A. Fischer, R. Illig, H. P. Kind, S. Bernasconi, A. Fanconi, and A. Prader. 1978. Pseudohypoparathyroidism and idiopathic hypoparathyroidism: Relationship between serum calcium and parathyroid hormone levels and urinary cyclic adenosine-3′,5′-monophosphate response to parathyroid extract. *J. Clin. Endocrinol. Metab.* **46**:872–879.

180. Drezner, M. K., R. A. Neelon, M. Haussler, H. T. McPherson, and H. E. Lebovitz. 1976. 1,25-Dihydroxycholecalciferol deficiency: The probable cause of hypocalcemia and metabolic bone disease in pseudohypoparathyroidism. *J. Clin. Endocrinol. Metab.* **42**:621–628.

181. Kidd, G. S., M. Schaaf, R. A. Adler, M. N. Lasman, and H. L. Wray. 1980. Skeletal responsiveness in pseudohypoparathyroidism: A spectrum of clinical disease. *Am. J. Med.* **68**:772–781.

182. Drezner, M. K., F. A. Neelon, and H. E. Lebovitz. 1973. Pseudohypoparathyroidism type II: A possible defect in the reception of the cyclic AMP signal. *N. Engl. J. Med.* **289**:1056–1060.

183. Rodriguez, H. J., H. Villarreal, S. Klahr, and E. Slatopolsky. 1974. Pseudohypoparathyroidism type II: Restoration of normal renal responsiveness to parathyroid hormone by calcium administration. *J. Clin. Endocrinol. Metab.* **39**:693–701.

184. Bell, N. H., S. Avery, T. Sinha, L. C. Clark, Jr., D. O. Allen, and C. Johnston, Jr. 1972. Effects of dibutyryl cyclic adenosine 3′-5′-monophosphate and parathyroid extract on calcium and phosphorus metabolism in hypoparathyroidism and pseudohypoparathyroidism. *J. Clin. Invest.* **51**:816–823.

185. Wertznian, R., and F. Murad. 1973. Effects of aminophylline, chlorpropamide and parathyroid extract on plasma and urinary cyclic AMP in pseudohypoparathyroidism. *Clin. Res.* **21**:89.

186. Marcus, R., J. F. Wilber, and G. D. Aurbach. 1971. Parathyroid hormone-sensitive adenyl cyclase from the renal cortex of a patient with pseudohypoparathyroidism. *J. Clin. Endocrinol. Metab.* **53**:537–541.

187. Drezner, M. K., and W. M. Burch, Jr. 1978. Altered activity of the nucleotide regulatory site in the parathyroid hormone-sensitive adenylate cyclase from the renal cortex of a patient with pseudohypoparathyroidism. *J. Clin. Invest.* **62**:1222–1227.

188. Ross, E. M., and A. G. Gilman. 1980. Biochemical properties of hormone-sensitive adenylate cyclase. *Annu. Rev. Biochem.* **49**:533–564.

189. Farfel, Z., A. S. Brickman, H. R. Kaslow, V. M. Brothers, and H. R. Bourne. 1980. Defect of receptor–cyclase coupling protein in pseudohypoparathyroidism. *N. Engl. J. Med.* **303**:237–242.

190. Farfel, Z., V. M. Brothers, A. S. Brickman, F. Conte, R. Neer, and H. R. Bourne. 1981. Pseudohypoparathyroidism: Inheritance of deficient receptor–cyclase coupling activity. *Proc. Natl. Acad. Sci. USA* **78**:3098–3102.

191. Levine, M. D., R. W. Downs, Jr., M. Singer, S. J. Marx, G. D. Aurbach, and A. M. Spiegel. 1980. Deficient activity of guanine nucleotide regulatory protein in erythrocytes from patients with pseudohypoparathyroidism. *Biochem. Biophys. Res. Commun.* **94**:1319–1324.

192. Farfel, Z., and H. R. Bourne. 1980. Deficient activity of receptor–cyclase coupling protein in platelets of patients with pseudohypoparathyroidism. *J. Clin. Endocrinol. Metab.* **51**:1202–1204.

193. Bourne, H. R., H. R. Kaslow, A. S. Brickman, and Z. Farfel. 1981. Fibroblast defect in pseudohypoparathyroidism, type I: Reduced activity of receptor–cyclase coupling protein. *J. Clin. Endocrinol. Metab.* **53**:636–640.

194. Farfel, Z., M. Abood, and H. R. Bourne. 1982. Deficient activity of receptor–cyclase coupling protein in virus-transformed lymphoblasts of patients with pseudohypoparathyroidism. *J. Clin. Endocrinol. Metab.* **55**:113–117.

195. Fischer, J. A., 1981. Biologically inactive parathyroid hormone in pseudohypoparathyroidism-type I (PHP-I). In: *Hormonal Control of Calcium Metabolism: Proceedings of the Seventh International Conference on Calcium Regulating Hormones.* D. V. Cohn, R. V. Talmage, and L. J. Matthews, eds. Excerpta Medica, Amsterdam.

196. Sinha, T. K., D. O. Allen, S. F. Queener, and N. H. Bell. 1977. Effects of acetazolamide on the renal excretion of phosphate in hypoparathyroidism and pseudohypoparathyroidism. *J. Lab. Clin. Med.* **89**:1188–1197.

197. McDonald, K. M. 1972. Responsiveness of bone to parathyroid extract in siblings with pseudohypoparathyroidism. *Metab. Clin. Exp.* **21**:521–531.

198. Bell, N. H., E. S. Gerard, and F. C. Bartter. 1963. Pseudohypoparathyroidism with osteitis fibrosa cystica and impaired absorption of calcium. *J. Clin. Endocrinol. Metab.* **23**:759–772.

199. Lambert, P. W., B. W. Hollis, N. H. Bell, and S. Epstein. 1980. Demonstration of a lack of change in serum 1α,25-dihydroxyvitamin D in response to parathyroid extract in pseudohypoparathyroidism. *J. Clin. Invest.* **66**:782–791.

200. Moses, A. M., N. Breslau, and R. Coulson. 1976. Renal responses to PTH in patients with hormone-resistant (pseudo) hypoparathyroidism. *Am. J. Med.* **61**:184.

201. Kolb, F. O., and H. L. Steinbach. 1962. Pseudohypoparathyroidism with secondary hyperparathyroidism and osteitis fibrosa. *J. Clin. Endocrinol. Metab.* **22**:59.

202. Singleton, E. B., and C. T. Teng. 1962. Pseudohypoparathyroidism with bone changes simulating hyperparathyroidism: Report of a case. *Radiology* **78**:388.

203. Bell, N. H., E. S. Gerard, and F. C. Bartter. 1963. Pseudohypoparathyroidism with osteitis fibrosa cystica and impaired absorption of calcium. *J. Clin. Endocrinol. Metab.* **23**:759.

204. Frame, B., C. A. Hanson, H. M. Frost, M. Block, and A. R. Arnstein. 1972. Renal resistance to parathyroid hormone with osteitis fibrosa: "Pseudohypohyperparathyroidism." *Am. J. Med.* **52**:311.

205. Wilson, J. D., and D. R. Hadden. 1980. Pseudohypoparathyroidism presenting with rickets. *J. Clin. Endocrinol. Metab.* **51**:1184–1189.

206. Sinha, T. K., H. F. DeLuca, and N. H. Bell. 1977. Evidence for a defect in the formation of 1,25-dihydroxyvitamin D in pseudohypoparathyroidism. *Metab. Clin. Exp.* **26**:731–738.

207. Metz, S. A., D. J. Baylink, M. R. Hughes, M. R. Hauser, and R. P. Robertson. 1977. Selective deficiency of 1,25 dihydroxycholecalciferol: A cause of isolated skeletal resistance to parathyroid hormone. *N. Engl. J. Med.* **97**:1084–1090.

208. Mason, R. S., D. Lissner, and S. Posen. 1980. Parathyroid hormone effect on 1,25-dihydroxyvitamin D in hypoparathyroidism. *Ann. Intern. Med.* **92**:260.

209. Aksnes, L., and D. Aarskog. 1980. Effect of parathyroid hormone on 1,25-dihydroxyvitamin D formation in type 1 pseudohypoparathyroidism. *J. Clin. Endocrinol. Metab.* **51**:1223.

210. Stogmann, W., and J. A. Fischer. 1975. Pseudohypoparathyroidism: Disappearance of the resistance of parathyroid extract during treatment with vitamin D. *Am. J. Med.* **59**:140.

211. Kind, H. P., D. K. Parkinson, S. M. Suh, D. Fraser, and S. W. Kooh. 1973. Parathyroid hormone response and effects of vitamin D in hypoparathyroidism and pseudohypoparathyroidism. *Abstr. Endocr. Soc.* **55**:164.

212. Kooh, S. W., D. Fraser, H. F. DeLuca, M. F. Holick, R. E. Belsey, M. B. Clark, and T. M. Murray. 1975. Treatment of hypoparathyroidism and pseudohypoparathyroidism with metabolites of vitamin D: Evidence for impaired conversion of 25-hydroxyvitamin D to 1,25-dihydroxyvitamin D. *N. Engl. J. Med.* **293**:840.

213. Werder, W. A., H. P. Kind, F. Egert, J. A. Fischer, and A. Prader. 1976. Effective long term treatment of pseudohypoparathyroidism with oral 1α-hydroxy- and 1,25-dihydroxyvitamin D in pseudohypoparathyroidism. *J. Pediatr.* **89**:266.

214. Davies, M., L. F. Hill, C. M. Taylor, and S. W. Stanbury. 1977. 1,25-Dihydroxycholecalciferol in hypoparathyroidism. *Lancet* **1**:55.

216. Salassa, R. M., J. Jowsey, and C. D. Arnaud. 1970. Hypophosphatemic osteomalacia associated with nonendocrine tumors. *N. Engl. J. Med.* **283**:65–70.

217. Evans, D. J., and J. G. Azzopardi. 1972. Distinctive tumors of bone and soft tissue causing acquired vitamin D-resistant osteomalacia. *Lancet* **1**:353–354.

218. Prader, A., R. Illig, E. Uehlinger, and G. Stadler, 1959. Rachitis in Folge Knochentumors. *Helv. Paediatr. Acta* **14**:554–565.

219. Dent, C. E., and M. Friedman. 1964. Hypophosphatemic osteomalacia with complete recovery. *Br. Med. J.* **1**:1676–1679.

220. Hauge, B. N. 1956. Vitamin D-resistant osteomalacia. *Acta Med. Scand.* **153**:271–282.

221. Yoshikawa, S., M. Kawabata, Y. Hatsuyama, O. Hosokawa, and T. Fujita. 1964. Atypical vitamin D-resistant osteomalacia. *J. Bone Jt. Surg. Am. Vol.* **46**:998–1107.

222. Case Records of the Massachusetts General Hospital (Case 38-1965). 1964. *N. Engl. J. Med.* **273**:494–504.

223. Drezner, M. K., and M. N. Feinglos. 1977. Osteomalacia due to 1,25-dihydroxycholecalciferol deficiency: Association with a giant cell tumor of bone. *J. Clin. Invest.* **60**:1046–1053.

224. Olefsky, J., R. Kempson, H. Jones, and G. Reaven. 1972. "Tertiary" hyperparathyroidism and apparent "cure" of vitamin-D-resistant rickets after removal of an ossifying mesenchymal tumor of the pharynx. *N. Engl. J. Med.* **286**:740–745.

225. Wyman, A. L., F. J. Paradinas, and J. R. Daly. 1977. Hypophosphataemic osteomalacia associated with a malignant tumor of the tibia: Report of a case. *J. Clin. Pathol.* **30**:328–335.

226. McCance, R. A. 1947. Osteomalacia with Looser's nodes (Milkman's syndrome) due to a raised resistance to vitamin D acquired about the age of 15 years. *Q. J. Med.* **16**:33–50.

227. Pollack, J. A., A. L. Schiller, and J. D. Crawford. 1973. Rickets and myopathy cured by removal of a nonossifying fibroma of bone. *Pediatrics* **52**:364–371.

228. Linovitz, R. J., D. Resnick, P. Keissling, J. J. Kondon, B. Sehler, R. J. Wejdl, J. H. Rowe, and L. J. Deftos. 1976. Tumor-induced osteomalacia and rickets: A surgically curable syndrome. *J. Bone Jt. Surg. Am. Vol.* **58A**:419–423.

229. Wilhoite, D. R. 1975. Acquired rickets and solitary bone tumor: The question of a causal relationship. *Clin. Orthop. Relat. Res.* **109**:210–211.

230. Sweet, R. A., J. L. Males, A. J. Hamstra, and H. F. DeLuca. 1980. Vitamin D metabolite levels in oncogenic osteomalacia. *Ann. Int. Med.* **93**:279–280.

231. Lyles, K., W. R. Berry, M. Haussler, J. M. Harrelson, and M. K. Drezner. 1980. Hypophosphatemic osteomalacia: Association with prostatic carcinoma. *Ann. Intern. Med.* **93**:275–278.

232. Tanaka, Y., and H. F. DeLuca. 1974. Role of 1,25-dihydroxyvitamin D_3 in maintaining serum phosphorus and curing rickets. *Proc. Natl. Acad. Sci. USA* **71**:1040–1044.

233. Braithar, N., M. Yaron, and J. W. Coburn. 1977. Interactions between parathyroid hormone and 1,25(OH)$_2$-vitamin D_3 on renal handling of phosphate by the dog. *Proceedings of the Third Workshop on Vitamin D.* de Gruyter, Berlin.

234. Stewart, A. F., R. Horst, L. J. Deftos, E. C. Cadman, R. Lang, and A. E. Broadus. 1980. Biochemical evaluation of patients with cancer associated hypercalcemia: Evidence for humoral and non-humoral groups. *N. Engl. J. Med.* **303**:1377–1383.

235. Stewart, A. F. 1982. Is there a role for parathyroid hormone in humoral hypercalcemia of malignancy? *Miner. Electrolyte Metab.* **8**:215–226.

236. Stewart, A. F., K. L. Insogna, D. Goltzman, and A. E. Broadus. 1983. Identification of adenylate cyclase-stimulating activity and cytochemical glucose-6-phosphate dehydrogenase-stimulating activity in extracts of tumors from patients with humoral hypercalcemia of malignancy. *Proc. Natl. Acad. Sci. USA* **80**:1454–1458.

237. Massry, S. G., R. M. Friedler, and J. W. Coburn. 1973. Excretion of phosphate and calcium: Physiology of their renal handling and relation to clinical medicine. *Arch. Intern. Med.* **131**:828–859.

238. Ikkos, D., R. Lufft, and C. A. Grinzel. 1959. The effect of human growth hormone in man. *Acta Endocrinol. (Copenhagen)* **32**:341–361.

239. Henneman, P. H., A. P. Forbes, A. Moldawer, E. F. Dempsy, and E. L. Carrol. 1960. Effects of human growth hormone in man. *J. Clin. Invest.* **39**:1223–1238.

240. Beck, J. C. 1960. Primate growth hormone studies in man. *Metabolism* **9**:699–737.

241. Corvilain, J., and M. Abramov. 1962. Some effects of human growth hormone on renal hemodynamics and on tubular phosphate transport in man. *J. Clin. Invest.* **41**:1230–1235.

242. Lambert, P. P., and J. Corvilain. 1963. Site of action of parathyroid hormone and role of growth hormone in phosphate excretion. In: *Hormones and the Kidney.* P. C. Williams, ed. Academic Press, New York. pp. 139–147.

243. Corvilain, J., and M. Abramov. 1964. Effect of growth hormone on tubular transport of phosphate in normal and parathyroidectomized dogs. *J. Clin. Invest.* **43**:1608.

244. Durand, D., M. Prelot, and Y. Raoul. 1976. Effets precoces de l'hormone de croissance sur les phosphates seriques et urinaires du rat. *Experientia* **32**:120.

245. Bijvoet, O. L. M., J. van der Sluys Veer, H. Greven, and A. M. P. Scheliekens. 1972. Influence of calcitonin on renal excretion of sodium and calcium. In: *Parathyroid Hormone and the Calcitonin.* R. V. Talmage and P. L. Munson, eds. Excerpta Medica, Amsterdam. pp. 284–298.

246. Hammerman, M. R., I. E. Carl, and K. A. Hruska. 1980. Regulation of canine renal vesicle P_i transport by growth hormone and parathyroid hormone. *Biochim. Biophys. Acta* **603**:322–325.

247. Bijvoet, O. L. M., J. Van der Sluys Veer, H. R. de Vries, and A. T. J. van Koppen. 1971. Natriuretic effect of calcitonin in man. *N. Engl. J. Med.* **284**:681–688.

248. Singer, F. R., N. J. Woodhouse, D. K. Parkinson, and G. F. Joplin. 1969. Some acute effects of administered porcine calcitonin in man. *Clin. Sci.* **37**:181–190.

249. Ardaillou, R. J., J. P. Faillastre, G. Milhaud, R. Rousselet, R. Delaunay, and G. Richet. 1969. Renal excretion of phosphate, calcium and sodium during and after a prolonged thyrocalcitonin infusion in man. *Proc. Soc. Exp. Biol. Med.* **131**:56–60.

250. Ardaillou, R., P. Vaugnat, G. Milhaud, and G. Richet. 1967. Effects de la thyrocalcitonine sur l'excretion renale des phosphates, du calcium et des ions H$^+$ chez l'homme. *Nephron* **4**:298–314.

251. Paillard, F., R. Ardaillou, H. Malendin, J. P. Fillastre and S. Prier. 1972. Renal effects of salmon calcitonin in man. *J. Lab. Clin. Med.* **80**:200–216.

252. Haas, H. G., M. A. Dambacher, J. Guncaga, and T. Lauffenburger. 1971. Renal effects of calcitonin and parathyroid extract in man. *J. Clin. Invest.* **50**:2689.

253. Sorensen, O. H., and I. Hindberg. 1972. The acute and prolonged effect of porcine calcitonin on urine electrolyte excretion in intact and parathyroidectomized rats. *Acta Endocrinol. (Copenhagen)* **70**:295.

254. Robinson, C. J., T. J. Martin, and I. McIntyre. 1966. Phosphaturic effect of thyrocalcitonin. *Lancet* **1**:83.

255. Milhaud, G., M. L. Mouhktar, G. Cherian, and A. M. Perault. 1966. Effect de l'administration de thyrocalcitonin sur les principaux parametres du metabolisme du calcium du rat normal et du rat thyroparathyroidectomise. *C. R. Acad. Sci.* **262**:511.

256. Marx, S. J., and G. D. Aurbach. 1975. Renal receptors for calcitonin: Coordinate occurrence with calcitonin-activated adenylate cyclase. *Endocrinology* **97**:448.

257. Marx, S. J., C. J. Woodard, and G. Aurbach. 1972. Calcitonin receptors of kidney and bone. *Science* **178**:999.

258. Dousa, T. P. 1974. Effects of hormones on cyclic AMP formation in kidneys of nonmammalian vertebrates. *Am. J. Phyisol.* **226**:1193.

259. Ardaillou, R. 1975. Kidney and calcitonin.. *Nephron* **15**:250.

260. Heersche, J. N. M., R. Marcus, and G. D. Aurbach. 1974. Calcitonin and the formation of 3',5'-AMP in bone and kidney. *Endocrinology* **94**:241.

261. Kurokawa, K., N. Nagata, M., Sasaki, and K. Nakane. 1974. Effects of calcitonin on the concentration of cyclic adenosine 3',5'-monophosphate in rat kidney *in vivo* and *in vitro*. *Endocrinology.* **94**:1514.

262. Gekle, D. 1972. Der Einfluss von synthetischem Calcitonin auf die proximal-tubulare Phosphatresorption. *Klin. Wochenschr.* **50**:527.

263. Deftos, L. J., and J. G. Parthemore. 1974. Secretion of parathyroid hormone in patients with medullary thyroid carcinoma. *J. Clin. Invest.* **54**:416–420.

264. Perlzweig, W. A., E. Latham, and C. S. Keefer. 1923. The behavior of inorganic phosphate in the blood and urine of normal and diabetic subjects during carbohydrate metabolism. *Proc. Soc. Exp. Biol. Med.* **21**:33.

265. Allen, F. N., B. R. Dickson, and J. Markowitz. 1925. The rela-

tionship of phosphate and carbohydrate metabolism. *Am. J. Physiol.* **70**:333.

266. Nishimoto, H. 1929. On relation between carbohydrate metabolism and inorganic phosphates. *Jpn. J. Exp. Med.* **7**:207.

267. Morey, E. R., and A. D. Kenny. 1964. Effects of catecholamines on urinary calcium and phosphorus in intact and parathyroidectomized rats. *Endocrinology* **75**:18.

268. Dousa, T. P., and H. Valtin. 1976. Cellular action of vasopressin in the mammalian kidney. *Kidney Int.* **10**:46.

269. Kurokawa, K., and S. G. Massry. 1973. Evidence for stimulation of renal gluconeogenesis by catecholamines. *J. Clin. Invest.* **52**:961.

270. Massara, F., and F. Camanni. 1970. Propranolol block of adrenalin-induced hypophosphataemia in man. *Clin. Sci.* **38**:245.

271. Body, J. J., P. E. Cryer, K. O. Offord, and H. Heath. 1983. Epinephrine is a hypophosphatemic hormone in man. *J. Clin. Invest.* **71**:572–578.

272. Ellis, S. 1956. The metabolic effects of epinephrine and related amines. *Pharmacol. Rev.* **8**:485.

273. Levitan, B. A. 1951. Effect in normal man of hyperglycemia and glycosuria on excretion and reabsorption of phosphate. *J. Appl. Physiol.* **4**:224–226.

274. Huffman, E. R., C. J. Hlad, Jr., N. E. Whipple, and H. Elrick. 1958. The influence of blood glucose on the renal clearance of phosphate. *J. Clin. Invest.* **37**:369–379.

275. Fox, M., S. Thier, and L. Rosenberg. 1964. Impaired renal tubular function induced by sugar infusion in man. *J. Clin. Endocrinol. Metab.* **24**:1318–1327.

276. Ginsburg, J. M. 1972. Effect of glucose and free fatty acid on phosphate transport in dog kidney. *Am. J. Physiol.* **222**:1153–1160.

277. Pitts, R. F., and R. S. Alexander. 1944. The renal reabsorptive mechanism for inorganic phosphate in normal and acidotic dogs. *Am. J. Physiol.* **142**:648–662.

278. Cohen, J., F. Berglund, and W. D. Lotspeich. 1957. Interactions during renal tubular reabsorption in the dog among several anions showing a sensitivity to glucose and phlorizin. *Am. J. Physiol.* **189**:331–338.

279. Sohdey, S. S., and F. N. Allan. 1924. The relationship of phosphates to carbohydrate metabolism. I. Time relationship of the changes in phosphate excretion caused by insulin and sugar. *Biochem. J.* **18**:1170–1184.

280. Eggleton, M. G., and S. Shuster. 1954. Glucose and phosphate excretion in the cat. *J. Physiol. (London)* **124**:613–622.

281. DeFronzo, R. A., C. R. Cooke, R. Andres, and G. R. Faloona. 1975. The effect of insulin on renal handling of sodium, potassium, calcium, and phosphate in man. *J. Clin. Invest.* **55**:845–855.

282. DeFronzo, R. A., M. Goldberg, and Z. Agus. 1976. The effects of glucose and insulin on renal electrolyte transport. *J. Clin. Invest.* **58**:83–90.

283. Roy, D. R. and J. F. Seeley. 1981. Effect of glucose on renal excretion of electrolytes in the rat. *Am. J. Physiol.* **240**:F17–F24.

284. Nizet, A., P. Lefebvre, A. Luyckx, and J. Crabbe. 1977. Hormone and non-specific humoral factors in the interferences between sodium, glucose and phosphate handling by the dog kidney. *Curr. Prob. Clin. Biochem.* **6**:262–272.

285. Anderson, J., and J. B. Forster. 1959. Effect of cortisone on urinary phosphate excretion in man, *Clin. Sci.* **18**:437.

286. Ingbar, S. H., E. H. Kass, C. H. Burnett, A. S. Reiman, B. A. Burrows, and J. H. Sisson. 1951. The effects of ACTH and cortisone on the renal tubular transport of uric acid, phosphorus, and electrolytes in patients with normal renal and adrenal function. *J. Lab. Clin. Med.* **38**:533.

287. Wajchenberg, B. L., E. R. Quintao, B. Liberman, and A. B. U. Cintra. 1965. Antagonism between adrenal steroids and parathyroid hormone. *J. Clin. Endocrinol. Metab.* **25**:1677.

288. Roberts, K. E., and R. F. Pitts. 1953. The effects of cortisone and desoxycorticosterone on the renal tubular reabsorption of phosphate and the excretion of titratable acid and potassium in dogs. *Endocrinology.* **52**:324.

289. Laron, Z., J. D. Crawford, and R. Klein. 1957. Phosphaturic effect of cortisone in normal and parathyroidectomized rats. *Proc. Soc. Exp. Biol. Med.* **96**:649.

290. Thorn, G. W., D. Jenkins, J. C. Laidlaw, F. C. Goetz, J. F. Dingman, W. L. Arons, D. H. P. Streeten, and B. H. McCracken. 1953. Pharmacologic aspects of adrenocortical steroids and ACTH in man. *N. Engl. J. Med.* **248**:141.

291. Rastegar, A., Z. Agus, T. B. Connor, and M. Goldberg. 1972. Renal handling of calcium and phosphate during mineralocorticoid "escape" in man. *Kidney Int.* **2**:279.

292. Pfaundler, M. 1900. Über ein Verfohien zur Bestemmung des Aminosäurenstickstoffs im Harne. *Z. Physiol. Chem.* **30**:75–89.

293. Thier. S. O. 1974. Amino acid transport in the renal tubule cell. In: *Recent advances in Renal Physiology and Pharmacology.* L. G. Wesson and G. M. Fanelli, Jr., eds. University Park Press, Baltimore. pp. 39–51.

294. Brodehl, J., and K. Gellissen. 1968. Endogenous renal transport of free amino acids in infancy and childhood. *Pediatrics* **42**:395–404.

295. Bergeron, M., and F. Morel. 1969. Amino acid transport in rat renal tubules. *Am. J. Physiol.* **216**:1139–1149.

296. Barfuss, D. W., and J. A. Schafer. 1979. Active amino acid absorption by proximal convoluted and proximal straight tubules. *Am. J. Physiol.* **236**:F149–F162.

297. Chan, A. W. K., H. B. Burch, T. R. Alvey, and O. H. Lowry. 1975. A quantitative histochemical approach to renal transport. I. Aspartate and glutamate. *Am. J. Physiol.* **229**:1034–1044.

298. Wedeen, R. P., and S. O. Thier. 1971. Intrarenal distribution of nonmetabolized amino acids *in vivo*. *Am. J. Physiol.* **220**:507–512.

299. Pitts, R. F. 1943. A renal reabsorptive mechanism in the dog common to glycine and creatinine. *Am. J. Physiol.* **140**:156–168.

300. Beyer, K. H., L. D. Wright, H. F. Russo, H. R. Skeggs, and E. A. Patch. 1946. The renal clearance of essential amino acids: Tryptophan, leucine, isoleucine and valine. *Am. J. Physiol.* **146**:330–335.

301. Beyer, K. H., L. D. Wright, H. R. Skeggs, H. F. Russo, and G. A. Shaner. 1947. Renal clearance of essential amino acids: Their competition for reabsorption by the renal tubules. *Am. J. Physiol.* **151**:202–210.

302. Wright, L. D., H. F. Russo, H. R. Skeggs, E. A. Patch, and K. H. Beyer. 1947. The renal clearance of essential amino acids: Arginine, histidine, lysine and methionine. *Am. J. Physiol.* **149**:130–134.

303. Kamin, H., and P. Handler. 1951. Effect of infusion of single amino acids upon excretion of other amino acids. *Am. J. Physiol.* **164**:654–661.

304. Gerok, W., and J. Gayer. 1961. Investigations on the reabsorption of amino acids in the kidney of the dog. *Proc. 1st Int. Congr. Nephrol.* pp. 720–724.

305. Gerok, W., and J. Gayer. 1961. Die Tubulare Rückresorption der L-Aminosäuren in der Niere des Hundes: Transport-maxima und Competitive Hemmung. *Klin. Wochenschr.* **39**:540–546.

306. Gerok, W., and J. Gayer. 1964. Investigations on the reabsorption of L-amino acids in the kidney of the dog with special reference to diaminomonocarbonic acids. *Proc. 2nd Int. Congr. Nephrol.* pp. 572–574.

307. Webber, S. W. 1962. Interactions of neutral and acidic amino acids in renal tubular transport. *Am. J. Physiol.* **202**:577–583.

308. Webber, W. A. 1963. Characteristics of acidic amino acid transport in mammalian kidney. *Can. J. Biochem. Physiol.* **41**:131–137.

309. Webber, W. A., J. L. Brown, and R. F. Pitts. 1961. Interactions of amino acids in renal tubular transport. *Am. J. Physiol.* **200**:380–386.

310. Scriver, C. R., M. L. Efron, and J. A. Schafer. 1964. Renal tubular transport of proline, hydroxyproline, and glycine in health and in familial hyperprolinemia. *J. Clin. Invest.* **43**:374–385.

311. Rosenberg, L. E., A. Blair, and S. Segal. 1961. The transport of amino acids by rat kidney cortex slices. *Biochim. Biophys. Acta* **54**:479–488.

312. Fox, M., S. O. Thier, L. E. Rosenberg, and S. Segal. 1964. Ionic requirements for amino acid transport in the rat kidney cortex slice. I. Influence of extracellular ions. *Biochim. Biophys. Acta* **79**:167–176.

313. Thier, S. O., A. Blair, M. Fox, and S. Segal. 1967. The effect of extracellular sodium on the kinetics of AIB transport in rat kidney cortex slice. *Biochim. Biophys. Acta* **135**:300–305.

314. Wedeen, R. P., and S. O. Thier. 1971. Intrarenal distribution of 512.

315. Wedeen, R. P., and B. Weiner. 1973. The distribution of p-aminohippuric acid in rat kidney slices. I. Tubular localization. *Kidney Int.* **3**:205–213.

316. Wedeen, R. P., and B. Weiner. 1973. The distribution of p-aminohippuric acid in rat kidney slices. II. Depth of uptake. *Kidney Int.* **3**:214–221.

317. Barfuss, D. W., J. M. Mays, and J. A. Schafer. 1980. Peritubular uptake and transepithelial transport of glycine in isolated proximal tubules. *Am. J. Physiol.* **238**:F224–F333.

318. Ausiello, D. A., S. Segal, and S. O. Thier. 1972. Cellular accumulation of L-lysine in rat kidney cortex *in vivo*. *Am. J. Physiol.* **222**:1473–1478.

319. Greth, W. E., S. O. Thier, and S. Segal. 1973. Cellular accumulation of L-cystine in rat kidney cortex *in vivo*. *J. Clin. Invest.* **52**:454–462.

320. Foulkes, E. C. 1972. Cellular localization of amino acid carriers in renal tubules. *Proc. Soc. Exp. Biol. Med.* **129**:1032–1033.

321. Bergeron, M., and M. Vandeboncoeur. 1971. Antiluminal transport of L-arginine and L-leucine following microinjections in peritubular capillaries of the rat. *Nephron* **8**:355–366.

322. Sacktor, B. 1977. Transport in membrane vesicles isolated from the mammalian kidney and intestine. *Curr. Top. Bioenerg.* **6**:39–81.

323. Reynolds, R. A., H. Wald, P. D. McNamara, and S. Segal. 1980. An improved method for isolation of basolateral membranes from rat kidney. *Biochim. Biophys. Acta* **601**:92–100.

324. McNamara, P. D., B. Ozegovic, L. M. Pepe, and S. Segal. 1976. Proline and glycine uptake by renal brushborder membrane vesicles. *Proc. Natl. Acad. Sci. USA* **73**:4521–4525.

325. Hammerman, M. R., and B. Sacktor. 1977. Transport of amino acids in renal brush border membrane vesicles. *J. Biol. Chem.* **252**:591–595.

326. McNamara, P. D., L. M. Pepe, and S. Segal. 1979. Sodium gradient dependence of proline and glycine uptake in rat renal brush-border membrane vesicles. *Biochim. Biophys. Acta* **556**:151–160.

327. Foulkes, E. C., and T. Gieske. 1973. Specificity and metal sensitivity of renal amino acid transport. *Biochim. Biophys. Acta* **318**:439–445.

328. Jepson, J. B. 1972. Hartnup disease. In: *The Metabolic Basis of Inherited Disease*, 3rd ed. J. B. Stanbury, J. B. Wyngaarden, and D. S. Fredrickson, eds. McGraw–Hill, New York. pp. 1486–1503.

329. Baron, D. N., C. E. Dent, H. Harris, E. W. Hart, and J. B. Jepson. 1956. Hereditary pellagra-like skin rash with temporary cerebellar ataxia, constant renal aminoaciduria, and other bizarre biochemical features. *Lancet* **1**:421–428.

330. Milne, M. D., M. A. Crawford, C. B. Girao, and L. Loughridge. 1960. The metabolic disorder in Hartnup disease. *Am. J. Med.* **29**:407–412.

331. Scriver, C. R. 1965. Hartnup disease: A genetic modification of intestinal and renal transport of certain neutral alpha-amino acids. *N. Engl. J. Med.* **273**:530–532.

332. Shih, V. E., E. M. Bixby, D. H. Alpers, C. S. Bartsocas, and S. O. Thier. 1971. Studies of intestinal transport defect in Hartnup disease. *Gastroenterology* **61**:445–453.

333. Navab, F., and A. M. Asatoor. 1970. Studies on intestinal absorption of amino acids and a dipeptide in a case of Hartnup disease. *Gut* **11**:373–379.

334. Winter, C. G., and H. N. Christensen. 1965. Contrasts in neutral amino acid transport by rabbit erythrocytes and reticulocytes. *J. Biol. Chem.* **240**:3594–3600.

335. Rosenberg, L. E., and C. R. Scriver. 1974. Disorders of amino acid metabolism. In: *Duncan's Diseases of Metabolism: Genetics and Metabolism*. P. K. Bondy and L. E. Rosenberg, eds. Saunders, Philadelphia. pp. 465–654.

336. Milne, M. D. 1972. Renal tubular dysfunction. In: *Diseases of the Kidney*, 2nd ed. M. B. Strauss and L. G. Welt, eds. Little, Brown, Boston. pp. 1071–1138.

337. Halvorsen, K., and S. Halvorsen. 1963. Hartnup disease. *Pediatrics* **31**:29–38.

338. Hooft, C., J. Timmenons, J. Snoeck, I. Anterer, W. Ouaert, and C. H. Vanden-Hends. 1964. Methionine malabsorption in a mentally defective child. *Lancet* **2**:20.

339. Drummond, K. N., A. F. Michael, R. A. Ulstrom, and R. A. Good. 1964. The blue diaper syndrome: Familial hypercalcemia with nephrocalcinosis and indicanaemia. *Am. J. Med.* **37**:928–948.

340. Sabater, J., C. Ferre, M. Puliol, and A. Maya. 1976. Histidinuria: A renal and intestinal histidine transport deficiency found in two mentally retarded children. *Clin. Genet.* **9**:117–124.

341. Thier, S. O., and S. Segal. 1972. Cystinuria. In: *The Metabolic Basis of Inherited Disease*, 3rd ed. J. B. Stanbury, J. B. Wyngaarden, and D. S. Fredrickson, eds. McGraw–Hill, New York. p. 1504.

342. Garrod, A. E. 1908. The Croonian lectures. *Lancet* **2**:1, 73, 142, 214.

343. Dent, C. E., J. G. Heathcote, and G. E. Joron. 1954. The pathogenesis of cystinuria. I. Chromatographic and microbiological studies of the metabolism of sulphur-containing amino acids. *J. Clin. Invest.* **33**:1210–1215.

344. Dent, C. E., B. Senior, and J. M. Walshe. 1954. The pathogenesis of cystinuria. II. Polarographic studies of the metabolism of sulphur-containing amino acids. *J. Clin. Invest.* **33**:1216–1226.

345. Crawhall, J. C., E. F. Scowen, C. J. Thompson, and R. W. E. Watts. 1967. The renal clearance of amino acids in cystinuria. *J. Clin. Invest.* **46**:1162–1171.

346. Rosenberg, L., S. Downing, and S. Segal. 1962. Competitive inhibition of dibasic amino acid transport in rat kidney. *J. Biol. Chem.* **237**:2265–2270.

347. Fox, M., S. O. Thier, L. E. Rosenberg, W. Kiser, and S. Segal. 1964. Evidence against a single renal transport defect in cystinuria. *N. Engl. J. Med.* **270**:556–561.

348. Foreman, J. W., S. M. Hwang, and S. Segal. 1980. Transport interactions of cystine and dibasic amino acids in isolated rat renal tubules. *Metabolism* **29**:53–61.

349. Segal, S., P. D. McNamara, and L. M. Pepe. 1977. Transport interaction of cystine and dibasic amino acids in renal brush border vesicles. *Science* **197**:169–171.

350. McNamara, P. D., L. M. Pepe, and S. Segal. 1981. Cystine uptake by rat renal brush-border vesicles. *Biochem. J.* **194**:443–449.

351. Rosenberg, L. E., and S. J. Downing. 1965. Transport of neutral and dibasic amino acids by human leukocytes: Absence of defect in cystinuria. *J. Clin. Invest.* **44**:1382–1393.

352. Crawhall, J. C., and S. Segal. 1966. The intracellular cysteine/cystine ratio in kidney cortex. *Biochem. J.* **99**:19c–20c.

353. Crawhall, J. C., and S. Segal. 1967. Intracellular ratio of cystine and cysteine in various tissues. *Biochem. J.* **105**:891–896.

354. Schwartzman, L., A. Blair, and S. Segal. 1966. A common renal transport system for lysine, ornithine, arginine and cysteine. *Biochem. Biophys. Res. Commun.* **23**:220–226.

355. Milne, M. D., A. M. Asatoor, K. D. G. Edwards, and L. W. Loughridge. 1961. The intestinal absorption defect in cystinuria. *Gut* **2**:323–337.

356. Asatoor, A. M., B. W. Lacey, D. R. London, and M. D. Milne. 1962. Amino acid metabolism in cystinuria. *Clin. Sci.* **23**:285–304.

357. Thier, S., M. Fox, S. Segal, and L. E. Rosenberg. 1964. Cystinuria: *In vitro* demonstration of an intestinal transport defect. *Science* **143**:482–484.

358. Thier, S. O., S. Segal, M. Fox, A. Blair, and L. E. Rosenberg.

1965. Cystinuria: Defective intestinal transport of dibasic amino acids and cystine. *J. Clin. Invest.* **44**:442–448.

359. McCarthy, C. F., J. L. Borland, H. J. Lynch, E. E. Owen, and M. P. Tyor. 1964. Defective uptake of basic amino acids and L-cystine by intestinal mucosa of patients with cystinuria. *J. Clin. Invest.* **43**:1518–1524.

360. Alpers, D. H., and S. O. Thier. 1972. Role of the free amino acid pool of the intestine in protein synthesis. *Biochim. Biophys. Acta* **262**:535–545.

361. Harris, H., U. Mittwoch, E. B. Robson, and F. L. Warren. 1955. Pattern of amino acid excretion in cystinuria. *Ann. Hum. Genet.* **12**:195–208.

362. Harris, H., U. Mittwoch, E. B. Robson, and F. L. Warren. 1955. Phenotypes and genotypes in cystinuria. *Ann. Hum. Genet.* **20**:57–91.

363. Rosenberg, L. E., S. J. Downing, J. L. Durant, and S. Segal. 1966. Cystinuria: Biochemical evidence for three genetically distinct diseases. *J. Clin. Invest.* **45**:365–371.

364. Rosenberg, L. E. 1967. Genetic heterogeneity in cystinuria. In: *Amino Acid Metabolism and Genetic Variation*. W. L. Nyhan, ed. McGraw–Hill, New York. p. 341.

365. Hershko, C., E. Ban-Ami, J. Paciorkovski, and N. Levin. 1965. Alleomorphism in cystinuria. *Proc. Tel-Hashomer Hosp.* **4**:21.

366. Crawhall, J. C., E. F. Scowen, and R. W. E. Watts. 1963. Effect of penicillamine on cystinuria. *Br. Med. J.* **1**:588–590.

367. King, J. S. 1968. Treatment of cystinuria with alpha-mercaptopropionylglycine: A preliminary report. *Proc. Soc. Exp. Biol. Med.* **129**:927–932.

368. Kinoshita, K., S. Yachiku, T. Kotake, M. Takeuchi, and T. Sonoda. 1972. Treatment of cystinuria with 2-mercaptopropionylglycine (MPG). *Proc. 2nd Int. Symp. Thiola.* Osaka, Santen Pharmaceutical Co. p. 50.

369. Sonoda, T., K. Kinoshita, T. Kotake, S. Yachiku, and M. Takeuchi. 1970. Effect of thiola on cystinuria. *Proc. Int. Symp. Thiola.* Osaka, Santen Pharmaceutical Co. p. 231.

370. Nishimura, R., T. Ishido, and S. Takai. 1972. Studies on cystinuria. *Proc. 2nd. Int. Symp. Thiola.* Osaka, Santen Pharmaceutical Co. p. 47.

371. Hautmann, R., B. Terhorst, H. W. Stuhlsatz, and W. Lutzeyer. 1977. Mercaptopropionylgylcine: A process in cystine stone therapy. *J. Urol.* **117**:628–630.

372. Kallistratos, G., I. Mita, and V. Vadaloyka-Kalfakakou. 1979. Management of cystinuric disorders with sulfhydryl drugs. In: *The Management of Genetic Disorders*. C. J. Papadatos and C. S. Bartsocas, eds. Liss, New York. pp. 255–263.

373. Lucky, P. A., F. Skovby, and S. O. Thier. 1983. Pemphigus foliaceus and proteinuria induced by alpha-mercaptopropionylglycine. *J. Am. Acad. Dermatol.* **8**:667–672.

374. Pak, C. Personal communication.

375. Miyagi, K., F. Nakada, and S. Ohshiro. 1979. Effect of glutamine on cystine excretion in a patient with cystinuria. *N. Engl. J. Med.* **301**:196–198.

376. Skovby, F., L. E. Rosenberg, and S. O. Thier. 1980. No effect of L-glutamine on cystinuria. [Letter to the Editor]. *N. Engl. J. Med.* **302**:236–237.

377. Joost, J., and E. Jarosch. 1981. Glutamine: A new anticystinuric drug? *Eur. Urol.* **7**:363–364.

378. Van Den Berg, C. J., J. D. Jones, D. M. Wilson, and L. H. Smith. 1980. Glutamine therapy of cystinuria. *Invest. Urol.* **18**:155–157.

379. Kekkomäki, M., J. K. Visakorpi, J. Perheentupa, and L. Saxen. 1967. Familial protein intolerance with deficient transport of basic amino acids: An analysis of ten patients. *Acta Paediatr. Scand.* **56**:617–630.

380. Whelan, D. T., and C. R. Scriver. 1968. Hyperdibasicaminoaciduria: An inherited disorder of amino acid transport. *Pediatr. Res.* **2**:525–534.

381. Brodehl, J., K. Gallissen, and S. Kowalewski. 1967. Isolated cystinuria (without lysine-ornithine-argininuria) in a family with hypocalcemic tetany. *Klin. Wochenschr.* **45**:38–40.

382. Oyanagi, K., H. Sogawa, R. Minami, T. Nakao, and T. Chiba.

1979. The mechanism of hyperammonemia in congenital lysinuria. *J. Pediatr.* **94**:255–256.

383. Desjeux, J.-F., J. Rajantie, O. Simell, A.-M. Dumontier, and J. Perheentupa. 1980. Lysine fluxes across the jejunal epithelium in lysinuric protein intolerance. *J. Clin. Invest.* **65**:1382–1387.

384. Simell, O., and J. Perheentupa. 1974. Renal handling of diamino acids in lysinuric protein intolerance. *J. Clin. Invest.* **54**:9–17.

385. Tada, K., T. Morikawa, T. Ando, T. Yoshida, and A. Minagawa. 1965. Prolinuria: A new renal tubular defect in transport of proline and glycine. *Tohoku J. Exp. Med.* **87**:133–143.

386. Scriver, C. R., and O. H. Wilson. 1967. Amino acid transport: Evidence for genetic control of two types in human kidney. *Science* **155**:1428–1430.

387. Hillman, R. E., and L. E. Rosenberg. 1969. Amino acid transport by isolated mammalian renal tubules. II. Transport systems for L-proline. *J. Biol. Chem.* **244**:4494–4498.

388. Mohyuddin, F., and C. R. Scriver. 1970. Amino acid transport in mammalian kidney: Identification and analysis of multiple systems for imino acids and glycine in rat kidney. *Am. J. Physiol.* **219**:1–8.

389. deVries, A., S. Kochwa, J. Lazebink, M. Frank, and M. Djaldetti. 1957. Glycinuria, a hereditary disorder associated with nephrolithiasis. *Am. J. Med.* **23**:408–415.

390. Greene, M. L., P. S. Lietman, L. E. Rosenberg, and J. E. Seegmiller. 1973. Familial hyperglycinuria: New defect in renal tubular transport of glycine and the imino acids. *Am. J. Med.* **54**:265–271.

391. Morikawa, T., K. Tada, T. Ando, T. Yoshida, Y. Tokoyama, and T. Arakawa. 1966. Prolinuria: Defect in intestinal absorption of imino acids and glycine. *Tohoku J. Exp. Med.* **90**:105–116.

392. Goodman, S. I., C. A. McIntyre, and D. O'Brien. 1967. Impaired intestinal transport of proline in a patient with familial iminoaciduria. *J. Pediatr.* **71**:246–249.

393. Teijema, H. L., H. H. van Gelderen, M. A. H. Giesberts, and S. L. Laurent de Angulo. 1974. Dicarboxylic aminoaciduria: An inborn error of glutamate and aspartate transport with metabolic implications, in combination with a hyperprolinemia. *Metabolism.* **23**:115.

394. Melancon, S. B., L. Dallaire, B. Lemieux, P. Robitaille, and M. Potier. 1977. Dicarboxylic aminoaciduria: An inborn error of amino acid conservation. *J. Pediatr.* **91**:422–427.

395. Thier, S. O. 1970. Genetic control of amino acid transport in gut and kidney. *Birth Defects: Orig. Artic. Ser.* **6**(3):20–21.

396. Scriver, C. R., S. Pueschel, and E. Davies. 1966. Hyper-beta-alaninemia associated with beta-aminoaciduria and gamma-aminobutyricaciduria, somnolence and seizures. *N. Engl. J. Med.* **274**:636–643.

397. Fanconi, G. 1936. Der Nephrotisch-glykosurische Zwergwuchs mit Hypophosphatamischer Rachitis. *Dtsch. Med. Wochenschr.* **62**:1169–1171.

398. Clay, R. D., E. M. Darmady, and M. Hawkins. 1953. The nature of the renal lesion in Fanconi syndrome. *J. Pathol. Bacteriol.* **65**:551–558.

399. Darmady, E. M., and F. Stranack. 1957. Microdissection of the nephron in disease. *Br. Med. Bull.* **13**:21–25.

400. Wallis, L. A., and R. L. Engle. 1957. Multiple myeloma and the adult Fanconi syndrome. *Am. J. Med.* **22**:5–12.

401. Costanza, D. J., and M. Smollen. 1963. Multiple myeloma with the Fanconi syndrome. *Am. J. Med.* **34**:125–133.

402. Jackson, J. D., F. G. Smith, N. N. Litman, C. L. Yuile, and H. Latta. 1962. The Fanconi syndrome with cystinosis. *Am. J. Med.* **33**:893–910.

403. Gentz, J., R. Jagenburg, and R. Zetterstrom. 1965. Tyrosinemia, an inborn error of tyrosine metabolism with cirrhosis of the liver and multiple renal tubular defects. *J. Pediatr.* **66**:670–696.

404. Cusworth, D. C., C. E. Dent, and F. V. Flynn. 1955. The aminoaciduria in galactosemia. *Arch. Dis. Child.* **30**:150–154.

405. Morris, R. C., L. Ueki, D. Loh, R. L. Eanes, and P. McLin. 1967. Absence of renal fructose-1-phosphate aldolase activity in hereditary fructose intolerance. *Nature (London)* **214**:920–921.

406. Morgan, H. G., W. K. Stewart, K. G. Lowe, J. M. Stowers, and J. H. Johnstone. 1962. Wilson's disease and the Fanconi syndrome. *Q. J. Med.* **31**:361–384.

407. Lee, D. B. N., J. P. Drinkard, V. J. Rosen, and H. C. Gonizk. 1972. The adult Fanconi syndrome. *Medicine (Baltimore)* **51**:107–138.

408. Chisolm, J. J., H. C. Harrison, W. R. Eberlein, and H. E. Harrison. 1955. Aminoaciduria, hypophosphatemia, and rickets in lead poisoning. *Am. J. Dis. Child.* **89**:159–168.

409. Morgan, J. M., M. W. Hartley, and R. E. Miller. 1966. Nephropathy in chronic lead poisoning. *Arch. Intern. Med.* **118**:17–29.

410. Worthen, H. G. 1963. Renal toxicity to maleic acid in the rat: Enzymatic and morphologic observations. *Lab. Invest.* **12**:791–801.

411. Bergeron, M., and P. Laporte. 1973. Effet membranaire du maleate au niveau du nephron proximal et distal. *Rev. Can. Biol.* **32**:275–279.

412. Kramer, J. H., and H. C. Gonick. 1970. Experimental Fanconi syndrome. I. Effect of maleic acid on renal cortical Na-K-ATPase activity and ATP-levels *J. Lab. Clin. Med.* **76**:799–808.

413. Bergeron, M., L. Dubord, and C. Hausser. 1976. Membrane permeability as a cause of transport defects in experimental Fanconi syndrome. *J. Clin. Invest.* **57**:1181–1189.

414. Rosenberg, L. E., and S. Segal. 1964. Maleic acid-induced inhibition of amino acid transport in rat kidney. *Biochem. J.* **92**:345–352.

415. Bovee, K. C., T. Joyce, R. Reynolds, and S. Segal. 1978. The Fanconi syndrome in basenji dogs: A new model for renal transport defects. *Science* **201**:1129–1131.

416. Omura, K., N. Yamanaka, N. Higami, O. Matsuoka, A. Fujimoto, G. Issiki, and K. Tada. 1976. Lysine malabsorption syndrome: A new type of transport defect. *Pediatrics* **57**:102–105.

CHAPTER 15

Pulmonary Edema

Aubrey E. Taylor and Bengt Rippe

1. Introduction

The formation of intraalveolar edema involves a complex series of events starting with increased fluid filtration by the pulmonary microvasculature and a heterogeneous accumulation of fluid within various portions of the lung parenchyma, which, if left unabated, culminates in accumulation of fluid within the airways. When intraalveolar edema forms, fluid must cross several barriers: (1) the endothelial cell barrier at the microvascular wall; (2) the basement membrane; (3) the tissue spaces; and (4) the epithelial barrier which lines the airways. Once fluid fills the airways, the ability of the lungs to oxygenate and remove carbon dioxide from the blood is compromised and the condition becomes life threatening.[1]

Prior to 1950, physiologists had scant information concerning the mechanisms responsible for edema formation in the lung. The classical study of Guyton and Lindsey[2] in 1959 provided the first quantitative assessment of edema formation. These authors introduced the measurement of tissue wet-to-dry weight ratios to assess interstitial water accumulation when pulmonary microvascular pressures were increased. Figure 1 is a schematic representation of their study. As left atrial pressure was increased, the lung did not gain a significant amount of fluid until some critical pressure was attained. At this critical left atrial pressure, the lung gained weight in direct proportion to the change in left atrial pressure. From these studies, it appeared that some mechanism(s) in lung tissue acted to oppose fluid accumulation at microvascular pressures less than the critical pressure. In Fig. 1, the dashed line was obtained when plasma proteins were diluted to approximately one-half normal. Since the critical left atrial pressure is less when plasma protein concentrations are decreased, the authors hypothesized that fluid accumulation in lung tissue was not only a function of the microvascular pressure tending to push fluid into the interstitium, but it was also a function of the plasma colloid osmotic pressure which tended to pull fluid into the microvascular system, i.e.,

Aubrey E. Taylor and Bengt Rippe • Department of Physiology, University of South Alabama College of Medicine, Mobile, Alabama 36688.

$$W \propto P_C - \pi_P \qquad (1)$$

where P_C is the microvascular hydrostatic pressure, π_P is the colloid osmotic pressure of the plasma proteins, and W is lung weight. Equation (1) is a definition of the Starling hypothesis in Guyton's studies: that fluid would not accumulate in significant amounts until $P_C > \pi_P$. This logic was the prevailing theory concerning the forces responsible for fluid accumulation in lung tissue until the mid-1960s.[3,4] At that time, Guyton and co-workers developed the concept of a subatmospheric interstitial fluid pressure which was responsive to interstitial fluid volume changes. Realizing that the concepts of lung fluid balance needed reassessing, Guyton and co-workers began to describe the forces responsible for fluid movement across the capillary walls in the following fashion[5-9]:

$$J_V = K_{F,C}[(P_C - P_T) - \sigma_d(\pi_P - \pi_T)] \qquad (2)$$

where J_V is the net movement of fluid across the microvascular walls, $K_{F,C}$ is the hydraulic conductance of the fluid-exchanging membrane, P_C is the microvascular pressure and is usually termed capillary pressure, P_T is the interstitial fluid hydrostatic pressure, and π_P and π_T are the oncotic pressures of total plasma proteins in plasma and tissue fluid, respectively. σ_d is the osmotic reflection coefficient and relates the measured colloid oncotic gradient to the effective oncotic pressure gradient operating across the microvessel wall. σ_d equals 1 if the molecule cannot cross the membrane, i.e., the molecule is totally reflected at the membrane, and σ_d is equal to zero if the membrane is freely permeable to the molecule, i.e., no reflection of the solute at the membrane.[10-15] σ_d, π_P, and π_T refer to values associated with total plasma proteins, not the individual components of plasma. In addition, the forces on the right-hand side of Eq. (2) were termed "Starling forces."[8,9]

Equation (2) is the Kedem and Katchalsky formulation[10] which describes volume flow across membranes in general, but it unmasked two important concepts which had been ignored for over 100 years in capillary physiology, although they had been discussed and even reviewed in the literature.[3,40] First, in most capillary beds, the sum of the forces on the right-hand side of Eq.

259

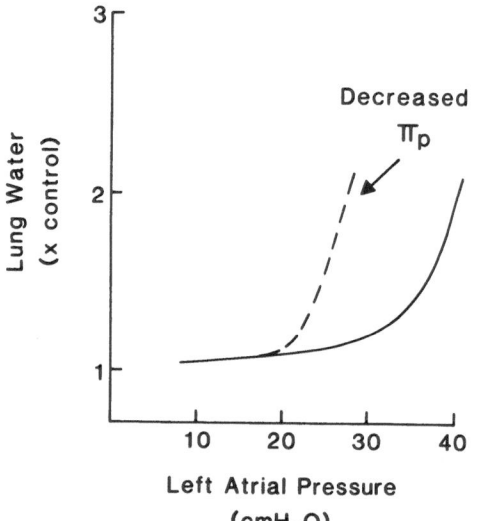

Fig. 1. Schematic representation of Guyton and Lindsey's classical study[2] on the effects of elevating left atrial pressure on lung water accumulation for normal (——) and decreased plasma colloid osmotic pressures (– – –). Note that the left atrial pressure at which the lung weight began to substantially change (critical capillary pressure) is decreased when plasma protein colloid osmotic pressure has been lowered.

(2) does not equal zero; rather, an imbalance exists in these forces which results in lymph formation. Second, P_T and π_T change as fluid accumulates in the tissues, such that J_V cannot be adequately described assuming that these forces are zero and/or constant.[16] This means of describing fluid fluxes in lung tissues provided the stimulus for several recent studies dealing with edema formation. The classical Starling[17] and Landis[4] papers were reevaluated and their findings used to design new experimental models and measuring techniques to evaluate the parameters in Eq. (2) in several different organs.[16]

The major stimulus for the recent surge of research activity in the area of pulmonary edema was provided by the experimental studies from Fishman's[18,19] and Guyton's[20,21] laboratories and the thorough review article by Staub.[22] In addition, the development of better lung experimental models and techniques for evaluating the various Starling forces (P_C, P_T, π_P, π_T) and flows (capillary filtration and lymph flow) in different experimental situations which cause pulmonary edema has produced a large literature of new information concerning the genesis of the edema process.

This chapter describes the present state of the art relative to the many factors which may cause pulmonary edema and the possible mechanisms responsible for either promoting or inhibiting fluid accumulation in lung tissue. To accomplish this, each Starling force will be discussed relative to the hydration state of lung tissues. Then, the complex pattern of the accumulation of lung interstitial fluid will be discussed relative to the different hydrostatic pressures which exist in different areas of the lung parenchyma. Finally, we will then postulate how the airways fill with fluid, especially in regard to the functional characteristics of the epithelia which line the airways. We hope that the reader will gain a greater appreciation for the complexities involved in the formation of tissue edema and be pointed in directions which will promote the development of experimental methods and models to provide the necessary data for describing the genesis

of pulmonary edema, and its absence in instances when it apparently should exist.

2. Starling Force Analysis

2.1. Pulmonary Microvascular Pressure (P_C)

The pulmonary microvascular pressure is a function of the pre- and postcapillary resistances and can be described as[21,23]

$$P_C = P_{PA} + \frac{r_V}{R_A + r_V}(P_{PA} - P_{LA}) \qquad (3)$$

where P_{LA} and P_{PA} are the pulmonary left atrial and arterial pressures, respectively, and r_A and r_V are the resistances up- and downstream from the filtration midpoint of the microvessels.

Because the lung's arterial and venous pressures are higher at the bottom of the lung than at the top, the corresponding capillary pressures at each lung height will be different. In addition, r_A and r_V will also vary from bottom to top of the lung but their ratios remain constant until P_{PA} is very close to alveolar pressure. As shown in Fig. 2, P_C decreases by about 1 cm H_2O per cm distance up the lung.[24] In addition, fluid filters in prealveolar (arteriolar) and postalveolar (venular) vessels, so the different pressures acting between P_{PA} and P_{LA} must also be considered when using single microvascular pressures to describe the capillary filtration in the lung. When discussing capillary (or microvascular) pressures in the lung, we must refer to some average or functional pressure which tends to push fluid out of the capillaries into the interstitium. Since left atrial pressure can change appreciably when the pumping ability of the heart is altered, increases in microvascular hydrostatic pressures are the most common causes of fluid accumulation in lung tissue. It is of interest to note that methods have been developed to measure this important determinant of fluid flux in intact animal and human lungs.[25] Therefore, the future should bring more information as to how this parameter changes in different pathological lung states.

Table I indicates how different compounds affect pulmonary microvascular pressure.[26,27] Only three compounds and

Table I. Effects of Vasoactive Substances on Pulmonary Vascular Pressures and Resistances[a,b]

	P_{PA}	P_V	P_C	r_A	r_V	r_T
		(cm H_2O)			(cm H_2O-min/100g per ml^{-1})	
Constant pressure						
Control	16.0	3.5	10.0	0.8	0.9	1.7 (26)
Histamine	16.3	3.6	11.7	3.6	5.5	9.1 (26)
Serotonin	15.4	3.2	9.5	4.2	4.6	8.8 (26)
Norepinephrine	15.3	3.0	8.8	6.2	5.0	11.2 (26)
Alveolar hypoxia	16.0	3.5	6.0	8.0	2.0	10.0 (27)
Constant flow						
Histamine	41.0	3.3	30.7	1.6	4.5	6.1 (26)
Serotonin	46.5	3.2	20.7	3.9	2.7	6.6 (26)
Norepinephrine	34.3	3.3	18.8	1.8	1.9	3.7 (26)
Alveolar hypoxia	31.0	3.5	9.0	8.0	2.0	10.0 (27)

[a]From the data of Rippe *et al.*[26] and Parker *et al.*[27].

[b]P_{PA}, P_V, and P_C refer to pulmonary arterial, venous outflow, and microvascular pressures, respectively; r_A, r_V, and r_T refer to pre-, post-, and total vascular resistances, respectively.

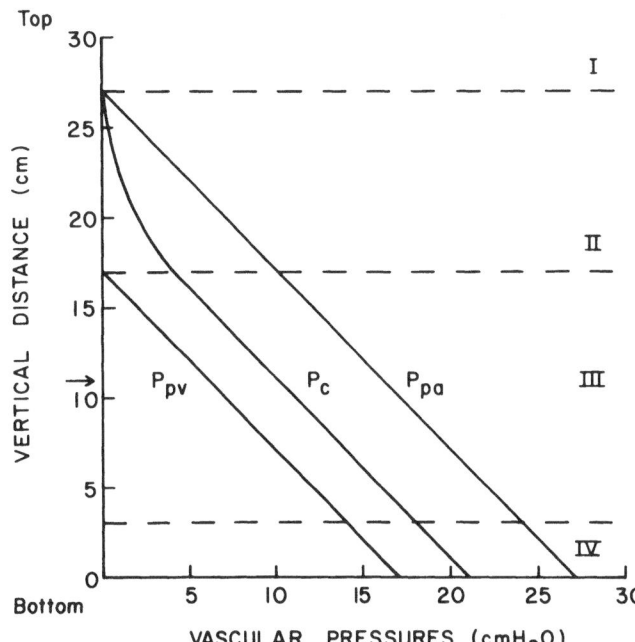

Fig. 2. Effect of lung vertical height on pulmonary arterial (P_{PA}), venous (P_{PV}), and capillary (P_C) pressures. I, II, and III refer to lung zones at which $P_{alveolar} > P_{PA} > P_{PV}$; $P_{PA} > P_{alveolar} > P_{PV}$; and $P_{PA} > P_{PV} > P_{alveolar}$. The arrow indicates heart level. Zone IV refers to the lowest portion of the lung and, for some reason, the vascular resistance increases in this portion of the lung. Reproduced with permission from Taylor and Parker.[1]

alveolar hypoxia are shown, since these are the only conditions for which P_C is known to any degree of certainty. The importance of these measurements is that some compounds increase P_{PA}, but r_A and r_V are affected differently. Serotonin and norepinephrine change both pre- and postcapillary resistances about equally and P_C is approximately doubled (see constant flow). However, for alveolar hypoxia the resistance change is mostly precapillary and P_C is unchanged from controls even when P_{PA} greatly increases.[27] Histamine also causes an increased P_{PA}, but this primarily reflects a large increase in postcapillary resistance and capillary pressure increases about threefold.[26] Table I indicates very clearly that P_C should be measured and not calculated using Eq. (3) in which r_A and r_V are assumed to not change with a particular challenge!

Another parameter which should be considered when evaluating pulmonary microvascular pressures is the effect of various endogenous vasoactive substances on vascular compliances and unstressed volumes (vascular capacity), since the portion of the lung at which filtration occurs also has the largest vascular capacity.[36] If a compound decreases the capacity of the vascular system, then any tendency for the lung to increase its blood volume, which occurs with elevated P_{LA} and systemic blood volume shifts, can significantly increase the capillary pressure.

Because of new methodology, P_C will be measured under many conditions in the future and perhaps many compounds which have been thought to cause pulmonary edema by mechanisms other than hydrostatic pressure effects may be correctly described, e.g., histamine increases pulmonary microvascular pressure and does not alter microvascular permeability as previously thought.[26,28–30] However, in peripheral tissues, histamine causes a dramatic and easily measured transient increase in microvascular permeability to macromolecules.[30]

2.2. Pulmonary Interstitial Fluid Pressure (P_T)

The interstitial fluid pressure in lung tissue changes as a function of vertical distance from heart level and is not the same within tissue spaces in different areas of the lung at the same horizontal height.[31,32] Figure 3A illustrates the interstitial fluid pressure changes at different lung heights. The interstitial pressure changes by about 0.6 cm H_2O per cm distance up or down the lung.[31] Also, the pressure in the fluid immediately surrounding the microvessels appears to be less negative than that surrounding perivascular spaces (Fig. 3B).[32,33] This gradient between microvessels, bronchioles, and the perivascular spaces is necessary to ensure that fluid will be propelled away from the thin gas-exchanging portions of the lungs toward the larger perivascular spaces where the lymphatics are located.[34–36]

When an average interstitial fluid pressure is used for lung tissue, it represents a very complex pressure which lumps together many different pressures acting in several lung regions and the entire lung behaves as if the average tissue fluid pressure is 4 to 5 mm Hg below atmospheric pressure.[37]

When microvascular pressures are increased, tissue fluid pressure increases.[38–40] Figure 3C shows the effect of increasing capillary pressure on interstitial fluid pressure as modified from studies of Parker et al.[38] Note that the pressure changes markedly when only small amounts of fluid enter the normal lung tissues. However, when the lung's interstitium expands appreciably, the tissue fluid pressure is no longer responsive to tissue hydration. Thus, it appears that tissue fluid pressure adjusts to oppose changes in microvascular pressure, especially around the range of normal pressures which are associated with dehydrated tissues. How much of this buffering effect is due to vascular volume expansion or capillary filtration into the tissues is not presently known.

2.3. Plasma Colloid Osmotic Pressure (π_P)

The plasma colloid osmotic pressure is the only Starling force which can be directly measured.[41] For many years, the concentration of total protein was measured in plasma and converted to osmotic pressures using equations which had been developed empirically by Landis and Pappenheimer for human

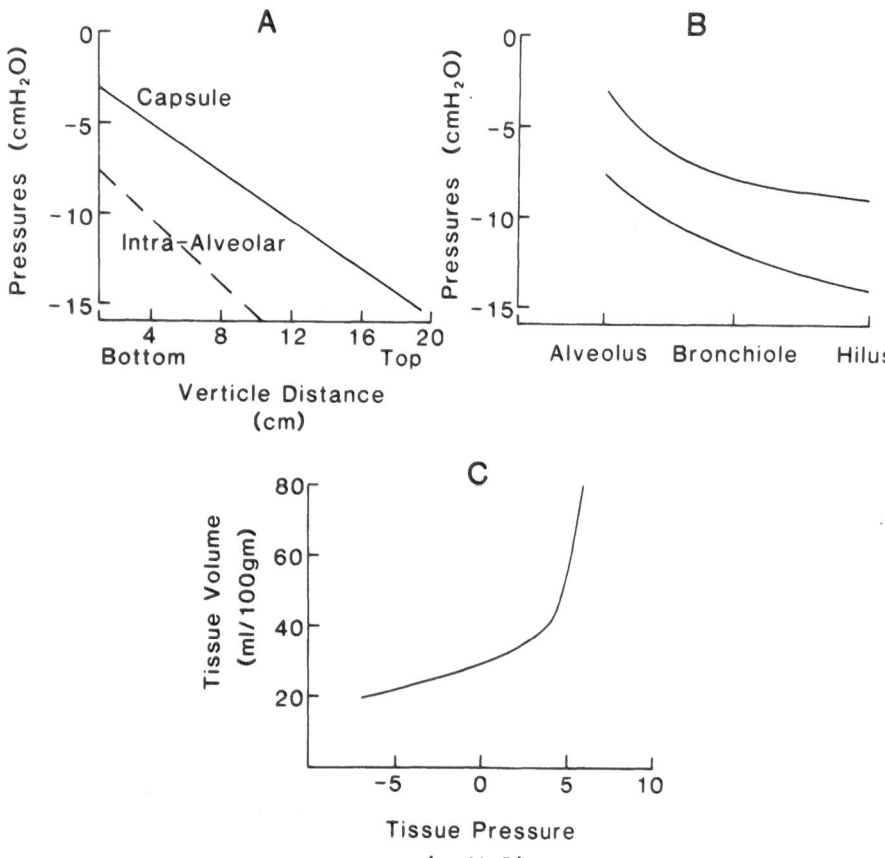

Fig. 3. (A) Effect of lung height on interstitial fluid pressures measured using fluid-filled alveoli (– – – –) or implanted plastic capsules (——). Note that the interstitial pressure is more negative at the top of the lung as compared to the bottom.[31,38] (B) Pressures measured in the tissues located in the alveolar, bronchiolar, and hilar regions of the lung. The lower curve represents the same measurements made with a larger transpulmonary pressure. The data clearly indicate that a pressure gradient exists between the small alveolar tissue spaces and the larger perivascular spaces and the pressures are also functions of transmural pressures. Redrawn from the data of Lai-Fook,[33] Inoue et al.,[32] and Mitzner and Robotham.[55] (C) A diagram showing how tissue volume is related to the average interstitial fluid pressure. Note that tissue pressure increases greatly for only small alterations in tissue volume (low interstitial compliance) and when tissue fluid pressure approaches alveolar pressures, the pressure changes are small while interstitial volume changes greatly (high interstitial compliance). Redrawn from the studies of Parker et al. [38]

plasma.[3] We now know that the relationship between plasma protein concentration (Cp) and colloid osmotic pressure is different between species.[42] And, more importantly, the albumin/globulin ratio in plasma is quite variable between individuals and even within the same individual at different times. Because of this complex makeup of plasma proteins in both normal and diseased states, the colloid osmotic pressure should be measured whenever possible and π_P calculated only if the albumin and globulin fractions of plasma are known. π_P is the only Starling force which can be considered a constant for all filtering lung microvessels, since only a small capillary filtration occurs relative to the large plasma flow in all areas of the lung's microcirculation.

2.4. Tissue Colloid Osmotic Pressure (π_T)

Tissue colloid osmotic pressure has not been directly measured in normally hydrated lung tissue. Usually, the lymph draining specific portions of the lung is used to assess tissue levels[43] or interstitial fluid is sampled during edema.[44] Lymph protein concentration can be greatly altered as it passes through nodes, so many estimates of π_T which were made in lung preparations utilizing postnodal lymph may not correctly represent lung tissue fluid.[45]

Two studies have been published in which lymph was collected from different vertical levels in upright lungs.[46,47] In one study, π_T varied by 0.1 cm H_2O per cm height up the lung, with π_T being higher at the apex of the lung as compared to the

base.[47] In the other study, π_T appeared to change by a larger amount.[46] However, both studies may not represent steady-state values of π_T and a particular lymphatic may contain tissue fluids draining different lung regions. The half-time for [131I]albumin equilibration in lung tissue is approximately 4 to 5 hr.[1] Certainly, experiments conducted for short time-frames can be far away from equilibrium and lymph may not reflect a steady-state π_T. From the tissue fluid pressure measurements, it is possible that π_T may change by 0.4 cm H_2O per cm lung height.[31,38]

π_T is affected by three different factors: (1) the rate of lymphatic protein removal from the interstitium; (2) the leakage of proteins from the vascular system into the interstitium, which is a function of both solute permeability (both charge and molecular configuration) and surface exchange area; and (3) the rate of capillary filtration.[1,8,22,43] Figure 4 illustrates how π_T changes when capillary pressure is elevated from normal to values which result in intraalveolar edema. π_T decreases because the filtration increases and the concentration of protein in the tissues decreases as also shown in Fig. 4 and as predicted by

$$C_L = \frac{C_p(1 - \sigma)}{(1 - \sigma e^{-x})} \qquad (4)$$

where $x = (1 - \sigma)J_L/PS$, PS being the permeability coefficient surface area product and J_L the lymph flow.[1,11,12,14,15]

The minimum value of π_T is a function of vascular permeability as shown by the dashed lines in Fig. 4. If the lung

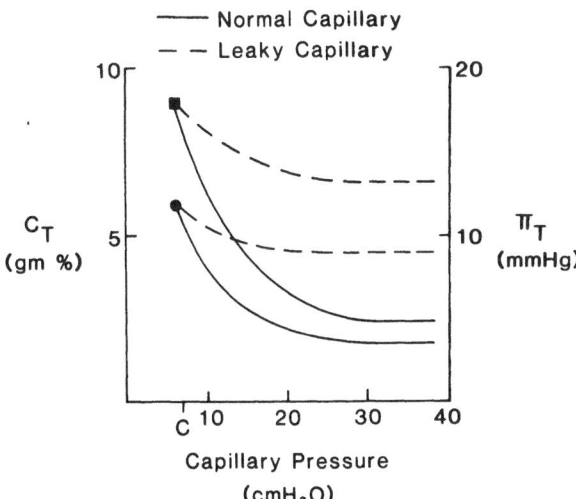

Fig. 4. Plot of total protein concentration (C_T) in the tissue (●) and colloid osmotic pressure (π_T) of the tissues (■) for normal (——) and abnormally leaky pulmonary capillaries (– – –) as a function of capillary pressure. Note that C_T is higher when the pulmonary capillaries are more leaky to plasma proteins and, consequently, π_T is higher and ($\pi_P - \pi_T$) lower.

capillaries are more leaky to plasma proteins than normal, the concentration of plasma proteins will be large even at high rates of capillary filtration. Therefore, π_T will be higher and the maximum osmotic gradient acting across the capillary wall will be lower (dashed lines). π_T is also a function of the albumin/globulin ratios of the plasma proteins and, in some species, the ratio is very variable, so calculated π_T's may be erroneous. Since π_T is a function of capillary filtration, a listing of literature values would not be too useful, but the ratio of the concentration of total protein in lymph (C_L) to that in plasma at normal vascular pressures is about 0.75 in many animal species and the minimal value obtained at high capillary filtration rate is approximately 0.25.[47] These ratios correspond to π_T's of approximately 18 and 4.6 mm Hg for control and minimum values when C_P is 7.5 g/100 ml ($\pi_P = 28.5$ mm Hg). These values correspond to transcapillary $\Delta\pi$'s of 10.5 and 23.9 mm Hg, respectively, at normal and absolute maximum capillary filtration states.

Finally, it must be emphasized that π_T measured from lymph samples represents some average tissue value which is weighted by the amount of lymph draining upper and lower portions of the lungs, the area of the lung's interstitium at which the particular lymphatic originates, and the permeability–surface area characteristics of the microcirculation drained by the particular lymphatic.[1]

Table II. Hydraulic Conductances ($K_{F,C}$) and Conductivities ($L_{P,C}$)[a]

	$K_{F,C}$ (ml/min-100 g-mm Hg)	$L_{P,C} \times 10^{11}$ (cm^3/dyne-sec)	Ref.
Pulmonary capillary walls			
Dog*	0.26	0.71	49
	0.15	0.40	49
	0.12	0.57	50
	0.33	0.88	51, 52
	0.17	0.44	53
	0.02–0.04	0.05–0.10	54
	0.99	2.60	55
	3.34	8.80	56
	0.12–0.38	0.31–1.00	57
	1.00	2.70	58
	30.20	90.10	56
Rabbit†	0.14	0.23	59
	0.63	1.00	60
	1.00	1.60	61
	2.20	3.50	62
	2.80	4.50	63, 64
Sheep*	0.017–0.027	0.025–0.038	65
Total alveolar-capillary membrane			
Dog*	0.05–0.09	0.13–0.25	66
	0.09	0.25	47, 57, 67, 68
	0.06	0.16	49
	0.07	0.19	1, 21, 49
Other capillary beds			
Skeletal muscle (cat)	0.015	2.2‡	23
Intestine (cat)	0.17	5.2§	69
Brain (dog)	0.004	—	70
Heart (rabbit)	0.40	—	71
Liver (dog)	0.30	—	75
Kidney (rat)	10.0	—	72
Cellular barrier			
Frog skin	—	0.40	73
Human erythrocytes	—	1.10	74
Alveolar membrane (maximum)			
Dog*	0.03	0.10	21, 22, 49

[a] *, †, ‡, and § calculated using surface areas of 3800, 6300, 70, and 700 cm^2/g, respectively.

2.5. The Hydraulic Conductance (or Filtration Coefficient) of the Microcirculation ($K_{F,C}$)

The filtration coefficient determines the rate at which fluid moves across the capillary wall for a given imbalance in Starling forces. Table II presents $K_{F,C}$ values determined for lung microvessels and for other membranes. The hydraulic conductance of lung tissue is high because of the tremendous capillary exchange surface; however, when the conductance values are converted to conductivity (L_{PC}), using published values for capillary wall surface area, the lung's hydraulic conductivity is actually smaller than several other capillary beds such as skeletal muscle and intestine and is an order of magnitude lower than the L_P for the alveolar membrane.

The hydraulic conductance measurements made in lung tissue are very variable. This variability appears to reflect different preparations, species, and techniques used for $K_{F,C}$ measurements. Again, as is true of the other parameters used in the Starling equation, $K_{F,C}$ is a composite of many different membrane parameters and filtration events occurring in both upper and lower portions of the lung's circulation. Basically, $K_{F,C}$ is a function of the microvascular wall porosity and surface area for exchange; however, experimental maneuvers used to measure this parameter may alter membrane selectivity, surface area, and that portion of the lung where fluid is filtered into the interstitium. $K_{F,C}$ has been shown to increase 8- to 10-fold when the capillary walls have been made more permeable to plasma proteins with α-naphthylthiourea[76] and 50–70% when plasma proteins are not present in the perfusates.[77]

2.6. The Osmotic Reflection Coefficient for Plasma Proteins (σ)

The osmotic reflection coefficient has been measured by using lymph,[1,22,48,84] osmotic transients,[52,53,61] or isotope movements[62] between lung tissue and blood. These approaches yield different σ's as shown in Table III. As compared to other capillary beds, the σ's for total proteins appear to be smaller in lungs with the exception of Kern's work (in which the σ for albumin was estimated to be 0.95). The reason for this difference is not clear, but it may be related to some type of

Table III. Estimation of the Osmotic Reflection Coefficients (σ_d) for Total Plasma Proteins in Different Capillary Beds

Tissue	σ_d	Ref.
Brain	1.0	70
Heart	0.9	78
Skeletal muscle	0.95	14
Subcutaneous	0.90	79
Stomach	0.78	80
Pancreas	0.85	—[a]
Small intestine	0.92	12
Colon	0.85	81
Liver	0.00	82
Lung		
Total protein	0.75 (lymph)	40, 48
Albumin	0.95 (isotope techniques)	62
Albumin	0.60 (osmotic transient)	52, 53
Albumin	0.30 (osmotic transient)	60, 61

[a]P. Kvietys (personal communication).

protein binding when using isotopically labeled albumin. It is difficult to see how σ can be greater than 0.8 in lung tissue unless all other techniques used to measure σ somehow increase vascular permeability. σ is only a function of membrane selectivity, so techniques using lymph sampled at minimal C_L (high capillary filtration rates) should provide the best estimates of the sieving characteristics of the pulmonary microvessel membrane since

$$C_L \cong C_P (1 - \sigma) \qquad (5)$$

It is important to note that Parker et al.[83] have obtained larger σ's by sampling lymph proteins when P_{LA} was elevated for 24 hr. This may reflect non-steady states for C_L in shorter duration lymphatic studies, extrapulmonary contribution to the lymph studied, nodal alterations in lymph, or some decrease in vascular permeability to macromolecules occurring with chronic elevations of P_C. The latter possibility has been postulated to explain the findings of Uhley et al. in animals with chronically elevated P_{LA}.[84] σ is known to change when lungs are challenged with caustic agents such as α-naphthylthiourea,[85] but many substances previously thought to alter vascular permeability are now known to increase vascular pressures or surface area when lymph is analyzed properly. Approximately 40 to 50 conditions have been postulated to cause pulmonary vascular permeability changes to macromolecules, but many conditions such as alveolar hypoxia, histamine, endotoxin, etc., fail to change σ's in dog lung capillaries.[1] Perhaps species differences exist, but the most likely explanation for the divergent results between species and in different preparations is that pulmonary vascular pressures are altered, not vascular permeability. The confusion may also reflect preparations in which lymph was sampled after passing through nodes or originating from extrapulmonary sources as discussed above.[1,13,14]

3. Safety Factors Associated with Hydrostatic Edemas

3.1. Tissue Fluid Pressure and Tissue Colloid Osmotic Pressure Changes

The ability of lung tissue to oppose edema formation has been described in terms of "tissue edema safety factors" which allow a "margin of safety" when pulmonary capillary pressure is elevated above normal values.[4,86] The upper panel in Fig. 5 shows the range of capillary pressures over which the edema safety factors normally operate. When capillary pressure is elevated from normal to approximately 25 mm Hg, a pressure differential of 18 mm Hg, three events occur in lung tissues which oppose the formation of edema fluid; (1) tissue colloid osmotic pressure decreases; (2) tissue fluid pressure increases; and (3) lymph flow draining the lung increases. Figures 3C and 4 demonstrate how the average tissue colloid osmotic pressure and fluid pressures change with increased capillary filtration. The contribution of these forces can easily be assessed by simply subtracting the control values from those measured after edema. This will give the maximum change in these tissue forces which can occur following elevations in microvascular pressure.

3.2. Lymphatic Safety Factor

The other tissue force, which is an actual volume removal factor, is lymph flow. We have defined the lymphatic safety factor as[9,87]

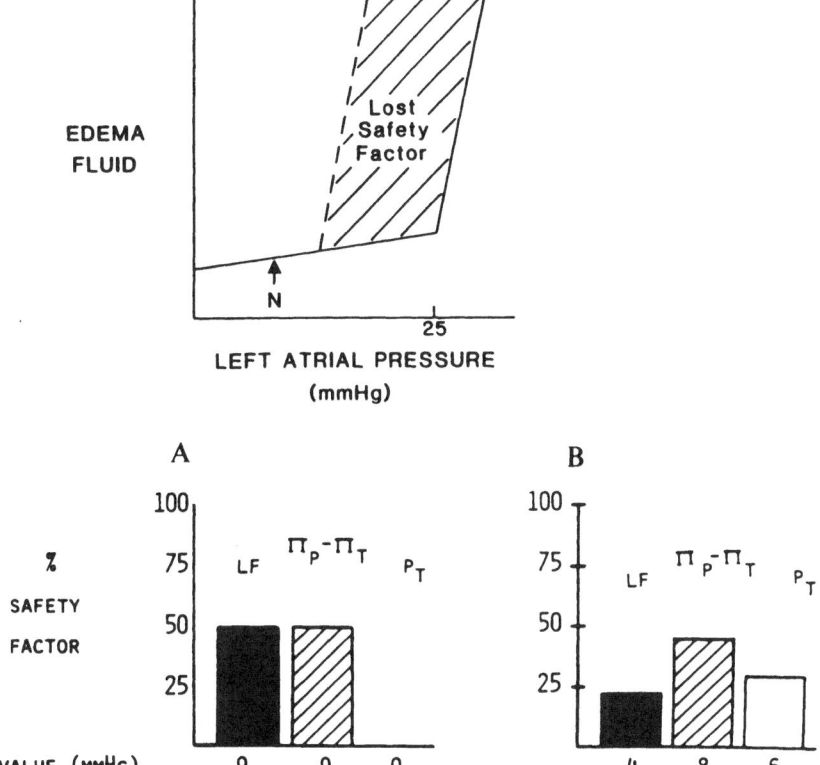

Fig. 5. (Top) Formation of edema fluid as a function of left atrial pressure. Between normal (N) and left atrial pressures of approximately 25 mm Hg, the safety factors change to oppose edema formation such that only a small amount of edema fluid accumulates. The striped area indicates how much this safety factor is reduced when the pulmonary capillaries become abnormally leaky to plasma proteins. From data of Rutili *et al.*[85] using α-naphthylthiourea. (Bottom) Safety factors calculated for lymph flow (LF), colloid osmotic pressure gradient ($\pi_P - \pi_T$) assuming $\sigma = 1$, and tissue fluid pressure (P_T) when capillary pressure was changed by 18 mm Hg. (A) Data from Erdmann and colleagues' study[65] in sheep lungs; (B) data from Drake's studies[39,40] in dog lungs. Numbers below each histogram are the actual changes in mm Hg for each force. Reproduced with permission from Taylor.[16]

$$\Delta P_{lymph} = \frac{\text{total capillary filtrate}}{K_{F,C}} \qquad (6)$$

where ΔP_{lymph} is the imbalance in the Starling forces. The total capillary filtrate is considered to be lymph flow; fluid is not entering the airways and the tissues are not gaining or losing interstitial fluid.[88]

3.3. Starling Force Changes Associated with Elevated P_{LA}

From Table II it is quite clear that we do not know the lymphatic factor to any degree of certainty because of the problems associated with estimating $K_{F,C}$. The lower panels in Fig. 5 depict the magnitude of ΔP_{lymph} (LF), the change in osmotic pressure ($\pi_P - \pi_T$) and tissue pressure (P_T) when capillary pressure was increased by 18 mm Hg. Panel A represents studies conducted in the chronically instrumented sheep preparation.[65] P_T was assumed to be zero, so the total safety factor was equally divided between the increased colloid osmotic pressure gradient and the lymph flow factor. Panel B depicts data obtained by Drake in isolated lungs[39,40] for which all Starling forces and lymph flow, with the exception of P_T, were measured. In this study, the change in tissue fluid pressure provided 30% of the safety factor, the colloid gradient change provided 50%, and the lymph flow factor another 20% of the change in capillary pressure.

3.4. Tissue Fluid Resistance Factor—The Correct Lymphatic Factor?

The differences in these two Starling force factors may actually be related to a species difference, but the assumptions made when assessing the magnitude of each factor may be questionable. For instance, the $K_{F,C}$ measured using total lymph flow divided by a calculated ΔP_{lymph} is 10–20 times smaller than $K_{F,C}$'s determined using weighing procedures. One might suspect that values which are different by orders of magnitude do not represent any real phenomena, but they could reflect that ΔP_{lymph} is a function of tissue volume conductance as well as $K_{F,C}$. Figure 6 shows how the tissue resistance may allow lymph formation to be a major edema safety factor. Consider that the small spaces surrounding the capillaries in the septal regions of the lung are connected to the larger perivascular spaces drained by lymphatics by a high-resistance pathway. The lower panel in Fig. 6 shows the pressure in these compartments. At normal capillary pressures, the pressure drop between septal and perivascular spaces is 7 mm Hg and this provides the force to move fluid into the perivascular spaces and fill the initial lymphatics which are located in this compartment. At capillary pressures greater than 25 mm Hg, this ΔP approaches zero because the tissue spaces have expanded and the tissue conductance has increased.[16,86] Therefore, the correct expression to describe ΔP_{lymph} in this system is

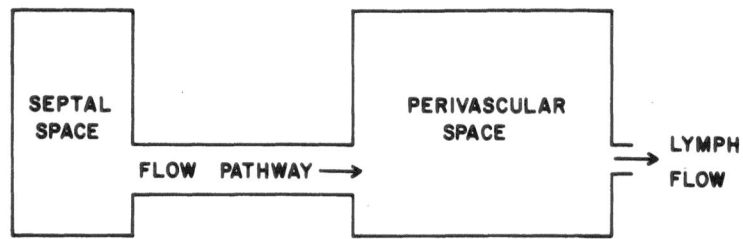

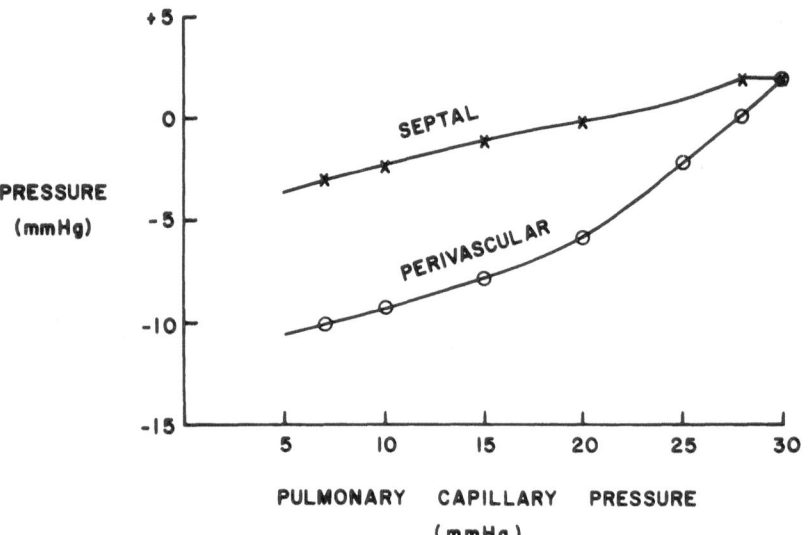

Fig. 6. Schematic representation of drainage patterns in the lung interstitium. The upper panel depicts the interstitial spaces surrounding alveolar capillary (septal resistance) as being connected to the larger perivascular spaces by a high-resistance pathway. Also, lymphatics only originate in the perivascular spaces. The model predicts that a pressure gradient exists between septal and perivascular spaces (lower panel) such that, at normal tissue hydrations, the conductance of the tissues is an important deterrent to edema formation. However, when the tissues swell, the flow pathway resistance will decrease and the pressure gradient will dissipate between the compartments. The tissue resistance safety factor decreases and lymph formation is now only a function of the capillary wall volume conductance ($K_{F,C}$).

$$\Delta P_{\text{lymph}} = \frac{\text{lymph flow}}{K_{F,\text{total}}} \qquad (7)$$

where

$$K_{F,\text{total}} = \frac{K_{FC}K_{FT}}{K_{FC} + K_{FT}} \qquad (8)$$

and K_{FT} is the volume conductance of the tissues. At normal hydration states, K_{FT} predominates because the tissues are very compact. When the tissues expand, K_{FT} becomes large and $K_{F,\text{total}} \cong K_{FC}$.

3.5. Summary of Starling Force Changes

Obviously, the interaction of $\Delta\pi$, P_T, and ΔP_{lymph} in providing a buffer against edema formation requires more precise measurements before quantitative values can be assigned to each of these forces when capillary filtration occurs in lungs. In fact, it is impossible to evaluate their importance in edema studies when only one or two forces are measured. It is desirable to measure each force if possible. This has been accomplished in some tissues at different capillary pressures.

Usually, $\Delta\pi$ is the major force which changes to oppose filtration in all organs studied with the exception of the very leaky (liver) and the very tight (brain) capillary beds. For the

former, $\sigma_d \cong 0$ such that $\sigma\Delta\pi$ is zero, and for the latter, virtually no proteins are in the tissues, so decreasing π_T is impossible. For most beds, P_T provides a substantial safety factor, especially in dependent areas such as the ankle, and the lymphatic factor provides from 50% to 0% in the liver and colon, respectively.[16]

Obviously, average Starling force studies leave much to be desired because of the complexities of the interactions of the various microforces and flows which are occurring in different sites within the lung's interstitium. However, these macrostudies do allow us to define the system in terms of measurable forces. In the future, the forces and flows will be dissected into their many components. For example, the pressure gradients postulated to exist between septal and perivascular spaces have been measured and behave in a fashion similar to that predicted by the model depicted in Fig. 6.[86]

4. Safety Factors Associated with Noncardiac Edema

4.1. Factors Which Are Thought to Alter Vascular Permeability

In a recent review, we discussed several factors which were thought to damage the lung's endothelium such that plasma proteins leaked more easily into the tissues.[1] However, when a careful analysis was applied to the published data, it was apparent that many conditions thought to cause capillary damage most likely only altered microvascular pressures and/or exchange surface area. Those which did appear to alter microvascular permeability such as emboli,[89–92] sepsis,[93–95] caustic agents such as α-naphthylthiourea,[85] oxygen toxicity caused by ventilating lungs with 100% O_2,[96,97] and high-altitude edema[98,99] were somehow related to the release of toxic O_2^- radicals by neutrophils or by damaged lung cells.[96,97] The oxygen anions combine with H_2O_2 to produce the extremely toxic hydroxyl radicals (OH·) which destroy cell membranes and disrupt interstitial glycosaminoglycans. Superoxide dismutase (SOD) catalyzes the reaction of O_2 to H_2O_2[100,101] and its concentration in tissues increases when lungs are challenged with low doses of 80% oxygen,[102] endotoxin,[103,104] hypoxia,[105] etc. In many forms of pathological lung states, it is thought that SOD is protective,[104,106,107] whereas in other forms of lung damage, SOD does not protect and H_2O_2 appears to be the damaging agent.[108,109] Many pathological changes can be avoided if animals are pretreated with SOD (increases O_2 conversion to H_2O_2), catalase (inactivates H_2O_2), and dimethylsulfoxide, mannitol, or ethyl alcohol (inactivate hydroxyl radicals). At the present time, it is not clear as to whether the toxic oxygen radicals are produced by neutrophils or tissue cells. In endothelial damage produced by microemboli, and endotoxin, decreasing the number of circulating neutrophils partially prevents the tissue damage[90,103] but O_2 toxicity and α-naphthylthiourea are not dependent on neutrophils for the production of the initial pulmonary damage.[109,111,112,113]

Many other systems appear to be involved in pathological lung states such as arachidonic acid products which produce O_2^- in the cyclooxygenase pathway and release leukotaxic substances in the lipoxygenase pathway which cause neutrophils to enter the tissue and release oxygen radicals. However, insufficient research has been conducted to define the interactions existing between the many systems postulated to be responsible for the various forms of lung damage. A symposium published in *The Physiologist* will introduce the reader to the more recent studies in oxygen radical research as applied to the lung and microcirculation in general.[106]

In our laboratory, we have found that SOD, ibuprofen, and dimethylsulfoxide protect the lung's endothelium against α-naphthylthiourea damage. Thus, this form of lung damage appears to be related to the generation of OH·.[110–112] Repine and Tate's studies, using a similar compound (thiourea) to pro-

duce the damage also indicate that OH· is partially responsible for the observed damage, although H_2O_2 appears to be responsible for many types of endothelial damage.[109]

4.2. Effects of Altering Microvascular Permeability on the Safety Factors

The normal safety factors require a rather low vascular permeability to plasma proteins in order for $\Delta\pi$ to increase and oppose filtration and for lymph flow to remove excess capillary filtrate. Figure 5A shows the range in which the edema safety factors operate when capillary pressures are elevated. The dashed line in the upper panel of Fig. 5 indicates how the safety factor is altered when the vascular permeability to plasma proteins is increased. The capillary pressure at which the edema forms is now much lower because $\sigma\Delta\pi$ is small.

Table IV shows how increased permeability affects the maximal effective $\Delta\pi$ which can be generated across the capillary wall. For normal capillaries, the maximal colloid osmotic gradient factor which can be generated in the pulmonary circulation is 10 mm Hg following left atrial pressure increase (see Fig. 5B). However, when the lung's endothelium is more leaky to plasma proteins, σ decreases to 0.4 and the amount of protein in the tissue fluids will also remain high, even at large capillary filtrations. Now, the maximal change in the colloid osmotic factor which can be generated when left atrial pressure is elevated is only 5.8 mm Hg. Thus, a difference in the ability of the lung to buffer (or oppose) increased capillary pressures has occurred such that the lung would gain excessive fluid at capillary pressures 7 mm Hg below that observed for the normal microcirculation.[111,114]

Increasing vascular permeability will also decrease the lung's lymphatic safety factor since its magnitude is inversely proportional to the filtration coefficient of the capillary wall. Since $K_{F,C}$ is a function of the pore radius through which filtration occurs to the fourth power, any large change in the capillary wall pore size should dramatically decrease the lymphatic safety factor. However, if the lung's interstitium is not expanded such that the tissue resistance factor is not changed, then this effect may be more apparent than real.[16,85]

4.3. Effects of Albumin Infusions on Withdrawing Lung Water in "Leaky Lung Syndromes"

In many forms of the adult respiratory distress syndrome (ARDS), it is evident that the lung's capillaries are abnormally

Table IV. Effects of Increased Microvascular Permeability on the Effective Colloid Osmotic Pressure Gradient Acting across the Capillary–Interstitial Barrier[a]

Condition	π_P	π_T	σ_d	$\sigma_d(\pi_P - \pi_T)$	Maximum change
Normal					
Control	28.0	18.0	0.75	7.5	
Increased P_{LA}	28.0	5.0	0.75	17.5	10.0
Leaky capillaries					
Control	28.0	22.0	0.40	2.4	
Increased P_{LA}	28.0	13.5	0.40	5.8	3.4
Protein added to "leaky" capillaries					2.8
Increased P_{LA}	38.0	17.0	0.40	8.0	

[a]All values are mm Hg.

leaky to plasma proteins. Many investigators have felt that pulmonary edema could be reversed by using colloid solutions (see Table IV). However, this is difficult to accomplish for two reasons: First, if π_p was increased by 10 mm Hg, then $\Delta\pi = 38.0 - 17.0 = 21$ mm Hg and the effective osmotic pressure ($\sigma\Delta\pi$) in the leaky lung state will only be 8.4 mm Hg. Although a 10 mm Hg change was imposed, because of the leakier capillaries, only a 3 mm Hg increase occurred at equilibrium! Usually, proteins pull more fluid out of extrapulmonary interstitial spaces and increase blood volume which tends to increase pulmonary capillary pressures. The increased P_C can easily be greater than the imposed change in $\sigma\Delta\pi$. Second, the half-time for albumin equilibration decreases to 1–1.5 hr in leaky lung states. Therefore, the protein placed into the circulation equilibrates rapidly in lung tissues, but more slowly with other vascular beds. Therefore, a condition may arise in which the net effect of changing colloids on lung fluid balance in ARDS patients is an increased fluid movement into the lung tissues, especially if P_C is increased. Thus, if colloids are increased in patients suspected of having ARDS, the pulmonary microvascular pressure must be carefully monitored to ensure that the small effective colloid osmotic pressure associated with the colloid change is not offset by increased capillary hydrostatic pressure.

5. Mechanism of Intraalveolar Edema Formation

Up to this point, we have discussed fluid accumulation in the lung's interstitium in a very general fashion. We will now present an analysis of the available information on how and where fluid accumulates in lung tissue and how the alveolar fluid leaves once the initiating cause has been eliminated, e.g., lowering of left atrial pressure.

5.1. Sequence and Pathways for Pulmonary Edema Formation

It now appears that fluid leaks from the small microvessels located in the alveolar region and percolates toward the hilus of the lung into the perivascular spaces.[115] Histologically, fluid accumulates first around these perivascular spaces in "fluid cuffs." The tissue regions surrounding the alveoli then begin to swell, especially in the thicker tissue regions around the alveolar structures. Finally, fluid fills the avleoli in an all-or-none fashion.

The reason that fluid accumulates first in the perivascular regions is that the pressure gradient between septal regions of the lung favors this route for fluid movement. In addition, the compliance of the perivascular spaces are high (i.e., their volumes can swell substantially without any significant pressure change) such that they are actually large potential spaces.[32,33]

A portion of the capillary filtrate[116–118] will enter the lymphatic system and will be propelled away from the tissues by either the lymphatic system's intrinsic pumping ability and/or the action of respiratory movements on the valved lymphatic system.[1,7,9,22,34,35,43]

An extravascular accumulation of lung interstitial fluid of approximately 35% above normal is the maximum fluid that the lung can accommodate before fluid begins to flood the alveoli. The reason that the alveoli flood is that the surface tension in the alveoli generates a fluid pressure within the alveoli of -2 to -3 mm Hg. Once tissue fluid pressure exceeds this intraalveolar pressure, a pressure gradient for alveolar filling occurs.[39,54,86,119] Figure 7 shows the magnitude of Starling forces at each vertical level of the lung. For many years, it was thought that edema fluid accumulated first in the most dependent area of the lung where P_C was higher. However, Fig. 7 clearly indicates that the same amount of fluid should filter at each lung

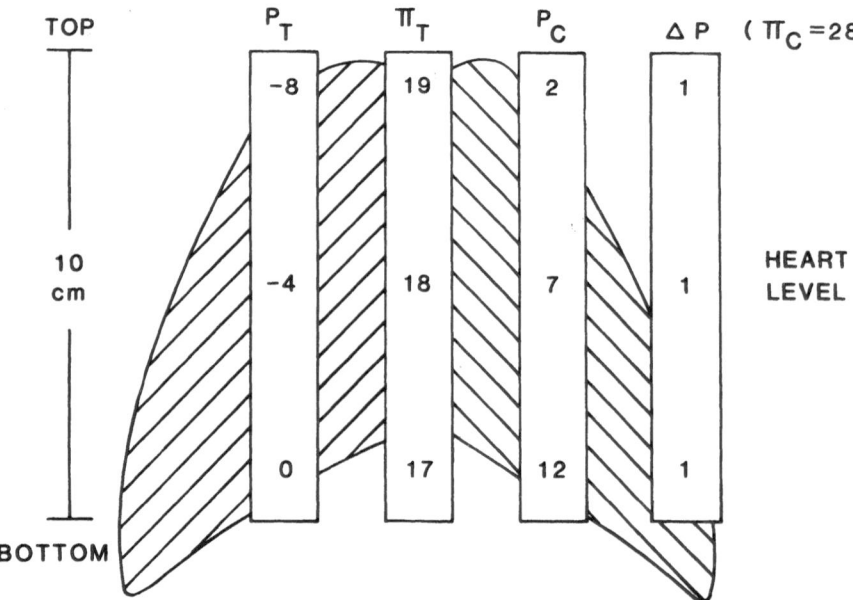

Fig. 7. Schematic representation of tissue fluid pressure (P_T), colloid osmotic pressure of the tissue fluids (π_T), and capillary pressure (P_C) at the bottom, midpoint, and top of an upright lung. Note that the imbalance in tissue pressure (ΔP) is the same at the top and bottom of the lung. The lung mass is considered to be 0.1 at bottom and top of the lungs and 0.25 at the lung midpoint. Therefore, more lymph will flow from the lung midpoint than from both the top and the bottom of the lung even at the same ΔP. However, it is certainly possible that the ability of the tissue forces to adjust when capillary pressures are elevated may be less in the dependent portions of the lung.

level since the sum of the forces is identical at each lung height. We now know that fluid does accumulate in all areas of the lung at the same rate initially, but the fluid either migrates from the upper portions of the lung to the dependent areas with time, or the safety factors are less in the dependent regions, i.e., they require less fluid to accumulate before being exhausted.[120]

5.2. Alveolar Filling

The route which fluid takes as it leaves the interstitium and floods the alveoli is presently not known. The fluid appears to fill the alveoli by bulk flow,[121–124] i.e., the fluid entering the alveoli has the same concentration of plasma proteins as tissue fluid. Since little or no selectivity is associated with alveolar flooding, the fluid certainly cannot cross the tight junctions between alveolar epithelial cells. The more probable route of alveolar flooding is through the small alveolar ducts which may be more permeable to large molecules than the alveolar walls. Since alveolar cells migrate toward the ducts as they age, alveolar epithelial cells may be displaced by pressure similar to that seen in the small intestine at the villus tips. The leakage of edema fluid into the alveoli may occur in a fashion similar to the "filtration secretion" observed in the small intestine following elevations of capillary pressures.[125–127] Only a small force is required to cause this phenomenon to occur in the small intestine (4 cm H_2O pressure gradient between serosa and mucosa) and gradients of this size could easily be generated in lung tissue during edema function because of the collapse tendency of the alveoli and the known pressure volume characteristics of the pulmonary tissue.

5.3. Absorption of Lung Intraalveolar Fluid

The manner in which lung fluid is absorbed once formed and the causative factor eliminated is not known. We know that electrolytes are transported into the larger airways and provide a fluid cushion to support the mucous layer and a low-resistance medium for ciliary motion.[128–130] There is now convincing evidence that the secretion transport process seen in large airways decreases in smaller airways and may actually reverse to absorption in the alveoli. A recent study by Matthay et al.[131] indicates that fluid is cleared from the airways against a considerable colloid osmotic gradient. The mechanism responsible for the fluid removal is not known, but it could be propelled by negative interstitial hydrostatic forces such that the fluid leaves the lung via the capillaries and/or crosses the visceral pleura space to be drained via the parietal lymphatics. In addition, an active transport mechanism may operate at the epithelial barrier of the alveoli and small airways to provide the osmotic force for water removal. The latter is the most likely process, since all epithelial cells transport electrolytes and water follows passively. One could ask the question as to why an active transport mechanism exists to move fluid out of the alveoli. Perhaps the physical force tending to pull fluid into the alveoli is much larger than we suspect; then the active transport of ions and, consequently, water would be necessary to prevent the filling of alveoli.

For years, physiologists felt that positive pressure ventilation should move fluid out of the alveoli. However, this is not the case since expansion of the lungs will simply decrease the pressure surrounding the perivascular spaces (as shown in Fig. 3B) and result in a greater accumulation of lung fluid. If the lung's microcirculation acts exactly as a "waterfall phenomenon,"

i.e., alveolar pressure collapses the blood vessels of the lungs, then positive pressure ventilation of the lung would be directly transmitted to the vascular system and no net pressure effect change for capillary reabsorption would occur. Positive pressure ventilation causes the alveolar fluid to layer in the alveoli into thin films and also opens up collapsed alveoli. These effects work in concert to increase the oxygen tension of blood flowing through the lung.[22] Also, since alveolar hypoxia causes increased pulmonary vascular resistance, oxygenating the alveoli will increase perfusion through the lungs. In pulmonary jargon, positive pressure ventilation may increase both alveolar ventilation and perfusion—both events resulting in better blood oxygenation and exchange of CO_2. The most fundamental steps in intraalveolar edema formation or resolution are presently not known. Further work must focus on this aspect of lung fluid balance before we can adequately describe how and where alveoli fill with fluid and what forces are responsible for its removal.

6. Summary

The various forces responsible for producing pulmonary edema have been presented in this chapter. Many areas of research associated with pulmonary edema formation are rapidly expanding and this chapter outlines several important concepts concerning how fluid balance is maintained in the lungs.

Hydrostatic edema: When left atrial pressures are elevated, the Starling forces, tissue fluid pressure, tissue colloid osmotic pressure, and lymph flow, change to oppose filtration. The lungs do not gain significant amounts of fluid until left atrial pressures are acutely elevated to levels exceeding 25–30 mm Hg. When left atrial pressures are chronically elevated, it appears that lymph flows can increase by an order of magnitude above that seen acutely. Therefore, patients will not develop pulmonary edema even with left atrial pressures of 40–50 mm Hg. The possibility exists that other Starling forces are also different in chronic conditions and this would be a most worthwhile area for future research.

Abnormally leaky lung capillaries: When the pulmonary capillary permeability to macromolecules is increased, intraalveolar edema will form at low left atrial pressures. Many compounds and pathological conditions are thought to alter vascular permeability and much more information is needed before we can quantitatively describe the mechanisms responsible for the various forms of lung damage. At the present time, it appears that the superoxide system is involved in many types of lung pathology and future research should focus on the relationship of superoxide radical formation to other physiological systems, such as prostaglandins, hormones, etc.

Intraalveolar edema: The alveoli appear to fill in an all-or-none fashion when tissue pressure exceeds alveolar pressure. This occurs after the lungs have gained considerable fluid. The fluid which enters the alveoli contains the same protein concentrations as the tissues, indicating that the fluid enters through holes which do not restrict plasma proteins, i.e., very large "pores." This is very different than the pulmonary capillary membrane which can be described by two pores of 60 and 200 Å[48] or the normal alveolar membrane which is composed of very small (5–8 Å) pores.[52] Therefore, either large pores open when edema forms, or pores have a favorable pressure gradient acting across their openings. This is an area of research which definitely needs to be explored.

Resolution of edema fluid: We presently do not know how fluid is removed from alveoli. Fluid obviously can go up the trachea, but it also appears that fluid removal may be related to either an active transport mechanism or a force operating between alveoli and tissue that is not presently known. Obviously, much work needs to be done in this area before we can explain the mechanisms responsible for removing edema fluid from alveoli.

Complexities of forces and flows within lung tissue: Throughout this chapter, we have described the forces and flows in terms of average capillary and tissue pressures, tissue and plasma colloid osmotic pressures, capillary filtrations, and lymph flows. We have also presented microanalysis of these forces in different regions of the lung's interstitium. Before we can expand our knowledge of how lung fluid balance is maintained, we must begin to measure the different forces and flows which occur in different areas of the lung and how they interact to oppose edema function.

It is obvious that many techniques presently utilized in membrane physiology need to be applied to lung fluid balance studies. Hopefully, this chapter will stimulate such research endeavors.

ACKNOWLEDGMENTS. I thank Sharon Miller and Sandy Worley for typing the manuscript and Penny Cook for preparing the figures.

References

1. Taylor, A. E., and J. C. Parker. 1985. The pulmonary interstitial spaces and lymphatics. In: *Handbook of Physiology: The Respiratory System*, Volume 1. A. P. Fishman and A. B. Fisher, eds. American Physiological Society, Bethesda Md., pp 167–230.
2. Guyton, A. C., and A. E. Lindsey. 1959. Effect of elevated left atrial pressure and decreased plasma protein concentration on the development of pulmonary edema. *Circ. Res.* **7**:649–657.
3. Landis, E. M., and J. R. Pappenheimer. 1963. Exchange of substances through the capillary walls. In: *Handbook of Physiology*, Volume 2. W. F. Hamilton and P. Dow, eds. American Physiological Society, Washington, D.C. pp. 961–1035.
4. Landis, E. M. 1934. Capillary pressure and capillary permeability. *Physiol. Rev.* **14**:404–481.
5. Guyton, A. C. 1963. A concept of negative interstitial pressure based on pressures in implanted perforated capsules. *Circ. Res.* **12**:399–414.
6. Guyton, A. C., B. J. Barber, and D. S. Moffatt. 1981. Theory of interstitial pressures. In: *Tissue Fluid Pressure and Composition*. A. Hargens, ed. Williams & Wilkins, Baltimore. pp. 11–20.
7. Guyton, A. C., A. E. Taylor, and R. A. Brace. 1976. A synthesis of interstitial fluid regulation and lymph flow. *Fed. Proc.* **35**:1881–1885.
8. Guyton, A. C., A. E. Taylor, and H. J. Granger. 1975. *Circulatory Physiology. II. Dynamics and Control of Body Fluids*. Saunders, Philadelphia. pp. 149–193.
9. Taylor, A. E., H. Gibson, H. J. Granger, and A. C. Guyton. 1973. The interaction between intracapillary and tissue forces in the overall regulation of interstitial fluid volume. *Lymphology* **6**: 192–208.
10. Kedem, O., and A. Katchalsky. 1958. Thermodynamic analysis of the permeability of biological membranes to non-electrolytes. *Biochim. Biophys. Acta* **27**:229–246.
11. Patlak, C. S., D. A. Goldstein, and J. F. Hoffman. 1963. The flow of solute and solvents across a two-membrane system. *J. Theor. Biol.* **5**:426–442.
12. Granger, D. N., and A. E. Taylor. 1980. Permeability of intestinal capillaries to endogenous macromolecules. *Am. J. Physiol.* **238**:H455–H464.
13. Taylor, A. E., J. C. Parker, D. N. Granger, N. A. Mortillaro, and G. Rutili. 1981. Assessment of capillary permeability using lymphatic protein flux: Estimation of the osmotic reflection coefficient. In: *The Microcirculation*. R. Effros, H. Schmid-Shonbein, and J. Ditzel, eds. Academic Press, New York. pp. 19–32.
14. Taylor, A. E., and D. N. Granger. 1984. Exchange of macromolecules across the circulation. In: *Handbook of Physiology*. E. M. Renkin and C. C. Michel, eds. *The Cardiovascular System* Volume IV Part 1, *Microcirculation*. American Physiological Society, Bethesda, Md., pp 867–520.
15. Taylor, A. E., and D. N. Granger. 1983. Equivalent pore modelling: Vesicles, channels and charge. *Fed. Proc.* **42**:2440–2445.
16. Taylor, A. E. 1981. Capillary fluid filtration: Starling forces and lymph flow. *Circ. Res.* **49**:557–575.
17. Starling, E. H. 1896. On the absorption of fluid from the connective tissue spaces. *J. Physiol. (London)* **19**:312–326.
18. Levine, O. R., R. B. Mellins, R. M. Senior, and A. P. Fishman. 1967. The application of Starling's law of capillary exchange to the lungs. *J. Clin. Invest.* **46**:934–944.
19. Mellins, R. B., O. R. Levine, R. Skalak, and A. P. Fishman. 1969. Interstitial pressure of the lung. *Circ. Res.* **24**:197–212.
20. Gaar, K. A., A. E. Taylor, L. J. Owens, and A. C. Guyton. 1967. Effect of capillary pressure and plasma proteins on the development of pulmonary edema. *Am. J. Physiol.* **213**:79–82.
21. Gaar, K. A., A. E. Taylor, L. J. Owens, and A. C. Guyton. 1967. Pulmonary capillary pressure and filtration coefficient in the isolated perfused lung. *Am. J. Physiol.* **213**:910–914.
22. Staub, N. C. 1974. Pulmonary edema. *Physiol. Rev.* **54**:687–811.
23. Pappenheimer, J. R., and A. Soto-Rivera. 1948. Effective osmotic pressure of the plasma proteins and the other quantities associated with the capillary circulation in the hindlimbs of cats and dogs. *Am. J. Physiol.* **152**:471–491.
24. Parker, J. C., R. E. Parker, D. N. Granger, and A. E. Taylor. 1979. Vertical gradient in regional vascular resistance and pre- to post-capillary resistance ratios in the dog lung. *Lymphology* **12**: 191–200.
25. Holloway, H., M. Perry, J. M. Downey, J. C. Parker, and A. E. Taylor. 1983. Estimation of effective pulmonary capillary pressure in intact lungs. *J. Appl. Physiol.* **54**:846–851.
26. Rippe, B., R. C. Allison, J. C. Parker, and A. E. Taylor. 1984. Effects of histamine, serotonin and norepinephrine on the circulation in dog lung. *J. Appl. Physiol.* **57**:223–232.
27. Parker, R. E., D. N. Granger, and A. E. Taylor. 1981. Estimates of isogravimetric capillary pressures during alveolar hypoxia. *Am. J. Physiol.* **241**:H732–H739.
28. Drake, R. E., and J. C. Gabel. 1980. Effects of histamine and alloxan on canine pulmonary vascular permeability. *Am. J. Physiol.* **239**:H96–H100.
29. Brigham, K., R. Bowers, and P. Owen. 1976. Effects of antihistamines on the lung vascular response to histamine in unanesthetized sheep: Diphenhydramine prevention of pulmonary edema and increased permeability. *J. Clin. Invest.* **58**:391–398.
30. Maron, M. B. 1982. Differential effect of histamine on protein permeability in dog lung and forelimb. *Am. J. Physiol.* **242**:F565–F572.
31. Parker, J. C., and A. E. Taylor. 1982. A comparison of capsular and intra-alveolar fluid pressure in lung. *J. Appl. Physiol.* **52**:1444–1452.
32. Inoue, H., C. Inoue, and J. Hildebrandt. 1982. Lobe weight gain and vascular, alveolar and peribronchial interstitial pressures. *J. Appl. Physiol.* **52**:173–183.
33. Lai-Fook, S. J. 1981. Interstitial pressure in the lung. In: *Tissue Fluid Pressure and Composition*. A. R. Hargens, ed. Williams & Wilkins, Baltimore. pp. 125–134.
34. Lauweryns, J. M. 1970. The juxta-alveolar lymphatics in the human adult lung. *Am. Rev. Respir. Dis.* **102**:877–885.
35. Lauweryns, J. M., and J. H. Baert. 1977. Alveolar clearance and the role of the pulmonary lymphatics. *Am. Rev. Respir. Dis.* **115**:625–683.
36. Weibel, E. R., and H. Bachofen. 1979. Structural design of the alveolar septum. In: *Pulmonary Edema*. A. P. Fishman and E. M.

Renkin, eds. American Physiological Society, Washington, D.C. pp. 1–20.

37. Meyer, B. J., A. Meyer, and A. C. Guyton. 1968. Interstitial fluid pressure. V. Negative pressure in the lung. *Circ. Res.* **22**:263–271.

38. Parker, J. C., A. C. Guyton, and A. E. Taylor. 1978. Pulmonary interstitial and capillary pressures estimated from intra-alveolar fluid pressures. *J. Appl. Physiol.* **44**:267–276.

39. Drake, R. E. 1975. Changes in Starling Forces During the Formation of Pulmonary Edema. Ph.D. dissertation. University of Mississippi.

40. Drake, R. E., and A. E. Taylor. 1976. Tissue and capillary force changes during the formation of intra-alveolar edema. In: *Progress in Lymphology.* R. C. Mayall and M. Witte, eds. Plenum Press, New York. pp. 13–17.

41. Prather, J. W., K. A. Gaar, and A. C. Guyton. 1968. Direct continuous recording of plasma colloid osmotic pressure of whole blood. *J. Appl. Physiol.* **24**:602–605.

42. Navar, P. C., and L. G. Navar. 1977. Relationship between colloid osmotic pressure and plasma protein concentration in the dog. *Am. J. Physiol.* **233**:H295–298.

43. Yoffey, J. M., and F. C. Courtice. 1970. *Lymphatics, Lymph and the Lymphomyeloid Complex.* Academic Press, New York.

44. Vreim, C. E., P. Snashall, and N. C. Staub. 1976. Protein composition of lung fluids in anesthetized dogs with acute cardiogenic edema. *Am. J. Physiol.* **231**:1466–1469.

45. Adair, T., D. S. Moffat, and A. C. Guyton. 1982. Lymph flow and composition is modified by the lymph node. *Am. J. Physiol.* **243**:H351–H359.

46. Albertine, K. H., E. L. Schultz, and N. C. Staub. 1983. Effect of lung height on protein concentration in sheep lung lymphatics. *Fed. Proc.* **42**:1272.

47. Taylor, A. E., J. C. Parker, P. R. Kvietys, and M. Perry. 1982. Pulmonary interstitium in capillary exchange. In: *Conference on Mechanisms of Lung Microvascular Injury.* A. B. Malik and N. C. Staub, eds. New York Academy of Science, New York. pp. 148–168.

48. Parker, J. C., R. E. Parker, D. N. Granger, and A. E. Taylor. 1981. Vascular permeability and transvascular fluid and protein transport in the dog lung. *Circ. Res.* **48**:545–561.

49. Drake, R., K. A. Gaar, and A. E. Taylor. 1978. Estimation of the filtration coefficient of pulmonary exchange vessels. *Am. J. Physiol.* **234**:H266–H274.

50. Weiser, P. C., and F. Grande. 1974. Estimation of fluid shifts and protein permeability during pulmonary edemagenesis. *Am. J. Physiol.* **227**:1008–1034.

51. Taylor, A. E., and K. A. Gaar. 1969. Calculation of equivalent pore radii of pulmonary capillary and alveolar membranes. *Rev. Argent. Angiol.* **111**:25–40.

52. Taylor, A. E., and K. A. Gaar, Jr. 1970. Estimation of equivalent pore radii of pulmonary capillary and alveolar membranes. *Am. J. Physiol.* **218**:1133–1140.

53. Perl, W. P., P. Chowdhury, and F. P. Chinard. 1975. Reflection coefficients of dog lung endothelium to small hydrophilic solutes. *Am. J. Physiol.* **228**:797–809.

54. Iliff, L. D. 1971. Extra-alveolar vessels and edema development in excised dog lungs. *Circ. Res.* **28**:524–532.

55. Mitzner, W., and J. L. Robotham. 1979. Distribution of interstitial compliance and filtration coefficient in canine lung. *Lymphology* **12**:140–148.

56. Oppenheimer, L., H. Unruh, C. Skoog, and H. S. Goldberg. 1983. Transvascular fluid flux measured from intravascular water concentration changes. *J. Appl. Physiol.* **54**:64–72.

57. Morriss, A. W., R. E. Drake, and J. C. Gabel. 1980. Comparison of microvascular filtration characteristics in isolated and intact dog lungs. *J. Appl. Physiol.* **48**:438–443.

58. Goldberg, H. S. 1980. Pulmonary interstitial compliance and filtration coefficients. *Am. J. Physiol.* **239**:H189–H198.

59. Lunde, P. K. M., and B. A. Waaler. 1969. Transvascular fluid balance in the lung. *J. Physiol. (London)* **205**:1–18.

60. Matalon, S. V., and O. D. Wangensteen. 1977. Pulmonary capillary filtration and reflection coefficients in the newborn rabbit. *Microvasc. Res.* **14**:99–110.

61. Wangensteen, O. D., E. Lysaker, and P. Savryn. 1977. Pulmonary capillary filtration and reflection coefficients in the adult rabbit. *Microvasc. Res.* **14**:81–97.

62. Kern, D. F. 1981. Pulmonary Capillary Permeabilities and Reflection Coefficients. Ph.D. thesis. University of Minnesota.

63. Nicolaysen, G., P. Aarseth, and B. A. Waaler. 1976. Transvascular fluid filtration and intravascular volume in isolated perfused lungs. *Acta Physiol. Scand.* **96**:25A–26A.

64. Perry, M. A. 1980. Capillary filtration and permeability coefficients calculated from measurements of interendothelial cell junctions in rabbit lung and skeletal muscle. *Microvasc. Res.* **19**:142–157.

65. Erdmann, A. J., T. R. Vaughan, K. L. Brigham, W. C. Woolverton, and N. C. Staub. 1975. Effect of increased vascular pressure on lung fluid balance in unanesthetized sheep. *Circ. Res.* **37**:271–284.

66. Levine, O. R., F. Rodriguez-Martinez, and R. B. Mellins. 1973. Fluid filtration in the lung of intact puppy. *J. Appl. Physiol.* **34**:683–686.

67. Drake, R. E., J. H. Smith, and J. C. Gabel. 1980. Estimation of the filtration coefficient in intact dog lungs. *Am. J. Physiol.* **238**:H430–H438.

68. Gabel, J. C., R. E. Drake, J. F. Arens, and A. E. Taylor. 1978. Unchanged pulmonary capillary filtration coefficients after *Escherichia coli* endotoxin infusion. *J. Surg. Res.* **25**:97–104.

69. Richardson, P. D. I., D. N. Granger, and A. E. Taylor. 1979. Capillary filtration coefficient: The technique and its application to the small intestine. *Cardiovasc. Res.* **13**:547–561.

70. Fenstermacher, J. D., and J. A. Johnson. 1966. Filtration and reflection coefficient of the rabbit blood brain barrier. *Am. J. Physiol.* **211**:341–346.

71. Vargas, F. F., and J. A. Johnson. 1964. An estimate of reflection coefficient from rabbit heart capillaries. *J. Gen. Physiol.* **47**:667–677.

72. Brenner, B. M., C. Baylis, and W. M. Deen. 1976. Transport of molecules across renal glomerular capillaries. *Physiol. Rev.* **65**:502–534.

73. Anderson, B., and H. H. Ussing. 1957. Solvent drag on non-electrolytes during osmotic flow through toad skin and its response to anti-diuretic hormone. *Acta Physiol. Scand.* **29**:228–239.

74. Sidel, V. W., and A. K. Solomon. 1957. Entrance of water into human red cells under an osmotic gradient. *J. Gen. Physiol.* **41**:243–257.

75. Greenway, C. V. 1981. Hepatic fluid exchange. In: *Hepatic Circulation in Health and Disease.* W. W. Lautt, ed. Raven Press, New York. pp. 153–167.

76. Allison, R. C., B. Rippe, J. C. Parker, and A. E. Taylor. 1983. ANTU increases vascular permeability in isolated dog lung. *Fed. Proc.* **42**:1108a.

77. Rippe, B., J. Parker, and A. E. Taylor. 1985. Effects of reduced plasma proteins on capillary filtration coefficients and isogravimetric capillary pressures in isolated dog lungs. *J. Appl. Physiol.* **26**:A8.

78. Laine, G. A., and H. J. Granger. 1980. Myocardial Starling forces and lymphatic dynamics during venous hypertension. *Microvasc. Res.* **20**:116.

79. Rutili, G., D. N. Granger, A. E. Taylor, J. C. Parker, and N. A. Mortillaro. 1982. Analysis of lymphatic protein data. IV. Comparison of the different methods used to estimate reflection coefficients and permeability–surface area products. *Microvasc. Res.* **23**:347–360.

80. Perry, M. A., and D. N. Granger. 1981. Permeability of gastric capillaries to small and large molecules. *Am. J. Physiol.* **241**:G478–G486.

81. Richardson, P. D. I., D. N. Granger, D. Mailman, and P. R. Kvietys. 1980. Permeability characteristics of colonic capillaries. *Am. J. Physiol.* **239**:G300–G305.

82. Granger, D. N., T. Miller, R. Allen, R. E. Parker, J. C. Parker, and A. E. Taylor. 1979. Permselectivity of the liver blood–lymph

barrier to endogenous macromolecules. *Gastroenterology* **77**: 103–109.

83. Parker, R. E., R. J. Rosseli, T. R. Harris, and K. L. Brigham. 1981. Effect of graded increases in pulmonary vascular pressures on lung fluid balance in unanesthetized sheep. *Circ. Res.* **49**:1164–1172.

84. Uhley, H. N., S. E. Leeds, J. J. Sampson, and M. Friedman. 1962. Role of pulmonary lymphatics in chronic pulmonary edema. *Circ. Res.* **11**:966–970.

85. Rutili, G., P. Kvietys, J. C. Parker, and A. E. Taylor. 1982. Increased microvascular permeability induced by ANTU. *J. Appl. Physiol.* **52**:1316–1323.

86. Guyton, A. C., A. E. Taylor, R. E. Drake, and J. C. Parker. 1976. Dynamics of subatmospheric pressure in the pulmonary interstitial fluid. In: *Lung Liquids.* Elsevier/Excerpta Medica, Amsterdam. pp. 77–100.

87. Taylor, A. E., and R. E. Drake. 1978. Fluid and protein exchange. In: *Lung Biology in Health and Disease,* Volume 9. N. C. Staub, ed. Dekker, New York. pp. 129–166.

88. Aukland, K., and G. Nicolaysen. 1981. Interstitial fluid volume: Local regulation mechanisms. *Physiol. Rev.* **61**:556–643.

89. Barie, P. S., F. L. Minnear, and A. B. Malik. 1981. Increased pulmonary vascular permeability after bone marrow injection in sheep. *Am. Rev. Respir. Dis.* **123**:648–653.

90. Flick, M. R., A. Perel, and N. C. Staub. 1981. Leukocytes are required for increased lung microvascular permeability after microembolization in sheep. *Circ. Res.* **48**:344–352.

91. Johnson, A., and A. B. Malik. 1982. Pulmonary edema after glass bead microembolization: Protective effect of granulocytopenia. *J. Appl. Physiol.* **52**:155–161.

92. Johnson, A., M. V. Tahamont, H. E. Kaplan, and A. B. Malik. 1982. Lung fluid balance after pulmonary embolization: Effects of thrombin vs. fibrin aggregates. *J. Appl. Physiol.* **52**:1565–1570.

93. Brigham, K. L., R. E. Bowers, and J. Haynes. 1979. Increased sheep lung vascular permeability caused by *Escherichia coli* endotoxin. *Circ. Res.* **45**:292–297.

94. Brigham, K. L., R. E. Bowers, and C. R. McKeen. 1981. Methylprednisolone prevention of increased lung vascular permeability following endotoxemia in sheep. *J. Clin. Invest.* **67**: 1103–1110.

95. Brigham, K. L., W. C. Woolverton, L. H. Blake, and N. C. Staub. 1974. Increased sheep lung vascular permeability caused by pseudomonas bacteremia. *J. Clin. Invest.* **54**:792–804.

96. Crapo, J. D., B. E. Barry, H. A. Foscue, and J. Shelburne. 1980. Structural and biochemical changes in rat lungs occurring during exposures to lethal and adaptive doses of oxygen. *Am. Rev. Respir. Dis.* **122**:123–143.

97. Clark, J. M., and C. J. Lambertsen. 1971. Pulmonary oxygen toxicity: A review. *Pharmacol. Rev.* **23**:38–117.

98. Scoggin, C. H., T. M. Hyers, J. T. Reeves, and R. F. Grove. 1977. High altitude pulmonary edema in children and young adults of Leadville, Colorado. *N. Eng. J. Med.* **297**:1269–1272.

99. Staub, N. C. 1980. Pulmonary edema—Hypoxia and overperfusion. *N. Eng. J. Med.* **302**:1086.

100. McCord, J. M. 1978. Free radicals and inflammation: Protection of synovial fluid by superoxide dismutase. *Science* **185**:529.

101. McCord, J. M., and I. Fridovich. 1978. The biology and pathology of oxygen radicals. *Ann. Intern. Med.* **89**:122–136.

102. Crapo, J. D., K. Sjostrom, and R. T. Drew. 1978. Tolerance and cross-tolerance using NO_2 and O_2. I. Toxicology and biochemistry. *J. Appl. Physiol.* **44**:364–369.

103. Frank, L., and R. J. Roberts. 1979. Endotoxin protection against oxygen-induced acute and chronic lung injury. *J. Appl. Physiol.* **47**:577–581.

104. Frank, L., J. Summerville, and D. Massaro. 1980. Protection from oxygen toxicity with endotoxin: Role of the endogenous antioxidant enzymes of the lung. *J. Clin. Invest.* **65**:1104–1110.

105. Rhodes, M. L., D. C. Zabala, and D. Brown. 1976. Hypoxic protection in paraquat poisoning. *Lab. Invest.* **35**:496–500.

106. Taylor, A. E. 1983. Oxygen radicals and the microcirculation. *Physiologist* **26**:152–182.

107. Flick, M. R., J. Hoeffel, and N. C. Staub. 1983. Superoxide dismutase with heparin prevents increased lung vascular permeability during air emboli in sheep. *J. Appl. Physiol* **55**:1284–1291.

108. Fox, R. B., J. R. Hoidal, D. N. Brown, and J. E. Repine. 1981. Pulmonary inflammation due to oxygen toxicity: Involvement of chemotactic factors and polymorphonuclear leukocytes. *Am. Rev. Respir. Dis.* **123**:521–523.

109. Repine, J., and R. M. Tate. 1983. Oxygen radicals and lung edema. *Physiologist* **26**:178–182.

110. Taylor, A. E., and D. Martin. 1983. Oxygen radicals and the microcirculation. *Physiologist* **26**:152–155.

111. Taylor, A. E., D. Martin, and J. C. Parker. 1983. The effects of oxygen radicals on pulmonary edema formation. *Surgery* **94**: 433–451.

112. Parker, J. C., D. J. Martin, G. Rutili, J. McCord, and A. E. Taylor. 1983. Prevention of free radical mediated vascular permeability increase in lung using superoxide dismutase. *Chest* **93**: 525–528.

113. Crapo, J. D., B. A. Freeman, B. E. Barry, J. F. Turrens, and S. L. Young. 1983. Mechanisms of hyperoxic injury to the pulmonary circulation. *Physiologist* **26**:171–178.

114. Taylor, A. E., J. C. Parker, and R. C. Allison. 1982. Capillary exchange of fluid and protein. In: *Critical Care, State-of-the Art.* W. C. Shoemaker and W. L. Thompson, eds. The Society of Critical Care Medicine, Fullerton, Calif. pp. B1–B15.

115. Staub, N. C., H. Nagano, and M. L. Pearce. 1967. Pulmonary edema in dogs: Especially the sequence of fluid accumulation in the lungs. *J. Appl. Physiol.* **22**:227–240.

116. Inoue, H., and J. Hildebrandt. 1980. Vascular and airway pressures and interstitial edema affect peribronchial fluid pressure. *J. Appl. Physiol.* **48**:177–185.

117. Lai-Fook, S. J. 1982. Perivascular interstitial fluid pressure measured by micropipettes in isolated dog lung. *J. Appl. Physiol.* **59**:9–15.

118. Lai-Fook, S. J., and K. C. Beck. 1982. Alveolar liquid pressures measured by micropipettes in isolated dog lung. *J. Appl. Physiol.* **53**:737–743.

119. Lai-Fook, S. J., and B. Toporoff. 1980. Pressure–volume behavior of perivascular interstitium measured in isolated dog lung. *J. Appl. Physiol.* **48**:939–946.

120. Hales, C. A., D. J. Kanarek, B. Ahluwalia, A. Latty, J. Erdmann, S. Javaheri, and H. Kazemi. 1981. Regional edema formation in isolated perfused dog lungs. *Circ. Res.* **48**:121–127.

121. Gee, M. H., and N. C. Staub. 1977. Role of bulk fluid flow and protein permeability of the dog lung alveolar membrane. *J. Appl. Physiol.* **42**:144–149.

122. Egan, E. A. 1982. Lung inflation, lung solute permeability, and alveolar edema. *J. Appl. Physiol.* **53**:121–125.

123. Staub, N. C. 1979. Pathways for fluid and solute fluxes in pulmonary edema. In: *Pulmonary Edema.* A. P. Fishman and E. M. Renkin, eds. American Physiological Society, Washington, D.C. pp. 113–124.

124. Vreim, C. E., P. Snashall, and N. C. Staub. 1976. Protein composition of lung fluids in anesthetized dogs with acute cardiogenic edema. *Am. J. Physiol.* **231**:1466–1469.

125. Duffy, P. A., D. N. Granger, and A. E. Taylor. 1978. Intestinal secretion induced by volume expansion in the dog. *Gastroenterology* **75**:413–418.

126. Yablonski, M. E., and N. Lifson. 1976. Mechanism of production of intestinal secretion by elevated venous pressure. *J. Clin. Invest.* **57**:904–915.

127. Mortillaro, N. A., and A. E. Taylor. 1976. Interaction of capillary and tissue forces in the cat intestine. *Circ. Res.* **39**:348–358.

128. Olver, R. E., and L. B. Strang. 1974. Ion fluxes across the pulmonary epithelium and the secretion of lung liquid in the fetal lamb. *J. Physiol. (London)* **241**:327–357.

129. Gatzy, J. T. 1967. Bioelectric properties of the isolated amphibian lung. *Am. J. Physiol.* **213**:425–431.

130. Crandall, E. D., and K. J. Kim. 1981. Transport of water and solutes across bullfrog alveolar epithelium. *J. Appl. Physiol.* **50**:1263–1271.

131. Matthay, M. A., C. C. Landolt, and N. C. Staub. 1982. Differential liquid and protein clearance from the alveoli of anesthetized sheep. *J. Appl. Physiol.* **53**:96–104.

Index